Cryptococcus neoformans

Cryptococcus neoformans

By

Arturo Casadevall
Albert Einstein College of Medicine
of Yeshiva University
Bronx, New York 10461-1602

and

John R. Perfect
Department of Microbiology
Duke University Medical Center
Durham, North Carolina 27710

ASM
PRESS

Washington, D.C.

Library of Congress Cataloging-in-Publication Data

Casadevall, Arturo, 1957–
Cryptococcus neoformans / by Arturo Casadevall and John R. Perfect.
    p. cm.
   Includes index.
   ISBN 1-5581-107-8
   1. Torulosis. 2. Cryptococcus neoformans. I. Perfect, John R.,
1949–   . II. Title.
   [DNLM: 1. Cryptococcus neoformans. 2. Cryptococcosis. QW
180.5.D3 C334c 1998]
RC186.T7C37 1998
616'.015—DC21
DNLM/DLC
for Library of Congress   98-8693
                                                                      CIP

CONTENTS

PREFACE

Cryptococcus neoformans has been a known human pathogen for over a century. The history of its epidemiology, ecology, taxonomy, molecular biology, pathobiology, diagnosis, and treatment strategies mirrors discoveries in the general field of medical mycology. This primary fungal pathogen, which can infect apparently normal hosts, has also become a major secondary fungal pathogen as it has been able to exploit immunosuppressive events of modern medicine such as human immunodeficiency virus and AIDS, cancer, and immunosuppressive therapies. As we approach the new millennium, the importance of this fungus and the infection it causes—cryptococcosis—has never been greater for clinical medicine. Infection with this encapsulated yeast occurs in all areas of the world, and clinicians at all levels of care-giving may be faced with its challenges.

We have been impressed with the knowledge base for cryptococcosis developed by many outstanding and dedicated investigators over the past 100 years. This book is a tribute to their scholarship and creative insights. In the 1950s, Littman and Zimmerman produced an excellent monograph on the understanding of cryptococcosis up to that time. In our present book, *Cryptococcus neoformans*, we attempt to chronicle the substantial progress in the understanding of this infection since their mid-20th-century treatise. With this comprehensive book, we have attempted to "draw a line in the sand" of knowledge regarding this fungus so that future generations can step over it. Our intention is that this book will be useful for both the student who is beginning to search the cryptococcal knowledge base, and the medical mycologist who wants in-depth coverage and references.

John E. Bennett coined the term "sugar-coated killer" for *C. neoformans* at a lecture to the Infectious Disease Society in the 1990s. This complex, killer yeast has become a model fungal pathogen for the study of pathobiology at all scientific levels. This book describes its history, assesses its current status, and provides an outlook on the future. There is much more to learn about this yeast's secrets. It is our hope that this book will stimulate scientists to pursue the understanding of this successfully evolving fungal pathogen.

Arturo Casadevall
John R. Perfect

ACKNOWLEDGMENT

We gratefully acknowledge Wiley Schell for his skillful help in photographing many of the color plates and for providing the photo for the cover for this book.

1 | Introduction to the Pathogen

HISTORY

1890s: The Discovery

Cryptococcus neoformans was identified as a human pathogen in 1894. The first description of cryptococcosis is generally attributed to two German physicians in Greifswald, Germany, Otto Busse and Abraham Buschke (reviewed in reference 106). Busse and Buschke described the case of a 31-year-old woman with a lesion on her tibia. Busse, a pathologist, described round corpuscles in the bone lesion and isolated a microorganism in culture that was thought to be a blastomycete. Busse named the organism *Saccharomyces hominis* and the infection saccharomycosis hominis. Buschke, a physician who cared for the patient, described the organism independently of Busse as a coccidium. Also in 1894 Sanfelice in Italy isolated an encapsulated yeast from fermenting peach juice and, in the following year, demonstrated that it was pathogenic in laboratory animals (167, 168). Sanfelice called his organism *Saccharomyces neoformans* because of its tendency to form tumorlike lesions in experimental animals (reviewed in reference 58). In addition, Sanfelice recovered a similar organism from the lymph node of an ox but called that isolate *Saccharomyces lithogenes* because of some differences from *S. neoformans* (58). A second case of human cryptococcosis was described in France by Curtis, who recovered a yeastlike fungus from a hip lesion in 1896 and named it *Saccharomyces subcutaneous tumefaciens* (reviewed in reference 58).

By the end of the 19th century three seminal observations had been made regarding *C. neoformans*. First, the organism had been recovered from lesions in humans and animals, establishing its potential to cause disease. Second, the organism had been recovered from the environment, establishing that it was free-living. Third, the organism was propagated in the laboratory and shown to cause disease in laboratory animals. Also apparent after the first decade was the use of multiple names for this organism, a phenomenon that was to persist for most of the 20th century and cause considerable confusion.

Knoke and Schwesinger (106) described several biographical details about Busse and Buschke and the events surrounding the discovery of *C. neoformans*. Otto Busse was born on December 6, 1867, in Guhlitz, Germany, and trained at the University of Greifswald. Busse subsequently became professor of pathology at

Greifswald and later continued his career in Posen and Zurich. He died in Zurich on February 3, 1922. Abraham Buschke was born on September 27, 1868, in Nakel, Germany, and trained as a physician in Breslau, Berlin, and Greifswald, specializing in surgery and dermatology. On November 11, 1942, he was deported to the concentration camp at Theresienstadt and died there on February 23, 1943.

In 1931 Freeman (81) suggested that the first description of cryptococcosis was made by Zenker in 1861. Despite the absence of a culture to accompany Zenker's case, Freeman argued that the clinical and pathological description of the Zenker case was that of cryptococcal meningitis (81). This may be true, but credit for the discovery of *C. neoformans* and for the description of cryptococcosis is given to Busse and Buschke, who provided the first detailed clinical, pathological, and mycologic description of the organism and saved it for comparative studies with subsequent isolates.

1900–1950: Clinical Descriptions and Consolidation of Its Status as a Pathogen

The first half of the 20th century was a period of consolidation in the status of *C. neoformans* as a pathogen.

In the early 1900s many case descriptions of *C. neoformans* infections produced an increased awareness of this fungus as a human pathogen (12, 23, 28, 31, 49, 73, 81, 82, 96, 101, 117, 126, 132, 133, 155, 157, 161, 171, 178, 184, 199). Cryptococcal infections were reported in various organs, including the brain, lungs, and skin. Several of the early case reports are noteworthy. In 1908 Brewer and Wood (34) described a case of localized cryptococcosis of the spine (which he called blastomycoses). This case was unusual in two ways: the lesion was a tumorlike lesion that appears to have been a large eosinophilic granuloma, and the patient recovered after surgery (34). The organism was cultured, shown to be a yeast, and used to infect guinea pigs. The authors noted a correlation between the ability of the guinea pig to mount a strong inflammatory response and long-term survival (34). In 1912 Rusk and Farnell (161) described two cases of apparent cryptococcal infection in the central nervous system and called the infection systemic oidiomycosis. Although the organism was not cultured in either case, the clinical and pathological descriptions provided by Rusk and Farnell are those of cryptococcal meningitis (161). These investigators provided a detailed description of the histology of infection and commented on "gelatinous masses" in the lungs, noting that the tissue reaction was a "chronic granulomatous process" distinct from that of tuberculosis (161).

The histopathology of cryptococcal infection in humans was described in detail with particular reference, and the absence of inflammatory response in some patients was noted. Freeman and Weidman (82) introduced the term "soapsuds appearance" to describe sharply demarcated cysts of cryptococci growing in brain tissue. The relative absence of polymorphonuclear infiltrates was commented on several early pathological reports (73, 155), but eosinophils were noted in some cases (34, 161). Granuloma formation was described as the classical tissue reaction of cryptococcosis in cases in which inflammation developed. Giant cells with multiple intracellular yeasts were described in various cases (12, 73, 182), indicating that this fungus could persist in both intracellular and extracellular spaces. The occurrence of atypical forms of cryptococci was also noted in the early literature (12, 168). Rappaport and Kaplan reported a case of disseminated cryptococcosis in which chain-like budding forms resembling a mycelium were noted in the kidney (cited in reference 22). The problems involved in diagnosing cryptococcosis were noted by several investigators who commented on the difficulty of distinguishing crypto-cocci from lymphocytes in cerebrospinal fluid (CSF) smears (73, 174).

The absence of effective therapy was obvious by the poor prognosis of infec-tion and the multiple ineffective treatments that were prescribed (105, 117, 157). With the exception of a case of localized cryptococcal infection in the spine, which was removed surgically, all patients in the early literature died from their infection (34). Among the therapies tried without success were serum therapy, iodides, sulfonamide, sodium thiosulfate, X-irradiation, gentian violet, and arsphenamine (13, 105, 157, 175). Daily spinal taps appeared to provide symptomatic relief in some cases.

Little progress was made in understanding the ecology of *C. neoformans*. The association of this organism with avian excreta was not made until the early 1950s (see below). Sources of infection were often assumed to be endogenous after the description of cryptococcus-like organisms in the gastrointestinal tract (4). Simi-larly, little progress was made in understanding the route of infection. Several investigators considered the lungs were to be a likely site for initial infection on the basis of associated lung lesions in patients with cryptococcal meningitis. However, most authorities were cautious in interpreting the relationship of lung lesions to meningitis, and most concluded that the route of infection was unknown. The gastrointestinal tract was considered to be a potential route for infection in some early cases (171).

Many of the early clinical reports of cryptococcosis included animal experi-ments with the organism isolated from patients. Experimental cryptococcal infec-tion was produced in guinea pigs (13, 28, 34, 49, 82, 126), mice (49, 105, 126), rats (105, 126), rabbits (13, 126, 175, 200), cats (15), and dogs (12, 13). Infection was induced in experimental animals by a variety of routes, including intratracheal (175), intraperitoneal (126), subcutaneous (82, 126), intranasal (82), and intracranial (15) inoculations. Animals were used to evaluate the potential efficacy of various drugs, but none were found effective (105, 190). Differences in animal susceptibility were noted: cats were more resistant than rats (15), and rats were more resistant than mice (126). The relative resistance of rabbits to experimental infection was appreciated, and an early review noted that "the rabbit is practically immune, while rats and mice are probably the most susceptible animals."

Descriptions of spontaneous cryptococcal infection in animals led to its recog-nition as a veterinary pathogen (84, 198). In 1902 Frothingham (84) isolated *C. neoformans* from a large tumorlike lesion in the lung of a horse and provided a detailed description of the pathological findings and the organism. The tumor was "as large again as the human head" and was located in the right lung of the horse (84). Histological examination revealed that a large number of "blastomyces em-bedded in a homogeneous, gelatinous mass." Frothingham (84) cultured the organ-ism and reproduced the disease in guinea pigs and rabbits. He did not name the organism but referred to it as a "blastomyces" and reviewed reports of pseudo-glanders in animals caused by similar organisms.

Confusion reigned in the terminology used to describe the pathogen and the clinical infection. Many names were used to refer to *C. neoformans* and cryptococ-cosis (Table 1). In 1901 Vuillemin reclassified the yeast recovered by Busse,

Table 1 Some names used to refer to *C. neoformans* and cyptococcosis in the literature

Name	Year or period	Notes
Saccharomyces hominis	1894	Name given by Busse for fungus recovered from human tibial lesion (for reviews, see references 106 and 115)
Saccharomyces neoformans	1895	Specific name given by Sanfelice to fungus isolated from peach juice, the same species as that of Busse and Buschke (17)
Saccharomycosis	1890s	Term used to refer to disease caused by a yeast infection. Used by Busse and Buschke. Later this term evolved to "blastomycosis" and "European blastomycosis" to distinguish it from coccidioidomycosis.
Saccharomyces tumefaciens	1895	Name given by Curtis. This isolate was thought by Benham (15) to have minor differences that suggested a related species. This isolate appears to belong to *C. neoformans* var. *gattii* (reviewed in reference 58).
Cryptococcus hominis	1901	Vuillemin used the name *Cryptococcus* to refer to the genus of the organism and gave the name *C. hominis* to the organism of Busse and Buschke (17).
European blastomycosis	Early 1900s	Term for cryptococcosis (17) intended to distinguish cryptococcosis from "American blastomycosis" caused by *Blastomyces dermatitidis*
Systemic oidiomycosis	1911	Name used by Rusk and Farnell (161) to refer to cryptococcosis.
Torula histolytica	1916-1950s	Common name for *C. neoformans* for over half a century. The name "histolytica" was given by Stoddard and Cutler (180) because empty spaces around yeast cells in tissue were assumed to represent tissue lysis. Subsequent work by Freeman (81) established that spaces were the result of capsule contraction during fixation, not tissue digestion.
Systemic blastomycosis	Early 1900s	Used in the description of some cases of *C. neoformans* infection (31, 34, 184, 200). The term "blastomycotic meningitis" was also used to describe cryptococcal meningitis (31). Some authors identified the organism as *Cryptococcus gilchristi* with a capsule (184). However, *C. gilchristi* was another name for *B. dermatitidis*.

Table 1 *(Continued)*

Name	Year or period	Notes
Torulosis	Early 1900s	Term used to refer to infection with *T. histolytica*
Debaryomyces hominis	1936	Todd and Herrmann (186) proposed this name after describing some spore-forming strains of *C. hominis*
Cryptococcus histolyticus	1936	Mook and Moore (143) argued against the use of the term *Torula* and proposed the name *C. histolyticus*
Cryptococcus neoformans	1950-present	Name supported by Benham (17, 18) on the basis of (i) the *Cryptococcus* name given by Vuillemin and (ii) the *neoformans* name given by Sanfelice. *C. hominis* and *T. histolytica* are considered synonyms. Two varieties of *C. neoformans* are recognized: var. *neoformans* and var. *gattii*.
Filobasidiella neoformans	1975-present	Name of the teleomorph. Two varieties of *F. neoformans* are recognized: var. *neoformans*, 1975, and var. *bacillispora* Kwon-Chung, 1982 (115).

Buschke, and Sanfelice to the genus *Cryptococcus* because these organisms did not form ascospores or fermented sugars like those in the genus *Saccharomyces*. In 1916 Stoddard and Cutler (180) described two cases of apparent cryptococcal infection and named the organism *Torula histolytica* when they misinterpreted the space around the capsule in tissue as host cell lysis. Stoddard and Cutler may have been misled into believing that they had identified a different organism from that of Busse and Buschke on the basis of Buschke's assertion (later retracted) that their organism formed ascospores (cited in reference 17). In 1916 Stoddard and Cutler published a monograph (180) that reviewed 10 cases of *C. neoformans* infection and described some animal studies. This monograph was widely cited in the early literature and provided a concise reference for the interpretation of clinical and pathological findings. Most important, these investigators clearly distinguished cryptococcosis from other mycotic infections, and their monograph brought some order and consistency to the literature.

The name *T. histolytica* led to the adoption of the name torulosis for cryptococcal infection. The name histolytica implied tissue lysis and led to the belief that *C. neoformans* infections were accompanied by tissue destruction through lytic digestion. In 1924 Weidman and Freeman (197) reported that the yeast cell capsule could be visualized by suspending cultures in a dilution of India ink. (There is a previous report by Swift and Bull (184) of India ink use to visualize the capsule.) This observation became the basis for the India ink stain that continues to be a useful diagnostic test, allowing rapid discrimination of *C. neoformans* from lymphocytes in CSF. In the introduction to their paper describing India ink staining, Weidman and Freedman wrote, "We came upon this method of study during an attempt to demonstrate in cultures the same invisible and unstainable mucinoid coating on *Torula histolytica* that we saw in tissue, and which was only inferred to be there on account of the halo which appeared between the yeast cells and the surrounding tissue" (197). Unfortunately, the idea that *C. neoformans* caused tissue lysis became widely accepted, and even as late as 1936 the finding of large collections of cryptococci in the brain was attributed to tissue lysis (136).

In 1931 Freeman (81) published a comprehensive review of all the clinical, pathological, and microbiological information available at the time. He noted the enormous range of histopathological findings in human cryptococcosis but emphasized that granulomatous inflammation was the usual host response. On the basis of his review of case reports and clinical specimens, Freeman proposed that the respiratory tract was the most likely portal of entry for human infection. The Freeman review is especially valuable for its detailed description of the histopathology of cryptococcal infection in the days before effective antifungal therapy was available.

Todd and Herrmann (186) proposed the name *Debaryomyces hominis* for *C. neoformans* after observing what appeared to be conjugation and spore formation within the zygote. In the 1930s Benham made major contributions to clarifying the classification of cryptococci and distinguishing *C. neoformans* from other pathogenic fungi (15–17). She studied many of the yeast isolates recovered from humans and grouped them as cryptococci (16), concluding that all the human isolates belonged to one species that she called *Cryptococcus hominis*. This name used the genus *Cryptococcus* assigned by the Vuillemin classification and the generic term *Cryptococcus* (Kutzing) em Vuillemin, commonly used by medical mycologists to refer to "fungi imperfecti which appear only as round or oval cells that reproduce by budding" (15). Later, Benham supported the use of the name *C. neoformans* (17). Nevertheless, she was aware of reports of ascospores in culture made by Todd and Herrmann (186) and others and acknowledged that a further name change may be necessary (17). Lodder (125) felt that the name *Cryptococcus* was "to be avoided in yeast taxonomy" and considered it "nomen dubium" and "nomen confusum." In Lodder's view, *Torulopsis* was the valid name for the asporogenous yeasts that included *C. neoformans* (125). In 1950 Skinner (177) revisited the controversial issue of nomenclature and provided support for the use of the name *Cryptococcus*, which Lodder subsequently accepted (cited in reference 18). Benham and Lodder are credited with helping to clarify the taxonomic relationship of *C. neoformans* to other yeasts (cited in reference 58).

Cox and Tolhurst (48) published the second monograph on *C. neoformans* in 1946. This fine study was rich in clinical description and summarized the state of the field in the mid-1940s. The monograph described 12 new clinical cases from Australia in great detail. Cox and Tolhurst attempted to achieve a coherent synthesis between the pathogenesis of human and experimental infection but were able to come to few firm conclusions. For example, analysis of available clinical information did not clearly indicate the route of infection or the mechanism or dissemination in the patients studied. Review of available therapies indicated that none had been proven to be effective. It is interesting that these authors used an autogenous therapeutic vaccine to stimulate the patient immune system and that vaccine use was associated with remission in two patients. The monograph included extensive descriptions of experimental infection in mice, guinea pigs, rabbits, monkeys, cats, and dogs.

By 1950 the term *C. neoformans* had come into use and was quickly adopted in both the mycology and medical literature. The salient features of cryptococcosis had been described in humans, and several investigators had described animal infection. Studies on the polysaccharide antigen had begun (70, 71, 95). The stage was set for an explosion of knowledge about *C. neoformans* ecology, biology, and immunology.

1951–1981: Hope and Many Major Discoveries

The decade of the 1950s witnessed several momentous developments in the field of cryptococcosis. This period is characterized by a major shift from the clinical description of cryptococcosis to laboratory investigation of the pathogen in controlled conditions. Significant advancements were made in the understanding of the biology of *C. neoformans*. The introduction of amphotericin B provided the first truly effective drug against *C. neoformans*.

In the 1950s Emmons (65, 66) recovered *C. neoformans* from soils, especially those contaminated with avian excreta, in particular, pigeon excreta. This observation led many to study the relationship between birds, *C. neoformans*, and cryptococcosis (153) (see chapter 3) and provided the foundation for the present concept that infection is acquired from the environment. Staib discovered that in bird seed agar *C. neoformans* colonies became pigmented, probably as a result of melanin synthesis (reviewed in reference 156). Pigmentation facilitated the identification of *C. neoformans* colonies, which together with the development of selective media (176) stimulated an explosion of ecological studies to define the habitat and ecology of this fungus (see chapter 3).

A rising number of cases of *C. neoformans* led to a greater appreciation of this organism as an opportunistic pathogen. In the early 1950s a firm association was made between lymphoproliferative disorders and increased risk for *C. neoformans* infection (104, 205). The introduction of cortisone therapy for a variety of ailments was associated with an increased risk for cryptococcosis (188). In 1956 Littman and Zimmerman (124) published the third monograph on *C. neoformans*. This beautifully illustrated book summarized the field in the mid-1950s and provided the standard authoritative reference on *C. neoformans* for several decades. The title of their monograph, *Cryptococcosis, Torulosis or European Blastomycosis* (124), is indicative of the confusion in terminology and nomenclature that plagued descriptions of this pathogen even in the mid-20th century.

Pathological studies by Baker and colleagues (9–11) in patients with or without meningeal cryptococcosis provided strong evidence for a pulmonary route of infection. The description of subpleural nodules, combined with the realization that *C. neoformans* was common in the environment and could be aerosolized, led to the current paradigm that cryptococcal infection is acquired through the pulmonary tree. This resulted in increased interest in the pathogenesis of pulmonary infection and suggested that asymptomatic cryptococcal pneumonia was probably a common infection in normal individuals (9–11, 122).

Amphotericin B was introduced in the late 1950s and provided the first effective therapy against cryptococcosis (64, 121, 179). Later, 5-fluorocytosine (flucytosine) was discovered and proved a useful adjunct for the therapy of cryptococcosis. The combination of amphotericin B therapy and flucytosine revolutionized the therapy of cryptococcal meningitis by reducing mortality from 100% to approximately 25 to 30%.

Studies on the antigenic characterization of *C. neoformans*, begun in the 1930s by Benham (16), were rigorously extended to the polysaccharide antigen and led to the description of three serotypes (A, B, and C) by Evans and collaborators (69–72, 121). The availability of serological reagents that could discriminate among strains provided essential tools for epidemiological studies (21, 114, 203) and the separation of *C. neoformans* strains into two varieties. A fourth serotype (D) was described with the use of adsorbed rabbit immune sera (203). Advances on serological studies led to the development of rapid antigen detection methods that greatly facilitated the diagnosis of cryptococcal infections (20, 30, 88). Serological studies of the polysaccharide antigen were paralleled by structural analysis, which led to proposals for the structure of the polysaccharide by Miyazaki and collaborators (138–140), Blandamer and Danishefsky (29), Bhattacharjee and collaborators (25–27), Merrifield and Stephen (134), and Cherniak and collaborators (44) (for reviews, see references 24 and 45).

Evidence accumulated rapidly that the polysaccharide capsule was a major determinant of virulence and that the polysaccharide antigen interfered with host defenses. In the early 1950s Drouhet and collaborators (58, 59) demonstrated that capsular polysaccharide could inhibit the migration of leukocytes and interfere with phagocytosis. In the late 1960s Bulmer and collaborators (36–38, 76) demonstrated that the capsule inhibited phagocytosis. In the 1970s polysaccharide was demonstrated to induce antibody unresponsiveness such that infection seldom elicited strong antibody responses (108, 109, 148).

Considerable progress was made in the taxonomy of this pathogen. Shadomy (172) studied a hypha-forming strain of *C. neoformans* and noted clamp connections, suggesting that this fungus was a basidiomycete. In 1970 Vanbreuseghem and Takashio (189) studied an atypical strain of *C. neoformans* that produced round and bacilliform cells; they proposed the existence of the variety *gattii*. However, the existence of variety *gattii* was not accepted until mating studies revealed two varieties on the basis of mating compatibility (115). In the mid-1970s Kwon-Chung (111–113) discovered two mating types of *C. neoformans* that produced fertile basidiospores under certain conditions. *C. neoformans* stains were subsequently separated into two varieties: var. *neoformans* (serotypes A and D) and var. *gattii* (serotypes B and C). Discovery of sexuality led to the proposal that the genus *Filobasidiella* accommodate this basidiomycete; this discovery ushered in a true revolution in epidemiology, classification, and application of genetic techniques to the study of *C. neoformans* (see chapters 2, 3, and 11).

A variety of sophisticated immunological studies were carried out to determine the effective host response mechanisms. Early studies by Gadebusch (85, 86) suggested the importance of antibody-mediated protection, but by the mid-1960s it was clear that cellular immune mechanisms were essential for containment of infection (22, 87). The importance of T cells in protection against *C. neoformans* was established in the 1970s (42, 91–93, 169). Similarly, studies of the interaction between the complement system and *C. neoformans* revealed the importance of complement-derived products for host defense (53, 54, 86, 130). Cryptococcal products were shown to elicit delayed-type hypersensitivity reactions (7, 94, 146, 165, 166), and considerable effort was devoted to the development of cryptococcal antigen preparations for delayed-type hypersensitivity testing. Skin testing with cryptococcin revealed immune defects in patients with cryptococcosis and the high likelihood of asymptomatic infection in pigeon handlers (151, 194).

By the late 1970s it was apparent that the incidence of cryptococcosis was rising rapidly. More cases were being described in the literature, and cryptococcosis was no longer a rare infection. Most cases were due to complications of immunosuppressive therapies for neoplastic diseases, autoimmune disorders, and organ transplantation. In recognition of this rising incidence of infection, in 1978 Kaufman and Blumer (103) termed cryptococcosis "the awakening giant" among mycologic infections, a prediction that would come true shortly with the recognition of the human immunodeficiency virus (HIV) epidemic.

1981-Present: *C. neoformans* Becomes a Major Pathogen

The medical importance of *C. neoformans* increased dramatically as a consequence of the AIDS epidemic. Before 1981 cryptococcosis was a rare infection, with the total number of cases in the United States being about 500 to 1,000. However, late-stage HIV infection (AIDS) was associated with a marked susceptibility to *C. neoformans* infection such that 5 to 10% of patients with AIDS developed cryptococcosis (see chapter 11). By the early 1990s cryptococcosis was epidemic in some regions. For example, in New York City alone there were over 1,200 cases of cryptococcal meningitis in 1991 (51). For comparison, the number of cases of *Haemophilus influenzae*, meningococcal, and aseptic (viral) meningitis at that time were 54, 30, and 426, respectively (5), making cryptococcal meningitis the most common infection of the central nervous system in New York City. The response to the high prevalence of cryptococcosis was a significant increase in the number of scientific and clinical studies of *C. neoformans*. For example, the number of MEDLINE citations to *C. neoformans* rose from an yearly average of 40 to 50 in the 1970s to approximately 150 by the mid-1990s.

In the setting of advanced HIV infection, cryptococcosis posed a major therapeutic challenge. In the mid-1980s Zuger and collaborators (206, 207) demonstrated that amphotericin B therapy was usually insufficient for curing cryptococcal infections in patients with AIDS. AIDS-associated cryptococcosis has high mortality, and patients who survived initial infection required lifelong antifungal therapy to prevent relapse of infection (46, 67, 206, 207). In the late 1980s the introduction of fluconazole provided an effective oral agent that greatly simplified the management of cryptococcosis (162, 183).

Molecular biology studies of *C. neoformans* began in earnest in the late 1980s and early 1990s. Many genes were identified, cloned, and characterized (see chapter 5). The development of transformation systems permitted the insertion and expression of DNA in cryptococcus, and molecular biology was applied to identify virulence-related genes (60, 187). Genes responsible for capsule production (43), melanin synthesis (202), mating type (144), biosynthetic enzymes (60, 154), ribosomal DNA (74, 75, 158), etc., were identified and characterized. The increased incidence of cryptococcosis made available hundreds of clinical isolates that were analyzed by DNA typing studies, revealing extensive genetic diversity among clinical strains (40, 57, 137, 191, 192). A major development in the understanding of the relationship between varieties, the environment, and pathogenesis came with the identification of eucalyptus trees as an ecological niche for *C. neoformans* var. *gattii* (63).

The association of cryptococcosis with the immunosuppression of HIV infection energized the study of the immunology and pathogenesis of cryptococcal infections (for reviews, see references 39, 107, 118, 119, 135, and 147). Extensive studies of polysaccharide capsule structure done by Cherniak and Sundstrom (45) revealed a multiplicity of structures among clinical strains. The polysaccharide antigen was associated with a variety of deleterious effects on host immune function, and much interest was focused on polysaccharide-host cell interactions (45, 55, 56).

Hence, by the late 1990s, cryptococcosis was a common infection in many parts of the world. The century following its discovery had seen this pathogen rise in importance from a rare infection to a common cause of life-threatening meningitis. The field had grown significantly, and three international congresses devoted solely to *C. neoformans* were held in Jerusalem, Milan, and Paris in 1989, 1993, and 1996, respectively. A fourth international congress on *C. neoformans* is planned for 1999 in London.

DESCRIPTION AND MORPHOLOGICAL CHARACTERISTICS

In clinical settings, *C. neoformans* is almost always found as a yeast. Yeast forms are seen in CSF, tissue sections, sputum, and culture media in the clinical microbiology laboratory. The yeast form is the asexual form of this organism and reproduces by budding. The sexual, or perfect, state is characterized by the presence of basidiospores, is observed only during mating, and has not been described in association with clinical specimens. Hyphal (or pseudohyphal) forms are produced by certain strains and are occasionally observed in tissue sections. There is practically no information available about the fungal form found in the environment. This gap in information is important, because environmental forms of *C. neoformans* are responsible for human and animal infections.

Morphological Characteristics

C. neoformans cells are spherical to oval. The yeast cell has a distinctive feature that sets *C. neoformans* apart from the other medically important yeasts: a polysaccharide capsule.

Capsule size

The size of the polysaccharide capsule in *C. neoformans* varies according to the strain and the culture conditions. Most *C. neoformans* isolates have medium-sized polysaccharide capsules that result in cell diameters ranging from 4 to 10 μm (50, 201). Some poorly encapsulated strains have diameters of only 2 to 5 μm (201). In heavily encapsulated strains recovered from infected tissues, the diameter of the cell can be as large as 80 μm (50). Cruickshank et al. (50) described an isolate that produced a giant capsule in tissue but had a normal appearance in culture. Love et al. (127) described another isolate with a diameter of 40 to 60 μm in vivo and 5 μm in vitro whose capsule production was induced by higher temperature.

In 1985 Bottone et al. (32) described 10 *C. neoformans* isolates, obtained from patients with AIDS, that had smaller than normal capsules and suggested that poorly encapsulated variants may be associated with cryptococcal infections in patients with AIDS. Nine of the 10 isolates produced small, dry, nonmucoid colonies after 48 h of incubation under CO_2 at 37°C (32). India ink examination showed that the diameter of the cells from these isolates ranged from 2.5 to 7.5 μm (median 4.9 μm) (32). In contrast, three isolates from patients without HIV infection had significantly larger capsules (33). The association between poorly encapsulated strains and HIV-associated cryptococcosis was challenged by several investigators, who cited examples of poorly encapsulated strains in patients without HIV infection (14, 19, 83, 131). There are several examples of *C. neoformans* strains with small capsules in patients without HIV infection (8, 77, 145). At this time the question of whether there is a consistent difference in size between isolates from patients with or without HIV infection is unresolved, since there have been no systematic large-scale studies to address this issue.

In summary, capsule size in clinical specimens can range from minute (capsule deficient) to 50 to 100 μm. Most strains produce larger capsules in tissue than in vitro, a phenomenon that may be related to higher CO_2 tension in tissue (90) and/or nutritional factors (120). The capsule is an important virulence factor that facilitates survival of the yeast in infection and can interfere with the host response to infection. All clinical specimens have capsules and produce soluble polysaccharide antigen, but the size of the capsule can be highly variable.

Perfect state

C. neoformans strains are capable of sexual reproduction. Sexual reproduction has been observed only in the laboratory, and its role in pathogenesis is uncertain. The teleomorph form, or perfect state, is characterized by the production of basidiospores, which are the product of sexual reproduction. The two varieties of *C. neoformans* are similar in the yeast form, or anamorph state. However, distinct phenotypic differences are evident in the teleomorph state: the basidiospores of the variety *neoformans* are spherical, oblong elliptical, or cylindrical, with finely roughened walls, whereas those of the variety *gattii* are bacilliform and smooth walled (62, 111–113, 116). For more extensive descriptions of mating, varieties, and the perfect state see chapters 2 and 3.

Light microscopy

Examination of *C. neoformans* by light microscopy reveals yeast-like cells that reproduce by budding and are of variable size. The yeast cells are usually oval or round, have a double contour, and may be highly refractile (117). The *C. neoformans* capsule is usually not visible by light microscopy in aqueous suspensions unless specific steps are taken to visualize it. The most common method for visualizing the capsule is to examine the organism in a dilute suspension of India ink. India ink particles are excluded by the capsule, which then appears as a clear area surrounding the yeast cell. The capsule can also be visualized by addition of capsule specific antibody, which binds to polysaccharide and produces a quellung-like phenomenon (47).

Early studies described several useful stains. Young cultures are gram-positive, but the staining is weaker for older cultures (182). In tissue the color of Gram staining is variable, ranging from intensely purple to shades of pink (117). Yeast cells can also be stained with Loeffler's methylene blue, toluidine blue, and

Wright's stain. Granules may be noted in the cytoplasm, and these can increase in number for cells of older cultures (49). Granules can be stained with Sudan III stain, indicating a lipid composition (157, 186). A chromium mercury modification of the basic India ink technique that permits visualization of internal and external structures by light microscopy has recently been described (204).

Electron microscopy

C. neoformans has been extensively studied by electron microscopy (2, 41, 61, 79, 141, 142, 163, 164, 181) (Fig. 1), which shows that *C. neoformans* cells have morphological features typical of eukaryotic cells. Cryptococcal cells have a single nucleus with a nucleolus (2, 61, 141). The nucleus is surrounded by a double-unit membrane, is irregularly shaped, and has a dimension of approximately 1.5 by 1.2 μm (61, 141). Pores are evident in the nuclear membrane (61, 141). The cytoplasm contains numerous organelles, including mitochondria, vacuoles, endoplasmic reticulum, and ribosomes (2, 61). Storage granules, which presumably contain lipid, are visible by both light and electron microscopy (61, 141). Glycogen granules have been demonstrated in the cytoplasm (2). Figure 2 shows an idealized diagram of *C. neoformans* cellular structures.

The ultrastructure of the cell wall and capsule has received considerable attention in electron microscopic studies. Defining the exact structural relationship of the cell to the capsule has been difficult because of problems in interpreting images after fixation and drying. Electron microscopy reveals that the capsule is composed of a fibrillar network of microfibrils (2, 61). The diameter reported for the microfibrils has ranged from 3 to 4 nm (2, 61) to 10 to 13 nm (164). The variation in diameter presumably reflects different fixative conditions. Early studies identified an "electron transparent" (or low-electron-density) zone separating the cell

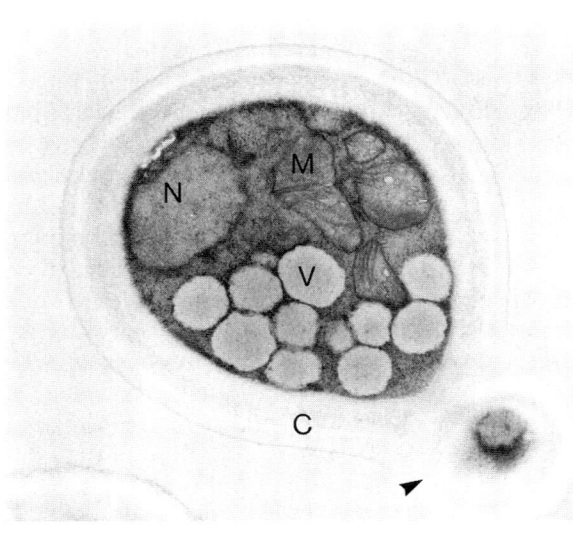

Figure 1 Transmission electron micrograph of *C. neoformans*. The cytoplasm shows typical eukaryotic cellular structures such as a nucleus, mitochondria, and various vesicles. Several vesicles contain low-electron-density material that may be lipid. Note the thick cell wall. N, nucleus; M, mitochondria; V, vacuole; C, cell wall. The arrow indicates a new bud. In this preparation the polysaccharide capsule is not apparent. Magnification, ×12,000. Reprinted with permission from A. Casadevall, Cryptococcosis: the case for immunotherapy. *Cliniguide to Fungal Infections* 4(4):2, 1993, Lawrence Della-Corte Publications, Inc.

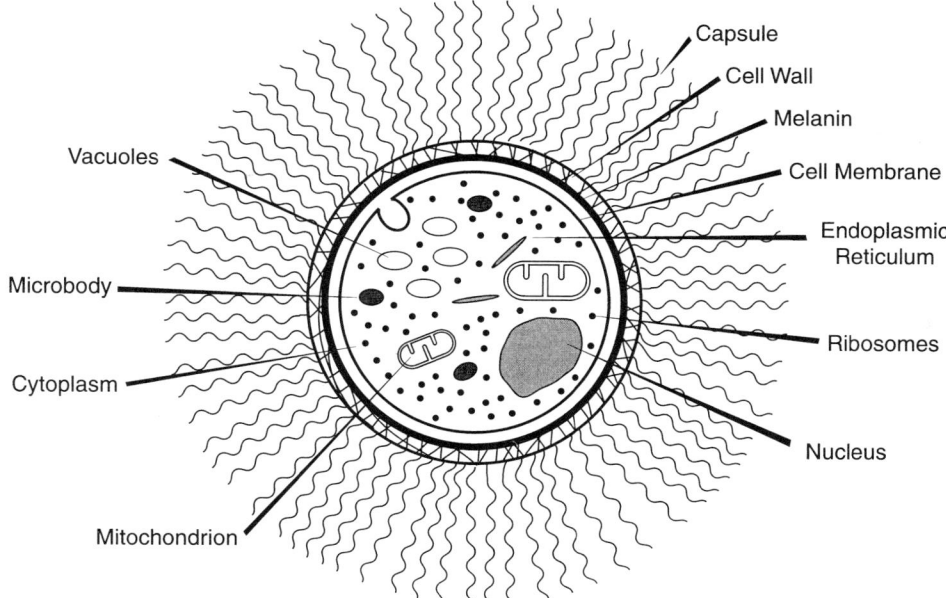

Figure 2 Idealized diagram of a *C. neoformans* cell based on electron micro-scopic studies. In cells grown with the appropriate substrate, melanin is found in the cell wall (see chapter 4).

wall from the microfibril network (2, 41, 61). This low-electron-density zone appeared as a white rim in electron micrographs (61) and was suggested to represent a site for active polysaccharide capsule synthesis (41). However, recent studies of the cell wall-capsule interface by quick freezing and deep-etching methods indicate that the capsule is attached to the cell wall and that the low-electron-density region observed in the early studies may have been an artifact of fixation (163, 164).

The ultrastructure of *C. neoformans* replication by budding and the mitotic apparatus has been described (41, 142). Budding involves the synthesis of new cell wall and the degradation of existing structures (41). The inner bud cell wall results from the synthesis of new material (41). Bud cell walls are consistently thinner than those in the mother cell, and bud scars occur at sites where daughter cells break off from the mother cell (2, 41). During interphase-prophase, a microtubule organizing center (or centriolelike organelle) is observed in close proximity to the nucleus. The microtubule organizing center in *C. neoformans* appears as an electron-dense structure composed of two globular elements and an intervening middle piece. In metaphase-anaphase, a spindle appears in the nucleus in the bud, the nuclear envelope becomes discontinuous, and the nucleus separates into a karyokinetic (bud) nucleus and a nucleolar nucleus (mother cell). In telophase-interphase, the separation of the two sets of chromosomes continues, and a nucleolus appears in the bud nuclei. This sequence of events is similar to that described for other basidiomycetes and is consistent with and supports the classification of *C. neoformans* as a basidiomycete. Analysis of mitochondrial kinetics during replication of *C. neoformans* revealed an increase in the number of mitochondria in cells containing interphase nuclei (141). During budding, mitochondria move through the isthmus of the budding cell. Branched mitochondria, but not giant mitochondria, were observed in *C. neoformans* cells (141).

There are significant differences in the ultrastructure of yeast cells in vitro and in vivo. Takeo et al. (185) carried out a comparative study of cellular structure by freeze-etching microscopy for yeast cells grown in vitro and in vivo (mouse brains). Freeze-etching microscopy of yeast forms grown in vitro revealed structural details that were significantly different from those found after growth in glycerol medium (185). The cell membrane of yeast cells grown in glycerol medium had many invaginations, whereas yeast cells in mouse tissue had larger cell bodies, cell walls, and capsules (185). Furthermore, in vivo forms were remarkable in that they had a marked increase in vacuoles, vesicles, storage organelles, and lipid deposits (185). Fusion of intracellular vesicles with cell membrane was observed, and yeast forms in mouse tissues had large numbers of vesicles, suggesting high secretory activity (185). Takeo and collaborators (185) suggested that capsular synthesis results from vesicles containing capsule precursors fusing with the cell wall. A recent study of *C. neoformans* by the quick-freezing and deep-etching method (164) confirmed and extended these findings. The outer layer of the cell wall of yeast cells grown in vitro and in vivo was found to be a granular structure that differed in thickness; the granules are presumably vesicles containing precursors for capsular synthesis (164).

Culture conditions can have a significant effect on the ultrastructure of *C. neoformans* yeast cells. Growth of *C. neoformans* in medium containing xanthine or urate resulted in the appearance of microbodies in the cytoplasm that stained for catalase and α-hydroxy oxidase (79). Alkaline phosphatase activity was also associated with microbodies. The induction of microbody formation in high xanthine medium may be related to the adaptation of *C. neoformans* for growth in avian excreta (79). Presumably, enzyme-loaded microbodies contribute to purine salvage in an otherwise nutritionally depleted milieu (79).

The ultrastructure of yeast cells in CSF of a patient with cryptococcal meningitis has been described (97). The patient had systemic lupus erythematosus, had received corticosteroids, and was treated with amphotericin B and flucytosine. Electron microscopy of yeast forms in the CSF revealed two forms of yeast cells: a majority with damaged and indistinct organelles and a minority with well-preserved organelles. Some yeast cells in CSF had vesicular membrane structures that have been called lomosomes in other fungi. After the commencement of therapy, yeast cells in the CSF became rapidly damaged with loss of plasma membrane and destruction of organelles. The ultrastructure of *C. neoformans* cells grown in fungal media differed from that observed in CSF by having fewer lomosomes.

Several investigators have studied the morphological features of *C. neoformans* cells in granulomas and after ingestion by phagocytic cells (102, 163, 164, 167). Kalina et al. (102) studied the interaction of *C. neoformans* with rabbit peritoneal macrophages and observed ring forms composed of neutrophils and monocytes around individual yeast cells (102). Phagocytic cells in the ring had pseudopods penetrating into the polysaccharide capsule, and hydrolytic enzymes were released into the yeast cell (102). Sakaguchi et al. (164) observed that occasional *C. neoformans* cells were ingested by phagocytes in mouse brain after intravenous infection. The cytoplasm of ingested cells was destroyed but the capsular fibrillar meshwork remained unchanged, suggesting resistance to digestion by enzymes in phagocytic cells (164). Macrophage pseudopodia were seen protruding into the tight structure of the capsule (163).

Hyphal forms, sporulation, and atypical variants

There are many reports in the literature of *C. neoformans* cells with germ tube-like structure, hyphal forms, pseudohyphae, and sporulation (52, 68, 80, 128, 129, 145, 149, 172, 173, 201). The formation of hyphal structures appears to be dependent on the *C. neoformans* strain and the growth conditions but has also been observed in clinical specimens (80, 201). The observation that some *C. neoformans* cells lack the classical yeast-like shape has been noted since the earliest descriptions of this organism. For example, Sanfelice (168) included diagrams of germ tubes and hyphal forms in his original description of *C. neoformans*.

In 1936 Todd and Herrmann (186) carried out detailed studies of the morphological characteristics of *C. neoformans* with culture age. In fresh young cultures the cells appeared globose to ovoid in shape, with an average diameter of 6 to 8 μm. However, in older cultures there were two morphological features: (i) round with thick walls and a single vacuole that stained with Sudan III, and (ii) osmic acid and ovoid cell forms with thin walls and granular cytoplasm. Occasional cells of both types had "tubes" of various lengths. Fusion of tubes from thick-and thin-walled cells was accompanied by the passing of contents of the thin-walled cells into the thick-walled cells. Fusion was followed by what appeared to be spore formation, and these investigators believed that they had described the perfect state of the organism (186).

Shadomy and collaborators (80, 128, 129, 172, 173) carried out detailed studies of a hypha-forming strain of *C. neoformans* known as the Coward strain. The Coward strain was isolated in 1957 from a human osteolytic lesion. Shadomy states that in 1964 and 1965 her work was "halted by what originally appeared to be a contaminant in the *C. neoformans* used for animal studies in our laboratory" (172). After attempting to separate the hypha-producing contaminant, she realized that the "contaminant" was a hyphal stage of *C. neoformans*. The Coward strain was unusual in that some colonies produced aerial hyphae in agar. The hyphae had clamp connection-like structures and appeared to be surrounded by a capsule when suspended in India ink (172). Mice inoculated with cells from the hyphal form developed meningitis, but only yeast cells were recovered from brain preparations. Examination of tissues of mice infected with hyphal forms at various times after inoculation of the Coward strain revealed various forms of transition from hyphal forms to the encapsulated yeast form (128). Histological analysis of lesions caused by inoculation of the Coward strain in mice revealed a strong inflammatory response to hyphal forms but weak inflammation around yeast cells with large polysaccharide capsules (173). Electron microscopic examination of the Coward strain hyphal forms revealed branching and true septa (129). Cutler and Erke (52) described a morphological variant strain of *C. neoformans* obtained from Shadomy that produced what appeared to be hyphae with clamplike structures by light microscopy. However, electron microscopy did not reveal fusion of the clamp structures with adjacent cells, suggesting that the septal structures were not true clamp connections.

Gordon and Devine (89) observed filamentation and sporulation in three of 10 *C. neoformans* strains grown in low concentrations of sodium deoxycholate (a component of amphotericin B). Filamentous forms were also observed in mouse tissues after experimental infection. After injection into the peritoneum of mice, recovery from peritoneal fluid, and agar plating, spore formation for one strain consisted of a large cell with four to eight internal spores, 2.5 to 6 μm in diameter (89). In some instances the mother cell developed an opening in the cell wall that resulted in spore release. Endosporulation was observed only in strains that had been recovered from the peritoneal cavity of mice. Some cells had tube-like projections that connected to adjacent cells, suggesting conjugation (89).

Erke and Schneidau (68) investigated the relationship of hypha-forming and standard strains of *C. neoformans,* comparing their nucleotide base composition ratio and DNA-DNA duplex formation. This analysis revealed no major differences in DNA composition among hypha-forming and standard strains, consistent with the notion that all strains belonged to one species.

Hyphal forms can be found in clinical specimens. Lurie and Shadomy (128) reviewed the autopsy cases at the Medical College of Virginia (Richmond) and identified two cases in which germ tubes and segmented hyphae were observed in tissue. Freed et al. (80) described a patient with leukemia who had hyphal forms in CSF preparations but only yeast forms in laboratory media. Williamson et al. (201) have described five cases with atypical forms of *C. neoformans* ranging from chains of budding yeasts, pseudohyphae, and germ tube-like structures. Anandi et al. (3) described a case of human *C. neoformans* with unusual yeast forms characterized by a "hand mirror" appearance and the formation of germ tube-like structures with blastoconidia at the tip. The occurrence of atypical non-yeast forms raises the possibility that *C. neoformans* can be mistaken for other yeasts in clinical specimens (201).

Pseudohyphal forms may represent variants that arise spontaneously and provide *C. neoformans* with a survival advantage in some conditions. When seven strains were fed to free-living amoebae, only pseudohyphal forms survived (150). Recovery of pseudohyphal forms from multiple strains suggested that the pseudohyphal forms were always present and were selected by the amoebae. Interestingly, pseudohyphal forms were less virulent than the parent strain but resulted in yeast forms when inoculated in mice (150). The investigators proposed that the ability of the yeast to form occasional pseudohyphae provided a "biological escape hatch" to survive certain environments.

In summary, atypical cells of *C. neoformans* with tubes, hyphae, and/or pseudohyphae can exist in both laboratory culture conditions and in clinical specimens but are nevertheless rare. In the overwhelming majority of clinical specimens, yeast forms predominate. The relationship of these atypical variants to the yeast forms and the significance of filamentous structures in the life cycle are uncertain. The role of atypical cells in pathogenesis, if any, is not understood. The ability of some strains of *C. neoformans* to produce atypical cells is important, because these cells can be confused with other yeasts in clinical situations where the organism is not recovered in culture (201).

Nutritional Requirements

C. neoformans grows well in most, if not all, types of fungal media. In fact, *C. neoformans* is probably an autotroph for most amino acids, sugars, lipids, and vitamins and can be grown in simple chemically defined media (193). In 1950 Schmidt and collaborators (170) demonstrated that *C. neoformans* could be grown successfully in a chemically defined medium containing glucose, ammonium, inorganic salts, and thiamine. Since then, many chemically defined formulations have been described. Many growth media formulations include thiamine because of early reports that it stimulated growth, but addition of this vitamin may be unnecessary for most, if not all, strains (35, 193). Vidotto and coworkers (35, 193) have described excellent growth of *C. neoformans* in simple synthetic media consisting only of glucose, ammonium sulfate, and inorganic salts with a pH of 4.5. Littman (120) investigated the ability of many sugars to support the growth of *C. neoformans* and published an extensive study of the effect of medium on capsule production. His analysis provides a comprehensive study of the carbon and nitrogen sources assimilated by *C. neoformans*. The ability of *C. neoformans* to utilize simple compounds as a source of carbon and nitrogen is consistent with the fact that it is a free-living organism adapted to survival in multiple ecological niches (see chapter 3).

Oxygen is a critical nutrient for the growth of *C. neoformans* (152). Growth and survival of *C. neoformans* are highly dependent on the oxygen supply (152). Noting large differences in growth rates for *C. neoformans* cultures placed in tissue culture plates that differed in well height, Odds et al. (152) concluded that aeration was a major variable in laboratory conditions. Addition of H_2O_2 to blood culture bottles can enhance isolation of *C. neoformans*, possibly by providing an oxygen source (99).

In summary, *C. neoformans* has simple requirements for growth: a carbon source, a nitrogen source, oxygen, and trace elements. Not all amino acids can serve as carbon sources, but *C. neoformans* can synthesize all necessary amino acids from ammonia (120). The simplicity of nutritional requirements facilitates laboratory maintenance of stocks, metabolic studies, and genetic studies involving auxotrophs.

Survival in Environmental Extremes

For an extensive discussion of the effect of various environmental conditions on the survival of *C. neoformans*, see chapter 3. Here, the effect of temperature, pH, sunlight, and moisture will be briefly considered.

Temperature

C. neoformans can grow well at 37°C, but higher temperatures inhibit growth and kill the fungus. Short incubations at 40 or 42°C have no significant effect on culture viability (159), but prolonged incubation is lethal: cultures are sterilized when incubated for 6 days at 40°C or 4 days at 41°C (105). The inability of *C. neoformans* to grow at temperatures above 40°C has been confirmed by others (123).

pH

C. neoformans tolerates a pH range of 4 to 7.5, but growth is significantly inhibited at a higher pH (98, 120). Alkalinization can be used to sterilize contaminated environmental sites (195).

Sunlight

Sunlight can sterilize sites contaminated with *C. neoformans* (100). UV radiation has been shown to kill *C. neoformans* (196).

Moisture

The level of moisture required for viability may depend on the medium used. *C. neoformans* can remain viable in desiccated pigeon excreta for more than 1 year (123). However, lack of moisture has been associated with a progressive decrease in viability of *C. neoformans* (160). (See chapter 3 for more extensive review of this subject.)

Approach to the *C. neoformans* Literature

For reviewing the early literature, three monographs (48, 124, 180) and several comprehensive reviews (49, 81, 117, 121) provide excellent summaries and starting points. In 1971 Al-Doory (1) performed a major service to the field by publishing a list of all papers on cryptococcosis up to that time. Since the mid-1960s, electronic indexing of the *C. neoformans* literature has been available in MEDLINE, and there are more than 2,000 citations listed under the name *C. neoformans*.

A major difficulty with the early literature is confusing nomenclature. Since its first description in 1894, *C. neoformans* has been referred to by multiple names (Table 1). As additional descriptions and observations were made, new terminology was coined, and this has resulted in considerable confusion. (For an excellent historical discussion, see reference 115.) Cryptococcosis has been referred to as European blastomycosis (as opposed to the American blastomycosis caused *B. dermatitidis*), torulopsis, systemic oidiomycosis, systemic blastomycosis (199), and yeast meningitis (186) (see above). The term blastomyces is found frequently in early reports of *C. neoformans* infection in refer to *C. neoformans*. Blastomycosis was a term applied to any infection in which the fungus was a rounded organism that resembled a yeast (15). Unfortunately, the term blastomycosis was used to describe many infections, including those now known as cryptococcosis, blastomycosis, and coccidioidomycosis. In recent years two terms have been consistently used: *C. neoformans,* for the asexual yeast form that is commonly found in clinical and environmental samples, and *Filobasidiella neoformans,* for the perfect state that has been observed only in laboratory conditions (115). The validity and accuracy of the term *Cryptococcus* to describe this fungus remains controversial among taxonomists (125). Nevertheless, the overwhelming majority of medical and scientific articles now use the term *Cryptococcus*. Well-reasoned arguments to preserve the name *Cryptococcus* have been put forward on the basis of historical usage and the need to maintain the continuity and coherence of the scientific literature (78). Considering the confusion that surrounded the terminology for this fungal infection in the first 100 years after its discovery, it would be a tremendous disservice to change the name of *C. neoformans,* even if more precise terminology exists.

REFERENCES

1. **Al-Doory, Y.** 1971. A bibliography of cryptococcosis. *Mycopathol. Mycol. Appl.* **45:**2–60.
2. **Al-Doory, Y.** 1971. The ultrastructure of *Cryptococcus neoformans. Sabouraudia* **9:**113–118.

3. **Anandi, V., P. G. Babu, and T. J. John.** 1991. Infection due to *Cryptococcus neoformans* of unusual morphology in a patient with AIDS. *Mycoses* **34:**377–379.

4. **Anderson, H. W.** 1917. Yeast-like fungi in the human intestinal tract. *J. Infect. Dis.* **21:**380–383.

5. **Anonymous.** 1996. Vital events and reportable diseases and conditions, 1976 to 1995. *City Health Informatd S. Shtokalko.* 1957. Cryptococcus meningitis arrested with Amphotericin B. *Ann. Intern. Med.* **47:**346–351.

7. **Atkinson, A. J., and J. E. Bennett.** 1968. Experience with a new skin test antigen prepared from *Cryptococcus neoformans. Am. Rev. Respir. Dis.* **97:**637–643.

8. **Attal, H. C., S. Grover, M. P. Bansai, B. S. Chaubey, and V. K. Joglekar.** 1983. Capsule deficient *Cryptococcus neoformans:* an unusual clinical presentation. *J. Assoc. Phys. India* **31:**49–51.

9. **Baker, R. D.** 1952. Resectable mycotic lesions and acutely fatal mycoses. *JAMA* **159:**1579–1581.

10. **Baker, R. D.** 1976. The primary pulmonary lymph node complex of cryptococcosis. *Am. J. Clin. Pathol.* **65:**83–92.

11. **Baker, R. D., and R. K. Haugen.** 1955. Tissue changes and tissue diagnosis in cryptococcosis. A study of 26 cases. *Am. J. Clin. Pathol.* **25:**14–24.

12. **Barlow, D. L.** 1923. Primary blasto-mycotic meningitis occurring in a child. *Med. J. Aust.* **2:**302–304.

13. **Beck, E. M., and G. Q. Voyles.** 1946. Systemic infection due to Torula Histolytica (Cryptococcus hominis). *Arch. Intern. Med.* **77:**516–525.

14. **Benbow, E. W., and R. W. Stoddardt.** 1985. Capsule-deficient *Cryptococcus neoformans* in AIDS patients. *Lancet* **i:**988–989.

15. **Benham, R. W.** 1934. The fungis of blastomycosis and coccidioidal granuloma. *Arch. Dermatol. Syph.* **30:**385–400.

16. **Benham, R. W.** 1935. Cryptococci—their identification by morphology and by serology. *J. Infect. Dis.* **57:**255–274.

17. **Benham, R. W.** 1950. Cryptococcosis and blastomycoses. *Ann. N.Y. Acad. Sci.* **50:**1299–1314.

18. **Benham, R. W.** 1956. The genus cryptococcus. *Bacteriol. Rev.* **20:**189–201.

19. **Benn, J. J., P. S. Monro, and G. Duncan.** 1985. Capsule-deficient *Cryptococcus neoformans* in AIDS patients. *Lancet* **i:**989.

20. **Bennett, J. E., and H. F. Hasenclever.** 1965. *Cryptococcus neoformans* polysaccharide: studies of serologic properties and role in infection. *J. Immunol.* **94:**916–920.

21. **Bennett, J. E., K. J. Kwon-Chung, and D. H. Howard.** 1977. Epidemiologic differences among serotypes of *Cryptococcus neoformans. Am. J. Epidemiol.* **105:**582–586.

22. **Bergman, F.** 1961. Pathology of experimental cryptococcosis. A study of course and tissue response in subcutaneously induced infection in mice. *Acta Pathol. Microbiol. Scand. Suppl.* **147:**1–163.

23. **Bettin, M. E.** 1924. Report of a case of Torula infection. *Calif. West. Med.* **22:**98–101.

24. **Bhattacharjee, A. K., J. E. Bennett, and C. P. J. Glaudemans.** 1984. Capsular polysaccharides of *Cryptococcus neoformans. Rev. Infect. Dis.* **6:**619–624.

25. **Bhattacharjee, A. K., K. J. Kwon-Chung, and C. P. J. Glaudemans.** 1979. The structure of the capsular polysaccharide from *Cryptococcus neoformans* serotype D. *Carbohydr. Res.* **73:**183–192.

26. **Bhattacharjee, A. K., K. J. Kwon-Chung, and C. P. J. Glaudemans.** 1980. Structural studies on the major, capsular polysaccharide from *Cryptococcus neoformans* serotype B. *Carbohydr. Res.* **82:**103–111.

27. **Bhattacharjee, A. K., K. J. Kwon-Chung, and C. P. J. Glaudemans.** 1981. Capsular polysaccharides from a parent strain and from a possible, mutant strain of *Cryptococcus neoformans* serotype A. *Carbohydr. Res.* **95:**237–248.

28. **Bindford, C. H.** 1941. Torulosis of the central nervous system. Review of recent literature and report of a case. *Am. J. Clin. Pathol.* **11**:242–251.

29. **Blandamer, A., and I. Danishefsky.** 1966. Investigations on the structure of the capsular polysaccharide from *Cryptococcus neoformans* type B. *Biochim. Biophys. Acta* **117**:305–313.

30. **Bloomfield, N., M. A. Gordon, and D. F. Elmendorf, Jr.** 1963. Detection of *Cryptococcus neoformans* antigen in body fluids by latex particle agglutination. *Proc. Soc. Exp. Biol. Med.* **114**:64–67.

31. **Bodham, C.** 1922. Preliminary report on the causal organism of blastomycotic meningitis in Australia. *Med. J. Aust.* **2**:385.

32. **Bottone, E. J., M. Toma, B. E. Johansson, and G. P. Wormser.** 1985. Capsule-deficient *Cryptococcus neoformans* in AIDS patients. *Lancet* **i**:400.

33. **Bottone, E. J., M. Toma, B. E. Johansson, and G. P. Wormser.** 1986. Poorly encapsulated *Cryptococcus neoformans* from patients with AIDS. I. Preliminary observations. *AIDS Res.* **2**:211–219.

34. **Brewer, G. E., and F. C. Wood.** 1908. Blastomycosis of the spine. *Ann. Surg.* **48**:889–896.

35. **Bruatto, M., V. Vidotto, and A. M. Maima.** 1992. Growth of *Cryptococcus neoformans* in a thiamine-free medium. *Mycopathologia* **119**:129–132.

36. **Bulmer, G. S., and M. D. Sans.** 1967. *Cryptococcus neoformans.* II. Phagocytosis by human leukocytes. *J. Bacteriol.* **94**:1480–1483.

37. **Bulmer, G. S., and M. D. Sans.** 1968. *Cryptococcus neoformans.* III. Inhibition of phagocytosis. *J. Bacteriol.* **95**:5–8.

38. **Bulmer, G. S., M. D. Sans, and C. M. Gunn.** 1967. *Cryptococcus neoformans.* I. Nonencapsulated mutants. *J. Bacteriol.* **94**:1475–1479.

39. **Casadevall, A.** 1995. Antibody immunity and invasive fungal infections. *Infect. Immun.* **63**:4211–4218.

40. **Casadevall, A., L. Freundlich, L. Marsh, and M. D. Scharff.** 1992. Extensive allelic variation in *Cryptococcus neoformans. J. Clin. Microbiol.* **30**:1080–1084.

41. **Cassone, A., N. Simonetti, and V. Strippoli.** 1974. Wall structure and bud formation on *Cryptococcus neoformans. Arch. Microbiol.* **95**:205–212.

42. **Cauley, L. K., and J. W. Murphy.** 1979. Response of congenitally athymic (nude) and phenotypically normal mice to *Cryptococcus neoformans* infection. *Infect. Immun.* **23**:644–651.

43. **Chang, Y. C., and K. J. Kwon-Chung.** 1994. Complementation of a capsule-deficient mutation of *Cryptococcus neoformans* restores its virulence. *Mol. Cell. Biol.* **14**:4912–4919.

44. **Cherniak, R., E. Reiss, M. E. Slodki, R. D. Plattner, and S. O. Blumer.** 1980. Structure and antigenic activity of the capsular polysaccharide of *Cryptococcus neoformans* serotype A. *Mol. Immunol.* **17**:1025–1032.

45. **Cherniak, R., and J. B. Sundstrom.** 1994. Polysaccharide antigens of the capsule of *Cryptococcus neoformans. Infect. Immun.* **62**:1507–1512.

46. **Chuck, S. L., and M. A. Sande.** 1989. Infections with *Cryptococcus neoformans* in the acquired immunodeficiency syndrome. *N. Engl. J. Med.* **321**:794–799.

47. **Cleare, W., S. Mukherjee, E. D. Spitzer, and A. Casadevall.** 1994. Prevalence in *Cryptococcus neoformans* strains of a polysaccharide epitope which can elicit protective antibodies. *Clin. Diagn. Lab. Immunol.* **1**:737–740.

48. **Cox, L. B., and J. C. Tolhurst.** 1946. *Human Torulosis.* Melbourne University Press, Melbourne, Australia.

49. **Crone, J. T., A. F. DeGroat, and J. G. Wahlin.** 1937. Torula infection. *Am. J. Pathol.* **13**:863–879.

50. **Cruickshank, J. G., R. Cavill, and M. Jelbert.** 1973. *Cryptococcus neoformans* of unusual morphology. *Appl. Microbiol.* **25**:309–312.

51. **Currie, B. P., and A. Casadevall.** 1994. Estimation of the prevalence of cryptococcal infection among HIV infected individuals in New York City. *Clin. Infect. Dis.* **19:**1029–1033.

52. **Cutler, J. E., and K. H. Erke.** 1971. Ultrastructural characteristics of *Coccidioides immitis,* a morphological variant of *Cryptococcus neoformans* and *Podossypha ravenelii. J. Bacteriol.* **105:**438–444.

53. **Diamond, R. D., J. E. May, M. Kane, M. M. Frank, and J. E. Bennett.** 1973. The role of late complement component and the alternate complement pathway in experimental cryptococcosis (37580). *Proc. Soc. Exp. Biol. Med.* **144:**312–315.

54. **Diamond, R. D., J. E. May, M. C. Kane, M. M. Frank, and J. E. Bennett.** 1974. The role of the classical and alternate complement pathways in host defenses against *Cryptococcus neoformans* infection. *J. Immunol.* **112:**2260–2270.

55. **Dong, Z. M., and J. W. Murphy.** 1996. Cryptococcal polysaccharides induce L-selectin shedding and tumor necrosis receptor loss from the surface of human neutrophils. *J. Clin. Invest.* **97:**689–698.

56. **Dong, Z. M., and J. W. Murphy.** 1997. Cryptococcal polysaccharide bind to CD18 on human neutrophils. *Infect. Immun.* **65:**557–563.

57. **Dromer, F., A. Varma, O. Ronin, S. Mathoulin, and B. Dupont.** 1994. Molecular typing of *Cryptococcus neoformans* serotype D clinical isolates. *J. Clin. Microbiol.* **32:**2364–2371.

58. **Drouhet, E.** 1997. Milestones in the history of *Cryptococcus* and cryptococcosis. *J. Mycol. Med.* **7:**10–27.

59. **Drouhet, E., and G. Segretain.** 1951. Inhibition de la migration leucocytaire *in vitro* par un polyoside capsulaire de *Torulopsis (Cryptococcus) neoformans. Ann. Inst. Pasteur* **81:**674.

60. **Edman, J. C., and K. J. Kwon-Chung.** 1990. Isolation of the *URA5* gene from *Cryptococcus neoformans* var *neoformans* and its use as a selective marker for transformation. *Mol. Cell. Biol.* **10:**4583–4544.

61. **Edwards, M. R., M. A. Gordon, E. W. Lapa, and W. C. Ghiorse.** 1967. Micromorphology of *Cryptococcus neoformans. J. Bacteriol.* **94:**766–777.

62. **Ellis, D., and T. J. Pfeiffer.** 1992. The ecology of *Cryptococcus neoformans. Eur. J. Epidemiol.* **8:**321–325.

63. **Ellis, D. H., and T. J. Pfeiffer.** 1990. Natural habitat of *Cryptococcus neoformans* var. *gattii. J. Clin. Microbiol.* **28:**1642–1644.

64. **Emanuael, B., E. Ching, A. D. Leiberman, and M. Goldin.** 1961. Cryptococcus meningitis in a child successfully treated with amphotericin B, with a review of the pediatric literature. *J. Pediatr.* **59:**577–591.

65. **Emmons, C. W.** 1951. Isolation of *Cryptococcus neoformans* from soil. *J. Bacteriol.* **62:**685–690.

66. **Emmons, C. W.** 1960. Prevalence of *Cryptococcus neoformans* in pigeon habitats. *Public Health Rep.* **75:**362–364.

67. **Eng, R. H. K., E. Bishburg, S. M. Smith, and R. Kapila.** 1986. Cryptococcal infections in patients with acquired immune deficiency syndrome. *Am. J. Med.* **81:**19–23.

68. **Erke, K. H. and J. D. Schneidau.** 1973. Relationship of some *Cryptococcus neoformans* hypha-forming strains to standard strains and to other species of yeasts as determined by deoxyribonucleic acid base ratios and homologies. *Infect. Immun.* **7:**941–948.

69. **Evans, E. D., and J. F. Kessel.** 1951. The antigenic composition of *Cryptococcus neoformans. J. Immunol.* **67:**109–114.

70. **Evans, E. E.** 1949. An immunologic comparison of twelve strains of *Cryptococcus neoformans (Torula histolytica). Proc. Soc. Exp. Biol. Med.* **71:**644–646.

71. **Evans, E. E.** 1950. The antigenic composition of *Cryptococcus neoformans.* I. A serologic

classification by means of the capsular and agglutination reactions. *J. Immunol.* **64**:423–430.

72. **Evans, E. E.** 1960. Capsular reactions of *Cryptococcus neoformans. Ann. N.Y. Acad. Sci.* **89**:184–192.

73. **Evans, N.** 1922. Torula infection. Report of two cases. *Calif. State J. Med.* **20**:383–385.

74. **Fan, M., L.-L. Chen, M. A. Ragan, R. R. Gutell, J. R. Warner, B. P. Currie, and A. Casadevall.** 1995. The 5S rRNA and the rDNA intergenic spacer of the two varieties of *Cryptococcus neoformans*: sequence, structure, and phylogenetic implications. *J. Med. Vet. Mycol.* **33**:215–221.

75. **Fan, M., B. P. Currie, R. R. Gutell, M. A. Ragan, and A. Casadevall.** 1994. The 16S-like, 5.8S, and 23S-like rRNAs of the two varieties of *Cryptococcus neoformans*: sequence, secondary structure, phylogenetic analysis, and restriction fragment polymorphisms. *J. Med. Vet. Mycol.* **32**:163–180.

76. **Farhi, F., G. S. Bulmer, and J. R. Tacker.** 1970. *Cryptococcus neoformans.* IV. The not-so-encapsulated yeast. *Infect. Immun.* **1**:526–531.

77. **Farmer, S. G., and R. A. Komorowski.** 1973. Histologic response to capsule-deficient *Cryptococcus neoformans. Arch. Pathol.* **96**:383–387.

78. **Fell, J. W., C. P. Kurtzman, and K. J. Kwon-Chung.** 1989. Proposal to conserve *Cryptococcus* (fungi). *Taxon* **38**:151–152.

79. **Fiskin, A. M., M. C. Zalles, and R. G. Garrison.** 1990. Electron cytochemical studies of *Cryptococcus neoformans* grown on uric acid and related sources of nitrogen. *J. Med. Vet. Mycol.* **38**:197–207.

80. **Freed, E. R., R. J. Duma, H. J. Shadomy, and J. P. Utz.** 1971. Meningoencephalitis due to hyphae-forming *Cryptococcus neoformans. Am. J. Clin. Pathol.* **51**:30–53.

81. **Freeman, W.** 1931. Torula infection of the central nervous system. *J. Psychol. Neurol.* **43**:236–345.

82. **Freeman, W., and F. D. Weidman.** 1923. Cystic blastomycosis of the cerebral gray matter caused by *Torula histolytica* Stoddard and Cutler. *Arch. Neurol. Psychiatr.* **9**:589–603.

83. **Fromtling, R. A., and G. S. Bulmer.** 1985. Capsule-deficient *Cryptococcus neoformans* in AIDS patients. *Lancet* **i**:988.

84. **Frothingham, L.** 1902. A tumor-like lesion in the lung of a horse caused by a blastomyces (torula*). J. Med. Res.* **8**:31–43.

85. **Gadebusch, H. H.** 1958. Passive immunization against *Cryptococcus neoformans. Proc. Soc. Exp. Biol. Med.* **98**:611–614.

86. **Gadebusch, H. H.** 1961. Natural host resistance to infection with *Cryptococcus neoformans.* The effect of the properdin system on the experimental disease. *J. Infect. Dis.* **109**:147–153.

87. **Gadebusch, H. H.** 1972. Mechanisms of native and acquired resistance to infection with *Cryptococcus neoformans. Crit. Rev. Microbiol.* **1**:311–320.

88. **Gordon, M.** 1981. Cryptococcal antigen test. *JAMA* **246**:1403.

89. **Gordon, M. A., and J. Devine.** 1970. Filamentation and endogenous sporulation in *Cryptococcus neoformans. Sabouraudia* **8**:227–234.

90. **Granger, D. L., J. R. Perfect, and D. T. Durack.** 1985. Virulence of *Cryptococcus neoformans.* Regulation of capsule synthesis by carbon dioxide. *J. Clin. Invest.* **76**:508–516.

91. **Graybill, J. R., and R. H. Alford.** 1974. Cell-mediated immunity in cryptococcosis. *Cell. Immunol.* **14**:12–21.

92. **Graybill, J. R., and D. J. Drutz.** 1978. Host defense in cryptococcosis. II. Cryptococcosis in the nude mouse. *Cell. Immunol.* **40**:263–274.

93. **Graybill, J. R., L. Mitchell, and D. J. Drutz.** 1979. Host defense in cryptococcosis. III. Protection of nude mice by thymus transplantation. *J. Infect. Dis.* **140**:546–552.

94. **Hay, R. J., and E. Reiss.** 1978. Delayed-type hypersensitivity responses in infected mice elicited by cytoplasmic fractions of *Cryptococcus neoformans*. *Infect. Immun.* **22:**72–79.

95. **Hehre, E. J., A. S. Carlson, and D. M. Hamilton.** 1949. Crystalline amylose from cultures of a pathogenic yeast (Torula histolytica). *J. Biol. Chem.* **177:**289–293.

96. **Hirsch, E. F., and G. H. Coleman.** 1929. Acute miliary torulosis of the lungs. *JAMA* **92:**437–438.

97. **Hiruma, M., and S. Kagawa.** 1985. Ultrastructure of *Cryptococcus neoformans* in the cerebrospinal fluid of a patient with cryptococcal meningitis. *Mycopathologia* **89:**5–12.

98. **Howard, D. H.** 1961. Some factors which affect the initiation of growth of *Cryptococcus neoformans*. *J. Bacteriol.* **82:**430–435.

99. **Huahua, T., J. Rudy, and C. M. Kunin.** 1991. Effect of hydrogen peroxide on growth of *Candida, Cryptococcus* and other yeasts in simulated blood culture bottles. *J. Clin. Microbiol.* **29:**328–332.

100. **Ishaq, C. M, G. S. Bulmer, and F. G. Felton.** 1968. An evaluation of various environmental factors affecting the propagation of *Cryptococcus neoformans*. *Mycopathol. Mycol. Appl.* **35:**81–90.

101. **Johns, F. M., and C. L. Attaway.** 1933. Torula meningitis. Report of a case and review of the literature. *Am. J. Clin. Pathol.* **3:**459–465.

102. **Kalina, M., Y. Kletter, and M. Aronson.** 1974. The interaction of phagocytes and the large-sized parasite *Cryptococcus neoformans*: cytochemical and ultrastructural study. *Cell Tissue Res.* **152:**165–174.

103. **Kaufman, L., and S. Blumer.** 1977. Cryptococcosis: the awakening giant, abstr. 176–182. *Proceedings of the Fourth International Conference on the Mycoses.* PAHO Scientific Publication No. 356.

104. **Keye, J. D., and W. E. Magee.** 1956. Fungal diseases in a general hospital. *Am. J. Clin. Pathol.* **26:**1235–1253.

105. **Kligman, A. M., and F. D. Weidman.** 1949. Experimental studies on treatment of human torulosis. *Arch. Dermatol. Syph.* **60:**726–741.

106. **Knoke, M., and G. Schwesinger.** 1994. One hundred years ago: the history of cryptococcosis in Greifswald. Medical mycology in the nineteenth century. *Mycoses* **37:**229–233.

107. **Kozel, T. R.** 1996. Activation of the complement system by pathogenic fungi. *Clin. Microbiol. Rev.* **9:**34–46.

108. **Kozel, T. R. and J. Cazin, Jr.** 1972. Immune response to *Cryptococcus neoformans* soluble polysaccharide. *Infect. Immun.* **5:**35–41.

109. **Kozel, T. R., W. F. Gulley, and J. Cazin, Jr.** 1977. Immune response to *Cryptococcus neoformans* soluble polysaccharide: immunological unresponsiveness. *Infect. Immun.* **18:**701–707.

110a. **Kozel, T. R. and T. G. McGaw.** 1979. Opsonization of *Cryptococcus neoformans* by human immunoglobulin G: role of immunoglobulin G in phagocytosis by macrophages. *Infect. Immun.* **25:**255–261.

111. **Kwon-Chung, K. J.** 1975. A new genus, *Filobasidiella*, the perfect state of *Cryptococcus neoformans*. *Mycologia* **67:**1197–1200.

112. **Kwon-Chung, K. J.** 1976. A new species of *Filobasidiella*, the sexual state of *Cryptococcus neoformans* B and C serotypes. *Mycologia* **68:**942–946.

113. **Kwon-Chung, K. J.** 1976. Morphogenesis of *Filobasidiella neoformans*, the sexual state of *Cryptococcus neoformans*. *Mycologia* **67:**821–833.

114. **Kwon-Chung, K. J., and J. E. Bennett.** 1984. Epidemiologic differences between the two varieties of *Cryptococcus neoformans*. *Am. J. Epidemiol.* **120:**123–130.

115. **Kwon-Chung, K. J., and J. E. Bennett.** 1992. Cryptococcosis, p. 397–446. *In* K. J. Kwon-Chung and J. E. Bennett. (ed.), *Medical Mycology*. Lea & Febiger, Philadelphia.

116. **Kwon-Chung, K. J., J. E. Bennett, and J. C. Rhodes.** 1982. Taxonomic studies on *Filobasidiella* species and their anamorphs. *Antonie van Leeuwenhoek* **48:**25–38.
117. **Levin, E. A.** 1937. Torula infection of the central nervous system. *Arch. Intern. Med.* **59:**667–684.
118. **Levitz, S. M.** 1992. Overview of host defenses in fungal infections. *Clin. Infect. Dis.* **14**(Suppl. 1)**:**S37–S42.
119. **Levitz, S. M., H. L. Mathews, and J. W. Murphy.** 1995. Direct antimicrobial activity by T cells. *Immunol. Today* **16:**387–391.
120. **Littman, M. L.** 1958. Capsule synthesis by *Cryptococcus neoformans. Trans. N.Y. Acad. Sci.* **20:**623–648.
121. **Littman, M. L.** 1959. Cryptococcosis (torulosis). *Am. J. Med.* **27:**976–998.
122. **Littman, M. L.** 1968. Cryptococcosis: current status. *Am. J. Med.* **45:**922–932.
123. **Littman, M. L., and R. Borok.** 1968. Relation of the pigeon to cryptococcosis: natural carrier state, heat resistance and survival of cryptococcus neoformans. *Mycopathol. Mycol. Appl.* **35:**329–345.
124. **Littman, M. L., and L. E. Zimmerman.** 1956. *Cryptococcosis, Torulosis or European Blastomycosis.* Grune & Stratton, New York.
125. **Lodder, J.** 1938. Torulopsis or Cryptococcus? *Mycopathologia* **1:**62–67.
126. **Longmire, W. P., and T. C. Goodwin.** 1939. Generalized torula infection. Case report and review with observations on pathogenesis. *Bull. Johns Hopkins Hosp.* **64:**22–43.
127. **Love, G. L., G. D. Boyd, and D. L. Greer.** 1985. Large *Cryptococcus neoformans* isolated from brain abscess. *J. Clin. Microbiol.* **22:**1068–1070.
128. **Lurie, H. I., and H. J. Shadomy.** 1971. Morphological variations of a hypha-forming strain of *Cryptococcus neoformans* (Coward strain) in tissues of mice. *Sabouraudia* **9:**10–14.
129. **Lurie, H. I., H. J. Shadomy, and W. J. S. Still.** 1971. An electron microscopic study of *Cryptococcus neoformans* (Coward strain). *Sabouraudia* **9:**15–16.
130. **Macher, A. M., J. E. Bennett, J. E. Gadek, and M. M. Frank.** 1978. Complement depletion in cryptococcal sepsis. *J. Immunol.* **120:**1686–1690.
131. **Mackensie, D. W. R., and R. J. Hay.** 1985. Capsule-deficient *Cryptococcus neoformans* in AIDS patients. *Lancet* **i:**642.
132. **Magruder, R. G.** 1939. A report of three cases of Torula infection of the central nervous system. *J. Lab. Clin. Med.* **24:**495–499.
133. **McKendree, C. A., and L. H. Cornwall.** 1926. Meningo-encephalitis due to torula. *Arch. Neurol. Psychiatr.* **16:**167–181.
134. **Merrifield, E. H., and A. M. Stephen.** 1980. Structural investigations of two capsular polysaccharides from *Cryptococcus neoformans. Carbohydr. Res.* **86:**69–76.
135. **Miller, G. P. G.** 1986. The immunology of cryptococcal disease. *Semin. Respir. Infect.* **1:**45–52.
136. **Mitchell, L. A.** 1936. Torulosis. *JAMA* **106:**450–452.
137. **Mitchell, T. G., E. Z. Freedman, T. J. White, and J. W. Taylor.** 1994. Unique oligonucleotide primers in PCR for identification of *Cryptococcus neoformans. J. Clin. Microbiol.* **32:**253–255.
138. **Miyazaki, T.** 1961. Studies on fungal polysaccharides. I. On the isolation and chemical properties of capsular polysaccharide from *Cryptococcus neoformans. Chem. Pharm. Bull.* **9:**715–718.
139. **Miyazaki, T.** 1961. Studies on fungal polysaccharides. II. On the componental sugars and partial hydrolysis of the capsular polysaccharides from *Cryptococcus neoformans. Chem. Pharm. Bull.* **9:**826–829.
140. **Miyazaki, T.** 1961. Studies on fungal polysaccharides. III. Chemical structure of the capsular polysaccharide from *Cryptococcus neoformans. Chem. Pharm. Bull.* **9:**829–833.
141. **Mochizuki, T., S. Tanaka, Y. Saito, and S. Watanabe.** 1988. Mitochondrial kinetics during mitosis in *Cryptococcus neoformans*—an ultrastructure study. *Mycoses* **32:**7–13.

142. **Mochizuki, T., S. Tanaka, and S. Watanabe.** 1987. Ultrastructure of the mitotic apparatus in *Cryptococcus neoformans. J. Med. Vet. Mycol.* **25**:223–233.

143. **Mook, W. H., and M. Moore.** 1936. Cutaneous torulosis. *Arch.Dermatol. Syph.* **33**:951–962.

144. **Moore, T. D. E., and J. C. Edman.** 1993. The alpha-mating type locus of *Cryptococcus neoformans* contains a peptide pheromone gene. *Mol. Cell. Biol.* **13**:1962–1970.

145. **Moser, S. A., L. Friedman, and A. R. Varraux.** 1978. Atypical isolate of *Cryptococcus neoformans* cultured from sputum of a patient with pulmonary cancer and blastomycosis. *J. Clin. Microbiol.* **7**:316–318.

146. **Muchmore, H. G., F. G. Felton, S. B. Salvin, and E. R. Rhoades.** 1968. Delayed hypersensitivity to cryptococcin in man. *Sabouraudia* **6**:285–288.

147. **Murphy, J. A.** 1991. Mechanisms of natural resistance to human pathogenic fungi. *Annu. Rev. Microbiol.* **45**:509–538.

148. **Murphy, J. W., and G. C. Cozad.** 1972. Immunological unresponsiveness induced by cryptococcal polysaccharide assayed by the hemolytic plaque technique. *Infect. Immun.* **5**:896–901.

149. **Neilson, J. B., R. A. Fromtling, and G. S. Bulmer.** 1981. Pseudohyphal forms of *Cryptococcus neoformans*: decreased survival in vivo. *Mycopathologia* **73**:57–59.

150. **Neilson, J. B., M. H. Ivey, and G. S. Bulmer.** 1978. *Cryptococcus neoformans*: pseudohyphal forms surviving culture with *Acanthamoeba polyphaga. Infect. Immun.* **20**:262–266.

151. **Newberry, W. M., J. Walter, J. W. Chandler, and F. E. Tosh.** 1967. Epidemiologic study of *Cryptococcus neoformans. Ann. Intern. Med.* **67**:727–732.

152. **Odds, F. C., T. De Backer, G. Dams, L. Vranckx, and F. Woestenborghs.** 1995. Oxygen as a limiting nutrient for growth of *Cryptococcus neoformans. J. Clin. Microbiol.* **33**:995–997.

153. **Partridge, B. M., and H. I. Winner.** 1965. *Cryptococcus neoformans* in bird droppings in London. *Lancet* **i**:1060–1061.

154. **Perfect, J. R., D. L. Toffaletti, and T. H. Rude.** 1993. The gene encoding phosphoribosylaminoimidazole carboxylase (ADE2) is essential for growth of *Cryptococcus neoformans* in cerebrospinal fluid. *Infect. Immun.* **61**:4446–4451.

155. **Pierson, P. H.** 1917. Torula in man. Report of a case with necropsy findings. *JAMA* **69**:2179–2181.

156. **Polacheck, I.** 1991. The discovery of melanin production in *Cryptococcus neoformans* and its impact on diagnosis and the study of virulence. *Zentralbl. Bakteriol.* **276**:120–123.

157. **Reeves, D. L., E. M. Butt, and R. W. Hammack.** 1941. Torula infection of the lungs and central nervous system. *Arch. Intern. Med.* **68**:57–79.

158. **Restrepo, B. I., and A. G. Barbour.** 1989. Cloning of the 18S and 25S rDNAs from the pathogenic fungus *Cryptococcus neoformans. J. Bacteriol.* **171**:5596–5600.

159. **Rosas, A. L., and A. Casadevall.** 1997. Melanization affects susceptibility of *Cryptococcus neoformans* to heat and cold. *FEMS Microbiol. Lett.* **153**:265–272.

160. **Ruiz, A., J. B. Neilson, and G. S. Bulmer.** 1982. A one year study on the viability of *Cryptococcus neoformans* in nature. *Mycopathologia* **77**:117–122.

161. **Rusk, G. Y., and F. J. Farnell.** 1912. Systemic oidiomycosis. A study of two cases developing terminal oidiomycetic meningitis. *Univ. Calif. Publ. Pathol.* **2**:47–58.

162. **Saag, M. S., and W. E. Dismukes.** 1988. Azole antifungal agents: emphasis on new triazoles. *Antimicrob. Agents Chemother.* **32**:1–8.

163. **Sakaguchi, N.** 1993. Ultrastructural study of hepatic granulomas induced by *Cryptococcus neoformans* by quick-freezing and deep-etching method. *Virchows Arch. (Cell. Pathol.)* **64**:57–66.

164. **Sakaguchi, N., T. Baba, M. Fukuzawa, and S. Ohno.** 1993. Ultrastructural study of

Cryptococcus neoformans by quick-freezing and deep-etching method. *Mycopathologia* **121**:133–141.

165. **Salvin, S. B.** 1959. Current concepts of diagnostic serology and skin hypersensitivity in the mycoses. *Am. J. Med.* **27**:97–114.

166. **Salvin, S. B., and R. F. Smith.** 1961. An antigen for detection of hypersensitivity to *Cryptococcus neoformans* (26977). *Proc. Soc. Exp. Biol. Med.* **108**:498–501.

167. **Sanfelice, F.** 1894. Contributo alla morfologia e biologia dei blastomiceti che si sviluppano nei succhi di alcuni frutti. *Ann. Igien.* **4**:463–495.

168. **Sanfelice, F.** 1895. Sull'azione patogena dei blastomiceti. *Ann. Inst. Igien. Univ. Roma* **5**:239–262.

169. **Schimpff, S. C., and J. E. Bennett.** 1975. Abnormalities in cell-mediated immunity in patients with *Cryptococcus neoformans* infection. *J. Allergy Clin. Immunol.* **55**:430–441.

170. **Schmidt, E. G., J. A. Alvarez-De Choudens, N. McElvain, F. J. Beardsley, and S. A. A. Tawab.** 1950. A microbiology study of *Cryptococcus neoformans*. *Arch. Biochem.* **26**:15–24.

171. **Semerak, C. B.** 1928. Torular leptomeningitis. *Arch. Pathol.* **6**:1142–1145.

172. **Shadomy, H. J.** 1970. Clamp connections in two strains of *Cryptococcus neoformans*, p. 67–72. *In* D. G. Ahearn (ed.), *Recent Trends in Yeast Research*. Georgia State University, Atlanta.

173. **Shadomy, H. J., and H. I. Lurie.** 1971. Histopathological observations in experimental cryptococcosis caused by hypha-producing strain of *Cryptococcus neoformans* in mice. *Sabouraudia* **9**:6–9.

174. **Shapiro, L. L., and J. B. Neal.** 1925. Torula meningitis. *Arch. Neurol. Psychiatr.* **13**:174–190.

175. **Sheppe, W. M.** 1924. Torula infection in man. *Am. J. Med. Sci.* **167**:91–108.

176. **Shields, A., and L. Ajello.** 1966. Medium for selective isolation of *Cryptococcus neoformans*. *Science* **151**:208–209.

177. **Skinner, C. E.** 1950. Generic name for imperfect yeasts, cryptococcus or torulopsis? *Am. Midland Nat.* **43**:242–250.

178. **Smith, F. B., and J. S. Crawford.** 1930. Fatal granulomatosis of the central nervous system due to a yeast (torula). *J. Pathol. Bacteriol.* **33**:291–296.

179. **Spickard, A., W. T. Butler, V. Andriole, and J. P. Utz.** 1963. The improved prognosis of cryptococcal meningitis with amphotericin B therapy. *Ann. Intern. Med.* **58**:66–83.

180. **Stoddard, J. L., and E. C. Cutler.** 1916. *Torula Infection in Man*. Waverly Press, The Williams and Wilkins Co., Baltimore.

181. **Stoetzner, H., and C. Kemmer.** 1971. The morphology of *Cryptococcus neoformans* in human cryptococcosis—a light-, phase contrast, and electron microscopic study. *Mycopathol. Mycol. Appl.* **45**:327–335.

182. **Stone, W. J., and B. F. Sturdivant.** 1929. Meningo-encephalitis due to Torula histolytica. *Arch. Intern. Med.* **44**:560–575.

183. **Sugar, A. M., and C. Saunders.** 1988. Oral fluconazole as suppressive therapy of disseminated cryptococcosis in patients with acquired immunodeficiency syndrome. *Am. J. Med.* **85**:481–489.

184. **Swift, H., and L. B. Bull.** 1917. Notes on a case of systemic blastomycosis-blastomycotic cerebrospinal meningitis. *Med. J. Aust.* **13**:265–267.

185. **Takeo, K., I. Uesaka, K. Uehira, and M. Nishiura.** 1973. Fine structure of *Cryptococcus neoformans* grown in vivo as observed by freeze-etching. *J. Bacteriol.* **113**:1449–1454.

186. **Todd, R. L., and W. W. Herrmann.** 1936. The life cycle of the organism causing yeast meningitis. *J. Bacteriol.* **32**:89–103.

187. **Toffaletti, D. L., T. H. Rude, S. A. Johnston, D. T. Durack, and J. R. Perfect.** 1993. Gene transfer in *Cryptococcus neoformans* by means of biolistic delivery of DNA. *J. Bacteriol.* **175**:1405–1411.

188. **Torack, R. M.** 1957. Fungus infections associated with antibiotic and steroid therapy. *Am. J. Med.* **22:**872–882.

189. **Vanbreuseghem, R., and M. Takashio.** 1970. An atypical strain of *Cryptococcus neoformans* (San Felice) Vuillemin 1894. *Ann. Soc. Belg. Med. Trop.* **50:**695–702.

190. **Van Cutsem, J., J. Fransen, F. Van Gerven, and P. A. J. Janssen.** 1986. Experimental cryptococcosis: dissemination of *Cryptococcus neoformans* and dermtropism in guinea pigs. *Mykosen* **29:**561–575.

191. **Varma, A., and K. J. Kwon-Chung.** 1989. Restriction fragment polymorphism in mitochondrial DNA of *Cryptococcus neoformans*. *J. Gen. Microbiol.* **135:**3353–3362.

192. **Varma, A., D. Swinne, F. Staib, J. E. Bennett, and K. J. Kwon-Chung.** 1995. Diversity of DNA fingerprints in *Cryptococcus neoformans*. *J. Clin. Microbiol.* **33:**1807–1814.

193. **Vidotto, V., S. Aoki, and G. Campanini.** 1996. A vitamin-free minimal synthetic medium for *Cryptococcus neoformans*. *Mycopathologia* **133:**139–142.

194. **Walter, J. E., and R. W. Atchison.** 1966. Epidemiological and immunological studies of *Cryptococcus neoformans*. *J. Bacteriol.* **92:**82–87.

195. **Walter, J. E., and E. G. Coffee.** 1968. Control of *Cryptococcus neoformans* in pigeon coops by alkalization. *Am. J. Epidemiol.* **87:**173–178.

196. **Wang, Y., and A. Casadevall.** 1994. Decreased susceptibility of melanized *Cryptococcus neoformans* to UV light. *Appl. Environ. Microbiol.* **60:**3864–3866.

197. **Weidman, F. D., and W. Freeman.** 1924. India ink in the microscopic study of yeast cells. *JAMA* **83:**1163–1164.

198. **Weidman, F. D., and H. L. Ratcliffe.** 1934. Extensive generalized torulosis in a chetah or hunting leopard (*Cynaelurus jubatus*). *Arch. Pathol.* **18:**362–369.

199. **Wilhelmj, C. M.** 1925. The primary meningeal form of systemic blastomycosis. *Am. J. Med. Sci.* **169:**712–721.

200. **Williams, J. R.** 1922. A case of systemic blastomycosis. *Med. J. Aust.* **9:**185–188.

201. **Williamson, J. D., J. F. Silverman, C. T. Mallak, and J. D. Christie.** 1996. Atypical cytomorphologic appearance of *Cryptococcus neoformans*. *Acta Cytol.* **40:**363–379.

202. **Williamson, P. R.** 1994. Biochemical and molecular characterization of the diphenol oxidase of *Cryptococcus neoformans*: identification as a laccase. *J. Bacteriol.* **176:**656–664.

203. **Wilson, D. E., J. E. Bennett, and J. W. Bailey.** 1968. Serologic grouping of *Cryptococcus neoformans*. *Proc. Soc. Exp. Biol. Med.* **127:**820–823.

204. **Zerpa, R., L. Huicho, and A. Guillen.** 1996. Modified india ink preparation for *Cryptococcus neoformans* in cerebrospinal fluid specimens. *J. Clin. Microbiol.* **34:**2290–2291.

205. **Zimmerman, L. E., and H. Rappaport.** 1954. Occurrence of cryptococcosis in patients with malignant disease of reticuloendothelial system. *Am. J. Clin. Pathol.* **24:**1050–1072.

206. **Zuger, A., E. Louie, R. S. Holzman, M. S. Simberkoff, and J. J. Rahal.** 1986. Cryptococcal disease in patients with the acquired immunodeficiency syndrome: diagnostic features and outcome of treatment. *Ann. Intern. Med.* **104:**234–240.

207. **Zuger, A., M. Shuster, M. S. Simberkoff, J. J. Rahal, and R. S. Holzman.** 1988. Maintenance amphotericin B for cryptococcal meningitis in the acquired immunodeficiency syndrome (AIDS). *Ann. Intern. Med.* **109:**592–593.

2 | Taxonomy

HISTORICAL NOMENCLATURE FOR CRYPTOCOCCOSIS

Cryptococcus neoformans (anamorph) has received a series of names in the last century. Rippon, Kwon-Chung, Bennett, and Drouhet give excellent accounts of the historical events leading to the present nomenclature for this pathogenic yeast (13, 36, 48a). In 1894, the German pathologist Busse observed the fungus in tissue from the original clinical case of sarcoma-like tibial lesions in a 31-year-old woman. He believed the round, encapsulated yeast to be *Saccharomyces* and named the disease *Saccharomycosis hominis* (7). The German surgeon Buschke reviewed the histopathological features of this same case and concluded that the etiologic agent was a coccidium (6). During this same period, the Italian scientist Sanfelice at the Hygiene Institute of University of Calgari (Sardinia) isolated a yeast from peach juice and in 1895 named it *Saccharomyces neoformans*, believing that it could produce cancerous tumors (50). This binomial established the taxonomic priority of the species epithet *neoformans*. In 1896, the French pathologist Curtis described a yeast infection in the groin of a man who later died with a central nervous system infection. Curtis named the yeast *Megalococcus myxoides* (6). Later, based on a series of animal experiments in which inoculation with this yeast produced tumor-like lesions and granulomatous reactions in multiple organs, Curtis named the isolate *Saccharomyces subcutaneous tumefaciens* (10). In 1901 Vuillemin renamed the isolates of Busse and Curtis as organism *Cryptococcus hominis* and the isolate of Sanfelice as *C. neoformans* to distinguish them from *Saccharomyces* species because they did not form ascospores and lacked the ability to ferment carbon sources (60). The genus *Cryptococcus*, from the Greek word *kryptos* meaning "hidden," was created by Kützing (31) in 1833 for a group of yeasts that lacked the ability to produce endospores. As more cases became recognized and pathological inspection of tissue was performed, Stoddard and Cutler (56) noticed by 1916 that yeast forms were surrounded by prominent areas of clearing. The scientists misinterpreted this finding as evidence of histolysis by the yeast, and they renamed the organism *Torula histolytica*. Therefore, by 1935 this yeast had been classified under at least three genera (*Saccharomyces, Cryptococcus, Torula*), which perpetuated confusion about the organism's name and possibly its clinical presentations as well. By this time there were more than 40 reported cases of cryptococcosis in humans and at least two cases in animals.

In 1935 Benham (4) conducted a critical study of a group of 22 yeast isolates. She carefully classified them into four groups, identifying each group by its fermentation, assimilation, colony morphology, and serology (4). Several of these groups likely represented *Torulopsis glabrata* and *Rhodotorula*. However, one of the yeast groups (group 3) contained *C. neoformans*. Benham was one of the first mycologists to recognize antigenic differences between strains within this group 3. She eventually accepted the name *C. hominis* for her group 3 yeasts; the group later was renamed *C. neoformans* based on the priority rule of nomenclature. The name *C. neoformans* for the yeast anamorph was considered the valid binomial based on priority because Sanfelice first proposed the species name in 1895. The genus *Cryptococcus* was first established by Kützing in 1833, and Vuillemin used the name only for this pathogenic yeast in 1901. Benham's proposal widened the classification for both pathogenic and nonpathogenic yeasts and ushered in the use of "cryptococcosis," rather than synonyms such as torulosis, European blastomycosis, and Busse-Buschke disease, as the standard designation for this infection. Phaff and Spencer (48) further delineated the genus to include only yeasts that can utilize inositol and lack the ability to ferment or produce pseudohyphae or true hyphae. Finally, following a proposal by Fell et al. (20) in 1989, the genus *Cryptococcus* was conserved and *C. neoformans* was designated as the type species.

As Benham (4) noted, there are different antigen types of *C. neoformans*, arising from antigenic differences in the polysaccharide capsule. *C. neoformans* var. *neoformans* contains those strains that possess serotype A and D capsular types, while *C. neoformans* var. *gattii* contains the serotypes B and C. These two varieties are not recognized as separate species because they can mate with each other to form fertile progeny. Further comparisons of similarities and differences between these varieties on a capsular, molecular biochemical, epidemiological, and clinical basis will be addressed in other chapters. The original Sanfelice isolate belonged to serotype D, and the Busse-Buschke isolate was serotype A. Thus, the original isolates were *C. neoformans* var. *neoformans*. However, Curtis's isolate, described in 1896, is similar to one from a case of meningitis in Africa reported by Gatti in 1970 (22). The isolate described by Gatti was the impetus for Vanbreuseghem and Takashio to rename these serotype B and C strains *Cryptococcus gattii* (59). *C. gattii* was reduced to varietal status in 1976, based on a better understanding of its life cycle (33).

In the 1960s Shadomy and coworkers (51, 52) made the interesting observation that some strains of *C. neoformans* produced hyphae with clamp connections. This finding was the impetus for finding a sexual stage for cryptococcus. In 1975–1976, Kwon-Chung et al. (32, 34) reported a major advance in our understanding of the life cycle of *C. neoformans*. Under certain conditions of nutritional starvation, yeasts of compatible mating types can develop into a filamentous, sexually reproducing teleomorph, which Kwon-Chung named *Filobasidiella neoformans*. Heterothallic mating in *C. neoformans* was thus confirmed. When subsequent mating experiments between isolates of *C. neoformans* var. *gattii* produced teleomorphs having bacilliform basidiospores, the binomial *Filobasidiella bacillispora* was proposed (33). However, when the ability of these two varieties to mate with each other was later discovered, *F. bacillispora* was reduced to synonymy with *F. neoformans*. The α-mating-type culture of *F. bacillispora* mated with an **a** tester strain of *F. neoformans*, producing both fertile progeny and segregation of phenotypic alleles. Furthermore, it was found that genetic relatedness between the two varieties ranged from 55 to 68% in hybridization studies for reassociation of similar DNA (1, 15). On the other hand, the highly conserved ribosomal DNA (rDNA) sequences have just a few nucleotide differences (18, 19) and a pyrimidine synthesis gene (*URA5*) has only an 8% nucleotide difference between the varieties (17). Thus, our present understanding of the life cycle of this species is that it has an asexual and a sexual stage. The yeast anamorph, *C. neoformans*, is composed of the two varieties *C. neoformans* var. *neoformans* and *C. neoformans* var. *gattii*. As further studies into the molecular biology and genetics of this yeast are made, the status of variety versus species for these two *C. neoformans* groups may be revisited. Boekhout et al. (5) emphasized that when a group of strains from these varieties is examined, there are substantial differences between the two varieties in their karyotypes, randomly amplified polymorphic DNA analysis, and killer toxin sensitivity patterns.

CRYPTOCOCCUS AND ITS SPECIES

Members of the genus *Cryptococcus* generally are considered to be yeasts that are nonfermentative, assimilate inositol, and produce urease. Besides the pathogenic *C. neoformans*, at least 38 other *Cryptococcus* species have been found in a wide variety of environmental locations (see Table 1) (3). Some species show striking environmental adaptation to extremely harsh exposures and widely variable climates. *Cryptococcus* species have been found in the extreme cold of Antarctica, at high elevations in the Himalayas, and in saline waters. *Cryptococcus* species such as *C. laurentii* and *C. humicolus* have been found in the food supply. Pork products such as country-cured hams and bacon may contain these yeasts, and it was suggested that colonization could be related to their lipid composition (49). Despite their ability to adapt to a variety of climates, most *Cryptococcus* species have not been able to survive in mammalian tissue because of the relatively high body temperatures and immune systems of these hosts. There are only rare reports of non-*neoformans* cryptococci such as *C. laurentii*, *C. curvatus*, and *C. albidus* causing infection in humans, and in most of these cases there is no histological proof of causality (9, 11, 12, 23, 24, 26–30, 38–41, 44, 46, 53, 57). Humans are frequently exposed to these species, which can be found in aqueous environments (58), soil, human skin such as toe webs (43), and bird guano (2), but for all practical purposes *C. neoformans* should be considered the only *Cryptococcus* species to consistently cause cryptococcosis.

FILOBASIDIELLA NEOFORMANS

In the 1970s, the taxonomy and the potential for understanding the phylogeny of *C. neoformans* received a major boost. Shadomy (51) discovered and reported the presence of clamp connections in isolates of *C. neoformans*. These clamp connections serve to maintain the dikaryotic condition during growth in most species of *Basidiomycota* and are highly distinctive for this phylum of fungi. Subsequently, Kwon-Chung et al. (32, 33) were able to induce the formation of basidia and basidiospores and named two teleomorphs: *F. neoformans* and *F. bacillispora*. *F. neoformans* was the teleomorph for the anamorph *C. neoformans* var. *neoformans*, and *F. bacillispora* was the teleomorph for *C. neoformans* var. *gattii*. However, with the

Table 1 *Cryptococcus* species (other than *C. neoformans*) and where they are found[a]

Species	Location
Cryptococcus albidus	Air, wine, soil, leaves, cheese, mammals
Cryptococcus amylolentus	Frass of bark beetles
Cryptococcus aquaticus	Scum on water
Cryptococcus asgardensis	Soil in Antarctica
Cryptococcus ater	Humans
Cryptococcus baldrensis	Soil in Antarctica
Cryptococcus bhutanensis	Soil in Himalayas
Cryptococcus consortionis	Soil in Antarctica
Cryptococcus curiosus	Frozen *Sillago japonica* in Japan
Cryptococcus curvatus	Mammals, sea
Cryptococcus dimerrae	Pasture
Cryptococcus elinovii	Soil
Cryptococcus feraegula	*Papio papio, Rhea americana*
Cryptococcus flavus	Air
Cryptococcus friedmannii	Rock fragment in Antarctica
Cryptococcus fuscescens	Soil in the former USSR
Cryptococcus gastricus	Soil, musk ox
Cryptococcus hempflingii	Soil in Antarctica
Cryptococcus heveanensis	Rubber
Cryptococcus himalayensis	Soil in Himalayas
Cryptococcus huempii	Rotten *Laurelia sempervirens* in Chile
Cryptococcus humicolus	Soil, tree, toadstools, mushrooms, water, pulp
Cryptococcus hungaricus	Soil, water, cereals, flowers
Cryptococcus kuetzingii	Fruit of medlar, air, humans
Cryptococcus laurentii	Wine, flowers, leaves, insect frass, water, fruit, beans, mail, soil, air, shrimp, mammals
Cryptococcus lupi	Gravel in Antarctica
Cryptococcus luteolus	Air, leaves
Cryptococcus macerans	Flax, flowers, deer
Cryptococcus magnus	Air
Cryptococcus marinus	Sea
Cryptococcus podzolieus	Soil
Cryptococcus skinneri	Insect frass in *Tsuga heterophylla*
Cryptococcus socialis	Soil in Antarctica
Cryptococcus terreus	Soil
Cryptococcus tsukubaensis	Flowers
Cryptococcus tyrolensis	Soil in Antarctica
Cryptococcus vishniacii	Soil in Antarctica
Cryptococcus wrightensis	Soil in Antarctica

[a]Table compiled from data in reference 3.

discovery of cross-fertility and intermediate genomic relatedness, as previously noted for the anamorph, the teleomorphs have been reduced to varietal status as *F. neoformans* var. *neoformans* and *F. neoformans* var. *bacillispora*.

Investigations have shown that the teleomorph *F. neoformans* is a bipolar heterothallic fungus. The anamorph *C. neoformans* exists as either an α- (alpha) or **a**-mating-type haploid yeast cell belonging to the genus *Cryptococcus* (see color plate 1, following p. 456). Under conditions of nitrogen starvation and relative desiccation, cells of these two mating types in physical proximity can conjugate and form a dikaryotic mycelium (heterokaryosis). Terminal basidia then form in which karyogamy and meiosis occur. Haploid basidiospores form in basipetal series from four loci on each basidium apex, resulting in chains of spores. Hyphae are regularly septate with dolipore septa and clamp connections at each septum. The basidiospores of *F. neoformans* var. *neoformans* are ovate to cylindrical, are typically about 2 to 3 μm in size, and with agitation can be released from the basidium onto culture medium. Within a period of as little as 6 h, each spore can germinate into an encapsulated haploid yeast cell that multiplies by asexual budding. The sexual stage has not yet been conclusively identified in nature although there has been some suggestion that structures of var. *gattii* might be found in flowers from eucalyptus trees (14). The α-mating-type bias for yeast isolates in nature (35) suggests that sexual reproduction may not be a common natural event. However, it has recently been reported that haploid strains with the α-mating locus have the ability under certain environmental stress conditions to form fruiting bodies, a process that has been termed monokaryotic, homokaryotic, or haploid fruiting (61). This vegetative ability to produce hyphae and haploid fruiting is a known feature of some basidiomycetes (16, 55). It is uncertain whether this ability of the α-mating locus to allow self-formation of these structures has any pathological consequences, since it remains unclear whether the infectious propagule is a small yeast or the basidiospore. It has been shown that α-mating-type strains occur in over 90% of clinical and environmental isolates (35). In fact, it has been difficult to find a MAT**a** strain of the clinically predominant serotype A group; it is possible that this serotype has evolved with the elimination of the opposite mating type. Therefore, the importance of the teleomorph stage for producing a potentially infectious propagule of *C. neoformans* may be limited. On the other hand, Madrenys et al. (42) have found that 96.8% of 195 clinical isolates could mate on sunflower agar. Also, the basidiospore is of ideal size for respiratory deposition, and monokaryotic fruiting could still be an important element in the infectious cycle of this yeast.

FILOBASIDIELLA DEPAUPERATA

With the discovery of the teleomorphs, it became possible to place *F. neoformans* (anamorph, *C. neoformans*) in the phylum Basidiomycota. By physical structures and subsequent molecular studies, the genus *Filobasidiella* has been found to contain two species, *F. neoformans* and *F. depauperata*. This genus is distinct among the Basidiomycota in that it (i) produces holobasidia with spores in four long chains; (ii) has identical basidial structure but with two distinct life cycles, physiology, and ecology; and (iii) contains different genomic locations of their 5S rRNA genes (37).

The differences between *F. neoformans* and *F. depauperata* are well defined. *F. depauperata* is homothallic with no yeast phase. This species of basidiospores develops into monokaryotic hyaline hyphae without clamp connections. Basidiospores of this species differ in that *F. neoformans* spores are subglobose to barrel-shaped to bacilliform, and *F. depauperata* spores are pentagonal. There are also substantial biochemical differences between the species. *F. depauperata* does not produce urease, phenoloxidase, or starch, assimilate inositol, or grow at 37°C, all of which are characteristics found in *F. neoformans*. An even greater contrast between the two species is their ecology. While the anamorph of *F. neoformans* has been found in many locales around the world and frequently causes mammalian disease, *F. depauperata* has been isolated only from dead insects and arachnids or their egg masses and has no known potential for mammalian pathogenicity (37).

Despite the differences in morphology, biochemistry, ecology, and pathogenicity, *F. neoformans* and *F. depauperata* are phylogenetically related and belong in the same genus based upon DNA homology. In a study of the 5.8S ribosomal sequence of the two species, Mitchell et al. (45) found that the sequences differ by one shared and another unique nucleotide substitution. Gueho et al. (25) used the most variable domain of the 26S rRNA to examine differences in *Filobasidiella* species and found remarkable similarities between the two species, with only nine nucleotide substitutions. On a molecular basis, these two fungi are genetically related and belong in the same genus. However, in DNA structures they have some differences. In general, basidiomycetes have their 5S RNA embedded in the nontranscribed spacer regions of the rDNA repeat unit, and except for *Coprinus* sp., transcription occurs in the same direction as in the other rRNA subunits. Unlike *F. neoformans*, which follows this standard pattern, *F. depauperata* 5S rRNA genes are found dispersed throughout the genome (37).

MOLECULAR PHYLOGENY

Although taxonomic relationships between *Cryptococcus* and *Tremella* have been based on fatty acid compositions and other phenotypic characteristics (54), molecular phylogeny studies will give us the most objective evidence for genus and species relationships. The molecular phylogeny of *Filobasidiella* in relationship to other closely related fungi has continued to be elucidated (Fig. 1) (21). The entire nucleotide sequence of the rDNA gene cassette has been identified for both varieties of *F. neoformans* (anamorph *C. neoformans*). Fan et al. (19) showed that there were only two nucleotide differences in the 16S-like rRNAs and 10 nucleotide substitutions in the 26S-like rRNAs between the two varieties of *C. neoformans*. Because rRNA is highly conserved, one would expect to find similarities. As more sequence of genome is made for the varieties, there will likely be sequence divergence at other alleles, as has already been noted with *URA5* (17). These findings support the taxonomic view that the varieties are very closely related, with only recent evolutionary divergence, and that there may be some slight evolutionary divergence between serotype A and the other serotypes (41). The rDNA primary sequence structure allows phylogenetic comparisons with other fungi and organisms (Fig. 2). A parsimony analysis of the small subunit of rDNA sequences found *C. neoformans* to be a basidiomycete and most closely related to species of *Tremella*, *Bullera*, and *Trichosporon* (2, 21). Using the conservative 16S-like rRNA, *C. neoformans* is also closely related to *Trichosporon beigelii*, with which it shares cross-reacting cell wall antigens (19). With the 26S-like RNA, *C. neoformans* is placed on a

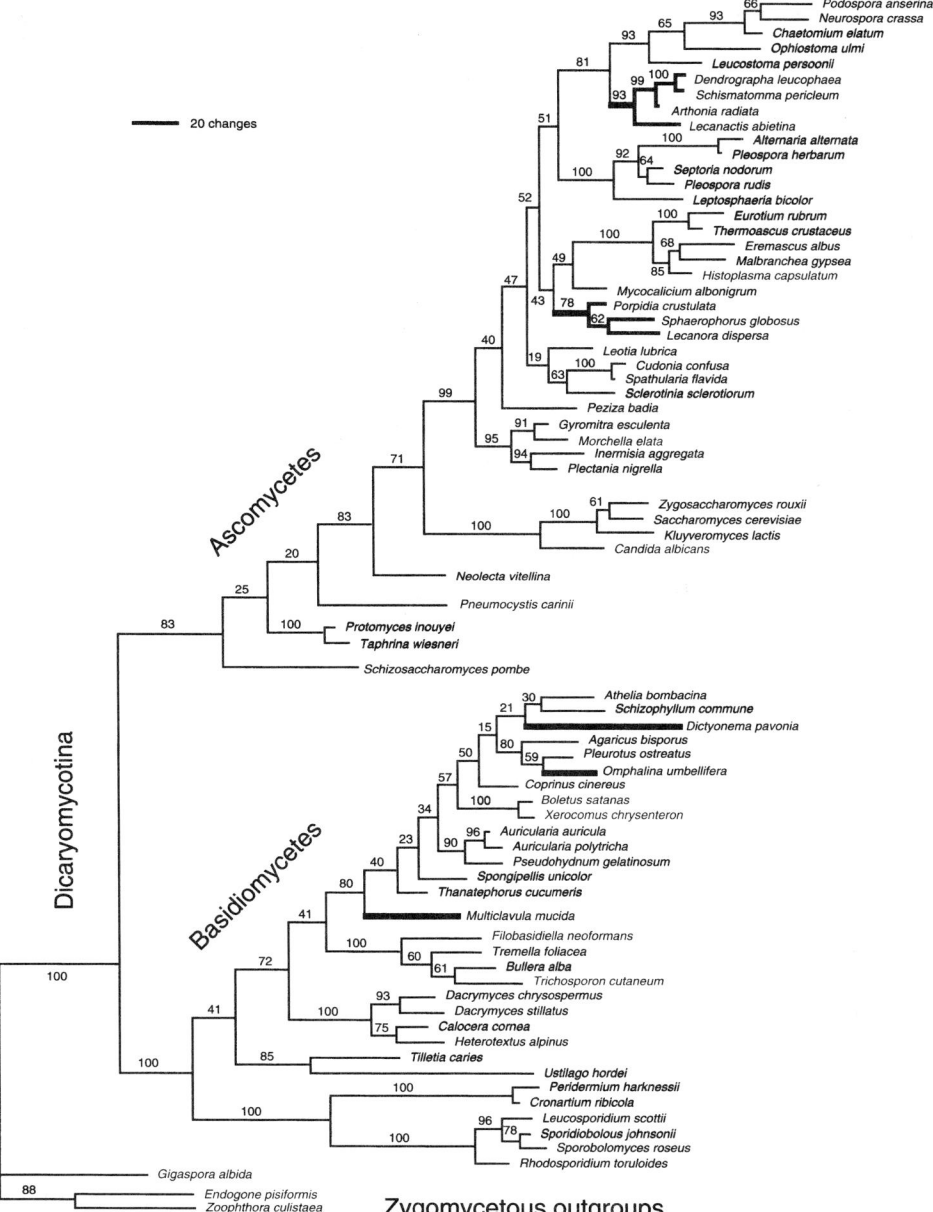

Figure 1 Phylogenetic relations within Amastigomycota as derived from parsimony analysis of 1,927 nucleotides of small subunit rDNA sequences. (From reference 21 with permission.)

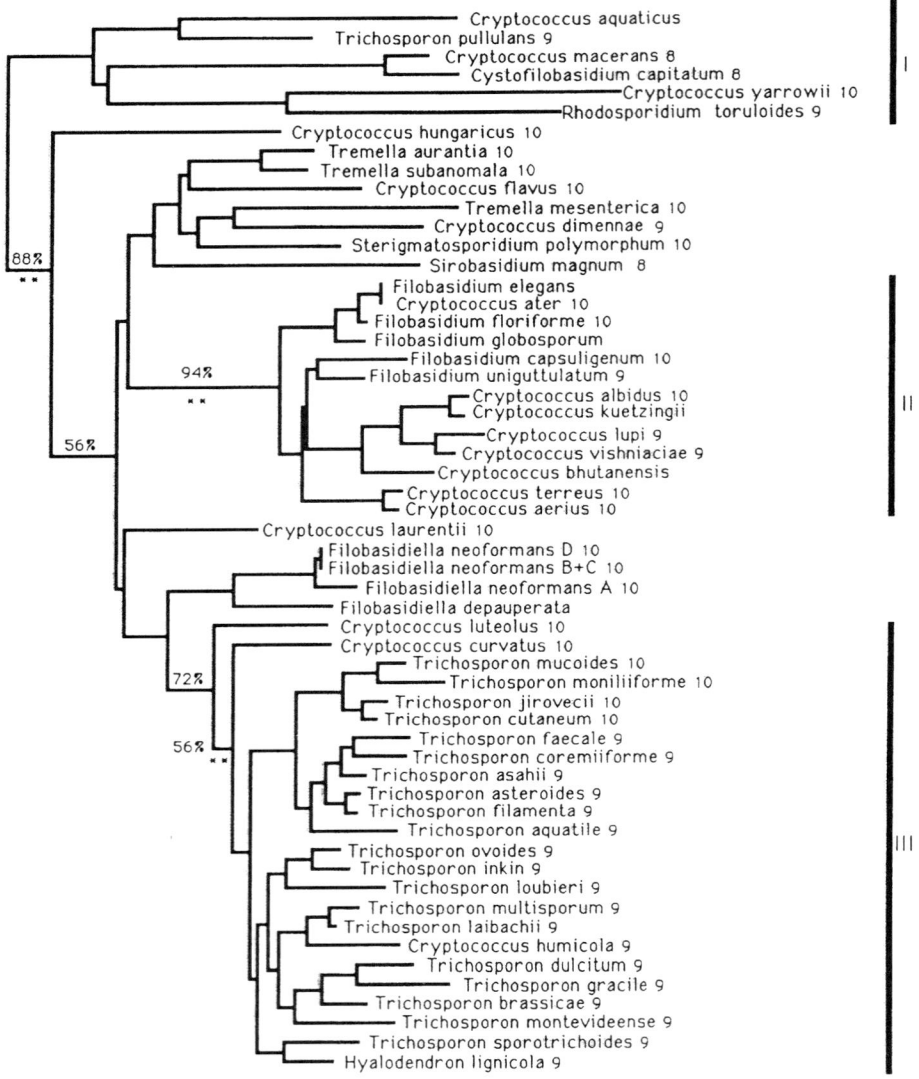

Figure 2   General phylogeny of *C. neoformans* and related species from partial sequence rDNA large subunit. (From reference 25 with permission.)

branch between *Mucor racemosus* and *Pneumocystis carinii/Schizosaccharomyces pombe* (19). As more genetic loci, such as the conserved actin gene (8), are sequenced and compared, the phylogenetic relationships on a molecular level can be interpreted further with regard to classical methods of taxonomy. This may lead to a better understanding of relationships between *F. neoformans* and other fungi.

The genera *Filobasidiella*, *Filobasidium*, and *Cystofilobasidium* form the family *Filobasidiaceae*. Studies on the conserved 5.8S rRNA show that *Filobasidiella* differs from *Filobasidium* and *Cystofilobasidium* by six and five nucleotides, respectively; this supports *Filobasidiella* as a separate but related genus. However, when the 16S rRNA gene of general heterobasidiomycetes was sequenced, and alignment was performed with construction of parsimonious trees, *Filobasidiella* species were more closely related to members of the family *Tremellaceaeae* than to the other members of the *Filobasidiaceae* (37). This closer phylogenetic relationship between the *Filobasidiella* and *Tremella* species has been supported by studies using the partial sequences of the 26S rRNA and 5S RNA genes (Fig. 2) (25). These studies begin to show a molecular divergence of serotype A *C. neoformans* var. *neoformans* from the other three serotypes. Investigations of molecular divergence of these related basidiomycetes at other genetic loci also have been studied in relationship to *F. neoformans*. When these basidiomycetes, such as *F. uniguttulatum*, *T. globospora*, *T. foliacea*, and *T. moriformis*, were examined for the presence of two *F. neoformans* virulence gene loci, *CAP59* and *CnLAC1*, by Southern analysis, these genes were found to be either absent or significantly less homologous (47). These findings suggest that the ancestral genes for these mammalian virulence factors and/or their sequences emerged and diverged early in evolution of these fungi, and likely represent different genetic responses as these related fungi adapted to different environmental challenges.

SUMMARY - CLASSIFICATION (*C. NEOFORMANS*)

Kingdom: Fungi
Phylum: Basidiomycota
Order: *Filobasidiales* (*Tremellales*)*
Family: *Filobasidiaceae* (*Tremellaceaeae*)*
Genus: *Filobasidiella*
*Molecular data suggest a close relationship to *Tremellaceaeae*.

Teleomorph	*Anamorph*
F. neoformans var. *neoformans* (serotypes A and D)	*C. neoformans* var. *neoformans*
F. neoformans var. *bacillispora* (serotypes B and C)	*C. neoformans* var. *gattii*

REFERENCES

1. **Aulakh, H. J., S. E. Straus, and K. J. Kwon-Chung.** 1981. Genetic relatedness of *Filobasidiella neoformans* (*Cryptococcus neoformans*) and *Filobasidiella bacillispora* (*Cryptococcus bacillisporus*) as determined by deoxyribonucleic acid base composition and sequence homology studies. *Int. J. Syst. Bacteriol.* **31:**97–103.

2. **Bangert, R. L., B. R. Cho, P. R. Widders, E. H. Stauber, and A. C. Ward.** 1988. A survey of aerobic bacteria and fungi in the feces of healthy psittacine birds. *Avian Dis.* **32:**46–52.

3. **Barnett, J. A., R. W. Payne, and D. Yarrow (ed.).** 1990. *Yeasts: Characteristics and Identification,* 2nd ed., p. 282–320. Cambridge University Press, Cambridge.

4. **Benham, R. W.** 1935. Cryptococci, their identification by morphology and serology. *J. Infect. Dis.* **57:**255–274.

5. **Boekhout, T., A. Van Belkum, A. C. Leenders, H. A. Verbrugh, P. Mukamurangwa, D. Swinne, and W. A. Scheffers.** 1997. Molecular typing of *Cryptococcus neoformans*: taxonomic and epidemiological aspects. *Int. J. Syst. Bacteriol.* **47:**432–442.

6. **Buschke, A.** 1895. Ueber eine durch Coccidien hemorgerufene Krankheit des Menschen. *Dtsch. Med. Wochenschr.* **21:**14.

7. **Busse, O.** 1894. Ueber parasitaere Zelleinschlusse und ihre Zuchtung. *Int. J. Med. Microbiol.* **16:**175–180.

8. **Cox, G. M., T. H. Rude, C. C. Dykstra, and J. R. Perfect.** 1995. The actin gene from *Cryptococcus neoformans*: structure and phylogenetic analysis. *J. Med. Vet. Mycol.* **33**:261–266.

9. **Cuna, T., and J. Lusins.** 1973. *Cryptococcus albidus* meningitis. *South. Med. J.* **66**:1230.

10. **Curtis, F.** 1895. Note sur un nouveau parasite humain, megalococcus myxoides, trouvé dans un néoplasme de la région inguino-crurole. *C.R. Soc. Biol.* (Paris) **1895**:715–718.

11. **Curtis, P. H., J. A. Haller, and E. de Juan.** 1995. An unusual case of cryptococcal endophthalmitis. *Retina* **15**:300–304.

12. **Dromer, F., A. Moulignier, B. Dupont, E. Gueho, M. Baudrimont, L. Improvisi, F. Provost, and G. Gonzalez-Cavali.** 1995. Myeloradiculitis due to *Cryptococcus curvatus* in AIDS. *AIDS* **9**:395–396.

13. **Drouhet, E.** 1997. Milestones in the history of cryptococcus and cryptococcosis. *J. Mycol. Med.* **7**:10–27.

14. **Ellis, D. H., and T. J. Pfeiffer.** 1990. Ecology, life cycle, and infectious propagule of *Cryptococcus neoformans*. *Lancet* **336**:923–925.

15. **Erke, K. H., and J. D. Schneidau.** 1973. Relationship of some *Cryptococcus neoformans* hypha-forming strains to standard strains and to other species of yeasts as determined by deoxyribonucleic acid base ratios and homologies. *Infect. Immun.* **7**:941–948.

16. **Esser, K., and F. Meinhardt.** 1977. A common genetic control of dikaryotic and monokaryotic fruiting in the basidiomycete *Agrocybe acgerita*. *Mol. Gen. Genet.* **155**:113–115.

17. **Fan, M., and A. Casadevall.** 1992. *URA5* gene of *Cryptococcus neoformans* var. *gattii* and *C. neoformans* var. *neoformans*. *J. Gen. Appl. Microbiol.* **38**:491–495.

18. **Fan, M., L. C. Chen, M. A. Ragan, R. R. Gutell, J. R. Warner, B. P. Currie, and A. Casadevall.** 1995. The 5S rRNA and rRNA intergenic spacer of the two varieties of *Cryptococcus neoformans*. *J. Med. Vet. Mycol.* **33**:215–221.

19. **Fan, M., B. P. Currie, R. R. Gutell, M. A. Ragen, and A. Casadevall.** 1994. The 16s-like, 5.8s and 23s-like rRNA of the two varieties of *Cryptococcus neoformans*: sequence secondary structure, phylogenetic analysis, and restriction fragment polymorphisms. *J. Med. Vet. Mycol.* **32**:163–180.

20. **Fell, J. W., C. P. Kurtzman, and K. J. Kwon-Chung.** 1989. Proposal to conserve cryptococcus. *Taxon* **38**:151–152.

21. **Gargas, A., P. T. DePriest, M. Grube, and A. Tehler.** 1995. Multiple origins of lichen symbioses in fungi: suggested by SSU rDNA phylogeny. *Science* **268**:1492–1495.

22. **Gatti, F., and H. Eeckels.** 1970. An atypical strain of *Cryptococcus neoformans* (Sanfelice) Vullemin. Part I. Description of the disease and of the strain. *Ann. Soc. Belg. Med. Trop.* **50**:689–694.

23. **Gluck, J. L., J. P. Myers, and L. M. Pass.** 1987. Cryptococcemia due to *Cryptococcus albidus*. *South. Med. J.* **80**:511–513.

24. **Gordon, M. A.** 1972. Pulmonary cryptococcosis. A case due to *Cryptococcus albidus*. *Am. Rev. Respir. Dis.* **106**:786–787.

25. **Gueho, E., L. Improvisi, R. Christen, and G. S. de Hoog.** 1993. Phylogenetic relationships of *Cryptococcus neoformans* and some related basidiomycetous yeasts determined from partial large subunit rRNA sequences. *Antonie van Leeuwenhoek* **63**:175–189.

26. **Horowitz, I. D., E. A. Blumberg, and L. Krevolin.** 1993. *Cryptococcus albidus* and mucormycosis empyema in a patient receiving hemodialysis. *South. Med. J.* **86**:1070–1072. (Abstract)

27. **Kamalam, A., and A. S. Thambiah.** 1976. A study of 3891 cases of mycoses in the tropics. *Sabouraudia* **14**:129–148.

28. **Kamalam, A., P. Yesudian, and A. S. Thambiah.** 1977. Cutaneous infection by *Cryptococcus laurentii*. *Br. J. Dermatol.* 97:221–223.

29. **Kromery, V., A. Kunova, and J. Mardiak.** 1997. Nosocomial *Cryptococcus laurentii* fungemia in a bone marrow transplant patient after prophylaxis with ketoconazole successfully treated with oral fluconazole. *Infection* **25:**130.

30. **Krumholz, R. A.** 1972. Pulmonary cryptococcosis. A case due to *Cryptococcus albidus. Am. Rev. Respir. Dis.* **105:**421–424.

31. **Kützing, F.** 1833. Systematische Zusammenstellung der niedern Algen-Gattungen und Arten. *Linneae* **8:**365.

32. **Kwon-Chung, K. J.** 1975. A new genus, *Filobasidiella,* the perfect state of *Cryptococcus neoformans. Mycologia* **67:**1197–1200.

33. **Kwon-Chung, K. J.** 1976. A new species of *Filobasidiella,* the sexual state of *Cryptococcus neoformans* B and C serotypes. *Mycologia* **68:**942–946.

34. **Kwon-Chung, K. J.** 1976. Morphogenesis of *Filobasidiella neoformans,* the sexual state of *Cryptococcus neoformans. Mycologia* **68:**821–833.

35. **Kwon-Chung, K. J., and J. E. Bennett.** 1978. Distribution of "alpha" and "a" mating types of *Cryptococcus neoformans* among natural and clinical isolates. *Am. J. Epidemiol.* **108:**337.

36. **Kwon-Chung, K. J., and J. E. Bennett.** 1992. Cryptococcosis, p. 426–429. *In* K. J. Kwon-Chung and J. E. Bennett (ed.), *Medical Mycolology.* Lea & Febiger, Philadelphia.

37. **Kwon-Chung, K. J., Y. C. Chang, R. Bauer, E. C. Swann, J. W. Taylor, and R. Goel.** 1995. The characteristics that differentiate *Filobasidiella depauperato* from *Filobasidiella neoformans,* p. 67–79. *In* T. Boekhout and A. Samson (ed.), *Heterobasidiomycetes: Systematic and Applied. Studies in Mycology.* CBS, Baarn, Switzerland.

38. **Lin, S. R., C. F. Peng, S. A. Yang, and H. S. Yu.** 1989. Isolation of *Cryptococcus albidus* var. *albidus* in patient with pemphigus foliaceous. *Kaohsiung J. Med. Sci.* **5:**126–128.

39. **Litherland, D. K.** 1972. Pulmonary cryptococcosis. A case due to *Cryptococcus albidus. Am. Rev. Respir. Dis.* **106:**786.

40. **Loison, J., J. P. Bouchara, E. Gueho, L. de Gentile, B. Cimon, J. M. Chennebault, and D. Chabasse.** 1996. First report of *Cryptococcus albidus* septicemia in an HIV patient. *J. Infect.* **33:**139–140.

41. **Lynch, J. P., III, D. R. Schaberg, D. G. Kissner, and C. A. Kauffman.** 1981. *Cryptococcus laurentii* lung abscess. *Am. Rev. Respir. Dis.* **123:**135–138.

42. **Madrenys, N., C. de Vroey, C. Raes-Wuytack, and J. M. Torres-Rodriquez.** 1993. Identification of the perfect state of *Cryptococcus neoformans* from 195 clinical isolates including 84 from AIDS patients. *Mycopathologia* **123:**65–68.

43. **McGinnis, M. R., M. G. Rinaldi, C. Halde, and A. E. Hilger.** 1975. Mycotic flora of the interdigital spaces of the human foot: a preliminary investigation. *Mycopathologia* **55:**47–52.

44. **Melo, J. C., S. Srinivason, M. L. Scott, and M. J. Raff.** 1980. *Cryptococcus albidus* meningitis. *J. Infect.* **2:**79–82.

45. **Mitchell, T. G., T. J. White, and J. W. Taylor.** 1992. Comparison of 5.8S ribosomal DNA sequences among the basidiomycetous yeast genera Cytofilobasidium, Filobasidium, Filobasidiella. *J. Med. Vet. Mycol.* **30:**207–218.

46. **Mocan, H., A. V. Murphy, T. J. Beattie, and T. A. McAllister.** 1989. Fungal peritonitis in children on continuous ambulatory peritoneal dialysis. *Scott. Med. J.* **34:**494–496.

47. **Petter, R., A. Varma, T. Boekhout, S. Salas, B. Davis, and K. J. Kwon-Chung.** 1996. Molecular divergence of the virulence factors of *Filobasidiella neoformans* found in other heterobasidiomycetous species, abstr. 2.2:36–38. 3rd International Conference on Cryptococcus and Cryptococcosis, Paris.

48. **Phaff, H. J., and J. F. T. Spencer.** 1966. Improved parameters in the separation of species in the genera Rhodotorula and Cryptococcus, p. 59–67. Proceedings of the Second International Symposium on Yeasts, Bratislava.

48a. **Rippon, J. W.** 1988. *Medical Mycology. The Pathogenic Fungi and the Pathogenic Actinomycetes*, 3rd ed., p. 582–609. W. B. Saunders, Philadelphia.

49. **Saldanha-da-Gama, A., M. Malfeito-Ferreira, and V. Loureiro.** 1997. Characterization of yeasts associated with Portuguese pork-based products. *Int. J. Food Microbiol.* **37:**201–207.

50. **Sanfelice, F.** 1894. Contributo alla morfologia e biolgia dei blastomiceti che si sviluppano nei succhi di alcuni frutti. *Ann. Ig.* **4:**463–495.

51. **Shadomy, H. J.** 1970. Clamp connections in two strains of *Cryptococcus neoformans*. Spectrum monograph series. Arts and Sciences, Georgia State University **1:**67–72.

52. **Shadomy, H. J., and J. P. Utz.** 1966. Preliminary studies on a hypha-forming mutant of *Cryptococcus neoformans*. *Mycologia* **58:**383–390.

53. **Sinnott, J. T., J. Rodnite, P. J. Emmanuel, and A. Campos.** 1989. *Cryptococcus laurentii* infection complicating peritoneal dialysis. *Pediatr. Infect. Dis. J.* **8:**803–805.

54. **Smit, E. J., J. L. F. Kock, J. P. J. Van der Westhuizen, and T. J. Britz.** 1988. Taxonomic relationships of *Cryptococcus* and *Tremella* based on fatty acid composition and other phenotypic characters. *J. Gen. Microbiol.* **134:**2849–2855.

55. **Stahl, U., and K. Esser.** 1976. Genetics of fruit body production in higher basidiomycetes. *Mol. Gen. Genet.* **148:**183–197.

56. **Stoddard, J. L., and E. C. Cutler.** 1916. *Torula Infection in Man*, p. 1–98. Monograph no. 6. Rockefeller Institute for Medical Research, New York.

57. **Taylor, G.D., M. Buchanan-Chell, T. Kirkland, M. McKenzie, and R. Wiens.** 1994. Trends and sources of nosocomial fungaemia. *Mycoses* **37:**187–190.

58. **Vadkertiova, R., and E. Slavikova.** 1995. Killer activity of yeasts isolated from the water environment. *Can. J. Microbiol.* **41:**759–766.

59. **Vanbreuseghem, R., and M. Takashio.** 1970. An atypical strain of *Cryptococcus neoformans* (Sanfelice) Vuillemin. Part II. *Cryptococcus neoformans* var. *gattii*. *Nov. Ann. Soc. Belg. Med. Trop.* **50:**695–702.

60. **Vuillemin, P.** 1901. Les blastomycetes pathogènes. *Rev. Gen. Sci. Pures Appl.* **12:**732–751.

61. **Wickes, B. L., M. E. Mayorga, J. Edman, and J. C. Edman.** 1996. Dimorphism and haploid fruiting in *Cryptococcus neoformans*: association with the alpha-mating type. *Proc. Natl. Acad. Sci. USA* **93:**7327–7331.

3 | Ecology of *Cryptococcus neoformans*

Cryptococcus neoformans is a cosmopolitan fungus, and cases of cryptococcosis have been reported from all regions of the world. The fungus is a free-living organism that can survive in a variety of environmental niches. The saprophytic nature of *C. neoformans* has been known since 1894 when Sanfelice cultured it from peach juice (172). *C. neoformans* is rarely isolated from healthy individuals (88, 157) and does not appear to be a common human commensal. The sporadic nature of human cryptococcosis, the extreme rarity of documented human-to-human transmission events (15, 80), and the high prevalence of *C. neoformans* in the environment (see below) indicate that human infection is acquired from environmental sources. Thus, knowledge of the ecology and the life cycle of this fungus is important for understanding of the epidemiology of human infection.

In the past half-century, considerable progress has been made in under-standing the ecology of *C. neoformans* and the epidemiology of cryptococcal infec-tions. Among the most significant observations were the division of *C. neoformans* strains into the varieties *neoformans* and *gattii* on the basis of phenotypic, biochemi-cal, and genetic differences; the association of *C. neoformans* var. *neoformans* with avian excreta; and the association of *C. neoformans* var. *gattii* with eucalyptus trees. Nevertheless, large gaps remain in our understanding of *C. neoformans* biology, ecology, and epidemiology. This chapter describes the distinctions between the two *C. neoformans* varieties and reviews the available information on the ecology of this fungal pathogen.

COMMENSALISM OF *C. NEOFORMANS*

Some of the early reports in the literature suggested that *C. neoformans* was an organism commonly found in human skin, mucous membranes, and the gastroin-testinal tract (10, 11, 69, 88). However, advances in mycology classification led to the conclusion that many of these isolates were not *C. neoformans* (88). Studies that use rigorous criteria for the identification of *C. neoformans* indicate that this fungus is rarely a commensal in healthy individuals. In 1973, Howard (88) reported the isolation of *C. neoformans* from only 3 of 561 sputum specimens and was unable to isolate *C. neoformans* from 102 saliva and 310 fecal specimens. Randhawa and Paliwal (157) confirmed this finding when they isolated *C. neoformans* from only 1 of 820 oropharyngeal washings and from the toes of only 6 of 723 persons. The presence of *C. neoformans* in these individuals was transient and without clinical evidence of invasive disease (157). An exception to the studies reporting a low prevalence of commensalism is the study by McGinnis et al. (131), who found *C. neoformans* in the foot interdigital spaces of 5 of 27 healthy men in San Francisco.

The likelihood of recovering *C. neoformans* from human samples is higher if patients are ill. Reiss and Szilagyi (160) reported *C. neoformans* in the throats of 6 of 92 patients with neoplastic disease. Randhawa and Mahendra (156) reported the isolation of *C. neoformans* from 3% of 469 patients with bronchopulmonary disorders. Tynes et al. (210) described the isolation of *C. neoformans* from 18 patients without pulmonary cryptococcosis, the majority of whom had preexisting pulmonary disease. A survey of patients with roentgenographic evidence of a pulmonary lesion and *C. neoformans* in their lungs revealed that many had lung carcinoma (54). These studies suggest that for some patients, isolation of *C. neoformans* from the respiratory tract may be an indication of underlying pulmonary disease (54).

In summary, the literature indicates that *C. neoformans* can, on occasion, be a transient human commensal. In chronically ill patients the likelihood of recovery of *C. neoformans* may be significantly greater than in healthy adults. Presumably, chronic illnesses weaken defense mechanisms and permit the establishment of *C. neoformans* on mucous membranes. The relationship between asymptomatic colonization and risk for invasive disease is unknown. However, the rarity of *C. neoformans* isolation from human specimens strongly suggests that a positive culture for *C. neoformans* in any patient should lead to a clinical evaluation for signs and symptoms of cryptococcosis. Furthermore, the low prevalence of human commensalism suggests that cryptococcosis is not the result of infection with endogenous flora but rather is caused by environmental infection.

TWO VARIETIES OF *C. NEOFORMANS*

C. neoformans strains have been divided into two varieties on the basis of phenotypic, serological, genetic, biochemical, and epidemiological criteria. Both varieties are pathogenic for humans. Variety *neoformans* is found worldwide, is the type invariably isolated from avian excreta, and represents the overwhelming majority of isolates recovered from patients with AIDS. Variety *gattii* is found in tropical and subtropical climates (108, 109), is associated with the eucalyptus tree (*Eucalyptus camaldulensis* and *Eucalyptus tereticornis*), and is rarely the cause of disseminated infection in patients with AIDS, even in endemic areas (164, 183, 184). To date there are fewer than 10 reports of *C. neoformans* var. *gattii* infection in patients with AIDS (2, 19, 32, 100, 163, 184, 186). The rarity of these infections in patients with AIDS is not understood.

Differences in the clinical presentation and outcome of cryptococcosis caused by strains of var. *neoformans* and var. *gattii* have been described (72, 183). A recent study from Australia revealed higher mortality for infections caused by var. *neoformans* (183). In contrast, infections caused by var. *gattii* had low mortality but were frequently complicated by neurological sequelae that required surgery and prolonged therapy (183). A need for frequent surgery and prolonged treatment in patients infected with var. *gattii* was also noted in an earlier study (72).

The rare occurrence of viable spores after matings between var. *neoformans* and var. *gattii* indicates that the differences between the varieties are not sufficiently great to warrant classification as two different species (110). Both varieties have been shown to undergo sexual and asexual reproduction. The yeast form (anamorph state) undergoes asexual reproduction by budding. Kwon-Chung (104–106) has described the perfect state for both varieties of *C. neoformans*. For *C. neoformans* var. *neoformans* the teleomorph (or perfect state) is *Filobasidiella neoformans* var. *neoformans*. For *C. neoformans* var. *gattii* the teleomorph state is *F. neoformans* var. *bacillispora*. The teleomorph state is characterized by the production of basidiospores, which are the product of sexual reproduction. The two varieties of *C. neoformans* are similar in the yeast form, or anamorph state, but distinct phenotypic differences are evident in the teleomorph state: the basidiospores of var. *neoformans* are spherical, oblong, elliptical, or cylindrical with finely roughened walls, whereas those of var. *gattii* are bacilliform and smooth-walled (56, 104–106, 110).

Many serological and biochemical differences have been described for the *C. neoformans* varieties (Table 1). Strains of serotypes A, D, and AD belong to *C. neoformans* var. *neoformans*. Strains of serotypes B and C belong to *C. neoformans* var. *gattii*. Serotype classification is done on the basis of absorbed rabbit sera and reflects antigenic differences in the main capsular polysaccharide, glucuronoxylomannan (40). The capsular polysaccharides of *C. neoformans* var. *gattii* are more complex than those of *C. neoformans* var. *neoformans* (40) (see also chapter 4). The two varieties differ in their ability to assimilate l-malic, fumaric, and succinic acids (13). The two varieties assimilate creatinine but differ in the regulation of creatinine metabolism (107, 149). EDTA inhibits the urease in *C. neoformans* var. *gattii* but not in *C. neoformans* var. *neoformans* (116). Strains of the two varieties also manifest differences in susceptibility to antifungal agents (176). On the average, strains of var. *gattii* appear to be less susceptible to flucytosine than are strains of var. *neoformans* (176).

Two colorimetric agar tests have been designed that distinguish between the *C. neoformans* varieties on the basis of biochemical differences (113, 171). The most widely used test is based on differential growth in canavanine–glycine–bromothymol blue (CGB) agar. CGB agar distinguishes among varieties because many *C. neoformans* var. *neoformans* strains are susceptible to l-canavanine and most cannot assimilate glycine as the sole source of nitrogen and carbon (132). *C. neoformans* var. *gattii* strains are able to grow in CGB agar and produce ammonia, which raises the pH and changes the pH indicator bromothymol blue to a blue-green color (132). *C. neoformans* var. *gattii* strains are naturally resistant to canavanine by a mechanism that metabolizes canavanine to a nontoxic product (150). *C. neoformans* var. *neoformans* strains do not grow in CGB and produce no color change. The second test was described by Salkin and Hurd (171) and is based on varietal differences in the resistance to low concentrations of cycloheximide and assimilation of glycine as a carbon source (171). However, some investigators have found that many isolates have intermediate results with the two colorimetric assays and hence "biochemical serotyping" may not completely replace the serological characterization (176). Differences in d-proline assimilation can also be used to group *C. neoformans* isolates according to varietal classification, but this method is less discriminatory than CGB agar (139). Differences in the ability of *C. neoformans* var. *neoformans*

Table 1 Differences between the two varieties of *C. neoformans*

Characteristic	References	*C. neoformans* var. *neoformans*	*C. neoformans* var. *gattii*
Geographic distribution	See text	Worldwide	Tropical and subtropical
Environmental sources	See text	Avian excreta	River and forest red gums (eucalyptus)
Serotypes	12, 109	A, D, and AD	B and C
Perfect state	104–106, 110	*Filobasidiella neoformans*	*Filobasidiella gattii*
Basidiospores	104–106, 110	Spherical or cylindrical, rough walls	Bacillary, smooth walls
CGB agar	113	No growth	Growth (blue-green color change)
Average chromosome number	223	12	13
L-Malic acid assimilation	13	Weak	Strong
Prevalence in patients with AIDS	20, 162, 178, 183	Common	Rare
D-Proline assimilation	139	No	Yes
Urease inhibition by EDTA	116	Yes	No
Repression of creatinine assimilation by NH_3	149	Yes	No
Isolation from AIDS patients	162, 178	Common	Rare

strains serotypes A and D to assimilate creatinine have been used to design a colorimetric agar test that discriminates between serotype A and D strains (93).

DNA sequence analysis indicates that the two varieties are closely related. Comparison of the *URA5* gene sequences revealed 8% nucleotide sequence divergence between *C. neoformans* var. *neoformans* and *C. neoformans* var. *gattii* strains (29). Considering that individual *C. neoformans* var. *neoformans* strains can differ by up to 5% in *URA5* sequence (30, 36, 75), the nucleotide differences between varieties for this gene are not very large. Analysis of the four ribosomal DNA (rDNA) genes revealed two differences in the 1,802-nucleotide 16S-like rDNA, 10 differences in the 3,358-nucleotide 23S-like rDNA, and no differences in the 5.8S and 5S rDNAs (67). For both varieties, the 5S rDNA gene is located within the 23S-16S rDNA repeat in the same orientation as the other rDNA genes (66, 111). Despite almost 100% identity in the rDNA coding sequences, the sequence homology in the rDNA intergenic regions between the varieties was only 78.5% (66). Nonhomology outside gene coding regions may account for the relatively low DNA relatedness values of 55 to 63% obtained by DNA hybridization studies (6). The close identity between the rDNA genes from the two varieties suggests recent evolutionary divergence (67).

DNA typing studies have revealed significant differences between *C. neoformans* var. *gattii* and *C. neoformans* var. *neoformans* strains. Analysis of mitochondrial DNA restriction fragment length polymorphisms (RFLPs) revealed intravarietal similarities and variety-specific RFLPs that can be useful in categorizing strains (212). Differences in the size of chromosomes among the varieties are apparent by electrophoretic karyotyping (223). The smallest chromosomes of var. *gattii* strains are 400 to 700 kb in size, whereas the smallest chromosome of var. *neoformans* strains is approximately 770 kb. The average number of chromosomes deduced by electrophoretic karyotyping for var. *gattii* and var. *neoformans* is 13 and 12, respectively (223). The CNRE-1 repetitive DNA element hybridizes strongly to DNA from *C. neoformans* var. *neoformans* but not from *C. neoformans* var. *gattii* (185).

In summary, the two varieties of *C. neoformans* are closely related but are different enough to suggest the beginnings of speciation. The adaptation of each variety to different ecological niches could provide the necessary separation for species divergence to occur. At this time, classification within one species is indicated by the successful mating of some var. *gattii* and var. *neoformans* strains in the laboratory (110).

C. NEOFORMANS VAR. *GATTII*

Geographic Distribution

The majority of clinical isolates of *C. neoformans* var. *gattii* have originated from tropical and subtropical regions (109). In the United States the majority of var. *gattii* isolates have originated from southern California (12, 109). In some parts of the world *C. neoformans* var. *gattii* is the predominant *C. neoformans* variety in clinical isolates (Table 2). In Papua, New Guinea, Currie et al. (45) found that five of six cases of cryptococcosis were caused by var. *gattii*. In Australia, Ellis (57) found that 50% of cases were due to var. *gattii*. In most studies, *C. neoformans* var. *neoformans* strains are significantly more common than *C. neoformans* var. *gattii* strains (Table

Table 2 Prevalence of the two C. *neoformans* varieties by location

Country	References	No. of isolates	Var. *neoformans* (% of isolates)	Var. *gattii* (% of isolates)
Australia	57	102	50	50
Canada	109	78	92.3	7.6
China	5	60	96.7	3.3
Brazil	109	31	64.5	35.5
	164	83	90	10
Germany	133	21	100	0
India	142	18	83	17
Kenya	7	20	100	0
New Guinea	45	6	17	83
Taiwan	89	21	95.2	4.8
Thailand	91	30	96.7	3.3
Thailand	92	26	100	0
Venezuela	214	27	70.3	29.6
United States (pre-AIDS)	12	272	85.7	14.3
United States (AIDS era)	162, 178	34	100	0

2). However, when interpreting varietal prevalence data, it is important to consider that *C. neoformans* var. *neoformans* has a predilection for patients with AIDS (20, 162, 178, 184). Since almost all cases of AIDS-related cryptococcosis are caused by *C. neoformans* var. *neoformans*, the high percentage of AIDS cases in many of the studies listed in Table 2 may contribute to the predominance of var. *neoformans* strains in the recent literature. In Zaire, the majority of isolates before 1970 were *C. neoformans* var. *gattii*, but only var. *neoformans* has been identified among clinical isolates after 1970 (200, 203). The cause for the switch in the type of varieties is not understood (203) but could reflect an increasing prevalence of human immunodeficiency virus (HIV)-associated cryptococcosis.

Rare cases of *C. neoformans* var. *gattii* infection have been reported in temperate areas (19, 50, 51, 77, 102, 120, 186). The source and route of infection for the occasional cases of cryptococcosis due to *C. neoformans* var. *gattii* in temperate areas is not understood. Some of these cases may represent reactivation of latent infection acquired in tropical and subtropical areas (19, 51, 120, 186). However, some cases could reflect undiscovered sources in the environment of temperate areas. In 1980, Muchmore et al. (135) analyzed 80 isolates from patients from Oklahoma and found 12 serotype B or C isolates of var. *gattii*. Half of these patients were long-term residents of that state and had no travel history to areas where var. *gattii* is prevalent. These investigators were unable to recover var. *gattii* from the soil in Oklahoma (135). Fromtling et al. (77) described three var. *gattii* isolates from patients who lived in Alabama, Louisiana, and Tennessee and who had never lived in or visited areas associated with *C. neoformans* var. *gattii* infections. Dromer et al. (49) described two cases of var. *gattii* infection in France among individuals who had never left the country. The source of the infection was unknown, and it was suggested that a potential source of infection was exposure to exotic imports (49).

Environmental Sources

A major advance in the understanding of the ecology of *C. neoformans* var. *gattii* was the association by Ellis and Pfeiffer (58) of this variety with *E. camaldulensis* (river red gum) trees in 1990. An extensive study of the vegetation in Barossa Valley, Australia, revealed a specific association between recovery of *C. neoformans* var. *gattii* from material associated with *E. camaldulensis* trees (58). *C. neoformans* var. *gattii* has also been recovered from the *E. camaldulensis* trees in San Francisco (147). Subsequent studies have documented an association of *C. neoformans* var. *gattii* with *E. tereticornis* (forest red gum), a close relative of *E. camaldulensis* (148). Var. *gattii* strains have been isolated from *E. camaldulensis* in Mexico City (122). Environmental isolates of *C. neoformans* var. *gattii* have been exclusively classified as serotype B (117). Serotype C has never been recovered from the environment.

C. neoformans var. *gattii* has been isolated from the environment in Australia (58, 148), Brazil (119), Mexico (122), Uruguay (79), and California (147). The only consistent environmental sources of *C. neoformans* var. *gattii* have been *E. camaldulensis* (58, 147) and *E. tereticornis* (122, 148). However, it is likely that there are other environmental sources of *C. neoformans* var. *gattii*. There is a report of isolation of *C. neoformans* var. *gattii* from bat guano found in an old house in Brazil (119). The significance of isolating this variety from bat guano is uncertain, since the site contained a variety of other animal and plant material (119). In Uruguay, an isolate of *C. neoformans* var. *gattii* was recovered from a nest of the wasp *Polybia occidentalis* during an investigation of environmental sources of *Polybia brasiliensis* (79). The site was in a gully heavily covered with subtropical vegetation, and it is unclear how the strain reached the wasp nest (79). Studies of eucalyptus trees in Africa (7) and South America (53) have not yielded *C. neoformans* var. *gattii* isolates to date. Currie et al. (45) have reported a high incidence of *C. neoformans* var. *gattii* infections in Port Moresby, New Guinea, despite the absence of *E. camaldulensis* in that environment. *C. neoformans* var. *gattii* isolates originating from *E. camaldulensis* trees have been shown to be virulent for mice and to have significant karyotypic differences from human isolates (117). DNA typing studies have identified a strain type in western Australia, where the two host eucalypti are not found (180, 181). This suggests the existence of other natural habitats for *C. neoformans* var. *gattii* apart from *E. camaldulensis* and *E. tereticornis* (180, 181).

Eucalyptus Trees and Cryptococcosis

The association between *C. neoformans* var. *gattii* and *E. camaldulensis* provides a plausible explanation for the epidemiology of *C. neoformans* var. *gattii* infections: the distribution of clinical var. *gattii* isolates approximates the geographic location of *E. camaldulensis* trees in the world (56, 58, 147, 148). Furthermore, the association of *C. neoformans* var. *gattii* with *E. camaldulensis* could explain the high prevalence of var. *gattii* infections among Australian aborigines who live in rural areas where these trees are common (64). Cryptococcosis in koala bears (17) may result from exposure to var. *gattii* in the environment, since koala bears inhabit *E. camaldulensis* trees (64). DNA typing of *C. neoformans* var. *gattii* isolates in Australia revealed the same strains in both clinical and environmental isolates, which is consistent with the hypothesis that human infection is acquired from exposure to eucalyptus trees (181).

C. NEOFORMANS VAR. *NEOFORMANS*

Geographic Distribution

C. neoformans var. *neoformans* is found in both tropical and temperate climate zones. *C. neoformans* var. *neoformans* is invariably isolated from soils and avian excreta (121). The isolation of *C. neoformans* var. *neoformans* from environmental sites has been reported from Africa (3, 159), Asia (5, 83, 89, 91, 92, 141, 142, 207, 225), Australia (76), Europe (14, 82, 133, 144, 145, 155, 182, 204, 205), North America (1, 103, 126, 130, 134, 153, 154), South America (169), the Caribbean (169), and the Pacific islands (3). Clinical infections with *C. neoformans* var. *neoformans* have also been reported from countries in all continents (Table 2). Hence, *C. neoformans* var. *neoformans* has a worldwide distribution. However, within continents there may be locations with a higher concentration of *C. neoformans*, resulting in a higher local prevalence of cryptococcosis. For example, in the United States it is generally believed that cryptococcosis is more common in the southeastern states than in the mountain states.

Environmental Sources

The first report of environmental isolation of *C. neoformans* by Sanfelice (172) in 1894 described the fungus in fermenting peach juice. Peaches and peach juice provide a good medium for the growth of *C. neoformans* (195). Interestingly, no additional environmental isolates were described until Emmons (60) described *C. neoformans* in soils contaminated with avian excreta in 1951. Since then, dozens of reports have appeared describing the isolation of *C. neoformans* var. *neoformans* from avian excreta, particularly pigeon (*Columba livia*) excreta all over the world. Less frequently, *C. neoformans* var. *neoformans* has been isolated from other sources (Table 3). Some of these sources will be considered in more detail.

Trees

There are several reports of the isolation of *C. neoformans* var. *neoformans* from trees and plant material. *C. neoformans* has been isolated from decaying wood forming hollows in living trees in Brazil (118). Tree species found to be positive for *C. neoformans* included pink shower, November shower, fig, java, munguba, coconut, and mango (118). *C. neoformans* has a laccase enzyme (224), and phenoloxidase activity in fungi has been associated with the ability to degrade wood lignin (55, 118). Although var. *gattii* is classically associated with eucalyptus trees, it is note-worthy that var. *neoformans* has also occasionally been isolated from eucalyptus

Table 3 Some environmental sources for *C. neoformans* var. *neoformans*

Source	Reference(s)
Air	177, 192, 205
Air conditioners	18, 123
Avian excreta	5, 9, 14, 47, 61–63, 76, 84, 86, 87, 92, 99, 124, 126, 130, 134, 141, 144, 145, 153–155, 158, 159, 166, 192, 204, 205, 207, 216, 221
Bagpipe	43
Bat guano	98, 118
Bat organs	44
Barns	47, 61, 129
Bat caves	8
Bird nests	9, 61, 99
Cockroaches	202
Fruits	76, 127, 194
Horse intestinal flora	211
Milk	28, 152
Pigeon excreta	5, 14, 31, 46, 47, 61, 76, 83, 84, 86, 90, 92, 99, 126, 129, 130, 134, 141, 144, 153–155, 159, 165, 166, 169, 205, 207, 216
Rabbit pens	47
Sawdust	9
Slime flux from mesquite trees	65
Soils	3, 4, 129, 182, 198
Trees and tree hollows	53, 118, 119
Water	205

trees (53). The reports of isolation of *C. neoformans* from wood (118, 119, 129) may reflect the ability of this fungus to use decaying wood as a habitat.

The reports linking var. *neoformans* to trees, together with the established association of var. *gattii* strains with eucalyptus trees and the close genetic relationship between the varieties, suggest that *C. neoformans* var. *neoformans* may also have an arboreal niche that remains to be discovered. In this regard, a flowering tree or grass has been formally suggested as the ecological niche of *C. neoformans* var. *neoformans* (59). When evaluating reports of isolation of *C. neoformans* var. *neoformans* from trees, one must consider that birds are common in trees and that the possibility of contamination with avian excreta may be a confounding variable.

In summary, *C. neoformans* var. *neoformans* can occasionally be isolated from trees. However, there is no firm connection between var. *neoformans* and any tree or plant species at this time. More research is needed to clarify the relationship of *C. neoformans* var. *neoformans* to arboreal sources.

Soils, dust, and cockroaches

Soil is often reported as an ecological site for *C. neoformans* (3, 4, 129, 182). In 1958, Ajello (4) reported the presence of *C. neoformans* in only 14 of 1,127 soil samples, 10 of which came from sites frequented by birds. The majority of reports describing *C. neoformans* in soil have used samples that originated from barns, aviaries, or areas potentially contaminated with avian excreta (3, 4, 47, 61). It has been suggested that soil is the source of infection for avian excreta, because it provides a rich medium for growth (1). However, it is unclear whether the *C. neoformans* in avian excreta originates from soil or whether *C. neoformans* in soil reflects contamination with avian excreta. There are occasional reports of *C. neoformans* isolation from soil contaminated with bat guano (8, 98, 119). The ecological relationship of *C. neoformans* colonization to mammals such as bats (44) and their guano (8, 98, 119) is uncertain and has received comparatively little attention. Other mammals, such as mice, have been shown to acquire infection when placed in cages containing soil experimentally infected with *C. neoformans* (179).

An epidemiological study in Burundi reported the isolation of *C. neoformans* from domestic dust in 54% of houses recently occupied by patients with HIV-related cryptococcosis and in 20% of randomly selected houses (201). Some dust samples had 10^6 colonies of *C. neoformans* per gram (201). Similar results were reported from Zaire, where *C. neoformans* was isolated from dust and air of the houses of patients with HIV-related cryptococcosis (202). Interestingly, that study also reported the isolation of *C. neoformans* from the digestive track of cockroaches (202). The role of cockroaches and other insects as vectors in the spread of *C. neoformans* within the environment is unknown (202).

Birds

The association of *C. neoformans* with avian excreta, particularly pigeon excreta, has been known since the 1950s and has been confirmed repeatedly by numerous investigators (Table 3). The *C. neoformans* variety found in pigeon excreta is invariably *C. neoformans* var. *neoformans*. For the remainder of this section the use of the name *C. neoformans* refers to var. *neoformans* unless stated otherwise. Most, if not all, *C. neoformans* strains recovered from pigeon excreta are virulent for mice (46, 87, 99).

C. neoformans can grow to densities of 30 to 60 million organisms per ml of either wet or desiccated pigeon excreta (124). Pigeon excreta allows long-term survival of *C. neoformans,* and cultures have been demonstrated to remain viable for almost 2 years when maintained on desiccated pigeon excreta (124). *C. neoformans* grown in pigeon excreta generally possesses small capsules (94), presumably because these conditions lack the stimuli found in host tissues that promote capsule growth. In fact, pigeon excreta have been shown to contain viable cryptococci 1 to 5.5 μm in diameter, compatible with alveolar deposition (153).

Pigeon excreta are the saprophytic source most commonly associated with *C. neoformans,* but the fungus has also been isolated from a variety of other avian excreta, including that of canaries (82, 188, 199), parrots (9, 127, 196), cuckoos (207), budgerigars (9), chickens (129, 202), and other species (84, 141, 194, 196, 221). *C. neoformans* was isolated from ambient air in the vicinity of a cage in the Berlin zoo in which a palm cockatoo was kept (192). Several studies of the comparative prevalence of *C. neoformans* in avian guano suggest that excreta of some bird species may be more likely to be contaminated with *C. neoformans* than others. Analysis of fresh bird excreta from 142 species of birds in the Berlin zoo revealed *C. neoformans* in the guano of only four species: tawny frogmouth, palm cockatoo, military macaw, and gray parrot (196). Bauwens et al. (9) sampled excreta from 261 aviaries at the Antwerp zoo, which housed 219 bird species, and recovered *C. neoformans* in only 13 samples, 7 of which came from aviaries inhabited by parrots and budgerigars. In a study of birds kept within human living areas, Weber and Schafer (221) found *C. neoformans* in 1.2% of droppings from psittacines, 1.7% of droppings from budgerigars, and 18.4% of droppings from carrier pigeons. *C. neoformans* was recovered from 26% of canary droppings in pet shops and homes in southern Italy, demonstrating a potential common source for human exposure from a popular pet bird (82). Pigeon excreta appear to be a better growth medium than chicken excreta (217). Growth of *C. neoformans* can be inhibited in chicken excreta by a combination of high pH and the presence of thermostable low-molecular-weight substances (217).

The fact that *C. neoformans* is highly prevalent in pigeon droppings is probably the result of several unique biochemical adaptations that allow its survival and growth in this ecological niche. Pigeon excreta could conceivably provide a suitable medium for *C. neoformans* mating since both α- and a-mating types can grow on pigeon manure agar and form the perfect state (191). Staib and coworkers (189, 193) obtained evidence that the association between *C. neoformans* and pigeon excreta may be due to the unique ability of *C. neoformans* to assimilate creatinine as a nitrogen source. Pigeon excreta also contain uric acid and purines, which *C. neoformans* can assimilate (101, 189, 193). Creatinine assimilation is a characteristic not common to most yeasts (101, 107, 189, 193). Thus, pigeon excreta may serve as a selective medium for *C. neoformans* because of its high creatinine content (107). The enzyme responsible for creatinine metabolism is a deiminase that catalyzes conversion to methylhydantoin and ammonia (107, 149). Growth of *C. neoformans* on uric acid or xanthine as the sole source of nitrogen results in the formation of enzyme-laden microbodies in the cytoplasm (73). The enzymes in microbodies are involved in purine salvage metabolism, and this phenomenon may reflect adaptation to growth in a milieu that is depleted of many nutrients (73). Hence, avian guano may be a natural reservoir for *C. neoformans,* because this variety has a specific regulatory mechanism for creatinine assimilation that is adapted to the special chemistry of avian guano (107).

It is interesting that *C. neoformans* var. *gattii* is not associated with avian excreta despite its also being capable of assimilating creatinine (107, 149). Creatinine deiminase is found in both varieties of *C. neoformans* (107, 149). However, the inability of *C. neoformans* var. *gattii* to colonize avian excreta may reflect differences in the regulation of the enzymatic pathway (107, 149). The deiminase responsible for creatinine metabolism in var. *neoformans* is repressed by ammonia, whereas that in var. *gattii* is not (107, 149). Regulation of deiminase by ammonia in var. *neoformans* has been proposed to help conserve energy and metabolic substances needed for survival in avian excreta (107, 149). Furthermore, since the metabolism of creatinine by var. *neoformans* results in the accumulation of ammonia and a rise in pH, inhibition of creatinine deiminase by ammonia in var. *neoformans* strains presumably prevents unchecked alkalinization of the environment, which would inhibit growth.

The mechanism by which guano becomes infected with *C. neoformans* is uncertain. Noninfected pigeon excreta have been shown to become infected when exposed to air that contains aerosolized *C. neoformans* cells (205), suggesting that avian guano can be infected by airborne spread. *C. neoformans* is not found in commercial bird feeds (21). This suggests that, for domesticated birds, ingestion of contaminated food is not likely to be the mechanism for infection of their excreta.

Survival of *C. neoformans* in avian excreta, soils contaminated by avian excreta, and other saprophytic sources can be influenced by several factors, including pH, humidity, temperature, sunlight, and biotic factors (90, 94, 124, 217). *C. neoformans* can survive on pigeon excreta for more than 1 year (124). Increased humidity can enhance the survival and proliferation of the fungus in soil at lower temperatures (4 to 26°C) but not at 37°C (94). Direct sunlight can significantly reduce the survival of cryptococci in soils (94) and may account for the fact that pigeon excreta from outdoor sites are less heavily contaminated than pigeon excreta from indoor sites (90, 124, 126). Denton and Di Salvo (47) demonstrated that sheltered environmental locations are more likely to be culture positive for *C. neoformans* than are locations exposed to sunlight. Maintenance of *C. neoformans* colonization in sites contaminated with pigeon excreta appears to be dependent on the moisture and the availability of fresh pigeon droppings at the site (168). Hubalek (90) studied the distribution of *C. neoformans* in a church tower, found a correlation between the presence of this fungus and the uric acid and creatinine content in pigeon excreta, and suggested that the nutritional value of excreta was a factor in the prevalence of the fungus.

Bulmer and coworkers (24, 165, 166, 168) carried out a series of longitudinal studies on the persistence and spread of *C. neoformans* in two building towers contaminated with pigeon excreta at a university in Oklahoma. These studies are significant because they were done at the same site over a number of years and provide some of the best studies available on the interplay between *C. neoformans*, pigeons, pigeon excreta, and human-made structures. *C. neoformans* was cultured from the air in the tower by exposing open petri dishes containing Sabouraud agar to air for 15 min. The air was analyzed with an Anderson air sampler, and a large percentage of the particles in the air were shown to be 0.65 to 4.7 μm in diameter (165, 166), a size compatible with alveolar deposition (153). Sweeping floors with brooms increased the number of aerosolized *C. neoformans* particles by 10-fold within 6 min (165). Analysis of 214 isolates recovered from the air in the towers revealed that all isolates were self-sterile, that 193 were α-mating type, and that none were **a**-mating type (97). The mating-type imbalance was interpreted to indicate that the aerosolized cells were yeast cells and not basidiospores, since basidiospores would have been expected to have a 1:1 distribution of α- and **a**-mating types (97). Chronic exposure of mice to air inside the towers did not result in fatal cryptococcosis, but several mice were naturally infected, as demonstrated by recovery of *C. neoformans* from internal organs. Analysis of air currents revealed that the concentration of *C. neoformans* in air was dependent on the wind direction, which provided evidence for a role of wind currents in the spread of *C. neoformans* in the environment. When the towers were sealed to prevent the entrance of pigeons, the density of *C. neoformans* slowly decreased over a 2-year period, indicating a need for infusion of fresh pigeon excreta to maintain the fungal population. Hence, it appears that *C. neoformans* can survive in pigeon excreta for a long time but not indefinitely without the infusion of new nutrients. These studies suggest that a safe and effective method for decontamination of human-made structures over time is to prevent access by birds.

The role of pigeons in the dissemination of cryptococcus in the environment is not fully understood despite considerable experimentation by many investigators. In 1955, Emmons (61) studied 20 pigeons for the presence of *C. neoformans* and found none that were infected. In the 1960s Littman and Borok (124) extensively studied the relationship of pigeons to *C. neoformans*. Of 94 feral pigeons caught in New York City, they found that *C. neoformans* could be cultured only from the feet and beaks of 7% of the birds (124). In contrast, Swinne-Desgain (204, 205) studied pigeons in Belgium, isolated *C. neoformans* from the crops of 50% of the birds, and concluded that pigeons were a reservoir for the fungus. Pigeons are resistant to *C. neoformans* infection by gastrointestinal (1, 124, 174) and intramuscular (190) inoculation, but infection can be induced by intracerebral (125) or ocular (175) inoculation. Resistance to infection in pigeons may be the result of a combination of high body temperature (41 to 43°C) and the presence of endogenous gut bacterial flora that are inhibitory to *C. neoformans*. Oral administration of *C. neoformans* to pigeons revealed that the fungus can survive passage through the pigeon intestinal tract despite the elevated body temperature (124, 174). *C. neoformans* can also survive passage through the canary alimentary tract (188). In some experiments, stool shedding after oral administration lasted for only 24 h, a phenomenon attributed to inhibition of *C. neoformans* by the pigeon intestinal bacterial flora (1). In summary, the available evidence suggests that, although pigeons may be infected with *C. neoformans* in their crops, they are resistant to disseminated infection. Pigeons may contribute to the propagation of *C. neoformans* by providing a rich culture medium in the form of pigeon excreta and could disseminate the fungus in the environment by carrying the fungus in their beaks and feet.

Relationship of Pigeons to Human Cryptococcosis

Because of the high concentration of *C. neoformans* yeasts in pigeon excreta, the high prevalence of pigeons in many urban areas, and the seriousness of cryptococcal infections in humans, there has been considerable interest in the relationship between the birds, their excreta, and human cryptococcosis (61, 124, 126, 144, 145, 153). There are many anecdotal reports in the literature linking human infection with exposure to *C. neoformans*-contaminated sites (48, 68, 83, 123, 154, 194). *C. neoformans* has been recovered from dust in an air conditioner contaminated with avian feces (74), raising the concern that mechanical ventilation may help to spread cryptococcus if the air intake is contaminated. Littman (123) described a medical student who developed cryptococcal infection after working in a hospital library near an air conditioner laden with pigeon excreta. In 1963, Muchmore et al. (134) reported three cases of cryptococcal meningitis in 1 year in a small community in Oklahoma. Analysis of bird droppings revealed *C. neoformans* in the immediate environment of each patient, leading to the speculation that infection occurred from exposure to contaminated bird droppings. Pigeon trapping and exposure to pigeon feces were associated with a case of fulminant cryptococcal hepatitis in a teenager (154). Gugnami et al. (83) isolated *C. neoformans* from an attendant of a pigeon house at the Delhi zoo who had pulmonary symptoms. More recently, two cases of cryptococcal meningitis were described in patients exposed to avian excreta (70). One individual developed cryptococcosis after using a chain saw to demolish an aviary that had been unused for 10 years, suggesting that *C. neoformans* can withstand prolonged drying (70). Cryptococcosis has been associated with pecking by a pigeon in a patient with HIV-negative lymphopenia (48).

Circumstantial evidence for subclinical infection after exposure to pigeon guano also comes from serological studies of pigeon fanciers (71, 215). Walter and Atchison (215) demonstrated that 22% of pigeon fanciers had complement-fixing anticryptococcal antibodies, compared with 3% of a control group composed of non-pigeon breeders (215). Tanphaichitra et al. (206) surveyed 101 individuals involved in pigeon racing or farming in Thailand and found that 4% had low-titer positive serum cryptococcal antigen tests, which suggested actual chronic infections. Finally, skin testing with cryptococcal antigen preparations (cryptococci) revealed that the prevalence of skin reactivity among pigeon fanciers was 32%, compared with 4% in a control group (137). The serological and skin reactivity studies strongly suggest that asymptomatic infection can occur among individuals with high exposure to *C. neoformans,* such as pigeon breeders and fanciers.

Several studies have attempted to clarify the relationship of environmental isolates primarily isolated from pigeon excreta to clinical cases by DNA typing. Currie et al (46) analyzed RFLPs of clinical and environmental isolates from New York City and identified two strains that were linked in both clinical and environmental samples. A study in Nagasaki using random amplified polymorphic DNA analysis demonstrated that the same strains could be found in clinical and environmental samples (225). In Belo Horizonte, Brazil, the same strains were recovered from patients and the environment (75). On the other hand, Varma et al. (213) compared clinical isolates with those obtained from the patient's immediate environment but were unable to demonstrate a connection by RFLP DNA typing studies. This negative result is not surprising considering the chronic nature of cryptococcal infection, the possibility that some cases represent reactivation of latent infection, the multitude of environmental sources, and the genetic heterogeneity of *C. neoformans* strains. Identifying a point source for cryptococcal infection for a given individual may be very difficult (46). Nevertheless, finding the same strains among environmental and clinical isolates in three studies (46, 75, 225) is consistent with the hypothesis that some human infections can originate from an immediate environmental reservoir.

The evidence suggesting pigeon excreta as a source of human infections is (i) anecdotal reports of cryptococcosis following exposure to contaminated sites (see above), (ii) identification of the same strains in pigeon excreta and clinical isolates by DNA typing in three independent studies (46, 75, 225), (iii) recovery of infectious *C. neoformans* particles from pigeon excreta of a size compatible with alveolar deposition (153), (iv) recovery of *C. neoformans* from the air above sites with contaminated avian guano (189, 205), (v) the virulence of strains isolated from pigeon excreta (87), and (vi) a high prevalence of anticryptococcal antibodies and positive skin tests in pigeon breeders (71). However, there are many unanswered questions about the ecology of *C. neoformans* and the pathogenesis of human infection that preclude a definitive association between the strains in pigeon excreta and those that cause human infection. For example, uncertainty as to whether human infections are the result of acute exposure or reactivation of latent infection makes it difficult to link cases to point sources in the environment. It is conceivable that the *C. neoformans* var. *neoformans* strains in pigeon excreta and human infection originate from a third, yet undiscovered, environmental source (59). In this regard, the discovery of the association of *C. neoformans* var. *gattii* with eucalyptus trees has led to the proposal that human *C. neoformans* var. *neoformans* infections are caused by basidiospores originating in a flowering plant, possibly a grass (59). Hence, the evidence linking some cases of cryptococcosis to acquisition of infection from pigeon excreta is strongly suggestive but not conclusive.

Pigeon Control and Decontamination of Environmental Sites

The case for pigeon control rests on the assumption that the high concentrations of *C. neoformans* found in pigeon excreta pose a health risk to human populations and, in particular, to immunocompromised individuals. Considering that cryptococcosis is usually a life-threatening infection in immunosuppressed patients, the question of what, if anything, to do about pigeons and pigeon excreta in urban environments is of potential importance.

The recommendation made by several authorities that immunosuppressed patients avoid birds and sites contaminated with bird excreta (41, 197) is reasonable and noncontroversial but at times impractical (see chapter 15). A study in Argentina described three HIV-infected patients who developed cryptococcosis after being admitted to the same ward, suggesting the possibility of in-hospital acquisition of infection (170). The isolation of *C. neoformans* from bird excreta in pet shops, bird hatcheries, and private homes raises the concern that pet birds can be environmental sources for exposure to cryptococcus (126, 194). Similarly, the recommendation to sterilize and clean up heavy deposits of weathered avian excreta in sites in close proximity to patient homes, hospitals, and nursing homes makes sense (145). The experience with the Oklahoma university towers shows that sealing a building to prevent penetration by pigeons results in long-term reduction of *C. neoformans*, even if the building is initially heavily contaminated (24). *C. neoformans* is sensitive to high pH, and contaminated sites can be sterilized by alkalinization with lime solution (216). Alkalinization with lime solution is useful in eradicating

C. neoformans from pigeon coops, because it appears to have no untoward effects on either pigeons or pigeon fanciers (216).

Proposals to control the pigeon population are more controversial (126, 173). An effective and humane method for pigeon control is to discourage pigeon feeding by citizens in public areas (85). In urban areas, pigeon populations are often over-crowded and are dependent on humans for sustenance (85). In Basel, a public information campaign to make citizens aware of the problems of pigeon overpopu-lation resulted in a 50% reduction in the number of pigeon flocks (85). However, it should be emphasized that pigeons are not the only possible vector for the spread and maintenance of *C. neoformans* in the environment. For example, in sub-Saharan Africa there are few pigeons but the prevalence of cryptococcosis is extremely high.

METHODOLOGY FOR THE ISOLATION OF *C. NEOFORMANS* FROM THE ENVIRONMENT

Many of the early studies of saprophytic sources of *C. neoformans* relied on inocu-lation of the sample in question into mice, since mice are highly susceptible to *C. neoformans* but resistant to other yeasts (4, 47, 61, 99, 129, 130). Later studies of the prevalence of *C. neoformans* in the environment have relied on selective media for the obvious reasons of cost and ease of study (46, 126, 144, 160). One study reported that mouse inoculation was superior to plating methods for recovery of *C. neofor-mans* from the environment (130).

Selective media are necessary for the isolation of *C. neoformans* from environ-mental sources, and several have been described (46, 158, 177, 192, 197). The media used should accomplish three goals: (i) inhibit bacteria, (ii) inhibit molds, (iii) permit the visual identification of putative *C. neoformans* colonies. In general, because *C. neoformans* grows more slowly than other fungi, it is important to use selective media that permit the growth of *C. neoformans* while simultaneously inhibiting other micro-organisms. Bacteria can be inhibited by the addition of antibiotics and the adjust-ment of agar pH to the acidic range (pH 4 to 5). Most molds can be inhibited by the combination of adding biphenyl to the agar (177) and incubating the plates at 37°C, a temperature that is nonpermissive for many molds. Visual identification of likely *C. neoformans* colonies can be facilitated by the addition of substrates suitable for the production of melanin. A widely used medium is *Guizotia abyssinica* (niger seed) creatinine agar supplemented with 0.1% biphenyl (192). Currie et al. (46) success-fully used a chemically defined medium supplemented with biphenyl and l-dopa for melanin induction for the recovery of pigeon guano.

The isolation of *C. neoformans* from environmental sources is more likely to be successful if the chosen site is shielded from direct sunlight. Direct sunlight can sterilize soil samples seeded with *C. neoformans* (94). Humidity enhances survival of *C. neoformans* in soils (94). Old, weathered pigeon excreta and the dirt and dust around the droppings are more likely to be positive for *C. neoformans* than are fresh droppings (160).

Biotic factors that influence the survival of *C. neoformans* in the environment include bacteria, amoebae, mites, and sow bugs (167). *Pseudomonas aeruginosa* and *Bacillus subtilis* isolated from pigeon excreta can inhibit *C. neoformans* (167, 208). Amoebae (*Acanthamoeba palestinensis*) isolated from pigeon excreta have been shown to phagocytose cryptococci and to inhibit their growth in vitro (167). Cryptococci have been called a "gastronomic delight" for soil amoebae (26). Interestingly, pseudohyphal forms of *C. neoformans* appear to be resistant to killing by amoebae, suggesting that these fungal forms may persist in sites inhabited by amoebae (136). Mites and sow bugs also appear to feed on *C. neoformans*, and yeast forms have been demonstrated in their guts (167). Other fungi may inhibit the growth of *C. neoformans* by elaborating killer toxins (16).

REPRODUCTION IN THE ENVIRONMENT: SEXUAL OR ASEXUAL?

C. neoformans strains are usually haploid and belong to either the **a**- or α-mating type. All *C. neoformans* strains are capable of asexual reproduction (budding), and some strains have been shown to be capable of sexual reproduction in the laboratory. Sexual reproduction results in the production of haploid basidiospores. The question of which mode of reproduction is employed by *C. neoformans* in the environment may be important for a complete understanding of its ecology and pathogenesis. For example, if sexual reproduction predominates in the environment, inhalation or ingestion of basidiospores (which are ideally sized at 1 to 2 μm for lung deposition) would be a plausible mode for human infection, as has been suggested (42, 59). Conversely, if *C. neoformans* reproduces asexually in the environment, then yeast forms are likely to be the infectious particles. Genetic analysis of other medically important fungi, notably *Coccidioides immitis* (27) and *Aspergillus fumigatus* (78), have provided evidence for sexual reproduction in the environment. Asexual reproduction would result in a clonal population structure, whereas sexual reproduction would involve chromosomal recombination and sorting.

Several lines of evidence suggest that the primary mode of *C. neoformans* reproduction is asexual. First, the ratio of α- to **a**-mating types among clinical and environmental isolates is 40:1. Since sexual reproduction results in a 1:1 mating-type ratio, the predominance of α-mating-type strains suggests strong selection for α-mating type, against **a**-mating type and/or infrequent mating. There are data that suggest that the α-mating-type locus is associated with virulence for *C. neoformans* (112), but this would not necessarily explain the imbalance in mating types in environmental isolates. First, at the very least, the 40:1 imbalance in mating types indicates a relative paucity of **a**-mating-type strains for mating. Second, the **a**-mating type appears to be very infrequent among serotype A strains. For example, a comprehensive mating study of 158 *C. neoformans* var. *neoformans* isolates from clinical and environmental isolates identified only two **a**-mating-type strains (128). Third, DNA typing of multiple isolates from single environmental sites in New York City revealed that the majority of isolates had the same CNRE-1 repetitive DNA RFLPs, consistent with clonal origin (46). Fourth, *C. neoformans* strains exhibit considerable heterogeneity in electrophoretic karyotype (52, 146, 223). Differences in the number and size of chromosomes may produce incompatible karyotypes for sexual reproduction or simply low spore viability (223). Fifth, sequence analysis of the *URA5* alleles combined with RFLP analysis (36, 75) and multilocus enzyme analysis (22, 209) suggests a clonal population structure for *C. neoformans* var. *neoformans*, which implies that the primary mode of replication for most strains in nature is asexual.

Recently, Wickes et al. (222) described haploid fruiting in *C. neoformans*. Under conditions of relative dryness and low ammonia concentration, several strains of *C. neoformans* formed hyphae and basidiospores. Hypha formation from haploid cells differed from that resulting from sexual reproduction in that asexual hyphae had only one nuclei and more blastospores than did sexual hyphae. The ability to undergo dimorphism and monokaryotic fruiting was associated with the α-mating locus. The majority of strains observed to undergo haploid fruiting in the laboratory were *MATα*, whereas none of the *MAT***a** strains produced hyphae. These observations suggest an explanation for the preponderance of α-mating-type strains among environmental and clinical strains, given the ability of this mating type to undergo haploid fruiting. Furthermore, haploid fruiting provides a possible mechanism for the generation of small infectious particles (basidiospores) that are capable of reaching the alveoli after inhalation without a requirement for sexual reproduction. Hence, the small airborne infectious particles of *C. neoformans* that are sometimes recovered from the air may be basidiospores arising from haploid fruiting of α-mating-type strains.

In summary, the relative importance of sexual and asexual replication of *C. neoformans* in the environment is poorly understood. Present evidence favors a clonal population structure for *C. neoformans* (75). A clonal population implies the predominance of asexual reproduction in the propagation of this fungus in nature (75, 209). Sexual reproduction for *C. neoformans* has been observed only in the laboratory and is difficult to achieve for most strains. However, it is not clear if the difficulty observed for mating *C. neoformans* in the laboratory is a result of forcing a vestigial reproductive mode on strains of *C. neoformans* or simply the inability to create optimal conditions for sexual reproduction to occur.

VIRULENCE FACTORS IN THE CONTEXT OF ENVIRONMENTAL SURVIVAL

Three classical virulence factors for *C. neoformans* are the polysaccharide capsule (25, 33, 115), the ability to grow at 37°C, and melanin synthesis (114, 151, 161). Since *C. neoformans* is not a known commensal for mammals and does not require animal infection for completion of its life cycle, it is likely that those phenotypes associated with virulence were selected because they provided a survival advantage in the environment. For example, the polysaccharide capsule is antiphagocytic and could provide protection against soil amoebae (167) and desiccation (140). The ability to survive over a broad temperature range (0 to 40°C) allows this fungus to survive in temperature extremes. The phenoloxidase activity, and its product melanin, may protect the yeast against solar radiation (218) and environmental oxidants (95, 96, 219). Furthermore, phenoloxidase activity has been associated with wood degradation (55, 220) and this enzyme may contribute to survival in decaying wood habitats.

Recently, other *C. neoformans* characteristics have been suggested to contribute to virulence, including mannitol production (35) and the elaboration of several enzymes that have the potential to degrade tissue, such as proteases (23, 37) and phospholipases (38, 39). Mannitol can serve an essential function in environmental survival of *C. neoformans* by protecting against osmotic stress (34) and serving as a nutritional reserve (138). Proteases, lipases, and phospholipases are likely to serve a nutritional role for *C. neoformans* by freeing extracellular nutrients.

In summary, most phenotypes associated with virulence are adaptations that are also useful for environmental survival. The ability of *C. neoformans* to cause disease means that it can successfully colonize and reproduce in the human host. Virulence (or pathogenic) factors represent characteristics that allow this free-living organism to survive in the ecological niche defined by the mammalian host. Mechanisms of virulence and traits that permit survival in the environment may be functionally linked for *C. neoformans*.

UNSOLVED PROBLEMS IN THE ECOLOGY OF *C. NEOFORMANS*

Much has been learned about the ecology of *C. neoformans* in the 4 decades since Emmons (61) isolated the fungus from soils contaminated with avian excreta. Major advances in recent years include the description of the perfect state of *C. neoformans* (104–106, 110), the separation of *C. neoformans* strains into distinct varieties, the association of var. *gattii* with eucalyptus trees (56, 58, 59, 147, 148), and the application of molecular tools to understanding the relationship between clinical and environmental isolates (46, 213, 225). However, fundamental questions of *C. neoformans* ecology and pathogenesis remain unanswered. The form of the infectious propagule for both varieties of *C. neoformans* is uncertain (42, 59). Many potential reservoirs for *C. neoformans* in the environment remain unexplored. In this regard, occasional reports of the isolation of var. *neoformans* from trees (65, 118) and of var. *gattii* from bat guano (119) and wasp nests (79) provide tantalizing hints of the existence of additional environmental sources and habitats for this fungus. In 1967, Pappagianis (143) reviewed aspects of *C. neoformans* epidemiology and stated that the "prefix *crypto* thus continues to be an appropriate one, since much of the epidemiology remains obscure." Thirty years later this statement remains true despite considerable progress in understanding the biology, ecology, and epidemiology of *C. neoformans*.

REFERENCES

1. **Abou-Gabal, M., and M. Atia.** 1978. Study of the role of pigeons in the dissemination of *Cryptococcus neoformans* in nature. *Sabouraudia* **16:**63–68.
2. **Abrahan, M., V. Mathews, A. Ganesh, J. John, and M. S. Mathews.** 1997. Infection caused by *Cryptococcus neoformans* var. *gattii* serotype B in an AIDS patient in India. *J. Med. Vet. Mycol.* **35:**283–284.
3. **Ajello, L.** 1956. Soil as natural reservoir for human pathogenic fungi. *Science* **123:**876–879.
4. **Ajello, L.** 1958. Occurrence of *Cryptococcus neoformans* in soils. *Am. J. Hyg.* **67:**72–77.
5. **Ansheng, L., K. Nishimura, H. Taguchi, R. Tanaka, W. Shaoxi, and M. Miyaji.** 1993. The isolation of *Cryptococcus neoformans* from pigeon droppings and serotyping of naturally and clinically sourced isolates in China. *Mycopathologia* **124:**1–5.
6. **Aulakh, H. S., S. E. Straus, and K. J. Kwon-Chung.** 1981. Genetic relatedness of *Filobasidiella neoformans* (*Cryptococcus neoformans*) and *Filobasidiella bacillispora* (*Cryptococcus bacillisporus*) as determined by deoxyribonucleic acid base composition and sequence homology studies. *Int. J. Syst. Bacteriol.* **31:**97–103.
7. **Batchelor, B. I. F., R. J. Brindle, and P. G. Waiyaki.** 1994. Clinical isolates of HIV-associated cryptococcosis in Kenya. *Trans. R. Soc. Trop. Med. Hyg.* **88:**85.
8. **Baum, G., and D. Artis.** 1966. Isolation of fungi from Judean soil. *Mycopathol. Mycol. Appl.* **29:**350–354.
9. **Bauwens, L., D. Swinne, C. De Vroey, and W. D. Meurichy.** 1986. Isolation of

Cryptococcus neoformans var *neoformans* in the aviaries of the Antwerp Zoological Gardens. *Mykosen* **29**:291–294.

10. **Benham, R. W.** 1935. Cryptococci—their identification by morphology and by serology. *J. Infect. Dis.* **57**:255–274.

11. **Benham, R. W., and A. M. Hopkins.** 1933. Yeastlike fungi found on the skin and in the intestines of normal subjects. *Arch. Dermatol. Syph.* **28**:532–543.

12. **Bennett, J. E., K. J. Kwon-Chung, and D. H. Howard.** 1977. Epidemiologic differences among serotypes of *Cryptococcus neoformans*. *Am. J. Epidemiol.* **105**:582–586.

13. **Bennett, J. E., K. J. Kwon-Chung, and T. S. Theodore.** 1978. Biochemical differences between serotypes of *Cryptococcus neoformans*. *Sabouraudia* **16**:167–174.

14. **Bergman, F.** 1963. Occurrence of *Cryptococcus neoformans* in Sweden. *Acta Med. Scand.* **174**:651–655.

15. **Beyt, B. E., and S. R. Waltman.** 1978. Cryptococcal endoththalmitis after corneal transplantation. *N. Engl. J. Med.* **298**:825–826.

16. **Boekhout, T., and G. Scorzetti.** 1997. Differential killer toxin sensitivity patterns of varieties of *Cryptococcus neoformans*. *J. Med. Vet. Mycol.* **35**:147–149.

17. **Bolliger, A., and E. S. Finckh.** 1962. The prevalence of cryptococcosis in the koala (Phascolarctos cinereus). *Med. J. Aust.* **49**:545–547.

18. **Botard, R. W., and D. C. Kelley.** 1969. A survey to determine the occurrence of *Histoplasma capsulatum* and *Cryptococcus neoformans* in air-conditioners. *Mycopathologia* **37**:372–376.

19. **Bottone, E. J., P. A. Kirschner, and I. F. Salkin.** 1986. Isolation of highly encapsulated *Cryptococcus neoformans* serotype B from a patient in New York City. *J. Clin. Microbiol.* **23**:186–188.

20. **Bottone, E. J., I. F. Salkin, N. J. Hurd, and G. P. Wormser.** 1987. Serogroup distribution of *Cryptococcus neoformans* in patients with AIDS. *J. Infect. Dis.* **156**:242.

21. **Brandsberg, J. W., and P. B. Kretchmer.** 1972. Cryptococci of commercial bird feeds. *Sabouraudia* **10**:43–46.

22. **Brandt, M., L. C. Hutwagner, L. A. Klug, W. S. Baughman, D. Rimland, E. A. Graviss, R. J. Hamill, C. Thomas, P. G. Pappas, A. L. Reingold, and R. W. Pinner.** 1996. Molecular subtype distribution of *Cryptococcus neoformans* in four areas of the United States. *J. Clin. Microbiol.* **34**:912–917.

23. **Brueske, C. H.** 1986. Proteolytic activity of a clinical isolate of *Cryptococcus neoformans*. *J. Clin. Microbiol.* **23**:631–633.

24. **Bulmer, G. S.** 1990. Twenty-five years with *Cryptococcus neoformans*. *Mycopathologia* **109**:111–122.

25. **Bulmer, G. S., M. D. Sans, and C. M. Gunn.** 1967. *Cryptococcus neoformans*. I. Nonencapsulated mutants. *J. Bacteriol.* **94**:1475–1479.

26. **Bunting, L. A., J. B. Neilson, and G. S. Bulmer.** 1979. *Cryptococcus neoformans*: gastronomic delight of a soil ameba. *Sabouraudia* **17**:225–232.

27. **Burt, A., D. A. Carter, G. L. Koenig, T. J. White, and J. W. Taylor.** 1996. Molecular markers reveal cryptic sex in the human pathogen *Coccidioides immitis*. *Proc. Natl. Acad. Sci. USA* **93**:770–773.

28. **Carter, H. S., and J. L. Young.** 1950. Note on the isolation of *Cryptococcus neoformans* from a sample of milk. *J. Pathol. Bacteriol.* **62**:271–273.

29. **Casadevall, A., and M. Fan.** 1992. *URA5* gene of *Cryptococcus neoformans* var. *gattii*. Evidence for a close phylogenetic relationship between *C. neoformans* var. *gattii* and *C. neoformans* var. *neoformans*. *J. Gen. Appl. Microbiol.* **38**:491–495.

30. **Casadevall, A., L. Freundlich, L. Marsh, and M. D. Scharff.** 1992. Extensive allelic variation in *Cryptococcus neoformans*. *J. Clin. Microbiol.* **30**:1080–1084.

31. **Castanon-Olivares, L. R., and R. Lopez-Martinez.** 1994. Isolation of *Cryptococcus neoformans* from pigeon (*Columba livia*) droppings in Mexico City. *Mycoses* **37**:325–327.

32. **Castanon-Olivares, L. R., R. Lopez-Martinez, G. Barriga-Angulo, and C. Rios-Rosas.** 1997. *Cryptococcus neoformans* var. *gattii* in an AIDS patient: first observation in Mexico. *J. Med. Vet. Mycol.* **35**:57–59.

33. **Chang, Y. C., and K. J. Kwon-Chung.** 1994. Complementation of a capsule-deficient mutation of *Cryptococcus neoformans* restores its virulence. *Mol. Cell. Biol.* **14**:4912–4919.

34. **Chatuverdi, V., T. Flynn, W. G. Niehaus, and B. Wong.** 1996. Stress tolerance and pathogenic potential of a mannitol mutant of *Cryptococcus neoformans*. *Microbiology* **142**:937–943.

35. **Chatuverdi, V., B. Wong, and S. L. Newman.** 1996. Oxidative killing of *Cryptococcus neoformans* by human leukocytes. Evidence that fungal mannitol protects by scavenging reactive oxygen intermediates. *J. Immunol.* **156**:3836–3840.

36. **Chen, F., B. P. Currie, L.-C. Chen, S. G. Spitzer, E. D. Spitzer, and A. Casadevall.** 1995. Genetic relatedness of *Cryptococcus neoformans* clinical isolates grouped with the repetitive DNA probe CNRE–1. *J. Clin. Microbiol.* **33**:2818–1822.

37. **Chen, L.-C., E. Blank, and A. Casadevall.** 1996. Extracellular proteinase activity of *Cryptococcus neoformans*. *Clin. Diagn. Lab. Immunol.* **3**:570–574.

38. **Chen, S. C. A., M. Muller, J. Z. Zhou, L. Wright, and T. C. Sorrell.** 1997. Phospholipase activity in *Cryptococcus neoformans*: a new virulence factor? *J. Infect. Dis.* **175**:414–420.

39. **Chen, S. C. A., L. C. Wright, R. T. Santangelo, M. Muller, V. R. Moran, P. W. Kuchel, and T. C. Sorrell.** 1997. Identification of extracellular phospholipase B, lysophospholipase, and acyltransferase produced by *Cryptococcus neoformans*. *Infect. Immun.* **65**:405–411.

40. **Cherniak, R., and J. B. Sundstrom.** 1994. Polysaccharide antigens of the capsule of *Cryptococcus neoformans*. *Infect. Immun.* **62**:1507–1512.

41. **Clark, R. A., D. Greer, W. Atkinson, G. T. Valianis, and N. Hyslop.** 1990. Spectrum of *Cryptococcus neoformans* infection on 68 patients infected with human immunodeficiency virus. *Rev. Infect. Dis.* **12**:768–777.

42. **Cohen, J., J. R. Perfect, and D. T. Durack.** 1982. Cryptococcosis and the basidiospore. *Lancet* **i**:1501.

43. **Corcroft, R., H. Kronenberg, and T. Wilkinson.** 1978. Cryptococcus in bagpipes. *Lancet* **i**:1308–1309.

44. **Crose, E., C. J. Marinkelle, and C. Striegel.** 1967. The use of tissue cultures in the identification of *Cryptococcus neoformans* isolated from Colombian bats. *Sabouraudia* **6**:127–132.

45. **Currie, B., T. Vigus, G. Leach, and B. Dwyer.** 1990. *Cryptococcus neoformans* var. *gattii*. *Lancet* **336**:1442.

46. **Currie, B. P., L. F. Freundlich, and A. Casadevall.** 1994. Restriction fragment length polymorphism analysis of *Cryptococcus neoformans* isolates from environmental (pigeon excreta) and clinical isolates in New York City. *J. Clin. Microbiol.* **32**:1188–1192.

47. **Denton, J. F., and A. F. Di Salvo.** 1968. The prevalence of *Cryptococcus neoformans* in various natural habitats. *Sabouraudia* **6**:213–217.

48. **Dev, D., G. S. Basran, D. Slater, P. Taylor, and M. Wood.** 1994. Consider HIV negative immunodeficiency in cryptococcosis. *Br. Med. J.* **308**:1436.

49. **Dromer, F., S. Mathoulin, B. Dupont, and A. Laporte.** 1996. Epidemiology of cryptococcosis in France: a 9-year survey (1985–1993). *Clin. Infect. Dis.* **23**:82–90.

50. **Dromer, F., S. Mathoulin, B. Dupont, L. Letenneur, and O. Ronin.** 1996. Individual and environmental factors associated with infection due to *Cryptococcus neoformans* serotype D. *Clin. Infect. Dis.* **23**:91–96.

51. **Dromer, F., O. Ronin, and B. Dupont.** 1992. Isolation of *Cryptococcus neoformans* var, *gattii* from an Asian patient in France: evidence for dormant infection in healthy subjects. *J. Med. Vet. Mycol.* **30**:395–397.

52. **Dromer, F., A. Varma, O. Ronin, S. Mathoulin, and B. Dupont.** 1994. Molecular typing of *Cryptococcus neoformans* serotype D clinical isolates. *J. Clin. Microbiol.* **32**:2364–2371.

53. **Duarte, A., N. Ordonez, and E. Castaneda.** 1994. Association de levaduras del genero *Cryptococcus* con especies de *Eucalyptus* en Santa Fe de Bogota. *Rev. Inst. Med. Trop. Sao Paulo* **36**:125–130.

54. **Duperval, R., P. E. Hermans, N. S. Brewer, and G. S. Roberts.** 1977. Cryptococcosis, with emphasis on the significance of isolation of *Cryptococcus neoformans* from the respiratory tract. *Chest* **72**:13–19.

55. **Eggert, C., U. temp, and K.-E. Ericksson.** 1996. The ligninolytic system of the white rot fungus *Pycnoporus cinnabarinus*: purification and characterization of the laccase. *Appl. Environ. Microbiol.* **62**:1151–1158.

56. **Ellis, D. and T. J. Pfeiffer.** 1992. The ecology of *Cryptococcus neoformans*. *Eur. J. Epidemiol.* **8**:321–325.

57. **Ellis, D. H.** 1987. *Cryptococcus neoformans* var. *gattii* in Australia. *J. Clin. Microbiol.* **25**:430–431.

58. **Ellis, D. H., and T. J. Pfeiffer.** 1990. Natural habitat of *Cryptococcus neoformans* var. *gattii*. *J. Clin. Microbiol.* **28**:1642–1644.

59. **Ellis, D. H., and T. J. Pfeiffer.** 1990. Ecology, life cycle, and infectious propagule of *Cryptococcus neoformans*. *Lancet* **336**:923–925.

60. **Emmons, C. W.** 1951. Isolation of *Cryptococcus neoformans* from soil. *J. Bacteriol.* **62**:685–690.

61. **Emmons, C. W.** 1955. Saprophytic sources of *Cryptococcus neoformans* associated with the pigeon (*Columba livia*). *Am. J. Hyg.* **62**:227–252.

62. **Emmons, C. W.** 1960. Prevalence of *Cryptococcus neoformans* in pigeon habitats. *Public Health Rep.* **75**:362–364.

63. **Emmons, C. W.** 1962. Natural occurrence of opportunistic fungi. *Lab. Invest.* **11**:1026–1032.

64. **Eng, R. H. K., E. Bishburg, S. M. Smith, and R. Kapila.** 1986. Cryptococcal infections in patients with acquired immune deficiency syndrome. *Am. J. Med.* **81**:19–23.

65. **Evenson, A., and J. W. Lamb.** 1964. Slime flux of mesquite as a new saprophytic source of *Cryptococcus neoformans*. *J. Bacteriol.* **88**:542.

66. **Fan, M., L.-L. Chen, M. A. Ragan, R. R. Gutell, J. R. Warner, B. P. Currie, and A. Casadevall.** 1995. The 5S rRNA and the rDNA intergenic spacer of the two varieties of *Cryptococcus neoformans*: sequence, structure, and phylogenetic implications. *J. Med. Vet. Mycol.* **33**:215–221.

67. **Fan, M., B. P. Currie, R. R. Gutell, M. A. Ragan, and A. Casadevall.** 1994. The 16S-like, 5.8S, and 23S-like rRNAs of the two varieties of *Cryptococcus neoformans*: sequence, secondary structure, phylogenetic analysis, and restriction fragment polymorphisms. *J. Med. Vet. Mycol.* **32**:163–180.

68. **Felkai-Voros, G.** 1967. Incidence of cryptococcus species in urban air. *Acta Microbiol. Acad. Sci. Hung.* **14**:305–308.

69. **Felsenfeld, O.** 1944. Yeast-like fungi in the intestinal tract of chronically institutionalized patients. *Am. J. Med.* **207**:60–62.

70. **Fessel, W. J.** 1993. Cryptococcal meningitis after unusual exposures to birds. *N. Engl. J. Med.* **328**:1354–1355.

71. **Fink, J. N., J. J. Barboriak, and L. Kaufman.** 1968. Cryptococcal antibodies in pigeon breeder's disease. *J. Allergy* **41**:297–301.

72. **Fisher, D., J. Burrow, D. Lo, and B. Currie.** 1993. *Cryptococcus neoformans* in tropical northern Australia: predominantly variant *gattii* with good outcomes. *Aust. N.Z. J. Med.* **23**:678–682.

73. **Fiskin, A. M., M. C. Zalles, and R. G. Garrison.** 1990. Electron cytochemical studies

of *Cryptococcus neoformans* grown on uric acid and related sources of nitrogen. *J. Med. Vet. Mycol.* **38**:197–207.

74. **Francisco, R., J. R. Durant, and R. A. Gams.** 1977. Demonstration of *Cryptococcus neoformans* in a stained bone marrow specimens. *Arch. Intern. Med.* **137**:688–690.

75. **Franzot, S. P., J. S. Hamdan, B. P. Currie, and A. Casadevall.** 1997. Molecular epidemiology of *Cryptococcus neoformans* in Brazil and the United States: evidence for both local genetic differences and a global clonal population structure. *J. Clin. Microbiol.* **35**:2243–2251.

76. **Frey, D., and E. B. Durie.** 1964. The isolation of *Cryptococcus neoformans (Torula histolytica)* from soil in New Guinea and pigeon droppings in Sydney, New South Wales. *Med. J. Aust.* **1**:947–949.

77. **Fromtling, R. A., S. Shadomy, J. Shadomy, and W. E. Dismukes.** 1982. Serotype B/C *Cryptococcus neoformans* isolated from patients in nonendemic areas. *J. Clin. Microbiol.* **16**:408–410.

78. **Geiser, D. M., M. L. Arnold, and W. E. Timberlake.** 1994. Sexual origins of British *Aspergillus nidulans* isolates. *Proc. Natl. Acad. Sci. USA* **91**:2349–2352.

79. **Gezuele, E., L. Calegari, D. Sanabria, G. Davel, and E. Civila.** 1993. Isolation in Uruguay of *Cryptococcus neoformans* var. *gattii* from a nest of the wasp *Polybia occidentalis*. *Rev. Iber. Micol.* **10**:5–6.

80. **Glaser, J. B., and A. Garden.** 1985. Inoculation of cryptococcosis without transmission of the acquired immunodeficiency syndrome. *N. Engl. J. Med.* **313**:266.

81. **Grillot, R., V. Portman-Coffin, and P. Ambroise-Thomas.** 1994. Growth inhibition of pathogenic yeasts by *Pseudomonas aeruginosa in vitro*: clinical implications in blood cultures. *Mycoses* **37**:343–347.

82. **Griseo, G., M. S. Bolignano, F. De Leo, and F. Staib.** 1995. Evidence of canary droppings as an important reservoir of *Cryptococcus neoformans*. *Zentralbl. Bakteriol.* **282**:244–254a.

83. **Gugnami, H. C., N. P. Gupta, and J. B. Shrivastav.** 1972. Prevalence of *Cryptococcus neoformans* in Delhi Zoological park and its recovery from the sputum of an employee. *Indian J. Med. Res.* **60**:182–185.

84. **Gustin, P. N., and D. C. Kelley.** 1971. A survey of zoo aviaries for the presence of *Histoplasma capsulatum* and *Cryptococcus neoformans*. *Mycopathol. Mycol. Appl.* **45**:93–102.

85. **Haag-Wackernagel, D.** 1993. Street pigeons in Basel. *Nature* **361**:200.

86. **Halde, C., and M. A. Fraher.** 1966. *Cryptococcus neoformans* in pigeon feces in San Francisco. *Calif. Med.* **104**:188–190.

87. **Hasenclever, H. F., and C. W. Emmons.** 1963. The prevalence and mouse virulence of *Cryptococcus neoformans* strains isolated from urban areas. *Am. J. Hyg.* **78**:227–231.

88. **Howard, D. H.** 1973. The commensalism of *Cryptococcus neoformans*. *Sabouraudia* **11**:171–174.

89. **Hsu, M. M., J.-C. Chang, K. Yokoyama, K. Nishimura, and M. Miyaji.** 1994. Serotypes and mating types of clinical strains of *Cryptococcus neoformans* in Taiwan. *Mycopathologia* **125**:77–81.

90. **Hubalek, Z.** 1975. Distribution of *Cryptococcus neoformans* in a pigeon habitat. *Folia Parasitol.* **22**:73–79.

91. **Imwidthaya, P.** 1994. One year's experience with *Cryptococcus neoformans* in Thailand. *Trans. R. Soc. Trop. Med. Hyg.* **88**:208.

92. **Imwidthaya, P., P. Dithaprasop, and C. Egtasaeng.** 1989. Clinical and environmental isolates of *Cryptococcus neoformans* in Bangkok (Thailand). *Mycopathologia* **108**:65–67.

93. **Irokannulo, E. A. O., C. O. Akueshi, and A. A. Makinde.** 1994. Differentiation of *Cryptococcus neoformans* serotypes A and D using creatine bromothymol blue thymine medium. *Br. J. Biomed. Sci.* **51**:100–103.

94. **Ishaq, C. M, G. S. Bulmer, and F. G. Felton.** 1968. An evaluation of various environmental factors affecting the propagation of *Cryptococcus neoformans. Mycopathol. Mycol. Appl.* **35:**81–90.

95. **Jacobson, E. S., and H. S. Emery.** 1991. Catecholamine uptake, melanization, and oxigen toxicity in *Cryptococcus neoformans. J. Bacteriol.* **173:**401–403.

96. **Jacobson, E. S., and S. B. Tinnell.** 1993. Antioxidant function of fungal melanin. *J. Bacteriol.* **175:**7102–7104.

97. **Jong, S. C., G. S. Bulmer, and A. Ruiz.** 1982. Serologic grouping and sexual compatibility of airborne *Cryptococcus neoformans. Mycopathologia* **79:**185–188.

98. **Kajihiro, E. S.** 1965. Occurrence of dermatophytes in fresh bat guano. *Appl. Microbiol.* **13:**720–724.

99. **Kao, C. J., and J. Schwarz.** 1957. The isolation of *Cryptococcus neoformans* from pigeon nests. *Am. J. Clin. Pathol.* **27:**652–663.

100. **Kapend'a, K., K. Komichelo, D. Swinne, and J. Vandepitte.** 1987. Meningitis due to *Cryptococcus neoformans* biovar *gattii* in a Zairean AIDS patient. *Eur. J. Clin. Microbiol.* **6:**320–321.

101. **Kelley, R. F., and M. P. O'Connell.** 1993. Thermodynamic analysis of an antibody functional epitope. *Biochemistry* **32:**6828–6835.

102. **Kohl, R.-H., H. Hof, A. Schrettenbrunner, H. P. R. Seeliger, and K. J. Kwon-Chung.** 1985. *Cryptococcus neoformans* var. *gattii* in Europe. *Lancet* **i:**1515.

103. **Kozel, T. R. and C. A. Hermerath.** 1984. Binding of cryptococcal polysaccharide to *Cryptococcus neoformans. Infect. Immun.* **43:**879–886.

104. **Kwon-Chung, K. J.** 1975. A new genus, *Filobasidiella*, the perfect state of *Cryptococcus neoformans. Mycologia* **67:**1197–1200.

105. **Kwon-Chung, K. J.** 1976. A new species of *Filobasidiella*, the sexual state of *Cryptococcus neoformans* B and C serotypes. *Mycologia* **68:**942–946.

106. **Kwon-Chung, K. J.** 1976. Morphogenesis of *Filobasidiella neoformans*, the sexual state of *Cryptococcus neoformans. Mycologia* **67:**821–833.

107. **Kwon-Chung, K. J.** 1991. The discovery of creatinine assimilation in *Cryptococcus neoformans,* and subsequent work on the characterization of the two varieties of *C. neoformans. Zentralbl. Bakteriol.* **275:**390–393.

108. **Kwon-Chung, K. J., and J. E. Bennett.** 1984. High prevalence of *Cryptococcus neoformans* var. *gattii* in tropical and subtropical regions. *Zentralbl. Bakteriol. Hyg. A* **257:**213–218.

109. **Kwon-Chung, K. J., and J. E. Bennett.** 1984. Epidemiologic differences between the two varieties of *Cryptococcus neoformans. Am. J. Epidemiol.* **120:**123–130.

110. **Kwon-Chung, K. J., J. E. Bennett, and J. C. Rhodes.** 1982. Taxonomic studies on *Filobasidiella* species and their anamorphs. *Antonie van Leeuwenhoek* **48:**25–38.

111. **Kwon-Chung, K. J., and Y. C. Chang.** 1994. Gene arrangement and sequence of the 5S rRNA in *Filobasidiella neoformans* (*Cryptococcus neoformans*) as a phylogenetic indicator. *Int. J. Syst. Bacteriol.* **44:**209–213.

112. **Kwon-Chung, K. J., J. C. Edman, and B. L. Wickes.** 1992. Genetic association of mating types and virulence in *Cryptococcus neoformans. Infect. Immun.* **60:**602–605.

113. **Kwon-Chung, K. J., I. Polacheck, and J. E. Bennett.** 1982. Improved diagnostic medium for separation of *Cryptococcus neoformans* var. *neoformans* (serotypes A and D) and *Cryptococcus neoformans* var. *gattii* (serotypes B and C). *J. Clin. Microbiol.* **15:**535–537.

114. **Kwon-Chung, K. J., I. Polacheck, and T. J. Popkin.** 1982. Melanin-lacking mutants of *Cryptococcus neoformans* and their virulence for mice. *J. Bacteriol.* **150:**1414–1421.

115. **Kwon-Chung, K. J., and J. C. Rhodes.** 1986. Encapsulation and melanin formation as indicators of virulence in *Cryptococcus neoformans. Infect. Immun.* **51:**218–223.

116. **Kwon-Chung, K. J., B. L. Wickes, J. L. Booth, H. S. Vishniac, and J. E. Bennett.** 1987.

Urease inhibition by EDTA in the two varieties of *Cryptococcus neoformans*. *Infect. Immun.* **55**:1751–1754.

117. **Kwon-Chung, K. J., B. L. Wickes, L. Stockman, G. D. Roberts, D. Ellis, and D. H. Howard.** 1992. Virulence, serotype, and molecular characteristics of environmental strains of *Cryptococcus neoformans* var. *gattii*. *Infect. Immun.* **60**:1869–1874.

118. **Lazera, M. S., F. D. A. Pires, L. Camillo-coura, M. M. Nishikawa, C. C. f. Bezerra, L. Trilles, and B. Wanke.** 1996. Natural habitat of *Cryptococcus neoformans* var. *neoformans* in decaying wood forming hollows in living trees. *J. Med. Vet. Mycol.* **34**:127–131.

119. **Lazera, M. S., B. Wanke, and M. M. Nishikawa.** 1993. Isolation of both varieties of *Cryptococcus neoformans* from saprophytic sources in the city of Rio de Janeiro, Brazil. *J. Med. Vet. Mycol.* **31**:449–454.

120. **Lehmann, P. F., R. J. Morgan, and E. H. Freimer.** 1984. Infection with *Cryptococcus neoformans* var. *gattii* leading to a pulmonary cryptococcoma and meningitis. *J. Infect.* **9**:301–306.

121. **Levitz, S. M.** 1991. The ecology of *Cryptococcus neoformans* and the epidemiology of cryptococcosis. *Rev. Infect. Dis.* **13**:1163–1169.

122. **Licea, B. A., D. G. Garza, and M. T. Zuniga.** 1996. Aislamiento de *Cryptococcus neoformans* var. *gattii* de *Eucalyptus tereticornis*. *Rev. Iber. Micol.* **13**:27–28.

123. **Littman, M. L.** 1959. Cryptococcosis (torulosis). *Am. J. Med.* **27**:976–998.

124. **Littman, M. L., and R. Borok.** 1968. Relation of the pigeon to cryptococcosis: natural carrier state, heat resistance and survival of cryptococcus neoformans. *Mycopathol. Mycol. Appl.* **35**:329–345.

125. **Littman, M. L., R. Borok, and T. J. Dalton.** 1965. Experimental avian cryptococcosis. *Am. J. Epidemiol.* **82**:197–207.

126. **Littman, M. L., and S. S. Schneierson.** 1959. *Cryptococcus neoformans* in pigeon excreta in New York City. *Am. J. Hyg.* **69**:49–59.

127. **Lopez-Martinez, R., and L. R. Castanon-Olivares.** 1995. Isolation of *Cryptococcus neoformans* var. *neoformans* from bird droppings, fruits and vegetables in Mexico City. *Mycopathologia* **129**:25–28.

128. **Madrenys, N., C. De Vroey, C. Raes-Wuytack, and J. M. Torres-Rodriguez.** 1993. Identification of the perfect state of *Cryptococcus neoformans* from 195 clinical isolates including 84 from AIDS patients. *Mycopathologia* **123**:65–68.

129. **McDonough, E. S., L. Ajello, R. J. Ausherman, A. Balows, J. T. McClellan, and S. Brinkman.** 1961. Human pathogenic fungi recovered from soil in an area endemic for North American blastomycosis. *Am. J. Hyg.* **73**:75–83.

130. **McDonough, E. S., A. L. Lewis, and L. A. Penn.** 1966. Relationship of *Cryptococcus neoformans* to pigeons in Milwaukee, Wisconsin. *Public Health Rep.* **81**:1119–1966.

131. **McGinnis, M. R., M. G. Rinaldi, C. Halde, and A. E. Hilger.** 1975. Mycotic flora of the interdigital spaces of the human foot: a preliminary investigation. *Mycopathologia* **55**:47–52.

132. **Min, K. H., and K. J. Kwon-Chung.** 1986. The biochemical basis for the distinction between the two *Cryptococcus neoformans* varieties with CGB medium. *Zentralbl. Bakteriol. Hyg. A* **261**:471–480.

133. **Mishra, S. K., F. Staib, U. Folkens, and R. A. Fromtling.** 1981. Serotypes of *Cryptococcus neoformans* strains isolated in Germany. *J. Clin. Microbiol.* **14**:106–107.

134. **Muchmore, H. G., E. R. Rhoades, G. E. Nix, F. G. Felton, and R. E. Carpenter.** 1963. Occurrence of *Cryptococcus neoformans* in the environment of three geographically associated cases of cryptococcal meningitis. *N. Engl. J. Med.* **268**:1112–1114.

135. **Muchmore, H. G., E. N. Scott, F. G. Felton, and R. A. Fromtling.** 1980. *Cryptococcus neoformans* serotype groups encountered in Oklahoma. *Am. J. Epidemiol.* **112**:32–38.

136. **Neilson, J. B., M. H. Ivey, and G. S. Bulmer.** 1978. *Cryptococcus neoformans*: pseudohyphal forms surviving culture with *Acanthamoeba polyphaga*. *Infect. Immun.* **20**:262–266.

137. **Newberry, W. M., J. Walter, J. W. Chandler, and F. E. Tosh.** 1967. Epidemiologic study of *Cryptococcus neoformans. Ann. Intern. Med.* **67:**727–732.

138. **Niehaus, W. G., and T. Flynn.** 1994. Regulation of mannitol biosynthesis and degradation by *Cryptococcus neoformans. J. Bacteriol.* **176:**651–655.

139. **Nishikawa, M. M., O. D. Sant'anna, M. S. Lazera, and B. Wanke.** 1996. Use of D-proline assimilation and CGB medium for screening Brazilian *Cryptococcus neoformans* isolates. *J. Med. Vet. Mycol.* **34:**365–366.

140. **Ophir, T., and D. L. Gutnick.** 1994. A role for exopolysaccharides in the protection of microorganisms from desiccation. *Appl. Environ. Microbiol.* **60:**740–745.

141. **Ostrowski, M., I. E. Salit, W. L. Gold, M. Sutton, M. L. Montpetit, D. Lepine, and T. Salas.** 1993. Idiopathic CD4+ T-lymphocytopenia in two patients. *Can. Med. Assoc. J.* **149:**1679–1683.

142. **Padhye, A. A., A. Chakrabarty, J. Chander, and L. Kaufman.** 1993. *Cryptococcus neoformans* var. *gattii* in India. *J. Med. Vet. Mycol.* **31:**165–168.

143. **Pappagianis, D.** 1967. Epidemiological aspects of respiratory mycotic infections. *Bacteriol. Rev.* **31:**25–34.

144. **Partridge, B. M., and H. I. Winner.** 1965. *Cryptococcus neoformans* in bird droppings in London. *Lancet* **i:**1060–1061.

145. **Partridge, B. M., and H. I. Winner.** 1966. *Cryptococcus neoformans* in bird droppings. *Lancet* **ii:**1251.

146. **Perfect, J. R., N. Ketabchi, G. M. Cox, C. W. Ingram, and C. L. Beiser.** 1993. Karyotyping of *Cryptococcus neoformans* as an epidemiological tool. *J. Clin. Microbiol.* **31:**3305–3309.

147. **Pfeiffer, T. and D. Ellis.** 1991. Environmental isolation of *Cryptococcus neoformans gattii* from California. *J. Infect. Dis.* **163:**929–930.

148. **Pfeiffer, T. J., and D. H. Ellis.** 1992. Environmental isolation of *Cryptococcus neoformans* var. *gattii* from *Eucalyptus tereticornis. J. Med. Vet. Mycol.* **30:**407–408.

149. **Polacheck, I., and K. J. Kwon-Chung.** 1980. Creatinine metabolism in *Cryptococcus neoformans* and *Cryptococcus bacillisporus. J. Bacteriol.* **142:**15–20.

150. **Polacheck, I., and K. J. Kwon-Chung.** 1985. Canavanine resistance in *Cryptococcus neoformans. Antimicrob. Agents Chemother.* **29:**468–473.

151. **Polacheck, I., and K. J. Kwon-Chung.** 1988. Melanogenesis in *Cryptococcus neoformans. J. Gen. Microbiol.* **134:**1034–1041.

152. **Pounden, W. D., J. M. Amberson, and R. F. Jaeger.** 1952. A severe mastitis problem associated with *Cryptococcus neoformans* in a large dairy herd. *Am. J. Vet. Res.* **13:**121–128.

153. **Powell, K. E., B. A. Dahl, R. J. Weeks, and F. E. Tosh.** 1972. Airborne *Cryptococcus neoformans*: particles from pigeon excreta compatible with alveolar deposition. *J. Infect. Dis.* **125:**412–415.

154. **Procknow, J. J., J. R. Benfield, J. W. Rippon, C. F. Diener, and F. L. Archer.** 1965. Cryptococcal hepatitis presenting as a surgical emergency. First isolation of *Cryptococcus neoformans* from point source in Chicago. *JAMA* **191:**93–98.

155. **Randhawa, H. S., Y. M. Clayton, and R. W. Riddel.** 1965. Isolation of *Cryptococcus neoformans* from pigeon habitats in London. *Nature* **208:**801.

156. **Randhawa, H. S., and P. Mahendra.** 1977. Occurrence and significance of *Cryptococcus neoformans* in the respiratory tract of patients with bronchopulmonary disorders. *J. Clin. Microbiol.* **5:**5–8.

157. **Randhawa, H. S., and D. K. Paliwal.** 1977. Occurrence and significance of *Cryptococcus neoformans* in the oropharynx and on the skin of a healthy human population. *J. Clin. Microbiol.* **6:**325–327.

158. **Randhawa, H. S., F. Staib, and A. Blisse.** 1973. Observations on the occurrence of *Cryptococcus neoformans* in an aviary, using niger-seed creatinine agar and membrane-filtration technique. *Zentralbl. Bakteriol.* **128:**795–799.

159. **Refal, M., M. Taha, A. Selim, F. Elshabourii, and H. H. Yousseff.** 1983. Isolation of *Cryptococcus neoformans, Candida albicans,* and other yeasts from pigeon droppings in Egypt. *Sabouraudia* **21**:163–165.

160. **Reiss, F., and G. Szilagyi.** 1965. Ecology of yeast-like fungi in a hospital population. Detailed investigation of *Cryptococcus neoformans. Arch. Dermatol.* **91**:611–614.

161. **Rhodes, J. C., I. Polacheck, and K. J. Kwon-Chung.** 1982. Phenoloxidase activity and virulence in isogenic strains of *Cryptococcus neoformans. Infect. Immun.* **36**:1175–1184.

162. **Rinaldi, M. G., D. J. Drutz, A. Howell, M. A. Sande, C. B. Wofsy, and W. K. Hadley.** 1986. Serotypes of *Cryptococcus neoformans* in patients with AIDS. *J. Infect. Dis.* **153**:642.

163. **Rozenbaum, R., A. J. Goncalves, B. Wanke, and W. Viera.** 1990. *Cryptococcus neoformans* var. *gattii* in a Brazilian AIDS patient. *Mycopathologia* **112**:33–34.

164. **Rozenbaum, R., A. J. Goncalves, B. Wanke, M. J. Caiuby, H. Clemente, M. D. S. Lazera, P. C. F. Monteiro, and A. T. Londero.** 1992. *Cryptococcus neoformans* varieties as agents of cryptococcosis in Brazil. *Mycopathologia* **119**:133–136.

165. **Ruiz, A., and G. S. Bulmer.** 1981. Particle size of airborne *Cryptococcus neoformans* in a tower. *Appl. Environ. Microbiol.* **41**:1225–1229.

166. **Ruiz, A., R. A. Fromtling, and G. S. Bulmer.** 1981. Distribution of *Cryptococcus neoformans* in a natural site. *Infect. Immun.* **31**:560–563.

167. **Ruiz, A., J. B. Neilson, and G. S. Bulmer.** 1982. Control of *Cryptococcus neoformans* in nature by biotic factors. *Sabouraudia* **20**:21–29.

168. **Ruiz, A., J. B. Neilson, and G. S. Bulmer.** 1982. A one year study on the viability of *Cryptococcus neoformans* in nature. *Mycopathologia* **77**:117–122.

169. **Ruiz, A., D. Velez, and R. A. Fromtling.** 1989. Isolation of saprophytic *Cryptococcus neoformans* from Puerto Rico: distribution and variety. *Mycopathologia* **106**:167–170.

170. **Rustan, M. E., H. R. Rubinstein, C. Siciliano, and D. T. Masih.** 1992. Possibility of in-hospital infection by *Cryptococcus neoformans* in patients with AIDS. *Rev. Inst. Med. Trop. Sao Paulo* **34**:383–387.

171. **Salkin, I. F., and N. J. Hurd.** 1982. New medium for differentiation of *Cryptococcus neoformans* serotype pairs. *J. Clin. Microbiol.* **15**:169–171.

172. **Sanfelice, F.,** 1894. Contributo alla morfologia e biologia dei blastomiceti che si sviluppano nei succhi di alcuni frutti. *Ann. Igien.* **4**:463–495.

173. **Schneidau, J. D.** 1964. Pigeons and cryptococcosis. *Science* **143**:525–526.

174. **Sethi, K. K., and H. S. Randhawa.** 1967. Survival of *Cryptococcus neoformans* in the gastrointestinal tract of pigeons ingestion of the organism. *J. Infect. Dis.* **118**:135–139.

175. **Sethi, K. K., and J. Schwarz.** 1966. Experimental ocular cryptococcosis in pigeons. *Am. J. Ophthalmol.* **62**:95–98.

176. **Shadomy, H. J., S. Wood-Helie, S. Shadomy, W. E. Dismukes, and R. Y. Chau.** 1987. Biochemical serogrouping of clinical isolates of *Cryptococcus neoformans. Diagn. Microbiol. Infect. Dis.* **6**:131–138.

177. **Shields, A., and L. Ajello.** 1966. Medium for selective isolation of *Cryptococcus neoformans. Science* **151**:208–209.

178. **Shimizu, R. Y., D. H. Howard, and M. N. Clancy.** 1986. The variety of *Cryptococcus neoformans* in patients with AIDS. *J. Infect. Dis.* **154**:1042.

179. **Smith, C. D., R. Ritter, H. W. Larsh, and M. L. Furcolow.** 1964. Infection of white swiss mice with airborne *Cryptococcus neoformans. J. Bacteriol.* **87**:1364–1368.

180. **Sorrell, T., A. G. Brownlee, P. Ruma, R. Malik, T. J. Pfeiffer, and D. H Ellis.** 1996. Natural environmental sources of *Cryptococcus neoformans* var. *gattii. J. Clin. Microbiol.* **34**:1261–1263.

181. **Sorrell, T. C., S. C. Chen, P. Ruma, W. Meyer, T. J. Pfeiffer, D. H. Ellis, and A. G. Brownlee.** 1996. Concordance of clinical and environmental isolates of *Cryptococcus*

neoformans var. *gattii* by random amplification of polymorphic DNA analysis and PCR fingerprinting. *J. Clin. Microbiol.* **34:**1253–1260.

182. **Sotgiu, G., A. Mazzoni, A. Mantovani, L. Ajello, and J. Palmer.** 1966. Survey of soils for human pathogenic fungi from the Emilia-Romagna region of Italy. II. Isolation of *Allescheria Boydii, Cryptococcus neoformans* and *Histoplasma capsulatum. Am. J. Epidemiol.* **88:**329–337.

183. **Speed, B., and D. Dunt.** 1995. Clinical and host differences between infections with the two varieties of *Cryptococcus neoformans. Clin. Infect. Dis.* **21:**28–34.

184. **Speed, B. R., L. Strawbridge, and D. H. Ellis.** 1993. *Cryptococcus neoformans* var. *gattii* meningitis in an Australian patient with AIDS. *J. Med. Vet. Mycol.* **31:**395–399.

185. **Spitzer, E. D., and S. G. Spitzer.** 1992. Use of a dispersed repetitive DNA element to distinguish clinical isolates of *Cryptococcus neoformans. J. Clin. Microbiol.* **30:**1094–1097.

186. **St.-Germain, G., G. Noel, and K. J. Kwon-Chung.** 1988. Disseminated cryptococcosis due to *Cryptococcus neoformans* variety *gattii* in a Canadian patient with AIDS. *Eur. J. Clin. Microbiol. Infect. Dis.* **7:**587–588.

188. **Staib, F.** 1962. *Cryptococcus neoformans* beim Kanarienvogel. *Zentralbl. Bakteriol.* **185:**129–134.

189. **Staib, F.** 1962. Vogelkot, ein Nahrsubstrat fur die Gattung Cryptococcus. *Zentralbl. Bakteriol. Hyg. Abt. 1 Orig. A* **186:**233–247.

190. **Staib, F.** 1962. *Cryptococcus neoformans* im Muskelgewebe. *Zentralbl. Bakteriol.* **185:**135–144.

191. **Staib, F.** 1981. The perfect state of *Cryptococcus neoformans, Filobasidiella neoformans,* on pigeon manure filtrate agar. *Zentralbl. Bakteriol. Hyg. Abt. 1 Orig. A* **248:**575–578.

192. **Staib, F.** 1985. Sampling and isolation of *Cryptococcus neoformans* from indoor air with the aids of the reuter centrifugal sampler (RCS) and guizotia abyssinica creatinine agar. A contribution to the mycological-epidemiological control of *Cr. neoformans* in the fecal matter of caged birds. *Zentralbl. Bakteriol. Hyg. Abt. 1 Orig. B* **180:**567–575.

193. **Staib, F., B. Grave, L. Altmann, S. K. Mishra, T. Abel, and A. Blisse.** 1976. Epidemiology of *Cryptococcus neoformans. Mycopathologia* **65:**73–76.

194. **Staib, F., and M. Haisenhuber.** 1989. *Cryptococcus neoformans* in bird droppings: a hygienic-epidemiological challenge. *AIDS-Forschung (AIFO)* **4:**649–655.

195. **Staib, F., H. S. Randhawa, M. Senska, A. Blisse, and R. Wulkow.** 1973. Peach and peach juice as a nutrient for *Cryptococcus neoformans* with comments on some observations of F. Sanfelice. *Zentralbl. Bakteriol. Hyg. Abt. 1 Orig. A* **224:**120–127.

196. **Staib, F. and J. Schulz-Dieterich.** 1984. *Cryptococcus neoformans* in fecal matter of birds kept in cages—control of *Cr. neoformans* habitats. *Zentralbl. Bakteriol. Hyg. Abt. 1 Orig. B* **179:**179–186.

197. **Staib, F., M. Seibold, E. Antweiler, B. Frohlich, S. Weber, and A. Blisse.** 1987. The brown color effect (BCE) of *Cryptococcus neoformans* in the diagnosis, control and epidemiology of *C. neoformans* infections in AIDS patients. *Zentralbl. Bakteriol. Hyg. A* **266:**167–177.

198. **Swatek, F. E., J. W. Wilson, and D. T. Omieczynski.** 1967. Direct plate isolation method for *Cryptococcus neoformans* from the soil. *Mycopathol. Mycol. Appl.* **32:**129–140.

199. **Swinne, D.** 1979. *Cryptococcus neoformans* and the epidemiology of cryptococcosis. *Ann. Soc. Belg. Med. Trop.* **59:**285–299.

200. **Swinne, D.** 1983. Study of *Cryptococcus neoformans* varieties. *Mykosen* **27:**137–141.

201. **Swinne, D., M. Deppner, S. Maniratunga, R. Laroche, J.-J. Floch, and P. Kadende.** 1991. AIDS-associated cryptococcosis in Bujumbura, Burundi: an epidemiological study. *J. Med. Vet. Mycol.* **29:**25–30.

202. **Swinne, D., K. Kayembe, and M. Niyimi.** 1986. Isolation of saprophytic *Cryptococcus neoformans* var. *neoformans* in Kinshasa, Zaire. *Ann. Soc. Belg. Med. Trop.* **66:**57–61.

203. **Swinne, D., J. B. Nkurikiyinfura, and T. L. Muyembe.** 1986. Clinical isolates of *Cryptococcus neoformans* from Zaire. *Eur. J. Clin. Microbiol.* **5:**50–51.

204. **Swinne-Desgain, D.** 1974. The pigeon as reservoir for *Cryptococcus neoformans. Lancet* **ii:**842–843.

205. **Swinne-Desgain, D.** 1975. *Cryptococcus neoformans* of saprophytic origin. *Sabouraudia* **3:**303–308.

206. **Tanphaichitra, D., S. Sahaphongs, and S. Srimuang.** 1988. Cryptococcal antigen survey among racing pigeon workers and patients with cryptococcosis, pythiosis, histoplasmosis, and penicilliosis. *Int. J. Clin. Pharmacol. Res.* **8:**433–439.

207. **Taylor, R. L., and C. Duangmani.** 1968. Occurrence of *Cryptococcus neoformans* in Thailand. *Am. J. Epidemiol.* **87:**318–322.

208. **Teoh-Chan, C. H., P. Y. Chau, M. H. Ng, and P. C. Wong.** 1975. Inhibition of *Cryptococcus neoformans* by *Pseudomonas aeruginosa. J. Med. Microbiol.* **8:**77–81.

209. **Tibayrenc, M., F. Kjellberg, J. Arnaud, B. Oury, S. F. Breniere, M.-L. Darde, and F. J. Ayala.** 1991. Are eukaryotic organisms clonal or sexual? A population genetics vantage. *Proc. Natl. Acad. Sci. USA* **88:**5129–5133.

210. **Tynes, B., K. Mason, A. E. Jennings, and J. E. Bennett.** 1968. Variant forms of pulmonary cryptococcosis. *Ann. Intern. Med.* **69:**1117–1125.

211. **Van Uden, N., L. D. C. Sousa, and M. Farinha.** 1958. On the intestinal yeast flora of horses, sheep, goats, and swine. *J. Gen. Microbiol.* **19:**435–445.

212. **Varma, A., and K. J. Kwon-Chung.** 1989. Restriction fragment polymorphism in mitochondrial DNA of *Cryptococcus neoformans. J. Gen. Microbiol.* **135:**3353–3362.

213. **Varma, A., D. Swinne, F. Staib, J. E. Bennett, and K. J. Kwon-Chung.** 1995. Diversity of DNA fingerprints in *Cryptococcus neoformans. J. Clin. Microbiol.* **33:**1807–1814.

214. **Villanueva, E., M. Mendoza, E. Torres, M. B. Albornoz, M. E. Cavazza, and G. Urbina.** 1989. Serotipificacion de 27 cepas de *Cryptococcus neoformans* aisladas en Venezuela. *Acta Cientifica Venezolana* **40:**151–154.

215. **Walter, J. E., and R. W. Atchison.** 1966. Epidemiological and immunological studies of *Cryptococcus neoformans. J. Bacteriol.* **92:**82–87.

216. **Walter, J. E., and E. G. Coffee.** 1968. Control of *Cryptococcus neoformans* in pigeon coops by alkalinization. *Am. J. Epidemiol.* **87:**173–178.

217. **Walter, J. E., and R. B. Yee.** 1968. Factors that determine the growth of *Cryptococcus neoformans* in avian excreta. *Am. J. Epidemiol.* **88:**445–450.

218. **Wang, Y., and A. Casadevall.** 1994. Decreased susceptibility of melanized *Cryptococcus neoformans* to UV light. *Appl. Environ. Microbiol.* **60:**3864–3866.

219. **Wang, Y., and A. Casadevall.** 1994. Susceptibility of melanized and nonmelanized *Cryptococcus neoformans* to nitrogen- and oxygen-derived oxidants. *Infect. Immun.* **64:**3004–3007.

220. **Wang, Z., T. chen, Y. Gao, C. Breuil, and Y. Hiratsuka.** 1995. Biological degradation of resin acids in wood chips by wood-inhabiting fungi. *Appl. Environ. Microbiol.* **61:**222–225.

221. **Weber, V. A., and R. Schafer.** 1991. Untersuchungen zum vorkommen von *Cryptococcus neoformans* in kopproven von im menschlichen Wohnbereich gehaltenen Vogeln. *Berl. Munch. Tierarztl. Wschr.* **104:**419–421.

222. **Wickes, B. L., M. E. Mayorga, U. Edman, and J. C. Edman.** 1996. Dimorphism and haploid fruiting in *Cryptococcus neoformans*: association with the alpha-mating type. *Proc. Natl. Acad. Sci. USA* **95:**7327–7331.

223. **Wickes, B. L., D. E. Moore, and K. J. Kwon-Chung.** 1994. Comparison of the electrophoretic karyotypes and chromosomal location of ten genes in the two varieties of *Cryptococcus neoformans. Microbiology* **140:**543–555.

224. **Williamson, P. R.** 1994. Biochemical and molecular characterization of the diphenol

oxidase of *Cryptococcus neoformans*: identification as a laccase. *J. Bacteriol.* **176:**656–664.

225. **Yamamoto, Y., S. Kohno, H. Koga, H. Kakeya, K. Tomono, M. Kaku, T. Yamazaki, M. Arisawa, and K. Hara.** 1995. Random amplified polymorphic DNA analysis of clinically and environmentally isolated *Cryptococcus neoformans* in Nagasaki. *J. Clin. Microbiol.* **33:**3328–3332.

4 | Biochemistry

INTRODUCTION

As a free-living eukaryotic microorganism, *Cryptococcus neoformans* is an enormously complex biochemical system. The fungus has several unusual traits that combine to confer on this organism a unique chemical profile. For example, the pathogen has a capsule composed of complex polysaccharides that function in virulence in a manner analogous to those of the classical encapsulated bacteria, such as *Streptococcus pneumoniae*, *Haemophilus influenzae*, and *Neisseria meningitidis* (173). *C. neoformans* has a phenoloxidase that can synthesize melanin and other pigments from a variety of phenolic precursors and a urease that catalyzes the hydrolysis of urea. *C. neoformans* produces a variety of extracellular products (Table 1), some of which have been associated with virulence.

Scientific interest in *C. neoformans* has been driven primarily by its ability to cause disease in humans. Hence, the biochemical studies of *C. neoformans* have been directed primarily at traits and pathways associated with virulence and drug mechanisms of action. Most biochemical studies on *C. neoformans* have focused on the polysaccharide capsule, the phenoloxidase activity, and the synthesis of melanin, mannitol, and sterols. As a consequence, there has never been a coordinated assault on the biochemistry of this organism, and the biochemical knowledge base is skewed toward traits linked to virulence. This chapter will survey the available information on *C. neoformans* biochemistry.

THE POLYSACCHARIDE CAPSULE

A distinctive feature of *C. neoformans* relative to other medically important yeasts is the presence of a polysaccharide capsule. The *C. neoformans* capsule is important for virulence and, as a result, the capsular polysaccharide has been studied extensively. The capsule can be visualized by a variety of techniques, including India ink staining (232, 243), capsular reactions resulting from the binding of antibody to acidic polysaccharides (64, 65, 152, 155), immunofluorescence (172), and electron microscopy (2, 192, 193, 207). Much is known about the structure, antigenicity, and biological properties of the soluble capsular polysaccharides of *C. neoformans*, but there is relatively little information available on the capsule itself.

Table 1 Substances described in *C. neoformans* culture supernatants

Substance	Component	Reference
Ethanol		See text
Glycoproteins	115-kDa and 38-kDa proteins	76, 78, 189
Polysaccharides	GXM	48
	GalXM	48
	MP	48
	Starch	80
	$(1\rightarrow)$-β-D-Glucan	165
Mannitol		241
Pigments		35–37
Enzymes	Proteinase	5, 25, 40
	Phospholipases	41, 42
3-Hydroxyanthranilic acid		164

Capsule Architecture

The capsule is composed primarily of polysaccharides but may also contain non-polysaccharide components such as enzymes and proteins (see below). The capsule is found immediately outside the cell wall and can vary in size from <1 μm to >50 μm, depending on the strain, environment, and growth conditions. The main polysaccharide component of the capsule is glucuronoxylomannan. The biochemical linkage responsible for holding the polysaccharide capsule at the cell wall is not known but is probably noncovalent (see below). Several studies have analyzed capsular architecture by electron microscopy (2, 58, 192, 193, 207). When evaluating electron micrographs of the capsule, it is important to consider that capsules are usually dehydrated for microscopy, and drying artifacts may occur. Some transmission electron studies have shown a low-electron-density layer at the junction between the capsule and the cell wall that has been referred to as a "halo," "white rim," or "electron transparent part" (32, 58, 204). This feature is probably an artifact of sample preparation for microscopy. Electron microscopy techniques that preserve the capsule and cell wall structure, such as the quick-freezing and deep-etching method, show these structures to be closely linked (192, 193). Scanning electron microscopy of the capsule reveals a loose fibrillar network (Fig. 1). The relevance of the capsular structural features observed in dehydrated cells by electron microscopy techniques to the normal hydrated state is unknown.

Studies of antibody binding to *C. neoformans* indicate that the architecture has a complex antigenic distribution. In 1966 Vogel (223) noted that the deep capsular layers in closest proximity to the cell wall had stronger fluorescence than the periphery of the capsule when stained with antibody. Using serotype absorbed sera and acid-treated *C. neoformans* cells, Vogel demonstrated that some antigens were in the periphery of the capsule whereas others were internal (223). Furthermore, the distribution of antigens differed among the A, B, and C serotypes. Different parts of the capsule manifested antigenic differences, presumably as a result of variation in polysaccharide structure and/or composition. Differences in the location of capsular polysaccharide antigens among strains have been confirmed by

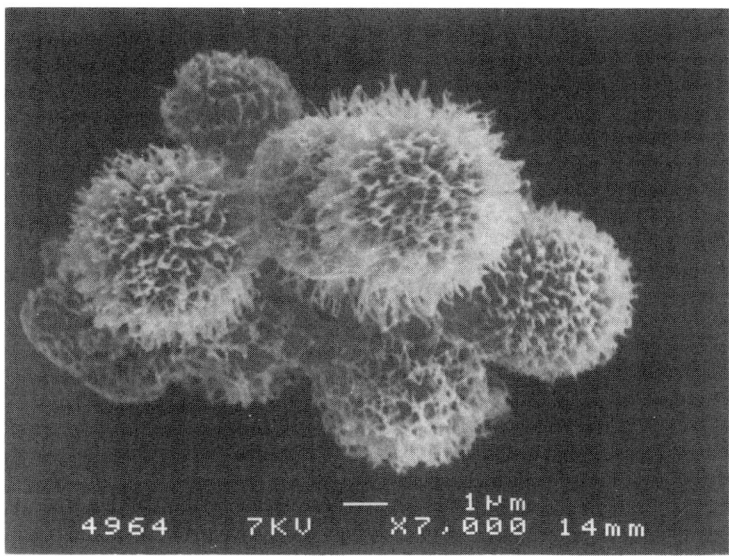

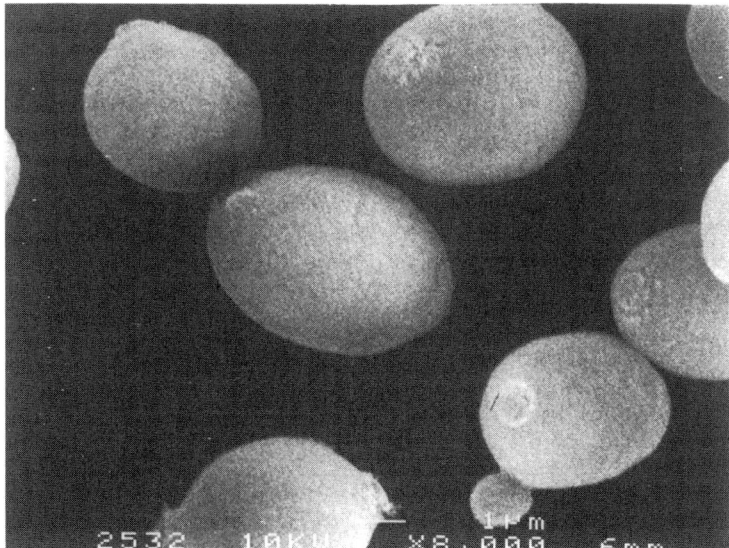

Figure 1 Scanning electron micrograph of encapsulated (top) and nonencapsulated (bottom) *C. neoformans* strains. On encapsulated strains the capsule appears as a loose fibrillar network (top). For the nonencapsulated strain Cap 67, the surface appears relatively smooth (bottom). Note that bud scars are apparent on some of the nonencapsulated yeast cells. Top micrograph provided by Wendy Cleare (Albert Einstein College of Medicine, Bronx, N.Y.).

studies with monoclonal antibodies (MAb). Dromer et al. (53) studied the binding of an immunoglobulin G1 (IgG1) MAb in clinical isolates assigned to the four serotypes and found differences in intensity, location, and pattern of antibody binding. Immunofluorescence studies with this MAb revealed that some strains produced speckled patterns whereas others had homogeneous patterns (53). The occurrence of speckled patterns would seem to imply geographic antigenic determinants within the capsule of *C. neoformans*. Two IgM MAbs have been described with specificity for different polysaccharide epitopes that produce either rim or punctate fluorescence patterns among strains of different serotypes (153, 163). Punctate immunofluorescence reactivity has been associated with the absence of antibody-mediated protection (153, 163). Hence, antibody binding studies provide evidence for structural and antigenic variation within the capsule of an individual *C. neoformans* strain. Additional evidence suggestive of structural heterogeneity within the *C. neoformans* capsule comes from complement activation studies that show that alternative complement activation occurs at a limited number of sites in the capsule, which then expand via alternative pathway amplification (119). Furthermore, complement component C3 deposition tends to occur near the periphery of the capsule in a location that permits its interaction with host effector cells bearing complement receptors (120). The results showing that antibody protection is dependent on the site of binding, that alternative complement activation begins at a few loci, and that C3 deposition occurs near the periphery suggest that the architecture of the capsule results in local differences in polysaccharide structure and antigen composition that are responsible for biological effects. The polysaccharide structures responsible for antigenic differences and complement activation foci are unknown.

In summary, the architecture of the *C. neoformans* capsule is difficult to study because electron microscopy techniques that employ dehydrated samples may significantly alter the native structure. Antibody binding and complement activation studies indicate that the architecture of the capsule is structurally complex. The important questions of how polysaccharide molecules are assembled into a capsule and how the capsule architecture contributes to the biological properties of the capsule remain unanswered.

Polysaccharide Components

The structure of the capsular polysaccharides have been inferred from studies of the polysaccharide material found in high concentration in culture filtrates, referred to as "exopolysaccharide." In 1949 Hehre et al. (80) reported the isolation of particles that stained blue with iodine from *C. neoformans* cultures treated with pentasol and demonstrated that these particles consisted of amylose (starch). This material could be converted to maltose by α- and β-amylase and was found to consist of 91% glucose (80). Additional studies of *C. neoformans* polysaccharide composition were done in the 1950s. Evans and Mehl (66) used filter paper chromatography to demonstrate xylose, galactose, glucose, mannose, and glucuronic acid in cryptococcal polysaccharide hydrolysate. In France, Drouhet and Segretain (55) carried out some of the earliest chemical studies on the cryptococcal polysaccharide and showed that it was biologically active by inhibiting leukocyte migration. Einbinder et al. (59) purified polysaccharides at low pH, removed protein by extraction with kaolin, and obtained a polysaccharide fraction that contained no starch or protein and had 47.5% reducing sugar and 6.7% hexuronic acid. Rebers et al. (182) studied the type-specific polysaccharides of *C. neoformans* (serotype A) and showed that they consisted of two fractions, one of which was rich in galactose. These investigators used cetyltrimethylammonium bromide precipitation to fractionate the cryptococcal polysaccharide into two fractions and demonstrated that both contained glucuronic acid and cross-reacted with antipneumococcal sera (182). In Japan, Miyazaki (147–149) carried out structural studies of the *C. neoformans* polysaccharide and proposed a structure consisting of a mannose backbone with mannose, xylose, and glucuronic acid residues. The Miyazaki structure has several significant differences from the presently accepted structure, including having (1→4)-mannose side chains from a basic (1→2)-mannose backbone (148). Hence, by the early 1960s it was known that cryptococcal polysaccharide (i) contained mannose, xylose, galactose, and glucuronic acid; (ii) consisted of several fractions; (iii) was biologically active; (iv) contained several antigenic determinants that permitted the grouping of strains by serotypes (11, 62, 63, 65, 203, 238); and (v) consisted of a mannose backbone with glucuronic acid and xylose residues (147–149).

In 1966 Blandamer and Danishefsky (19) revised the structure of the polysaccharide and reported a structure of *C. neoformans* polysaccharide consisting of mannan backbone with only xylose and glucuronic acid residues. Over the last three decades the *C. neoformans* capsular polysaccharides have been studied extensively by serological techniques, sugar composition, and nuclear magnetic resonance (NMR). Gradual improvements in the technology of polysaccharide purification led to the description of new polysaccharide fractions. Three polysaccharide components have now been described in detail in *C. neoformans* exopolysaccharide: glucuronoxylomannan (GXM), galactoxylomannan (GalXM), and mannoprotein (MP). GXM composes approximately 90% of the capsular polysaccharide. Differences in the GXM structure among strains produce antigenic differences that provide the basis for the classical separation of *C. neoformans* strains into four serotypes known as A, B, C, and D (48). Because some strains display characteristics of both A and D serotypes, AD may constitute a fifth serotype (86). An important theme in GXM structure studies is structural diversity among strains, even within strains assigned to a given serotype (48, 200, 201). The contribution of GalXM and MP to capsular structure, if any, is unknown; these components appear to have been more closely associated with the cell wall preparations than with the capsule itself (see below).

GXM composes most of the mass of the capsule and of the exopolysaccharide. The structure of GXM has been intensively studied by several groups (6, 15–18, 43–46, 146, 210, 213). There is consensus that GXM is composed of a (1→3)-linked linear α-D-mannopyranan with β-D-xylopyranosyl (Xyl$p$), β-D-glucupyranosyluronic acid (Glc$p$A), and 6-O-acetyl substituents (48). The GXM from all serotypes has been proposed to have a core repeating unit consisting of

$$\beta\text{-D-Glc}p\text{A}$$
$$1$$
$$\downarrow$$
$$2$$
$$\rightarrow\!3)\text{-}\alpha\text{-D-Man}p\text{-}(1\!\rightarrow\!3)\text{-}\alpha\text{-D-Man}p\text{-}(1\!\rightarrow\!3)\text{-}\alpha\text{-D-Man}p\text{-}(1\!\rightarrow$$

with Xyl*p* residues added to the mannose residues at *O*-2 and *O*-4 (48). Xyl*p* residues in the GXM of serotypes A and D are found primarily at *O*-2, whereas in serotypes B and C they are also found at *O*-4 (48). Molar ratios of xylose, mannose, and glucuronic acid residues vary depending on the serotype and are 1:3:1, 2:3:1, 3:3:1, and 4:3:1 for serotypes D, A, B, and C, respectively (48). However, there is considerable structural variation for the GXM of strains assigned to a serotype (45). For example, serotype A includes multiple GXM structures that differ in the degree of xylose, glucuronic acid, and *O*-acetyl substitution (45).

Cherniak and collaborators (48a) have recently proposed grouping cryptococcal strains on the basis of the occurrence of the minimum GXM repeating unit. The minimum GXM repeat unit is a triad composed of (1→3)-linked-α-D-mannosyl residues bearing various amounts of 2-*O*-β-D-glycopyranosyluronic acid, 2-*O*-β-D-xylopyranosyl, and 4-*O*-β-D-xylopyranosyl residues. Six triads have been defined (Fig. 2). Each triad produces a characteristic series of chemical shifts in the proton NMR spectrum of the polysaccharide. The primary structural assignment of a GXM is made by using characteristic chemical shifts of the anomeric protons and mannosyl residues. This is referred to as the structure-reporter group concept. The relative percentage of each triad in a particular polysaccharide is computed by using a computer-based neural network trained with many GXM proton NMR data sets. The strain is characterized on the basis of its triad signature (48a). This system could provide a new strain classification system based on the concept of "chemotype" defined by the quantitative distribution of the structural-reporter groups in a particular strain's GXM. A computer-based neural network analysis of proton NMR spectra can be expected to provide a rapid method for the identification of each strain, and the chemotyping system can be expected to be useful in clinical and epidemiological studies.

GalXM is a minor polysaccharide component of the capsule, the exopolysaccharide material, and the cell wall. GalXM is not covalently bound to the cell wall (105) and appears to be in a loose, poorly characterized association with the other yeast-cell components. GalXM is antigenically and chemically different from GXM (47). The molecular mass of GalXM estimated from column chromatography is 275,000 ± 24,000 Da (47). GalXM can be separated from GXM by concanavalin A column chromatography into several fractions that constitute a diverse group of complex polysaccharides composed of xylose, mannose, glucose, and galactose (105). For individual strains there are differences in sugar composition in the GalXM fractions, indicating that this polysaccharide is structurally heterogeneous (105).

MP is the second minor component of the capsular exopolysaccharide and is the least-characterized fraction from the structural point of view. MP and GalXM often fractionate together in polysaccharide preparations. When GalXM-containing material is further fractionated on a concanavalin A affinity chromatography column, three peaks are observed, of which peak III has 21% protein (214). Analysis of peak III material reveals that the predominant amino acids are serine, threonine,

Figure 2 Triads identified in GXMs of *C. neoformans* by proton NMR spectroscopy. Each strain contains a variable amount of each triad (range 0 to 100%). Figure courtesy of Robert Cherniak and reproduced with permission.

Serotype   Structural Reporter Group

D   M1
$$\begin{array}{cc}
\beta\text{-D-Glc}p\text{A} & \beta\text{-D-Xyl}p \\
1 & 1 \\
\downarrow & \downarrow \\
2 & 2 \\
\end{array}$$
$$\left[\rightarrow 3)\text{-}\alpha\text{-D-Man}p\text{-}(1\rightarrow 3)\text{-}\alpha\text{-D-Man}p\text{-}(1\rightarrow 3)\text{-}\alpha\text{-D-Man}p\text{-}(1\rightarrow\right]_n$$

A   M2
$$\begin{array}{ccc}
\beta\text{-D-Glc}p\text{A} & \beta\text{-D-Xyl}p & \beta\text{-D-Xyl}p \\
1 & 1 & 1 \\
\downarrow & \downarrow & \downarrow \\
2 & 2 & 2 \\
\end{array}$$
$$\left[\rightarrow 3)\text{-}\alpha\text{-D-Man}p\text{-}(1\rightarrow 3)\text{-}\alpha\text{-D-Man}p\text{-}(1\rightarrow 3)\text{-}\alpha\text{-D-Man}p\text{-}(1\rightarrow\right]_n$$

B   M3
$$\begin{array}{ccc}
\beta\text{-D-Glc}p\text{A} & \beta\text{-D-Xyl}p & \beta\text{-D-Xyl}p \\
1 & 1 & 1 \\
\downarrow & \downarrow & \downarrow \\
2 & 2 & 2 \\
\end{array}$$
$$\left[\rightarrow 3)\text{-}\alpha\text{-D-Man}p\text{-}(1\rightarrow 3)\text{-}\alpha\text{-D-Man}p\text{-}(1\rightarrow 3)\text{-}\alpha\text{-D-Man}p\text{-}(1\rightarrow\right.$$
$$\begin{array}{c}
4 \\
\uparrow \\
1 \\
\beta\text{-D-Xyl}p
\end{array}$$

C   M4
$$\begin{array}{ccc}
\beta\text{-D-Glc}p\text{A} & \beta\text{-D-Xyl}p & \beta\text{-D-Xyl}p \\
1 & 1 & 1 \\
\downarrow & \downarrow & \downarrow \\
2 & 2 & 2 \\
\end{array}$$
$$\left[\rightarrow 3)\text{-}\alpha\text{-D-Man}p\text{-}(1\rightarrow 3)\text{-}\alpha\text{-D-Man}p\text{-}(1\rightarrow 3)\text{-}\alpha\text{-D-Man}p\text{-}(1\rightarrow\right.$$
$$\begin{array}{cc}
4 & \qquad\qquad 4 \\
\uparrow & \qquad\qquad \uparrow \\
1 & \qquad\qquad 1 \\
\beta\text{-D-Xyl}p & \qquad\qquad \beta\text{-D-Xyl}p
\end{array}$$

M5
$$\begin{array}{cc}
\beta\text{-D-Glc}p\text{A} & \beta\text{-D-Xyl}p \\
1 & 1 \\
\downarrow & \downarrow \\
2 & 2 \\
\end{array}$$
$$\left[\rightarrow 3)\text{-}\alpha\text{-D-Man}p\text{-}(1\rightarrow 3)\text{-}\alpha\text{-D-Man}p\text{-}(1\rightarrow 3)\text{-}\alpha\text{-D-Man}p\text{-}(1\rightarrow\right.$$
$$\begin{array}{cc}
4 & \qquad\qquad 4 \\
\uparrow & \qquad\qquad \uparrow \\
1 & \qquad\qquad 1 \\
\beta\text{-D-Xyl}p & \qquad\qquad \beta\text{-D-Xyl}p
\end{array}$$

M6
$$\begin{array}{c}
\beta\text{-D-Glc}p\text{A} \\
1 \\
\downarrow \\
2 \\
\end{array}$$
$$\left[\rightarrow 3)\text{-}\alpha\text{-D-Man}p\text{-}(1\rightarrow 3)\text{-}\alpha\text{-D-Man}p\text{-}(1\rightarrow 3)\text{-}\alpha\text{-D-Man}p\text{-}(1\rightarrow\right]_n$$

and alanine (214). Peak III is the most serologically active fraction in the polysaccharide preparation and corresponds to MP (214). The nature of the protein-carbohydrate linkage has not been determined. MP is composed primarily of mannose but also has significant amounts of galactose and xylose residues (48). The molar ratio of the MP in peak III from concanavalin A column fractionation for xylose-mannose-galactose is 1:7.5:0.8 (214). MP is also released from cell wall preparations by mechanical disruption (47, 105, 214). MP appears to be the material responsible for the delayed-type hypersensitivity reactions elicited by culture filtrates in sensitized animals (154).

The capsular polysaccharide is responsible for the high negative charge of C. neoformans cells (117, 118, 161). Acapsular mutants of C. neoformans had cell charge measured by zeta potential of −3.08 mV (106). For encapsulated strains, the zeta potentials were in the range of −18 to −27 mV (106). Incubation of the acapsular mutant with purified polysaccharide restored the high negative cell charge (106), consistent with the reversible binding described by others (117, 199).

Capsule Synthesis

The structural complexity of the polysaccharide capsule and its components implies that it is the product of a complex enzymatic machinery. However, little information is available regarding the pathways and mechanisms of capsule synthesis. Several genes required for the capsular phenotype have now been identified (34, 94) but their function is not known. Jacobson and collaborators (94, 235) have studied several enzymatic reactions related to polysaccharide synthesis. The presence of xylosyltransferase and glucuronyltransferase activities has been demonstrated in microsomal preparations of C. neoformans (235). Analysis of side group addition by xylosyltransferase and glucuronyltransferase suggests that the order of sugar addition is mannosyl, acetyl, glucuronyl, and xylosyl (235). Several acapsular mutants were studied, but the enzymatic defects responsible for loss of capsular production have not been assigned (99).

The site of capsule synthesis is not known. Capsular polysaccharide could be synthesized in situ at the cell wall or exported from the cytoplasm. Some have suggested that the capsular polysaccharide is synthesized in the cytoplasm and then exported to the outer layer of the cell wall in vesicles for assembly into a capsule (193, 207). For MP there is evidence from immune electron microscopy studies for an origin in the cytoplasm (221). Several C. neoformans mutants that lack a polysaccharide capsule have been characterized by Jacobson and Tingler (101). These acapsular mutants lack GXM and produce rough colonies when grown on agar plates (101). The biochemical defects responsible for the acapsular phenotype in these strains are unknown, since no enzymatic defect has been identified after analysis of UDP-glucose dehydrogenase, UDP-glucuronate decarboxylase, UDP-glucuronyl:acceptor transferase, UDP-xylosyl:acceptor transferase, and lipid-linked oligosaccharide biosynthetic pathways (101).

An interesting feature of the C. neoformans capsular polysaccharide is its ability to be released from encapsulated cells and to be bound by nonencapsulated cells (113, 117a, 199). Little is known about the mechanism of polysaccharide release or attachment. The original finding of the binding phenomenon was based on the observation that the addition of soluble polysaccharide to nonencapsulated mutants made them resistant to phagocytosis (26). Kozel and collaborators (113, 117a) provided the first detailed description of polysaccharide binding by nonencapsulated cells. Incubation of nonencapsulated cells with soluble polysaccharide resulted in a dose-dependent attachment of polysaccharide to the yeast surface (113, 117a). Binding of cryptococcal polysaccharide to *C. neoformans* cells is a specific interaction since no binding occurs to other yeasts, and polysaccharide from other microorganisms does not bind to cryptococcus (113). However, polysaccharide from all serotypes can bind to a nonencapsulated mutant; hence, this phenomenon is not dependent on a match between the serotype of the soluble polysaccharide and the serotype of the wild-type cell from which the mutant was derived (113). Quantitative studies of polysaccharide binding indicate that about 30,000 molecules of polysaccharide per cell are required to inhibit phagocytosis (117a). De-*O*-acetylation, carboxyl reduction, and periodate oxidation of cryptococcal polysaccharide had no effect on the binding phenomenon (117a). However, the polyalcohol or Smith degradation products of cryptococcal polysaccharide bound poorly to nonencapsulated cells, as measured by inhibition of phagocytosis (117a). Small and Mitchell (199) also studied the phenomenon of attachment by adding radioiodinated capsular polysaccharides to acapsular mutants. Polysaccharide binding to yeast cells was rapid, specific, reversible, and saturable, and the reaction was consistent with a receptor-mediated mechanism (199). Treatment of capsular polysaccharide, but not the cells, with proteinase eliminated the binding phenomenon (199).

In summary, the biochemistry of the *C. neoformans* polysaccharide capsule remains, for the most part, unexplored territory. The location and machinery responsible for capsular polysaccharide synthesis remain largely uncharacterized. The interaction of the capsular polysaccharide with the fungal cell appears to be noncovalent but dependent upon proteinase-sensitive structures in the capsular polysaccharide. Detailed studies of capsule structure and synthesis would seem to be an important prerequisite for a better understanding of the function of the capsule and capsular polysaccharides in infection and pathogenesis.

Regulation of Capsule Synthesis

The most exhaustive study of the effect of culture conditions on the synthesis of the polysaccharide capsule was published by Littman (132) in 1958. Littman studied the effect of carbon sources, nitrogen sources, and vitamins on the growth of *C. neoformans* and its ability to make capsular polysaccharide. Thiamine was found to stimulate capsule production. The amino acids (and derivatives) D-glutamic acid, sodium glutamate, and L-proline strongly induced capsule formation. Many sugars were found to induce capsule formation, including D-mannose, D-glucose, D-fructose, D-xylose, and sucrose. On the basis of his observations, Littman (132) formulated a synthetic medium to elicit capsule production in *C. neoformans*. The medium consisted of mineral salts, thiamine, sodium glutamate, maltose, and sucrose (132). This preparation is often referred to as "Littman's capsule media." Capsule synthesis by *C. neoformans* is also responsive to the concentration of NaCl in medium (102). The production of capsular polysaccharide is suppressed by growth in 1 M NaCl (102). The suppressive effect was not observed in other hypertonic media, suggesting a NaCl-related effect (102). Interestingly, suspension of encapsulated cells in 1 M NaCl led to a physical contraction of the capsule consistent with a salt-induced physico-chemical contraction of the capsular gel (102).

Most, if not all, strains of *C. neoformans* manifest larger polysaccharide capsules during infection than under in vitro culture conditions. Transfer of *C. neoformans* between mouse tissues and Sabouraud dextrose agar can result in major changes in capsule size (14). In rats and mice, instillation of *C. neoformans* into the lungs has been shown to result in rapid release of capsular polysaccharide into tissue (68, 74). Hence, infection is accompanied by an increase in capsule size and a rapid release of capsular polysaccharide into tissues, where it almost certainly interferes with the host immune response. A potential explanation for the observation that *C. neoformans* makes a large capsule in vivo is suggested by the observation of Granger et al. (75) who showed that capsule synthesis was affected by CO_2 and HCO_3^- concentration. A link between the ability to make a large capsule and virulence was suggested by the observation that an encapsulated mutant of the serotype A strain H99, which did not respond to CO_2 by enlarging the capsule, was avirulent in steroid-treated rabbits (75).

Another important regulator of capsule production is iron (95, 219). Limitation of iron by chelators combined with high CO_2 concentrations in media produce additive effects on the production of capsular polysaccharide (219). Capsular expression is suppressed by growth of *C. neoformans* in high concentrations of ferric ion (95). These observations suggest that during infection the combination of high concentrations of CO_2 and low concentrations of iron in tissue may be responsible for the increase in capsule size associated with infection.

In summary, most biochemical and structural aspects of capsule synthesis by *C. neoformans* are poorly understood. Capsule synthesis is under complex regulation, and multiple factors, including nutrient availability, type of nutrient, pH, and CO_2, have been shown to affect the size of the *C. neoformans* capsule.

Immune Responses to Polysaccharide Antigens

GXM, GalXM, and MP, the three major components of the *C. neoformans* capsular exopolysaccharides, can each elicit antibody responses, but only the MP component elicits cell-mediated immunity as measured by delayed-type hypersensitivity reaction (154). GXM and GalXM are poorly immunogenic polysaccharides. The immune response to *C. neoformans* polysaccharide antigens is discussed in detail in chapter 8.

Serotype Classification and Immunological Reactivity of Capsular Polysaccharides

The serotype classification is based on antigenic differences between *C. neoformans* strains that allow their grouping into distinct classes with the use of antibody reagents. Cryptococcal strains have been divided into four major serotypes, known as A, B, C, and D, on the basis of antigenic differences resulting from variation in the polysaccharide structure (61, 62, 238). The original assignment of the A, B, and C serotypes was provided by Evans (62, 63) and Evans and Kessel (61) in the late 1940s and early 1950s using agglutination, precipitation, and capsular reactions with absorbed rabbit sera. In the 1960s, serotype D was described by Wilson et al.

(238). Serotype A and D compose *C. neoformans* var. *neoformans,* and serotypes B and C compose *C. neoformans* var. *gattii.* In the 1980s, serotype AD was described; AD constitutes a set of strains that have the same antigenic determinants found in the classical A and D serotypes (86).

The polysaccharide component for the antigenic differences leading to the serotype classification is GXM. Classical studies of *C. neoformans* serotype have utilized absorbed rabbit sera because the rabbit is the laboratory animal that most consistently makes a strong antibody response to cryptococcal polysaccharide preparations. In recent years a variety of mouse MAbs have been generated to *C. neoformans* capsular polysaccharide epitopes (29, 30, 54, 87, 174, 201, 210).

Despite considerable effort, relatively few serotype-specific serological reagents are available. A recurring theme in serological studies of *C. neoformans* is that many antigenic structures in the capsule are shared by most strains. An important contribution to the understanding of the serological relationships between the *C. neoformans* serotypes was made by Ikeda et al. (88), who devised an antigenic classification system based on eight antigenic factors (Table 2). In this scheme (88), the antigenic composition of a given *C. neoformans* strain and serotype is based upon the presence of one or more of eight antigenic factors defined by factor sera obtained by reciprocal absorption methods. Important support for the factor sera classification scheme comes from the generation of mouse MAbs with reactivity patterns for the *C. neoformans* serotypes that correspond closely to that of the factor sera (8, 54, 56, 87, 210). It is noteworthy that, of the dozens of MAbs generated, most react with more than one serotype (8, 29, 54, 56, 210). However, some MAbs have significantly higher reactivity with one serotype. The MAb E1 generated by Dromer et al. (53) reacts most strongly with serotype A strains and has been used for serotyping strains. Another MAb has been isolated that reacts primarily with serotype D and may correspond to Factor 8 serum in the factor sera classification scheme (87). The antigenic classification based on factor sera suggests that the explanation for the relative paucity of serotype-specific MAbs is that the serotype classification is a composite of several antigenic factors, most of which are shared by more than one serotype.

The serotype classification of *C. neoformans* strains has been useful for helping to understand the global epidemiology of cryptococcal infections and the serological relationship of clinical strains (12, 13, 121, 122, 127). Several methods for serotype discrimination have been described, including agglutination, capsular reactions, immunoprecipitation, and immunofluorescence (61–63, 65, 86–88, 109, 238). Serological reagents based on the factor sera developed by Ikeda et al. (88) are

Table 2 Antigenic patterns of *C. neoformans* serotypes[a]

Serotype	Factor sera reactivity							
	1	2	3	4	5	6	7	8
A	+	+	+	−	−	−	+	−
B	+	+	+	−	−	−	−	+
C	+	+	−	+	+	−	−	−
D	+	−	−	+	−	+	−	−

[a]Based on the factor sera classification of Ikeda et al. (88).

commercially available from Iatron Laboratories (Tokyo, Japan) for typing of *C. neoformans* strains. A method of serotyping utilizing factor sera or selected anti-*C. neoformans* polysaccharide MAbs has been described that employs a dot enzyme assay on polysaccharide absorbed onto nitrocellulose strips (9). However, serotype classification has been of little use in the clinical management of patients with cryptococcosis and is seldom used except in research laboratories.

Mixing of *C. neoformans* with specific antibody produces a capsular reaction analogous to the classical quellung reaction observed for antibody binding to the polysaccharide capsule of pneumococcus (65). A capsular reaction is a phenomenon whereby the capsule becomes visible after being mixed with antibody in aqueous solutions (65, 152). Capsular reactions in *C. neoformans* can also be produced by mixing yeast cells with a basic polygalactosamine polysaccharide isolated from *Aspergillus parasiticus* (64), presumably as a consequence of binding to the polyanions in the cryptococcal capsule. However, the serotype classification of *C. neoformans* strains can be deduced by capsular reactions with specific antisera, but not with the *Aspergillus* polysaccharide (64). Capsular reactions can also be produced by MAbs to the polysaccharide, a phenomenon that may be associated with a structural change in the capsule fibrils (152). The physical basis of the antibody-induced capsular reaction phenomenon is not fully understood. Evans (65) proposed that the capsular reaction effect was a result of a line of precipitation at the capsule surface that allowed visualization of the capsule outline.

Serological Assays for Capsular Polysaccharides

The detection and analysis of capsular polysaccharides in tissue remain dependent upon serological assays. Many assays have been described for the measurement of cryptococcal capsular polysaccharides based on the use of antibody reagents (Table 3). Currently, latex bead agglutination and enzyme-linked immunosorbent assay (ELISA) are the most commonly used methods for the detection of cryptococcal polysaccharide in clinical specimens.

Biological Effects of the Capsule and Capsular Polysaccharides

The polysaccharide capsule is an important determinant of virulence for *C. neoformans* (27, 34, 116). Polysaccharide is believed to contribute to virulence by protect-

Table 3 Assays for the serological detection of cryptococcal polysaccharide

Assay	Reference
Counterimmunoelectrophoresis	135
Dot blot	9, 10
ELISA	183, 196
Latex agglutination	21, 111
Staphylococcal agglutination	136
Immunohistochemistry	130
Immunoprecipitation	182

ing the yeast cell from phagocytosis and by interfering with immune responses after being shed into tissues.

Classically, the polysaccharide capsule has been thought to function by being antiphagocytic. The exact mechanism by which the capsule inhibits phagocytosis and the structural features of the polysaccharide responsible for this effect are not fully understood. It has been suggested that the capsule does not modulate phagocytosis directly but rather interferes with the phagocytic process by presenting a surface that is not recognized by phagocytic cells and blocking recognition of cell wall structures by host cell receptors (117). The capsule has been shown to mask the opsonic function of cell wall–bound IgG (142). Capsular polysaccharide confers a strong negative charge to *C. neoformans* cells (161), but a causal association between charge and inhibition of phagocytosis has not been demonstrated (118). Decarboxylation and de-O-acetylation of capsular polysaccharide has no effect on its phagocytosis-inhibiting properties (117). Similarly, chemical modification of the capsule to produce a more hydrophobic surface has no effect on its antiphagocytic properties (114). Hence, the antiphagocytic properties may be a combination of (i) masking surface structures that can be recognized by receptors in phagocytic cells and (ii) the absence of receptors in phagocytic cells that can recognize the capsule.

Although there is conclusive evidence that the capsule is antiphagocytic in vitro, the question of whether soluble or attached polysaccharide is more important for virulence is unclear. The capsule can activate the alternative pathway of the complement system, and complement-derived opsonins can promote the phagocytosis of encapsulated *C. neoformans* (115). Recently, it was demonstrated that encapsulated yeasts deposited in the alveoli were rapidly internalized by macrophages regardless of whether antibody opsonin was provided (68). This observation raises some question as to whether interference with phagocytosis is the primary function of the capsule in virulence. Soluble polysaccharide, such as that released from yeast cells during infection, has been demonstrated to produce a variety of potentially deleterious effects on host immune cells in vitro.

In summary, capsular polysaccharide appears to enhance virulence both when attached to the yeast cell and as an exopolysaccharide released from cells. Capsular polysaccharide probably contributes to virulence through multiple mechanisms. The relative contribution of the various effects is listed in chapter 9.

Antigenic Similarities of Capsular Polysaccharide to Other Microbial Products

The *C. neoformans* capsular polysaccharide has antigenic similarities to products of other microorganisms. Antibodies to the polysaccharides of types 2 and 14 *S. pneumoniae* can precipitate cryptococcal polysaccharide, indicating serological cross-reactivities (182). Cross-reactions have also been reported with *Cryptococcus* sp. (85, 136) and a *Klebsiella* strain (135). The polysaccharide of *Cryptococcus albidus* var. *albidus* has very similar antigenic determinants to those of *C. neoformans* serotype A (85). Antibodies to *C. neoformans* capsular polysaccharide also bind to a polysaccharide made by the fungus *Trichosporon beigelii* (52, 143, 145). The antigenic cross-reaction between *C. neoformans* and *Trichosporon* is clinically important because the latex antigen assay used for serological diagnosis of cryptococcosis also produces a positive reaction in cases of disseminated trichosporonosis (145). A false-positive result in the latex antigen assay has also been reported in a patient with infection with the bacterium DF-2 (233). The antigenic similarities between *C. neoformans* and other microbes introduce a potential confounding variable in the analysis of antibody titers in patients with or without cryptococcosis. For example, *C. neoformans* was initially implicated as the etiological agent in *Trichosporon* sp.-induced summer-type hypersensitivity pneumonitis because of antigenic similarities between these two fungi (160).

THE CELL WALL

Like other fungi, *C. neoformans* has a cell wall, a complex structure exterior to the cell membrane but interior to the polysaccharide capsule. The cell wall defines cell shape and provides support for the cell against osmotic stress. The fungal cell wall consists of cross-linked polysaccharides providing a protective "chain mail" type of armor for the cell. Several investigators have analyzed the *C. neoformans* cell wall by transmission electron microscopy (2, 32, 151, 204). Electron micrographs of *C. neoformans* reveal the cell wall to be composed of multiple parallel fibrillar layers of sheath-like plates that are more dense in the inner section. Fungal cell walls are dynamic structures that must accommodate cell changes such as growth and budding. Because of the essential nature of the cell wall for fungal survival and the absence of a comparable structure in mammalian cells, the biosynthetic pathways of the cell wall are an attractive target for antifungal drug design. Relatively few biochemical and structural studies have been done on the cell wall of *C. neoformans*.

An early study of the chemical composition of crude cell wall preparations from two isolates of *C. neoformans* revealed that it contained hexose (73 to 84%), hexosamine (0.23 to 0.43%), protein (3.2 to 3.5%), and lipid (10 to 18%) (49). However, the early studies utilizing encapsulated strains had difficulty in separating cell wall fractions from the polysaccharide capsule and cytoplasmic fractions. For *C. neoformans*, the analysis of the cell wall is complicated because of the difficulty in separating capsular polysaccharide from cell wall fractions. The strategy used by Cherniak and collaborators (106, 184) to study cell wall composition was to employ acapsular mutants that lack capsule GXM. Sequential extraction of cell wall preparations with detergent, dilute alkaline borohydride, hot dilute acetic acid, and alkaline borohydride from the acapsular mutant Cap 67 revealed that the cell wall was composed primarily of glucose (106, 184). The composition of the cell wall is 86% glucose, 7.3% hexosamine, 2.2% nitrogen, and 0.3% phosphate (106). Cell walls from 9-day cultures have more hexosamine but less glucose, nitrogen, and phosphate than do cell walls from 3-day cultures, indicating changes in cell wall chemistry with age (106). This detergent-alkaline-acid-alkaline extraction sequence preserved cell wall shape but removed all galactose, xylose, and mannose residues, indicating that neither GalXM nor MP was covalently linked to the cell wall (184, 221). Electron microscopy of cell wall preparations generated by this method revealed a structure with alternating electron-dense layers with a punched-out appearance (184).

The cell wall of *C. neoformans* has been divided into water-soluble and water-insoluble fractions (106). Analysis of the water-soluble fraction revealed polysaccharides composed primarily, if not exclusively, of (1→6)-β-D-glucupyranans (106). The (1→6)-β-D-glucupyranans were branched at *O*-3 (~10 to 12%) with β-D-glucupyranoside-(1→3)-β-D-glucupyranoside side chains (106). Analysis of the water-insoluble fraction revealed polysaccharides composed primarily of (1→3)-α-D-glu-cupyranans with no chain branching and a small fraction of (1→4) linkages (106). A detailed structure of the *C. neoformans* cell wall polysaccharides and their respective linkages is not available. However, there is evidence that some biosynthetic *C. neoformans* enzymes are regulated in a manner similar to those of *Saccharomyces cerevisiae*. For example, homologs of *S. cerevisiae* genes involved in yeast-cell morphogenesis have been identified in *C. neoformans* (141).

The *C. neoformans* cell wall contains an MP that is also found in culture supernatants (221). The MP is immunogenic and is probably responsible for the phenomenon of delayed-type hypersensitivity observed when culture filtrates are injected into infected animals (see chapter 8). Immune electron microscopy has located most of the MP in the *C. neoformans* cell at the inner aspect of the cell wall (221). The cell wall of *C. neoformans* stains with the fluorescent dye Calcofluor white, probably as a consequence of the presence of chitin.

The cell wall of *C. neoformans* provides a tough obstacle to biochemical and molecular studies of this organism. In general, many *C. neoformans* strains are difficult to lyse for the preparation of DNA, cytoplasmic extracts, etc. Enzymes suitable for the digestion of the cell wall of *S. cerevisiae* are not effective for generating *C. neoformans* protoplasts, presumably because of differences in cell wall chemistry (186). However, protoplasts of *C. neoformans* can be generated by using enzyme preparations from snails and fungi (171, 186, 218). The most commonly used preparation is Novozyme 234 (Novo Industries, Denmark), which originates from *Trichoderma harzianum* and contains α-1,3 and β-1,3 hydrolases (186). The generation of protoplasts and spheroplasts is an important first step in making intracellular preparations (DNA, RNA, etc.) and is usually accomplished with crude complex enzyme preparations including snail gut enzymes (171). Electron microscopy of the process of cell wall digestion with snail gut enzymes shows that protoplast-spheroplast formation is a two-stage process (171). First, the enzymes induce a hole in the equatorial region of the cell wall, through which the protoplast-spheroplast emerges from a cell wall "ghost" (171). Second, continued digestion of the cell wall leads to the disappearance of these structures (171). The process of protoplast-spheroplast generation is slow; it requires 5 h for the first stage and 20 h for the second stage (171). Similar events occur during protoplast-spheroplast generation in *Candida albicans* (171).

In many fungal infections, cell wall polysaccharides such as (1→3)-β-D-glucan are released into serum, and their detection can be useful in the diagnosis of invasive infection (165). (1→3)-β-D-Glucan polysaccharides produce positive results in the classical *Limulus* test used for the detection of endotoxin. The *Limulus* amebocyte reaction can be triggered by two pathways, one sensitive to endotoxin and the other sensitive to (1→3)-β-glucans (165). The reaction of cryptococcal glucans with a commercially available endotoxin assay based on amebocyte lysate from the Japanese horseshoe crab was more sensitive than that based on amebocyte lysates from the American horseshoe crab (82). This reaction can mimic endotoxin contamination in cryptococcal polysaccharide preparations. Attempts to use (1→3)-β-D-glucans for the diagnosis of *C. neoformans* infection have not proved successful because cryptococcal cells, unlike *Candida* and *Aspergillus* cells, shed only small amounts of (1→3)-β-D-glucans (150, 165).

In summary, the cell wall of *C. neoformans* is a structure that remains largely uncharacterized. There are significant differences between the *C. neoformans* cell wall and those of other fungi. For example, mannose is found in the cell wall of *S. cerevisiae* but not in that of *C. neoformans* (106). Differences in cell wall composition between *C. neoformans* and other medically significant fungi such as *C. albicans* and *Aspergillus fumigatus* are probably responsible for the relative resistance of *C. neoformans* to pneumocandin-type drugs that inhibit cell wall biosynthetic enzymes (1).

MELANIN AND OTHER PIGMENTS

One of the distinctive features of *C. neoformans* is its ability to synthesize dark pigments when grown in media with phenolic compounds. The discovery of pigment production by *C. neoformans* is attributed to Staib, who reported in 1962 that colonies in agar containing *Guizotia abyssinica* seed extracts were brown (cited in reference 175). *C. neoformans* produces several dark pigments when provided with suitable substrates, a characteristic used for diagnostic microbiology (112, 197, 205). Although other members of the *Cryptococcus* genus produce pigments when provided with suitable substrates, pigment formation in *C. neoformans* is faster, stronger, and can be elicited by compounds not active in other *Cryptococcus* spp. (37). The type of pigment made is a function of the substrate provided. In medium containing L-dopa, the black pigment has been shown by electron spin resonance spectroscopy criteria to be a type of melanin (226). Melanization results in additional negative charges to the *C. neoformans* cell (161). Pigment production is associated with virulence in *C. neoformans* (124, 187, 194), and there is considerable interest in this pathway from the viewpoint of understanding fungal pathogenesis and antifungal drug design.

Phenoloxidase

Melanin synthesis is catalyzed by a phenoloxidase (237). The phenoloxidase of *C. neoformans* has been categorized as a laccase on the basis of substrate specificity (237). The enzyme is a glycosylated copper-containing protein of 624 amino acids (237) that may exist as a dimeric species in cell extracts (89). The exact cellular location of the enzyme is uncertain, because two studies have provided discordant results. One study reported that most of the phenoloxidase activity was membrane bound (176), but the other localized most of the enzymatic activity in soluble fractions (89). This discrepancy may reflect differences in the protocols used to purify the enzyme (89), but it appears that the enzyme can be detected in both soluble and membrane fractions. The activity of phenoloxidase can be measured by the oxidation of L-epinephrine to melanin (185). Phenoloxidase activity is regulated by glucose concentration and temperature. More enzyme is present in conditions of glucose starvation (176) and when cultures are grown at lower temperature (e.g., 25°C in contrast to 37°C) (89). The reduction in phenoloxidase activity at 37°C may seem odd for a virulence factor that presumably functions at that temperature, but the effect of reduced activity may be minor; there was no difference in melanization as measured by electron spin resonance spectroscopy at 30°C and 37°C (227). The phenoloxidase produced at 25°C and 37°C showed differences in activity, electrophoretic mobility, and concanavalin A binding ability (84, 89, 97). Iron and copper ions increase phenoloxidase activity (95). Interestingly, there appears to be discordant regulation of the capsular polysaccharide and phenoloxidase activity in *C. neoformans*, since some stimuli that suppress capsular production, such as iron, tend to increase phenoloxidase activity (95).

A variety of diphenols can serve as substrates for pigment production by the *C. neoformans* phenoloxidase (36). This enzyme has broad specificity and can generate pigments from many structurally different compounds (36, 57, 125, 176, 179, 205, 225). Substrates consisting of *o*-diphenols (OH− groups at the 2,3- or 3,4-position of the phenyl ring) result in cellular pigment deposition (36). In contrast, substrates consisting of *p*-diphenols (OH− groups at the 1,4- or 2,5-position of the phenyl ring) result in the formation of pigments that diffuse into the culture medium (36). Monophenols such as tyrosine and phenol are poor substrates for the phenoloxidase (237). A wide variety of structurally distinct indole compounds are substrates for pigment production (125). The pigment color is dependent on the substrate. L-Dopa (3,4-dihydroxyphenylalanine) results in a black pigment, whereas other substrates can result in green-black, brown, or orange-red pigments (36). The ability of *C. neoformans* to make pigment is also dependent upon the nitrogen source for growth (36). The amino acids glutamine, asparagine, and glycine are strong inducers of pigment synthesis (36). *C. neoformans* can also make pigment from esculin (6-*β*-D-glucose-dihydroxycoumarin) by oxidizing the 6,7-dihydroxycoumarin component to a melanin-like pigment, and esculin agar has been suggested as a possible medium for the isolation and identification of *C. neoformans* (57). Biochemical differences in the activity of the phenoloxidases from the two varieties of *C. neoformans* were inferred by the ability of glutamine and $(NH_4)_2SO_4$ to affect enzyme activity (162). Glutamine was found to suppress phenoloxidase activity of all isolates except those assigned to serotype B, whereas $(NH_4)_2SO_4$ suppressed the activity of serotype A isolates (162).

Melanin

Melanin is formed by the action of phenoloxidase on 2,3- or 3,4-diphenol substrate and is deposited in the *C. neoformans* cell wall (197, 226). The melanin synthetic pathway in *C. neoformans* involves oxidation of exogenous dihydroxyphenols and is biochemically different from the polyketide pathway found in other melanotic fungal pathogens such as *Wangiella dermatitidis* (178, 234). In *C. neoformans*, melanin synthesis occurs only in the presence of exogenous dihydroxyphenols, because this fungus lacks an enzyme pathway capable of generating endogenous substrates. Melanin production from L-dopa has been suggested to proceed in a scheme adapted after the classical tyrosinase pathway, which produces melanin in mammals (162) (Fig. 3). Minor modifications of this scheme have been proposed for melanogenesis in *C. neoformans* (178). However, unlike the mammalian pathway, which can utilize only tyrosine and dihydroxyphenylalanine, the *C. neoformans* phenoloxidase can oxidize a wide variety of substrates to produce melanin (see above). The *C. neoformans* melanin synthetic pathway is believed to involve the oxidation of L-dopa (or the dihydroxyphenol) to the quinone derivative in the presence of O_2, and the product then spontaneously rearranges and undergoes autooxidation to melanin. The intermediates dopachrome and 5,6-dihydroxyindole have been identified in the *C. neoformans* melanogenesis (178). Decarboxy dopachrome has been demonstrated in vitro by the action of purified phenoloxi-

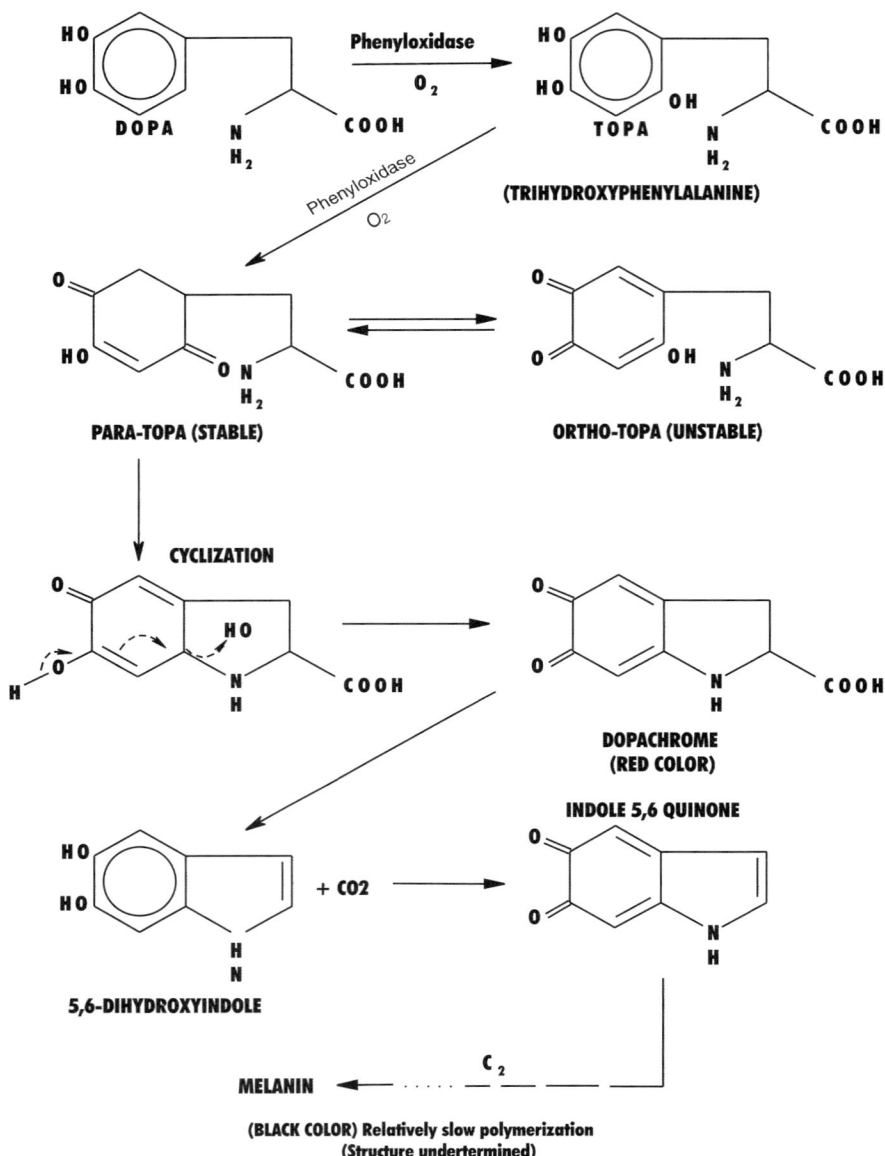

Figure 3   Proposed pathway for the oxidation of L-dopa to melanin (reprinted from reference 162). Some modifications to this pathway were suggested by Polacheck and Kwon-Chung (178).

dase enzyme on L-dopa (237). The structure of fungal melanin is unknown. The reaction of *C. neoformans* phenoloxidase with L-dopa results in consumption of oxygen (197). Although melanin synthesis requires only the presence of phenoloxidase, substrate, and oxygen, at least seven genes are involved in melanin synthesis in *C. neoformans* (212). In some melanin-negative mutants, the phenotype can be suppressed by the addition of Cu^{2+}, suggesting that melanogenesis and copper metabolism are closely linked (212).

Electron microscopy of melanized cells revealed electron-dense material in the inner aspect of the cell wall, leading to the suggestion that this was the location for melanin deposition (226). When *C. neoformans* cells are grown with L-dopa, melanization is progressive with time (227). During experiments to extract RNA from melanized cells, it was serendipitously found that melanin ghosts were generated by treatment with detergent followed by hydrolysis with hot concentrated HCl acid (227). Figure 4 shows melanin ghosts that appear to be melanin particles in the shape of the original cell wall. The availability of melanin ghosts has permitted some preliminary structural studies on melanin composition. In heavily melanized cells,

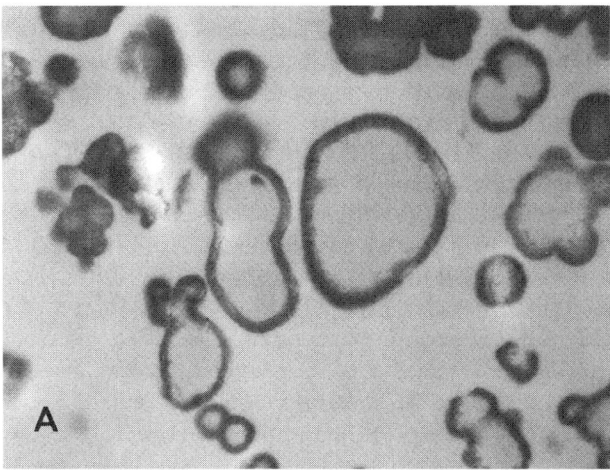

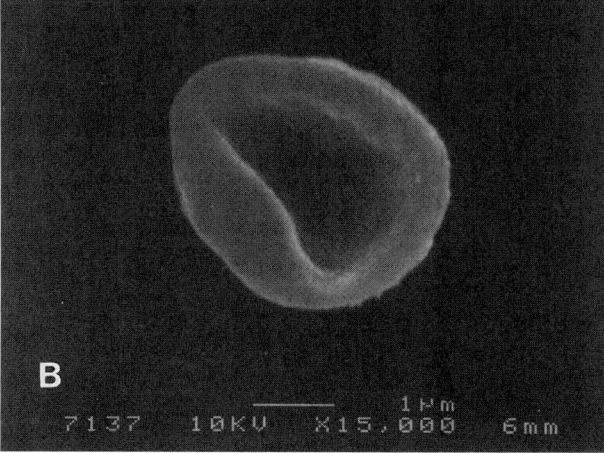

Figure 4 (A) Scanning electron micrograph of melanin ghosts ($\times$15,000). (B) Transmission electron micrograph of a melanin ghost ($\times$45,000). The structure represents melanin left after detergent treatment with guanidinium isothiocyanate and digestion with hot concentrated HCl acid. Reprinted from reference 227.

melanin accounted for 15.4% of the dry mass of the cells (227). The amount of melanin made per cell when *C. neoformans* is grown in medium with L-dopa varies for different strains (161, 227). The carbon-to-nitrogen ratio of L-dopa melanin "ghosts" is 12.5 (227), and melanin particles are found to be negatively charged (161).

Given that melanization is easily detectable in L-dopa and niger seed agar plates because colonies turn black or brown, one can easily identify melanin-deficient mutants because they lack pigments. Several melanin-negative mutants of *C. neoformans* have been generated and characterized (187, 212). Melanin-deficient mutants are less virulent than wild-type strains (124, 187, 194, 226). The contribution of the phenoloxidase to virulence has also been demonstrated by gene disruption techniques (194). Melanization in *C. neoformans* is associated with a variety of phenomena, some of which have been implicated in virulence (Table 4). Melanin is a powerful antioxidant with the capacity to protect the fungal cell against oxygen- and nitrogen-derived oxidants of the type made by host effector cells (96, 98, 103, 226, 228, 234). The antioxidant power of melanin is comparable to that of superoxide dismutase (SOD) (98). Relative to SOD, melanin has the added advantage of having greater chemical reactivity and could, in theory, protect against a wider variety of oxidants (98). Melanization may also contribute to virulence by inhibiting phagocytosis (226) and reducing the susceptibility of yeast cells to amphotericin B (229). *C. neoformans* melanin is negatively charged, and melanization imparts an additional negative charge to the fungal cell (161). Melanization is associated with increased susceptibility of *C. neoformans* to some melanin-binding drugs, such as trifluoperazine (231). Drugs that target melanin may be useful against *C. neoformans* provided that melanization occurs in vivo (231).

Melanization of *C. neoformans* cells is likely to occur in vivo during infection given the existence of melanin precursors such as L-dopa and epinephrine in tissue. However, no study to date has unequivocally shown melanin formation in vivo. Many studies cite silver staining of *C. neoformans* as evidence of in vivo melanization, but silver staining is not specific for melanin given the existence of reducing substances in the fungal cell wall (123). The ability of *C. neoformans* to metabolize catecholamines to melanin has been suggested as an explanation for the remarkable neurotropism of this pathogen (176). The increased concentration of catecho-

Table 4 Biological phenomena associated with and attributed to melanin production in *C. neoformans*

Phenomenon or effect	Reference
Protection against oxidants	60, 96, 103, 179, 226, 230
Cell wall integrity	227
Resistance to amphotericin B	229
Increased susceptibility to neutropic drugs	231
Protection against UV light	230
Interference with inflammatory responses	83
Inhibition of phagocytosis	226
Resistance to heat and cold	188
$Fe^{3+} \rightarrow Fe^{2+}$ reduction: role in Fe uptake?	164
Increase in negative cell charge	161

lamines in the brain may provide the necessary substrate for melanization, which in turn could translate into an increased survival advantage for *C. neoformans*. Consistent with this hypothesis, areas of the brain rich in catecholamines (such as the basal ganglia) are often invaded by *C. neoformans* in cryptococcal meningoencephalitis (131). Darkly pigmented *C. neoformans* cells have been described in the brain, and the pigment in these cells may be melanin (131).

In summary, *C. neoformans* has a phenoloxidase enzyme system that catalyzes pigment production from structurally diverse phenolic substrates. The pigments generated from L-dopa and epinephrine have been shown to be melanin. The biochemistry of melanin and the assembly of melanin on the cell wall remain poorly understood. Melanogenesis is interesting because of its association with virulence and because it is a potential target for antifungal drug design.

Other Pigments

C. neoformans and other *Cryptococcus* species produce a variety of colorful pigments when grown in medium containing tryptophan (35). Approximately 30% of *C. neoformans* strains produce a pink to purple extracellular pigment and a brown intracellular pigment when grown in the dark (35). Furthermore, yeast-cell autofluorescence was observed for *C. neoformans* strains grown in tryptophan medium (35). The biochemical nature of these pigments and their relationship to the *C. neoformans* phenoloxidase activity is unknown.

EXTRACELLULAR PRODUCTS

Polysaccharides and MP

The major extracellular polysaccharides of *C. neoformans* are GXM, GalXM, and MP. The relative amounts of these polysaccharides in culture supernatants are GXM > GalXM > MP (184). Both GXM and GalXM are shed during infection and can be detected in body tissues. The structure, antigenicity, and biological properties of the *C. neoformans* polysaccharides are described above.

Mannitol

C. neoformans produces large amounts of hexitol mannitol in culture and in tissues of experimentally infected rabbits (241). Mannitol biosynthetic enzymes are not constitutively expressed but are induced in response to glucose (156). Glucose, fructose, and mannose are converted to mannitol, but maltose and xylose are not (156). A NAD- and NADP-linked mannitol dehydrogenase found in *C. neoformans* cytoplasmic extracts is induced by mannitol (169). During attempts to express this enzyme in *S. cerevisiae*, a *C. neoformans* regulatory gene was discovered that enhanced the activity of the endogenous yeast mannitol dehydrogenase (169).

In rabbits, the concentration of mannitol in cerebrospinal fluid (CSF) is proportional to the severity of infection (241). A *C. neoformans* mutant deficient in mannitol production was less virulent than the parent strain (38). In nature, mannitol may serve as a reservoir for the storage of carbohydrate and reducing equivalents (156). In infection, mannitol may contribute to virulence by promoting osmotic and heat stress tolerance, scavenging hydroxyl radicals, and promoting brain edema through osmotic effects (38, 39). Mannitol has been demonstrated in the CSF of patients with AIDS suffering from cryptococcal meningitis. However, unlike the studies in rabbits, there was no correlation between mannitol concentration in human CSF and the severity of infection (144). The explanation for the discordance between the mannitol concentration and the severity of infection in humans and rabbits may lie in interspecies differences in CSF synthesis and resorption (144).

Enzymes and Protein Antigens

A variety of proteins and enzyme activities have been detected in *C. neoformans* cells and culture supernatants (Table 5). These include protein antigens that elicit immune reactions and many enzymes that have been associated with virulence for other microorganisms, including proteases, lipases, and SOD. In the environment the fungus presumably elaborates these enzymes to obtain nutrients. However, when secreted during infection, these enzymes could destroy tissues, interfere with host defense mechanisms, and contribute to virulence.

C. *neoformans* produces an extracellular DNase (33). This enzyme probably contributes to the difficulties encountered in generating high-molecular-weight DNA from cryptococcus for molecular studies (33). Many investigators have resorted to high concentrations of EDTA in the preparation of DNA to inhibit DNase (28, 202, 218). The DNase could participate in virulence by degrading host DNA and supplying the fungus with nucleotides. No detailed biochemical or structural information on the *C. neoformans* DNase is available. A survey of several yeast species, including *C. neoformans*, suggests a correlation between urease activity and extracellular DNase production (33).

A glycoprotein has been isolated from the supernatants of *C. neoformans* cultures (189). This material was found with the major polysaccharide fraction when analyzed by anion exchange chromatography (189). Amino acid composition analysis revealed that aspartate/asparagine, glutamate/glutamine, serine, and threonine were the most common amino acids. The relationship of this protein component to the polysaccharide or to proteins that have been subsequently described is unknown.

Two SODs have been described in *C. neoformans* (77, 208). One is a Cu,Zn SOD with a molecular mass of 145 kDa and a pI of 7.5 that is inhibited by KCN. Partial N-terminal sequence data reveal some amino acid differences between the SOD of var. *neoformans* and var. *gattii* (77). Biochemical differences between the Cu,Zn SOD from the two varieties were suggested by the fact that diethyldithiocarbamate and sodium azide inhibited only the SOD from var. *gattii* (77). The second SOD has significant homology to SODs from other organisms that contain Mn but no homology to the Cu,Zn SOD of *C. neoformans* (208). The Mn SOD has a molecular mass of 19.5 kDa and a pI of 6.6 and is inhibited by sodium dodecyl sulfate, sodium azide, *o*-phenanthrolene, and EDTA (208). In its native conformation, the Mn SOD exists as an 80-kDa tetrameric structure (208). SOD could contribute to virulence by protecting against superoxide generated by host immune effector cells. Melanin and SOD may provide complementary defenses for the *C. neoformans* cells against oxidative damage. SOD production is regulated by temperature such that more is made at 37°C than at 25°C (98). Increased production

Table 5 Some enzymes and enzyme activities and their cellular location in *C. neoformans*

Enzyme activity	Location[a]	Reference
Acid phosphatase	Capsule and lysosomes	31, 140
Acyltransferase	Extracellular	42
Alkaline phosphatase	Endoplasmic reticulum	69
Aminopeptidase	Lysosome, microbody	140
Catalase	Lysosome	140
	Microbody	69
Creatinine deiminase	Cell associated	177
DHFR	Intracellular	198
DNase	Extracellular	33
Esterase lipase	Cell associated	31
Glucose-6-phosphate dehydrogenase	Intracellular	157
Glucose-phosphate isomerase	Intracellular	191
α-Glucosidase	Cell associated	31
β-Glucosidase	Cell associated	31
β-Glucuronidase	Cell associated	31
Glucuronyltransferase	Intracellular	235
α-Hydroxy oxidase	Microbody	69
Imidizole glycerol phosphate dehydratase	Intracellular	168
Laccase	Intracellular	89, 176, 237
Lysophopholipase	Extracellular	42
Lysophospholipase and lysophospholipase-transcylase	Extracellular	41
Malate dehydrogenase	Intracellular	137
Malic dehydrogenase	Mitochondria	140
Mannitol dehydrogenase	Intracellular	169
α-Mannosidase	Cell associated	31
α-Mannosyl transferase	Intracellular	236
Myristoyl-CoA:protein *N*-myristoyltransferase	Intracellular	134
Phosphoaminase	Cell associated	31
Phosphoglucomutase	Intracellular	191
6-Phosphogluconate dehydrogenase	Intracellular	158, 159
Phospholipase B	Extracellular	41, 42
Protease	Extracellular	5, 25, 40
SOD	Extracellular	77, 208
Thymidylate synthase	Intracellular	133
UDP-glucose dehydrogenase	Intracellular	94
Urease	Intracellular	244
Xylotransferase	Intracellular	235

[a]Location refers to site of isolation of enzyme or enzyme activity according to the reference cited.

of SOD at mammalian temperatures may protect the fungus against oxidants produced by host effector cells (98).

Phospholipases have been associated with virulence for many microorganisms and presumably function by destroying cell membrane components and promoting attachment to host tissues. *C. neoformans* culture supernatants contain phospholipase B, lysophospholipase, and acyltransferase (41, 42). A correlation between phospholipase activity and fungal burden in mouse tissues for a small number of strains suggests that phospholipase production may be important for virulence in *C. neoformans* (41).

Protease activity in *C. neoformans* cultures has been reported by several investigators (5, 25, 40, 67). Proteases presumably contribute to virulence by destroying host proteins, facilitating tissue invasion, and interfering with host defense mechanisms. The protease activity in *C. neoformans* culture supernatants is weak (40). However, this activity appears to be sufficient to permit the organism to grow in medium containing immunoglobulin or complement as the sole source of carbon and nitrogen (40). Protein substrate gels have identified three proteins with proteolytic activity in *C. neoformans* supernatants (40).

Hamilton and Goodley (76) have described a 115-kDa antigen in culture supernatants that is recognized by sera from patients with cryptococcosis and by the mouse MAb 7C9 (76). This antigen is produced in late-log-phase cultures, and more is made at 25°C than at 37°C (76). The 115-kDa antigen is a glycoprotein with N-linked oligosaccharides composed, in part, of mannose (76). The function of this 115-kDa antigen is unknown, and it appears to be antigenically different from the extracellular MP of *C. neoformans* (76). Another, smaller glycoprotein of 34 to 38 kDa is also present in culture supernatants; it appears to originate from the cytoplasm and cell wall (78). This smaller antigen is glycosylated, defines a species-specific epitope, is also unrelated to the MP, and has an unknown function (78).

Ethanol

Ethanol is the end product of glucose fermentation by yeast cells. There were several studies of ethanol production by *C. neoformans* in the 1950s and 1960s, but the subject has not been studied recently. The earlier studies used the term alcohol (51, 215), whereas the later studies specifically referred to ethanol (167). The original suggestion that *C. neoformans* synthesized alcohol was made by Tyler (215) in 1956. He based this suggestion on the reasoning that all yeasts produce ethanol during fermentation and on a positive dichromate reduction reaction in *C. neoformans* supernatants. Tyler reported alcohol in the spinal fluid of patients with cryptococcal meningitis but not in those with bacterial meningitis; thus, he suggested its potential efficacy as a diagnostic test for fungal meningitis (215). Dawson and Taghavy (51) subsequently confirmed this finding by using a more sensitive test based on the reduction of diphosphopyridine nucleotide by alcohol in the presence of alcohol dehydrogenase. They also measured alcohol in a patient with presumed cryptococcal meningitis and demonstrated alcohol production in CSF inoculated with *C. neoformans* (51). Pappagianis and Marovitz (167) used gas chromatography and an enzymatic technique to detect ethanol in *C. neoformans* culture supernatants and compared ethanol production by cryptococcus to that of *S. cerevisiae* and *C. albicans* (167). These investigators confirmed the presence of ethanol but found that the ethanol concentration in *C. neoformans* supernatants was less than 1% of the theoretical yield, based on fermentation of glucose to ethanol and carbon dioxide (167). In contrast, both *S. cerevisiae* and *C. albicans* produced ethanol concentrations in excess of half the theoretical yield (167). On the basis of low production, Pappagianis and Marovitz (167) suggested that measurement of CSF alcohol concentration had limited diagnostic value. The question of the utility of CSF ethanol as a diagnostic tool in cryptococcal meningitis was then reevaluated in a large study involving 31 patients with cryptococcal meningitis, 1 patient with candidal meningitis, 1 patient with coccidioidal meningitis, and 37 patients without meningitis (239). No correlation was found between cryptococcal meningitis and CSF ethanol concentration (239). Small amounts of ethanol were detected in patients without meningitis. It was suggested that this originated from normal tissue metabolism and contamination from skin disinfection with antiseptic preparations containing ethanol (239).

In summary, there is evidence that small amounts of ethanol can be found in the supernatants of *C. neoformans* cultures. Ethanol is presumably a product of fungal growth in culture, but the synthesis pathway has not been elucidated. Ethanol production in vivo has not been convincingly demonstrated. The question of whether *C. neoformans* makes ethanol in vivo and whether or not this contributes to virulence or pathogenesis is unknown.

Other Extracellular Fungal Products

Apart from the fungal products listed above, it is likely that there are other substances produced by *C. neoformans* that have not been identified. Gas chromatography of fungal cultures revealed many products that are also found in the CSF of patients with cryptococcal meningitis (195).

BIOSYNTHETIC AND PROTEIN MODIFICATION PATHWAYS

Like many free-living microorganisms, *C. neoformans* can synthesize most organic molecules necessary for growth from simple carbon- and nitrogen-precursor compounds. In the laboratory, *C. neoformans* can grow in simple, chemically defined medium containing glycine (simplest amino acid), glucose, and salts. Autotrophy implies the existence of complex biochemical biosynthetic pathways for synthesizing amino acids, sugars, lipids, nucleotides, and polysaccharides. Table 5 lists enzymes and enzymatic activities described for *C. neoformans*.

Information on biosynthetic pathways is important for antifungal drug design. The ability of *C. neoformans* to make many of the complex organic compounds may also contribute to virulence by allowing the fungus to survive in host tissues with minimal nutritional requirements. In this regard, *C. neoformans* mutants defective in adenine synthesis have been shown to have reduced virulence relative to the wild-type strain (170). Biochemical differences between fungal and host enzymatic pathways provide the basis for rational drug design. Lysine is an essential amino acid that is obtained from the diet in humans but is synthesized in fungi via the α-aminoadipate pathway (72). Differences in the metabolism of lysine between humans and fungi suggest that inhibition of this pathway may be a target for new antifungal drugs. Five enzyme activities associated with the lysine biosynthetic pathway have been described in *C. neoformans* extracts: homocitrate synthase, homoisocitrate dehydrogenase, aminoadipate reductase, saccharopine reductase, and saccharopine dehydrogenase (72). Interestingly, saccharopine reductase was detected in two *C. neoformans* var. *neoformans* strains but not in one *C. neoformans* var. *gattii* strain (72). The absence of this enzyme in the *C. neoformans* var. *gattii* strain is interesting because the organism grew in the absence of lysine (72).

For adaptation to growth in avian excreta, *C. neoformans* has enzymatic pathways capable of metabolizing creatinine, xanthines, and uric acid (see also chapter 3). Most *C. neoformans* strains have strong urease activity (126, 224, 244), although occasional urease-negative strains have been described in clinical isolates (7, 190, 244). Urease activity in *C. neoformans* is influenced by metal ions, and differences in urease activity of the two varieties can be demonstrated by adding metal-chelating compounds such as EDTA (126). *C. neoformans* can utilize creatinine as a source of nitrogen but not as a source of carbon (177). The fungus has a creatinine deiminase that cleaves creatinine to ammonia and methylhydantoin (177). Creatinine deiminase synthesis is repressed by ammonia in *C. neoformans* var. *neoformans* but not in *C. neoformans* var. *gattii* (177). This biochemical difference may contribute to the ability of var. *neoformans* to colonize avian excreta, since enzyme inhibition could prevent the uncontrolled alkalinization that would inhibit growth.

Several biosynthetic enzymes have been studied in detail, including dihydrofolate reductase (DHFR) (198) and thymidylate synthase (133). In general, the characterization of these *C. neoformans* enzymes has revealed biosynthetic proteins similar to those described for other organisms. The *C. neoformans* DHFR has been expressed in *Escherichia coli* and has been characterized (198). The cryptococcal DHFR has significant homology to DHFRs from other organisms and is inhibited by classical inhibitors of this enzyme, including methotrexate, trimethoprim, and pyrimethamine (198). Comparison of the *C. neoformans* thymidylate synthase to that of other organisms revealed conservation of all catalytic and protein structural elements with some differences in nonconserved residues (133). Niehaus and collaborators (137, 158, 159) have characterized several *C. neoformans* metabolic enzymes, including 6-phosphogluconate dehydrogenase and malate dehydrogenase. A *C. neoformans* malate dehydrogenase dimer of 35-kDa subunits (probably of mitochondrial origin) has been described with several unusual biochemical characteristics, including inhibition by Zn ion and by heparin (137).

Protein myristoylation is another enzymatic reaction that is a potential target for antifungal drug design. *C. neoformans* has a myristoyl-CoA:protein *N*-myristoyltransferase that catalyzes the transfer of myristate from coenzyme A to the amino-terminal glycine residue (107, 134). *C. neoformans* produces *N*-myristoyl proteins, and inhibition of myristoylation by the myristic acid analog 4-oxatetradecanoic acid is fungicidal for this organism (129).

GLYCOGEN

Several electron microscopy studies of *C. neoformans* have suggested that *C. neoformans* cells contain glycogen, on the basis of distinctive glycogen granules in the cytoplasm (2, 58, 69, 193).

LIPIDS AND STEROLS

Several studies have addressed issues of lipid and sterol composition in *C. neoformans*. Lipid bodies in the cytoplasm of *C. neoformans* have been observed by light and electron microscopy, but their chemical composition is unknown (2, 58, 151, 204, 207). (Poly-*β*-hydroxybutyrate was suggested as a component [58].) Comparative analysis of fatty acid composition profiles of several yeasts, including *Candida* sp., *Torulopsis glabrata*, and *C. neoformans*, by gas-liquid chromatography indicate that each has a distinctive fatty acid composition that can be used to discriminate and identify the species (139). Together with the finding of pyrophosphatidic acid (see below), this suggests that *C. neoformans* has significant qualitative and quantitative differences from other pathogenic yeasts in its lipid and sterol biochemistry.

Rawat and colleagues (181) studied the lipid composition of *C. neoformans* after growth in Sabouraud dextrose broth using several biochemical techniques. The analysis revealed that *C. neoformans* cells contain only 0.96% lipid per dry weight (181). This very low lipid composition may be an artifact of the large mass of the polysaccharide capsule, since fungal cells with larger capsules have lower lipid contents (216). The percentage of lipid composition per cell was substantially lower than that reported for other fungi (181). The lipid composition was 86.1% nonpolar lipid, 3.4% phospholipid, and 10.5% glycolipids and pigments (181). The phospholipid composition was similar to that of other yeasts, with the predominant phospholipids being phosphatidylinositol (11.5%), lysophosphatidyl ethanolamine (10.9%), and cardiolipin (10.1%) (181). A comparison of three strains of *C. neoformans* with different degrees of virulence revealed no major difference in phospholipid content (216).

A novel phospholipid, pyrophosphatidic acid [*P*,*P'*-*bis*-(1,2-diacyl-*sn*-glycero-3)-pyrophosphate], was first described in *C. neoformans* (91–93). This lipid appears to be a dimer of phosphatidic acid and has a 1:1 molar ratio of diglyceride to phosphorus (91). A survey of the distribution of pyrophosphatidic acid in nature documented this lipid in yeasts of the genera *Rhodotorula*, *Kloeckera*, and *Trichosporon*. Pyrophosphatidic acid was not detected in *S. cerevisiae*, *Schizosaccharomyces pombe*, or other ascosporogenous yeasts. Pyrophosphatidic acid has been proposed to serve as a stable cellular pool of diglyceride and phosphatidic acid (93).

Like other fungi, *C. neoformans* has ergosterol in its cell membranes. The presence of ergosterol and related sterols in the cell membrane and their biosynthetic pathways are the target for two of the most effective antifungal drugs: amphotericin B and the azoles. Amphotericin B is believed to mediate antifungal activity by binding to cell wall sterols and promoting cell damage (24). However, there is considerable evidence that amphotericin B also functions by enhancing host macrophage function and affecting other aspects of fungal metabolism (24, 211, 240). The azole class of antifungal agents is believed to inhibit sterol synthesis through effects on the P450-dependent 14*α*-demethylase and associated enzymes. It is noteworthy that most studies of the mechanism of action of amphotericin B and the azoles have been done with fungi other than *C. neoformans*. Extrapolating mechanisms of action to *C. neoformans* from studies with other fungi may or may not be appropriate, considering the many biochemical differences among fungal species.

The sterol composition of *C. neoformans* varies with the strain. Kim et al. (110) studied two strains of *C. neoformans* by gas liquid chromatography and found that the major sterols were epifungissterol and ergosterol. Analysis of several mutants resistant to nystatin revealed loss of ergosterol and alterations of sterol content (110). However, mutants resistant to amphotericin B had normal ergosterol content, suggesting a mechanism of amphotericin resistance by means other than membrane sterol alteration (110). Vanden Bossche et al. (217) studied one strain of *C. neoformans* and reported that ergosterol constituted 74.5% of the total sterol content. Treatment of this strain with itraconazole resulted in inhibition of ergosterol synthesis with accumulation of obtusifolione and eburicol (217).

Franzot and Hamdam (70, 71) analyzed the lipid and sterol composition of several *C. neoformans* isolates before and after exposure to amphotericin B. For five strains the total lipid content ranged from 4.38 to 7.00 mg/100 ml, and the sterol content ranged from 0.48 to 0.96 mg/100 ml (71). These values for lipid and sterol content are in general agreement with the typical yeast composition range from 7 to 15 mg/100 ml and 0.1 to 1.0 mg/100 ml, respectively (71). Sterol fractionation for the five strains revealed ergosterol and squalene and the absence of lanosterol (71). Incubation of *C. neoformans* with a concentration of amphotericin B corresponding to one-half the MIC produced a significant reduction in lipid and sterol content relative to control cells (71). The mechanism by which amphotericin B alters *C. neoformans* lipid and sterol composition is not known. Exposure of *C. neoformans* to the azole compounds ketoconazole, fluconazole, and itraconazole at concentrations below their MIC also produced major changes in lipid and sterol content and composition (70). The effectiveness of these drugs for inhibiting *C. neoformans* sterol synthesis was itraconazole > fluconazole > ketoconazole (70).

Ghannoum et al. (73) studied 13 strains of *C. neoformans* and also found considerable interstrain variation in the amount and type of sterols present. The sterol content of *C. neoformans* ranged from 0.3 to 5.9% of cell dry weight depending on the strain (73). Obtusifoliol and ergosterol were the predominant sterols in nine and three isolates, respectively (73). Table 6 lists the average sterol composition measured for the 13 *C. neoformans* strains in the presence or absence of fluconazole (73). By comparing sterol content in the presence and absence of fluconazole, it was suggested that this drug also inhibited the conversion of 24-methylenedihydrolanosterol to obtusifoliol and 4,14-dimethylzymosterol to zymosterol (73). The sterol content of *C. neoformans* has been found to be affected by passage in mice, suggesting that infection can result in selection of phenotypically stable variants with altered sterol composition (50). Mouse passage was associated with increased ergosterol content and reduced obtusifoliol content in four of five strains (50). Furthermore, mouse passage of environmental *C. neoformans* isolates produced isolates with higher amphotericin B and fluconazole MICs than before passage, suggesting that sterols and antifungal drug resistance could be altered by mammalian infection without a need for exposure to antifungal drugs (50).

Vincent and Klig (222) studied inositol metabolism in *C. neoformans* and its effect on phospholipid metabolism. *C. neoformans* has the unusual capacity to grow while using inositol as the sole source of carbon (222). This six-carbon sugar is a precursor for phosphatidylinositol, a component of the cell membrane that also serves as a lipid anchor for some membrane proteins. *C. neoformans* can synthesize inositol, and inositol was identified in three fungal lipids: ceramide-(P-inosi-

Table 6 Sterol composition of 13 *C. neoformans* isolates in the presence or absence of fluconazole[a]

	% of total sterols		
Sterol	Control	Fluconazole	*P* value
Squalene	7.9 ± 5.1	12.0 ± 8.8	1.24
24-Methylenedihydrolanosterol	2.0 ± 3.3	0.0 ± 0.0	0.43
Lanosterol	4.6 ± 8.0	10.7 ± 8.2	0.742
4,14-Dimethylzymosterol	5.2 ± 6.0	34.5 ± 15.4	0.001
Zymosterol	4.4 ± 9.3	0.4 ± 0.8	0.23
Obtusifoliol	44.2 ± 14.1	16.1 ± 8.9	<0.001
Calciferol	5.7 ± 4.6	9.3 ± 4.5	1.26
Ergosterol	27.4 ± 15.6	18.6 ± 13.4	1.23

[a]Adopted from reference 73. Values are means ± standard deviations. *P* values are obtained by *t*-test and adjusted for the Bonferroni correction.

tol)$_2$mannose, ceramide-P-inositol-mannose, and ceramide-P-inositol. However, growth of *C. neoformans* in medium with inositol had no effect on membrane lipid composition. *C. neoformans* (unlike other yeasts, such as *S. cerevisiae*) is able to maintain a stable lipid composition despite the presence of inositol in the growth medium. It was speculated that the ability of *C. neoformans* to maintain stable lipids regardless of exogenous inositol concentration may contribute to its neurotropism, considering the existence of high inositol concentrations in the brain (222).

In summary, *C. neoformans* strains contain common fungal lipids and sterols but display great variation in terms of sterol content and composition. However, *C. neoformans* has some unique characteristics with regard to lipid synthesis and regulation, as exemplified by the inositol studies (222). The lipid and sterol composition profile of *C. neoformans* can be altered by polyene- and azole-type antifungal drugs and animal passage. The results indicate great plasticity in the lipid and sterol content of *C. neoformans* strains, which appears to be a result of genetic and environmental factors.

VACUOLAR AND LYSOSOMAL ENZYMES

Electron microscopy reveals that *C. neoformans* cells contain many vacuolar and organelle-like structures (140). Cytochemical and biochemical studies of enzymatic activities in internal *C. neoformans* structures have been described, and the existence of lysosomes has been proposed on the basis of acid phosphatase activity in membrane-bound organelles (140). Staining of *C. neoformans* cells with acridine orange, a lysosomal indicator, revealed reddish fluorescence that localized to individual organelles in the cell cytoplasm (140). Lead staining followed by electron microscopy revealed lead deposition consistent with acid phosphatase activity in the vacuoles, cell wall, and polysaccharide capsule (138) (Fig. 5). Lysosomal function in *C. neoformans* may include internal digestion of broken-down cell membranes, lipids, etc., for recycling and further metabolic use (140). The finding of acid phosphatase in the capsule suggests that lysosomes may function in enzyme secretion by discharging their contents to the outside through exocytosis. Phos-

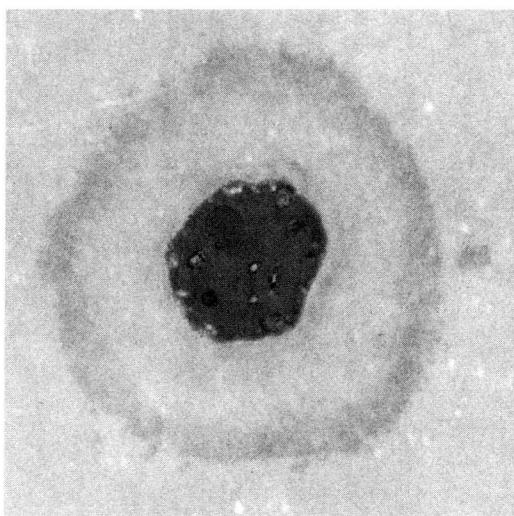

Figure 5   Acid phosphatase in *C. neoformans* cells. Electron-dense deposits inside the cell are due to lead phosphate staining, which indicates the presence of acid phosphatase ($\times$4,500). Note also staining at the edge of the capsule. Figure courtesy of Marta Feldmesser and Phyllis Novikoff (Albert Einstein College of Medicine, Bronx, N.Y.).

phatases have been demonstrated both in yeast cells grown in culture and in granulomas, and they include both acid and alkaline phosphatases (242).

When *C. neoformans* is grown in medium containing uric acid or xanthine as the sole source of nitrogen, the fungal cells contain numerous membrane-bound microbodies (69). These microbodies contain catalase, urate oxidase activity, and alkaline phosphatase activity (69). Energy dispersive X-ray spectra of *C. neoformans* microbodies suggested that xanthine oxidase is a Mb-containing enzyme (69). The increased number of microbodies containing enzymes in minimal media supplemented with either uric acid or xanthine was believed to be a result of induction of biosynthetic pathways to grow in a nutritionally poor environment (69). These results imply that induction of microbody formation may be a survival strategy for growth in the nutrient-poor (but uric acid- and xanthine-rich) environment of avian excreta (69).

IRON METABOLISM

Iron is an essential nutrient for *C. neoformans*. Some microorganisms acquire iron from the environment by elaborating substances that chelate iron, such as hydroxamates or phenolates (siderophores). Other microorganisms have iron uptake mechanisms that utilize transport proteins and are siderophore-independent. Iron metabolism in *C. neoformans* is important because it could represent a target for drug design and there is evidence that it is involved in the regulation of virulence factors such as capsule synthesis and melanin formation. Furthermore, the host uses iron sequestration by serum proteins as a nonspecific mechanism for protection against *C. neoformans* (81, 206).

Analysis of *C. neoformans* culture medium for the presence of siderophores revealed no hydroxamates, organic acids, or other iron chelators, suggesting that *C. neoformans* obtains iron by a siderophore-independent mechanism (100). However, growth inhibition caused by the iron chelator ethylenediamine-di(*o*-hydroxyphenylacetic acid) was reversed by addition of deferoxamine, raising the possibility of the presence of a hydroxamate-dependent uptake mechanism (100). This result suggested that *C. neoformans* may have a hydroxamate-dependent iron uptake mechanism without making its own hydroxamate (100). Subsequent studies have established that iron is reduced before uptake (104). *C. neoformans* can acquire iron by a mechanism present in the fungal surface that reduces Fe^{3+} and internalizes it (220). Nyhus et al. (164) have described three systems that may be involved in extracellular reduction of Fe^{3+} to Fe^{2+}: a cell-associated ferric reductase, secretion of 3-hydroxyanthranilic acid, and melanin. The relative efficacy of the three Fe^{3+} reduction systems is ferric reductase > 3-hydroxyanthranilic acid > melanin. The peak activity of each reduction system varies with the stage of growth: ferric reductase, 3-hydroxyanthranilic acid, and melanin are most active during exponential growth, late exponential growth, and the stationary phase, respectively (164). A transferrin-like 80-kDa membrane protein that binds Fe^{3+} and reacts with polyclonal sera raised against ovotransferrin has been demonstrated in *C. neoformans* (209).

In summary, *C. neoformans* has efficient mechanisms for iron uptake that are siderophile-independent but require ferric ion reduction. Iron uptake may involve binding and transport by transferrin-like proteins. The enzymes and transport proteins responsible for iron acquisition in *C. neoformans* have not been characterized at the molecular level.

BIOCHEMISTRY OF ANTIMICROBIAL DRUG RESISTANCE

For *C. neoformans* the emergence of antifungal drug resistance has not been a major problem to date. Nevertheless, several *C. neoformans* isolates with reduced susceptibility to antifungal drugs have been described. For some isolates, resistance to antifungal agents has been associated with changes in virulence and yeast-cell biochemistry. In 1969 Bodenhoff (22) described strains of *C. neoformans* resistant to amphotericin B and polymyxin B. Although the biochemical mechanism for resistance was not characterized, the resistant strains were shown to have reduced virulence for mice (22).

In recent years several drug-resistant *C. neoformans* mutants have been isolated from patients treated with antifungal agents. Powderly et al. (180) described an amphotericin B-resistant isolate from a patient who had been treated initially with fluconazole. This isolate had reduced sterol content relative to an earlier isolate from the patient and was shown to have a defect in δ-8→isomerase that resulted in depletion of ergosterol and accumulation of aberrant δ-8 double-bonded ergosterol precursors (4, 79). Fluorescent membrane probes were used to study the mechanism of amphotericin B activity in this resistant mutant and in a susceptible strain (79). The results showed a two-phase process involving the formation of amphotericin-ergosterol complexes followed by their aggregation in the cell membrane and pore formation (79).

The biochemical basis for reduced susceptibility to azole drugs in *C. neoformans* has been investigated (128). Azole-type drugs mediate their antifungal effects by inhibiting 14α-demethylase, a crucial enzyme in the production of ergosterol and other sterols. Analysis of four strains from patients who failed fluconazole therapy revealed that a 10-fold-higher concentration of fluconazole was required to inhibit sterol 14α-demethylation in these strains relative to azole-sensitive strains (128). Although the molecular basis for the differences in 14α-demethylase activity was not determined, it was suggested that changes in enzyme concentration or mutations may be responsible for reduced susceptibility to azole drugs in some *C. neoformans* strains (23).

C. neoformans mutants that are cross-resistant to both azole-type drugs and amphotericin B have been described (108). Cross-resistance was found at a frequency of 10^{-8}, consistent with the occurrence of a single mutation (108). Mutations that confer both azole and amphotericin B resistance would produce isolates resistant to the two major classes of antifungal drugs and hence pose a major challenge for therapy. The phenomenon of azole and amphotericin B cross-resistance has been described in other yeasts, such as *S. cerevisiae,* but the mechanism of cross-resistance in *C. neoformans* appears to be different (108).

Flucytosine is a nucleotide analog that is an effective antifungal drug. A problem with the use of flucytosine in patients with cryptococcal meningitis has been the emergence of resistance during therapy (20). Resistance to flucytosine appears to result from a single mutation, and rates of mutation leading to drug resistance in the range of 1.2×10^{-7} to 4.8×10^{-7} per cell division have been reported. Characterization of several of these isolates suggests mutations in the uridine-5'-monophosphate pyrophosphorylase and/or the cytosine-specific permease.

REFERENCES

1. **Abruzzo, G. K., A. M. Flattery, C. J. Gill, L. Kong, J. G. Smith, D. Krupa, V. B. Pikounis, H. Kropp, and K. Bartizal.** 1995. Evaluation of water soluble pneumocandin analogs L-733560, L-705589, and L-731373 with mouse models of disseminated aspergillosis, candidiasis, and cryptococcosis. *Antimicrob. Agents Chemother.* **39:**1077–1081.

2. **Al-Doory, Y.** 1971. The ultrastructure of *Cryptococcus neoformans. Sabouraudia* **9:**113–118.

3. **Anderson, D. M., and M. A. Dykstra.** 1984. Pulmonary cell response in mice following intranasal instillation with *Cryptococcus neoformans. Mycopathologia* **86:**179–184.

4. **Anonymous.** 1980. Garlic in cryptococcal meningitis. A preliminary report of 21 cases. *Chin. Med. J.* **93:**123–126.

5. **Aoki, S.** 1993. Extracellular proteolytic activity of *Cryptococcus neoformans,* abstr. P1–4. 2nd International Conference on Cryptococcus and Cryptococcosis, Milan.

6. **Bacon, B. E., and R. Cherniak.** 1995. Structure of the *O*-deacetylated glucuronoxylomannan from *Cryptococcus neoformans* serotype C as determined by 2D H NMR spectroscopy. *Carbohydr. Res.* **276:**365–386.

7. **Bava, A. J., R. Negroni, and M. Bianchi.** 1993. Cryptococcosis produced by a urease negative strain of *Cryptococcus neoformans. J. Med. Vet. Mycol.* **31:**87–89.

8. **Belay, T., R. Cherniak, T. R. Kozel, and A. Casadevall.** 1997. Reactivity patterns and epitope specificities of anti-*Cryptococcus neoformans* monoclonal antibodies by enzyme-linked immunosorbent assay and dot enzyme assay. *Infect. Immun.* **65:**718–728.

9. **Belay, T., R. Cherniak, E. B. O'Neill, and T. R. Kozel.** 1996. Serotyping of *Cryptococcus neoformans* by dot enzyme assay. *J. Clin. Microbiol.* **34:**466–470.

10. **Belay, T., R. Cherniak, and T. Shinoda.** 1993. Specificity of *Cryptococcus neoformans* factor sera determined by enzyme-linked immuosorbed assay and dot enzyme assay. *Infect. Immun.* **61:**2879–2885.

11. **Benham, R. W.** 1935. Cryptococci—their identification by morphology and by serology. *J. Infect. Dis.* **57**:255–274.

12. **Bennett, J. E., K. J. Kwon-Chung, and D. H. Howard.** 1977. Epidemiologic differences among serotypes of *Cryptococcus neoformans*. *Am. J. Epidemiol.* **105**:582–586.

13. **Bennett, J. E., K. J. Kwon-Chung, and T. S. Theodore.** 1978. Biochemical differences between serotypes of *Cryptococcus neoformans*. *Sabouraudia* **16**:167–174.

14. **Bergman, F.** 1965. Studies on capsule synthesis of *Cryptococcus neoformans*. *Sabouraudia* **4**:23–31.

15. **Bhattacharjee, A. K., J. E. Bennett, and C. P. J. Glaudemans.** 1984. Capsular polysaccharides of *Cryptococcus neoformans*. *Rev. Infect. Dis.* **6**:619–624.

16. **Bhattacharjee, A. K., K. J. Kwon-Chung, and C. P. J. Glaudemans.** 1979. On the structure of the capsular polysaccharide from *Cryptococcus neoformans* serotype C – II. *Mol. Immunol.* **16**:531–532.

17. **Bhattacharjee, A. K., K. J. Kwon-Chung, and C. P. J. Glaudemans.** 1980. Structural studies on the major, capsular polysaccharide from *Cryptococcus neoformans* serotype B. *Carbohydr. Res.* **82**:103–111.

18. **Bhattacharjee, A. K., K. J. Kwon-Chung, and C. P. J. Glaudemans.** 1981. Capsular polysaccharides from a parent strain and from a possible, mutant strain of *Cryptococcus neoformans* serotype A. *Carbohydr. Res.* **95**:237–248.

19. **Blandamer, A., and I. Danishefsky.** 1966. Investigations on the structure of the capsular polysaccharide from *Cryptococcus neoformans* type B. *Biochim. Biophys. Acta* **117**:305–313.

20. **Block, E. R., A. E. Jennings, and J. E. Bennett.** 1973. 5-Fluorocytosine resistance in *Cryptococcus neoformans*. *Antimicrob. Agents Chemother.* **3**:649–656.

21. **Bloomfield, N., M. A. Gordon, and D. F. Elmendorf, Jr.** 1963. Detection of *Cryptococcus neoformans* antigen in body fluids by latex particle agglutination. *Proc. Soc. Exp. Biol. Med.* **114**:64–67.

22. **Bodenhoff, J.** 1969. Alteration in virulence in strains of *Crytococcus neoformans* resistant to amphotericin b and polymyxin b. *Acta Pathol. Microbiol. Scand.* **75**:153–168.

23. **Bozette, S. A., R. A. Larsen, J. Chie, M. A. Leal, J. Jacobsen, P. Rothman, P. Robinson, G. Gilbert, J. A. McCutchan, J. Tilles, J. M. Leedom, and D. D. Richman.** 1991. A placebo-controlled trial of maintenance therapy with fluconazole after treatment of cryptococcal meningitis in the acquired immunodeficiency syndrome. *N. Engl. J. Med.* **324**:580–584.

24. **Brajtburg, J., W. G. Powderly, G. Kobayashi, and G. Medoff.** 1993. Amphotericin B: current understanding of mechanisms of action. *Antimicrob. Agents Chemother.* **34**:183–188.

25. **Brueske, C. H.** 1986. Proteolytic activity of a clinical isolate of *Cryptococcus neoformans*. *J. Clin. Microbiol.* **23**:631–633.

26. **Bulmer, G. S., and M. D. Sans.** 1968. *Cryptococcus neoformans* III. Inhibition of phagocytosis. *J. Bacteriol.* **95**:5–8.

27. **Bulmer, G. S., M. D. Sans, and C. M. Gunn.** 1967. *Cryptococcus neoformans*. I. Nonencapsulated mutants. *J. Bacteriol.* **94**:1475–1479.

28. **Casadevall, A., L. Freundlich, L. Marsh, and M. D. Scharff.** 1992. Extensive allelic variation in *Cryptococcus neoformans*. *J. Clin. Microbiol.* **30**:1080–1084.

29. **Casadevall, A., J. Mukherjee, S. J. N. Devi, R. Schneerson, J. B. Robbins, and M. D. Scharff.** 1992. Antibodies elicited by a *Cryptococcus neoformans* glucuronoxylomannan-tetanus toxoid conjugate vaccine have the same specificity as those elicited in infection. *J. Infect. Dis.* **65**:1086–1093.

30. **Casadevall, A., and M. D. Scharff.** 1991. The mouse antibody response to infection with *Cryptococcus neoformans*: V_H and V_L usage in polysaccharide binding antibodies. *J. Exp. Med.* **174**:151–160.

31. **Casal, M., and M. J. Linares.** 1983. Contribution to the study of the enzymatic profiles of yeast organisms with medical interest. *Mycopathologia* **81:**155–159.

32. **Cassone, A., N. Simonetti, and V. Strippoli.** 1974. Wall structure and bud formation on *Cryptococcus neoformans*. Arch. Microbiol. **95:**205–212.

33. **Cazin, J., Jr., T. R. Kozel, D. M. Lupan, and W. R. Burt.** 1969. Extracellular deoxyribonuclease production by yeasts. *J. Bacteriol.* **100:**760–762.

34. **Chang, Y. C., and K. J. Kwon-Chung.** 1994. Complementation of a capsule-deficient mutation of *Cryptococcus neoformans* restores its virulence. *Mol. Cell. Biol.* **14:**4912–4919.

35. **Chaskes, S., and R. Tyndall.** 1978. Pigmentation and autofluorescence of cryptococcus species after growth on tryptophan and anthranilic acid media. *Mycopathologia* **64:**105–112.

36. **Chaskes, S., and R. L. Tyndall.** 1975. Pigment production by *Cryptococcus neoformans* from *para-* and *ortho-*diphenols: effect of the nitrogen source. *J. Clin. Microbiol.* **1:**509–514.

37. **Chaskes, S., and R. L. Tyndall.** 1978. Pigment production by *Cryptococcus neoformans* and other *Cryptococcus* species from aminophenols and diaminobenzenes. *J. Clin. Microbiol.* **7:**146–152.

38. **Chaturvedi, V., T. Flynn, W. G. Niehaus, and B. Wong.** 1996. Stress tolerance and pathogenic potential of a mannitol mutant of *Cryptococcus neoformans*. *Microbiology* **142:**937–943.

39. **Chaturvedi, V., B. Wong, and S. L. Newman.** 1996. Oxidative killing of *Cryptococcus neoformans* by human leukocytes. Evidence that fungal mannitol protects by scavenging reactive oxygen intermediates. *J. Immunol.* **156:**3836–3840.

40. **Chen, L.-C., E. Blank, and A. Casadevall.** 1996. Extracellular proteinase activity of *Cryptococcus neoformans*. *Clin. Diagn. Lab. Immunol.* **3:**570–574.

41. **Chen, S. C. A., M. Muller, J. Z. Zhou, L. Wright, and T. C. Sorrell.** 1997. Phospholipase activity in *Cryptococcus neoformans*: a new virulence factor? *J. Infect. Dis.* **175:**414–420.

42. **Chen, S. C. A., L. C. Wright, R. T. Santangelo, M. Muller, V. R. Moran, P. W. Kuchel, and T. C. Sorrell.** 1997. Identification of extracellular phospholipase B, lysophospholipase, and acyltransferase produced by *Cryptococcus neoformans*. *Infect. Immun.* **65:**405–411.

43. **Cherniak, R., R. G. Jones, and E. Reiss.** 1988. Structure determination of *Cryptococcus neoformans* serotype A-variant glucuronoxylomannan by 13C-NMR spectroscopy. *Carbohydr. Res.* **172:**113–138.

44. **Cherniak, R., R. G. Jones, and M. E. Slodki.** 1988. Type-specific polysaccharides of *Cryptococcus neoformans*. NMR-spectral study of a glucuronoxylomannan chemically derived from a *Tremella mesenterica* exopolysaccharide. *Carbohydr. Res.* **182:**227–239.

45. **Cherniak, R., L. C. Morris, B. C. Anderson, and S. A. Meyer.** 1991. Facilitated isolation, purification, and analysis of glucuronoxylomannan of *Cryptococcus neoformans*. *Infect. Immun.* **59:**59–64.

46. **Cherniak, R., E. Reiss, M. E. Slodki, R. D. Plattner, and S. O. Blumer.** 1980. Structure and antigenic activity of the capsular polysaccharide of *Cryptococcus neoformans* serotype A. *Mol. Immunol.* **17:**1025–1032.

47. **Cherniak, R., E. Reiss, and S. Turner.** 1982. A galactoxylomannan antigen of *Cryptococcus neoformans* serotype A. *Carbohydr. Res.* **103:**239–250.

48. **Cherniak, R., and J. B. Sundstrom.** 1994. Polysaccharide antigens of the capsule of *Cryptococcus neoformans*. *Infect. Immun.* **62:**1507–1512.

48a. **Cherniak, R., H. Valafor, L. C. Morris, and S. Valafor.** *Cryptococcus neoformans* neotyping by quantitative analysis of H-NMR of glucuronoxylomannan with a computer simulated artificial neural network. *Clin. Diagn. Lab. Immunol.* **5:**146–159.

49. **Cook, W. L., F. G. Felton, H. G. Muchmore, and E. R. Rhoades.** 1970. Cell wall

differences of patient and soil isolates of *Cryptococcus neoformans*. *Sabouraudia* **7**:257–260.

50. **Currie, B., H. Sanati, A. S. Ibrahim, J. E. Edwards, A. Casadevall, and M. A. Ghannoum.** 1995. Sterol compositions and susceptibilities to amphotericin B of environmental *Cryptococcus neoformans* isolates are changed by murine passage. *Antimicrob. Agents Chemother.* **39**:1934–1937.

51. **Dawson, D. M., and A. Taghavy.** 1963. A test for spinal-fluid alcohol in torula meningitis. *N. Engl. J. Med.* **269**:1424–1425.

52. **Devi, S. J. N., P. G. Reddy, C. A. Lyman, T. J. Walsh, C. E. Frasch, and A. C. Bush.** 1996. Immunohistochemical properties of a polysaccharide antigen of *Trichosporon beigelii* that cross-reacts with the capsular polysaccharide of *Cryptococcus neoformans*. *Immunol. Infect. Dis.* **6**:87–92.

53. **Dromer, F., E. Gueho, O. Ronin, and B. Dupont.** 1993. Serotyping of *Cryptococcus neoformans* by using a monoclonal antibody specific for capsular polysaccharide. *J. Clin. Microbiol.* **31**:359–363.

54. **Dromer, F., J. Salamero, A. Contrepois, C. Carbon, and P. Yeni.** 1987. Production, characterization, and antibody specificity of a mouse monoclonal antibody reactive with *Cryptococcus neoformans* capsular polysaccharide. *Infect. Immun.* **55**:742–748.

55. **Drouhet, E., and G. Segretain.** 1951. Inhibition de la migration leucocytaire *in vitro* par un polyoside capsulaire de *Torulopsis (Cryptococcus) neoformans*. *Ann. Inst. Pasteur* **81**:674.

56. **Eckert, T. F., and T. R. Kozel.** 1987. Production and characterization of monoclonal antibodies specific for *Cryptococcus neoformans* capsular polysaccharide. *Infect. Immun.* **55**:1895–1899.

57. **Edberg, S. C., S. J. Chaskes, E. Alture-Werber, and J. M. Singer.** 1980. Esculin-based medium for isolation and identification of *Cryptococcus neoforman*. *J. Clin. Microbiol.* **12**:332–335.

58. **Edwards, M. R., M. A. Gordon, E. W. Lapa, and W. C. Ghiorse.** 1967. Micromorphology of *Cryptococcus neoformans*. *J. Bacteriol.* **94**:766–777.

59. **Einbinder, J. M., R. W. Benham, and C. T. Nelson.** 1954. Chemical analysis of the capsular substance of *Cryptococcus neoformans*. *J. Invest. Dermatol.* **22**:279–283.

60. **Emery, H. S., C. P. Shelburne, J. P. Bowman, P. G. Fallon, C. A. Schulz, and E. S. Jacobson.** 1995. Genetic study of oxygen resistance and melanization in *Cryptococcus neoformans*. *Infect. Immun.* **62**:5694–5697.

61. **Evans, E. D., and J. F. Kessel.** 1951. The antigenic composition of *Cryptococcus neoformans*. *J. Immunol.* **67**:109–114.

62. **Evans, E. E.** 1949. An immunologic comparison of twelve strains of *Cryptococcus neoformans (Torula histolytica)*. *Proc. Soc. Exp. Biol. Med.* **71**:644–646.

63. **Evans, E. E.** 1950. The antigenic composition of *Cryptococcus neoformans*. I. A serologic classification by means of the capsular and agglutination reactions. *J. Immunol.* **64**:423–430.

64. **Evans, E. E.** 1959. Reaction of an Aspergillus polysaccharide with Cryptococcus capsules and various acidic polysaccharides. *Proc. Soc. Exp. Biol. Med.* **101**:733–739.

65. **Evans, E. E.** 1960. Capsular reactions of *Cryptococcus neoformans*. *Ann. N.Y. Acad. Sci.* **89**:184–192.

66. **Evans, E. E., and J. W. Mehl.** 1951. A qualitative analysis of capsular polysaccharides from *Cryptococcus neoformans* by filter paper chromatography. *Science* **114**:10–11.

67. **Federici, F.** 1982. A note on milk clotting ability in the yeast genera *Cryptococcus* and *Rhodotorula*. *J. Appl. Bacteriol.* **52**:293–296.

68. **Feldmesser, M., and A. Casadevall.** 1997. Effect of serum IgG1 against murine pulmonary infection with *Cryptococcus neoformans*. *J. Immunol.* **158**:790–799.

69. **Fiskin, A. M., M. C. Zalles, and R. G. Garrison.** 1990. Electron cytochemical studies

of *Cryptococcus neoformans* grown on uric acid and related sources of nitrogen. *J. Med. Vet. Mycol.* **38**:197–207.

70. **Franzot, S. P., and J. S. Hamdan.** 1994. Effects of three azole derivatives on the lipids of different strains of *Cryptococcus neoformans. Mycoses* **38**:183–189.

71. **Franzot, S. P., and J. S. Hamdan.** 1994. Effects of amphotericin B on the lipids of five different strains of *Cryptococcus neoformans. Mycopathologia* **128**:85–89.

72. **Garrad, R. C., and J. K. Battacharjee.** 1992. Lysine biosynthesis in selected pathogenic fungi: characterization of lysine auxotrophs and the clones *LYS1* gene of *Candida albicans. J. Bacteriol.* **174**:7379–7384.

73. **Ghannoum, M. A., B. J. Spellberg, A. S. Ibrahim, J. A. Ritchie, B. Currie, E. D. Spitzer, J. E. Edwards, and A. Casadevall.** 1994. Sterol composition of *Cryptococcus neoformans* in the presence and absence of fluconazole. *Antimicrob. Agents Chemother.* **38**:2029–2033.

74. **Goldman, D., S. C. Lee, and A. Casadevall.** 1994. Pathogenesis of pulmonary *Cryptococcus neoformans* infection in the rat. *Infect. Immun.* **62**:4755–4761.

75. **Granger, D. L., J. R. Perfect, and D. T. Durack.** 1985. Virulence of *Cryptococcus neoformans.* Regulation of capsule synthesis by carbon dioxide. *J. Clin. Invest.* **76**:508–516.

76. **Hamilton, A. J., and J. Goodley.** 1993. Purification of the 115-kilodalton exoantigen of *Cryptococcus neoformans* and its recognition by immune sera. *J. Clin. Microbiol.* **31**:335–339.

77. **Hamilton, A. J., and M. D. Holdom.** 1997. Biochemical composition of the Cu,Zn superoxide dismutases of *Cryptococcus neoformans* var. *neoformans* and *Cryptococcus neoformans* var. *gattii. Infect. Immun.* **65**:488–494.

78. **Hamilton, A. J., L. Jeavons, P. Hobby, and R. J. Hay.** 1992. A 34- to 38-kilodalton *Cryptococcus neoformans* glycoprotein produced as an exoantigen bearing a glycosylated species-specific epitope. *Infect. Immun.* **60**:143–149.

79. **Haynes, M. P., P. L.-C. Chong, H. R. Buckley, and R. A. Pieringer.** 1996. Fluorescence studies on the molecular action of amphotericin b on susceptible and resistant fungal cells. *Biochemistry* **35**:7983–7992.

80. **Hehre, E. J., A. S. Carlson, and D. M. Hamilton.** 1949. Crystalline amylose from cultures of a pathogenic yeast (Torula histolytica). *J. Biol. Chem.* **177**:289–293.

81. **Hendry, A. T., and A. Bakerspigel.** 1969. Factors affecting serum inhibited growth of *Candida albicans* and *Cryptococcus neoformans. Sabouraudia* **7**:219–229.

82. **Hodes, D. S., D. Heon, A. Hass, A. C. Hyatt, and H. Hodes.** 1987. Reaction of fungal products with amebocyte lysates of the Japanese horseshoe crab, *Tachypleus tridentatus. J. Clin. Microbiol.* **25**:1701–1704.

83. **Huffnagle, G. B., G.-H. Chen, J. L. Curtis, R. A. McDonald, R. M. Strieter, and G. B. Toews.** 1995. Down-regulation of the afferent phase of T cell-mediated pulmonary inflammation and immunity by a high melanin-producing strain of *Cryptococcus neoformans. J. Immunol.* **155**:3507–3516.

84. **Ikeda, R., and E. S. Jacobson.** 1992. Heterogeneity of phenol oxidases in *Cryptococcus neoformans. Infect. Immun.* **60**:3552–3555.

85. **Ikeda, R., H. Matsuyama, A. Nishikawa, T. Shinoda, and Y. Fukazawa.** 1991. Comparison of serological and chemical characteristics of capsular polysaccharides of *Cryptococcus neoformans* var. *neoformans* serotype A and *Cryptococcus albidus* var. *albidus. Microbiol. Immunol.* **35**:125–138.

86. **Ikeda, R., A. Nishikawa, T. Shinoda, and Y. Fukazawa.** 1985. Chemical characterization of capsular polysaccharide from *Cryptococcus neoformans* serotype A-D. *Microbiol. Immunol.* **29**:981–991.

87. **Ikeda, R., S. Nishimura, A. Nishikawa, and T. Shinoda.** 1996. Production of agglutinating monoclonal antibody against antigen 8 specific for *Cryptococcus neoformans* serotype D. *Clin. Diagn. Lab. Immunol.* **3**:89–92.

88. **Ikeda, R., T. Shinoda, Y. Fukuzawa, and L. Kaufman.** 1982. Antigenic characterization of *Cryptococcus neoformans* serotypes and its application to serotyping of clinical isolates. *J. Clin. Microbiol.* **36:**22–29.

89. **Ikeda, R., T. Shinoda, T. Morita, and E. S. Jacobson.** 1993. Characterization of phenol oxidase from *Cryptococcus neoformans* var. *neoformans. Microbiol. Immunol.* **37:**759–764.

91. **Itoh, T., and H. Kaneko.** 1974. Pyrophosphatidic acid. A new phospholipid from *Cryptococcus neoformans. J. Biochem.* **75:**1291–1300.

92. **Itoh, T., and H. Kaneko.** 1975. Positional specificity of fatty acids in pyrophosphatidic acid from *Cryptococcus neoformans. J. Biochem.* **77:**777–781.

93. **Itoh, T., and H. Kaneko.** 1975. The distribution of pyrophosphatidic acid in nature. *J. Biochem.* **78:**817–820.

94. **Jacobson, E. S.** 1987. Cryptococcal UDP-glucose dehydrogenase: enzymatic control of capsular biosynthesis. *J. Med. Vet. Mycol.* **25:**131–135.

95. **Jacobson, E. S. and G. M. Compton.** 1996. Discordant regulation of phenoloxidase and capsular polysaccharide in *Cryptococcus neoformans. J. Med. Vet. Mycol.* **34:**289–291.

96. **Jacobson, E. S., and H. S. Emery.** 1991. Catecholamine uptake, melanization, and oxygen toxicity in *Cryptococcus neoformans. J. Bacteriol.* **173:**401–403.

97. **Jacobson, E. S., and H. S. Emery.** 1991. Temperature regulation of the cryptococcal phenoloxidase. *J. Med. Vet. Mycol.* **29:**121–124.

98. **Jacobson, E. S., N. D. Jenkins, and J. M. Todd.** 1994. Relationship between superoxide dismutase and melanin in a pathogenic fungus. *Infect. Immun.* **62:**4085–4086.

99. **Jacobson, E. S., and W. R. Payne.** 1981. UDP glucuronate decarboxylase and synthesis of capsular polysaccharide in *Cryptococcus neoformans. J. Bacteriol.* **152:**932–934.

100. **Jacobson, E. S., and M. J. Petro.** 1987. Extracellular iron chelation in *Cryptococcus neoformans. J. Med. Vet. Mycol.* **25:**415–418.

101. **Jacobson, E. S., and M. J. Tingler.** 1994. Strains of *Cryptococcus neoformans* with defined capsular phenotypes. *J. Med. Vet. Mycol.* **32:**401–404.

102. **Jacobson, E. S., M. J. Tingler, and L. Quynn.** 1989. Effect of hypertonic solutes upon the polysaccharide capsule in *Cryptococcus neoformans. Mycoses* **32:**14–23.

103. **Jacobson, E. S., and S. B. Tinnell.** 1993. Antioxidant function of fungal melanin. *J. Bacteriol.* **175:**7102–7104.

104. **Jacobson, E. S., and S. E. Vartivarian.** 1992. Iron assimilation in *Cryptococcus neoformans. J. Med. Vet. Mycol.* **30:**443–450.

105. **James, P. D., and R. Cherniak.** 1992. Galactoxylomannans of *Cryptococcus neoformans. Infect. Immun.* **60:**1084–1088.

106. **James, P. G., R. Cherniak, R. G. Jones, and C. A Stortz.** 1998. Cell-wall glucans of *Cryptococcus neoformans* CAP 67. *Carbohydr. Res.* **198:**23–28.

107. **Jong, S. C., G. S. Bulmer, and A. Ruiz.** 1982. Serologic grouping and sexual compatibility of airborne *Cryptococcus neoformans. Mycopathologia* **79:**185–188.

108. **Joseph-Horne, T., D. Hollomon, R. S. T. Loeffler, and S. L. Kelly.** 1995. Cross-resistance to polyene and azole drugs in *Cryptococcus neoformans. Antimicrob. Agents Chemother.* **39:**1526–1529.

109. **Kaplan, W., S. L. Bragg, S. Crane, and D. G. Ahearn.** 1981. Serotyping *Cryptococcus neoformans* by immunofluorescence. *J. Clin. Microbiol.* **14:**313–317.

110. **Kim, S. J., K. J. Kwon-Chung, G. W. A. Milne, W. B. Hill, and G. Patterson.** 1975. Relationship between polyene resistance and sterol compositions in *Cryptococcus neoformans. Antimicrob. Agents Chemother.* **7:**99–106.

111. **Kiska, D. L., D. R. Orkieszewski, D. Howell, and P. H. Gilligan.** 1994. Evaluation of new monoclonal antibody-based latex agglutination test for detection of cryptococcal polysaccharide antigen in serum and cerebrospinal fluid. *J. Clin. Microbiol.* **32:**2309–2311.

112. **Korth, H., and G. Pulverer.** 1971. Pigment formation for differentiating *Cryptococcus neoformans* from *Candida albicans. Appl. Microbiol.* **21:**541–542.

113. **Kozel, T. R.** 1977. Non-encapsulated variant of *Cryptococcus neoformans.* II. Surface receptors for cryptococcal polysaccharide and their role in inhibition of phagocytosis by polysaccharide. *Infect. Immun.* **16:**99–106.

114. **Kozel, T. R.** 1983. Dissociation of a hydrophobic surface from phagocytosis of encapsulated and non-encapsulated *Cryptococcus neoformans. Infect. Immun.* **39:**1214–1219.

115. **Kozel, T. R.** 1996. Activation of the complement system by pathogenic fungi. *Clin. Microbiol. Rev.* **9:**34–46.

116. **Kozel, T. R., and J. Cazin, Jr.** 1971. Nonencapsulated variant of *Cryptococcus neoformans. Infect. Immun.* **3:**287–294.

117. **Kozel, T. R., and E. Gotschlich.** 1982. The capsule of *Cryptococcus neoformans* passively inhibits phagocytosis of the yeast by macrophages. *J. Immunol.* **129:**1675–1680.

117a. **Kozel, T. R., and C. A. Hermerath.** 1984. Binding of cryptococcal polysaccharide to *Cryptococcus neoformans. Infect. Immun.* **43:**879–886.

118. **Kozel, T. R., E. Reiss, and R. Cherniak.** 1980. Concomitant but not causal association between surface charge and inhibition of phagocytosis by cryptococcal polysaccharide. *Infect. Immun.* **29:**295–300.

119. **Kozel, T. R., M. A. Wilson, and J. W. Murphy.** 1991. Early events in initiation of alternative complement pathway activation by the capsule of *Cryptococcus neoformans. Infect. Immun.* **59:**3101–3110.

120. **Kozel, T. R., M. A. Wilson, G. S. Pfrommer, and A. M. Schlagetter.** 1989. Activation and binding of opsonic fragments of C3 on encapsulated *Cryptococcus neoformans* by using an alternative complement pathway reconstituted from six isolated patients. *Infect. Immun.* **57:**1922–1927.

121. **Kwon-Chung, K. J., and J. E. Bennett.** 1984. High prevalence of *Cryptococcus neoformans* var. *gattii* in tropical and subtropical regions. *Zentralbl. Bakteriol. Hyg. A* **257:**213–218.

122. **Kwon-Chung, K. J., and J. E. Bennett.** 1984. Epidemiologic differences between the two varieties of *Cryptococcus neoformans. Am. J. Epidemiol.* **120:**123–130.

123. **Kwon-Chung, K. J., W. B. Hill, and J. E. Bennett.** 1981. New, special stain for histopathological diagnosis of cryptococcosis. *J. Clin. Microbiol.* **13:**383–387.

124. **Kwon-Chung, K. J., and J. C. Rhodes.** 1986. Encapsulation and melanin formation as indicators of virulence in *Cryptococcus neoformans. Infect. Immun.* **51:**218–223.

125. **Kwon-Chung, K. J., W. K. Tom, and J. L. Costa.** 1983. Utilization of indole compounds by *Cryptococcus neoformans* to produce a melanin-like pigment. *J. Clin. Microbiol.* **18:**1419–1421.

126. **Kwon-Chung, K. J., B. L. Wickes, J. L. Booth, H. S. Vishniac, and J. E. Bennett.** 1987. Urease inhibition by EDTA in the two varieties of *Cryptococcus neoformans. Infect. Immun.* **55:**1751–1754.

127. **Kwon-Chung, K. J., B. L. Wickes, L. Stockman, G. D. Roberts, D. Ellis, and D. H. Howard.** 1992. Virulence, serotype, and molecular characteristics of environmental strains of *Cryptococcus neoformans* var. *gattii. Infect. Immun.* **60:**1869–1874.

128. **Lamb, D. C., A. Corran, B. C. Baldwin, K. J. Kwon-Chung, and S. L. Kelly.** 1995. Resistant P45051A1 activity in azole antifungal tolerant *Cryptococcus neoformans* from AIDS patients. *FEBS Lett.* **368:**326–330.

129. **Langner, C. A., J. K. Lodge, S. J. Travis, J. E. Caldwell, T. Lu, Q. Li, M. L. Bryant, B. Devadas, G. W. Gokel, G. S. Kobayashi, and J. I. Gordon.** 1992. 4-Oxatetradecanoic acid is fungicidal for *Cryptococcus neoformans* and inhibits replication of human immunodeficiency virus I. *J. Biol. Chem.* **267:**17159–17169.

130. **Lee, S. C., A. Casadevall, and D. W. Dickson.** 1996. Immunohistochemical localization of capsular polysaccharide antigen in the central nervous system cells in cryptococcal meningoencephalitis. *Am. J. Pathol.* **148:**1267–1274.

131. **Lee, S. C., D. W. Dickson, and A. Casadevall.** 1996. Pathology of cryptococcal meningoencephalitis: analysis of 27 patients with pathonenetic implications. *Hum. Pathol.* **27**:839–847.

132. **Littman, M. L.** 1958. Capsule synthesis by *Cryptococcus neoformans. Trans. N.Y. Acad. Sci.* **20**:623–648.

133. **Livi, L. L., U. Edman, G. P. Schneider, P. J. Greene, and D. V. Santi.** 1994. Cloning, expression and characterization of thymidylate synthase from *Cryptococcus neoformans. Gene* **150**:221–226.

134. **Lodge, J. K., R. L. Johnson, R. A. Weinberg, and J. I. Gordon.** 1994. Comparison of myristoyl-CoA:protein N-myristoyltransferases from three pathogenic fungi: *Cryptococcus neoformans, Histoplasma capsulatum,* and *Candida albicans. J. Biol. Chem.* **269**:2996–3009.

135. **Maccani, J. E.** 1977. Detection of cryptococcal polysaccharide using counterimmunoelectrophoresis. *Am. J. Clin. Pathol.* **68**:39–44.

136. **Maccani, J. E.** 1981. Rapid presumptive identification of *Cryptococcus neoformans* by staphylococcal coagglutination. *J. Clin. Microbiol.* **13**:828–832.

137. **Mahmoud, Y. A.-G., S. M. Abu El Souod, and W. G. Niehaus.** 1995. Purification and characterization of malate dehydrogenase from *Cryptococcus neoformans. Arch. Biochem. Biophys.* **322**:69–75.

138. **Mahvi, T. A., S. S. Spicer, and N. J. Wright.** 1974. Cytochemistry of acid mucosubstance and acid phosphatase in *Cryptococcus neoformans. Can. J. Microbiol.* **20**:833–838.

139. **Marumo, K., and Y. Aoki.** 1990. Discriminant analysis of cellular fatty acids of *Candida* species, *Torulopsis glabrata,* and *Cryptococcus neoformans* determined by gas-liquid chromatography. *J. Clin. Microbiol.* **28**:1509–1513.

140. **Mason, D. L., and C. L. Wilson.** 1979. Cytochemical and biochemical identification of lysosomes in *Cryptococcus neoformans. Mycopathologia* **68**:183–190.

141. **Mazur, P., and W. Baginsky.** 1996. In vitro activity of 1,3-β-D-glucan synthase requires the GTP-binding protein Rho1. *J. Biol. Chem.* **271**:14604–14609.

142. **McGaw, T., and T. R. Kozel.** 1979. Opsonization of *Cryptococcus neoformans* by human immunoglobulin G: masking of immunoglobulin G by cryptococcal polysaccharide. *Infect. Immun.* **25**:262–267.

143. **McManus, E. J., and J. M. Jones.** 1985. Detection of a *Trichosporon beigelii* antigen cross-reactive with *Cryptococcus neoformans* capsular polysaccharide in serum from a patient with disseminated *Trichosporon* infection. *J. Clin. Microbiol.* **21**:681–685.

144. **Megson, G. M., D. A. Stevens, J. R. Hamilton, and D. W. Denning.** 1996. d-Mannitol in cerebrospinal fluid of patients with AIDS and cryptococcal meningitis. *J. Clin. Microbiol.* **34**:218–221.

145. **Melcher, G. P., K. D. Reed, M. G. Rinaldi, J. W. Lee, P. A. Pizzo, and T. J. Walsh.** 1991. Demonstration of a cell wall antigen cross-reacting with cryptococcal polysaccharide in experimental disseminated trichosporonosis. *J. Clin. Microbiol.* **29**:192–196.

146. **Merrifield, E. H., and A. M. Stephen.** 1980. Structural investigations of two capsular polysaccharides from *Cryptococcus neoformans. Carbohydr. Res.* **86**:69–76.

147. **Miyazaki, T.** 1961. Studies on fungal polysaccharides. I. On the isolation and chemical properties of capsular polysaccharide from *Cryptococcus neoformans. Chem. Pharm. Bull.* **9**:715–718.

148. **Miyazaki, T.** 1961. Studies on fungal polysaccharides. III. Chemical structure of the capsular polysaccharide from *Cryptococcus neoformans. Chem. Pharm. Bull.* **9**:829–833.

149. **Miyazaki, T.** 1961. Studies on fungal polysaccharides. II. On the componental sugars and partial hydrolysis of the capsular polysaccharides from *Cryptococcus neoformans. Chem. Pharm. Bull.* **9**:826–829.

150. **Miyazaki, T., S. Kohno, K. Mitsutake, S. Maesaki, K.-I. Tanaka, N. Ishikawa, and**

K. Hara. 1995. Plasma $(1\rightarrow3)$-β-d-glucan and fungal antigenemia in patients with candidemia, aspergillosis, and cryptococcosis. *J. Clin. Microbiol.* **33**:3113–3118.

151. **Mochizuki, T., S. Tanaka, Y. Saito, and S. Watanabe.** 1988. Mitochondrial kinetics during mitosis in *Cryptococcus neoformans*—an ultrastructure study. *Mycoses* **32**:7–13.

152. **Mukherjee, J., W. Cleare, and A. Casadevall.** 1995. Monoclonal antibody mediated capsular reactions (quellung) in *Cryptococcus neoformans*. *J. Immunol. Methods* **184**:139–143.

153. **Mukherjee, J., G. Nussbaum, M. D. Scharff, and A. Casadevall.** 1995. Protective and non-protective monoclonal antibodies to *Cryptococcus neoformans* originating from one B-cell. *J. Exp. Med.* **181**:405–409.

154. **Murphy, J. W., R. L. Mosley, R. Cherniak, G. H. Reyes, T. R. Kozel, and E. Reiss.** 1988. Serological, electrophoretic, and biological properties of *Cryptococcus neoformans* antigens. *Infect. Immun.* **56**:424–431.

155. **Neill, J. M., C. G. Castillo, R. H. Smith, and C. E. Kapros.** 1949. Capsular reactions and soluble antigens of Torula histolytica and Sporotrichum schenckii. *J. Exp. Med.* **89**:93–106.

156. **Niehaus, W. G., and T. Flynn.** 1994. Regulation of mannitol biosynthesis and degradation by *Cryptococcus neoformans*. *J. Bacteriol.* **176**:651–655.

157. **Niehaus, W. G., and T. C. Mallett.** 1994. Purification and characterization of glucose-6-phosphate dehydrogenase from *Cryptococcus neoformans*: identification as "nothing dehydrogenase." *Arch. Biochem. Biophys.* **313**:304–309.

158. **Niehaus, W. G., S. B. Richardson, and R. L. Wolz.** 1996. Slow binding inhibition of 6-phosphogluconate dehydrogenase by zinc ion. *Arch. Biochem. Biophys.* **333**:333–337.

159. **Niehaus, W. G., R. H. White, S. B. Richardson, A. Bourne, and W. K. Ray.** 1995. Polyethylene sulfonate: a tight-binding inhibitor of 6-phosphogluconate dehydrogenase of *Cryptococcus neoformans*. *Arch. Biochem. Biophys.* **324**:325–330.

160. **Nishiura, Y., K. Nakagawa-Yoshida, M. Suga, T. Shinoda, E. Gueho, and M. Ando.** 1997. Assignment and serotyping of *Trichosporon* species: the causative agents of summer-type hypersensitivity pneumonitis. *J. Med. Vet. Mycol.* **35**:45–52.

161. **Nosanchuk, J. D., and A. Casadevall.** 1997. Cellular charge of *Cryptococcus neoformans*: contributions from the capsular polysaccharide, melanin, and monoclonal antibody binding. *Infect. Immun.* **65**:1836–1841.

162. **Nurudeen, T., and D. G. Ahearn.** 1979. Regulation of melanin production by *Cryptococcus neoformans*. *J. Clin. Microbiol.* **10**:724–729.

163. **Nussbaum, G., W. Cleare, A. Casadevall, M. D. Scharff, and P. Valadon.** 1997. Epitope location in the *Cryptococcus neoformans* capsule is a determinant of antibody efficacy. *J. Exp. Med.* **185**:685–697.

164. **Nyhus, K. J., A. T. Wilborn, and E. S. Jacobson.** 1997. Ferric iron reduction by *Cryptococcus neoformans*. *Infect. Immun.* **65**:434–438.

165. **Obayashi, T., M. Yoshida, T. Mori, H. Goto, A. Yasuoka, H. Iwasaki, H. Teshima, S. Kohno, A. Horiuchi, A. Ito, H. Yamaguchi, K. Shimada, and T. Kawai.** 1995. Plasma $(1\rightarrow3)$-β-D-glucan measurement in diagnosis of invasive deep mycosis and fungal febrile episodes. *Lancet* **345**:17–20.

167. **Pappagianis, D., and R. Marovitz.** 1966. Studies on ethanol production by *Cryptococcus neoformans*. *Sabouraudia* **4**:250–255.

168. **Parker, A. R., T. D. E. Moore, J. C. Edman, J. M. Schwab, and V. J. Davidson.** 1994. Cloning, sequence analysis and expression of the gene encoding imidazole glycerol phosphate dehydrogenase in *Cryptococcus neoformans*. *Gene* **145**:135–138.

169. **Perfect, J. R., T. H. Rude, B. Wong, T. Flynn, V. Chatuverdi, and W. Niehaus.** 1996. Identification of a *Cryptococcus neoformans* gene that directs expression of a cryptic *Saccharomyces cerevisiae* mannitol dehydrogenase gene. *J. Bacteriol.* **178**:5257–5262.

170. **Perfect, J. R., D. L. Toffaletti, and T. H. Rude.** 1993. The gene encoding phosphoribo-

sylaminoimidazole carboxylase (ADE2) is essential for growth of *Cryptococcus neoformans* in cerebrospinal fluid. *Infect. Immun.* **61**:4446–4451.

171. **Peterson, E., R. J. Hawley, and R. A. Calderone.** 1976. An ultrastructural analysis of protoplast-spheroplast induction in *Cryptococcus neoformans*. *Can. J. Microbiol.* **22**:1518–1521.

172. **Pidcoe, V., and L. Kaufman.** 1968. Fluorescent-antibody reagent for the identification of *Cryptococcus neoformans*. *Appl. Microbiol.* **16**:271–275.

173. **Pirofski, L., and A. Casadevall.** 1996. Antibody immunity to *Cryptococcus neoformans*: paradigm for antibody immunity to the fungi? *Zentralbl. Bakteriol.* **284**:475–495.

174. **Pirofski, L., R. Lui, M. DeShaw, A. B. Kressel, and Z. Zhong.** 1995. Analysis of human monoclonal antibodies elicited by vaccination with a *Cryptococcus neoformans* glucuronoxylomannan capsular polysaccharide vaccine. *Infect. Immun.* **63**:3005–3014.

175. **Polacheck, I.** 1991. The discovery of melanin production in *Cryptococcus neoformans* and its impact on diagnosis and the study of virulence. *Zentralbl. Bakteriol.* **276**:120–123.

176. **Polacheck, I., V. J. Hearing, and K. J. Kwon-Chung.** 1982. Biochemical studies of phenoloxidase and utilization of catecholmines in *Cryptococcus neoformans*. *J. Bacteriol.* **150**:1212–1220.

177. **Polacheck, I., and K. J. Kwon-Chung.** 1980. Creatinine metabolism in *Cryptococcus neoformans* and *Cryptococcus bacillisporus*. *J. Bacteriol.* **142**:15–20.

178. **Polacheck, I., and K. J. Kwon-Chung.** 1988. Melanogenesis in *Cryptococcus neoformans*. *J. Gen. Microbiol.* **134**:1034–1041.

179. **Polacheck, I., Y. Platt, and J. Aronovitch.** 1990. Catecholamines and virulence of *Cryptococcus neoformans*. *Infect. Immun.* **58**:2919–2922.

180. **Powderly, W. G., E. J. Keath, M. Sokol-Anderson, D. Kitz, J. Russell Little, and G. Kobayashi.** 1992. Amphotericin B-resistant *Cryptococcus neoformans* in a patient with AIDS. *Infect. Dis. Clin. Pract.* **1**:314–316.

181. **Rawat, D. S., H. B. Upreti, and S. K. Das.** 1984. Lipid composition of *Cryptococcus neoformans*. *Microbiologica* **7**:299–307.

182. **Rebers, P. A., S. A. Barker, M. Heidelberger, Z. Dische, and E. E. Evans.** 1958. Precipitation of the specific polysaccharide of *Cryptcococcus neoformans* A by types II and XIV antipneumococcal antisera. *J. Am. Chem. Soc.* **80**:1135–1137.

183. **Reiss, E., R. Cherniak, R. Eby, and L. Kaufman.** 1984. Enzyme immunoassay detection of IgM to galactoxylomannan of *Cryptococcus neoformans*. *Diagn. Immunol.* **2**:109–115.

184. **Reiss, E., E. H. White, R. Cherniak, and J. E. Dix.** 1986. Ultrastructure of acapsular mutant *Cryptococcus neoformans* cap 67 and monosaccharide composition of cell extracts. *Mycopathologia* **93**:45–54.

185. **Rhodes, J. C.** 1986. A simplified assay for cryptococcal phenoloxidase. *Mycologia* **78**:867–868.

186. **Rhodes, J. C., and K. J. Kwon-Chung.** 1985. Production and regeneration of protoplasts from *Cryptococcus*. *Sabouraudia* **23**:77–80.

187. **Rhodes, J. C., I. Polacheck, and K. J. Kwon-Chung.** 1982. Phenoloxidase activity and virulence in isogenic strains of *Cryptococcus neoformans*. *Infect. Immun.* **36**:1175–1184.

188. **Rosas, A. L., and A. Casadevall.** 1997. Melanization affects susceptibility of *Cryptococcus neoformans* to heat and cold. *FEMS Microbiol. Lett.* **153**:265–272.

189. **Ross, A., and I. E. P. Taylor.** 1981. Extracellular glycoprotein from virulent and avirulent *Cryptococcus* species. *Infect. Immun.* **31**:911–918.

190. **Ruane, P. J., L. J. Walker, and W. L. George.** 1988. Disseminated infection caused by urease-negative *Cryptococcus neoformans*. *J. Clin. Microbiol.* **26**:2224–2225.

191. **Safrin, R. E., L. A. Lancaster, C. E. Davis, and A. I. Braude.** 1986. Differentiation of *Cryptococcus neoformans* serotypes by isoenzyme electrophoresis. *Am. J. Clin. Pathol.* **86**:204–208.

192. **Sakaguchi, N.** 1993. Ultrastructural study of hepatic granulomas induced by *Crypto-coccus neoformans* by quick-freezing and deep-etching method. *Virchows Arch. (Cell. Pathol.)* **64:**57–66.

193. **Sakaguchi, N., T. Baba, M. Fukuzawa, and S. Ohno.** 1993. Ultrastructural study of *Cryptococcus neoformans* by quick-freezing and deep-etching method. *Mycopathologia* **121:**133–141.

194. **Salas, S. D., J. E. Bennett, K. J. Kwon-Chung, J. R. Perfect, and P. R. Williamson.** 1996. Effect of the laccase gene, *CNLAC1*, on virulence of *Cryptococcus neoformans. J. Exp. Med.* **184:**377–386.

195. **Schlossberg, D., J. B. Brooks, and J. A. Shulman.** 1976. Possibility of diagnosing meningitis by gas chromatography: cryptococcal meningitis. *J. Clin. Microbiol.* **3:**239–245.

196. **Scott, E. N., H. G. Muchmore, and F. G. Felton.** 1980. Comparison of enzyme immunoassay and latex agglutination methods for detection of *Cryptococcus neoformans* antigen. *Am. J. Clin. Pathol.* **73:**790–794.

197. **Shaw, C. E., and L. Kapica.** 1972. Production of diagnostic pigment by phenoloxidase activity of *Cryptococcus neoformans. Appl. Microbiol.* **24:**824–830.

198. **Sirawaraporn, W., M. Cao, D. V. Santi, and J. C. Edman.** 1993. Cloning, expression, and characterization of *Cryptococcus neoformans* dihydrofolate reductase. *J. Biol. Chem.* **268:**8888–8892.

199. **Small, J. M., and T. G. Mitchell.** 1986. Binding of purified and radioiodinated capsular polysaccharides from *Cryptococcus neoformans* serotype A to capsule-free mutants. *Infect. Immun.* **54:**742–750.

200. **Small, J. M., and T. G. Mitchell.** 1989. Strain variation in antiphagocytic activity of capsular polysaccharides from *Cryptococcus neoformans* serotype A. *Infect. Immun.* **57:**3751–3756.

201. **Spiropulu, C., R. A. Eppard, E. Otteson, and T. R. Kozel.** 1989. Antigenic variation within serotypes of *Cryptococcus neoformans* detected by monoclonal antibodies specific for the capsular polysaccharide. *Infect. Immun.* **57:**3240–3242.

202. **Spitzer, E. D., and S. G. Spitzer.** 1992. Use of a dispersed repetitive DNA element to distinguish clinical isolates of *Cryptococcus neoformans. J. Clin. Microbiol.* **30:**1094–1097.

203. **Stacey, M., and S. A. Barker.** 1960. *Polysaccharides of Micro-Organisms*, p. 198–201. Clarendon Press, Oxford.

204. **Stoetzner, H., and C. Kemmer.** 1971. The morphology of *Cryptococcus neoformans* in human cryptococcosis: a light–, phase contrast, and electron microscopic study. *Mycopathol. Mycol. Appl.* **45:**327–335.

205. **Strachan, A. A., R. J. Yu, and F. Blank.** 1971. Pigment production of *Cryptococcus neoformans* grown with extracts of *Guizotia abyssinica. Appl. Microbiol.* **22:**478–479.

206. **Szilagyi, G., F. Reiss, and J. C. Smith.** 1966. The anticryptococcal factor of blood serum. *J. Invest. Dermatol.* **46:**306–308.

207. **Takeo, K., I. Uesaka, K. Uehira, and M. Nishiura.** 1973. Fine structure of *Cryptococcus neoformans* grown in vivo as observed by freeze-etching. *J. Bacteriol.* **113:**1449–1454.

208. **Tesfa-Selase, F., and R. J. Hay.** 1995. Superoxide dismutase of *Cryptococcus neoformans*: purification and characterization. *J. Med. Vet. Mycol.* **33:**253–259.

209. **Tesfa-Selase, F., and R. J. Hay.** 1996. Partial characterization and identification of a transferrin-like molecule of pathogenic yeast *Cryptococcus neoformans. J. Gen. Appl. Microbiol.* **42:**61–70.

210. **Todaro-Luck, F., E. Reiss, R. Cherniak, and L. Kaufman.** 1989. Characterization of *Cryptococcus neoformans* capsular glucuronoxylomannan polysaccharide with monoclonal antibodies. *Infect. Immun.* **57:**3882–3887.

211. **Tohyama, M., K. Kawakami, and A. Saito.** 1996. Anticryptococcal effect of amphotericin B is mediated through macrophage production of nitric oxide. *Antimicrob. Agents Chemother.* **40:**1919–1923.

212. **Torres-Guerrero, H., and J. C. Edman.** 1994. Melanin-deficient mutants of *Cryptococcus neoformans*. *J. Med. Vet. Mycol.* **32:**303–313.

213. **Turner, S. H., and R. Cherniak.** 1990. Multiplicity in the structure of the glucuronoxylomannan of *Cryptococcus neoformans*, p. 123–142. *In* J. P. Latge and D. Boucias (ed.), *Fungal Cell Wall and Immune Response.* Springer-Verlag, Berlin.

214. **Turner, S. H., R. Cherniak, and E. Reiss.** 1984. Fractionation and characterization of galactoxylomannan from *Cryptococcus neoformans*. *Carbohydr. Res.* **125:**343–349.

215. **Tyler, R.** 1956. Spinal fluid alcohol in yeast meningitis. *Am. J. Med. Sci.* **232:**560–561.

216. **Upreti, H. B., D. S. Rawat, and S. K. Das.** 1984. Virulence, capsule size and lipid composition interrelation of *Cryptococcus neoformans*. *Microbiologica* **7:**371–374.

217. **Vanden Bossche, H., H. Marichal, L. Le Jeune, M. Coene, J. Gorrens, and W. Cools.** 1993. Effects of itraconazole on cytochrome P-450-dependent sterol 14-alpha-demethylation and reduction of 3-ketosteroids in *Cryptococcus neoformans*. *Antimicrob. Agents Chemother.* **37:**2101–2105.

218. **Varma, A., and K. J. Kwon-Chung.** 1991. Rapid method to extract DNA from *Cryptococcus neoformans*. *J. Clin. Microbiol.* **29:**810–812.

219. **Vartivarian, S. E., E. J. Anaissie, R. E. Cowart, H. A. Sprigg, M. J. Tingler, and E. S. Jacobson.** 1993. Regulation of cryptococcal capsular polysaccharide by iron. *J. Infect. Dis.* **167:**186–190.

220. **Vartivarian, S. E., R. E. Cowart, E. J. Anaissie, T. Tashiro, and H. A. Sprigg.** 1995. Iron acquisition by *Cryptococcus neoformans*. *J. Med. Vet. Mycol.* **33:**151–156.

221. **Vartivarian, S. E., G. H. Reyes, E. S. Jacobson, P. G. James, R. Cherniak, V. R. Mumaw, and M. J. Tingler.** 1989. Localization of mannoprotein in *Cryptococcus neoformans*. *J. Bacteriol.* **171:**6850–6852.

222. **Vincent, V. L., and L. S. Klig.** 1995. Unusual effect of myo-inositol on phospholipid biosynthesis in *Cryptococcus neoformans*. *Microbiology* **141:**1829–1837.

223. **Vogel, R. A.** 1966. The indirect fluorescent antibody test for the detection of antibody in human cryptococcal disease. *J. Infect. Dis.* **116:**573–580.

224. **Vogel, R. A.** 1969. Primary isolation medium for *Cryptococcus neoformans*. *Appl. Microbiol.* **18:**1100.

225. **Wang, H. S., R. T. Zeimis, and G. D. Roberts.** 1977. Evaluation of a caffeic acid-ferric citrate test for rapid identification of *Cryptococcus neoformans*. *J. Clin. Microbiol.* **6:**445–449.

226. **Wang, Y., P. Aisen, and A. Casadevall.** 1995. *Cryptococcus neoformans* melanin and virulence: mechanism of action. *Infect. Immun.* **63:**3131–3136.

227. **Wang, Y., P. Aisen, and A. Casadevall.** 1996. Melanin, melanin "ghosts," and melanin composition in *Cryptococcus neoformans*. *Infect. Immun.* **64:**2420–2424.

228. **Wang, Y., and A. Casadevall.** 1994. Susceptibility of melanized and nonmelanized *Cryptococcus neoformans* to nitrogen- and oxygen-derived oxidants. *Infect. Immun.* **64:**3004–3007.

229. **Wang, Y., and A. Casadevall.** 1994. Growth of *Cryptococcus neoformans* in presence of L-dopa decreases its susceptibility to amphotericin B. *Antimicrob. Agents Chemother.* **38:**2646–2650.

230. **Wang, Y., and A. Casadevall.** 1994. Decreased susceptibility of melanized *Cryptococcus neoformans* to UV light. *Appl. Environ. Microbiol.* **60:**3864–3866.

231. **Wang, Y., and A. Casadevall.** 1996. Susceptibility of melanized and non-melanized *Cryptococcus neoformans* to the melanin-binding compounds trifluoperazine and chloroquine. *Antimicrob. Agents Chemother.* **40:**541–545.

232. **Weidman, F. D., and W. Freeman.** 1924. India ink in the microscopic study of yeast cells. *JAMA* **83:**1163–1164.

233. **Westerink, M. A., D. Amsterdam, R. J. Petell, M. N. Stram, and M. A. Apicella.** 1987. Septicemia due to DF-2. Cause of a false-positive cryptococcal latex agglutination result. *Am. J. Med.* **83:**155–158.

234. **Wheeler, M. H., and A. A. Bell.** 1988. Melanins and their importance in pathogenic fungi. *Curr. Top. Med. Mycol.* **2:**338–387.

235. **White, C. W., R. Cherniak, and E. S. Jacobson.** 1990. Side group addition by xylosyl-transferase and glycuronyltransferase in biosynthesis of capsular polysaccharide in *Cryptococcus neoformans. J. Med. Vet. Mycol.* **28:**289–301.

236. **White, C. W., and E. S. Jacobson.** 1992. Mannosyl transfer in *Cryptococcus neoformans. Can. J. Microbiol.* **38:**129–133.

237. **Williamson, P. R.** 1994. Biochemical and molecular characterization of the diphenol oxidase of *Cryptococcus neoformans*: identification as a laccase. *J. Bacteriol.* **176:**656–664.

238. **Wilson, D. E., J. E. Bennett, and J. W. Bailey.** 1968. Serologic grouping of *Cryptococcus neoformans. Proc. Soc. Exp. Biol. Med.* **127:**820–823.

239. **Wilson, D. E., T. W. Willimans, and J. E. Bennett.** 1966. Further experience with the alcohol test for cryptococcal meningitis. *Am. J. Med. Sci.* **252:**62–66.

240. **Wolf, J. E., and S. E. Massof.** 1990. In vivo activation of macrophage oxidative burst activity by cytokines and amphotericin B. *Infect. Immun.* **58:**1296–1200.

241. **Wong, B., J. R. Perfect, S. Beggs, and K. A. Wright.** 1990. Production of the hexitol d-mannitol by *Cryptococcus neoformans* in vitro and in rabbits with experimental meningitis. *Infect. Immun.* **58:**1664–1670.

242. **Yen-Fang, L.** 1959. Preliminary observations on the phosphatases of *Cryptococcus neoformans. Chin. Med. J.* **79:**138–142.

243. **Zerpa, R., L. Huicho, and A. Guillen.** 1996. Modified India ink preparation for *Cryptococcus neoformans* in cerebrospinal fluid specimens. *J. Clin. Microbiol.* **34:**2290–2291.

244. **Zimmer, B. L., and G. D. Roberts.** 1979. Rapid selective urease test for presumptive identification of *Cryptococcus neoformans. J. Clin. Microbiol.* **10:**380–381.

5 | Molecular Biology

INTRODUCTION

The understanding of the epidemiology and pathogenic features for *Cryptococcus neoformans* infections can now be aided by the use of molecular biological techniques. In the last three decades, model systems for eukaryotic molecular biology have developed around certain fungi, such as *Saccharomyces cerevisiae*, *Schizosaccharomyces pombe*, *Neurospora crassa*, and *Aspergillus nidulans*. Using the experience with these molecular model systems for fungi, investigations have now advanced in the molecular biology of *C. neoformans* (46). *C. neoformans* has the potential to become a molecular model system for the study of pathogenic fungi because of its clinical importance, well-studied pathophysiology, genetics, molecular biological foundation, and relevant animal models. Particularly, the use of new molecular tools will allow direct insights into pathogenesis that can lead to drug target validation and/or vaccine epitopes (71).

The first molecular studies of *C. neoformans* were performed in the 1960s and 1970s and were focused on determining G+C content of cryptococcal DNA and the DNA relatedness of various *Cryptococcus* species to each other. The average G+C content of the *C. neoformans* genome varied from 43 to 45.9 mol% for several strains (26), 51.5 mol% for a single strain (95), and 49 to 50 mol% for two strains (66). In early hybridization studies Erke and Schneidau (26) convincingly showed through DNA-DNA reassociation that *C. neoformans* clearly represented a species separate from the other *Cryptococcus* species. In their studies using reannealing DNA as a marker, 60% homology was found among several *C. neoformans* strains compared to 7.5% for *Cryptococcus laurentii* and 3.2% for *Cryptococcus uniguttulatus*. In 1981, Aulakh et al. (2) confirmed this genetic relatedness and extended it to *C. neoformans* var. *gattii*. They found that *C. neoformans* showed between 55 and 63% reassociation between var. *neoformans* and var. *gattii*. In fact, within the two varieties, isolates were found to have an 88 to 94% reassociation, which reaffirmed their relatedness. When G+C content was measured, *C. neoformans* var. *gattii* had a slightly higher content (56 to 57.3 mol%) than did var. *neoformans* (53 to 56 mol%). With the use of ^{32}P-labeling and two-dimensional thin-layer chromatography, it was shown that there is considerable variation among strains in both DNA nucleotide composition and nucleotide adducts, which are modified nucleotides formed through methylation, alkylation, oxidation, and reduction (79a).

The first cryptococcal gene to receive molecular attention was actually cloned not from *C. neoformans* but from *Cryptococcus albidus*. This gene encoded the xylanase of *C. albidus,* which is a xylan-degrading enzyme, and has now received molecular attention for the past decade (4, 63–65).

In the late 1980s, studies for the present foundation of molecular biology in *C. neoformans* began. The isolation of *C. neoformans* DNA by several spheroplast techniques was refined (106) (see chapter 12). Optimal conditions for spheroplasting this yeast and obtaining large amounts of DNA include careful attention to treating yeast cells with cell wall-degrading enzymes in mid- to late logarithmic phase of yeast cell growth. The cell wall of *C. neoformans* can be degraded with several enzyme preparations, including Mureinase, Novozyme 234, and Lysing enzyme. For instance, Lysing enzyme is a product from *Trichoderma harzianum* and contains cellulase, protease, and chitinase. All these enzymes have been used successfully to produce spheroplasts in *C. neoformans* strains. Cell wall-degrading enzymes used for other yeasts, such as glusulase and zymolyase, are not as effective in making spheroplasts. Although *C. neoformans* can release large amounts of nucleases, this fact is less of a technical problem than the massive contamination of the DNA/RNA preparations with cryptococcal capsular polysaccharide, which generally precipitates with the nucleic acid purification steps. For PCR and some restriction enzyme digestions, ethanol-precipitated DNA can be used. However, the use of hexadecyltrimethylammonium bromide or even cesium chloride gradient separation of DNA may be needed to reduce the massive polysaccharide contamination of DNA that occurs with standard DNA isolation procedures. This purification of DNA may be necessary for work with certain restriction enzymes and particularly DNA ligases. In most cases, *Taq* polymerases function well in the presence of polysaccharide contamination, but occasionally the presence of large amounts of polysaccharide will alter the fidelity of the PCR, and this interaction may need to be taken into account. Once these specific unique factors for *C. neoformans* are addressed, cryptococcal DNA work becomes part of standard molecular biological practice. However, occasionally, some cryptococcal DNA sequences will inhibit the growth of certain strains of *Escherichia coli* used for cloning.

RNA isolation procedures can be performed with standard techniques for breaking open cells, such as glass-beading or the use of a French press on cells to release RNA. Typically, RNA yields per cell tend to be lower in *C. neoformans* than in unencapsulated yeasts, such as *S. cerevisiae*. These findings may reflect the difficulty of releasing RNA through the capsule or simply polysaccharide contamination of the RNA preparations, which makes it more difficult to obtain purified preparations as measured by spectrophotometric assays. However, there are a series of RNA isolation kits that will help with purifying RNA preparations (see chapter 12).

MOLECULAR EPIDEMIOLOGY OF *C. NEOFORMANS*

The early development of molecular biological techniques for studying a pathogen generally involves questions of epidemiology in relationship to pathogenicity. Investigators in the late 1980s and the 1990s have used the newly acquired ability to type individual strains of *C. neoformans* by molecular techniques. Several molecular typing systems have been developed and used for *C. neoformans* infections. These typing systems can identify unique strains from both clinical and environmental sources and in some instances are used to determine relapse from true reinfection isolates. These molecular epidemiological tests include multilocus enzyme typing (5, 6); electrophoretic karyotyping with pulsed-field gel electrophoresis (48, 72, 73, 81, 118); restriction fragment length polymorphisms (RFLPs); DNA fingerprinting by hybridization of restricted DNA with genomic repetitive DNA fragments (20, 80, 88, 107, 109), mitochondrial DNA probes (105), and oligonucleotide probes to microsatellite sequences; and random amplified polymorphic DNA (RAPD) analysis and PCR fingerprinting (19, 55, 56, 58, 87).

Pulsed-field gel electrophoresis using a CHEF (contour-clamped homogeneous electric field) apparatus has been used to separate chromosomes of *C. neoformans*. A consistent finding in these studies was the significant polymorphic variation in chromosome sizes between strains (72, 73). This finding suggests that *C. neoformans* strains have not clonally expanded from one isolate but that there has been some evolutionary pressure for karyotype instability. It appears that these yeasts have developed a mechanism(s) that allows for chromosomal rearrangement. Despite the heterogeneity of these chromosomal patterns between various strains, it appears that some strains can maintain a stable pattern during repeated passage in vitro for many years and in vivo in animals during short-term experiments (weeks) (73). Therefore, the karyotypic analysis of strains may be used in epidemiological studies to separate different strains (72). However, recent work has cast some doubt on the use of this technique for epidemiological purposes because of the rapidity of chromosome instability in some strains. There have been findings in which colonies from an isolate passed through mice had karyotype patterns different from those in the original isolate (33). It has also been shown that karyotype instability in vitro can be shown by having a single well-characterized *C. neoformans* strain (ATCC 24067) passed in a variety of laboratories over years. Small differences in chromosome sizes were noted with pulsed-field gel electrophoresis of several isolates from different laboratories (32). These observations serve to emphasize the flexibility of genome rearrangement in these wild-type *C. neoformans* strains. In some respects, this polymorphism in *C. neoformans* is not surprising. For instance, on examination of serial isolates from patients, some changes in karyotypes between isolates from the same patient have been noted (5), and microevolution in vivo for *C. neoformans* has been noted by other DNA typing methods (90, 96). These results emphasize that investigators need to maintain their stock yeast cultures carefully under minimal passage and periodically these cultures need to be checked to ensure chromosome stability.

A variety of karyotype studies have found between 8 and 13 chromosomes in *C. neoformans* strains, with its predicted genome size of approximately 23 Mb. The genome size appears to be larger than that of *S. cerevisiae* or *Candida albicans*. In *C. neoformans* var. *neoformans*, there are generally between 8 and 12 chromosomes; the largest chromosome is ≥2.2 Mb and the smallest is about 770 kb, but smaller chromosomes have been noted (72, 73, 81, 118). Similarly, *C. neoformans* var. *gattii* strains most commonly have between 11 and 13 chromosomes, with several small chromosomes in the range of 400 to 700 kb (48, 73). These general differences in chromosome sizes and numbers can allow a reasonable prediction of the strain's variety status from its initial karyotype pattern.

The known genetic map of *C. neoformans* is extremely small, and it is not extremely precise. For example, hypocapsular mutants that had their genes mapped to a similar linkage group by meiotic recombination (93) were later found to be located on completely different chromosomes (11). Furthermore, the genome has not been sequenced, and the chromosome locations of only a few genes have been determined and reported (17, 73, 75, 118). Specific genes have been located on most chromosomes (118), and repetitive element(s) or cDNA clones have also been identified for all chromosomes (88), so there are potential markers for chromosomal recombination studies. Within the two *C. neoformans* varieties, var. *neoformans* and var. *gattii*, the homologous genes are generally located on similar-sized chromosomes, although these locations vary more within var. *gattii* (118). However, between the two varieties, homologous genes may be present on significantly different-sized chromosomes (118). Another difference between the two varieties and their karyotypes is that environmental and clinical isolates of var. *neoformans* have widely different karyotypes. However, Kwon-Chung et al. (48) have found that *C. neoformans* var. *gattii* strains appear to have developed a clonal pattern in their environmental isolates, even when isolates from widely divergent environmental areas are compared (48). It is hypothesized that, although there is a clonal strain of var. *gattii* that is associated with eucalyptus trees, karyotypes will change as strains are passed through animals, humans, or other environmental stresses.

Karyotypic analysis can also be used to identify progeny after matings between different strains. However, it is clear that the exchange of chromosomes during meiosis can be complex, with first-generation progeny possessing a variety of new-sized chromosomes. This complexity of chromosomal alignment, together with karyotypically stable progeny, is particularly difficult to follow when serotypes and/or varieties are crossed. Chromosome-length polymorphisms are common in other fungi, including pathogens like *Candida* species. Frequent chromosomal deletions and rearrangements have been associated not only with fungi but also with protozoa such as *Plasmodium* spp. (malaria) during their life cycle. These findings suggest that *C. neoformans* possesses the ability under stress or certain selection pressures to produce chromosomal breaks and/or rearrangements. There is no insight into whether this instability is deleterious or helpful to the yeast's survival, but it could help explain the diversity of karyotypes in *C. neoformans*. Studies have now shown that up to 90% of clinical isolates of *C. neoformans* var. *neoformans* have unique karyotypes (72). However, it is clear that with the present knowledge of microevolution in this yeast, epidemiologists will need to exercise some caution in using karyotyping as the only tool for distinguishing different strains when there are only minor differences in electrophoretic karyotypes. The mechanisms and evolutionary importance of the chromosome-size polymorphisms and variability in gene locations between varieties represent important questions for future research in this area of *C. neoformans* pathobiology.

The well-described molecular epidemiological strategy of using repetitive elements to probe for RFLPs in many pathogenic yeast species has also been developed for epidemiological studies with *C. neoformans*. First, the use of repetitive, multiple-copy ribosomal DNA (rDNA) genes of *C. neoformans* has been explored. The cryptococcal rDNA gene complex has been cloned (82) and sequenced

(27, 28). The *C. neoformans* rDNA repetitive fragment (approximately 8 kb) contains genes in a 16S, 5.8S, and 23S arrangement, and these genes are transcribed in the same direction. The rDNA complex is an attractive target for RFLPs, because it consists of a tandem array of many repeating segments and thus a high copy number that does not require Southern probes. However, the conserved nature of the rDNA complex in *C. neoformans*, in which the nucleotide sequences of the 16S-like, 5.8S, and 23S-like rRNAs of var. *neoformans* and var. *gattii* differ in only 15 nucleotides in 5,656 bases (28), does not allow for much discrimination between strains.

RFLPs with other repetitive elements can distinguish between strains of *C. neoformans*. For example, Polacheck et al. (80) identified middle-repetitive DNA sequences from *C. neoformans* (CND1.4 and CND1.7) that can be used to distinguish varieties and strains for epidemiological typing. Although certain hybridization patterns were found to be more common in AIDS versus non-AIDS isolates with the use of these repetitive probes, none of the molecular typing methods have conclusively proved that there is a unique genotype(s) from isolates that is AIDS-specific. In our opinion, it remains unlikely that repetitive probes will identify an AIDS-specific genotype.

The CNRE-1 probe contains a member of a dispersed family of repetitive elements (20, 88, 90, 91). Spitzer and Spitzer (91) identified this repetitive element and showed that *C. neoformans* possesses multiple copies of CNRE-1 that are located on at least seven chromosomes in some strains. There are at least 10 to 20 members of this family with a minimum length of 5 to 10 kb. The function of CNRE-1 is not known, but it does not possess features typical of a transposable element (91). The origin of the distinct CNRE-1 hybridization patterns in *C. neoformans* is not known, but the RFLP differences could arise from chromosomal rearrangements, sexual recombination, or simply genetic drift. Since more than 70% of isolates within a geographically constrained population had an RFLP pattern different from that of CNRE-1, this probe has some discriminating power to identify different clinical strains. Fortunately, the probe also does not appear to change hybridization patterns during infection or with antifungal treatment (20, 90). It has been used in concert with specific gene sequences from the *URA5* gene to help identify clonal populations of *C. neoformans* within a geographically restricted environment (20) and to compare the genetic diversity of isolates from Brazil with those from New York (31). In the first study (20), isolates within the same CNRE-1 group showed fewer differences in nucleotide sequence of the *URA5* gene than did those from different CNRE-1 groups. In the second study (31), there were differences in discriminating power between CNRE-1 and *URA5* gene sequence and karyotypes, but these molecular studies showed that there were differences in local diversity between the two sites but that some isolates were closely related to each other, suggesting global dispersal of some pathogenic strains.

Varma and Kwon-Chung (107) identified another DNA probe (UT-4p) that has also been used to examine the diversity of *C. neoformans* strains (42, 107, 109). This 7-kb linear plasmid contains the cryptococcal *URA5* gene, and telomere-like sequences (TTAGGGGG) have been added to the ends of the plasmid. The acquisition of repetitive DNA sequences that are present on each chromosome and polymorphisms within the *URA5* gene made this probe useful for strain identification. Initial work with the probe found that 21 of 26 strains had unique DNA fingerprints and that the fingerprints of isolates within a serotype group were more similar to each other than to those from another serotype (107). The fingerprint pattern with *Acc*1-digested genomic DNA was stable with in vitro passage of strains in animals and from different sites within the same patient (107). Although there were no unique patterns in AIDS or non-AIDS patients in 156 isolates, the patterns could be grouped into 9 to 12 distinct fingerprint patterns depending on the *C. neoformans* variety (109). In a study of 40 serotype D strains from France by Dromer et al. (22), at least 10 groups were identified with the UT-4p probe and RFLPs. There was some suggestion that certain risk groups, such as drug addicts and homosexuals, had isolates from the same fingerprint group; also, five of seven isolates from cryptococcal pneumonia patients had the same fingerprint group. These findings suggest that this probe might identify characteristics of the genotype that predict body localization or even risk factors for infection, but further studies are needed to confirm this finding. The DNA RFLPs for UT-4p in serotype D strains were also found to be stable during human infection and its treatment.

Synthetic oligonucleotide probes homologous to microsatellite sequences have been used to identify genotypes of isolates. Although a series of oligonucleotide probes, including $5'-(GT)_8-3'$, $5'-(GTG)_5-3'$, $5'-(GATA)_4-3'$, and $5'-(GACA)_4-3'$, have been effectively used (35, 57, 58), Meyer and coworkers (57, 58) found that the oligonucleotide $5'-(GGAT)_4-3'$ yielded the most consistent and reproducible hybridization signals. With this highly discriminatory probe, all patients were found to have different isolates by RFLP analysis. This technique has also been used to discriminate relapse from reinfection isolates (35). On the other hand, while it appears that the banding patterns are stable within clinical and environmental isolates during in vitro growth, *C. neoformans* may become more genetically polymorphic under certain environmental stresses, such as in vivo. Furthermore, certain strains may have a more inherently unstable genotype than others. This rapid genetic polymorphism has now been displayed both at the microsatellite level and with larger DNA sequences such as chromosomes (96). The concept of microevolution occurring in *C. neoformans* strains as they pass through a variety of environmental and host stresses should be taken into account when very sensitive genotyping methods are used.

Varma and Kwon-Chung (105) have isolated mitochondrial DNA of *C. neoformans* by its affinity to bisbenzimide, which binds AT-rich sequences. The 40- to 50-kb DNA of this organelle was found to have some sequence polymorphisms that allowed RFLP differences between strains to be detected. The differences were found in seven patterns in 20 *C. neoformans* isolates by probing the RFLPs with the *S. cerevisiae* cytochrome oxidase gene. There were also some intravarietal specific patterns, and hybridization of some cloned mitochondrial DNA fragments to total DNA revealed polymorphisms as well as variety-specific patterns of homology. This study emphasized the rapid evaluation of *C. neoformans* mitochondrial DNA by its high intraspecific sequence divergence and, as in other mitochondrial genomes, this represents an area of study for recombination and repair. On the other hand, mitochondrial DNA is not particularly discriminating for epidemiological studies, and it is likely that more sensitive molecular typing will generally be used. In fact, little work has yet been performed on the genetics and molecular biology of mitochondrial DNA in *C. neoformans*. In at least one serotype A strain (H99), mitochondrial function appears essential. For instance, respiration-deficient (petite) mutants have not yet been produced by standard ethidium bromide treatments. In preliminary studies this yeast also does not appear to have a functional alternative oxidase pathway, as is found in plants and some fungi; it generally uses the standard cytochrome oxidative system (71a). Finally, a *C. neoformans* mitochondrial gene in this pathway, cytochrome oxidase C subunit 1 (*Cox1*) has been cloned from serotype A strain H99 (78), and it appears to be regulated in its expression by certain stress conditions, such as temperature changes and in vivo growth. Further work on the importance of mitochondria to this aerobic yeast as it attempts to survive a host immune attack is likely to be extremely helpful in our understanding of its pathogenesis.

The use of PCR technology combined with repetitive primers (arbitrary primed PCR) has allowed investigators to identify the DNA fingerprints of different fungal strains rapidly, specifically, and with a minimal number of DNA purifications. These procedures have been adapted and refined for use in *C. neoformans* (19, 35, 55–59, 61). When performed in a laboratory with experience in this technology, and with proper genotyping controls, a PCR fingerprint can distinguish one strain from another in more than 95% of isolates. The first PCR fingerprint studies used a variety of RAPD fragments within a strain. This work reflects the use of arbitrary single primers to identify the known polymorphic sequence structure of DNA from individual yeast strains. However, another step was to refine this PCR technique and use oligonucleotides that are specific to minisatellite or simple repetitive sequences designed for conventional hybridization fingerprinting. This PCR fingerprinting technique has been used to amplify hypervariable repetitive DNA sequences in *C. neoformans*. PCR fingerprinting with the oligonucleotide primers $(GTG)_5$, $(GACA)_4$, and the phage M13 core sequence (GAGGGTGGXGGXTCT), but not with $(CA)_8$-generated polymorphic bands, has been successful in all *C. neoformans* strains tested. Strains of serotype A, B/C, or D could also be classified to serotype by their PCR fingerprint pattern. Thus, this method has the sensitivity to detect both inter- and intravarietal differences. When carefully performed and controlled, this strategy is a reproducible and rapid method for distinguishing specific *C. neoformans* strains. In vitro the pattern of an isolate appears to remain the same, but in vivo, such as in cerebrospinal fluid from a patient with meningitis, changes or actual microevolution can be detected within a single isolate (71a). Therefore, it is important to interpret each banding pattern to be sure that true strain differences are detected. In at least one case of human cryptococcal meningitis, there is reasonable evidence that by using this technique a true reinfection with another strain was documented (35).

By these PCR methods, some amplified bands have been sequenced, and it appears that they generally represent inter-repeat sequences that are likely to be located in the noncoding regions of the *C. neoformans* genome. The high degree of homology among some of these minisatellite sequences and the primers produces such strong binding for PCR fingerprinting that these sequences may be more reliable and discriminating than standard RAPD fragments, which may use shorter primers (6 to 10 nucleotides) and arbitrary sequences (51, 83). For instance, RAPD analysis using five primers separated 344 *C. neoformans* isolates into 19 subtypes but could not distinguish isolates of *C. neoformans* var. *gattii* from *C. neoformans* var. *neoformans* (19). On the other hand, RAPD analysis with 12- to 22-mer pairs and PCR fingerprinting with a single primer derived from the microsatellite core sequence of the wild-type phage M13 found only three major genetic profiles in isolates of *C. neoformans* var. *gattii* (87). This high level of genetic concordance between the majority of clinical and environmental isolates of var. *gattii* is consistent with the hypothesis that human disease can derive from exposure to eucalyptus trees. Only one study by RAPD analysis has suggested that *C. neoformans* isolates from AIDS patients and other immunocompromised hosts may have a common genetic profile, and this result awaits confirmation (14). Finally, a word of caution for investigators who use these very sensitive PCR technologies for strain identification in *C. neoformans*. It is essential to control carefully for differences in techniques when first starting to use them. Always use a known standard with specifically identified bands to ensure that the banding patterns are reproducible and comparable in each run.

Although the method is more time-consuming, specific gene sequences have been used to detect *C. neoformans* polymorphisms and thus strain differences (13, 20). It is clear that there is some allelic variation in sequence between *C. neoformans* strains. For example, the single-copy *URA5* gene has shown extensive base pair substitutions between individual strains and the two varieties that reach up to 6% of the nucleotides encoding the gene (8). These base pair substitutions are generally found in the third codon position or within introns, so protein structures are not changed (8). We have also made similar findings of sequence gene diversity in the *C. neoformans* recombinational gene for topoisomerase I, in which sequence differences between a serotype A strain and a serotype D strain were 3.6%. The apparent ability of *C. neoformans* to undergo frequent extensive gene conversion and/or recombination events has been explored further. For example, it was known that individual CNRE-1 elements from one isolate exhibit numerous nucleotide differences (91). In a complementary investigation of multigene families in *C. neoformans*, two ubiquitin-encoding genes, *UBI1* and *UBI4*, located on separate chromosomes were identified and sequenced (89). Extensive nucleotide differences (approximately 15%) in the ubiquitin repeats of both genes were found, along with differences in the location and number of introns. The mechanism(s) for these frequent rearrangements within duplicated sequences has not been elucidated. In fact, it is also not known if and to what extent methylation, which can affect gene expression and/or change RFLPs, occurs in *C. neoformans*. However, it is clear from sequence data that *C. neoformans* can undergo extensive gene conversion and/or recombination events. This fact can be exploited as a sensitive measure to determine strain identification at the level of gene sequences.

In the final analysis, it may be wise for investigators to test *C. neoformans* strains by several different molecular techniques. This strategy was elegantly followed by Brandt et al. and the Cryptococcal Disease Active Surveillance Group (7) when they typed 33 strains with the use of multilocus enzyme electrophoresis, electrophoretic karyotyping, RAPD analysis, and the CNRE-1 DNA probe. There was reasonable correlation between methods and serial isolates. This strategy allowed differences in one method to be interpreted in comparison to the other methods. It should be emphasized that there must be a balance between methods that can detect actual microevolution during infection and those that lack discriminating power to detect strain differences. Together, these multiple techniques can probably define and distinguish *C. neoformans* strains from each other and provide precise measurements for epidemiological studies of *C. neoformans*.

MOLECULAR PHYLOGENY FOR *C. NEOFORMANS*

As the genome era continues to mature, the use of DNA sequence information will further document the phylogenetic relationships among fungi. There have been several studies of the molecular phylogeny of *C. neoformans*, and some of this information has been highlighted in chapter 2. The highly conserved rRNA and rDNAs are generally the first emphasis for this molecular analysis, and *C. neoformans* has been studied and compared for these conserved genes (27, 28, 34). In *C. neoformans* the 8-kb repetitive DNA fragment contains the 16S, 5.8S, and 23S gene arrangements, which are transcribed in the same direction. This particular genetic arrangement is similar to that of other basidiomycetes, such as *Schizophyllum commune*, *Agaricus bisporus*, and *Coprinus cinereus*. As previously noted, there is high conservation of rDNA sequences between the two varieties of *C. neoformans*. The sequences of the 5.8S gene and the intergenic areas around this gene have been used to support the classification of *C. neoformans* as a separate species and to identify its similarity to *Tremella mesenterica* (27). Through this phylogenetic analysis, the teleomorph *Filobasidiella neoformans* is also related to *Filobasidiella depauperata* but is clearly separate from the genus *Filobasidium*. Further studies have developed the molecular phylogeny of *C. neoformans* in relationship to other fungi. Fan et al. (28) sequenced both the 16S-like rRNA and 23S-like rRNA of *C. neoformans* and used the sequence to create a distance tree with other fungi (Fig. 1 and 2) (28). In fact, a further reduction of relationships among these yeasts by Gueho et al. (35) used partial sequences of the most variable domain of the large subunit rRNA of genera related to Cryptococcus and of other *Cryptococcus* species (34). The general topology of these relationships is shown in chapter 2 of this volume. In this phylogenetic tree, it is noted that *F. neoformans* serotype A is evolutionarily slightly divergent from the other three serotypes, and it is hypothesized that it may be evolving at a different rate. In fact, there may be support for this hypothesis. Serotype A strains may have lost a functional *MAT***a** locus since there is no convincing evidence that a *MAT***a** strain from serotype A has been identified, unlike that in serotype D and the *C. neoformans* var. *gattii* strains.

Recent work in molecular phylogeny of *C. neoformans* has focused on other genes and their ancestral lineages. For example, Cox et al. (17) have examined the highly conserved actin gene of *C. neoformans* for its phylogenetic relationships. Although *C. neoformans* can be grouped with this gene sequence to a discrete evolutionary fungal branch, more actin genes from other fungi need to be sequenced to determine relationships between various subgroups and to make further comparisons with the much larger rRNA database.

Another interesting avenue of molecular evolutionary studies with *C. neoformans* is to determine the divergence of specific genetic virulence factors among related fungal species. *C. neoformans*, the anamorph of *F. neoformans*, is the only zoopathogenic species among members of the family *Filobasidiaceae* and the major human pathogen among basidiomycetes. Since we are now making progress in identifying virulence factors at the molecular level in this species, the potential now exists to examine their evolution in response to challenges posed by various ecological niches. For example, recent work in this area suggests that a capsule gene (*CAP59*) and a laccase gene (*CnLAC1*) of *C. neoformans* were not detected when probed with genomes of other heterobasidiomycetes (79). There was no

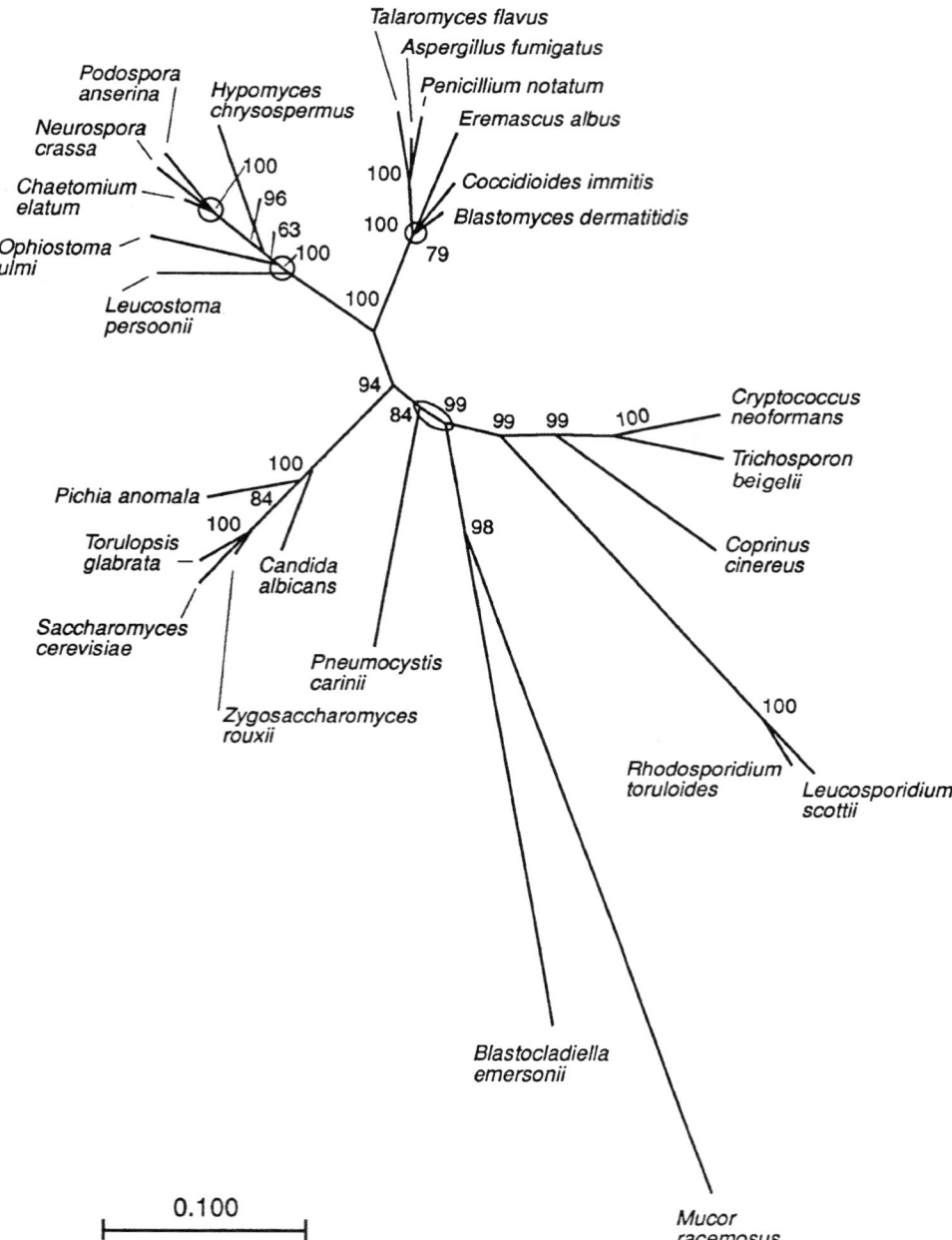

Figure 1 Fitch-Margoliash distance tree inferred from the more conservative 16S-like rRNA sequence matrix. (From reference 28 with permission.)

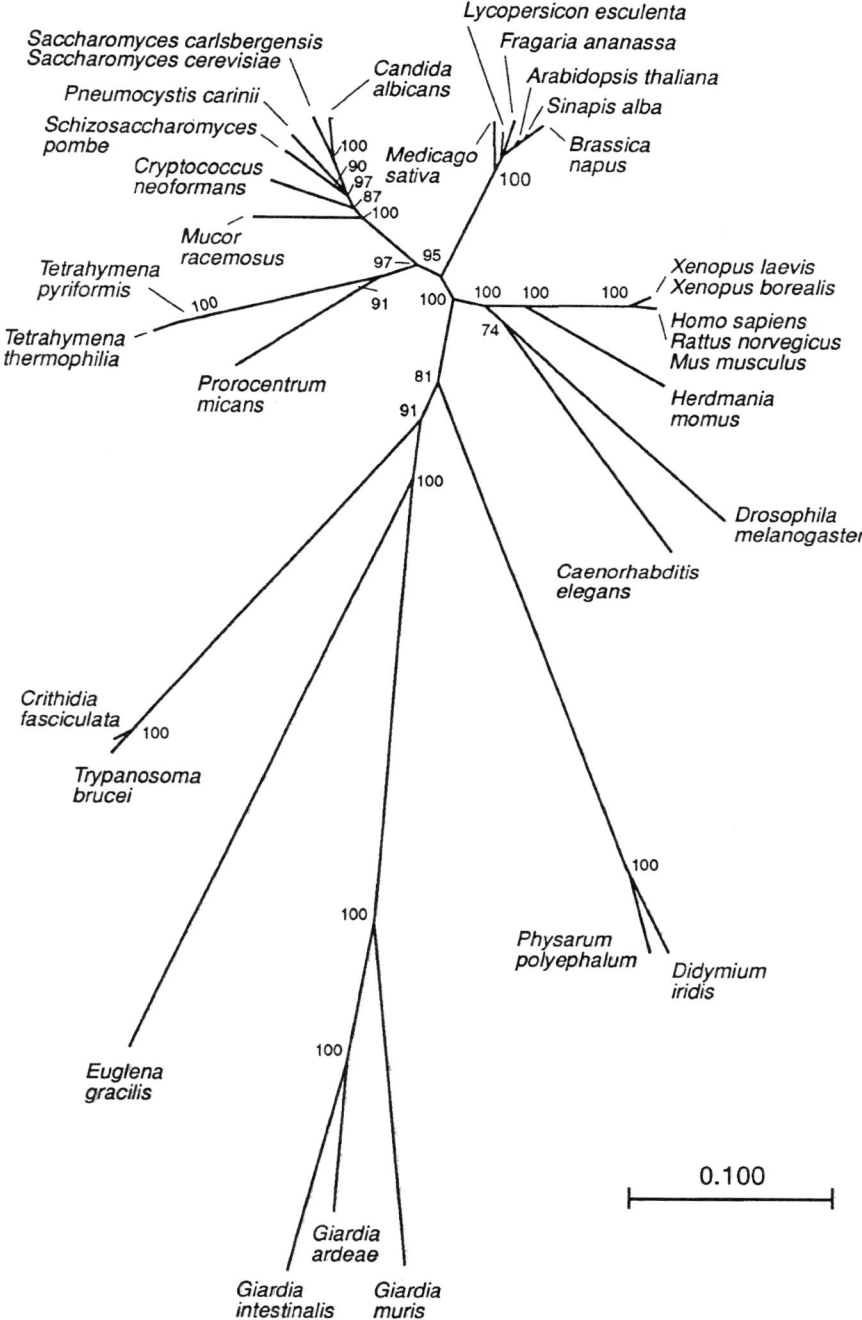

Figure 2 Fitch-Margoliash distance tree inferred from the more conservative 28S-like rRNA sequence matrix. (From reference 28 with permission.)

apparent homologous *CAP59* gene in capsule-producing species such as *Filobasidium uniguttulatum*, several *Tremella* species, or *C. laurentii*. Similarly, when the *CnLAC1* gene for melanin production was used as a probe for several other heterobasidiomycetes, it did not identify a homolog in these species. These results suggest that the ancestral genes for capsule and laccase sequences either emerged early in evolution or profoundly diverged in response to different ecological niches among a group of related fungi. Further understanding of relationships between other heterobasidiomycetous fungi and the pathogen *C. neoformans* for the evolution of these virulence factors can be studied as the virulence genes are identified and compared with those in other related fungal species.

Finally, Boekhout et al. (3) applied a series of molecular typing methods to the two varieties of *C. neoformans* in environmental, veterinary, and clinical isolates. *C. neoformans* var. *neoformans* and *C. neoformans* var. *gattii* differed significantly in chromosome numbers and sizes, RAPD patterns, and killer toxin sensitivity. With karyotyping and RAPD analysis, there was also some suggestion of clonal expansion within restricted geographic areas. These differences made Boekhout et al. (3) suggest that the two varieties actually represent different species. Although debate still continues on variety versus species status for var. *neoformans* and var. *gattii*, it is clear that their genotypes are significantly different and that further molecular studies of genetic crosses between these varieties are probably warranted.

THE GENETICS OF *C. NEOFORMANS*

A major discovery of the sexual, or perfect, state of *C. neoformans* was made in 1976 by Kwon-Chung (43), who identified this yeast as a bipolar heterothallic basidiomycete with two mating types, **a** and α (*MAT***a** and *MAT*α). She further showed that the serotype B/C strains also possessed a sexual stage (44). *C. neoformans* generally grows in the host and probably in nature in the imperfect yeast stage. Under nutritionally deprived conditions, such as in the laboratory on minimal media, V-8 juice, or hayseed agar, signals such as pheromones are induced through signal transduction pathways, including a G protein (CnGPA1) that regulates intracellular cyclic AMP (1). Under nutrient-starved conditions, the perfect state of *F. neoformans* forms when *MAT***a** and *MAT*α strains are physically mixed together on the agar plate. The perfect state is formed through transient dikaryotic hyphae with clamp connections that create basidia at the terminal ends of the hyphae. In these basidia, karyogamy and meiosis occur and form basidiospores, which through postmeiotic replication develop four long chains of these spores on the surface of the basidium. These meiotic products of specific postmeiotic nuclei are channeled into a particular basidiospore chain (41). However, these long chains of spores are rarely seen in routine matings, and there may only be one or two spores per basidium. The hyphae form as a mold-like growth around the edge of the yeast streak on an agar plate (see chapter 2). When these haploid spores are allowed to regrow on culture medium, they convert to their yeast forms within 5 to 7 h of growth on agar plates. Heterokaryon analysis with marked strains can be used to detect dominant or recessive mutations. Whelan (112, 113) identified the use of heterokaryons for complementation studies. A heterokaryon was identified as a dry white colony with abundant hyphae bearing clamp connections. Odom et al. (67) used this technique to identify both recessive and dominant drug mutations. The heterokaryon stage of the C. neoformans life cycle is accessible for studies of complementation, cytoplasmic inheritance, and diploidy.

The meiotic products of a single genetic event are unfortunately not contained in asci, as occurs with S. cerevisiae. Detailed genetic linkage maps are difficult to make and not very accurate, since random spore analysis is the major method for detection of segregants. Random spore analysis of meiotic products has demonstrated 1:1 segregation of several alleles, including mating types, in laboratory crosses (10, 25). With serotype D strains, backcrosses have helped to eliminate genetic defects in growth of transformed strains and have allowed specific allele (laccase) studies of virulence (84). Finally, auxotrophic mutants in C. neoformans have been made, and crosses for complementation have been reported (40, 41). It is obvious that this life cycle has the potential to be used as a powerful tool for genetic analysis in the study of C. neoformans.

The ability to perform genetic crosses and use meiosis strengthens the usefulness of this organism as a molecular model for fungal pathogenesis. However, there remain several concerns about the genetics of this yeast. First, single dissection of small (1- to 2-μm) basidiospores, which many times are not in chains, from a single meiotic event allows only a general random spore analysis of segregants, and as previously noted it is not always easy to determine gene linkage groups. Second, there is the recent finding that some α-mating strains do possess the ability to form fruiting body-like structures in the vegetative phase, which has been called monokaryotic, homokaryotic, or haploid fruiting (117). This life cycle form has been described for other fungi. In C. neoformans it appears to be controlled by the α-mating locus and is stimulated by nutritional starvation. If not specifically looked for in control plates with genetic markers, this characteristic has the potential to confuse genetic crosses. Third, at present the original MAT**a** and MATα strains (3501 and 3502) (43) and congenic pairs JEC 20 and JEC 21 are serotype D (62). It is extremely useful that congenic pairs of the original strains, except at the mating locus of these strains, have been made (45). On the other hand, serotype A strains are the most common serotype to cause clinical disease worldwide and in almost all AIDS patients. Unfortunately, a serotype A MAT**a** strain has been extremely difficult to identify. The reason for this possible evolutionary deletion remains unexplained, and with further investigations a strain may still be found. However, it is essential that investigators who appear to find a serotype A MAT**a** strain be sure that it does not contain the α-mating-type locus and that mating is not simply related to haploid fruiting. It would definitely be helpful to have a congenic serotype A MAT**a**/α pair for pathogenesis studies. The fourth concern about the genetics of this yeast is that not all wild-type strains are haploid (99, 114); a ploidy analysis was performed on a group of 34 clinical isolates, and approximately one-third were found to be diploid and 20% asexual (97). Approximately two thirds of the strains were heterothallic, haploid, α-mating strains. Finally, the exact biological relevance of the mating-type locus and surrounding sequences to the ecology and pathogenesis of this yeast remains unclear. For instance, the MATα locus was originally believed by Edman and coworkers (45, 62) to be a 45-kb DNA region unique to α-mating cells, but recent studies in that laboratory suggest that it is larger, with a minimum size of 60 kb, and that it contains both a pheromone gene (62) and at least two genes involved in signal transduction events (STE11-α and possibly STE12-α) (24). It is also possible that the α-mating locus is different for serotype D and A strains. Further work is needed to understand the evolutionary and functional importance of this locus. Also, there have been no reports on the structure of the **a**-mating locus, but it is interesting to note that *MAT***a** strains do not possess the transactivating gene (*STE12-α*). This gene is apparently not essential for viability and appears to be *MAT*α specific (116).

Despite many uncertainties about the genetics of *C. neoformans*, genetic crosses can be used in some strains of *C. neoformans* to identify alleles, ensure gene disruptions, and, most important, eliminate unwanted ectopic effects of transformation on the total virulence composite during the study of specific site-directed mutagenic effects. These advantages make this sexual life cycle a very useful tool for investigators in molecular pathogenesis and may also be important to the pathobiology of this yeast.

MOLECULAR TECHNOLOGY FOR THE CLINICAL LABORATORY

The use of molecular technology for identification and diagnosis has been developed in *C. neoformans* (see chapter 12). For molecular diagnosis, a commercial probe with excellent sensitivity and specificity has been developed for the rapid identification of *C. neoformans* from a yeast colony (39, 60, 94). For the potential identification of the presence of *C. neoformans* in clinical specimens, PCR techniques have been used to amplify a portion of the cryptococcal rDNA from the clinical specimen. This amplicon can then be digested with a series of restriction enzymes to distinguish it from the amplicons of other yeasts (29, 38, 53, 60, 61, 98, 111). It is clear that the identification and diagnosis of *C. neoformans* are readily amenable to molecular techniques in clinics. As clinical microbiological laboratories adapt more automated molecular methods, *C. neoformans* will easily be converted to these strategies. On the other hand, the use of excellent histological staining methods for the capsule, its relatively facile growth and identification features, and an excellent serological test make this yeast less likely to be an initial focus for molecular diagnostics. Molecular techniques will simply be incorporated into diagnostic strategies for *C. neoformans* when nonculture methods are routine practice.

C. NEOFORMANS GENE ISOLATIONS

The identification, cloning, and sequencing of cryptococcal genes have begun, and a series of these genes have now been reported. Table 1 lists more than 30 genes that have been identified. It is hoped that this is only a partial catalog that will rapidly expand in the next few years. This list can grow particularly rapidly if more investigators study this model fungal organism. Another giant leap in the molecular studies with this organism will occur when the genome is finally sequenced and available for public use. In fact, a database of even partial sequences of expressed genes would be extremely helpful. Spitzer and Spitzer (92) have reported between 35 and 44 partial sequences of cDNAs from *C. neoformans*. Although this is a small beginning, a larger-scale sequencing project of partially sequenced expression tags entered into a database would advance the field.

The breadth of techniques that have been used to clone *C. neoformans* genes indicates the significant advances over the last 6 years within its molecular biology foundation (Table 2). Genes have been cloned by the following strategies: (i) het-

Table 1 Cloned and identified genes of *C. neoformans*

Genes	References
rDNA (*23S, 16S, 5.8S*)	82
CL27 (L27 ribosomal protein-encoding gene)	12
TOPOI (topoisomerase)	21
TOPOII (topoisomerase)	70
Actin	17
β-Tubulin	21
UBI1 and *UBI4* (ubiquitin)	89
URA5 (orotidine, monophosphate, pyrophosphorylase)	8, 25
URA3 (orotidine-5′-phosphate decarboxylase)	118
ADE1 (phosphoribosylaminoimidazole-succinocarboxamide synthetase)	118
ADE2 (phosphoribosylaminoimidazole carboxylase)	77
LEU2 (3-isopropyl malate dehydrogenase)	118
HIS3 (imidazole glycerolphosphate dehydratase)	69
TRP-1 (phosphoribosylanthranilate isomerase)	75
CAP59, CAP60, CAP64, CAP110	9–11, 78
CnLAC1 (laccase)	119
CNA1 (calcineurin)	68
NMT (*N*-myristoyl transferase)	50
ARF (ADP-ribosylation factor)	50
MATα locus	62
MATα locus	62
STE11 (*S. cerevisiae* homologs)	24
STE12 (*S. cerevisiae* homologs)	116
PRE1 (proteasome subunit)	11
Cox1 (cytochrome oxidase subunit 1)	78
Urease (*URE1*)	16
EF3 (elongation factor 3)	36
Dihydrofolate reductase	86
Thymidylate synthase	49
GPA1 (GTP-binding protein α subunit)	103
Mtl-1 (regulator of mannitol dehydrogenase in *S. cerevisiae*)	76
CneMDRI (multidrug resistance protein)	100
GAL7 (UOP glucose-D-galactase-1-phosphate uridylyltransferase)	115

Table 2 Strategies for gene isolations in *C. neoformans*

Strategy	Example	References
Heterologous probes	Actin gene	17
Complementation bacterial mutants with cDNA	*ADE2, URA5*	8, 25, 77
Expression or complementation in *S. cerevisiae*	*NMT-1, Mtl-1, TRP-1*	50, 75, 76
Amino acid sequence of protein	*CnLAC1*	119
Plasmid gene rescue	*CAP59, CAP64*	10, 11, 120
PCR with degenerate oligonucleotides from conserved regions	*CNA1, TOPOI*	21, 68, 70
Screening lambda libraries with antibody	Immunosuppressive protein	54
Differential gene expression screens	*Cox1*	78

erologous gene probes to genomic libraries (e.g., actin), (ii) cDNA libraries used to complement bacterial mutants with recovery of the transcript (e.g., *URA3, URA5, ADE2, ADE1*), (iii) identification of mutations in *S. cerevisiae* by expression and complementation (e.g., *NMT-1, Mtl-1*), (iv) identification of partial amino acid sequence from the protein, with the use of oligonucleotides to identify the gene from *C. neoformans* libraries (e.g., *CnLAC1*), (v) complementation of a known *C. neoformans* mutation with a library, followed by a plasmid gene rescue technique (e.g., *CAP59, CAP64*), (vi) use of degenerate oligonucleotides from conserved regions of identified genes from other species for PCR amplification (*TOPOI, CNA1*), (vii) lambda expression libraries with clones identified by specific antibodies (e.g., immunosuppressive protein), and (viii) screening for differential gene expression under specific environmental conditions (e.g., *Cox1*).

The first method of gene isolation is the use of heterologous probes with the entire gene from other species. This method has been useful when genes are particularly well conserved and when low-stringency conditions are used. The rDNA genes of *C. neoformans* were first identified by cross-hybridization with the *S. cerevisiae* rDNA gene (82), and the cryptococcal actin gene was identified with the use of the *C. albicans* actin gene (17).

The second method of gene isolation is to use bacterial mutants and their powerful genetic background and known functional mutations to identify *C. neoformans* transcripts. This strategy has been used to clone multiple metabolic genes in *C. neoformans*, such as *URA5, URA3, LEU2, ADE1, ADE2*, and *HIS3* (8, 25, 77). If the *C. neoformans* gene of interest has a bacterial mutant, this method of complementation using a cryptococcal cDNA library under a bacterial promoter can be very successful. It should be stressed, however, that some *C. neoformans* genes can be toxic to *E. coli*, and this may be a problem at times in general cloning strategies. Sometimes, simply reorienting the gene in the opposite direction within the plasmid may help the gene to be amplified.

A third method for cloning is to use *S. cerevisiae* as a surrogate. This eukaryotic system with its shuttle vectors and well-described genetics is ideal for cloning genes of other species with a less-developed molecular biology foundation. It has been shown that *S. cerevisiae* can express *C. neoformans* genes, and complementation of certain mutants can occur. Several genes, including *C. neoformans TRP-1* (75), *NMT-1* (50), and *Mtl-1* (76), have been cloned by their expression in *S. cerevisiae*. However, experience suggests that only some *C. neoformans* genes will be expressed and complement in *S. cerevisiae* with the use of *C. neoformans* genomic libraries. It is possible that this fact is related to the tendency for *C. neoformans* to possess frequent intronic structures, which are much less common in *S. cerevisiae*. For instance, two *C. neoformans* genes (*TRP-1* and *Mtl-1*) that were cloned directly by genomic libraries in *S. cerevisiae* do not possess introns. A higher efficiency of cloning *C. neoformans* genes in *S. cerevisiae* will probably occur if the libraries are made with *C. neoformans* cDNA and inserted into vectors that contain high-level expression promoters for *S. cerevisiae*. We recommend this approach if *S. cerevisiae* is to be used as a surrogate for *C. neoformans* gene cloning.

The fourth approach allows the cloning of a gene when the encoded protein can be purified and a partial amino acid sequence can be obtained by microsequencing (119). Although more tedious than the other methods, this may be the only strategy available if there is little information on the genetics or molecular biology of the gene or protein under study. For instance, the interesting and important *C. neoformans CnLAC1* (laccase) gene was extremely difficult to identify by classical cloning methods. However, through careful biochemical separations, the enzyme was purified and a portion of the amino-terminal end sequence was obtained. With the use of degenerate oligonucleotides to the obtained sequences, the entire gene was then cloned and sequenced (119).

Fifth, with the ability to perform genetic analysis and the development of the molecular biology of *C. neoformans*, the cloning of genes has matured to the stage where complementation of defined mutations and recovery of genes by plasmid or PCR rescue techniques can now be performed. This strategy has elegantly been used to clone a series of genes (*CAP10*, *CAP59*, *CAP60*, *CAP64*) that are essential for capsular synthesis in *C. neoformans* (10, 11, 120). This strategy uses a high-efficiency vector pCnTel plasmid developed by Edman (23). It contains *C. neoformans* telomeres for extrachromosomal replication, but care must be taken in its maintenance, or the *C. neoformans* inserts can be lost during amplification.

With the use of PCR technology and available databases on gene structures and conservation of gene arrangements, a particularly useful technique in the genome era has been to identify several *C. neoformans* genes through amplification of a partial fragment of the gene between conserved sequences (21, 42, 70, 91, 108, 109). For example, areas of sequence conservation in the gene of interest from other species may be identified through database searches. These areas are used to construct a series of degenerate PCR primers, and low-stringency conditions are used to amplify a part of the *C. neoformans* gene. This amplified fragment can then be sequenced to ensure that it is part of the gene of interest by homology and then used as a probe to identify the entire gene from a genomic library. This strategy has been used to isolate many *C. neoformans* genes, such as the genes for urease 1, topoisomerases I and II, and calcineurin A. Of course, this strategy requires a knowledge of the structure of the gene from several other organisms and identification of some areas of conservation within the gene. At times, this information is simply not available.

The last method of gene identification used in *C. neoformans* is differential gene expression under certain conditions. Using this strategy, investigators have cloned several genes that appear to be up- or down-regulated at the site of a central nervous system infection in rabbits (74, 120). There are now several molecular methods for analyzing gene expression by capturing and quantifying cellular transcripts. These methods have become particularly important as the genome projects convert from structural to functional studies, and they can now be adapted to *C. neoformans*. cDNA libraries from cells under various conditions have been used as probes for differential hybridization to genomic libraries for gene identification in *C. neoformans* (74). cDNA library subtraction techniques could be used with the isolation of small amounts of RNA from organisms at the site of infection compared to in vitro-grown cells. A popular technique for identifying transcripts with limited mRNA, differential-display reverse transcription (RT)-PCR, has been adapted for *C. neoformans* (74, 120). As the genome era has progressed, new techniques for quantitative gene expression have been developed and could be adapted for *C. neoformans*. These techniques include serial analysis of gene expression (110) and quantitative monitoring of gene expression patterns with a complementary DNA microassay (85). This latter method will become more feasible as genome sequencing technology reaches *C. neoformans* and as microchip technology develops. This strategy may become particularly useful for rapid analysis of gene expression.

In *C. neoformans*, it is also possible to consider in vivo expression technology, which has been used successfully in bacteria (52). In this system, in vivo-expressed promoters are captured by their ability to turn on an essential gene in purine metabolism and thus allow viability in vivo but not in vitro. Since the *ADE2* gene has been shown to be essential for *C. neoformans* in vivo (77), a similar strategy can be used to identify in vivo regulated promoters and their genes in *C. neoformans*. An advantage of using the *ADE2* system is that the turned-off gene in vitro produces a detectable phenotype: the colony turns pink. A possibly more sensitive technique for in vivo work would be the use of the recorder gene for green fluorescent protein. By fluorescence-activated cell sorter (FACS) analysis, cells containing turned-on or turned-off genes under a variety of conditions could be easily identified and separated. This recorder gene has been elegantly optimized by Cormack et al. (15) for expression in *C. albicans* and can also be used in *C. neoformans* (20a).

When these differential gene strategies are used, it is important to identify the relevant environmental signals to be studied, or it may become quite difficult to decide which genes are actually important to the pathophysiology of infection. When studying in vivo gene expression, it seems reasonable to divide the pathophysiological state of *C. neoformans* into at least five stages, which may require differential gene expressions and proteins for each stage of infection. All these genes and their proteins are essential for survival of *C. neoformans* and its ability to produce disease. The five stages during infection are (i) initiation, (ii) dormancy, (iii) reactivation, (iv) dissemination, and (v) proliferation. Although these stages, as noted in chapter 6, are arbitrary and there is no proof that they are correlated with gene expression, future investigators are advised to consider linking pathophysiological understanding with molecular gene expression studies to reduce the complexity of their work. It will also be important to observe gene expression as infection progresses, even at the same site of infection. For instance, the pattern of gene expression in *C. neoformans* in the rabbit model of cryptococcal meningitis will dramatically vary by differential-display RT-PCR from day 2 to day 7 of infection (120).

GENE STRUCTURES AND REGULATION IN *C. NEOFORMANS*

Investigations into *C. neoformans* gene structures and the understanding of gene regulation in this yeast are in their infancy. However, certain features have been identified. First, it has been noted that there appears to be allelic variation in sequences between various *C. neoformans* strains and serotypes. The *URA5* gene shows approximately 6% base pair divergence between strains (8), and the *TOPOI* gene has 3.6% differences in nucleotide sequences between serotypes A and D. These allelic variations generally occur in the third codon position and within introns and do not affect the primary amino acid sequence. This sequence divergence is expected to become a common theme as more alleles are carefully examined and used for evolutionary and epidemiological studies. Further work on genome structural divergence is needed to identify various gene characteristics between the two varieties and their implications for pathogenesis. For example, it has been noted that *C. neoformans* var. *gattii* shows more polymorphism in gene locations on chromosomes than does *C. neoformans* var. *neoformans*. On the other hand, strains from var. *gattii* in Australia appear to be more clonal in origin than those in var. *neoformans* from the United States (118).

As further studies are published, the actual gene arrangements of *C. neoformans* will be likely to have a series of general structural features but have some unique variation compared to other microorganisms. First, it appears that *C. neoformans* uses typical intronic sequence structures, but many introns per gene may be a particularly common features of *C. neoformans* genes. The putative intron splice sites closely follow consensus 5' and 3' sequences of GTNNGY and YAG, respectively, with the exception that the fourth nucleotide of the 5' sequence could be either a purine or a pyrimidine. Several sequenced genes have more than six introns, and even the highly conserved actin gene in *C. neoformans* has five introns (17), which is the most of any fungus reported so far. *C. neoformans* introns can be small, some being less than 55 bp (100). The 5' end of *C. neoformans* genes, which includes the promoter or upstream regulatory elements, may or may not have consensus transcription regulatory sequences such as TATA and/or CAAT boxes to identify promoter motifs and these upstream activating sequences. The consensus signals for polyadenylation in *C. neoformans* have not yet been defined. Regarding the function-structure organization of genes, *C. neoformans* has been shown to differ from other microorganisms in its composition. *ADE2* appears to be bifunctional and can complement both *purE* and *purK* in bacteria (30). On the other hand, the *TRP-1* gene appears to encode only phosphoribosylanthranilate isomerase activity; this arrangement is different from that of some filamentous fungi in which *TRP-1* is trifunctional (75). There are also enough submitted *C. neoformans* gene sequences in databases to make a rudimentary table of codon bias use in *C. neoformans* that may be helpful in strategies to make degenerate primers.

Studies on gene regulation in *C. neoformans* have just begun. However, several heterologous recorder gene constructs have been shown to be functional in *C. neoformans*. For example, bacterial enzymes such as β-galactosidase (101), β-glucuronidase (115), and hygromycin B (18) have been expressed as fusion proteins in *C. neofor-*

mans. A study using a *C. neoformans* actin–β-galactosidase fusion has emphasized that even the constitutively produced actin gene used by many investigators as a control gene for gene expression can be regulated by temperature and growth phase in *C. neoformans* (101), and these findings may need to be considered in comparative gene regulation studies. Another molecular advance for *C. neoformans* is the identification of a classically regulated promoter, *GAL7*, by galactose/glucose conditions (115). This system allows for genes that need to be differentially expressed in vitro, and it has already been used to drive phenotypical expression in vitro for phero-mone (62) and capsule (12). These reports suggest a tightly regulated galactose promoter that does not produce transcripts under glucose conditions and has at least 500-fold greater expression in galactose than in glucose. However, in our experience, *GAL7* can be used as a regulated promoter under these conditions, but it can still drive low expression of certain gene products under repressed conditions, and this gene expression will be detected by RT-PCR (20a).

The most progress in understanding transcriptional regulations in *C. neofor-mans* has been made through the studies on *CnLAC1* by Williamson and Zhang (120, 121). This gene, which has its own natural recorder system in which it produces the easily identifiable and measurable melanin pigment, may be ideal for the study of *C. neoformans* gene regulation. Several features of gene regulation have already been identified for this gene (120). The upstream area has both TATA and CAAT boxes at position -539 and -503, respectively, and also a canonical en-hancer Sp1-like site at position -1727. This gene's expression has also been shown to be regulated by glucose concentration in the medium. Further studies on iden-tification of nuclear proteins that bind to regulatory elements of the promoter are in progress. However, protocols for sequential lysis of cytoplasmic and nuclear membranes were found to require additional protease inhibitors and a lower concentration of glycerol for stability of *C. neoformans* nuclear proteins. EGTA was also required with EDTA to inhibit the high amount of nuclease activity found in nuclear extracts of *C. neoformans*.

Further studies on regulatory genes and circuits will be important. At least one *C. neoformans* gene, *Mtl-1*, has actually crossed fungal species in its ability to regulate gene activities. *Mtl-1* can regulate expression of mannitol dehydrogenase in an *S. cerevisiae* strain (76), but its function in *C. neoformans* remains undefined. Several genes related to the mating pathway and signal transduction have identified viru-lence circuits through environmental cues. CnGPA1, a G protein, is regulated by various environmental factors, including glucose, nitrogen, and iron starvation. Although the receptors (genes) that regulate CnGPA1 are not known, CnGPA1 regulates cyclic AMP levels through adenylate cyclase, with resulting control of capsule production, melanin synthesis, and mating (1). A second signal transduction gene for *C. neoformans*, *CNA1*, which encodes the catalytic portion of the calcineurin protein, is also regulated uniquely by a series of important environmental factors. This gene has been found to be essential for *C. neoformans* growth at host factors such as temperatures of 37 to 39°C, pCO_2 of 5%, and pH of 7.3 to 7.4 (68). Regulation of these genes, including their central regulatory genes and circuits, may be complex but it undoubtedly exists. With the use of two-hybrid technology and gene disrup-tion protocols, genes involved in these virulence pathways can now be identified and the pathways further elucidated. Other regulons with central regulatory genes should also be pursued in *C. neoformans*. These systems may be complex, but they have been shown to exist in nonpathogenic fungi such as *S. cerevisiae* and *A. nidulans* as well as in plant pathogens such as *Ustilago maydis* and *Cyphronectria parasitica*. The basic genetic system and the foundation in molecular biology for *C. neoformans* will support its use as a model for human pathogenic fungi.

TRANSFORMATION SYSTEMS FOR *C. NEOFORMANS*

The development of two transformation systems for *C. neoformans* has been a significant step in the molecular study of its pathogenesis (25, 102). The ability to introduce, disrupt, and/or retrieve genes from an organism is an essential cornerstone to the study of its molecular pathobiology. In *C. neoformans* there are now several selectable markers and two methods for introduction of DNA. The first system, developed by Edman and Kwon-Chung (25), used complementation of *URA5* auxotrophs in the *C. neoformans* serotype D background. The *URA5* gene was introduced through electroporation delivery of DNA. The *URA5* auxotrophs were initially selected by spontaneous 5-fluoroorotic acid resistance, and it was found that uracil auxotrophs in this serotype D strain were generally *URA5* mutants, in comparison to a var. *gattii* strain in which half of the uracil auxotrophs were *URA3* (47). This transformation system was initially a low-efficiency method with both stable ectopic integrative events and many unstable transformants in which the vector DNA appeared to be maintained extrachromosomally. Since it was discovered that most transformation events were actually extrachromosomal, a further analysis of these extrachromosomal events was made by recovery of the introduced plasmids. These unstable transformants showed the presence of an autonomously replicating plasmid that was similar in size to or smaller than the transforming plasmid (104). It was maintained in a linear form and had acquired sequences with homology to a sequence on all chromosomes. It was then shown with exonuclease digestion that the extrachromosomal fragments were actually linear. Sequencing of these linear plasmids showed that the ends contained repetitive sequences consistent with telomeres (23). These findings demonstrated that *C. neoformans*, like another pathogenic fungus, *Histoplasma capsulatum*, possesses very active telomerase activity. This activity appears to act on foreign DNA when it is introduced into the yeast by placing telomere sequences on the ends of the plasmid. These findings allowed Edman (23) to use the telomeres on vectors to develop a high-efficiency transformation system. A pCnTel plasmid was constructed and shown to possess transformation frequencies of 200 colonies per μg of supercoiled DNA and 90,000 colonies per μg of linearized DNA. With the introduction of certain restriction enzyme sites, selectable markers, and cloning of genomic DNA, this plasmid has been used to screen *C. neoformans* genomic libraries in mutants and then to recover the complementing plasmid for gene identification (10, 11). However, present vectors are generally unstable and do not stably replicate extrachromosomally in *C. neoformans*. The identification of autonomous replication sequences for high-level expression of genes will be helpful. Also, more-stable plasmids are being created with the use of sequences from a minichromosome (108). These sequences act more like centromeres and may give more stability to extrachromosomal plasmids. Stable extrachromosomal vectors in *C. neoformans* will be extremely useful molecular tools. Unlike *S. cerevisiae*, no known natural plasmid has been found in *C. neoformans* strains analyzed so far.

One serious drawback to this transformation system has been the frequency of homologous recombination events, reported to be between 0.1 and 0.01% (10, 11, 119). It is not known whether this low frequency represents factors in the strain, DNA delivery, or alleles used for disruption. However, with known phenotypes, a disruption strategy can be performed. To further help in complete gene displacement, Chang and coworkers (10, 11) have elegantly used a positive-negative selection that forces double crossover events to be detected at the site of disruption.

A second transformation system has also been developed (102). It uses complementation of the *ADE2* locus through DNA delivery by a biolistic apparatus. This system uses a serotype A strain (H99) that has a mutation in the *ADE2* gene, rather than a serotype D strain. The mutants of this serotype A strain, M001 and M049, were made with UV and gamma irradiation, respectively, and display a typical pink color when grown on fully supplemented medium. This transformation protocol (see chapter 12) produces high-level transformation frequencies that can reach thousands of colonies per plate. Both extrachromosomal and integrative transformation events occur with this method, and there is little difference in the frequency or location of these events after the introduction of linear or circular plasmids. Between 15 and 40% of transformants are stable, and all stable transformants have either single or multiple integrative events. Approximately one fourth of the integrative transformants produce homologous recombination at the *ADE2* locus, but the selectable *ADE2* gene was transformed from a serotype D strain into a serotype A recipient in an attempt to reduce the frequency of this event (102). This system has now been used to produce site-directed gene disruptants or replacements of the *NMT1, CNA1, URE1, TOPOI,* and *GPA1* genes in *C. neoformans* (1, 50, 68, 108). In these loci, the frequency of gene replacement ranges from 3 and 10%. However, the ability to observe homologous recombination and complete gene replacements may be dependent on the allele. For instance, another genetic allele, *URA5*, appears to be even less frequently replaced by this method. Frequencies of homologous integrations in fungi have been generally reported to be 100% in *S. cerevisiae*, 80% in *A. nidulans*, and 1 to 5% in *N. crassa* and *C. cinereus*. It is likely that for most *C. neoformans* strains, homologous recombination will generally fall in the lower range of frequencies. However, it should be emphasized that, although *C. neoformans* appears to lack an efficient gene replacement mechanism, it does occur and can be exploited for production of site-directed mutants. In our experience, there are transformants in which the phenotype has been altered without a corresponding change in genotype by PCR or Southern analysis. Thus, there may also be both ectopic and homologous integration with actual excision of the DNA. It should be emphasized that introduced naked DNA possesses mutagenic properties and serves to stimulate abnormal recombination. It is likely that *C. neoformans* with its integrative and nonintegrative events, including frequent multiple and tandem insertions, will occasionally produce ectopic events that can unexpectedly change the phenotype of the yeast. For example, transformation procedures with *C. neoformans* can change the position of chromosomes in transformants or actually develop new minichromosomes (108). Also, the chromosomal site and/or the number of copies integrated may influence the fate of these sequences through inactivation by methylation. Methylation has been shown in other fungi to cause duplicated gene inactivation, and it might also be active in *C. neoformans* transformation protocols.

Another transformation system, developed by Cox et al. (18), uses a fusion gene of the *C. neoformans* actin gene and the bacterial hygromycin B resistance gene. This is an efficient dominant selection transformation system (hundreds per plate) that allows movement of DNA into many wild-type strains of *C. neoformans*, including isolates from both AIDS and non-AIDS patients. Although the present construct is very inefficient with *C. neoformans* var. *gattii,* it has been used to obtain low transformation frequencies in this variety. It is likely that var. *gattii* will require the use of specific promoters from a var. *gattii* strain to improve transformation frequencies. This selectable marker, which allows the selection of transformants on plates containing hygromycin B, has been used to disrupt the urease gene directly in a wild-type strain (108). Similar to the complementation of auxotrophic strains, this system produces a substantial number of unstable and stable ectopically integrated transformants. However, this feature of random integrative events can also be used as a positive factor. For instance, it is now possible to develop libraries of signature-tagged mutants in *C. neoformans* (17a) modeled after the strategy of Hensel and coworkers in salmonella (37). In these ongoing studies with libraries of signature-tagged mutants, most transformants appear to represent random events and sometimes multiple events, and there do appear to be some recombinational "hot spots" in the genome by Southern analysis. A series of phenotypic mutants have been identified, and the identification of genes is in progress. This dominant vector system also allows investigators to insert a second gene back into previously transformed cells. The replacement of the gene for CnGPA1 into the site-directed mutant of the allele was accomplished with this strategy (1). Another possible selectable marker for *C. neoformans* is phleomycin resistance, and a similar dominant system can be developed that will further help in the molecular domestication of this pathogenic yeast.

REFERENCES

1. **Alspaugh, J. A., J. R. Perfect, and J. Heitman.** 1997. *Cryptococcus neoformans* mating and virulence are regulated by the G-protein alpha subunit GPA1 and cAMP. *Genes Dev.* **11**:3206–3217.

2. **Aulakh, H. J., S. E. Straus, and K. J. Kwon-Chung.** 1981. Genetic relatedness of *Filobasidiella neoformans* (*Cryptococcus neoformans*) and *Filobasidiella bacillospora* (*Cryptococcus bacillosporus*) as determined by deoxyribonucleic acid base composition and sequence homology studies. *Int. J. Syst. Bacteriol.* **31**:97–101.

3. **Boekhout, T., A. Van Belkum, A. C. Leenders, H. A. Verbrugh, P. Mukamurangwa, D. Swinne, and W. A. Scheffers.** 1997. Molecular typing of *Cryptococcus neoformans*: taxonomic and epidemiological aspects. *Int. J. Syst. Bacteriol.* **47**:432–442.

4. **Boucher, F., R. Morosoli, and S. Durand.** 1988. Complete nucleotide sequence of the xylanase gene from the yeast *Cryptococcus albidus. Nucleic Acids Res.* **16**:9874.

5. **Brandt, M. E., L. C. Hutwagner, L. A. Klug, W. S. Baughman, D. Rimland, E. D. Graviss, R. J. Hamill, C. Thomas, P. G. Pappas, A. L. Reingold, R. W. Pinner, and Cryptococcal Disease Active Surveillance Group.** 1996. Molecular subtypes distribution of *Cryptococcus neoformans* in four areas of the United States. *J. Clin. Microbiol.* **34**:912–917.

6. **Brandt, M. E., L. C. Hutwagner, R. J. Kuykendall, R. W. Pinner, and The Cryptococcal Disease Active Surveillance Group.** 1995. Comparison of multilocus enzyme electrophoresis and random amplified polymorphic DNA analysis for molecular subtyping of *Cryptococcus neoformans. J. Clin. Microbiol.* **33**:1890–1895.

7. **Brandt, M. E., M. A. Pfaller, R. A. Hajjeh, E. A. Graviss, J. Rees, E. D. Spitzer, R. W. Pinner, L. W. Mayer, and Cryptococcal Disease Active Surveillance Group.** 1996. Molecular subtypes and antifungal susceptibilities of serial *Cryptococcus neoformans* isolates in human immunodeficiency virus-associated cryptococcosis. *J. Infect. Dis.* **174:**812–820.

8. **Casadevall, A., L. F. Freundlich, L. Marsh, and M. Scharff.** 1992. Extensive allelic variation in *Cryptococcus neoformans. J. Clin. Microbiol.* **30:**1080–1084.

9. **Chang, Y. C., R. Cherniak, T. R. Kozel, D. L. Granger, L. C. Morris, L. C. Weinhold, and K. J. Kwon-Chung.** 1997. Structure and biological activities of acapsular *Cryptococcus neoformans* 602 complemented with CAP64 gene. *Infect. Immun.* **65:**1584–1592.

10. **Chang, Y. C., and K. J. Kwon-Chung.** 1994. Complementation of a capsule-deficiency mutation of *Cryptococcus neoformans* restores its virulence. *Mol. Cell. Biol.* **14:**4912–4919.

11. **Chang, Y. C., L. A. Penoyer, and K. J. Kwon-Chung.** 1996. The second capsule gene of *Cryptococcus neoformans,* Cap64, is essential for virulence. *Infect. Immun.* **64:**1977–1983.

12. **Chang, Y. C., B. L. Wickes, and K. J. Kwon-Chung.** 1995. Further analysis of the Cap59 locus of *Cryptococcus neoformans*: structure defined by forced expression and description of a new ribosomal protein-encoding gene. *Gene* **167:**179–183.

13. **Chen, F., B. P. Currie, L. C. Chen, S. G. Spitzer, E. D. Spitzer, and A. Casadevall.** 1995. Genetic relatedness of *Cryptococcus neoformans* clinical isolates grouped with the repetitive DNA probe CRE–1. *J. Clin. Microbiol.* 2818–2822.

14. **Chen, S. C. A., A.G. Brownlee, T. C. Sorrell, P. Ruma, D. H. Ellis, T. Pfeiffer, B. R. Speed, and G. Nimmo.** 1996. Identification by random amplification of polymorphic DNA of a common molecular type of *Cryptococcus neoformans* in patients with AIDS or other immunosuppressive conditions. *J. Infect. Dis.* **173:**754–758.

15. **Cormack, B. P., G. Bertram, M. Egerton, N. A. R. Gow, S. Falkow, and A. J. P. Brown.** 1992. Yeast-enhanced green fluorescent protein (yGFP): a reporter of gene expression in *Candida albicans. Microbiology* **143:**303–311.

16. **Cox, G. M., G. T. Cole, and J. R. Perfect.** 1996. Identification and disruption of the *Cryptococcus neoformans* urease gene, abstr. 1.6:150. 3rd International Conference on Cryptococcus and Cryptococcosis, Paris.

17. **Cox, G. M., C. Dykstra, T. H. Rude, and J. R. Perfect.** 1995. *Cryptococcus neoformans* actin gene: characterization and its use as a phylogenetic marker. *J. Med. Vet. Mycol.* **33:**261–266.

17a. **Cox, G. M., and K. A. Haynes.** Unpublished data.

18. **Cox, G. M., D. L. Toffaletti, and J. R. Perfect.** 1996. Dominant selection system for use in Cryptococcus neoformans. *J. Med. Vet. Mycol.* **34:**385–391.

19. **Crampin, A. C., R. C. Matthews, D. Hall, and E. G. V. Evans.** 1993. PCR fingerprinting *Cryptococcus neoformans* by random amplification of polymorphic DNA. *J. Med. Vet. Mycol.* **31:**463–465.

20. **Currie, B. P., L. F. Freundlich, and A. Casadevall.** 1994. Restriction fragment length polymorphism analysis of *Cryptococcus neoformans* isolates from environmental (pigeon excreta) and clinical sources in New York City. *J. Clin. Microbiol.* **32:**1188–1192.

20a. **Del Poeta, M., and J. R. Perfect.** Unpublished data.

21. **Del Poeta, M., T. Rude, and J. R. Perfect.** 1996. Identification of the topoisomerase I gene of *C. neoformans,* abstr. F33. 36th International Conference of Antimicrobial Agents Chemotherapy (ICAAC), New Orleans. American Society for Microbiology, Washington, D.C.

22. **Dromer, F., A. Varma, O. Ronin, S. Mathoulin, and B. Dupont.** 1994. Molecular typing of *Cryptococcus neoformans* serotype D clinical isolates. *J. Clin. Microbiol.* **32:**2364–2371.

23. **Edman, J. C.** 1992. Isolation of telomerelike sequences from *Cryptococcus neoformans* and their use in high-frequency transformation. *Mol. Cell. Biol.* **12:**2777–2783.

24. **Edman, J. C.** 1996. Paradoxes of mating in *Cryptococcus neoformans,* abstr. 1.1:7. 3rd International Conference on Cryptococcus and Cryptococcosis, Paris.

25. **Edman, J. C., and K. J. Kwon-Chung.** 1990. Isolation of the *URA5* gene from *Cryptococcus neoformans* var. *neoformans* and its use as a selective marker for transformation. *Mol. Cell. Biol.* **10:**4538–4544.

26. **Erke, K. H., and J. D. Schneidau.** 1973. Relationship of some *Cryptococcus neoformans* hypha-forming strains to standard strains and to other species of yeasts as determined by deoxyribonucleic acid base ratios and homologies. *Infect. Immun.* **7:**941–948.

27. **Fan, M., L. C. Chen, M. A. Ragan, R. R. Gutell, J. R. Warner, B. P. Currie, and A. Casadevall.** 1995. The 5S rRNA and rRNA intergenic spacer of the two varieties of *Cryptococcus neoformans. J. Med. Vet. Mycol.* **33:**215–221.

28. **Fan, M., B. P. Currie, R. R. Gutell, M. A. Ragen, and A. Casadevall.** 1994. The 16s-like, 5.8s and 23s-like rRNA of the two varieties of *Cryptococcus neoformans*: sequence secondary structure, phylogenetic analysis, and restriction fragment polymorphisms. *J. Med. Vet. Mycol.* **32:**163–180.

29. **Fell, J. W.** 1995. rDNA targeted oligonucleotide primers for the identification of pathogenic yeasts in a polymerase chain reaction. *J. Industr. Microbiol.* **14:**475–477.

30. **Firestine, S. M., S. Misialek, D. Toffaletti, T. J. Klem, J. R. Perfect, and V. J. Davisson.** 1998. Biochemical role of the *Cryptococcus neoformans* ADE2 protein in fungal de novo purine biosynthesis. *Arch. Biochem. Biophys.* **351:**123–134.

31. **Franzot, S. P., J. S. Hamdan, B. P. Currie, and A. Casadevall.** 1997. Molecular epidemiology of *Cryptococcus neoformans* in Brazil and the United States: evidence for both local genetic differences and a global clonal population structure. *J. Clin. Microbiol.* **35:**2243–2251.

32. **Franzot, S. P., J. Mukherjee, R. Cherniak, L. Chen, J. S. Hamdan, and A. Casadevall.** 1998. Microevolution of a standard strain of *Cryptococcus neoformans* resulting in differences in virulence and other phenotypes. *Infect. Immun.* **66:**89–97.

33. **Fries, B. C., F. Chen, B. P. Currie, and A. Casadevall.** 1996. Karyotype instability in *Cryptococcus neoformans* infection. *J. Clin. Microbiol.* **34:**1531–1534.

34. **Gueho, E., L. Improvisi, R. Christen, and G. S. de Hoog.** 1993. Phylogenetic relationships of *Cryptococcus neoformans* and some related basidiomycetous yeasts determined from partial large subunit rRNA sequences. *Antonie van Leeuwenhoek* **63:**175–189.

35. **Haynes, K. A., D. J. Sullivan, D. C. Coleman, J. C. K. Clarke, R. Emilianus, J. Atkinson, and K. J. Cann.** 1995. Involvement of multiple *Cryptococcus neoformans* strains in a single episode of cryptococcosis and reinfection with novel strains in recurrent infection demonstrated by random amplification of polymorphic DNA and DNA fingerprinting. *J. Clin. Microbiol.* **33:**99–102.

36. **Hekman, J., and P. R. Williamson.** 1996. Structural comparison of EF-3 between *Cryptococcus neoformans* and ascomycetes, abstr. 1.9. 3rd International Conference on Cryptococcus and Cryptococcosis, Paris.

37. **Hensel, M., J. E. Shea, C. Gleeson, M. D. Jones, E. Dalton, and D. W. Holden.** 1995. Simultaneous identification of bacterial virulence genes by negative selection. *Science* **269:**400–403.

38. **Hopfer, R. L., P. Walden, S. Setterquist, and W. E. Highsmith.** 1993. Detection and differentiation of fungi in clinical specimens using polymerase chain reaction (PCR) amplification and restriction enzyme analysis. *J. Med. Vet. Mycol.* **31:**65–75.

39. **Huffnagle, K. E., and R. M. Gander.** 1993. Evaluation of Gen-probe's *Histoplasma capsulatum* and *Cryptococcus neoformans* ACCU probes. *J. Clin. Microbiol.* **31:**419–421.

40. **Jacobson, E. S., and D. J. Ayers.** 1979. Auxotrophic mutants of *Cryptococcus neoformans. J. Bacteriol.* **139:**318–319.

41. **Kline, M. T., and E. S. Jacobson.** 1979. Auxotrophic markers and the cytogenetics of *Filobasidiella neoformans. Mycopathologia* **139:**318–319.

42. **Kohno, S., S. Varma, K. J. Kwon-Chung, and K. Hara.** 1994. Epidemiology studies of clinical isolates of *Cryptococcus neoformans* of Japan by restriction fragment length polymorphism. *J. Jpn. Assoc. Infect. Dis.* **68:**1512–1517.

43. **Kwon-Chung, K. J.** 1976. Morphogenesis of *Filobasidiella neoformans*, the sexual state of *Cryptococcus neoformans. Mycologia* **68:**821–833.

44. **Kwon-Chung, K. J.** 1976. A new species of *Filobasidiella,* the sexual state of *Cryptococcus neoformans* B and C serotypes. *Mycologia* **68:**942–946.

45. **Kwon-Chung, K. J., J. C. Edman, and B. L. Wickes.** 1992. Genetic association of mating types and virulence in *Cryptococcus neoformans. Infect. Immun.* **60:**602–605.

46. **Kwon-Chung, K. J., T. Pfeiffer, Y. C. Chang, B. L. Wickes, D. Mitchell, and J. J. Stern.** 1994. Molecular biology of *Cryptococcus neoformans* and therapy of cryptococcosis. *J. Med. Vet. Mycol.* **32:**407–415.

47. **Kwon-Chung, K. J., A. Varma, J. C. Edman, and J. E. Bennett.** 1992. Selection of Ura5 and Ura3 mutants from the two varieties of *Cryptococcus neoformans* on 5-fluoroorotic acid medium. *J. Med. Vet. Mycol.* **30:**61–69.

48. **Kwon-Chung, K. J., B. L. Wickes, L. Stockman, G. D. Roberts, D. Ellis, and D. H. Howard.** 1992. Virulence, serotype, and molecular characteristics of environmental strains of *Cryptococcus neoformans* var. *gattii. Infect. Immun.* **60:**1869–1874.

49. **Livi, L. L., U. Edman, G. P. Schneider, P. J. Greene, and D. V. Santi.** 1994. Cloning, expression and characterization of thymidylate synthase from *Cryptococcus neoformans. Gene* **150:**221–226.

50. **Lodge, J. K., R. L. Johnson, R. A. Weinberg, and J. I. Gordon.** 1994. Comparison of myristoyl-coA: protein N-myristoyl transferases from three pathogenic fungi: *Cryptococcus neoformans, Histoplasma capsulatum,* and *Candida albicans. J. Biol. Chem.* **269:**2996–3010.

51. **Lo Passo, C., I. Pernice, M. Gallo, C. Barbara, F. T. Luck, G. Criseo, and A. Pernice.** 1997. Genetic relatedness and diversity of *Cryptococcus neoformans* strains in the Maltese Islands. *J. Clin. Microbiol.* **35:**751–755.

52. **Mahan, M. J., J. M. Slauch, and J. J. Mekalanos.** 1993. Selection of bacterial virulence genes that are specifically induced in host tissues. *Science* **259:**686–688.

53. **Maiwald, M., R. Kappe, and H. G. Sontag.** 1994. Rapid presumptive identification of medically relevant yeasts to the species level by polymerase chain reaction and restrictive enzyme analysis. *J. Med. Vet. Mycol.* **32:**115–122.

54. **Martinez-Marino, B., and B. Bolanos.** 1995. Characterization of a cytoplasmic protein that inhibits phagocytosis, abstr. F9. Abstr. 95th Annu. Meet. Am. Soc. Microbiol. 1995. American Society for Microbiology, Washington, D.C.

55. **Meyer, W., E. Lieckfeldt, K. Kuhls, E. Z. Freedman, T. Borner, and T. G. Mitchell.** 1993. DNA- and PCR-fingerprinting in fungi, p. 311–320. *In* S. D. J. Pena, R. Chakraborty, J. T. Epplen, and A. J. Jeffreys (ed.), *DNA Fingerprinting: State of the Science.* Birkhauser Verlag, Basel.

56. **Meyer, W., and T. G. Mitchell.** 1994. PCR fingerprinting to distinguish species and strains of yeasts, p. 293–302. *In* B. Maresca and G. S. Kobayashi (ed.), *Molecular Biology of Pathogenic Fungi: A Laboratory Manual.* Telos Press, New York.

57. **Meyer, W., and T. G. Mitchell.** 1995. Polymerase chain reaction fingerprinting in fungi using single primers specific to minisatellites and simple repetitive DNA sequences: strain variation in *Cryptococcus neoformans. Electrophoresis* **16:**1648–1656.

58. **Meyer, W., T. G. Mitchell, E. Z. Freedman, and R. Vilgalys.** 1993. Hybridization probes for conventional DNA fingerprinting used as single primers in the polymerase chain reaction to distinguish strains of *Cryptococcus neoformans. J. Clin. Microbiol.* **31:**2274–2280.

59. **Mitchell, T. G., E. Z. Freedman, W. Meyer, T. J. White, and J. W. Taylor.** 1993. PCR identification of *Cryptococcus neoformans*, p. 431–436. *In* D. H. Persing, T. F. Smith, F. C. Tenover, and T. J. White (ed.), *Diagnostic Molecular Microbiology. Principles and Applications.* American Society for Microbiology, Washington, D.C.

60. **Mitchell, T. G., E. Z. Freedman, T. J. White, and J. W. Taylor.** 1994. Unique oligonucleotide primers in PCR for identification of *Cryptococcus neoformans. J. Clin. Microbiol.* **32:**253–255.

61. **Mitchell, T. G., R. L. Sandin, B. H. Bowman, W. Meyer, and W. G. Merz.** 1994. Molecular mycology: DNA probes and applications of PCR technology. *J. Med. Vet. Mycol.* **32:**351–366.

62. **Moore, T. D., and J. C. Edman.** 1993. The alpha-mating type locus of *Cryptococcus neoformans* contains a peptide pheromone gene. *Mol. Cell. Biol.* **13:**1962–1970.

63. **Moreau, A., S. Durand, and R. Morosoli.** 1992. Secretion of a *Cryptococcus albidus* xylanase in *Saccharomyces cerevisiae. Gene* **116:**109–113.

64. **Morosoli, R.** 1985. Molecular expression of xylanase gene in *Cryptococcus albidus. Biochim. Biophys. Acta* **826:**202–207.

65. **Morosoli, R., S. Durand, and A. Moreau.** 1992. Cloning and expression in *Escherichia coli* of a xylanase-encoding gene from the yeast *Cryptococcus albidus. Gene* **117:**145–150.

66. **Nakase, T., and K. Komagata.** 1971. Significance of DNA base composition in the classification of yeast genera *Cryptococcus* and *Rhodotorula. J. Gen. Appl. Microbiol.* **17:**121–130.

67. **Odom, A., M. Del Poeta, J. Perfect, and J. Heitman.** 1997. The immunosuppressant FK506 and its non-immunosuppressive analog L-685,818 are toxic to *Cryptococcus neoformans* by inhibition of a common target protein. *Antimicrob. Agents Chemother.* **41:**156–161.

68. **Odom, A., S. Muir, E. Lim, D. L. Toffaletti, J. R. Perfect, and J. Heitman.** 1997. Calcineurin is required for virulence of *Cryptococcus neoformans. EMBO J.* **16:**2576–2589.

69. **Parker, A. R., T. D. E. Moore, J. C. Edman, J. M. Schwab, and V. J. Davisson.** 1995. Cloning, sequence analysis and expression of the gene encoding imidazole glycerol phosphate dehydratase in *Cryptococcus neoformans. Gene* **145:**135–138.

70. **Pazin, J. G., T. H. Rude, C. C. Dykstra, and J. R. Perfect.** 1995. Cloning of the topoisomerase II gene in *Cryptococcus neoformans,* abstr. 345. 33rd Infectious Disease Society Meetings, San Francisco.

71. **Perfect, J. R.** 1996. Fungal virulence genes as targets for antifungal chemotherapy. *Antimicrob. Agents Chemother.* **40:**1577–1583.

71a. **Perfect, J. R.** Unpublished data.

72. **Perfect, J. R., N. Ketabchi, G. M. Cox, C. I. Ingram, and C. Beiser.** 1993. Karyotyping of *Cryptococcus neoformans* as an epidemiological tool. *J. Clin. Microbiol.* **31:**3305–3309.

73. **Perfect, J. R., B. B. Magee, and P. T. Magee.** 1989. Separation of chromosomes of *Cryptococcus neoformans* by pulsed field gel electrophoresis. *Infect. Immun.* **57:**2624–2627.

74. **Perfect, J. R., and T. H. Rude.** 1993. Identifying *Cryptococcus neoformans* by differential expression at the site of infection, abstr. L3, p. 24. 2nd International Conference on Cryptococcus and Cryptococcosis, Milan.

75. **Perfect, J. R., T. H. Rude, L. M. Penning, and S. A. Johnston.** 1992. Cloning of *Cryptococcus neoformans* TRP1 gene by complementation in *Saccharomyces cerevisiae. Gene* **122:**213–217.

76. **Perfect, J. R., T. H. Rude, B. Wong, T. Flynn, V. Chaturvedi, and W. Niehaus.** 1996. Identification of a *Cryptococcus neoformans* gene that directs expression of the cryptic *Saccharomyces cerevisiae* mannitol dehydrogenase gene. *J. Bacteriol.* **178:**5257–5262.

77. **Perfect, J. R., D. L. Toffaletti, and T. H. Rude.** 1993. The gene encoding for phosphori-

bosylaminoimidazole carboxylase (ADE2) is essential for growth of *Cryptococcus neoformans* in cerebrospinal fluid. *Infect. Immun.* **61:**4446–4451.

78. **Perfect, J. R., B. Wong, Y. Chang, K. J. Kwon-Chung, and P. R. Williamson.** 1998. *Cryptococcus neoformans*: virulence and host defenses. *Med. Mycol.* **36,** in press.

79. **Petter, R., A. Varma, T. Boekhout, S. Salas, B. Davis, and K. J. Kwon-Chung.** 1996. Molecular divergence of the virulence factors of *Filobasidiella neoformans* found in other heterobasidiomycetous species, abstr. 2.2:36–38. 3rd International Conference on Cryptococcus and Cryptococcosis, Paris.

79a. **Podwall, D., E. Yador, S. Miller, J. Pena, S. P. Franzot, J. Lipetz, A. Casadevall, and J. J. Steinberg.** 1998. Interstrain variation in the deoxynucleotide composition of *Cryptococcus neoformans*: nucleotide composition of *Cryptococcus neoformans*. *Med. Mycol.* **36:**1–5.

80. **Polacheck, I., G. Lebens, and J. B. Hicks.** 1992. Development of DNA probes for early diagnosis and epidemiological study of cryptococcosis in AIDS patients. *J. Clin. Microbiol.* **30:**925–930.

81. **Polacheck, I., and G. A. Lebens.** 1989. Electrophoretic karyotype of the pathogenic yeast *Cryptococcus neoformans*. *J. Gen. Microbiol.* **135:**67–71.

82. **Restrepo, B. I., and A. G. Barbour.** 1989. Cloning of the 18S and 25S ribosomal DNA from the pathogenic fungus *Cryptococcus neoformans*. *J. Bacteriol.* **171:**5596–5600.

83. **Ruma, P., S. C. A. Chen, T. C. Sorrell, and A. G. Brownlee.** 1996. Characterization of *Cryptococcus neoformans* by random DNA amplification. *Lett. Appl. Microbiol.* **23:**312–316.

84. **Salas, S. D., J. E. Bennett, K. J. Kwon-Chung, J. R. Perfect, and P. R. Williamson.** 1996. Effect of the laccase gene, CNLAC1, on virulence of *Cryptococcus neoformans*. *J. Exp. Med.* **184:**377–386.

85. **Schena, M., D. Shalon, R. W. Davis, and P. O. Brown.** 1995. Quantitative monitoring of gene expression patterns with a complementary DNA microassay. *Science* **270:**467–470.

86. **Sirawaraporn, W., C. Ming, D. V. Santi, and J. C. Edman.** 1993. Cloning, expression, and characterization of *Cryptococcus neoformans* dihydrofolate reductase. *J. Biol. Chem.* **268:**8888–8892.

87. **Sorrell, T. C., S. C. A. Chen, P. Ruma, N. Meyer, T. J. Pfeiffer, D. H. Ellis, and A. G. Brownlee.** 1996. Concordance of clinical and environmental isolates of *Cryptococcus neoformans* var. *gattii* by random amplification of polymorphic DNA analysis and PCR fingerprinting. *J. Clin. Microbiol.* **34:**1253–1260.

88. **Spitzer, E. D., and S. G. Spitzer.** 1992. Use of a dispersed repetitive DNA element to distinquish clinical isolates of *Cryptococcus neoformans*. *J. Clin. Microbiol.* **30:**1094–1097.

89. **Spitzer, E. D., and S. G. Spitzer.** 1995. Structure of the ubiquitin-encoding genes of *Cryptococcus neoformans*. *Gene* **161:**113–117.

90. **Spitzer, E. D., S. G. Spitzer, L. F. Freundlich, and A. Casadevall.** 1993. Persistence of initial infection in recurrent *Cryptococcus neoformans* meningitis. *Lancet* **341:**595–596.

91. **Spitzer, S. G., and E. D. Spitzer.** 1994. Characterization of the CNRE-1 family of repetitive DNA elements in *Cryptococcus neoformans*. *Gene* **144:**103–106.

92. **Spitzer, S. G., and E. D. Spitzer.** 1997. Isolation of *Cryptococcus neoformans* chromosome-specific probes using expressed sequence tags. *J. Med. Vet. Mycol.* **315:**257–262.

93. **Still, C. N., and E. S. Jacobson.** 1983. Recombinational mapping of capsule mutations in *Cryptococcus neoformans*. *J. Bacteriol.* **156:**460–462.

94. **Stockman, L., K. A. Clark, J. M. Hunt, and G. D. Roberts.** 1993. Evaluation of commercially available acridinium ester-labeled chemiluminescent DNA probes for culture identification of *Blastomyces dermatitidis, Coccidioides immitis, Cryptococcus neoformans, Histoplasma capsulatum. J. Clin. Microbiol.* **31:**845–850.

95. **Storck, R., C. J. Alexopoulos, and H. J. Phaff.** 1969. Nucleotide composition of

deoxyribonucleic acid of some species of *Cryptococcus, Rhodotorula* and *Sporobolomyces. J. Bacteriol.* **98**:1069–1072.

96. **Sullivan, D., K. Haynes, G. Moran, D. Shanley, and D. Coleman.** 1996. Persistence, replacement, and microevolution of *Cryptococcus neoformans* strains in recurrent meningitis in AIDS patients. *J. Clin. Microbiol.* **34**:1739–1744.

97. **Takeo, K., R. Tanaka, H. Taguchi, and K. Nishimura.** 1993. Analysis of ploidy and sexual characteristics of natural isolates of *Cryptococcus neoformans. Can. J. Microbiol.* **39**:958–963.

98. **Tanaka, K., T. Miyazaki, S. Maesaki, K. Mitsutake, H. Kakeya, Y. Yamamoto, K. Yanagihara, M.A. Hossain, T. Tashiro, and S. Kohno.** 1996. Detection of *Cryptococcus neoformans* gene in patients with pulmonary Cryptococcosis. *J. Clin. Microbiol.* **34**:2826–2828.

99. **Tanaka, R., H. Taguchi, K. Takeo, M. Miyaji, and K. Nishimura.** 1996. Determination of ploidy in Cryptococcus neoformans by flow cytometry. *J. Med. Vet. Mycol.* **34**:299–301.

100. **Thornwell, S. J., R. B. Peery, and P. C. Skatrud.** 1997. Cloning and characterization of CneMDR1: a *Cryptococcus neoformans* gene encoding a protein related to multidrug resistance proteins. *Gene* **201**:21–29.

101. **Toffaletti, D. L., and J. R. Perfect.** 1997. Study of *Cryptococcus neoformans* actin gene regulation with a beta-galactosidase-actin fusion. *J. Med. Vet. Mycol.* **35**:313–320.

102. **Toffaletti, D. L., T. H. Rude, S. A. Johnston, D. T. Durack, and J. R. Perfect.** 1993. Gene transfer in *Cryptococcus neoformans* by use of biolistic delivery of DNA. *J. Bacteriol.* **175**:1405–1411.

103. **Tolkacheva, T., P. McNamara, E. Piekarz, and W. Courchesne.** 1994. Cloning of a *Cryptococcus neoformans* gene, *GPA1*, encoding a G-protein α-subunit homolog. *Infect. Immun.* **62**:2849–2856.

104. **Varma, A., J. C. Edman, and K. J. Kwon-Chung.** 1992. Molecular and genetic analysis of *URA5* transformants of *Cryptococcus neoformans. Infect. Immun.* **60**:1101–1108.

105. **Varma, A., and K. J. Kwon-Chung.** 1989. Restriction fragment polymorphism in mitochondrial DNA of *Cryptococcus neoformans. J. Gen. Microbiol.* **135**:3353–3362.

106. **Varma, A., and K. J. Kwon-Chung.** 1991. Rapid method to extract DNA from *Cryptococcus neoformans. J. Clin. Microbiol.* **29**:810–812.

107. **Varma, A., and K. J. Kwon-Chung.** 1992. DNA probe for strain typing of *Cryptococcus neoformans. J. Clin. Microbiol.* **30**:2960–2967.

108. Varma, A., and K. J. Kwon-Chung. 1994. Formation of a minichromosome in *Cryptococcus neoformans* as a result of electroporative transformation. *Curr. Genet.* **26**:54–61.

109. **Varma, A., D. Swinne, F. Staib, J. E. Bennett, and K. J. Kwon-Chung.** 1995. Diversity of DNA fingerprints in *Cryptococcus neoformans. J. Clin. Microbiol.* **33**:1807–1814.

110. **Velculescu, V. E., L. Zhang, B. Vogelstein, and K. W. Kinzler.** 1995. Serial analysis of gene expression. *Science* **270**:484–487.

111. **Vilgalys, R., and M. Hester.** 1990. Rapid genetic identification and mapping of enzymatically amplified ribosomal DNA from several *Cryptococcus* species. *J. Bacteriol.* **172**:4238–4246.

112. **Whelan, W. L.** 1987. The genetic basis of resistance to 5-fluorocytosine in *Candida* species and *Cryptococcus neoformans. Crit. Rev. Microbiol.* **15**:45–56.

113. **Whelan, W. L.** 1987. The genetics of medically important fungi. *Crit. Rev. Microbiol.* **14**:99–170.

114. **White, C. W., and E. S. Jacobson.** 1985. Occurrence of diploid strains of *Cryptococcus neoformans. J. Bacteriol.* **161**:1231–1232.

115. **Wickes, B. L., and J. C. Edman.** 1995. The *Cryptococcus neoformans* Gal7 gene and its use as an inducible promoter. *Mol. Microbiol.* **16**:1099–1109.

116. **Wickes, B. L., U. Edman, and J. C. Edman.** 1997. The *Cryptococcus neoformans* STE12

gene: a putative *Saccharomyces cerevisiae* STE12 homologue that is mating type specific. *Mol. Microbiol.* **26:**951–960.

117. **Wickes, B. L., M. E. Mayorga, U. Edman, and J. C. Edman.** 1996. Dimorphism and haploid fruiting in *Cryptococcus neoformans* association with the alpha-mating type. *Proc. Natl. Acad. Sci. USA* **93:**7327–7331.

118. **Wickes, B. L., T. D. E. Moore, and K. J. Kwon-Chung.** 1994. Comparison of the electrophoretic karyotypes and chromosomal location of ten genes in the two varieties of *Cryptococcus neoformans*. *Microbiology* **140:**543–550.

119. **Williamson, P. R.** 1994. Biochemical and molecular characterization of the diphenol oxidase of *Cryptococcus neoformans*: identification as a laccase. *J. Bacteriol.* **176:**656–664.

120. **Williamson, P. R., and S. Zhang.** 1996. Transcriptional regulators of CNLAC1, abstr. 1–4:16–17. 3rd International Conference on Cryptococcus and Cryptococcosis, Paris.

121. **Zhang, S., and P. R. Williamson.** 1998. Structural and functional analysis of the promoter region of *CnLAC1* in *Cryptococcus neoformans*, abstr. F-68. Abstr. 98th Annu. Meet. Am. Soc. Microbiol. 1998. American Society for Microbiology, Washington, D.C.

6 | Virulence Factors

DEFINITION OF VIRULENCE IN FUNGI

A common term in the description of pathogens is *virulence factor*. In its simplest definition, it describes the relative infectiousness of a microorganism or its ability to overcome the natural defenses of a host. For bacterial pathogens, there are elegant studies describing a variety of virulence factors, such as adherence factors, invasive properties, and toxins. Progress in the area of bacterial virulence is now at the level of understanding genes, and even some of the regulatory circuits and their genes have been identified. On the other hand, in fungi, the identification of certain virulence factors and the corresponding genes may be more complex and even less well defined than for bacteria. For instance, as Steele (139) has emphasized, these eukaryotic pathogens are generally not passed from person to person, and their evolution has not necessarily attained specific abilities to infect or invade the human host. A fungal infection is generally an accidental encounter in the life cycle of the fungus. It is likely that there are many subtle virulence factors in fungi, such as the simple ability to survive and grow within the host, that are important to their ability to be pathogens (139). The host for a fungal infection may or may not be in an immunodepressed state, depending on the particular fungus and/or inoculum of fungal exposure. A further complexity in understanding fungal virulence is the need to take into account genomic variability and differing virulence properties among fungal strains within the same species. In fact, Cutler (26) proposed a virulence-set hypothesis. In this hypothesis he suggested that, within a given isolate, virulence traits belong to a set of genes expressing a finite number or subset of traits to make the composite virulence phenotype of that particular strain. Cutler further suggested that not all genes within the set are necessary, but that a critical number of genes must act in concert to cause disease. If correct, these concepts will force us to broaden our scope for both defining and investigating fungal virulence factors and may help explain differences between strains.

A second important feature in the study of fungal virulence factors is that we now have the potential to study them at the genetic and molecular biological levels. Therefore, it is important that we use stringent criteria in our studies for assigning virulence genes in fungi. Falkow (38) has made the cogent argument that there needs to be a fulfilling of the "molecular Koch's postulates" before determining that a particular property contributes to the virulence of a microorganism. These criteria are (i) the property is associated with the pathogenic microorganism, (ii) inactivation of the gene responsible for the property results in decreased virulence, and (iii) restoration of the impact gene through gene reversion or replacement is associated with a return to wild-type virulence. It is important that this strict identification of virulence genes be considered, and it will be even more important when *Cryptococcus neoformans*, which has a high rate of ectopic integration events, is studied.

A third feature of fungal virulence factors is to understand the genes' implications for the virulence composite. For example, there are at least three types of virulence genes: (i) those essential to fungal growth on living hosts, (ii) those that produce disease symptoms, and (iii) those that determine host ranges (45). The first group of genes may simply allow the fungus to grow at 37°C or may produce a dimorphic transition so that the specific parasitic form is allowed to grow in the host. The second group of genes requires expression to produce disease symptoms, and expression may be dependent on the site of infection (e.g., lung versus central nervous system). Finally, the third set of genes may be specifically required to cause infection in certain species and, even more important with fungi, in specific immunosuppressed hosts such as AIDS patients or transplant recipients. Investigators in this area must understand that these virulence gene types exist and that there may be no genetic linkage between the groups in the composite virulence phenotype.

Most virulence factor studies are initiated with specific animal model systems and with proper controls, and this investigative strategy is encouraged. However, the measurement of endpoints in these model systems can be quite broad and yet still include certain virulence features. For instance, the measurements of virulence may be lethality for the host, changes in tissue burdens or specific tissue tropisms, influences on histopathology, and/or changes of infection rate from acute to chronic.

In many respects, most fungi live a saprophytic existence. However, under all these properties of saprobes, there remains a subgroup of pathogenic fungi that possess the ability to cause acute and chronic infections in humans. In fact, the dimorphic fungi can even change their shape for the host environment, and some fungi in certain clinical settings can infect essentially every organ of the mammalian host and eventually lead to its death. Thus, there is a fungal composite virulence gene set for these human pathogens. It is likely to be complex, interrelated, and at times specific for the fungal genus, species, or strain. The molecular tools are now available to dissect and thus understand the circuitry for virulence regulation in the pathogenic fungi (109).

VIRULENCE FACTORS

In the area of fungal virulence studies, *C. neoformans* has become an attractive fungal pathogen for study (Table 1), and there have been at least two comprehensive reviews of virulence factors in this yeast (55, 76). First, *C. neoformans* is a primary fungal pathogen that causes invasive disease in both normal and immunocompromised patients. The early molecular studies on virulence with two major opportunistic (secondary) pathogens, *Candida* spp. and *Aspergillus* spp., are difficult to understand. For example, it appears that in a major human fungal colonizer,

Table 1 Advantages and disadvantages of *C. neoformans* as a model yeast for the study of virulence

Advantages	Disadvantages
Important pathogen	No genome sequence
Genetic system	Lack of critical mass of investigators
Some molecular biology foundation	Need for continued development of molecular
Known virulence phenotypes	biology
Well-studied pathophysiology	Not most common deep-seated mycoses
Excellent animal model targets	Virulence factors may not translate into drug
	targets

Candida albicans, most specific gene disruptions result in an attenuation of virulence features. In contrast, specific gene null mutants in *Aspergillus* have had no, or infrequent, apparent impact on fungal virulence of the fungus in the animal models used. The reason for this wide range of outcomes remains unclear. The results could be explained by characteristics of the animal models or simply by the complex nature of the virulence in these secondary pathogens. On the other hand, *C. neoformans* bridges the gap between primary and secondary pathogens. It can produce infections in humans and animals (see chapters 10 and 13) and in both normal and immunosuppressed hosts; thus, the present animal models are very robust in their ability to detect differences in strain virulence, and there are several defined endpoints to measure.

A second feature that makes *C. neoformans* attractive for virulence studies at a molecular level is that there already are known important phenotypic characteristics of *C. neoformans* that are associated with the organism's ability to produce disease. Some of these phenotypes are supported by genetic studies. For example, of all the *Cryptococcus* species, only *C. neoformans* can consistently grow well at mammalian host temperatures of 37°C and above. This virulence phenotype of growth at host body temperatures in *C. neoformans* appears to be requisite for survival in the mammalian host and a general feature of fungal pathogens. Furthermore, the temperature phenotype for production of cryptococcosis has been shown to be under genetic control (128). A recent molecular study has identified a signaling pathway, through the use of calcineurin, as essential for *C. neoformans* growth at 37°C (108). While this temperature-growth virulence phenotype under genetic control in *C. neoformans* is a central theme in all pathogenic fungi, there are two virulence phenotypes that are particularly unique to *C. neoformans* and not shared by most human fungal pathogens. These two virulence phenotypes are the ability to produce a large polysaccharide capsule (11) and to form melanin (128). The following discussion will expand on these virulence factors and on several other potential virulence factors identified in *C. neoformans* pathobiology (Table 2).

Capsule

The capsule of *C. neoformans* is composed of long, unbranched polymers of α-1,3-mannan with monosaccharide branches of xylose and glucuronic acid (see chapter 4). In speculation about the capsule's value for *C. neoformans* in nature, it is reason-

Table 2 Phenotypes and pathways: association with virulence in *C. neoformans* studied on a genetic, molecular, or biochemical basis

Phenotype	Association with virulence	References
Capsule	Proven	11, 13
Melanin synthesis	Proven	133
Matα	Probable	80
Growth at 37°C	Proven	108
Mannitol synthesis	Probable	15
Proteinase secretion	Possible	2, 7, 17
Phospholipase	Possible	19, 20
Myristoylation	Proven	91
Urease	Proven	24
Auxotrophy (purine/pyrimidine)	Proven	114, 147
Signal transduction (through calcineurin and α subunit of a G protein	Proven	1, 108

able to hypothesize that it may protect the yeast from desiccation or reduce its ability to be ingested and destroyed by soil amebae (104). Whatever the reason for its evolutionary existence, the capsule is clearly a major virulence factor for *C. neoformans,* and its importance for host immunity has been discussed in detail (see chapter 7). Table 3 summarizes the broad range of its potential effects on host immunity. For instance, in experimental cryptococcosis, acapsular mutants that are produced by nonspecific methods are less virulent than capsular wild-type strains in animal models (44). Human cases of cryptococcosis attributable to acapsular or hypocapsular isolates are not common, and in apparently normal hosts they induce a stronger host response with a greater degree of inflammation and less severe disease (39). In vitro experiments show that capsule-free yeast cells are more easily ingested than are encapsulated strains or acapsular strains to which polysaccharide has been added (8, 77, 79, 137, 138), although, paradoxically, the unencapsulated strains appear to be less readily killed by monocytes and macrophages (87, 100). Genetic studies have also supported the concept that genes controlling capsule production are associated with virulence (83).

Through a series of elegant genetic and molecular biology strategies, Chang and Kwon-Chung (11) provided proof for the essential nature of the capsule in virulence of *C. neoformans*. With the use of a directed strategy to determine capsular genes' importance for virulence, they attempted to complete Falkow's "molecular Koch's postulates" for the capsule phenotype. They used genetically defined acapsular mutants (140) for gene complementation with *C. neoformans* libraries in the pCnTel plasmid from Edman (35), followed by rescue of a plasmid(s) containing a gene(s) that restored complete encapsulation of the acapsular mutants. The first capsule gene isolated with this strategy was *CAP59* (11). It was cloned and specifically disrupted, and the gene was then replaced back into the null mutant. The acapsular mutant of *CAP59* was less virulent than the parental strain, with the use of mortality in mice as an endpoint, and virulence was completely restored when the wild-type gene was reintroduced back into the mutant. These studies represent the first convincing molecular evidence, with identification of a specific isolated

Table 3 Capsular polysaccharide effects which may contribute to and function for virulence

Effect	Reference
Antiphagocytosis	52, 77, 79
Complement depletion	93
Antibody unresponsiveness	78, 103
Inhibition of leukocyte migration	30, 32
Dysregulation of cytokine secretion	126, 149, 150
Brain edema	51–54
Enhancement of human immunodeficiency virus infection	117, 118
Interference with antigen presentation	22
L-selectin and tumor necrosis factor receptor loss	31
High negative charge of cells	105

gene, that the capsule of *C. neoformans* is necessary for maximal pathogenicity. The study defines an ideal molecular strategy for identifying fungal virulence genes and confirming their importance in *C. neoformans*.

Since classical recombinational genetic analysis (140) showed that acapsular mutants involving multiple genes could be produced by nonspecific mutagenesis, it was theorized that other genes necessary for capsular synthesis could be cloned by similar methods. Chang and Kwon-Chung (11, 12) and Chang et al. (13) then extended their work to the identification of several other capsular genes. *CAP64*, *CAP60*, and *CAP10* have also been cloned. All four genes, *CAP59*, *CAP64*, *CAP60*, and *CAP10,* are essential to capsule formation and to the virulence of *C. neoformans* in mice (11, 12, 12a, 13). *CAP59* and *CAP60* genes are located on the same chromosome, but *CAP64* and *CAP10* are unlinked on different-sized chromosomes (115). It appears that *CAP59*, *CAP60*, and *CAP64* are closely linked to a convergently transcribed gene. For example, *CAP59* is linked to a putative mitochondrial ribosomal I-27 gene (11). *CAP64* is linked to a putative proteasome subunit gene, *PRE1* (13). Chang et al. (10) designed an experiment that showed that the *CAP64* gene of a serotype D strain could complement an acapsular clinical isolate, 602, which has been widely used in studies of virulence and host-parasite interactions. These results confirmed that 602 was a serotype A strain when capsule was made. Restoration of capsule structure in 602 improved the yeast's ability to mask induction of tumor necrosis factor alpha and nitric oxide synthase production in primed macrophage-like cells and enabled the newly created, encapsulated 602 strain to cause a fatal infection in mice when acapsular 602 strain was avirulent (10). The *CAP60* gene is also linked to a gene that shares high homology with a cellulose growth-specific gene of *Agaricus bisporus* (115). The functional relevance to other genes of this linkage for capsular genes is not yet clear. Although these identified genes are essential for capsule formation, the biochemical function of these genes within the pathway of capsule synthesis remains uncertain. For instance, analysis of primary DNA sequence for the genes in databases does not reveal any significant homology with known genes or identify specific motifs that may help to understand their function. However, a detailed molecular analysis of *CAP59* showed a possible transmembrane domain at the N-terminal end that is required for complementation, and there is a glycine residue in the middle of the gene that is essential for its function. Further progress in this area of gene function would be aided by further basic understanding of the biochemistry of capsular synthesis, identification of the other capsule genes, and possibly the cellular locations of the functional gene products for capsule production within this yeast.

Although the body of experimental evidence confirms that the presence of a capsule in *C. neoformans* is required for optimal pathogenicity, it is not necessarily sufficient to produce clinical disease. For instance, there have been clinical reports that hypocapsular or acapsular strains of *C. neoformans* have produced human infections (5, 6, 39, 129), although it is not absolutely certain that these strains possessed no evidence of capsule through detailed electromicrographic studies of the clinical isolates. It has also clearly been shown that there is no direct correlation between the size of the capsule and the virulence of the strains (34, 90), so the mere presence or size of the capsule does not ensure *C. neoformans* pathogenic features. In fact, there are *Cryptococcus* species that produce capsule but cannot produce clinical disease. However, within *C. neoformans* there are also differences in the quality of the capsules for different strains. The *C. neoformans* capsule should be viewed as a dynamic structure with its architecture loosely woven around the cell and extruding into the local environment. This architectural structure, glucuronoxylomannan, has been shown to inhibit phagocytosis, activate the alternative complement pathway that leads to deposition of C3b on and within the capsule, and inhibit antibody production to itself (see chapter 7). Its production is even regulated by exposure to environmental cues such as high pCO_2 (50) and low ferric iron concentrations (148) in the media or the host. When a strain loses its ability to respond to these signals and becomes hypocapsular under physiological conditions, it can become less virulent than its parental strain. Granger et al. (50) identified such a strain, which no longer induced capsular production under elevated environmental pCO_2 conditions. Recently, Alspaugh et al. (1) also illustrated the dynamic molecular interaction between environmental cues and capsule production. Capsule synthesis is mediated by a specific α GTP-binding (G) protein encoded by the *GPA1* gene in *C. neoformans* (1). This signaling pathway participates in the regulation of capsule size both in vitro and in vivo. Further support for the capsule as an important part, but only one complex part, of the virulence composite is found within capsular strains of the same serotype or different serotypes. In these strains, there can be wide ranges of virulence potential when compared in controlled studies with experimental animal models (59, 85).

Other important features of the capsule, such as its regulation and response to environmental signals and linkage to other virulence phenotypes, need further study. Alspaugh et al. (1) found a linkage with other virulence factors, such as melanin production and mating through the α G protein. Jacobson and Compton (64) found discordant regulation of capsular polysaccharide and phenoloxidase production. For instance, exogenous iron increased phenoloxidase activity but decreased capsular synthesis, and high temperature (37°C) decreased phenoloxidase production but had no effect on capsule synthesis. Furthermore, specific differences in capsular structures between strains may play important roles in the pathobiology of this yeast. The capsule structure is so dynamic that it can even change its composition within a single strain during infection (21). Thus, there has been significant progress at a molecular level to anchor further studies on this most prominent virulence factor for *C. neoformans*, but further investigations are necessary to fully understand this complex virulence structure and its relationship to the composite virulence phenotype.

Melanin

Williamson (161) has recently provided an excellent review of our present knowledge concerning *C. neoformans* laccase and melanin production in relationship to the pathogenesis of this yeast. Among the *Cryptococcus* species, *C. neoformans* is relatively unique in its possession of an enzyme system that allows it to metabolize a variety of catechol precursors to melanin, such as dopa, dopamine, norepinephrine, and epinephrine (119). Occasionally, a *Cryptococcus laurentii* strain has been reported to produce pigment when placed on medium with diphenolic substrates, but the laccase gene of *C. neoformans* did not identify a homologous gene in *C. laurentii* by Southern hybridization (116). Biochemistry studies initially suggested that *C. neoformans* possessed a single phenoloxidase enzyme (laccase) that converted diphenolic compounds through a series of autooxidation reactions to the final pigmented, antioxidant melanin (68, 120, 121). Melanin production by this yeast occurs in the presence of dihydroxyphenolic or polyaminobenzene compounds and molecular oxygen and under glucose-limiting conditions. Unlike many other fungi, *C. neoformans* lacks a tyrosinase enzyme capable of endogenous dihydroxyphenol production but possesses a phenoloxidase that can oxidize the dihydroxyphenols and generally follows the Mason-Roper model as modified by Ito (60). The dihydroxyphenol is oxidized to its corresponding quinone, which will spontaneously rearrange through sequential polymerization and undergo autooxidation to form melanochrome and then melanin (142). This property of *C. neoformans* has been used as a diagnostic test for identification of *C. neoformans* (135) (see chapter 4). However, this feature of melanin production is also present in other pathogenic fungi, and as a group these pathogenic dematiaceous fungi have produced infections classified as phaeohyphomycoses. Although there has been some suggestion through studies of albino mutants that production of melanin is a virulence factor for these fungi (29, 48, 123), the link of melanization with pathogenesis is still most fully developed in *C. neoformans*.

Melanin structure and function remain poorly understood. However, it contains stable populations of free radicals, and with electron spin resonance spectroscopy, melanin appears to be concentrated in the inner aspects of the cell wall (155). Both Ikeda et al. (58) and Williamson (162) purified the laccase from *C. neoformans*, which was helpful in the cloning of the gene. It should be emphasized that laccases have broad-range specificity in oxidizing a series of substrates. For instance, in brain tissue there are significant quantities of dopa, norepinephrine, epinephrine, and dopamine that could be oxidized by the *C. neoformans* laccase. This lack of enzyme specificity could yield a mixture of laccase products that in vivo may yield a heterogeneous and unpredictable melanin structure against which the host may interact. Therefore, more study and understanding of melanin structure in vivo is necessary to define its importance to the composite virulence status of *C. neoformans*.

In the early 1980s Rhodes et al. (128), Polacheck et al. (122), and Kwon-Chung et al. (82, 83) performed the first studies that directly identified melanin's impact on *C. neoformans* virulence. They used classical genetics and mutants of the melanin phenotype. In these elegant experiments, albino mutants (Mel$^-$) produced by nonspecific mutagenesis were found to be less virulent, as measured by survival times, than the wild-type parental strain (Mel$^+$). Importantly, when animals receiving the Mel$^-$ strain died, cultures of the brain showed that 50% of CFU were revertants to Mel$^+$ phenotype. These mutants not only regained melanin production but also were restored to the virulence state of the parental strains (128). These genetic and animal studies were convincing in showing that melanin must play a role in the ability of *C. neoformans* to produce disease but did not identify the mechanism for protection of the yeast.

In vitro studies were then performed to examine the pathobiological significance of melanin. Studies were focused on melanin's role as an antioxidant for fungal protection against oxidative damage by professional phagocytes (68). For example, it was shown that an albino mutant (Mel$^-$) but not the wild-type parent was killed in vitro by an epinephrine oxidative system that used epinephrine as the electron donor in the presence of a catalytic transition metal ion, Fe^{3+}, and hydrogen peroxide as an electron acceptor (122). This work suggested that melanin could act as an antioxidant and, in particular, could neutralize the harmful effects of catecholamines under certain conditions. Further genetic work has supported the focus on antioxidants. There has been confirmation of a strong link between the production of oxygen-sensitive mutants and their development of albino phenotypes. However, some melanin-positive revertants may still retain their sensitivity to hyperoxia (65). Similarly, a series of cellular studies supported the hypothesis that melanin may play a part in protecting the yeast from both oxygen- and nitrogen-based oxidative damage from the host cells (154, 157). Melanized cells also show protection from macrophage killing and reduced antibody-mediated phagocytosis by macrophages. Melanized cells are more resistant to killing by chemically generated oxygen- and nitrogen-derived oxidants, amphotericin B, trifluoperazine, and UV light (158). However, a concern about melanin's antioxidant properties for virulence of *C. neoformans* is that the production of phenoloxidase appears to be down-regulated at host temperatures of 37°C (66). On the other hand, Jacobson et al. (66) have shown that there may be compensatory changes in antioxidants with temperature elevation so that, at 37°C, as phenoloxidase activity decreases, superoxide dismutase activity increases (66). Furthermore, Shaw and Kapica (135) have shown that phenoloxidase enzyme activity of cell extracts is maximal at 37 to 40°C. Although there has been some work that suggests that oxidants may not be primary effectors against *C. neoformans* (75), it remains reasonable to hypothesize that melanin helps the yeast as an antioxidant against the host effector systems.

A second hypothesis for the value of melanin to *C. neoformans* is that it acts as structural support for the cell. With the use of electron spin resonance spectroscopy, it was shown that antioxidant melanin was contained within the cell wall and could potentially contribute to fungal survival against host damage by promoting cell wall integrity. Despite the fact that the phenoloxidase enzyme production is greater at 30 than at 37°C, similar amounts of melanin were detected in cells grown at both temperatures with prolonged incubation. It was also noted that *C. neoformans* strains can vary up to eightfold in their melanin content (155). Another possible mechanism for melanin as a virulence factor is its ability to change the charge of the cell wall so that the combination of the polysaccharide and melanin significantly increases the negative charge of the yeast (105). Finally, melanin prevents T-cell response and cytokine secretion (56), reduces antibody-mediated phagocytosis (105), or simply makes the organism more resistant to treatments such as amphotericin B (158).

As a virulence factor in mammalian hosts, melanin was probably coopted from its use by *C. neoformans* in its natural environment. It is possible that *C. neoformans* in its environmental niche has used melanin to protect itself from oxidative damage from UV irradiation (156) or extremes in temperature (130) and/or to provide structural integrity of the cell wall against osmotic challenges in the environment. However, the role of melanin in the natural habitat may also be related to its ability to break down polyphenolic compounds during wood cutting (74). *C. neoformans,* with its possible ecological niche in decaying trees (86), could use laccase in degradation of wood pulp lignin. Laccase has been found in related lignolytic mushrooms.

Since *C. neoformans* has a unique predilection for the brain, with its high dopaminergic areas that produce large amounts of other catechol precursors, the presence of substrates for the laccase enzyme might help to explain the location of infection (121). Although the use of a Fontana-Masson stain has suggested that cells within the central nervous system are melanized (81), this stain is not specific for melanin, and definitive studies on the quantity and quality of melanin production in host tissue, particularly the brain with its variety of substrates, remain to be performed. On the other hand, recent work has shown that despite the relatively high temperatures in vivo (approximately 39°C), the *C. neoformans* laccase gene (*CnLAC1*) transcript can be detected in cells in the subarachnoid space by reverse transcription-PCR during experimental cryptococcal meningitis in rabbits (133). Therefore, the focus of research has now centered on genes in this pathway of melanin synthesis to further understand melanin's importance.

There is likely a pathway of proteins and their genes leading up to or regulating the oxidation of diphenolic compounds by *C. neoformans*. This concept is supported by genetic studies in which albino strains were easily produced with nonspecific mutagenesis strategies and documentation of at least seven separate complementing mutant groups (142). Those melanin-deficient mutants that have been identified demonstrate that multiple genes participate in effective melanogenesis in *C. neoformans*. In fact, the ability to suppress some of these mutations with copper suggests that there is a signaling pathway with copper-containing or transport enzymes that helps to regulate melanin expression. Recent work on delivery of copper into yeast cells may be useful to help understand its potential link to melanogenesis (145). At least one other signaling pathway that utilizes the α G protein encoded by the *GPA1* gene, which regulates cyclic AMP (cAMP) levels in *C. neoformans,* has already been identified as important for optimal melanin expression (1). Despite a lack of complete understanding of the pathway(s) for melanin production and its regulation, it has been conclusively shown that there is one crucial entry enzyme (phenoloxidase or laccase) encoded by a single gene (*CnLAC1*) in this pathway.

Laccases are part of the family of blue copper-containing oxidases, and the cryptococcal laccase is believed to catalyze the oxidation of catecholamines into quinones, which then autooxidize and polymerize into melanin or melanin-like compounds. Definitive studies in this area for the role of melanin in virulence have specifically required the cloning and disruption of this gene for the central enzyme. The *C. neoformans* laccase gene (*CnLAC1*) was first identified by Williamson (162). It contained a single open reading frame encoding a polypeptide of 624 amino acids in length, with 14 introns. It included an essential copper-binding site. In fact, site-directed mutagenesis of a single base in a histidine copper-binding site of *CnLAC1* abrogated laccase activity without preventing transcription of *CnLAC1*. The amino acid sequence of *CnLAC1* showed a large amount of sequence diversity, typical of fungal laccases with homology to them only in the copper-binding regions. No apparent membrane-binding domains were found in the laccase gene, but since little enzyme was found either extracellularly or in the cytoplasm, it was postulated that the product was secreted into the periplasmic space or bound to the cell wall. As a potential virulence factor, this gene had somewhat unusual regulation by *C. neoformans*, according to the initial studies. The enzyme was suppressed by glucose, incubation at 37°C, and logarithmic growth. On the other hand, within the central nervous system, low glucose levels and stationary-phase growth characteristics within the subarachnoid spaces of hosts may actually be ideal for the expression of laccase, and as previously noted, transcripts of *CnLAC1* were detected by reverse transcription-PCR of yeast cells during cryptococcal meningitis in rabbits.

The study of gene regulation of *CnLAC1* has begun. It is a particularly attractive pathway for the study of genetic regulation of virulence-associated genes because of its simplicity and easily measured phenotype. The transcriptional regulation of *CnLAC1* is being mapped (163). The gene appears to be repressed by elevated glucose concentrations and during the sexual or filamentous state of yeast. It has a canonical TATA box and also contains an Spl-like enhancer site, but functional activities and protein binding domains are still being investigated (see chapter 5).

Definitive studies of the role of melanin in the virulence of *C. neoformans* were hindered by the lack of melanin-negative mutants characterized at the molecular level. Therefore, with the cloned *CnLAC1* gene, a melanin-negative mutant was created by site-directed mutagenesis. It was an infrequent event (<1 in 10,000), and the stable transformant containing the disrupted gene also was missing a substantial amount of upstream sequence from the 5' end of the gene. The melanin-negative mutant disrupted at the 5' end was backcrossed with a parental strain to remove unintended genetic defects introduced by the transformation process, and the melanin-negative progeny were found to be much less virulent than the melanin-positive progeny. Finally, complementation of a melanin-negative mutant with the *CnLAC1* gene restored the melanin-positive phenotype and increased virulence (133). These molecular pathogenesis studies support the concept that *CnLAC1* from a serotype D strain has a direct role in virulence of *C. neoformans* in mice.

The sum of all the present studies supports the concept that melanin has a significant role in the composite virulence phenotype of *C. neoformans* (Table 4). However, it should be emphasized that the melanin-negative phenotype that is so strongly linked to *C. neoformans* virulence may also be controlled by other genes besides *CnLAC1*, such as those associated with diphenolic transport, earlier metabolic pathways, and various signaling and regulatory pathway genes. It would also be helpful to prove the existence of melanin in the yeast cells within the host.

Table 4 Possible molecular mechanism(s) for melanin in pathogenesis of cryptococcosis

Effect	References
Antioxidant	65, 68, 122, 154, 157
Cell wall support or integrity	155
Alterations in cell wall charge	105
Interference with T-cell response	56
Reduction of susceptibility to antifungal agents	158
Abrogation of antibody-mediated phagocytosis	105
Protection from extreme temperatures	130

The focus on *C. neoformans* and its melanin production has two potential benefits. First, it may us help to understand the regulation of virulence of this infection with a well-defined phenotype that is easy to measure. Lessons from *C. neoformans* may also be extrapolated to other fungi that produce melanin, such as those that produce phaeohyphomycoses. Second, with further understanding of the molecular biology and biochemistry of melanin, it could become a unique and selective target for antifungal drugs against *C. neoformans* and other dematiaceous fungi.

Metabolic Targets

Virulence factors in this area of investigation need to be defined broadly as those that are essential for survival and proliferation in the host but do not simply represent growth defects on nonliving substrates. Although this runs the risk of identifying certain housekeeping genes, studies in bacteria with screens using in vivo expression technology suggest that these basic metabolic pathways are essential in vivo and actually may become better antifungal targets than specific pathogen-based virulence factors.

Pyrimidine and purine metabolism

Initial investigations in *C. neoformans* with the *URA5* gene encoding orotidine monophosphate pyrophosphorylase showed that 5-fluoroorotic-resistant mutants in this gene were less virulent than the wild-type parent in mice (37, 84). Replacement of the *URA5* gene into these mutants of one strain gave variable responses to infection, and restoration of prototrophy did not always recover a fully virulent strain (147). Similarly, the *ADE2* gene encoding phosphoribosylaminoimidazole decarboxylase was found to be essential for the establishment of a *C. neoformans* central nervous system infection in rabbits (114). A block at this site in purine metabolism could simply not be recovered by the yeast from the host adenine pool. Mutants with blocks in this pathway were avirulent in an animal model and were completely eliminated several days after inoculation in immunosuppressed rabbits. This avirulence was not related to in vitro growth rate with fully supplemented media, and virulence could be fully reconstituted by replacing the gene through transformation back into the adenine auxotroph. Along with *C. albicans*, which also requires a functioning *ADE2* gene for full virulence in vivo (73), the importance of this pathway for pathogenic fungi may be essential. This area of purine metabolism could become an area for antifungal target development if fungal enzymes in the purine pathway have differences when compared to mammalian enzymes. In fact, a structure-function analysis of the *C. neoformans ADE2* gene has been completed; it shows that this gene encodes a two-domain, bifunctional protein that produces a different intermediate (N^5-carboxyamino-imidazole ribonucleotide) in its carboxylation pathway compared to the avian pathway (40). This novel biochemical variation in a pathway essential for virulence of this yeast but not found in the host could potentially be exploited for drug development.

As important as endogenous purine and pyrimidine metabolism appears to be for *C. neoformans* to produce disease, auxotrophic requirements for specific amino acids for this yeast in the host will likely require study with each auxotrophy. At least one lysine auxotroph of a *C. neoformans* strain (H99) had no effect on virulence when compared with the wild-type parental strain in the rabbit model of meningitis, but an arginine mutant was avirulent (110a). Rhodes and Howard (127) similarly found that arginine (arginosuccinate lyase) mutants produced by UV irradiation were avirulent but that one of these mutants had abnormal budding. When genetic crosses were performed to help eliminate unknown gene mutations, another arginine mutant was found to be fully virulent in a mouse model. Therefore, it will be important to eliminate the alteration of other potentially important pathways when auxotrophic mutants are studied for pathogenesis.

Mannitol

Another possible metabolic target that has received some attention in *C. neoformans* is the mannitol pathway. Many fungi produce and accumulate large amounts of acyclic polyols, which appear to function as energy stores (reducing equivalent osmolytes) and stress protectants. Mannitol, which is both produced and utilized by *C. neoformans*, can be secreted extracellularly by *C. neoformans* in substantial amounts at the site of infection (65, 96). The amount of mannitol correlates with the numbers of organisms and antigen present at the site of infection in animals (164), but not in humans (96). It has been hypothesized that mannitol could act as either an osmoticum or an antioxidant to promote cryptococcal infection. Clinically, in the presence of a high concentration of yeasts in cerebrospinal fluid, a syndrome of elevated intracranial pressure has been identified (27). Extracellular mannitol might contribute to the osmotic pressure in both the subarachnoid space and the brain tissue and thus add to the acute elevation of intracranial pressure in these heavily infected patients.

Mannitol, like melanin, has been shown to scavenger host oxidants. Chaturvedi et al. (16) isolated one mutant with hypoproduction of mannitol (CnMLP) that was found to be more susceptible to polymorphonuclear leukocyte killing and was specifically sensitive to the generation of a distal reactive oxygen intermediate such as OH^-. These investigators were also able to link this mutant with osmotolerance by expressing a bacterial mannitol-1-PO_4 dehydrogenase under the cryptococcal *GAL7* promoter in *C. neoformans*, thereby increasing both its mannitol production and improving its ability to grow on hypertonic media (2 M NaCl) (14). Finally, in an intravenous mouse model, the CnMLP mutant was found to be less virulent than the parental strain (15). Unfortunately, definitive proof of the importance for this pathway in virulence will require isolation of genes within the mannitol pathway and production of a specific gene disruption to make isogenic pairs.

The mannitol pathway in *C. neoformans* is used for both synthesis and utilization of mannitol. It links up with glycolysis and could be an essential part of the general stress response. *C. neoformans* possesses mannitol dehydrogenase activity with both NADP and NAD cofactors, and the inability to find a null mutant for mannitol production suggests either a significant redundancy in this pathway or the fact that it is an essential factor for yeast survival. The first cloned *C. neoformans* gene (*Mtl-1*) related to mannitol synthesis was found to regulate the *Saccharomyces cerevisiae* mannitol dehydrogenase gene (113). Although *Mtl-1* regulates expression of mannitol dehydrogenase in *S. cerevisiae*, its function in *C. neoformans* mannitol metabolism still remains uncertain.

The genetic, biochemical, and virulence studies on mannitol in *C. neoformans* suggest that mannitol functions both as an intracellular osmolyte and an oxidative stress protectant. The capabilities of wild-type biosynthetic mannitol for both production and utilization of this yeast are probably required for the wild-type composite virulence of the strain. However, further genetic and biochemical work is needed to fully understand this pathway's importance to virulence.

Inositol

Inositol is a key cellular metabolite that participates in formation of membrane lipids and signal transduction mechanisms. Inositol can also act as a major intracellular osmolyte during stress and has a role in cell development and growth. Interestingly, *C. neoformans* has the capacity to catabolize inositol and use it as a sole carbon and energy source, but it also can synthesize inositol. Inositol appears to be involved in regulating phospholipid biosynthesis in *C. neoformans* (153). Previous studies have shown that *C. neoformans* has a typical eukaryotic phospholipid composition (61, 62). However, one group found that common lipids in other yeasts, such as phosphatide serine, sphingolipids, and cerebrosides, were absent in *C. neoformans* (124). On the other hand, a direct correlation with any quantitative measures of various lipids and virulence has yet to be established (143).

This dual capacity of inositol suggests that it has the potential to be an important factor in regulating and facilitating balanced growth of *C. neoformans* and could be particularly important in the inositol-rich central nervous system of the host. Therefore, further investigations into inositol and its phospholipid pathways in *C. neoformans* may find unique features that are exploited by this yeast to produce disease.

Myristoylation

Myristoylation of certain proteins is an essential part of eukaryotic cellular biochemistry, and without this function cells will die. Lodge et al. (92) have isolated the *C. neoformans* N-myristoyl transferase (NMT) gene of *C. neoformans* and purified its protein. This particular protein catalyzes the myristoylation of a series of small proteins, such as the adenosine ribosylation factor, for activating cellular functions. Yeast cells with a temperature-sensitive mutation in the allele can recover from a defect in this enzyme in vitro if the medium is supplied with myristic acid. Lodge et al. (91) were also able to obtain site-directed mutants of the NMT gene in *C. neoformans*. In their study, the NMT gene was replaced with an NMT gene that would perform normally at 30°C but would not function at 37°C. This replacement was necessary because the gene was likely to be essential for in vitro growth. It was then determined that a completely functional protein is also essential for survival of *C. neoformans* in the immunosuppressed rabbit model of cryptococcal meningitis. A fully virulent *C. neoformans* strain that now contained a temperature-sensitive mutation in the NMT gene was unable to establish infection when directly inoculated into the subarachnoid space of rabbits. In fact, unlike the parental wild-type strain, the temperature-sensitive mutant was rapidly killed by the immunocompromised host. These experiments clearly demonstrate that a properly functioning myristoylation system in *C. neoformans* is necessary to produce disease. The myristoylation pathway is considered a housekeeping function for growth on nonliving substrates and does not represent a classic virulence factor. On the other hand, these experiments confirmed that a functional NMT gene is also essential in vivo, and that no other redundant or scavenger pathways could suppress this lethal defect in the host environment. Furthermore, since there are known differences in substrate specificity between the mammalian and the *C. neoformans* NMT proteins (91), selective inhibitors may now be devised for use in antifungal therapy.

Iron utilization
Iron is essential for all microorganisms' growth, and *C. neoformans* is particularly susceptible to changes in extracellular iron. Iron has profound effects on yeast growth and its virulence phenotypes, such as capsular production (148). However, unlike many fungi, *C. neoformans* does not produce siderophores to help in its iron utilization scheme (67). Because only ferrous iron is taken up by *C. neoformans*, and ferric iron is generally present in the environment or host (69), it is likely that there are important mechanisms on the surface of *C. neoformans* for ferric iron reduction. Physiological experiments reveal that *C. neoformans* does regulate its iron-reducing capacity in response to environmental conditions (106). It is likely that there is a ferric reductase(s) and other enzymes involved in iron metabolism, and these enzymes will be extremely important to this yeast's ability to grow in the host's hostile environment for iron acquisition. There are likely to be virulence factors for *C. neoformans* that are directly related to iron metabolism.

Mating Locus
An interesting and still incompletely explained observation is that more than 95% of in vivo and in vitro isolates of the heterothallic *C. neoformans* are of the α-mating phenotype (136). A finding by Wickes et al. (160) that only the α-mating-type strains possess the ability to produce hyphae in the vegetative phase with its monokaryotic or haploid fruiting may explain some of this bias. This finding suggests the hypothesis that this haploid fruiting structure and its basidiospores will make mating-type ratios biased toward the α-mating type. For instance, the strains with this mating locus could make infectious propagule basidiospores that are produced on low-nitrogen substrates. Also, investigators have reasoned that this locus or areas linked to it may be directly related to virulence of the yeast which may explain its epidemiology. With multiple genetic backcrosses, Kwon-Chung et al. (80) made congenic pairs of two serotype D strains that differed primarily in their mating-type locus. Virulence studies performed in mice confirmed that the congenic strain containing the α-mating-type locus was more

virulent in mice than its opposite congenic pair possessing the **a**-mating-type locus. Using a difference cloning method for the two congenic pairs, an approximately 45-kb region present only in the *MAT*α strain was cloned, and a 2.1-kb fragment containing an α-mating-factor gene, which likely encodes a pheromone precursor, was identified (102).

The molecular mechanisms that may add to the virulence of *C. neoformans* through the α-mating-type locus still remain unclear. Recent experiments have further examined the α-mating locus; with more than 90% of it sequenced, its minimum size appears to be at least 60 to 75 kb. With further sequence information, unexpected findings have now been revealed. Two genes for a putative signal transduction pathway, *STE11*-α and *STE12*-α, appear to be found only in α-mating-type cells (36, 159). These two proteins are part of a conserved *MAP* kinase cascade involved in signal transduction events for mating that function in both mating types within *S. cerevisiae*. The *C. neoformans STE12*-α displays substantial similarity in sequence identity to other fungal transcriptional activator genes and includes a highly conserved homeodomain. Overexpression of this gene results in poor growth, altered cell shape, presence of hyphal projections, induction of a phoromone, and induction of *CnLAC1* (159). These genes and others only found in the α-mating-type locus remain to be further studied for their influence on virulence. In *C. neoformans*, it is likely that these genes and their signaling pathway are involved in haploid fruiting and possibly virulence (13a). However, it does appear that *C. neoformans* has a unique genetic arrangement compared with the ascomycete *S. cerevisiae* and other fungi. It will also be important to examine organizational differences among the four serotypes at the mating-type locus; there is little information on the **a** locus, but its importance to virulence of the yeast is not apparent.

Another line of investigation into the mating genes of *C. neoformans* and their relationship to virulence has been to examine certain G proteins that signal mating responses in other fungi, such as *S. cerevisiae* and *Ustilago maydis*. A *C. neoformans GPA1* gene that encodes a GTP-binding protein α subunit (G protein) has been cloned (141). The protein encoded by this gene has some homology to the G protein α subunits that function in the mating-response pathways in *S. cerevisiae* and *Schizosaccharomyces pombe,* and *GPA1* transcriptional regulation is consistent with a role in mating of *C. neoformans*. Generally, these proteins involved in mating processes will also reduce growth through down-regulation of the cell cycle. Early studies to determine the pathobiological importance of this G protein used the nonspecific mutagenesis of G proteins by selection of *C. neoformans* mutants growing in the presence of aluminum trifluoride (23). These mutants were shown to have poor or absent mating and displayed reduced virulence in mice, compared with the parental wild-type strain. Further linkage of this protein and the mating process to virulence requires specific site disruption of the gene and comparison of isogenic sets of mutants. These experiments were performed by Alspaugh et al. (1) to confirm this pathway's importance to mating response and virulence in *C. neoformans*. Further discussion of these experiments can be found in the section on signal transduction.

The potential links between signal transduction and mating-type locus or identification of the influence of hormonal factors, such as estrogens, on *C. neoformans* growth are the most compelling molecular and biochemical links for yeast sex factors to the fungal virulence composite for *C. neoformans*. However, both mating-type and hormonal links may also be recurrent themes in fungal pathogenesis, because virulence in *Coccidioides immitis, Paracoccidioides braziliensis,* and *C. albicans* have all been suggested to be associated with certain hormonal factors. *C. albicans* even has a putative estrogen-binding gene that has already been identified (93a). Similarly, the mating-type bias within *C. neoformans* strains is also known to occur in certain pathogenic dermatophytes. Sex and mating factors should continue to be explored in fungi for their relevance to virulence.

Proteinase

Extracellular proteinase production has been implicated as a potential virulence factor in the pathogenesis of several fungi, but its importance has not been fully explored in *C. neoformans*. A series of studies have now identified the presence of extracellular proteolytic activity by *C. neoformans* (2, 7, 17), and data suggest that this activity may be produced by several proteinase enzymes secreted by *C. neoformans*. These enzymes are hypothesized to contribute to microbial virulence by degrading host tissues and/or destroying immunologically important proteins. It has been generally considered that free-living organisms such as *C. neoformans* developed these extracellular proteinases to obtain nutrients from the environment. Although proteolytic production by *C. neoformans* is low, it is likely that these enzymes digest immunoglobulins and complement at the site of infection. A series of other extracellular enzymes have been identified, including esterases, β-glucosidase, phosphohydrolase, and acid phosphatase (18). The secretion of these enzymes is dependent on strain and growth conditions, and the host can recognize some of these extracellular proteins during infection by developing antibodies against them (18). Since histopathological examination of *C. neoformans* infections will occasionally show tissue necrosis and extracellular matrix degradation, the study of these proteinases as contributing virulence factors is warranted, but further gene and protein identifications and manipulations will be required.

Signal Transduction

Signal transduction pathways serve as critical links that enable eukaryotic organisms to respond to a variety of extracellular events. Understanding the pathways used by fungi in response to environmental cues, and thus the activation of certain virulence traits, may help us to devise strategies to interrupt these pathways in fungal pathogens. Both the good and bad feature for study of these pathways as a target(s) for antifungal development is their conserved nature. Conservation of structures allows us to study these pathways rapidly, but with too much conservation it is difficult to find unique differential targets between the yeast and host. Recently, significant progress has been made in the study of signal transduction in *C. neoformans*. The signaling pathways for the three major virulence phenotypes of temperature, capsule, and melanin have been identified. Variations in the use of these conserved pathways has been elucidated between several fungi and *C. neoformans*.

Calcineurin

The first work in *C. neoformans* that specifically identified a link between a gene, environmental signals, and virulence was based upon signal transduction studies, which had previously focused on T-cell activation. Odom et al. (107) discovered that the immunosuppressive agents cyclosporine and tacrolimus, which were known to specifically inhibit a certain signal transduction pathway at the calmodulin-activated serine/threonine-specific phosphatase, calcineurin, killed *C. neoformans* in a temperature-dependent manner. These drugs had no activity on the growth of *C. neoformans* at 24°C, but at 37°C they were fungicidal in vitro (107). Since it was known that cyclosporine and tacrolimus complex with the immunophilins cyclophilin A and FKBP12, respectively, to inactivate the Ca^{2+}-regulated calcineurin, Odom et al. (108) performed a series of experiments in which they cloned, sequenced, and disrupted the calcineurin A catalytic subunit in *C. neoformans* and determined the mutant's predicted phenotype. The site-specific calcineurin disruptant strain was found to be completely viable in vitro at 24°C in air and a medium pH of 7.0, but it did not survive in a series of in vitro conditions that mimicked the host environment. For example, the mutant did not grow in temperatures of 37 to 39°C, as predicted by the immunophilin drug studies. The calcineurin A mutant did not grow in the presence of 5% CO_2 concentrations in the environment or at medium pH conditions of 7.3 to 7.4, whereas it had normal growth in air and neutral pH conditions. Both environmental features were found to be toxic to the *C. neoformans* strain that had an interrupted calcineurin pathway. The importance of the calcineurin gene in signal transduction within the host was further confirmed in animal experiments in which the null mutant for calcineurin A was completely avirulent in the rabbit meningitis model and, by reintroducing the wild-type gene into the mutant, virulence was partially restored to the wild-type pathogenic level (108).

There are several messages in these experimental observations. First, these findings identify a conserved signaling pathway within *C. neoformans* that is used for its regulated survival response to multiple host environmental signals (temperature, CO_2, pH). This is a theme that will likely be repeated in most pathogenic fungi, although the specific details may vary among species. Second, although the identification of this conserved pathway has been used to study signal transduction biology in many systems, the functional use of this particular pathway by *C. neoformans* could not have been predicted by other fungal models such as *S. cerevisiae*, *S. pombe*, or *Aspergillus nidulans*. For instance, in *S. cerevisiae*, calcineurin is required for survival during cationic stress and recovery of pheromone arrest. In *S. pombe*, it is induced by nitrogen starvation and is required for growth at low temperatures, mating, and proper cytokinesis. Finally, in *A. nidulans*, calcineurin is essential for regulating cell cycle progression. Third, these observations prove the importance of calcineurin gene function in pathobiology and demonstrate that this intact signaling pathway is required for the expression of virulence by *C. neoformans*. Through its signaling of multiple host environmental stress responses and its essential features with inactivation, calcineurin's potential as an antifungal target for therapeutic interventions is revived. Agents such as cyclosporine and tacrolimus have too powerful immunosuppressive activity on the mammalian pathway to be used as antifungal agents, and this fact has been documented in some animals (111). On the other hand, selective nonimmunosuppressive analogs that block this site and not the mammalian target can possess anticryptococcal activity (107). Further structural understanding of the cryptococcal calcineurin protein might be used to model and develop even more potent and selective inhibitors of this protein. Since it is also clear that this is a major signaling pathway for *C. neoformans*, further work on understanding the receptors upstream of calcineurin and the genes regulated downstream will be important to understand the molecular circuitry of this important virulence pathway.

G proteins

A second major signaling pathway for *C. neoformans* and its virulence phenotype has been identified. The heterotrimeric GTP-binding (G) proteins have been found to serve as critical links in signal pathways in a variety of eukaryotic organisms. Two plant fungal pathogens, *Cryphonectria parasitica* and *U. maydis*, have been shown to have their virulence attenuated when an α subunit of a G protein is inactivated (125). In *C. parasitica*, disruption of the gene *CPG-1*, which encodes an α subunit G protein, resulted in reduced fungal growth rates, mating and sporulation changes, inhibition of laccase, and loss of virulence. In *U. maydis*, in which four genes have been identified (*UmGPA1*, *UmGPA2*, *UmGPA3*, and *UmGPA4*) for encoding α subunits of G proteins, *UmGPA3* has been found to play an active role in transmission of the pheromone signal and is also required for pathogenesis. With these plant pathogens using this pathway for virulence, it has been interesting to speculate that human fungal pathogens might also conserve function and similarly use this signaling pathway in their virulence schema.

Tolkacheva et al. (141) cloned the first G-protein gene of *C. neoformans* (*CnGPA1*). As previously reviewed, nonspecific mutants defective in this gene produced mating defects and hypovirulence in mice (23). It is interesting that *CnGPA1* was most closely related to *UmGPA3* (75% identity), the pathogenic gene for *U. maydis*. Recently, Alspaugh et al. (1) used site-directed mutagenesis to disrupt *CnGPA1* in *C. neoformans* and observed the phenotypes. The null mutant had predicted defects in mating and was hypovirulent in the rabbit meningitis model. However, other virulence phenotypes were also noted. Capsule synthesis was not induced by low iron conditions in vitro or in vivo. Melanin synthesis was depressed in the presence of diphenolic compounds, and *CnLAC1* was found to be down-regulated. These results demonstrate that *CnGPA1* is a central member of a signaling pathway in *C. neoformans* that controls three major virulence factors for *C. neoformans*: capsule synthesis, melanin production, and mating. It was also found that *CnGPA1* mediates its effects specifically through its control of adenylate cyclase. For instance, addition of cAMP to the medium can influence the development of the three phenotypes in *C. neoformans*, i.e., it suppresses the negative phenotypes of capsule, melanin, and mating found in the null mutant (47). *CnGPA1* appears to up-regulate adenylate cyclase-producing cAMP, which is a positive switch to turn on the machinery for capsule production, melanin synthesis, and mating. Further work on the circuitry of this pathway, from its sensing of nutrient deprivation to its expression of major virulence structures and proteins, will be essential in understanding the molecular pathogenesis of this yeast. It is clear that there are different receptors for this pathway, such as low glucose levels for melanin production, low iron concentrations for capsule synthesis, and nutrient starvation for mating. These signals do not appear to overlap until they reach *CnGPA1*. In fact, investigators have even found that the regulation of capsule and phenoloxidase can be discordant under certain conditions (64).

From these early studies in signaling pathways for *C. neoformans*, and with the recent identification of *STE11*-α and *STE12*-α genes in the *MAP* kinase cascade, it is clear that these conserved genes and their circuitry are closely intertwined for the composite virulence of this pathogenic yeast. Although studies in other model organisms may be helpful as general road maps for strategies, specific studies with *C. neoformans* are necessary to understand how these genes fit into the regulation of virulence for this pathogenic fungus. Both the receptors for these environmental cues and many of the downstream effectors, such as the genes for the various kinases and the terminal phenotypic enzymes, remain to be elucidated. However, we now have a focus for some of the central pathways for expression of virulence in *C. neoformans*. These pathways have probably been used by the organism to respond to the environmental conditions and nutrient limitations in its ecological niches, but they have been coopted for their use as circuits for pathogenesis.

Yeast Growth and Responses In Vivo

Growth rates and the genes controlling them are extremely important to the composite virulence of a *C. neoformans* strain. The importance of in vivo growth rate in a strain has been demonstrated in several studies. First, in studies on the null mutant of *CnLAC1*, its importance to virulence was demonstrated only after the gene knockout strain was backcrossed to remove growth rate gene defects produced by the transformation events (133). In gene replacement experiments with the calcineurin A disruptant, the wild-type gene was placed back into the mutant ectopically. Although the phenotype of temperature tolerant growth was recovered at 37°C, under higher temperatures (39°C) the growth rates were substantially slower in the mutant with the ectopic calcineurin gene compared with the wild-type parental strain. In the animal model at 39°C, the partially reconstituted mutant was not able to survive the peak host inflammatory response at 1 week within the central nervous system, as the parental strain did (108). These examples are important considerations as specific genetic factors of virulence are studied. *C. neoformans* strains survive best when the strains have optimal growth rates under all conditions.

One of the most important factors for pathogenic fungal growth is temperature. Only 250 to 300 fungal species have been reported to cause human disease despite over 20,000 known fungal species, and this is largely due to the single fact that most fungi do not grow at the elevated temperatures of the mammalian host. For instance, only rare strains of other *Cryptococcus* species are capable of growth at 37°C. Even *C. neoformans* strains are particularly sensitive to environmental temperatures above 39°C and demonstrate a significant decrease in growth rates at those temperatures, with the development of unusual budding patterns, pseudo-hyphal growth, and prominent vacuolization within the yeast. These structural changes demonstrate a serious stress reaction within the yeast and its replication machinery. Despite these known environmental stresses, *C. neoformans* strains can adapt to a mammalian host and cause disease in animals, such as rabbits, with normal body temperatures of 39.5°C and in humans with fevers over 39°C if certain immunosuppressive conditions are met. It has already been shown that there is at least some genetic control of temperature sensitivity and virulence in *C. neoformans* (83) and that the signaling for this stress response involves calcineurin (108). This adaptability to conditions in the mammalian host environment, such as high temperature, is clearly available to this pathogenic yeast.

This environmental or host-infectious site paradigm should be a cornerstone of future research into the study of virulence factors in *C. neoformans*. One strategy for the identification of virulence genes is to use differential gene expressions so that the importance of certain genes can be determined during infection. There are a variety of molecular techniques for the identification of both up- and down-regulated genes. These include differential hybridization (72, 132), cDNA library subtraction (33), differential display PCR (89, 110), serial analysis of gene expression (151), complementary cDNA microassay (134), and promoter traps with the use of in vivo expression technology (94).

With these molecular techniques, it should be cautioned that the environmental cues within the host and the regulated *C. neoformans* virulence genes may be complex, and investigators will need to set parameters for their investigations. For instance, *C. neoformans* has at least five yeast stages during host infection, with potentially different gene expressions at each stage. These stages of infection are (i) initiation, (ii) dormancy, (iii) reactivation, (iv) dissemination, and (v) proliferation. Studies on virulence genes identified through environmental or infection site-specific regulation will need to take into account these stages of infection.

In the rabbit meningitis model, both differential hybridization and differential display PCR have been used to identify genes that are regulated within the central nervous system of an immunosuppressed host (110). When compared with cells grown under certain defined in vitro conditions, there are genes that are both up- and down-regulated in vivo. The pattern of gene expression changes as infection progresses, and gene patterns of expression early in infection are different from those as infection progresses in an experimental meningitis model. Some of these genes have now been sequenced, and one gene that has been identified as highly up-regulated within the subarachnoid space of a rabbit is the cytochrome C oxidase subunit 1 (COX1). The up-regulation of this gene may be due to the elevated temperatures signaling increased expression, since this gene appears to be up-regulated when yeasts are moved from 30°C to 37°C (115). These observations support the potential importance of mitochondrial function for *C. neoformans* during the stress responses of infection.

It should be emphasized that the work on differentially expressed genes does not ensure their importance in virulence. First, their regulations need to be confirmed by other expression methods, such as Northern blotting or quantitative PCR. Second, site-directed mutants will need to be evaluated for the magnitude of the gene's importance in virulence within animal models. However, in this era of genome analysis, the ability to capture and identify the genes being regulated by *C. neoformans* has great potential in further understanding the virulence traits of this yeast. When the genome of *C. neoformans* is sequenced, these studies using molecular expression to determine gene importance in virulence will be greatly aided.

Other Determinants of Pathogenicity

Site tropism and adherence

The neurotropism of *C. neoformans* has not been conclusively understood despite its impressive clinical presentation, which suggests that this yeast has a particular predilection for the central nervous system. There have been suggestions that high concentrations of natural catecholamines as substrates within the brain allow for vigorous melanogenesis and thus yeast protection from host factors. However, this hypothesis remains unproved. The second possibility is that the yeast possesses specific ligands that bind to receptors present only on neuronal cells and/or choroid plexus cells. There are some data that suggest there is adherence to both glial and lung cells by *C. neoformans* that is trypsin sensitive and inhibitable by aminosugars and disaccharides (99). Merkel et al. (97) have further studied the interaction between adherence to glial cells and the encapsulation of *C. neoformans*. In these studies both encapsulated and nonencapsulated cells appeared to use similar adherent mechanisms, but the nonencapsulated strain was three times more adherent. There have also been *C. neoformans* adherence mutants produced by several nonspecific mutagenic events. Several of these mutants had growth defects, but one mutant without a growth defect was compared with the parental wild-type strain in rats and displayed attenuation of virulence compared to the wild-type strain (98). Furthermore, studies by Ibrahim et al. (57) showed that with the use of an isogenic *C. neoformans* pair consisting of an acapsular strain and its encapsulated parent strain, endothelial cells are prevented from phagocytosis of these *C. neoformans* cells if the cells contain a capsule. These studies suggest that the capsule does have some influence on adherence of the yeast within blood vessels. It is possible that poor encapsulation may be the best form for *C. neoformans* to escape into the vasculature for dissemination. It is also likely that other protein adhesins are produced by the yeast to influence attachment and initiate infection. This area of attachment to cells and specific tissue tropism may be extremely important in the virulence of fungi. For instance, there are also *C. neoformans* strains known to be rhinotropic (28, 42) or dermatotrophic (146) in animal models. The reasons for this tissue tropism within specific strains remain relatively unexplored.

Subtle and complex areas of virulence studies with different strains

It is clear that within and between serotypes of clinical strains of *C. neoformans*, there may be variation in virulence within the same animal models (43, 59, 70, 112). Furthermore, environmental and clinical isolates of *C. neoformans* may not possess the same virulence traits for animals. For instance, several studies have examined dozens of isolates from an environmental source for virulence in mice and have found that multiple isolates from the environment were less virulent than clinical isolates of *C. neoformans* susceptible mouse models (43, 71, 101). It is also clear that prolonged in vitro passage can attenuate strain virulence, but, as in bacteria, passage through animals can restore virulence to an attenuated strain (110a). The plasticity of the *C. neoformans* genome is recognized in studies in which a single isolate demonstrated microevolution in several laboratories as it was being studied (41). These molecular events also translate into changes in virulence phenotypes. This emphasizes the need for investigators to store their isolates for virulence studies carefully. These observations indicate that there are subtle virulence characteristics that are not identified by the major virulence phenotypes in *C. neoformans*. Finally, several other factors may be associated with virulence of *C. neoformans*. There is some suggestion that drug-resistant *C. neoformans* strains may be less virulent. Iwata et al. (63) found that itraconazole- and ketoconazole-resistant mutants of virulent strains were found to be avirulent in mice. However, these studies remain inconclusive until specific gene disruption and replacement experiments can be performed on drug resistance genes. Also, as noted in chapter 14, drug-resistant strains of *C. neoformans* can cause disease in severely immunocompromised hosts. Between the two varieties of *C. neoformans*, virulence potential can vary; specifically, the sensitivity to killer toxins is different in the two varieties (4). Carefully controlled studies of specific phenotypes and genotypes within these strains are needed to determine why the composite virulence phenotypes are different between strains and varieties. We have not yet begun to address these subtle virulence differences at a molecular or biochemical level.

The area of intracellular survival and dormancy is particularly relevant to *C. neoformans*. There has been very little work on the biochemical response of the yeast as it encounters the host cells. Several proteins have been identified that regulate an immune response to the yeast (46, 95). Identification of these proteins and their genes could lead to further studies on how they affect the pathobiology of the yeast and thus contribute to its virulence. With the understanding of eukaryotic biochemistry and the foundation of molecular biology in *C. neoformans*, targeting the biochemical responses of the yeast to the host cell's attack has the potential to lead to important discoveries in the virulence of these eukaryotic pathogens.

Extracellular proteins (enzymes)

Identification and characterization of extracellular proteins in *C. neoformans* are important because proteinases, esterases, and lipases have been associated with virulence in other pathogens, and these proteins may also elicit strong immune responses from the host. Recently, Chen et al. (18) demonstrated that *C. neoformans* can excrete multiple proteins besides proteinases extracellularly and that there are different protein profiles depending on the strains and the growth conditions for each strain. A series of enzymes such as esterases, lipases, β-glucosidases, and acid phosphatase have been identified, and some proteins elicit antibody responses. Although production of killer toxin activity has been found in some *Cryptococcus* species (144), it has not yet been identified in any *C. neoformans* strains. It is likely that some extracellular proteins have profound effects on host immunity and others may have effects on the intrinsic virulence of the pathogen. It will be important to dissect each enzyme's contribution to the virulence profile, and it will not be surprising if multiple extracellular enzymes in *C. neoformans* are found that may make them redundant for pathogenesis in some circumstances. Several of these extracellular proteins in have now been investigated for their impact on virulence in *C. neoformans*.

Urease Urease is a nickel metalloenzyme that catalyzes the hydrolysis of urea to ammonia and carbamate. This reaction allows the environment to rise in pH, and there are media such as Christensen's agar that can detect this environmental pH alteration with a color change. *C. neoformans* appears to invest a significant amount of energy in its urease production, and as noted in chapter 12, it has been used as a rapid diagnostic test for *C. neoformans* since yeasts like *Candida* species do not produce urease.

The potential importance of urease in the virulence composite, besides its use as a scavenger enzyme in its environmental niche during low-nitrogen states, could involve its ability to change local pH in the host, such as when the yeast resides in a phagolysosome. Urease has already been shown to be a significant virulence factor for the gastrointestinal pathogen *Helicobacter pylori*. Furthermore, preliminary clinical data have shown that only rarely does a urease-negative *C. neoformans* strain cause disease; there have been two reported cases of meningitis in AIDS patients with this type of strain (3, 131). Also, there has been no difference in the quantitative urease production between clinical and environmental isolates (9), and there has been only one known environmental urease-negative *C. neoformans* isolate (88). However, to determine the potential importance of urease in the composite virulence phenotype, the most direct experiments include the creation of an isogenic parent and a null mutant in urease production.

Cox et al. (24) have cloned, sequenced, and site-directed the disruption of the urease gene of *C. neoformans*. The urease gene (*CnURE1*) is a single-copy gene with 55 to 60% homology to the sequences of the structural urease genes from jack bean and *H. pylori*, with conservation of all the histidine residues thought to be involved in binding to the nickel metallocenter, in binding to the substrate, and in catalysis.

The observations on the pathogenic effect of *C. neoformans* urease through the use of isogenic wild-type strains and a null mutant represented important lessons in molecular fungal pathogenesis. These studies emphasized the importance of infection site, model used, and influence of this product on histopathology. When the null mutant of *CnURE1* was compared with the wild-type strain in the immunosuppressed rabbit model of experimental meningitis, there was no difference in the numbers of cryptococci found in the cerebrospinal fluid with either strain over a 2-week period of infection. This finding suggested that urease was not important to the growth of *C. neoformans* within the subarachnoid space (24). If the evaluation had stopped with this model, the conclusion would be that urease is probably not a virulence factor. However, in a second set of experiments, the isogenic pair were inoculated intratracheally into a mouse pulmonary model. It was hypothesized that urease may be important during infection in the lung, since many of the other fungal pathogens that produce urease primarily infect the lung. For example, *C. immitis*, *Histoplasma capsulatum*, and *Aspergillus* species produce urease, whereas *Candida* species, which rarely infect the lung, do not produce urease. When the isogenic pair was used in the mouse pulmonary model, there was a dramatic attenuation of virulence during infection with the null mutant. Animals infected with the parental strain had significantly shorter lives, and brain counts of cryptococci were higher. Furthermore, histological examination of lung tissue revealed differences in the host inflammatory response. Infection with the null mutant was associated with more collections of lymphocytes in the lung than was infection with the parental strain (23a). This finding, that an isogenic mutant in urease production might influence histopathological responses, is similar to the concept proposed by Curtis et al. (25). These results suggest that (i) urease activity is a virulence factor for *C. neoformans*, (ii) different animal models and different sites of infection may be necessary to determine the importance of certain virulence traits, and (iii) urease activity appears to directly influence the type of host inflammatory response to infection. These molecular studies emphasize the potential importance of the animal model, site of infection, influence on host factors, and virulence endpoints examined when virulence gene identifications are made. It is possible from this work that urease inhibitors or vaccines could be explored for the management of cryptococcal disease.

Phospholipase Extracellular phospholipase activity has been shown to be a potential virulence factor in *C. albicans*. Chen et al. (19) have also identified phospholipase activity in *C. neoformans* strains. Production of the phospholipases appears to be extracellular, and at least one enzymatic component appears to be phospholipase B. It has been shown that strains of *C. neoformans* vary in their production of phospholipase. When strains are divided into groups of high or low producers of phospholipase, the high-producer strains, when inoculated intravenously into mice, produce larger numbers of cryptococci in brain and lung tissues than do the low-producer strains (20). Vidotto et al. (152) have linked phospholipase production to the size of the capsule and the strain's virulence. These preliminary findings suggest that phospholipase may have a role in the virulence of *C. neoformans*. Recently, a *C. neoformans* phospholipase B gene has been cloned (49). It is hoped that this advance will allow the production of isogenic null mutants of phospholipase, to further help in determining the importance of extracellular phospholipase in the composite virulence of this yeast. It is interesting to hypothesize that phospholipase might be particularly important in the initial pulmonary infection, when high concentrations of phospholipid substrates are present.

REFERENCES

1. **Alspaugh, J. A., J. R. Perfect, and J. Heitman.** 1997. *Cryptococcus neoformans* mating and virulence are regulated by the G-protein alpha subunit GPA1 and cAMP. *Genes Dev.* **11:**3206–3217.
2. **Aoki, S., S. Ito-kowa, K. Nakamura, J. Kato, K. Ninomiya, and V. Vidotto.** 1994. Extracellular proteolytic activity of *Cryptococcus neoformans*. *Mycopathologia* **128:**143–150.
3. **Bava, A. J., R. Negroni, and M. Bianchi.** 1993. Cryptococcosis produced by a urease negative strain of *Cryptococcus neoformans*. *J. Med. Vet. Mycol.* **31:**87–89.
4. **Boekhout, T., and G. Scorzetti.** 1997. Differential killer toxin sensitivity patterns of varieties of *Cryptococcus neoformans*. *J. Med. Vet. Mycol.* **35:**147–149.
5. **Bottone, E. J., B. E. Johansson, M. Toma, and G. P. Wormser.** 1986. Poorly encapsulated *Cryptococcus neoformans* from patients with AIDS. I. Preliminary observations. *AIDS Res. Hum. Retroviruses* **2:**221.
6. **Bottone, E. J., M. Toma, B. E. Johansson, and G. P. Wormser.** 1985. Capsule-deficient *Cryptococcus neoformans* in AIDS patients. *Lancet* **ii:**553.
7. **Brueske, C. H.** 1986. Proteolytic activity of a clinical isolate of *Cryptococcus neoformans*. *J. Clin. Microbiol.* **23:**631.
8. **Bulmer, G. S., and J. R. Tacker.** 1975. Phagocytosis of *Cryptococcus neoformans* by alveolar macrophages. *Infect. Immun.* **11:**73–79.
9. **Canteros, C., C. Rivas, L. Rodero, and G. Davel.** 1996. Quantitative urease detection of clinical and environmental strains of *Cryptococcus neoformans*, abstr. 1.5:150. 3rd International Conference on Cryptococcus and Cryptococcosis, Paris.
10. **Chang, Y. C., R. Cherniak, T. R. Kozel, D. L. Granger, L. C. Morris, L. C. Weinhold, and K. J. Kwon-Chung.** 1997. Structure and biological activities of acapsular *Cryptococcus neoformans* 602 complemented with CAP64 gene. *Infect. Immun.* **65:**1584–1592.
11. **Chang, Y. C., and K. J. Kwon-Chung.** 1994. Complementation of a capsule-deficiency mutation of *Cryptococcus neoformans* restores its virulence. *Mol. Cell. Biol.* **14:**4912–4919.
12. **Chang, Y. C., and K. J. Kwon-Chung.** 1996. Molecular dissection of capsule formation in *Cryptococcus neoformans*, abstr 7.2:132–133. 3rd International Conference on Cryptococcus and Cryptococcosis, Paris.

12a. **Chang, Y. C., and K. J. Kwon-Chung.** 1998. Isolation of the third capsule-associated gene, *CAP60*, required for virulence in *Cryptococcus neoformans*. *Infect. Immun.* **66**:2230–2236.

13. **Chang, Y. C., L. A. Penoyer, and K. J. Kwon-Chung.** 1996. The second capsule gene of *Cryptococcus neoformans*, Cap64, is essential for virulence. *Infect. Immun.* **64**:1977–1983.

13a. **Chang, Y., B. Wickes, and K. J. Kwon-Chung.** 1998. The *STE12α* gene of *Crytococcus neoformans*, abstr. F-63. Abstr. 98th Annu. Meet. Am. Soc. Microbiol. 1998. American Society for Microbiology, Washington, D.C.

14. **Chaturvedi, V., A. Bartiss, and B. Wong.** 1997. Induced expression of bacterial mtlD supplements mannitol production in a mannitol mutant of *Cryptococcus neoformans*, abstr. 019:81. 13th Congress of International Society for Human and Animal Mycology, Italy.

15. **Chaturvedi, V. P., T. Flynn, W. G. Niehaus, and B. Wong.** 1996. Stress tolerance and pathogenic potential of a mannitol-mutant of *Cryptococcus neoformans*. *Microbiology* **142**:937–943.

16. **Chaturvedi, V. P., B. Wong, and S. L. Newman.** 1996. Oxidative killing of *Cryptococcus neoformans* by human neutrophils: evidence that fungal mannitol protects by scavengering reactive oxygen intermediates. *J. Immunol.* **156**:3836–3840.

17. **Chen, L. C., E. S. Blank, and A. Casadevall.** 1996. Extracellular proteinase activity of *Cryptococcus neoformans*. *Clin. Diagn. Lab. Immunol.* **3**:570–574.

18. **Chen, L. C., L. A. Pirofski, and A. Casadevall.** 1997. Extracellular proteins of *Cryptococcus neoformans* and host antibody response. *Infect. Immun.* **65**:2599–2605.

19. **Chen, S. C., L. C. Wright, R. T. Santangelo, M. Muller, V. R. Moran, P. W. Kuchel, and T. C. Sorrell.** 1997. Identification of extracellular phospholipase B, lysophospholipase, acyltransferase produced by *Cryptococcus neoformans*. *Infect. Immun.* **65**:405–411.

20. **Chen, S. C. A., M. Muller, J. Z. Zhou, L. C. Wright, and T. C. Sorrell.** 1997. Phospholipase activity in *Cryptococcus neoformans*: a new virulence factor? *J. Infect. Dis.* **175**:414–420.

21. **Cherniak, R. L., L. C. Morris, T. Belay, E. D. Spitzer, and A. Casadevall.** 1995. Variation in the structure of glucuronoxylomannan in isolates from patients with recurrent cryptococcal meningitis. *Infect. Immun.* **63**:1899–1905.

22. **Collins, H. L., and G. J. Bancroft.** 1991. Encapsulation of *Cryptococcus neoformans* impairs antigen-specific T-cell responses. *Infect. Immun.* **59**:3883–3888.

23. **Courchesne, W., C. Savoy, and D. M. Lupan.** 1996. Mutants of *C. neoformans* defective in mating and virulence, abstr 1.8:151. 3rd International Conference on Cryptococcus and Cryptococcosis, Paris.

23a. **Cox, G. M., et al.** Unpublished data.

24. **Cox, G. M., G. T. Cole, and J. R. Perfect.** 1996. Identification and disruption of the *Cryptococcus neoformans* urease gene, abstr. 1.6:150. 3rd International Conference on Cryptococcus and Cryptococcosis, Paris.

25. **Curtis, J. L., G. B. Huffnagle, G.-H. Chen, M. L. Warnock, M. R. Gyetko, R. A. McDonald, P. J. Scott, and G. B. Toews.** 1994. Experimental murine pulmonary cryptococcosis: differences in pulmonary inflammation and lymphocyte recruitment induced by two encapsulated strains of *Cryptococcus neoformans*. *Lab. Invest.* **71**:113–126.

26. **Cutler, J. E.** 1991. Putative virulence factors of *Candida albicans*. *Annu. Rev. Microbiol.* **45**:187–218.

27. **Denning, D. W., R. W. Armstrong, B. H. Lewis, and D. A. Stevens.** 1991. Elevated cerebrospinal fluid pressures in patients with cryptococcal meningitis and acquired immunodeficiency syndrome. *Am. J. Med.* **91**:267–272.

28. **Dixon, D. M., and A. Polak.** 1986. *In vivo* and *in vitro* studies with an atypical, rhinotrophic isolate of *Cryptococcus neoformans. Mycopathologia* **96:**33–40.

29. **Dixon, D. M., A. Polak, and P. J. Szaniszlo.** 1987. Pathogenicity and virulence of wild-type and melanin-deficient *Wangiella dermatitidis. J. Med. Vet. Mycol.* **25:**97–106.

30. **Dong, Z. M., and J. W. Murphy.** 1995. Effects of the two varieties of *Cryptococcus neoformans* cells and culture filtrate antigens on neutrophil locomotion. *Infect. Immun.* **63:**2632–2644.

31. **Dong, Z. M., and J. W. Murphy.** 1996. Cryptococcal polysaccharides induce L-selectin shedding and tumor necrosis factor receptor loss from the surface of human neutrophils. *J. Clin. Invest.* **97:**689–698.

32. **Drouhet, E., and G. Segretain.** 1951. Inhibition de la migration leucocytaire *in vitro* par un polyoside capsulaire de *Torulopsis (Cryptococcus) neoformans. Ann. Inst. Pasteur* **81:**674.

33. **Duguid, J. R., R. G. Rohwer, and B. Seed.** 1988. Isolation of cDNA of scrapie-modulated RNAs by substractive hybridization of cDNA library. *Proc. Natl. Acad. Sci. USA* **85:**5738–5742.

34. **Dykstra, M. A., L. Friedman, and J. W. Murphy.** 1977. Capsule size of *Cryptococcus neoformans:* control and relationship to virulence. *Infect. Immun.* **16:**129.

35. **Edman, J. C.** 1992. Isolation of telomerelike sequences from *Cryptococcus neoformans* and their use in high-frequency transformation. *Mol. Cell. Biol.* **12:**2777–2783.

36. **Edman, J. C.** 1996. Paradoxes of mating in *Cryptococcus neoformans,* abstr. 1.1:7. 3rd International Conference on Cryptococcus and Cryptococcosis, Paris.

37. **Edman, J. C., and K. J. Kwon-Chung.** 1990. Isolation of the *URA5* gene from *Cryptococcus neoformans* var. *neoformans* and its use as a selective marker for transformation. *Mol. Cell. Biol.* **10:**4538–4544.

38. **Falkow, S.** 1988. Molecular Koch's postulates applied to microbial pathogenicity. *Rev. Infect. Dis.* **10**(Suppl. 2)**:**5274–5276.

39. **Farmer, S. G., and R. A. Komorowski.** 1973. Histologic response to capsule-deficient *Cryptococcus neoformans. Arch. Pathol.* **96:**383–387.

40. **Firestine, S. M., S. Misialek, D. Toffaletti, T. J. Klem, J. R. Perfect, and V. J. Davisson.** 1998. Biochemical role of the *Cryptococcus neoformans* ADE2 protein in fungal de novo purine biosynthesis. *Arch. Biochem. Biophys.* **351:**123–134.

41. **Franzot, S. P., J. Mukherjee, R. Cherniak, L. Chen, J. S. Hamdan, and A. Casadevall.** 1998. Microevolution of a standard strain of *Cryptococcus neoformans* resulting in differences in virulence and other phenotypes. *Infect. Immun.* **66:**89–97.

42. **Fromtling, R. A., G. K. Abruzzo, and A. Ruiz.** 1988. *Cryptococcus neoformans:* a central nervous system isolate from an AIDS patient that is rhinotropic in a normal mouse model. *Mycopathologia* **102:**72–86.

43. **Fromtling, R. A., G. K. Abruzzo, and A. Ruiz.** 1989. Virulence and antifungal susceptibility of environmental and clinical isolates of *Cryptococcus neoformans* from Puerto Rico. *Mycopathologia* **106:**163–166.

44. **Fromtling, R. A., H. J. Shadomy, and E. S. Jacobson.** 1982. Decreased virulence in stable, acapsular mutants of *Cryptococcus neoformans. Mycopathologia* **79:**23–29.

45. **Gabriel, D. W.** 1986. Specificity and gene function in plant-pathogen interactions. *ASM News* **52:**19–25.

46. **Galgiani, J. N.** 1996. Identification and cloning of a 20KDA protein from *Cryptococcus neoformans* that stimulates murine delayed-type hypersensitivity, abstr. I7. 3rd International Conference on Cryptococcus and Cryptococcosis, Paris.

47. **Gao, S., and D. L. Nuss.** 1996. Distinct roles for two G-protein alpha subunits in fungal virulence, morphology and reproduction revealed by targeted gene disruption. *Proc. Natl. Acad. Sci. USA* **93:**14122–14127.

48. **Goldman, D., S. C. Lee, and A. Casadevall.** 1994. Pathogenesis of pulmonary *Cryptococcus neoformans* infection in the rat. *Infect. Immun.* **62:**4755–4761.

49. **Gottfredsson, M., G. M. Cox, M. Ghannoum, and J. R. Perfect.** 1998. Molecular cloning of the *Cryptococcus neoformans* phospholipase B gene, a putative virulence factor, abstr. F-71. Abstr. 98th Annu. Meet. Am. Soc. Microbiol. 1998. American Society for Microbiology, Washington, D.C.

50. **Granger, D. L., J. R. Perfect, and D. T. Durack.** 1985. Virulence of *Cryptococcus neoformans*: regulation of capsule synthesis by carbon dioxide. *J. Clin. Invest.* **76:**508–516.

51. **Hirano, A., H. M. Zimmerman, and S. Levine.** 1965. The fine structure of cerebral fluid accumulation. VII. Reactions of astrocytes to cryptococcal polysaccharide implantation. *J. Neurol.* **24:**386–396.

52. **Hirano, A., H. M. Zimmerman, and S. Levine.** 1964. The fine structure of cerebral fluid accumulation. III. Extracellular spread of cryptococcal polysaccharides in the acute stage. *Am. J. Pathol.* **46:**1–11.

53. **Hirano, A., H. M. Zimmerman, and S. Levine.** 1964. Fine structure of cerebral fluid accumulation. V. Transfer of fluid from extracellular compartments in acute phase of cryptococcal polysaccharide lesions. *Arch. Neurol.* **11:**632–641.

54. **Hirano, A., H. M. Zimmerman, and S. Levine.** 1965. Fine structure of cerebral fluid accumulation. VI. Intracellular accumulation of fluid and cryptococcal polysaccharide in oligodendria. *Arch. Neurol.* **12:**189–196.

55. **Hogan, L. H., B. S. Klein, and S. M. Levitz.** 1996. Virulence factors of medically important fungi. *Clin. Microbiol. Rev.* **9:**469–488.

56. **Huffnagle, G. B., G. H. Chen, J. L. Curtis, R. A. McDonald, R. M. Streiter, and G. B. Toews.** 1995. Down-regulation of the different phase of T cell-mediated pulmonary inflammation and immunity by a high melanin-producing strain of *Cryptococcus neoformans*. *J. Immunol.* **155:**3507–3516.

57. **Ibrahim, A. S., S. G. Filler, M. S. Alcouloumre, T. R. Kozel, J. E. Edwards, and M. A. Ghannoum.** 1995. Adherence to and damage of endothelial cells by *Cryptococcus neoformans* in vitro: role of the capsule. *Infect. Immun.* **63:**4368–4374.

58. **Ideka, R., T. Shinoda, T. Morita, and E. S. Jacobson.** 1993. Characterization of a phenoloxidase from *Cryptococcus neoformans* var. *neoformans*. *Microbiol. Immunol.* **37:**759–764.

59. **Irokanulo, E. A., and C. O. Akveshi.** 1995. Virulence of *Cryptococcus neoformans* serotypes A,B,C, and D for four mouse strains. *J. Med. Microbiol.* **43:**289–293.

60. **Ito, S.** 1993. Biochemistry and physiology of melanin, p. 33–59. *In* N. Levine (ed.), *Pigmentation and Pigmentory Disorders.* CRC Press, Ann Arbor, Mich.

61. **Itoh, T., and H. Kaneko.** 1977. The *in vivo* incorporation of ^{32}P-labeled orthophosphate into pyrophosphatidic acid and other phospholipids of *Cryptococcus neoformans* through cell growth. *Lipids* **12:**804–813.

62. **Itoh, T., H. Waki, and H. Kaneko.** 1975. Changes of lipid composition with growth phase of *Cryptococcus neoformans*. *Agric. Biol. Chem.* **39:**2365–2371.

63. **Iwata, K., T. Yamashita, M. Ohsumi, M. Baba, N. Naito, A. Taki, and N. Yamada.** 1990. Comparative morphological and biological studies on itraconazole- and ketoconazole-resistant mutants of *Cryptococcus neoformans*. *J. Med. Vet. Mycol.* **28:**77–90.

64. **Jacobson, E. S., and G. M. Compton.** 1996. Discordant regulation of phenoloxidase and capsular polysaccharide in *Cryptococcus neoformans*. *J. Med. Vet. Mycol.* **34:**289–291.

65. **Jacobson, E. S., and H. S. Emery.** 1991. Catecholamine uptake, melanization, and oxygen toxicity in *Cryptococcus neoformans*. *J. Bacteriol.* **173:**401–403.

66. **Jacobson, E. S., N. D. Jenkins, and J. M. Todd.** 1994. Relationship between superoxide dismutase and melanin in a pathogenic fungus. *Infect. Immun.* **62:**4085–4086.

67. **Jacobson, E. S., and M. J. Petro.** 1987. Extracellular iron chelation in *Cryptococcus neoformans. J. Med. Vet. Mycol.* **25:**415–418.

68. **Jacobson, E. S., and S. B. Tinnell.** 1993. Antioxidant function of fungal melanin. *J. Bacteriol.* **175:**7102–7104.

69. **Jacobson, E. S., and S. E. Vartivarian.** 1992. Iron assimilation in *Cryptococcus neoformans. J. Med. Vet. Mycol.* **30:**443–450.

70. **Kagaya, K., T. Yamada, Y. Miyakawa, Y. Fukazawa, and S. Saito.** 1985. Characterization of pathogenic constituents of Cryptococcus neoformans strains. *Microbiol. Immunol.* **29:**517–532.

71. **Kao, C. J., and J. Schwartz.** 1957. The isolation of *Cryptococcus neoformans* from pigeon nests. *Am. J. Pathol.* **27:**652–663.

72. **Keath, E. J., A. A. Painter, G. S. Kobayashi, and G. Medoff.** 1989. Variable expression of a yeast-phase-specific gene in *Histoplasma capsulatum* strains differing in thermotolerance and virulence. *Infect. Immun.* **57:**1384–1390.

73. **Kirsch, D. R., and R. R. Whitney.** 1991. Pathogenicity of *Candida albicans* auxotrophic mutants in experimental infections. *Infect. Immun.* **59:**3297–3300.

74. **Kojima, Y., Y. Tsukuda, K. Kawai, A. Tsukamoto, J. Sugiura, M. Sakaino, and Y. Kita.** 1990. Cloning, sequence analysis and expression of lignolytic phenoloxidase genes of the white-rot basidiomycete coriolus hirsutus. *J. Biol. Chem.* **265:**15224–15230.

75. **Kovacs, J. A., A. A. Kovacs, M. Polis, W. C. Wright, V. J. Gill, C. U. Tuazon, E. P. Gelmann, H. C. Lane, R. Longfield, G. Overturf, A. M. Macher, A. S. Fauci, J. E. Parrillo, J. E. Bennett, and H. Masur.** 1985. Cryptococcosis in the acquired immunodeficiency syndrome. *Ann. Intern. Med.* **103:**533–538.

76. **Kozel, T. R.** 1995. Virulence factors of *Cryptococcus neoformans. Trends Microbiol.* **3:**295–299.

77. **Kozel, T. R., and E. C. Gotschlich.** 1982. The capsule of *Cryptococcus neoformans* passively inhibits phagocytosis of the yeast by macrophages. *J. Immunol.* **129:**1675–1680.

78. **Kozel, T. R., W. F. Gulley, and J. Cazin, Jr.** 1977. Immune response to *Cryptococcus neoformans* soluble polysaccharide: immunological unresponsiveness. *Infect. Immun.* **18:**701–707.

79. **Kozel, T. R., G. S. T. Pfrommer, A. S. Guerlain, B. A. Highison, and G. J. Highison.** 1988. Role of the capsule in phagocytosis of *Cryptococcus neoformans. Rev. Infect. Dis.* **10:**S436–S439.

80. **Kwon-Chung, K. J., J. C. Edman, and B. L. Wickes.** 1992. Genetic association of mating types and virulence in *Cryptococcus neoformans. Infect. Immun.* **60:**602–605.

81. **Kwon-Chung, K. J., W. B. Hill, and J. E. Bennett.** 1981. New, special stain for histopathological diagnosis of cryptococcosis. *J. Clin. Microbiol.* **13:**383–387.

82. **Kwon-Chung, K. J., I. Polacheck, and T. J. Popkin.** 1982. Melanin-lacking mutants of *Cryptococcus neoformans* and their virulence for mice. *J. Bacteriol.* **150:**1414–1421.

83. **Kwon-Chung, K. J., and J. C. Rhodes.** 1986. Encapsulation and melanin formation as indicators of virulence in *Cryptococcus neoformans. Infect. Immun.* **51:**218–223.

84. **Kwon-Chung, K. J., A. Varma, J. C. Edman, and J. E. Bennett.** 1992. Selection of Ura5 and Ura3 mutants from the two varieties of *Cryptococcus neoformans* on 5-fluoroorotic acid medium. *J. Med. Vet. Mycol.* **30:**61–69.

85. **Kwon-Chung, K. J., B. L. Wickes, L. Stockman, G. D. Roberts, D. Ellis, and D. H. Howard.** 1992. Virulence, serotype and molecular characteristics of environmental strains of *Cryptococcus neoformans* var *gattii. Infect. Immun.* **60:**1869–1874.

86. **Lazera, M. S., F. D. A. Pires, L. Camillo-Coura, M. M. Nishikawa, C. C. F. Bezerra,**

L. Trilles, and B. Wanke. 1996. Natural habitat of *Cryptococcus neoformans* var. *neoformans* in decaying wood forming hollows in living trees. *J. Med. Vet. Mycol.* **34**:127–131.

87. Levitz, S. M., and D. J. DiBenedetto. 1989. Paradoxical role of capsule in murine bronchoalveolar macrophage-mediated killing of *Cryptococcus neoformans. J. Immunol.* **142**:659–665.

88. Li, A., and S. Wu. 1992. An urease negative *Cryptococcus neoformans. Acta Microbiol. Sin.* **32**:68–71.

89. Liang, P., and A. B. Pardee. 1992. Differential display of eucaryotic messenger RNA by means of the polymerase chain reaction. *Science* **257**:967–971.

90. Littman, M. L., and E. Tsubura. 1959. Effect of degree of encapsulation upon virulence of *Cryptococcus neoformans. Proc. Soc. Exp. Biol. Med.* **101**:773–776.

91. Lodge, J. K., E. Jackson-Machelski, D. L. Toffaletti, J. R. Perfect, and J. I. Gordon. 1994. Targeted gene replacement demonstrates that myristoyl CoA: protein N-myristoyl-transferase is essential for the viability of *Cryptococcus neoformans. Proc. Natl. Acad. Sci. USA* **91**:12008–12012.

92. Lodge, J. K., R. L. Johnson, R. A. Weinberg, and J. I. Gordon. 1994. Comparison of myristoyl-CoA: protein N-myristoyl transferases from three pathogenic fungi: *Cryptococcus neoformans, Histoplasma capsulatum,* and *Candida albicans. J. Biol. Chem.* **269**:2996–3010.

93. Macher, A. M., J. E. Bennett, J. E. Gadek, and M. M. Frank. 1978. Complement depletion in cryptococcal sepsis. *J. Immunol.* **120**:1686–1690.

93a. Madani, N. D., P. J. Malloy, P. Rodriquez-Pombo, A. V. Krishnan, and D. Feldman. 1994. *Candida albicans* estrogen-binding protein gene encodes an oxidoreductase that is inhibited by estradiol. *Proc. Natl. Acad. Sci. USA* **91**:922–926.

94. Mahan, M. J., J. M. Slauch, and J. J. Mekalanos. 1993. Selection of bacterial virulence genes that are specifically induced in host tissues. *Science* **259**:686–688.

95. Martinez-Marino, B., and B. Bolanos. 1995. Characterization of a cytoplasmic protein that inhibits phagocytosis, abstr. F9. Abstr. 95th Annu. Meet. Am. Soc. Microbiol. 1995. American Society for Microbiology, Washington, DC.

96. Megson, G. M., D. A. Stevens, J. R. Hamilton, and D. W. Denning. 1996. D-Mannitol in cerebrospinal fluid of patients with AIDS and cryptococcal meningitis. *J. Clin. Microbiol.* **34**:218–221.

97. Merkel, G., B. Scofield, F. Rescoria, and R. Yang. 1994. Comparison between *in vitro* glial cell adherence and internalization of nonencapsulated strains of *Cryptococcus neoformans. J. Med. Vet. Mycol.* **32**:361–372.

98. Merkel, G., B. Scofield, F. Rescoria, R. Yang, and J. Grosfeld. 1995. Reduced recovery of *Cryptococcus neoformans* adherent mutant from a rat model of cryptococcosis. *Can. J. Microbiol.* **41**:428–432.

99. Merkel, G. J., and B. A. Scofield. 1993. Conditions affecting the adherence of *Cryptococcus neoformans* to rat glial and lung cells *in vitro. J. Med. Vet. Mycol.* **31**:55–64.

100. Miller, M. F., and T. G. Mitchell. 1991. Killing of *Cryptococcus neoformans* strains by human neutrophils and monocytes. *Infect. Immun.* **59**:24–28.

101. Montero-Gei, F., and F. Avardo. 1978. Clinical and epidemiological aspects of cryptococcosis in Costa Rica. The black and white yeasts. Proceedings of the 9th International Conference on Mycoses. Pan American Health Organization, Washington, D.C.

102. Moore, T. D. E., and J. D. Edman. 1993. The alpha mating type locus of *Cryptococcus neoformans* contains a peptide pheromone gene. *Mol. Cell. Biol.* **13**:1962–1970.

103. Murphy, J. W., and G. C. Cozad. 1972. Immunological unresponsiveness induced by cryptococcal polysaccharide assayed by the hemolytic plaque technique. *Infect. Immun.* **5**:896–901.

104. Neilson, J. B., M. H. Ivey, and G. S. Bulmer. 1978. *Cryptococcus neoformans:* pseudo-

hyphal forms surviving culture with *Acanthamoeba polyphaga*. *Infect. Immun.* **20:**262–266.

105. **Nosanchuk, J. D., and A. Casadevall.** 1997. Cellular charge of *Cryptococcus neoformans*: contributions from the capsular polysaccharide, melanin, and monoclonal antibody binding. *Infect. Immun.* **65:**1836–1841.

106. **Nyhus, K. J., and E. S. Jacobson.** 1996. Reduction of ferric iron by *Cryptococcus neoformans*, abstr. 1.4:148. 3rd International Conference on Cryptococcus and Cryptococcosis, Paris.

107. **Odom, A., M. Del Poeta, J. Perfect, and J. Heitman.** 1997. The immunosuppressant FK506 and its non-immunosuppressive analog L-685,818 are toxic to *Cryptococcus neoformans* by inhibition of a common target protein. *Antimicrob. Agents Chemother.* **41:**156–161.

108. **Odom, A., S. Muir, E. Lim, D. L. Toffaletti, J. R. Perfect, and J. Heitman.** 1997. Calcineurin is required for virulence of *Cryptococcus neoformans*. *EMBO J.* **16:**2576–2589.

109. **Perfect, J. R.** 1996. Fungal virulence genes as targets for antifungal chemotherapy. *Antimicrob. Agents Chemother.* **40:**1577–1583.

110. **Perfect, J. R.** 1997. Differential expressions and virulence, abstr. S125. 13th Congress of International Society for Human and Animal Mycology, Italy.

111. **Perfect, J. R., and D. T. Durack.** 1985. Effects of cyclosporine in experimental cryptococcal meningitis. *Infect. Immun.* **50:**22–26.

112. **Perfect, J. R., S. D. R. Lang, and D. T. Durack.** 1980. Chronic cryptococcal meningitis: a new experimental model in rabbits. *Am. J. Pathol.* **101:**177–194.

113. **Perfect, J. R., T. H. Rude, B. Wong, T. Flynn, V. Chaturvedi, and W. Niehaus.** 1996. Identification of a *Cryptococcus neoformans* gene that directs expression of the cryptic *Saccharomyces cerevisiae* mannitol dehydrogenase gene. *J. Bacteriol.* **178:**5257–5262.

114. **Perfect, J. R., D. L. Toffaletti, and T. H. Rude.** 1993. The gene encoding for phosphoribosylaminoimidazole carboxylase (ADE2) is essential for growth of *Cryptococcus neoformans* in cerebrospinal fluid. *Infect. Immun.* **61:**4446–4451.

115. **Perfect, J. R., B. Wong, Y. Chang, K. J. Kwon-Chung, and P. R. Williamson.** 1998. *Cryptococcus neoformans*: virulence and host defenses. *Med. Mycol.* **36,** in press.

116. **Petter, R., A. Varma, T. Boekhout, S. Salas, B. Davis, and K. J. Kwon-Chung.** 1996. Molecular divergence of the virulence factors of *Filobasidiella neoformans* found in other heterobasidiomycetous species, abstr. 2.2:36–38. 3rd International Conference on Cryptococcus and Cryptococcosis, Paris.

117. **Pettoello-Mantovani, M., A. Casadevall, T. R. Kollman, A. Rubinstein, and H. Goldstein.** 1992. Enhancement of HIV-1 infection by the capsular polysaccharide of *Cryptococcus neoformans*. *Lancet* **339:**21–23.

118. **Pettoello-Mantovani, M., A. Casadevall, P. Smarnworawong, and H. Goldstein.** 1994. Enhancement of HIV Type 1 infectivity *in vitro* by capsular polysaccharide of *Cryptococcus neoformans* and *Haemophilus influenzae*. *AIDS Res. Hum. Retroviruses* **10:**1079–1087.

119. **Polacheck, I.** 1991. The discovery of melanin production in *C. neoformans* and its impact on diagnosis and the study of virulence. *Int. J. Med. Microbiol.* **276:**120–123.

120. **Polacheck, I., V. J. Hearing, and K. J. Kwon-Chung.** 1982. Biochemical studies of phenoloxidase and utilization of catechalamines in *Cryptococcus neoformans*. *J. Bacteriol.* **150:**1212–1220.

121. **Polacheck, I., and K. J. Kwon-Chung.** 1988. Melanogenesis in *Cryptococcus neoformans*. *J. Gen. Microbiol.* **134:**1037–1041.

122. **Polacheck, I., Y. Platt, and J. Aronovitch.** 1990. Catecholamines and virulence of *Cryptococcus neoformans*. *Infect. Immun.* **58:**2919–2922.

123. **Polak, A.** 1990. Melanin as a virulence factor in pathogenic fungi. *Mycoses* **33:**215–224.

124. **Rawat, D. S., H. B. Upreti, and S. K. Das.** 1984. Lipid composition of *Cryptococcus neoformans. Microbiologica* **7:**299–307.

125. **Regenfelder, E., T. Spellig, A. Hartman, S. Lavenstein, M. Bolker, and R. Kahmann.** 1996. G-proteins in *Ustilago maydis.* Transmission of multiple signals. *EMBO J.* **16:**1934–1942.

126. **Retini, C., A. Vecchiarelli, C. Monari, C. Tascini, F. Bistoni, and T. R. Kozel.** 1996. Capsular polysaccharide of *Cryptococcus neoformans* induces proinflammatory cytokine release by human neutrophils. *Infect. Immun.* **64:**2897–2903.

127. **Rhodes, J. C., and D. H. Howard.** 1980. Isolation and characterization of arginine auxotrophs of *Cryptococcus neoformans. Infect. Immun.* **27:**910–914.

128. **Rhodes, J. C., I. Polacheck, and K. J. Kwon-Chung.** 1982. Phenoloxidase activity and virulence in isogenic strains of *Cryptococcus neoformans. Infect. Immun.* **36:**1175–1184.

129. **Ro, J. Y., S. S. Lee, and A. G. Ayala.** 1987. Advantage of Fontana-Masson stain in capsule-deficient cryptococcal infection. *Arch. Pathol. Lab. Med.* **111:**53–57.

130. **Rosas, A. L., and A. Casadevall.** 1997. Melanization affects susceptibility of *Cryptococcus neoformans* to heat and cold. *FEMS Microbiol. Lett.* **153:**265–272.

131. **Ruane, P. J., L. J. Walker, and W. L. George.** 1988. Disseminated infection caused by urease-negative *Cryptococcus neoformans. J. Clin. Microbiol.* **26:**2224–2226.

132. **Rude, T. H., and J. R. Perfect.** 1994. Isolation and identification of several site-specific genes of *Cryptococcus neoformans* differentially expressed in the CNS, abstr. F49. Abstr. 94th Annu. Meet. Am. Soc. Microbiol. 1994. American Society for Microbiology, Washington, D.C.

133. **Salas, S. D., J. E. Bennett, K. J. Kwon-Chung, J. R. Perfect, and P. R. Williamson.** 1996. Effect of the laccase gene, CNLAC1, on virulence of *Cryptococcus neoformans. J. Exp. Med.* **184:**377–386.

134. **Schena, M., D. Shalon, R. W. Davis, and P. O. Brown.** 1995. Quantitative monitoring of gene expression patterns with a complementary DNA microassay. *Science* **270:**467–470.

135. **Shaw, C. E., and L. Kapica.** 1972. Production of diagnosis pigment by phenoloxidase activity of *Cryptococcus neoformans. Appl. Environ. Microbiol.* **24:**824–830.

136. **Sherman, F., G. R. Fink, and J. B. Hicks.** 1987. *Laboratory Course Manual for Methods in Yeast Genetics.* Cold Spring Harbor Laboratory, Cold Spring Harbor, N.Y.

137. **Small, J. M., and T. G. Mitchell.** 1986. Binding of purified and radioiodinated capsular polysaccharides from *Cryptococcus neoformans* serotype A strains to capsule-free mutants. *Infect. Immun.* **54:**742–750.

138. **Small, J. M., and T. G. Mitchell.** 1989. Strain variation in the antiphagocytic activity of capsular polysaccharides from *Cryptococcus neoformans* serotype A. *Infect. Immun.* **57:**3751–3756.

139. **Steele, P. E.** 1991. Current concepts of fungal virulence. *Adv. Pathol. Lab. Med.* **4:**107–119.

140. **Still, C. N., and E. S. Jacobson.** 1983. Recombinational mapping of capsule mutations in *Cryptococcus neoformans. J. Bacteriol.* **156:**460–462.

141. **Tolkacheva, T., T. P. McNamara, E. Piekarz, and W. Courchesne.** 1994. Cloning of a *Cryptococcus neoformans* gene, GPA1, encoding a G-protein α-subunit homolog. *Infect. Immun.* **62:**2849–2856.

142. **Torres-Guererro, H., and J. C. Edman.** 1994. Melanin-deficient mutants of *Cryptococcus neoformans. J. Med. Vet. Mycol.* **32:**303–313.

143. **Upreti, H. B., D. S. Rawat, and S. K. Das.** 1984. Virulence, capsule size and lipid composition interrelation of *Cryptococcus neoformans. Microbiologica* **7:**371–374.

144. **Vadkertiova, R., and E. Slavikova.** 1995. Killer activity of yeasts isolated from the water environment. *Can. J. Microbiol.* **41:**759–766.

145. **Valentine, J. S., and E. B. Gralla.** 1997. Delivering copper inside yeast and human cells. *Science* **278**:817–818.

146. **Van Cutsem, J., J. Fransen, F. Van Gerven, and P. A. J. Janssen.** 1986. Experimental cryptococcosis: dissemination of *Cryptococcus neoformans* and dermatropism in guinea pigs. *Mykosen* **29**:561–575.

147. **Varma, A., J. C. Edman, and K. J. Kwon-Chung.** 1992. Molecular and genetic analysis of *URA5* transformants of *Cryptococcus neoformans*. *Infect. Immun.* **60**:1101–1108.

148. **Vartivarian, S. E., E. J. Anaissie, R. E. Cowart, H. A. Sprigg, M. J. Tingler, and E. S. Jacobson.** 1993. Regulation of cryptococcal capsular polysaccharide by iron. *J. Infect. Dis.* **167**:186–190.

149. **Vecchiarelli, A., C. Retini, C. Monari, C. Tascini, F. Bistoni, and T. R. Kozel.** 1996. Purified capsular polysaccharide of *Cryptococcus neoformans* induces interleukin-10 secretion by human monocytes. *Infect. Immun.* **64**:2846–2849.

150. **Vecchiarelli, A., C. Retini, D. Pietrella, C. Monari, C. Tascini, T. Beccari, and T. R. Kozel.** 1995. Down-regulation by cryptococcal polysaccharide of tumor necrosis factor alpha and interleukin-1 beta secretion from human monocytes. *Infect. Immun.* **63**:2919–2923.

151. **Velculescu, V. E., L. Zhang, B. Vogelstein, and K. W. Kinzler.** 1995. Serial analysis of gene expression. *Science* **270**:484–487.

152. **Vidotto, V., A. Sinicco, D. Di Fraia, S. Cardaropoli, S. Aoki, and S. Ito-Kuwa.** 1997. Phospholipase activity in *Cryptococcus neoformans*. *Mycopathologia* **136**:119–123.

153. **Vincent, V. L., and L. S. Klig.** 1995. Unusual effect of myo-inositol on phospholipid biosynthesis in *Cryptococcus neoformans*. *Microbiology* **141**:1829–1837.

154. **Wang, Y., P. Aisen, and A. Casadevall.** 1995. *Cryptococcus neoformans* melanin and virulence: mechanism of action. *Infect. Immun.* **63**:3131–3136.

155. **Wang, Y., P. Aisen, and A. Casadevall.** 1996. Melanin, melanin "ghosts," and melanin composition in *Cryptococcus neoformans*. *Infect. Immun.* **64**:2420–2424.

156. **Wang, Y., and A. Casadevall.** 1994. Decreased susceptibility of melanized *Cryptococcus neoformans* to UV light. *Appl. Environ. Microbiol.* **60**:3864–3866.

157. **Wang, Y., and A. Casadevall.** 1994. Susceptibility of melanized and nonmelanized *Cryptococcus neoformans* to nitrogen- and oxygen-derived oxidants. *Infect. Immun.* **64**:3004–3007.

158. **Wang, Y., and A. Casadevall.** 1994. Growth of *Cryptococcus neoformans* in presence of L-dopa decreases its susceptibility to amphotericin B. *Antimicrob. Agents Chemother.* **38**:2648–2650.

159. **Wickes, B. L., U. Edman, and J. C. Edman.** 1997. The *Cryptococcus neoformans* STE12 gene: a putative *Saccharomyces cerevisiae* STE12 homologue that is mating type specific. *Mol. Microbiol.* **26**:951–960.

160. **Wickes, B. L., M. E. Mayorga, U. Edman, and J. C. Edman.** 1996. Dimorphism and haploid fruiting in *Cryptococcus neoformans* association with the alpha-mating type. *Proc. Natl. Acad. Sci. USA* **93**:7327–7331.

161. **Williamson, P.** 1997. Laccase and melanin in the pathogenesis of *Cryptococcus neoformans*. *Front. Biosci.* **2**:99–107.

162. **Williamson, P. R.** 1994. Biochemical and molecular characterization of the diphenol oxidase of *Cryptococcus neoformans*: identification as a laccase. *J. Bacteriol.* **176**:656–664.

163. **Williamson, P. R., and S. Zhang.** 1996. Transcriptional regulators of CNLAC1, abstr. 1-4:16–17. 3rd International Conference on Cryptococcus and Cryptococcosis, Paris.

164. **Wong, B., J. R. Perfect, S. Beggs, and K. A. Wright.** 1990. Production of the hexitol D-mannitol by *Cryptococcus neoformans* in vitro and in rabbits with experimental meningitis. *Infect. Immun.* **58**:1664–1670.

7 | Physical Defenses and Nonspecific Immunity

A human cryptococcal infection can be considered the successful establishment of *Cryptococcus neoformans* in the ecological niche defined by a mammalian host. Mammalian tissues present a difficult and hostile environment for *C. neoformans*, even in the absence of specific anticryptococcal immunity. The fungus must survive in conditions of elevated body temperature, slightly alkaline pH, nutrient depletion, and constant attack by host effector cells. Hence, mammalian hosts are initially protected by a combination of physical barriers, unfavorable conditions for fungal growth, and a multitude of effector cells capable of killing or inhibiting *C. neoformans*. The low frequency of symptomatic cryptococcal infections despite the presumably high frequency of exposure suggests that physical barriers and nonspecific immunity provide adequate defenses to protect the host against infection. This chapter describes physical barriers and humoral and cellular nonspecific defense mechanisms that are believed to provide the first line of defense against cryptococcal infection.

PHYSICAL FACTORS

Temperature

Mammalian temperatures are usually significantly higher than environmental temperatures. *C. neoformans* can grow at normal human temperature, but the rate of growth at 37°C is generally slower than that at 25 or 30°C. At temperatures above 37°C, *C. neoformans* grows poorly (117). Hence, normal mammalian temperatures in the range of 37 to 38°C may provide some protection against infection by slowing fungal growth. At temperatures above 40°C, *C. neoformans* ceases to grow and slowly loses viability (117). These observations suggest that fever could provide some protective effects against *C. neoformans*. Several animal experiments support the notion that the high physiological temperature of mammals provides protection against *C. neoformans*. Mice infected with *C. neoformans* live longer when maintained at temperatures of 35 to 36°C than do infected mice maintained at lower temperatures (117, 118). Chick embryos inoculated with *C. neoformans* are more likely to survive if incubated at temperatures above 39°C (103). The higher body temperature of rabbits (39.5°C) may contribute to their resistance to infection (117). Experimental infection in rabbits revealed higher *C. neoformans* growth in testis than in the abdomen, a finding that probably reflected the lower temperature of the scrotal sac (9). However, other factors than temperature alone must contribute to the resistance of rabbits to *C. neoformans*, since corticosteroid administration makes this species susceptible to infection (183). *C. neoformans* extracts are pyrogenic in rabbits. Rabbits respond to *C. neoformans* infection with fever (40 to 41°C), which may further inhibit fungal replication (104). In contrast to rabbits, mice and humans do not consistently mount febrile responses when infected with *C. neoformans* (103).

An anecdotal case report supports a role for fever in defense against *C. neoformans*. A patient with simultaneous malaria and cryptococcosis was found to have a disappearance of yeast cells in the cerebrospinal fluid during febrile episodes caused by relapsing malaria (104). In 1949 Kligman and Weidman (104) suggested that fever therapy, possibly induced with malaria, should be considered for therapy of cryptococcosis, given the hopeless prognosis of cryptococcosis at that time. On the basis of results with chick embryo experiments, Kligman et al. (103) calculated that the cure for human cryptococcosis would require prolonged hyperpyrexia of 40°C (104°F) for 8 days. It is not known whether fever therapy was ever used for therapy of *C. neoformans* infections.

Some heat-intolerant strains of *C. neoformans* have been described that are rhinotrophic (44, 52). These strains grow poorly or are killed at mammalian core temperatures. Rhinotrophism presumably reflects a preference for growth in the cooler areas of the body (44, 52).

pH

C. neoformans is susceptible to killing and growth inhibition by alkaline pH. In 1961 Howard (88) showed that the growth of *C. neoformans* was partially inhibited at pH 7.3. The slightly alkaline pH of human serum (pH 7.4) may provide an unfavorable environment for the growth of some strains of *C. neoformans*.

PHYSICAL BARRIERS

Skin

The intact skin provides an effective barrier to many fungal infections (218). The rarity of primary cutaneous cryptococcal infections is a testament to the effectiveness of the skin in preventing *C. neoformans* invasion. Cutaneous involvement in human cryptococcal infections is common, but it is almost always a result of systemic dissemination (198). Human primary cutaneous cryptococcal infection is usually the result of injury to the skin with a contaminated object. Two unequivocal human cases of primary cutaneous cryptococcosis have resulted from accidental needlestick injuries with needles contaminated with *C. neoformans* (27, 59). The inoculum was low in one accident and high in the other, yet both developed local infection (27, 59).

Nasal Passages

The nasal and upper respiratory airways may be sites for the deposition of *C. neoformans* cells inhaled through the nares. Air turbulence in the nasal cavity results in the deposition of particles within the nasal mucosa by direct physical impingement or by gravity settling (6, 229). Mucociliary flow results in transport of trapped particles posteriorly, where they are presumably swallowed (6). Despite the fact that the nasal epithelium is likely to be a pivotal site for the initial contact of *C. neoformans* and host, little is known about nasal defenses against *C. neoformans*. Susceptibility to nasal cryptococcal infection varies with the animal species studied. Nasal cryptococcosis is frequent in some animals but rare in humans. For example, cats and dogs frequently present with nasal and facial cryptococcosis (see chapter 10). Asymptomatic carriage of *C. neoformans* in the nasal epithelium was demonstrated in 14% of 56 dogs and 7% of 45 cats in a study from Australia (149). It has been suggested that the nasal epithelium in animals is a common site for initial infection (149).

Among laboratory animals, mice and rats can be readily infected by nasal inoculation of *C. neoformans*, but guinea pigs are resistant (119). Intranasal instillation of *C. neoformans* in Swiss Webster mice results in persistent infection that disseminates after 14 to 28 days (3). A comparative study of the relative invasiveness of *C. neoformans* and pathogenic bacteria (*Staphylococcus aureus*, group B *Streptococcus*) and fungi (*Aspergillus fumigatus*, *Blastomyces dermatitidis*, *Histoplasma capsulatum*, *Coccidioides immitis*, and *Candida albicans*) after intranasal instillation revealed that only *C. neoformans* attached and invaded the nasal epithelium of mice and rats (119).

In guinea pigs the nasal mucosa is an effective barrier against *C. neoformans*. Intranasal instillation of *C. neoformans* in guinea pigs with normal olfactory mucosa results in rapid clearance of fungal cells without infiltration by immune cells and without histological evidence of dissemination (143). However, damage to the nasal epithelium with zinc sulfate results in invasion of yeast into subepithelial tissues (143).

In humans *C. neoformans* can cause a rhinophyma-like condition (150). *C. neoformans* has been reported in the subarachnoid space around olfactory nerves in human immunodeficiency virus (HIV)-infected patients with meningeal cryptococcosis (142). This presumably results from drainage of infected cerebrospinal fluid into the nose rather than direct extension from the nasal cavity (142). A cerebrospinal fluid drainage connection that can carry red cell-sized particles from the subarachnoid space to the nasal mucosa has been demonstrated in humans (147).

Two strains of *C. neoformans* have been described that are rhinotrophic in mice (44, 52). One strain produces gross nasal enlargement with granulomatous and ulcerated lesions in the nasal epithelium (44). The other strain also produces granulomatous lesions in nasal tissues as well as periocular tissues, ears, feet, and tail (52). Both strains grow best at temperatures lower than 37°C; the rhinotrophism exhibited by this strain may be a result of cooler temperatures in the nares and nasal epithelium of the mouse (44).

In summary, *C. neoformans* has the capacity to invade the nasal epithelium in some animal species. The existence of a connection between the nasal cavity and the subarachnoid space indicates that contiguous extension is a possible route for *C. neoformans* entry into the central nervous system (CNS). Nasal passages may also become secondarily infected from a subarachnoid infection by contiguous extension. The extent to which a nasal route of infection contributes to the establishment and dissemination of *C. neoformans* infections in humans is unclear.

Eye

Primary *C. neoformans* involvement of eye structures is rare but can occur (10, 224). In contrast, secondary involvement of the eye is not an infrequent event in patients with disseminated cryptococcal infections (101). Eye involvement is a relatively uncommon event in patients with AIDS and cryptococcal meningitis (145). The eye can be involved by direct extension of infection from the brain along the meningeal sheaths or by hematogenous dissemination (84). Catastrophic visual loss can occur, but this is usually the result of optic nerve involvement rather than infection of the orbit and the eye (57, 101, 178, 189). The eye was implicated as the portal of entry for *C. neoformans* in a rare case of cryptococcosis following transplantation with an infected cornea (10). The mechanisms of host defense in the eye against *C. neoformans* are unknown, but granulomatous chorioretinitis appears to be a common histological finding (84). The rarity of primary eye infections strongly suggests that ocular defenses are highly effective.

Lung

Human infection is generally believed to occur following inhalation of aerosolized *C. neoformans*. The evidence that the lungs are the portal of entry for *C. neoformans* is largely circumstantial but a pulmonary route of infection is widely accepted (see chapter 11). Physical protection against pulmonary infection includes upper airway turbulence that can result in the deposition of larger dust particles in the airways and nasal passages such that these do not reach the alveoli (6, 229). Most particles with a size of 5 to 10 μm are retained in the upper airways. Since the diameter of *C. neoformans* yeast cells ranges from 4 to 20 μm, it is unlikely that the normal-sized yeast cell with an enlarged capsule can reach the alveolar space (45). However, air sampling above contaminated sites in the environment has revealed airborne *C. neoformans* particles of 1 to 5.5 μm, a size compatible with alveolar deposition (185) (see chapter 3). When an encapsulated strain was grown in soil for several weeks, the average diameter of the cells significantly decreased, with some yeast cells being in the 0.65- to 2.0-μm range (176). Hence, small *C. neoformans* particles capable of alveolar penetration may exist in soils, and soils may actually promote the growth of small-sized variants.

It is possible that the infective form of *C. neoformans* is not a yeast cell but rather the basidiospore (30, 45). Basidiospores range from 1.8 to 2.5 μm in diameter and are produced as a result of sexual reproduction by the perfect form of *C. neoformans*, *Filobasidiella neoformans* (30). Germination of basidiospores produces typical *C. neoformans* yeast cultures, and air-sampling techniques would not distinguish between small yeast cells and basidiospores (30). An argument against the basidiospore being the infecting form is that sexual reproduction should result in equal numbers of α- and **a**-mating types, yet clinical and environmental isolates are primarily α-mating type (120). However, basidiospores may still be the infecting form if monothallic reproduction with spore formation by *C. neoformans* occurs in the environment. Asexual reproduction with spore formation has recently been described for several α-mating-type strains (225) (see chapter 3).

Airway turbulence and ciliary function of the bronchial epithelium are likely to prevent most inhaled yeast forms from reaching the alveoli, except for small hypocapsular forms and basidiospores. However, penetration of *C. neoformans* to the level of alveoli may not actually be required for infection. In the rat model of experimental infection, a histopathological study of intratracheal infection showed that yeast cells invaded at the level of the bronchiole, suggesting that penetration to the alveoli may not be a requirement for *C. neoformans* infection (60). Macrophages resident in the alveolar space presumably provide defense against infectious yeast forms that reach the alveolar space (see below).

Gastrointestinal Tract

A gastrointestinal route of infection has not been completely excluded by epidemiological or animal experimentation studies. *C. neoformans* could reach the gastrointestinal tract through two routes: (i) swallowing of mucous secretions containing yeasts removed from inhaled air and/or (ii) ingestion of contaminated foods. *C. neoformans* was originally isolated from peach juice and has been found in a variety of fruits (see chapter 3). Several studies have isolated *C. neoformans* from locations in the human gastrointestinal tract. In 1917 Anderson (4) showed the presence of various species of cryptococci in the feces of normal individuals. In 1944 Felsenfeld (49) reported the identification of *Cryptococcus* species in the stools of 24% of patients admitted to a Chicago hospital. Although the isolates in these early studies were not rigorously identified as *C. neoformans* by present standards, recent studies suggest that *C. neoformans* can be occasionally found in the gastrointestinal tract. One study demonstrated *C. neoformans* in throat swabs of 6 of 92 patients with cancer (187).

Relatively little is known about gastrointestinal defense mechanisms against this fungus. Gastrointestinal infection with *C. neoformans* can be established in laboratory animals (68, 197, 208). Disseminated cryptococcosis was observed in monkeys fed banana sandwiches infected with *C. neoformans* (208). Sethi (202) fed *C. neoformans*-infected mouse brains to hamsters and induced infection in 4 of 14 animals (202). In two hamsters, *C. neoformans* was recovered from spleen and liver without pulmonary infection, suggesting a gastrointestinal route for systemic infection. In mice, gastric inoculation of Swiss White outbred and C3H/HE mice with *C. neoformans* resulted in persistent stool shedding of cryptococci (68). *C. neoformans* was viable in mouse stool pellets for at least 6 months (68). Among Swiss White outbred and C3H/HE mice, the percentage of mice with persistent stool shedding was 15.4% and 44.4%, respectively, and some developed fatal disseminated infection (68). In another study, oral infection of congenitally immunodeficient athymic nude (*nu/nu*) BALB/c mice resulted in colonization of the alimentary tract, with cerebral dissemination after 3 to 4 weeks (197). Immunocompetent heterozygous (*nu/+*) BALB/c mice were more resistant to gastrointestinal tract colonization, suggesting an important role for T-cell immunity in mucosal defense (197).

Several cases of human gastrointestinal cryptococcosis have been described in the absence of meningeal disease (18, 36, 90, 94). *C. neoformans* has been shown histologically to invade the gastric, duodenal, and colonic mucosa (18). Aphthous ulcers caused by *C. neoformans* have been described in patients with AIDS (51). Two patients with hyper-immunoglobulin E (IgE) syndrome have been reported to have localized gastrointestinal *C. neoformans* infection (90, 94). One patient had an erosive esophageal cryptococcoma and presented with massive hematemesis (94). The other patient had granulomatous colitis resembling Crohn's disease and a chronic perirectal abscess caused by *C. neoformans* (90). These cases illustrate the potential of *C. neoformans* to invade the human gastrointestinal tract.

The observation that disseminated infection can follow ingestion of *C. neoformans* in mice and monkeys suggests that the gastrointestinal tract could serve as a portal of entry for this fungal pathogen. Although rare, human cases of gastrointestinal cryptococcosis also suggest the possibility of acquisition of infection through a gastrointestinal portal of entry. At this time, the frequency and importance of a gastrointestinal route of infection in humans are unknown.

ATTACHMENT TO HOST TISSUES

Attachment to host tissues is usually a necessary event for successful microbial infection. Despite the importance of attachment interactions for microbial pathogenesis and virulence, relatively little is known about the mechanisms by which *C. neoformans* attaches to host cells. Microbial structures by which microbes attach themselves to host cells are often called adhesins. For *C. neoformans*, no adhesins have been described thus far. The paucity of knowledge about *C. neoformans* adhesins is in marked contrast to that of *C. albicans*, in which the study of adhesins has received considerable attention (23). *C. neoformans*, like *C. albicans*, probably has specific structures that allow fungal cells to adhere to and colonize mammalian epithelial surfaces. Oral and nasal administration of *C. neoformans* in mice results in mucosal colonization (3, 197), a finding consistent with and indicative of attachment of yeast cells to mucosal surfaces.

Incubation of *C. neoformans* with primary rat lung cultures results in attachment and internalization of some yeast forms by epithelial cells (152). Similarly, yeast cell attachment occurs after incubation of *C. neoformans* with glial cells (153). Adherence of yeast cells was influenced by culture age, glucose concentration, and growth temperature (152). The best condition for attachment was culture growth at 37°C for 48 h. Treatment of yeast cells with proteinase inhibited adherence, suggesting the existence of a protein adhesin on *C. neoformans* (152). Addition of several carbohydrates also inhibited attachment, suggesting the involvement of carbohydrate receptors (153). Comparison of the adherence of encapsulated and nonencapsulated strains to glial cells revealed that the nonencapsulated strains attached better than the encapsulated strains (154). A *C. neoformans* hypoadherence mutant has been described that is less virulent in rats, as indicated by increased clearance from lungs after pulmonary infection (155).

C. neoformans cells can bind to glycosphingolipid lactosylceramide (Galβ1-4Glcβ1-1Cer), a compound widely distributed in epithelial tissues that has been suggested as a possible attachment receptor in human tissues (96). The fungal structure responsible for binding lactosylceramide is unknown. Since interference with attachment of *C. neoformans* to host tissues could prevent infection, the identification and characterization of the fungal adhesin molecules may be an important investigation for vaccine design.

NONSPECIFIC HUMORAL DEFENSES

Antifungal Substances in Serum and Saliva

Human serum can inhibit the growth of *C. neoformans* in vitro (7, 8, 76, 88, 92, 173, 206, 207). Sera from several mammalian species can also inhibit the growth of *C. neoformans,* but avian sera appear to lack this activity and may even promote fungal growth (188). The growth inhibitory factor is destroyed by trypsin and is nondialyzable (92). Growth inhibition by human sera requires the addition of at least 13% serum to fungal media and is associated with the alpha-2-globulin and gamma-2-globulin protein fractions (206). Another study associated serum growth inhibitory activity with the beta-2-globulin fraction (92). Human cord blood has lower growth inhibitory activity than does normal serum (92). The growth inhibitory effect of human serum does not appear to be related to specific *C. neoformans* antibodies (92, 206). Heating to 56°C has little or no effect on the inhibitory action of serum; this argues against a complement-mediated effect (8, 92, 206). However, heating to 70°C produces protein precipitation and abolishes the fungistatic property of serum (92). Nassar et al. (173) have confirmed that the inhibitory effect of human serum is donor independent, heat stable, and not due to albumin or globulin (173). These investigators demonstrated two components in human serum that have activity against *C. neoformans*: a macromolecular component that is inhibitory and a low-molecular-mass (,000 Da) substance that can enhance the efficacy of the antifungal drug fluconazole (173). Murine and bovine sera appear to lack *C. neoformans* growth inhibitory activity (173).

Addition of iron as ferrous ammonium chloride or ferric chloride has been demonstrated to reverse the growth inhibitory effects of human serum in some studies (76, 206) but not in others (71). These differences may reflect differences in experimental conditions. Iron is an essential nutrient for *C. neoformans* (214), and growth inhibition can be expected in conditions where iron is a limiting nutrient. Part of the fungistatic effect of serum could involve sequestration of iron by serum proteins, with consequent depletion of this essential nutrient. Additional evidence for the importance of iron in the growth of *C. neoformans* comes from the observation that deferoxamine can exacerbate experimental infection in guinea pigs (14). However, serum-mediated fungistasis has also been demonstrated after iron supplementation, suggesting the existence of other, yet uncharacterized, antifungal factors in human sera (71).

As noted above, cryptococcal infection of the oral cavity in humans is extremely rare. This may be due, in part, to parotid saliva, which contains a fungicidal substance for *C. neoformans* (92). Unlike the fungistatic serum component, the fungicidal substance in saliva is dialyzable and is stable after heating for 2 h at 72°C (92). This fungicidal substance in saliva has been proposed to provide protection against infection (92).

The exact role of human serum and saliva humoral factors in protection against *C. neoformans* is uncertain, but it is likely that they contribute to host defense against cryptococcosis. In contrast to serum and saliva, human cerebrospinal fluid has no inhibitory activity against *C. neoformans* (92). The inability of human cerebrospinal fluid to inhibit *C. neoformans* growth may contribute to the marked propensity of this fungus to proliferate within the subarachnoid space.

Surfactant and Collectins

Antimicrobial substances in surfactant may contribute to innate resistance against inhaled *C. neoformans*. Primitive opsonins, termed collectins, which are found in surfactant and serum, have been shown to bind to acapsular variants of *C. neoformans* but not to encapsulated cells (200). One collectin, pulmonary surfactant protein SP-D, agglutinates acapsular *C. neoformans* (200). The role of surfactant and collectins in innate resistance to *C. neoformans* is presently unknown. However, the demonstration of collectin binding to some forms of *C. neoformans* raises the possibility that these proteins play a role in the early antifungal defense mechanisms of the lung (200).

Complement

The complement system is a nonspecific humoral defense system that can provide opsonins for phagocytosis and chemotactic factors for the recruitment of inflammatory cells. Complement system activation can occur via two pathways known as the classical and the alternative pathways. Activation of the classical pathway is triggered by antigen-antibody complexes and requires the presence of specific antibody. Activation of the alternative pathway results from the interaction of complement proteins with microbial surfaces and does not require the presence of specific antibody. Both pathways lead to the activation of complement component C3, the binding ligand for complement receptors, which promote phagocytosis.

Many fungi, including *C. neoformans*, are potent activators of the complement system (106). The ability of *C. neoformans* to activate complement was shown in the late 1960s when Goren and Warren (62) demonstrated C3 binding on the cryptococcal capsule by immunofluorescence studies. Since then, the interaction of *C. neoformans* with the complement system has been extensively studied. (For a comprehensive review of the interactions of the complement system with fungi, see reference 106.)

Evidence for the importance of complement in protection

The importance of the complement system in host defense against *C. neoformans* was established by demonstrating that complement deficiency is associated with increased mortality in laboratory animals. Two animal species with congenital complement deficiency have been studied: C4-deficient guinea pigs and C5-deficient mice. C4 deficiency in guinea pigs was not associated with enhanced susceptibility to infection, but depletion of treatment of C3 by cobra venom factor resulted in significantly increased mortality (40, 41). Presumably, C4 deficiency has no effect on susceptibility to infection because C4 is involved in classical pathway activation, and the experiments used nonimmune guinea pigs that lacked specific antibody. In contrast, C3 depletion was associated with markedly depressed neutrophil function, which reflects the crucial role of this complement component in opsonization (41). These results were suggestive of, and consistent with, complement activation in *C. neoformans* infection through the alternative pathway.

Congenital C5 deficiency (C5$^-$) in mice has been associated with enhanced susceptibility to *C. neoformans* (191). Intravenous *C. neoformans* infection in C5$^-$ mice results in an acute pneumonia characterized by sequestration of cryptococci in the lung and pulmonary edema (190). Histopathological examination of lung tissue from C5$^-$ and C5$^+$ mice given intravenous infection revealed larger numbers of polymorphonuclear cells and mononuclear cells in the lungs of C5$^+$ mice at early times after infection, consistent with a chemotactic defect resulting from C5 deficiency (190). In vitro complement activation by the *C. neoformans* capsule generates a chemotactic response by neutrophils (121) that is presumably absent in C5$^+$ mice. As in guinea pigs, cobra toxin-mediated C3 depletion reduced survival in DBA/1J, C3H/HeJ, and BALB/c mice infected with *C. neoformans* (65). Neutralization of C3 by administration of a C3-binding monoclonal antibody (MAb) has also been demonstrated to significantly shorten average survival times in mice (35).

The complement system may also contribute to defense against *C. neoformans* by enabling antibody function. The efficacy of rabbit immunoglobulin in mediating protection in mice is dependent on an intact complement system, indicating a role for complement in antibody-mediated protection (65). Hence, the complement system appears to have a central role in host protection by contributing opsonins and chemotactic factors and enhancing the efficacy of anticryptococcal antibody.

Complement activation by *C. neoformans*

The interaction of *C. neoformans* with the complement system is complex. The pathway of complement activation by *C. neoformans* is dependent on the presence of specific antibody to fungal antigens and the state of encapsulation of the yeast cell (105). Normal human sera usually lack sufficient capsule-binding antibody to activate the complement system by the classical pathway (87). In contrast, human sera commonly contain antibodies to glucans, which bind to the *C. neoformans* cell wall and can activate the classical pathway (100). However, in encapsulated cells the polysaccharide capsule blocks access of these antibodies to the yeast cell surface (227). Hence, classical pathway activation is unlikely to occur for encapsulated *C. neoformans* in nonimmune hosts, because capsule-binding antibodies are absent and the polysaccharide capsule blocks access of naturally occurring antibodies to cell wall epitopes.

Two lines of evidence indicate sole reliance on the alternative complement pathway for complement activation by encapsulated cells. First, comparison of the kinetics for the binding of ^{125}I-labeled C3 to the *C. neoformans* capsule revealed no differences between normal serum and a reconstituted alternative complement pathway consisting of purified factors B, D, H, I, C3, and properdin (115). This result demonstrates that C3 deposition on the capsule can be accomplished entirely by the alternative complement pathway (115). Second, complement deposition following incubation of *C. neoformans* in nonimmune human serum is a relatively slow process consistent with alternative pathway activation (see below).

The comparison of complement activation by encapsulated and nonencapsulated cells has provided important insights into the interaction of *C. neoformans* with the complement system (114, 116, 184, 227). C3 deposition on encapsulated and nonencapsulated *C. neoformans* strains is qualitatively and quantitatively different. Nonencapsulated cells activate both the classical and alternative complement pathways, leading to C3 deposition on the yeast cell surface (114). Classical pathway activation by nonencapsulated cells is triggered by naturally occurring antibodies to fungal wall components (100). Such complement activation by the classical pathway occurs rapidly and results in the deposition of C3 throughout the fungal cell surface (114). In contrast, complement activation by the alternative pathway is slower, being characterized by a lag of 6 to 8 min before significant amounts of C3 are deposited on the fungal surface (114). More C3 molecules bind to encapsulated cells than to nonencapsulated cells. In nonencapsulated cells, C3 is deposited on the surface of the yeast cell, whereas in encapsulated cells C3 binds to the capsule. Furthermore, the pattern of complement activation is markedly different depending on whether a capsule is present. In nonencapsulated strains, C3 deposition occurs rapidly as a result of simultaneous initiation by antibodies to the cell wall binding to multiple sites on the surface of the yeast cell. In encapsulated strains C3 binding begins in discrete foci, which then expand to form large patches of C3 deposition. Incubation of encapsulated *C. neoformans* cells in normal human serum results in the deposition of an average of 1×10^7 to 5×10^7 molecules of C3 (or C3 fragments) on each yeast cell (105). C3 deposition occurs throughout the capsule and is particularly heavy at the capsular surface, which places it in position to interact with the C3 complement receptor CR3 in phagocytic cells (108) (Fig. 1). There is a positive correlation between capsular volume and the maximal amount of C3 that can be deposited in the capsule (233).

The kinetics of conversion of C3b to iC3b are different depending on whether or not the *C. neoformans* cell has a capsule (184) (Fig. 2). In encapsulated cells there

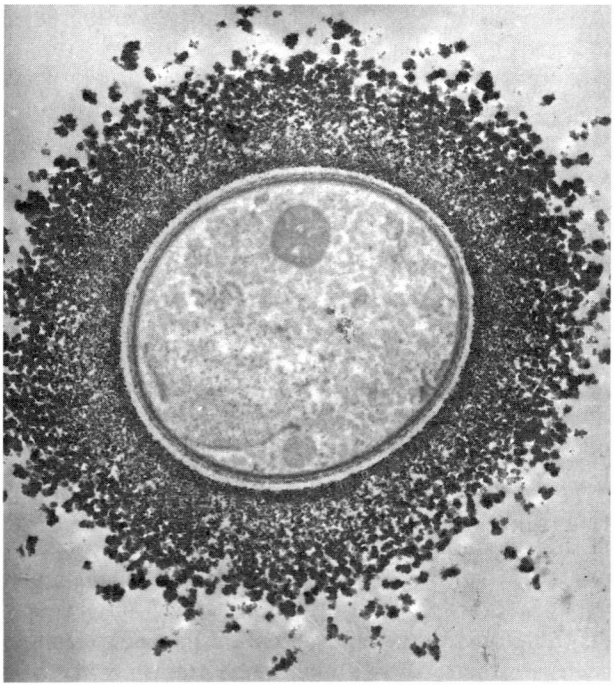

Figure 1 Ultrastructural localization of C3 fragments bound to encapsulated *C. neoformans*. Yeast cells were incubated with normal human serum, and the site of C3 binding was identified by immunoperoxidase staining. Sites of C3 deposition are seen as dense staining throughout the capsule, particularly at the capsular surface. Photograph kindly provided by T. Kozel and adapted from reference 106.

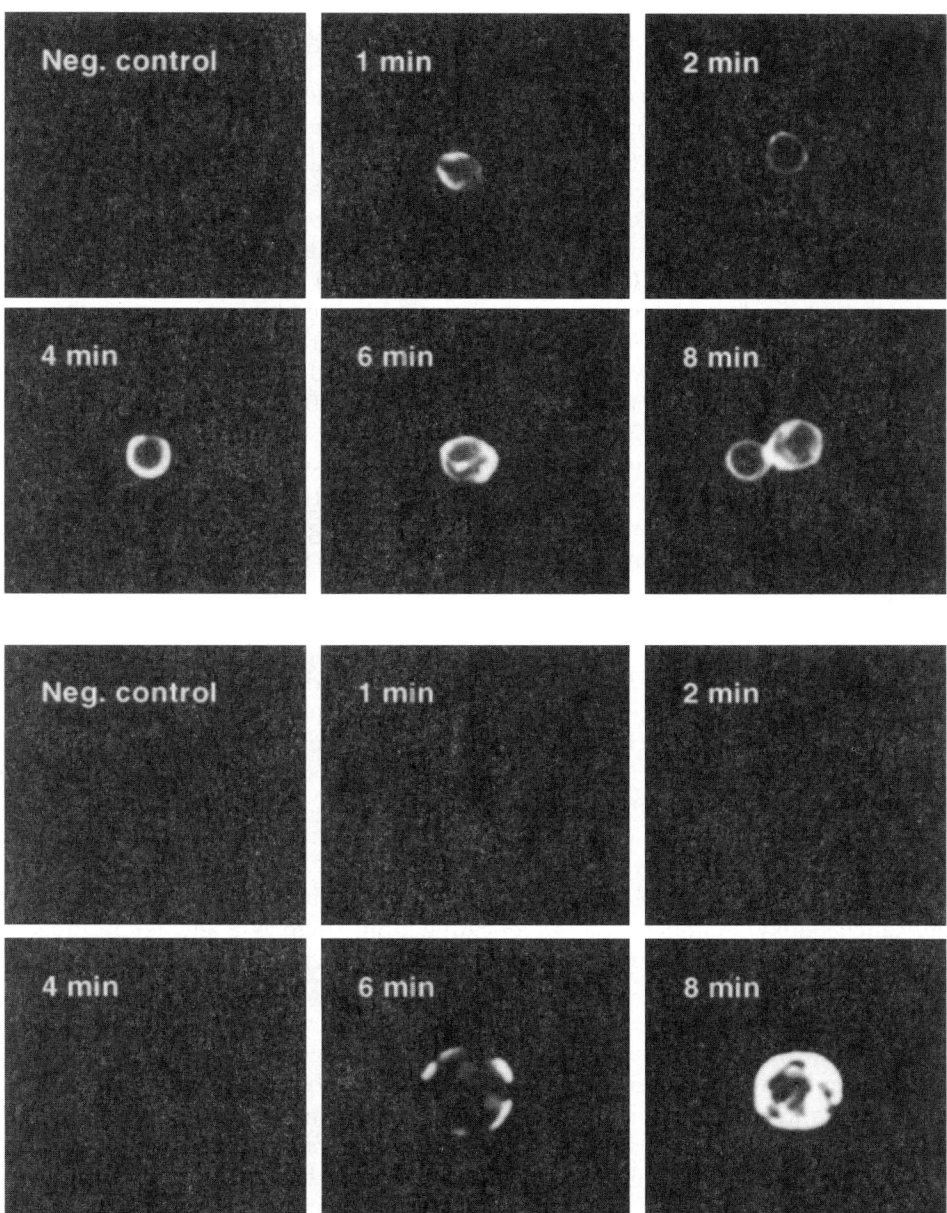

Figure 2 Localization of C3 by immunofluorescence of cellular sites for early deposition of C3 fragments on non-encapsulated (top two rows) and encapsulated (bottom two rows) *C. neoformans*. Yeast cells were incubated with heat-inactivated serum (negative control) or incubated for 1, 2, 4, 6, or 8 min with normal serum and stained with fluorescein-labeled antiserum specific for human C3. Early synchronous activation and binding of C3 to nonencapsulated cryptococci is mediated by the classical complement pathway, which is initiated by ubiquitous antiglucan IgG antibody found in normal human serum (100, 227). The delayed, focal activation and binding of C3 to encapsulated cryptococci is mediated solely by the action of the alternative pathway (100, 114, 115, 227). Figure kindly provided by T. Kozel and adapted from reference 115.

is a high rate of conversion of C3b to iC3b by complement factors H and I, whereas C3b bound to the surface of nonencapsulated cells is relatively resistant to conversion (109, 184). This suggests that C3b and iC3b are the more important opsonic ligands for serum-opsonized nonencapsulated and encapsulated cells, respectively (184). The differences in C3 deposition observed in encapsulated and nonencapsulated strains may be physiologically important if initial infection is believed to occur with poorly encapsulated yeast cells. The critical role of the capsule in the *C. neoformans*–complement interaction is also highlighted by comparative studies done with *C. albicans* (107). Incubation of *C. albicans* with serum results in maximal deposition of C3b on the yeast surface after 2.5 to 5.0 min. In contrast, maximal binding of C3b to *C. neoformans* requires 10 to 20 min (107). For *C. albicans*, the rapid activation of complement deposition appears to be the result of activation of the classical complement pathway by antibody to fungal surface antigens (113).

There are differences in complement activation between the two varieties of *C. neoformans* that are potentially important for the pathogenesis of infection and host defense (222, 233). For example, Washburn et al. (222) reported that *C. neoformans* var. *gattii* isolates (serotypes B and C) bind approximately half the number of complement component molecules as do *C. neoformans* var. *neoformans* isolates (serotypes A and D). However, Young and Kozel (233) reported no differences in the amount of C3 deposited within capsules of strains differing in serotype despite major differences in the rate of C3 accumulation. Accumulation of C3 in cells of *C. neoformans* var. *neoformans* was more rapid than that in cells of *C. neoformans* var. *gattii* (233). The discrepancies between these studies most likely reflect differences in incubation time and/or experimental design (222, 233). Comparison of the efficiency of C3 deposition on strains from the four serotypes also revealed greater binding to *C. neoformans* var. *gattii* strains than to *C. neoformans* var. *neoformans* strains (105, 193). C3 deposition correlated with the number of xylose residues in the glucuronoxylomannan (GXM) repeating unit; there is a greater number of xylose residues in *C. neoformans* var. *gattii* strains (105, 193).

Complement-mediated opsonization and leukotaxis

It has been known for many years that addition of fresh serum to mixtures of *C. neoformans* and phagocytic cells promotes attachment and phagocytosis (16, 22, 37, 41, 42, 69). In nonimmune sera, incubation of *C. neoformans* results in activation of the alternative complement pathway, which leads to deposition of C3 in the capsule; this is sufficient to opsonize the yeast cells for phagocytosis by macrophages and neutrophils (37, 41, 108, 121). Incubation of *C. neoformans* cells in serum produces a phenomenon similar to antibody-mediated capsular swelling in which the capsule becomes visible by light microscopy (41). Heat inactivation of complement system components abrogates the serum-mediated phagocytosis (108, 111).

Complement-mediated phagocytosis is a complex process that is dependent on many variables, including strain type, capsule size, presence of antibody to capsular polysaccharide, and the state of activation of the phagocytic cell. Kozel (105) has defined three criteria for the occurrence of complement-mediated opsonization: (i) activation of the complement pathway by *C. neoformans*, leading to deposition of iC3b at the capsular surface; (ii) the presence of cytokines that up-regulate the efficiency of complement-mediated phagocytosis; and (iii) the availability of phagocytic cells with complement receptors. Hence, complement-mediated opsonization is dependent on both host and *C. neoformans* factors.

Effective complement-mediated opsonization requires the presence of complement receptors in phagocytic cells. Activation of neutrophils with agents that enhance complement receptor expression results in a significant increase in complement-mediated phagocytosis (111). On the other hand, blockage of any of three C3 complement receptors (CR1, CR3, and CR4) can significantly inhibit the binding of serum-opsonized *C. neoformans* cells to human monocytes (138). The interaction of individual complement receptors with serum-opsonized *C. neoformans* has been dissected by using CHO cells stably transfected with human CR1, CR3, and CR4 receptors (140). Each receptor was shown to independently contribute to binding of serum-opsonized *C. neoformans* with the relative efficacy of CR3 > CR1 > CR4 (140).

Unlike the immunoglobulin receptor, the complement receptors are not constitutively expressed in phagocytic cells, and their expression is dependent on the activation state of the cell. Several cytokines have been shown to enhance complement-mediated *C. neoformans* phagocytosis, presumably through their effects on complement receptor expression. Collins and Bancroft (32) found that addition of granulocyte-macrophage colony-stimulating factor (GM-CSF) and tumor necrosis factor (TNF)-α enhanced complement-mediated phagocytosis by murine peritoneal macrophages. Individually, GM-CSF and TNF-α increased phagocytosis by 2- to 3-fold, but in combination the enhancement was 5- to 10-fold (32). In the absence of cytokine stimulation, only 70% of macrophages ingest cryptococci, and only 1 or 2 yeast cells are taken up by each macrophage (35). Cytokine stimulation increases the percentage of cells that are phagocytic and promotes the ingestion of 6 to 8 yeast cells per macrophage (35). GM-CSF and TNF-α probably increase both CR3 and the affinity of CR3 for C3 ligand (35). Interleukin-4 (IL-4) and gamma interferon (IFN-γ) have no effect on phagocytosis (32). These observations indicate that binding of complement components to the capsule is a necessary, but not sufficient, condition for effective phagocytosis (32, 105).

C1q, a subunit of C1, has been shown to enhance phagocytosis and killing of *C. neoformans* cells by human monocytes (13). C1q can augment the phagocytosis of yeast cells coated with either capsule-specific IgG or complement (13). It has been suggested that elaboration of C1q by human monocytes at the site of infection could facilitate ingestion and killing of *C. neoformans* and potentiate the opsonic efficacy of complement and capsule-specific antibody (13).

Ikeda et al. (93) studied complement opsonization for weakly and heavily encapsulated *C. neoformans* strains and showed that the efficacy of complement-mediated phagocytosis was dependent on the degree of yeast cell encapsulation. Weakly encapsulated cells were effectively opsonized by the alternative complement pathway, but heavily encapsulated strains were ingested only in the presence of specific antibody (93). These investigators suggested that the alternative pathway plays an important role in phagocytosis of weakly encapsulated strains but that phagocytosis of heavily encapsulated strains requires an immune response in which the classical pathway is activated through specific antibody (93).

The relationship between capsule size, phagocytosis, and complement deposition was recently investigated by Kozel et al. (112). *C. neoformans* cells with large capsules often bind less efficiently to phagocytic cells than do those with small capsules in the presence of serum. These investigators further analyzed C3 deposition on yeast cells grown in conditions conducive to production of either large or small capsules. Optimal deposition of C3 on cells with large capsules required a reduction in the number of yeast cells per milliliter of serum, suggesting an overactivation of the complement system by these cells, with a consequent reduction in the density of C3 bound to the capsule. These results suggest that in infections involving large numbers of yeast cells, there may not be sufficient C3 in serum for optimal complement-mediated opsonization. However, in conditions optimized for maximal C3 deposition on yeast with large capsules, i.e., fewer yeast cells per milliliter of serum, there was no difference in either the concentration of capsule-bound C3 or the binding of complement-opsonized yeast cells to human monocytes. Analysis of the interaction of complement with phagocytosis-resistant and phagocytosis-sensitive strains revealed no major differences in the amount, molecular form, or capsular location of bound C3 (110), suggesting that strain factors other than complement activation are responsible for the differences in phagocytosis.

   Activation of the complement system by *C. neoformans* can generate chemotactic factors that are probably important for the inflammatory response to infection in vivo. Incubation of *C. neoformans* with fresh serum induces a chemotactic response by neutrophils and monocytes (39, 121). This phenomenon was observed with both encapsulated and nonencapsulated *C. neoformans* cells and was abrogated by heat inactivation of serum (121).

Complement depletion in cryptococcal infection

Disseminated cryptococcal infections result in an acquired complement-deficient state (148). Complement depletion presumably results from direct complement activation by *C. neoformans* cells, resulting in consumption of complement factors (148). The problems inherent in achieving sufficient C3 deposition in large encapsulated cells may be compounded by a deficiency of complement components in late-stage infections. Gadebusch (53) noted that injection of cryptococcal polysaccharide resulted in a transient reduction in properdin levels and that mice given experimental infection had significantly depressed properdin serum levels.

Complement binding to *C. neoformans* in tissue

The presence of complement components in the *C. neoformans* capsule has been demonstrated in infected tissues. Deposition of C3 in the *C. neoformans* capsule has been demonstrated by immunohistochemistry for yeast cells in the alveolar space after intratracheal infection in mice (47). Hence, it appears that complement interacts with *C. neoformans* early in the course of infection. Immunofluorescence studies have shown C3 staining in the capsules of yeast cells in lung and liver tissue (213). However, in brain tissue, *C. neoformans* cells had weak or absent C3 staining, a finding that may reflect the relative absence of complement system factors in the CNS (213). Absence of complement in the CNS could impair the efficacy of host defense mechanisms and thus indirectly contribute to the remarkable neurotropism of this fungal pathogen.

   In summary, there is conclusive evidence that the complement system contributes to host defense against *C. neoformans*. The complement system performs a variety of functions that help the host contain *C. neoformans*, including providing opsonins and chemotactic factors and enabling antibody function. The interaction of complement and *C. neoformans* is complex and only partially understood.

NONSPECIFIC CELLULAR DEFENSES

PMNs

Polymorphonuclear cells (PMNs), or neutrophils, are professional phagocytic cells that originate from pluripotent stem cells in the bone marrow (199). PMNs are highly effective antimicrobial cells, and neutropenia is associated with a variety of severe and life-threatening infections (199). Although neutropenia is not classically associated with cryptococcosis, there is convincing in vitro evidence that PMNs can phagocytose and kill *C. neoformans* (42, 156, 157, 207). PMNs are often present in the inflammatory response to *C. neoformans* infection and can be found in close association with cryptococci in tissue (46, 131, 224). For example, in experimental ocular infection in rabbits, PMNs and monocytes have been shown to form distinctive rosette-like structures surrounding individual yeast cells (224). PMNs are also the predominant phagocytic cells found in early histiocytic ring formation after experimental peritoneal infection in laboratory animals (see Macrophages, below, for a full discussion of this phenomenon) (5, 98, 201, 203). The PMN response to experimental *C. neoformans* infections in mice is variable; it has been observed in some types of infections and not in others (46) (see chapter 9). Subcutaneous infection results in a strong inflammatory response consisting predominantly of neutrophils (131). In contrast, few neutrophils were observed in cryptococcal lesions resulting from brain and intraperitoneal infection (131). Administration of a 300-mg/kg dose of cyclophosphamide to BALB/c mice, followed by infection with *C. neoformans*, produced a transient reduction in blood PMNs, which was associated with reduced survival (67). However, this experiment, although suggestive of and consistent with a role for PMNs in defense against *C. neoformans*, is not conclusive since cyclophosphamide can also affect lymphocytes (67). Another suggestive line of evidence for the importance of PMNs in host resistance is the finding that mice given recombinant granulocyte colony-stimulating factor had a lower tissue fungal burden when subsequently challenged with *C. neoformans* (66). PMNs are a prominent cell type during the early stages of experimental cryptococcal meningitis in rabbits (183), possibly as a result of chemotactic factors that attract these cells to the subarachnoid space (180).

PMNs have been shown to be more effective at killing *C. neoformans* than are monocytes in vitro (156, 157). PMNs appear capable of killing *C. neoformans* by both intracellular and extracellular mechanisms. Killing of *C. neoformans* by human PMNs has been documented by using quantitative colony counts (42, 207). Phase-contrast microscopy of *C. neoformans* ingested by PMNs shows that yeast cells lose their defining characteristics, consistent with intracellular killing (42). A limiting factor in the efficacy of PMNs for killing *C. neoformans* is their ability to phagocytose yeast cells (42). Phagocytosis of *C. neoformans* appears to be strain dependent, since some strains are efficiently phagocytosed by PMNs whereas others resist phagocytosis (97). Killing by PMNs is most effective in the presence of complement opsonins (37, 42) or capsule-binding antibody (156). The complement system can provide adequate opsonins for PMN-mediated killing in the absence of specific antibody (37, 42, 157). Complement-mediated phagocytosis is proportional to the amount of CR3 complement receptor on the cell surface (111). Since the capsule can activate the alternative complement pathway directly, killing can be demonstrated by incubating human PMN and *C. neoformans* in medium with fresh serum (42, 157). However, PMNs are capable of some fungicidal activity even in the absence of serum (207).

The mechanism by which PMNs kill *C. neoformans* includes both oxidative and nonoxidative microbicidal mechanisms. Human PMNs require the generation of H_2O_2 and myeloperoxidase activity for maximal activity (42). Distal reactive oxygen intermediates (ROIs) produced by PMNs, such as hydroxyl radical and HOCl, are important effector molecules against *C. neoformans* (29). PMNs from patients with chronic granulomatous disease failed to kill *C. neoformans* cells, consistent with a requirement for ROIs in fungicidal action (42). Nonoxidative mechanisms are also likely to contribute to fungicidal activity. PMN granule fractions contain multiple components that can kill *C. neoformans*, including cationic proteins (129). Human PMNs contain potent low-molecular-mass (<3,500 Da) antimicrobial peptides (defensins) in cytoplasmic granules (130), which have been shown to kill *C. neoformans* in vitro (56). PMNs also contain calcium-binding leukocyte L1 protein (also known as calprotectin), which has been demonstrated to inhibit *C. neoformans* (205).

In summary, PMNs are potent effector cells that are capable of killing *C. neoformans* in vitro, although cryptococcal strains may differ in their susceptibility to killing by these cells (97, 157). Despite their apparent efficacy against *C. neoformans*, PMNs are not reliably present at sites of cryptococcal infection, and neutropenia has not been associated with cryptococcal infection. However, the lack of association between neutropenia and cryptococcosis should not be used as an argument to rule out an important role for the neutrophil in protection against *C. neoformans*. In clinical practice, neutropenia is usually a result of antineoplastic therapy. Neutropenic episodes tend to be of short duration and are followed by recovery of neutrophil function. If cryptococcosis is a result of acute infection, then neutropenia secondary to chemotherapy may not be associated with cryptococcosis because of lack of exposure to *C. neoformans* in the hospital environment and adequate host resistance in the presence of normal monocyte/macrophage function. Furthermore, cryptococcosis is usually a chronic infection with few symptoms at the early stages, when PMNs are presumably most important in host defense. The recovery of neutrophil function after chemotherapy could conceivably eradicate new or reactivated infections without any clinical manifestations.

NK Cells

Natural killer (NK) cells constitute a population of lymphocytes that kill malignant and virus-infected cells (212). NK cells have been defined as large granular lymphocytes that are negative for cell surface expression of CD3 and T-cell-receptor antigens and positive for cell surface expression of CD16 and NKH-1 antigens in humans and NK1.1/NK2.1 antigens in mice. NK cells can function as effector cells against bacteria and parasites (212). Several studies provide evidence that NK cells are involved in protection against *C. neoformans*.

Murphy and her collaborators (77–82, 168–172) have characterized the *C. neoformans*–NK cell interaction extensively. In 1982 this group demonstrated that the anticryptococcal activity of murine splenic lymphocyte preparations correlated with NK activity in vitro, and they suggested a role for NK cells in defense against *C. neoformans* (169). The process of NK cell-mediated fungistasis is similar to that of NK cell-mediated tumor cell lysis and requires binding to the yeast cell. NK cell binding to *C. neoformans* is temperature independent but Mg^{2+} dependent (80). Binding is followed by release of NK cell cytolytic components, resulting in yeast cell death (82, 168). After damaging or killing a *C. neoformans* cell, NK cells may recycle to another target cell. However, there are differences between NK cell action against tumor cells and against *C. neoformans* cells. Murine NK cells bind to *C. neoformans* by microvilli that penetrate the capsule, and they bind to tumor cells by cell membrane interactions (77). NK cells lyse YAC-1 tumor cells after 5 to 10 min, whereas up to 4 h is required for killing of *C. neoformans* (168). Rabbit antibody to the *C. neoformans* capsule enhances mouse (172) and human (158) NK cell activity against *C. neoformans*. Comparison of the fungistatic effect of human NK cells with that of human monocytes indicated that NK cells are significantly less active against *C. neoformans* than are monocytes (158). The mechanism of murine NK cell action against *C. neoformans* appears to involve direct toxicity to the yeast cell, possibly through soluble cytotoxic factors (82). NK cell granules contain cytolysin, which can kill *C. neoformans*. Furthermore, NK cells secrete a variety of cytokines, including IFN-γ, GM-CSF, and TNF (212). Cytokine release by NK cells may contribute to protection against *C. neoformans* by activating the effector cells and promoting effective inflammatory responses.

Mouse studies suggest a role for NK cells in resistance to *C. neoformans* infection. Beige mice (C57BL/6 *bg/bg*), which have defects in lymphocyte and neutrophil function and low NK cell activity, are more susceptible to infection than are heterozygous littermates (*bg/+*) despite comparable phagocytic cell function (79). However, the presence of simultaneous defects in the function of NK cells, neutrophils, and lymphocytes in beige mice does not permit unequivocal conclusions regarding the role of NK cells in beige mice. Histological studies of cryptococcal infection in C57BL/6N *bg/bg* and *bg/+* mice showed altered inflammatory responses in the *bg/bg* mice and suggested that the marked susceptibility of this strain to *C. neoformans* was associated with defective neutrophil and macrophage function (195). Increased susceptibility of beige mice relative to the parental strain was confirmed by another group (234).

Studies from three research groups have shown that depletion of NK cells can alter the course of intravenous *C. neoformans* infection in mice. Depletion of NK cells in C57BL/6N *bg/+* mice with an anti-NK cell MAb (anti-NK-1.1) resulted in increased lung CFU in mice relative to controls at early times after intravenous infection with *C. neoformans* (196). However, the administration of the MAb to NK-1.1 had no effect on brain, kidney, liver, or spleen CFU, and by day 14 of infection there was also no difference in lung CFU (196). Treatment of CBA/J mice with cyclophosphamide to deplete NK cell activity reduced in vivo clearance of *C. neoformans* from tissues (78). Passive transfer of splenic nylon wool-nonadherent cells to cyclophosphamide-treated CBA/J mice restored NK cell activity, which correlated with enhanced clearance of *C. neoformans* from organ tissues (78). Evidence that this effect was mediated by NK cells comes from the fact that depletion of NK cells before the adoptive transfer of spleen cells with anti-asialo GM_1 and complement abrogated the ability of the cell preparation to mediate protection (78).

Administration of anti-asialo GM$_1$ antiserum to C57BL/6 mice temporarily reduced NK activity and was associated with higher lung CFU in treated mice given
intravenous infection (144). However, CFU in liver, spleen, kidney, and brain were
not significantly different in control mice and in mice treated with anti-asialo GM$_1$
antiserum (144). Similarly, the administration of the NK cell-specific MAb (anti-NK
1.1) resulted in significantly greater numbers of lung CFU in treated mice after
intravenous infection but had no effect on CFU in other organs or on survival (144).
In contrast to the mouse studies using intravenous infection, administration of
anti-NK 1.1 to mice given intratracheal infection had no effect on lung, brain, or
spleen CFU relative to control mice (144).

   Horn and Washburn (86) documented the fact that the anticryptococcal activity of NK cells from both early asymptomatic and late-stage HIV-infected patients
is markedly impaired (86). They restored the anticryptococcal activity of NK cells
from HIV-infected patients by addition of exogenous IL-12, indicating that the
defect is reversible. These observations suggest that an NK functional deficit may
contribute to the susceptibility of patients with AIDS to *C. neoformans* infections.

   In summary, the mouse studies consistently show that NK cell depletion can
result in increased lung CFU at early times after intravenous infection, but the
effect is temporary and has little or no effect on in the overall course of infection.
In vitro, NK cells are capable of inhibiting *C.* neoformans growth. Overall, the
results of the NK cell studies indicate a role for NK cells in protection against *C.*
neoformans infection, but the relative importance of this cell type in host defense
mechanisms is uncertain.

Macrophages

Macrophages have a central role in host defense against *C. neoformans*. In contrast
to other cells, histological studies of *C. neoformans*-infected tissue consistently show
macrophages at sites of infection, often in intimate association with yeast cells.
Experimental evidence for an important role of macrophages in protection against
C. neoformans infection comes from mouse experiments that show that the state of
macrophage function affects susceptibility to infection. Administration of silica to
mice destroys macrophages and markedly decreases resistance to *C. neoformans*
infection (162). Conversely, activation of macrophage function in mice by *Mycobacterium bovis* BCG vaccination is associated with increased resistance to *C. neoformans* (58, 64, 162).

   Perhaps the strongest evidence that macrophages play a crucial role in host
defense against *C.* neoformans is the observation that successful host responses to
C. neoformans infection almost always involve granuloma formation (see chapter
8). In granulomas, macrophages form giant cells that contain ingested cryptococci.
Potential roles of macrophages in protection against *C. neoformans* include phagocytosis and killing of yeast cells, granuloma formation, antigen presentation, cytokine release, and sequestration of polysaccharide antigen.

Phagocytosis and killing

Macrophage phagocytosis of *C. neoformans* can occur through antibody (136, 165,
167), complement (138), mannose (17, 34), and β-glucan (34) receptors. Efficient
phagocytosis of *C. neoformans* by macrophages requires the presence of opsonins

and the expression of the appropriate receptor on the macrophage cell. Antibody-mediated phagocytosis occurs through Fc receptors, which are constitutively expressed in macrophages, but it should be emphasized that opsonic capsule-binding antibody is not consistently present during human and animal infections. Complement-mediated phagocytosis occurs through C3 receptors, the expression of which may depend on cytokine stimulation (32, 34). In human macrophages, complement-mediated binding is an energy-dependent process that requires three C3 complement receptors (CR1, CR3, and CR4) and actin but does not necessarily lead to phagocytosis (138). Mannose receptors are constitutively expressed, but phagocytosis by this route appears to be less efficient, as shown by suboptimal phagocytosis in the absence of serum opsonins (17). On the whole, macrophages are efficient phagocytic cells in conditions in which opsonins are present and cells are stimulated by cytokines (32, 34, 35, 135). In vitro macrophage phagocytosis of C. neoformans is greatly dependent on the surface to which the cells are attached (135, 141). For example, culture on surfaces coated with mannose-binding protein (141) or coincubation with endothelial cells (192) can significantly enhance macrophage phagocytic function.

Phagocytosis of C. neoformans by macrophages in vitro is not necessarily accompanied by killing of the ingested yeast cells. Some studies with human, rat, mouse, and guinea pig macrophages showed little or no killing of C. neoformans after phagocytosis (22, 38, 157, 159) (Table 1). This led to some controversy as to the effectiveness of macrophages as an effector cell against C. neoformans. However, it is now clear that the antifungal efficiency of macrophages is dependent on the degree of activation of the macrophage cell, and there is convincing evidence that, with appropriate stimulation, macrophages can be efficient killers of C. neoformans. In recent years, many studies have demonstrated that human and murine macrophages are effective at mediating fungistatic and fungicidal activity against C. neoformans (Table 1).

Studies of murine alveolar and peritoneal macrophages from C. neoformans-infected animals reveal that the phagocytic and fungicidal efficacy of macrophages is dependent on the timing of macrophage harvest after respiratory infection (99). Peritoneal macrophages display maximal phagocytic efficacy between days 21 and 28 of infection (99). Alveolar macrophages exhibit a fungistatic effect at day 14, and peritoneal macrophages are fungicidal for C. neoformans at day 28 (99). For both alveolar and peritoneal macrophages, phagocytosis and antifungal efficacy decline after 28 days, suggesting the development of an immunosuppressive state during progressive cryptococcal infection (99). In contrast, rat alveolar macrophages are phagocytic and fungicidal for C. neoformans in the absence of opsonins or exogenous cytokines (17). These findings are consistent with differences in macrophage function depending on the degree of activation. The divergent experimental observations on the role of macrophages may also result from differences in macrophage origin, anatomic site, activation, culture conditions, ratio of macrophages to yeast cells, and the form of opsonization used. Differences between mouse strains in anticryptococcal activity of murine macrophages have been described: macrophages from SW, ICR, BALB/c, C57BL/5, and DBA/2J mice were fungistatic without IFN-γ and lipopolysaccharide (LPS), whereas macrophages from CD-1 mice lacked fungistatic activity (21). However, activation of CD-1 macrophages with IFN-γ and LPS resulted in fungistatic activity against C. neoformans (21).

Table 1 Studies demonstrating macrophage efficacy against C. *neoformans*

Study	Species	Source	Opsonin	E:T ratio[a]	Stimulation	Activity
Gentry and Remington (58)	Mouse	Peritoneum	Serum	1:10	Intracellular pathogens	Fungistatic?
Mitchell and Friedman (159)	Rat	Peritoneum	Serum	1:5	None	Minimal
Diamond and Bennett (38)	Human	Blood	Serum	25:1 to 1:2,500	None	None
Bulmer and Tacker (22)	Guinea pig	Lung	Complement	1:4	Endotoxin	None
Miller and Kohl (156)	Human	Blood	Anti-GXM Ig	1:1 to 100:1	None	Fungicidal
Granger et al. (64)	Mouse (BCG infected)	Peritoneum	Serum	20:1	Endotoxin	Fungistatic
Weinberg et al. (223)	Human	Lung	Serum	30:1	None	Fungicidal
Perfect et al. (181)	Mouse (BCG infected)	Peritoneum	Serum	10:1	IFN-γ and LPS	Fungistatic
Bolanos and Mitchell (15)	Rat	Lung	Serum	1:1	None	Fungicidal
Flesch et al. (50)	Mouse	Bone marrow	None	50:1	IFN-γ	Fungicidal

Reference	Species	Source	Opsonin	E:T ratio[a]	Activation	Result
Levitz and DiBenedetto (133)	Mouse	Lung	Serum	20:1	None	Fungicidal
Levitz and Farrell (135)	Human	Blood	Serum		None	Fungistatic
Cameron et al. (25)	Human	Lung, peritoneum	Serum	10:1 to 20:1	None	Fungistatic
Miller and Mitchell (157)	Human	Blood	Serum	1:1	None	Fungicidal
Mody et al. (161)	Rat	Lung	Serum		IFN-γ and LPS	Fungicidal
Levitz et al. (136)	Human	Blood	Serum and anti-GXM Ig	400:1	Killed *C. neoformans*	Fungicidal
Brummer and Stevens (21)	Mouse	Peritoneum	Serum	1:2	None or IFN-γ and LPS	Fungistatic
Mukherjee et al. (167)	Mouse	J774.16 cell line	MAbs to GXM	1:1	IFN-γ, LPS	Fungicidal

[a]E:T ratio, effector-to-target ratio. In these assays the "effector" is the macrophage and the target cells are the cryptococci.

Macrophage-mediated fungistasis was demonstrated with murine peritoneal macrophages harvested from BCG-infected mice after in vitro stimulation with endotoxin (64). Murine bone marrow macrophages stimulated with recombinant IFN-γ were fungicidal to *C. neoformans* in the absence of complement or antibody opsonins (50). The antifungal activity of bone marrow macrophages was also enhanced by bacterial LPS and occurred extracellularly (50). Macrophage fungicidal activity against *C. neoformans* has been reported to be independent of oxygen radicals but was associated with secreted proteins with an apparent molecular mass of 15 to 30 kDa (50). Rat alveolar macrophages activated by recombinant IFN-γ were significantly more effective in reducing CFU in vitro than were nonactivated cells (161). Addition of LPS did not enhance the effect of optimal IFN-γ concentrations but did enhance the effect of suboptimal stimulation (161). However, neither cytokine nor endotoxin stimulation appears to be necessary for stimulating alveolar macrophages in some conditions (70). Primary rat alveolar macrophages phagocytosed complement-opsonized *C. neoformans*, resulting in phagosomal acidification and respiratory burst (70). This suggests that alveolar macrophages may have intrinsic antifungal properties suited for defense against aerosolized *C. neoformans* (70).

Macrophages are able to inhibit and kill *C. neoformans* in both intra- and extracellular compartments. Human alveolar macrophages can inhibit *C. neoformans* through two mechanisms: one is serum independent and occurs extracellularly and the other is serum dependent (i.e., opsonin dependent) and occurs intracellularly (223). Similar results have been observed with murine macrophage-like cells (165, 167). The antifungal efficacy of macrophages against intracellular yeast appears to be significantly greater than that against extracellular organisms (165, 167, 223). On the other hand, *C. neoformans* is also capable of intracellular replication in human macrophages (38) and microglial cells (127) in vitro, suggesting that phagocytosis without killing could provide a niche for the survival and dissemination of cryptococci inside macrophages.

Coincubation of murine macrophages and *C. neoformans* in the presence of capsule-binding antibody can significantly enhance macrophage antifungal efficacy. IFN-γ and LPS-stimulated cells of the murine macrophage-like cell line J774.16 were fungicidal to *C. neoformans* in the presence of murine MAbs (167). Among the murine IgG subclasses, IgG1 was most effective in enhancing the antifungal activity of J774.16 cells against *C. neoformans* (167). Electron microscopy of *C. neoformans* cells engulfed during antibody-mediated phagocytosis revealed damaged and degenerating yeast forms consistent with intracellular killing (167).

Two studies have shown that human monocytes kill *C. neoformans* in the presence of high-titer rabbit antibody to the capsular polysaccharide (136, 156). Levitz (132) showed that the antifungal activity of human peripheral blood mononuclear cells is dependent on cytokine activation, and they demonstrated fungicidal activity in these cells after IL-2 and GM-CSF activation (132). Human monocyte antifungal efficacy in vitro is influenced by cell culture conditions, cytokines, opsonins, and the encapsulation of the fungal cell (186).

Electron microscopy of hepatic granulomas in rats given experimental infection has shown that the cytoplasm of ingested yeast forms is usually destroyed but that the cell wall and capsule remains (194). This suggests that mammalian macrophages may lack the enzymes necessary for the digestion and elimination of yeast polysaccharide components (194). Macrophage pseudopodia extend into the polysaccharide capsule of the yeast, and actin filaments are seen in close association with pseudopodia, suggesting involvement of actin in this process (194).

Histiocytic rings and granuloma formation

During infection, some *C. neoformans* strains can produce very large capsules with a diameter exceeding 50 μm. These cells are larger than phagocytic cells and cannot be engulfed by single macrophages. A mechanism for containment and killing of *C. neoformans* in the peritoneal cavity is formation of histiocytic rings around yeast cells (5, 98, 201, 203). Intraperitoneal infection in mice revealed that macrophages form histiocyte rings around *C. neoformans* cells and may fuse to form giant cells (201). Early (24-h) histiocytic rings contain primarily PMNs, but macrophages predominate at later times (120 h) (98). Histiocyte ring formation in the peritoneal cavity is enhanced by the presence of rabbit antibody to the capsule polysaccharide (201). In the presence of capsule-specific antibody, the number of contacts between macrophages and cryptococcal cells and the number of macrophages surrounding the yeast cells were significantly increased (179). In contrast to mice, peritoneal infection in rabbits resulted in histiocytic ring formation without the need for exogenous antibody, an observation that may reflect the fact that rabbits are good antibody responders to capsular polysaccharide (5, 98, 201, 203). Electron microscopy and cytochemical studies of histiocytic rings revealed that phagocytic cells formed pseudopodia that penetrated into the *C. neoformans* capsule (98). Furthermore, several hydrolytic enzymes, including acid phosphatase, β-glucuronidase, and an esterase were released by phagocytic cells into the yeast cell (98). Histiocytic ring formation resulted in initial fungistasis followed by slow killing of the yeast cell. Histiocytic ring formation is dependent on the presence of opsonins and may be a precursor to granuloma formation.

In experimental pulmonary *C. neoformans* infection in rats and mice, a decline in lung tissue CFU paralleled the development of granuloma formation by lung macrophages (60, 85). In mice, the decline in pulmonary CFU has been associated with a T-cell-dependent response that results in the formation of multinucleated giant cells around *C. neoformans* (201). Multinucleated giant cells probably develop as a result of macrophage cell fusion and serve to engulf large particles, such as encapsulated yeast cells (85). These observations strongly suggest that macrophages contribute to protection against *C. neoformans* by forming giant cells and granulomas, which serve to localize the infection and prevent dissemination.

Antigen presentation

Human alveolar macrophages containing phagocytosed *C. neoformans* can induce proliferation of autologous T lymphocytes (216). The mechanism appears to involve antigen presentation through HLA class II DR molecules and secretion of IL-1 (216). Antigen presentation by HLA class II molecules requires phagocytosis of yeast cells and antigen processing. Since the capsule of *C. neoformans* is antiphagocytic and interferes with efficient antigen presentation (31), opsonins may be required for efficient antigen presentation by macrophages. The finding that nonstimulated human alveolar macrophages were not fungicidal to *C. neoformans* after 6-h incubation in vitro but promoted T-lymphocyte proliferation has led to the suggestion that the primary function of alveolar macrophages in infection is antigen presentation (216, 217). However, addition of IFN-γ to human alveolar macrophages induced fungicidal activity against *C. neoformans* (217). Again, these observations suggest that the role of alveolar macrophages in defense against *C. neoformans* may depend on their state of activation.

Cytokine production

Macrophages are likely to be one of the first inflammatory cell types to come into contact with *C. neoformans*, and their function includes the production of cytokines for the recruitment and activation of the host inflammatory response. Macrophages can produce a variety of cytokines after ingestion of *C. neoformans*. Human monocytes secrete IL-1 (216) and TNF-α (139) in response to ingestion of *C. neoformans* (216). Ingestion of acapsular *C. neoformans* by murine macrophages occurs via mannose receptors and results in secretion of pro-inflammatory cytokines, including TNF-α, IL-1, macrophage inflammatory protein-1-α, and GM-CSF, which activate macrophages for efficient phagocytosis of encapsulated strains (34). Secretion of pro-inflammatory cytokines by alveolar macrophages shortly after contact with *C. neoformans* may provide early stimuli to recruit additional effector cells to the alveolar space (34, 60). Furthermore, macrophages respond to cytokine signals at the site of infection. Chemotactic factors in the subarachnoid space are important in stimulating an inflammatory response, which includes monocytes and macrophages, in the meninges (180). This chemotactic response can be abrogated by the administration of corticosteroid to rabbits (180).

Polysaccharide antigen sequestration

Immunostaining of mouse, rat, and human tissues infected with *C. neoformans* reveals that cryptococcal polysaccharide GXM antigen is frequently found inside macrophages (60, 61, 122, 124, 166). The mechanism of GXM uptake has not been defined but may involve direct phagocytosis via the mannose receptor or the Fc receptor if antibody is present (61). The effect of intracellular GXM on macrophage function is unknown. Since soluble GXM has been associated with a variety of deleterious effects on host immune function, macrophage uptake and sequestration of soluble GXM could theoretically benefit the host by removing a potentially immunosuppressive fungal product.

Macrophage function in patients with AIDS and *C. neoformans* infection

Human blood monocytes and peritoneal macrophages have been shown to have reduced anticryptococcal activity in vitro after infection with some strains of HIV-1 (24, 72). In contrast, human alveolar macrophages retain normal anticryptococcal activity after HIV-1 infection, suggesting that the effect of viral infection may vary with the anatomical source of the macrophage (24). Defective macrophage function in patients with AIDS could contribute to their marked susceptibility for *C. neoformans* infections.

Helper T cells (CD4+ lymphocytes) stimulate macrophages for efficient microbicidal activity, and CD4+ T-cell depletion in the late stages of HIV infection is likely to have profound effects on macrophage function. HIV infection of human macrophages by macrophage-tropic strains of HIV could also contribute to defective macrophage function. Furthermore, soluble recombinant HIV-1 surface glycoprotein (gp120) has been shown to interfere with phagocytosis and fungistasis of

C. neoformans by human bronchoalveolar macrophages in vitro (219). A comparison of monocyte interactions with *C. neoformans* using cells from HIV-seropositive and HIV-seronegative individuals revealed no differences in phagocytosis of yeast cells but did find reduced H_2O_2 synthesis and release of β-glucuronidase in monocytes from HIV-seropositive individuals (74). The monocyte cytokine response to *C. neoformans* is also abnormal in cells from patients with AIDS (73). Human monocytes from HIV-infected individuals produce significantly less IFN-γ when challenged with *C. neoformans* than do monocytes from seronegative donors (73). Monocyte IL-12 production is also deficient in patients with HIV infection but can be restored when the cells are primed with IFN-γ (73). Dysregulation in macrophage cytokine synthesis in patients with AIDS could contribute to their marked susceptibility for *C. neoformans* infection, since proper cytokine expression is a critical element for inflammatory cell recruitment and activation.

The question of whether macrophage defects in patients with HIV infection are the result of macrophage infection by HIV or lack of T-cell stimulation due to the CD4$^+$ lymphopenia has been explored in rhesus monkeys (20). Infection of rhesus monkey macrophages in vitro or in vivo with simian immunodeficiency virus did not result in reduced anticryptococcal activity. However, alveolar macrophages from symptomatic monkeys with AIDS had major deficits in fungicidal activity, regardless of whether the cells themselves were infected. These observations suggest that the macrophage defect that accompanies AIDS is a consequence of CD4$^+$ T cell depletion.

The interaction of *C. neoformans* with HIV-infected monocytes may also promote the activation of HIV infection. Incubation of *C. neoformans* with a latently infected cell line stimulated HIV reverse transcriptase activity through TNF-α and NF-κB-dependent mechanisms (75). Addition of antibody to the capsule to cocultures of *C. neoformans* and HIV-infected cells resulted in a further stimulation of reverse transcriptase activity, possibly as a result of Fc-mediated phagocytosis (75). Activation of HIV infection in latently infected macrophages suggests a mechanism by which *C. neoformans* and HIV infection could synergistically degrade host immune function.

In summary, there is considerable evidence that macrophages from patients with AIDS are less effective against *C. neoformans* than are those from HIV-seronegative individuals. This effect is probably the result of inadequate macrophage activation as a consequence of CD4$^+$ T-cell depletion. HIV infection of macrophage cells may contribute to macrophage dysfunction, and the interaction of *C. neoformans* with infected macrophages may stimulate HIV infection. Defects in macrophage function may contribute to the marked susceptibility of patients with AIDS to *C. neoformans* infection.

Mechanisms of macrophage anticryptococcal activity

Macrophages have a variety of microbicidal mechanisms that are potentially active against *C. neoformans*. Oxidative microbicidal mechanisms include the generation of ROIs and nitrogen-derived oxidants. Nonoxidative mechanisms include phagosomal acidification, microbicidal proteins, lysosomal fusion, amino acid-degrading enzymes, and iron-binding proteins.

Several studies suggest that macrophages mediate anticryptococcal activity through both oxidative and nonoxidative microbicidal mechanisms (Table 2). The

Table 2 Antimicrobial mechanisms of phagocytic cells against *C. neoformans*

Mechanism	Activity for *C. neoformans* (references)	Comments
Oxidative		
Oxygen-derived microbicidal molecules		
H_2O_2	Yes (42, 95, 221)	Neutrophils from patients with chronic granulomatous disease are defective in *C. neoformans* killing (42).
Myeloperoxidase-halide system	Yes (95)	Myeloperoxidase-halide killing is more effective than H_2O_2.
Polyamine oxidase-polyamine system	Yes (134)	Two products, hydrogen peroxide and acrolein, kill *C. neoformans*.
Nitrogen-derived microbicidal molecules		
Nitric oxide	Yes (2, 63, 123)	Nitrogen-derived microbicidal molecules are important for murine macrophage anti-*C. neoformans* activity. Role in human macrophages is uncertain. Human astrocytes inhibit *C. neoformans* by a nitric oxide-related mechanism. Nitric oxide is fungistatic at low concentration and fungicidal at high concentration (2).
Other reactive nitrogen intermediates	Yes (232)	

Nonoxidative		
Phagolysosomal acidification	Probably not	Low pH (< 4.0) may inhibit intracellular replication. Macrophage activity against *C. albicans* includes phagolysosome acidification and microbicidal proteins. However, for *C. neoformans* there is indirect evidence that phagosome acidification may contribute to intracellular survival (137).
Microbicidal proteins and peptides		
Murine microbicidal proteins	Yes (83)	Murine microbicidal proteins are histone-like proteins with broad antimicrobial spectrum found in murine macrophages (83).
Defensins	Yes (56)	Defensins are found in human, rat, and rabbit neutrophils (130). Murine neutrophils lack defensins (130).
Lactoferrin	?	Iron chelator. Human lactoferrin is fungicidal tor *C. albicans* (177).
Lysosome	Yes (54)	Budding cells are more susceptible (54). Mechanism of action against *C. neoformans* is unknown. Toxicity to *C. albicans* involves cell wall damage and cytoplasmic membrane injury.
Proteases	?	Proteases include cathepsin G, azurocidin (26), and elastase. Azurocidin is active against *C. albicans* (26).
Calcium-binding protein	?	Calcium-binding proteins are effective inhibitors of *C. albicans* and *A. fumigatus* growth (204). Human leukocyte calcium-binding protein L1 has been shown to be fungicidal for *C. neoformans* (205).
Amino acid-degrading enzymes (e.g., L-arginase)	?	Unlikely to be a major mechanism against *C. neoformans* since this fungus can grow on minimal media.

fungistatic activity of murine macrophages is associated with the oxidation of L-arginine to L-citrulline and nitrate, which presumably results in the production of several reactive nitrogen species, including nitric oxide (2, 63). Methylated L-arginine analogs (such as N^G-monomethyl-L-arginine) inhibit the fungistatic activity of murine macrophages (63). The role of nitrogen-related oxidants in human macrophage fungistasis is uncertain (174). There is evidence that human macrophages mediate anticryptococcal activity by mechanisms other than the production of reactive nitrogen species (25, 217). ROIs kill *C. neoformans* in vitro, but they have not been directly implicated in human macrophage-mediated anticryptococcal activity. Coincubation of *C. neoformans* and murine macrophages results in fungistasis without O_2^- release (50). However, a role for ROIs in macrophage fungistasis has not been rigorously ruled out, and the defective killing of *C. neoformans* by neutrophils of patients with chronic granulomatous disease (42) suggests that ROIs are important anticryptococcal molecules for some phagocytic cells. Macrophages have polyamine oxidases that can catalyze the enzymatic degradation of spermine and spermidine, which can subsequently be metabolized to produce reactive aminoaldehydes, ammonia, and H_2O_2 (160). The products of the polyamine oxidase-polyamine system can be fungicidal to *C. neoformans*, and this antimicrobial system could participate in macrophage-mediated antifungal activity (160). Cationic microbicidal proteins are also likely to be an important mechanism of anticryptococcal activity for both murine (50, 83) and human (217) macrophages.

Macrophages and antifungal drugs
Some antifungal drugs may mediate their effects through host effector cells such as macrophages. Amphotericin B can activate macrophages to produce oxygen- and nitrogen-derived oxidants (146, 210, 226, 228). Amphotericin B accumulates inside human monocytic cells (151). Azoles such as ketoconazole and itraconazole also bind to macrophages (181). Accumulation of azoles by macrophages (azole-loading) may contribute to the efficacy of these drugs against *C. neoformans*, because it exposes intracellular yeasts to high concentrations of the drug (181). The ability of macrophages to bind antifungal drugs, combined with their phagocytic capacity, could translate into increased drug efficacy by placing the fungal cells in a drug-rich environment. Chloroquine has been shown to increase human macrophage killing of intracellular *C. neoformans*, possibly by increasing phagolysosomal pH (137).

In summary, the macrophage is a versatile effector cell against *C. neoformans*. Macrophages are reliably present at sites of cryptococcal infection, but their function against *C. neoformans* cells may vary during the course of infection. Early in infection, macrophages may function to present antigen to lymphocytes and to secrete cytokines that stimulate the inflammatory process. The influx of lymphocytes triggered by macrophage-derived inflammatory signals could then activate the macrophage for more effective, direct antifungal action. This action, together with the presence of opsonins in the form of complement and/or antibody, can convert macrophages into efficient antifungal effector cells. Later in infection, macrophages are involved in granuloma formation, which serves to localize and control *C. neoformans* infection, particularly in the lung. The available evidence indicates that the macrophage is a critical effector cell against *C. neoformans*, but its efficacy may depend on lymphocytes to provide activating cytokines and on the availability of opsonins to enhance phagocytosis of yeast cells.

Microglia and Astrocytes

Despite the affinity of *C. neoformans* for CNS tissues, relatively little is known about immune defense mechanisms against cryptococcus in the brain, and few studies have been done with brain effector cells. Perfect et al. (182) studied the functional characteristics of cerebrospinal fluid macrophages in the course of experimental intracisternal infection in rabbits. Rabbit cerebrospinal fluid macrophages become activated during cryptococcal infection, as measured by their ability to produce H_2O_2 and acquire tumoricidal activity (182). However, activation did not correlate with the ability to kill *C. neoformans* in vitro (182).

Several studies have suggested that microglia and astrocytes may be important anticryptococcal effector cells in the CNS. Microglia are resident macrophage-like cells in the CNS (43). Blasi et al. (11) demonstrated that the mouse microglial cell line BV-2 is phagocytic and fungistatic for *C. neoformans*. Phagocytosis and fungistasis of *C. neoformans* by BV-2 cells were enhanced by serum derived (complement) opsonins. Treatment of BV-2 cells with IFN-γ and LPS increased their anticryptococcal activity through a nitric oxide-related mechanism (12).

Human primary microglial cells can ingest *C. neoformans* when supplied with antibody to GXM, and phagocytosis results in inhibition of fungal growth (126) (Fig. 3). Antibody-mediated fungistasis by human microglia is not affected by inhibitors of nitric oxide synthase or scavengers of ROIs (126). However, in the absence of cytokine stimulation, *C. neoformans* was observed to replicate intracellularly (127). Intracellular replication was associated with the formation of large spacious vacuoles filled with polysaccharide material, and fungal proliferation ultimately caused the cells to burst (127). The physiological significance of this in vitro phenomenon is not known, but the result establishes a precedent for *C. neoformans*-mediated cytotoxicity through unchecked intracellular replication. Whether intracellular replication can be prevented by cytokine stimulation is not known.

Astrocytes are star-shaped cells in the CNS that presumably perform a supportive role for neurons and may be involved in a variety of infectious diseases (164). *C. neoformans* involvement of the brain parenchyma places yeast cells in close proximity to astrocytes. IL-1β and IFN-γ stimulation of human astrocytes results in nitric oxide production (125) which can inhibit *C. neoformans* growth extracellularly in vitro (123). Although the physiological significance of this observation is uncertain, the finding raises the tantalizing possibility that astrocytes may contribute to host defense in the brain parenchyma (123).

Eosinophils and Mast Cells

Eosinophils are granulocytic cells that are prominent in the response to allergens. Although eosinophils are classically associated with the host response to helminthic infections, their role in protection against infectious diseases is uncertain. The inflammatory response to *C. neoformans* often contains eosinophils, but little is known about the efficacy of eosinophils as antifungal effector cells. Eosinophils are

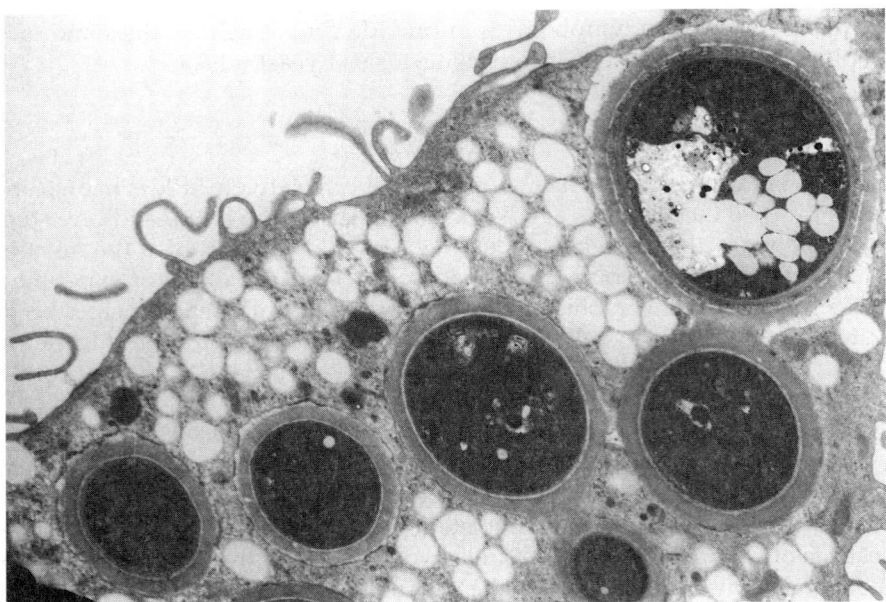

Figure 3 Electron micrograph of human microglia containing numerous *C. neoformans* cells. Micrograph obtained at a magnification of ×12,000. In the presence of capsule-specific antibody, human microglia avidly phagocytose yeast cells and mediate fungistatic effects in vitro (126). Micrograph kindly provided by S. Lee (Bronx, N.Y.).

occasionally found in human inflammatory infiltrates (19) and are common in the inflammatory response of mice and rats to *C. neoformans* pulmonary infection (48, 60, 89). Intratracheal cryptococcal infection in C57BL/6 and A/J mice produces an eosinophilic pneumonia (48, 89). Electron microscopy studies of A/J pulmonary infiltrates to *C. neoformans* infection reveal eosinophils in close proximity to yeast cells (97). Yeast cells in close proximity to eosinophils in lung were often coated with electron-dense material that may represent discharged eosinophil granule contents (48) (Fig. 4). In vitro rat eosinophils phagocytose *C. neoformans* cells in the presence of capsule-binding IgG and IgE MAbs. However, ingestion of *C. neoformans* by eosinophils is seldom observed in vivo.

The role of mast cells in host defense against *C. neoformans* infection is not known. Mast cells are versatile immune effector cells that can produce a variety of cytokines and mediators of inflammation (55). In vitro, mast cells have been shown to degranulate and to occasionally ingest *C. neoformans* cells in the presence of a capsule-specific IgE MAb (48).

In summary, eosinophils are commonly observed in rodent inflammatory infiltrates but are seldom reported in human inflammation. In humans, the overwhelming majority of histopathological information comes from autopsies and material from immunocompromised patients. Hence, the differences between rodents and humans in regard to eosinophils in inflammatory infiltrates could reflect species differences, differences in the response between immunocompetent animals and immunocompromised humans, or possibly differences in the timing of tissue sampling during the course of cryptococcal infection. The function of eosi-

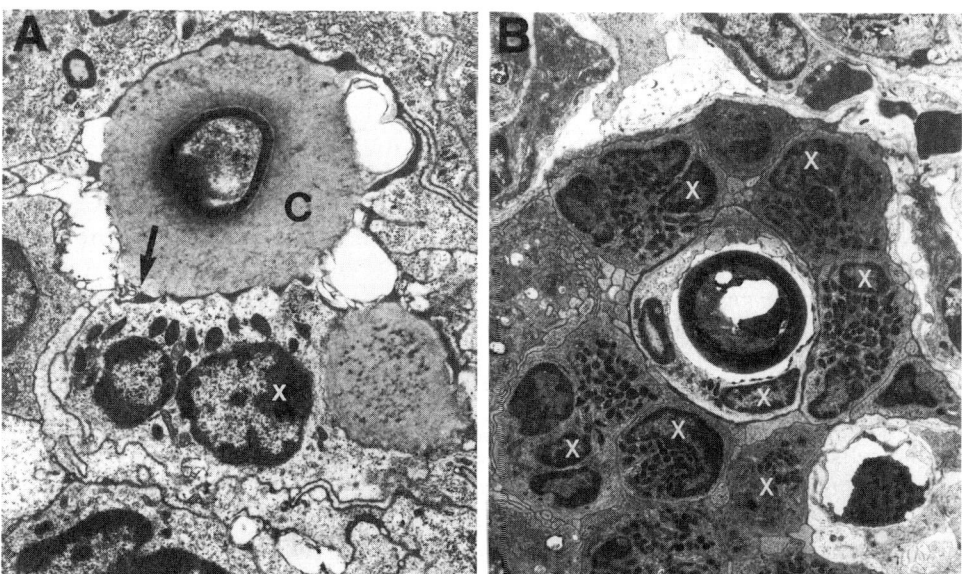

Figure 4 (A) Electron micrograph of a section of mouse lung 14 days after infection, showing an eosinophil in direct contact with a *C. neoformans* cell. The arrow indicates electron-dense material that may originate from eosinophil granule contents discharged in response to eosinophil–*C. neoformans* contact. Magnification, ×10,000. (B) Multiple eosinophils (denoted by white X's) surround a *C. neoformans* cell in mouse lung. Micrograph obtained at a magnification of ×3,000. Figure adapted from reference 48.

nophils in protection against *C. neoformans* or in the pathogenesis of cryptococcal infection is not known. However, the intimate association of yeast cells and eosinophils in rodent infection indicates that eosinophils are an active participant in the host response, suggesting a role for this cell type in host defense.

Platelets

The role of platelets in the pathogenesis of and defense against *C. neoformans* infection has not been investigated. Platelets may be involved in defense against other fungal pathogens, such as *C. albicans* (231). Platelets contain platelet microbicidal protein, which is active against a variety of pathogens, including *C. neoformans*. Platelet microbicidal protein is more active against *C. albicans* than against *C. neoformans*, but incubation of cryptococci with this protein results in fungal killing (231).

Endothelial Cells

Endothelial cells line the lumen of blood vessels and are capable of secreting a variety of cytokines in response to microbes. Since *C. neoformans* is often recovered from blood, and since the endothelial cell layer must be crossed for the organism to produce metastatic infections, the endothelial cells are likely to have an important function in the defense and pathogenesis of disseminated cryptococcal infection. Human endothelial cells have no activity against *C. neoformans* in vitro, but coincubation of human PMNs with endothelial cells significantly increases phagocytosis and antifungal activity by PMNs (192). Endothelial cells are able to qualitatively modify the PMN–*C. neoformans* interaction in vitro: in the presence of endothelial cells, cryptococci are often surrounded by clusters of five or more PMNs (192). These results suggest that endothelial cells may have an important role in protecting against metastatic infection by enhancing leukocyte function against *C. neoformans*.

Intravascular granulomas have been described in the veins of rats infected intravenously with *C. neoformans* (230). The intravascular granulomas were covered by endothelial cells and reassembled mural thrombi. The pathogenesis of such lesions is unclear but may involve an initial attachment of *C. neoformans* to endothelial cells. Analysis of the cellular adhesion molecules ICAM and β_2-integrins in infected rats showed higher levels in endothelial cells than in noninfected controls. This suggests that infection may lead to expression of adhesion molecules, which in turn may promote the adherence of inflammatory cells to the vessel wall and lead eventually to the formation of a granuloma.

Interactions between endothelial cells and *C. neoformans* may be important for the escape of fungal cells from the vascular space into organ tissues. Ibrahim et al. (91) have shown that acapsular *C. neoformans* adhere better to endothelial cells and cause more injury than do encapsulated cells in vitro. Endothelial cells are able to phagocytose *C. neoformans* through a serum-dependent process that is inhibited by cytochalasin D (91). These observations suggest that poorly encapsulated forms may be taken up by endothelial cells and that this phenomenon may contribute to *C. neoformans* dissemination (91).

OVERVIEW OF CELLULAR MICROBICIDAL MECHANISMS

Oxygen-Derived Oxidants

Neutrophils and macrophages produce a variety of microbicidal ROIs (102). Oxygen-derived oxidants can be fungicidal for *C. neoformans* (42, 134, 221). Hydrogen peroxide kills *C. neoformans* in vitro at concentrations of 0.1 to 1.0 mM, which is within the concentration range for H_2O_2 in human neutrophils (42). However, the antifungal activity of H_2O_2 may be lower than that of other oxidative substances produced by effector cells. Another study reported that the concentration of H_2O_2 necessary to kill *C. neoformans* was 7.7 ± 4.0 mM, or 100 times the concentration of hypochlorite required for fungicidal activity (95). The myeloperoxidase enzyme system catalyzes the conversion of H_2O_2 to hypochlorite in the presence of chloride. The observation that neutrophils from patients with chronic granulomatous disease (a defect in H_2O_2 generation) are defective in killing *C. neoformans* (42) suggests that oxygen-derived oxidants are important antimicrobial molecules against this fungal pathogen. The polyamine oxidase-polyamine system (PAO) is a potent antimicrobial oxidative system found in macrophages. Products of the PAO system include H_2O_2, reactive aminoaldehydes, and ammonia. Of these products, H_2O_2 and acrolein have been shown to kill *C. neoformans* at concentrations attainable for the PAO system (134).

C. neoformans strains vary significantly in their susceptibility to killing by products of the halide-H_2O_2 reaction in vitro (97). Fungal factors that may influence susceptibility to ROIs include enzymatic activities such as catalase and superoxide dismutase (209), melanin (95, 220, 221), and possibly the polysaccharide capsule. Although the capsule has not been directly implicated in this phenomenon, it is noteworthy that mycobacterial lipids and polysaccharides can scavenge oxygen radicals (28) and that the cryptococcal polysaccharide may serve a similar function. Differences between strains in susceptibility to killing by oxygen-derived oxidants may contribute to differences in virulence and to in vitro variability in experiments studying *C. neoformans*–effector cell interactions.

Nitrogen-Derived Oxidants

Nitrogen-derived oxidants are powerful antimicrobial molecules and are produced by several mammalian cell types (174, 175). L-Arginine is required for fungistasis against *C. neoformans* by murine macrophages (63). L-Arginine is converted into nitrite, nitrate, and L-citrulline by an enzymatic reaction catalyzed by nitrogen synthase (174, 175). This reaction generates a variety of reactive nitrogen species (including nitric oxide) that are directly toxic to *C. neoformans* (2). The presence of specific antibody to *C. neoformans* polysaccharide enhances nitric acid production by murine macrophages incubated with cryptococcus or cryptococcal polysaccharide (163, 215). Nitric oxide and nitrogen-derived oxidants have been shown to be important antimicrobial effector molecules against *C. neoformans* in murine macrophages, but a role for nitric oxide in human macrophages remains to be established. L-Arginine-dependent generation of reactive nitrogen species does not appear to be involved in *C. neoformans* fungistasis by human alveolar and peritoneal macrophages (25). Similarly, antibody-dependent fungistatic and fungicidal activity of human microglia occurs without evidence of nitrite production and is not affected by inhibitors of nitrogen synthase (126). However, human astrocytes inhibit *C. neoformans* replication extracellularly by a nitric oxide-related mechanism (123). In summary, nitric oxide and related products have been shown to be important effector molecules against *C. neoformans* in murine macrophages and human astrocytes, but their role in human macrophage function is uncertain.

NONOXIDATIVE ANTIMICROBIAL MECHANISMS

Phagocytic effector cells produce antimicrobial proteins and peptides that are capable of nonoxidative killing of *C. neoformans*. Murine macrophages produce histone-like murine microbicidal proteins that are fungicidal for *C. neoformans* (83). IFN-γ stimulation of murine macrophages enhances the production of murine microbicidal proteins (83). Azurophil granules in human neutrophils contain small antimicrobial peptides known as human neutrophil peptides or defensins (130). Defensins are arginine-rich polypeptides of 29 to 34 amino acids with molecular sizes of 3.5 to 4.0 kDa (130). Defensins constitute 5 to 7% of the total protein in human neutrophils and 30 to 50% of the total protein in azurophil granules (56, 128, 130). *C. neoformans* cells are rapidly killed by human defensins (56). Synthesis of defensins appears to be both species and tissue specific (130). Defensins are found in human, rabbit, rat, and guinea pig neutrophils but not in mouse neutrophils (130). However, in the mouse, defensin mRNA has been detected in gut epithelial (Paneth) cells and lamina propria macrophages (130). In rabbit, defensin production is developmentally regulated; their alveolar macrophages lack defensins at birth but achieve adult levels after a few weeks of age (130). Humans and the laboratory animals rabbits, rats, and guinea pigs are relatively resistant to cryptococcal infection, whereas mice are highly susceptible to infection. The causes for the species differences in susceptibility to infection are not known. However, the absence of defensins in murine neutrophils could conceivably contribute to the marked susceptibility of this species to cryptococcal infections.

There are other nonoxidative antimicrobial mechanisms available to effector cells that could help contain *C. neoformans*. Acidification following phagocytosis is an important nonspecific antimicrobial mechanism. Phagolysosome acidification has been demonstrated for rat alveolar macrophages (70) and human monocyte-derived macrophages (137) after ingestion of *C. neoformans*. However, it is unclear whether the pH in phagolysosomes is sufficiently acidic to damage *C. neoformans*. The pH of neutrophil phagolysosomes is about 6.0, which is well tolerated by *C. neoformans*. In fact, Levitz and collaborators (137) have shown that chloroquine can enhance the antifungal efficacy of *C. neoformans* against intracellular cryptococci; the mechanism appears to involve an inhibition of phagolysosomal acidification. Since *C. neoformans* grows better in acidic environments than in neutral or slightly alkaline environments, phagolysosomal acidification may not be an important mechanism for controlling intracellular growth.

Iron-free lactoferrin is fungicidal against *C. albicans* (177) and may also be active against *C. neoformans*. Cationic proteins in tissue have also been shown to inhibit *C. neoformans* (54). Human calcium-binding leukocyte L1 protein (calprotectin) is found in neutrophils, monocytes, and epithelial cells and is fungicidal for *C. neoformans* (205).

In summary, multiple components of the antimicrobial arsenal of host effector cells have been shown to have anti-*C. neoformans* activity. The relative importance of oxidative and nonoxidative antimicrobial mechanisms of host effector cells in inhibiting and killing *C. neoformans* is uncertain. In vivo, it is likely that the oxidative and nonoxidative antimicrobial mechanisms cooperate to produce the antifungal effects described for host effector cells.

REFERENCES

1. **Aguirre, K. M., and L. L. Johnson.** 1997. A role for B cells in resistance to *Cryptococcus neoformans* in mice. *Infect. Immun.* **65:**525–530.
2. **Alspaugh, J. A., and D. L. Granger.** 1991. Inhibition of *Cryptococcus neoformans* replication by nitrogen oxide supports the role of these molecules as effectors of macrophage-mediated cystostasis. *Infect. Immun.* **59:**2291–2296.
3. **Anderson, D. A., and H. M. Sagha.** 1988. Persistence of infection in mice inoculated intranasally with *Cryptococcus neoformans*. *Mycopathologia* **104:**163–169.
4. **Anderson, H. W.** 1917. Yeast-like fungi in the human intestinal tract. *J. Infect. Dis.* **21:**380–383.
5. **Aronson, M., and J. Kletter.** 1973. Aspects of the defense against a large-sized parasite, the yeast, *Cryptococcus neoformans*. *Isr. J. Med. Sci.* **1:**132–162.

6. **Bang, F. B.** 1961. Mucociliary function as protective mechanism in upper respiratory tract. *Bacteriol. Rev.* **25:**228–236.

7. **Baum, G. L., and D. Artis.** 1961. Growth inhibition of *Cryptococcus neoformans* by cell free human serum. *Am. J. Med. Sci.* **98:**613–616.

8. **Baum, G. L., and D. Artis.** 1963. Characterization of the growth inhibition factor for *Cryptococcus neoformans* (GIFc) in human sera. *Am. J. Med. Sci.* **90:**87–91.

9. **Bergman, F.** 1966. Effect of temperature on intratesticular cryptococcal infection in rabbits. *Sabouraudia* **5:**54–58.

10. **Beyt, B. E., and S. R. Waltman.** 1978. Cryptococcal endophthalmitis after corneal transplantation. *N. Engl. J. Med.* **298:**825–826.

11. **Blasi, E., R. Barluzzi, R. Mazzola, P. Mosci, and F. Bistoni.** 1992. Experimental model of intracerebral infection with *Cryptococcus neoformans*: roles of phagocytes and opsonization. *Infect. Immun.* 60:3682–3688.

12. **Blasi, E., R. Barluzzi, R. Mazzola, B. Tancini, S. Saleppico, M. Puliti, L. Pitzurra, and F. Bistoni.** 1995. Role of nitric oxide and melanogenesis in the accomplishment of anticryptococcal activity by the BV-2 microglial cell line. *J. Neuroimmunol.* **58:**111–116.

13. **Bobak, D. A., R. G. Washburn, and M. M. Frank.** 1988. C1q enhances the phagocytosis of *Cryptococcus neoformans* blastospores by human monocytes. *J. Immunol.* **141:**592–597.

14. **Boelaert, J. R., M. de Locht, J. Van Cutsem, V. Kerrels, B. Cantinieaux, A. Verdonck, H. W. Van Landuyt, and Y.-J. Schneider.** 1993. Mucormycosis during deferoxamine therapy is a siderophore mediated infection. In vitro and in vivo animal studies. *J. Clin. Invest.* **91:**1979–1976.

15. **Bolanos, B., and T. G. Mitchell.** 1989. Killing of *Cryptococcus neoformans* by rat alveolar macrophages. *J. Med. Vet. Mycol.* 27:219–228.

16. **Bolanos, B., and T. G. Mitchell.** 1989. Phagocytosis of *Cryptococcus neoformans* by rat alveolar macrophages. *J. Med. Vet. Mycol.* **27:**203–217.

17. **Bolanos, B., and T. G. Mitchell.** 1989. Phagocytosis and killing of *Cryptococcus neoformans* by rat alveolar macrophages in the absence of serum. *J. Leukocyte Biol.* **48:**521–528.

18. **Bonacini, M., J. Nussbaum, and C. Ahluwalia.** 1990. Gastrointestinal, hepatic, and pancreatic involvement with *Cryptococcus neoformans* in AIDS. *J. Clin. Gastroenterol.* **12:**296–297.

19. **Brewer, G. E., and F. C. Wood.** 1908. Blastomycosis of the spine. *Ann. Surg.* **48:**889–896.

20. **Brodie, S. J., V. G. Sasseville, K. A. Reinmann, M. A. Simon, P. K. Sehgal, and D. J. Ringler.** 1994. Macrophage function in simian AIDS. Killing defects in vivo are independent of macrophage function, associated with alterations in Th phenotype, and reversible with IFN-gamma. *J. Immunol.* **153:**5790.

21. **Brummer, E., and D. A. Stevens.** 1994. Anticryptococcal activity of macrophages: role of mouse strain, C5, contact, phagocytosis and L-arginine. *Cell. Immunol.* **157:**1–10.

22. **Bulmer, G. S., and J. R. Tacker.** 1975. Phagocytosis of *Cryptococcus neoformans* by alveolar macrophages. *Infect. Immun.* **11:**73–79.

23. **Calderone, R. A., and P. C. Braun.** 1991. Adherence and receptor relationships of *Candida albicans*. *Microbiol. Rev.* **55:**1–20.

24. **Cameron, M. L., D. L. Granger, T. J. Matthews, and J. B. Weinberg.** 1994. Human immunodeficiency virus (HIV)-infected human blood monocytes and peritoneal macrophages have reduced anticryptococcal activity whereas HIV-infected alveolar macrophages retain normal activity. *J. Infect. Dis.* **170:**60–70.

25. **Cameron, M. L., D. L. Granger, J. B. Weinberg, W. J. Kozumbo, and H. S. Koren.** 1990. Human alveolar and peritoneal macrophages mediate fungistasis independently of L-arginine oxidation to nitrate or nitrate. *Am. Rev. Respir. Dis.* **142:**1313–1319.

26. **Campanelli, D., P. A. Detmers, C. F. Nathan, and J. E. Gabay.** 1990. Azurocidin and a homologous serine protease from neutrophils. *J. Clin. Invest.* **85:**904–915.

27. **Casadevall, A., J. Mukherjee, Y. RuiRong, and J. Perfect.** 1994. Management of *Cryptococcus neoformans* contaminated needle injuries. *Clin. Infect. Dis.* **19:**951–953.

28. **Chan, J., T. Fujiwara, P. Brennan, M. McNeil, S. J. Turco, J.-C. Sibille, M. Snapper, P. Aisen, and B. R. Bloom.** 1989. Microbial glycolipids: possible virulence factors that scavenge oxygen radicals. *Proc. Natl. Acad. Sci. USA* **86:**2453–2457.

29. **Chatuverdi, V., B. Wong, and S. L. Newman.** 1996. Oxidative killing of *Cryptococcus neoformans* by human leukocytes. Evidence that fungal mannitol protects by scavenging reactive oxygen intermediates. *J. Immunol.* **156:**3836–3840.

30. **Cohen, J., J. R. Perfect, and D. T. Durack.** 1982. Cryptococcosis and the basidiospore. *Lancet* **i:**1501.

31. **Collins, H. L., and G. J. Bancroft.** 1991. Encapsulation of *Cryptococcus neoformans* impairs antigen-specific T-cell responses. *Infect. Immun.* **59:**3883–3888.

32. **Collins, H. L., and G. J. Bancroft.** 1992. Cytokine enhancement of complement-dependent phagocytosis by macrophages: synergy of tumor necrosis-alpha and granulocyte-macrophage colony stimulating factor for phagocytosis of *Cryptococcus neoformans*. *Eur. J. Immunol.* **22:**1447–1454.

33. **Cox, R. A., R. M. Pope, and D. A. Stevens.** 1982. Immune complexes in coccidioidomycosis. *Am. Rev. Respir. Dis.* **126:**439–443.

34. **Cross, C. E., and G. J. Bancroft.** 1995. Ingestion of acapsular *Cryptococcus neoformans* occurs via mannose and beta-glucan receptors, resulting in cytokine production and increased phagocytosis of the encapsulated form. *Infect. Immun.* **63:**2604–2611.

35. **Cross, C. E., H. L. Collins, and G. J. Bancroft.** 1997. CR3-dependent phagocytosis by murine macrophages: different cytokines regulate ingestion of a defined CR3 ligand and complement-opsonized *Cryptococcus neoformans*. *Immunology* **91:**289–296.

36. **Daly, J. S., K. A. Porter, F. K. Chong, and R. J. Robillard.** 1990. Disseminated, nonmeningeal gastrointestinal cryptococcal infection in an HIV-negative patient. *Am. J. Gastroenterol.* **85:**1421–1424.

37. **Davies, S. F., D. P. Clifford, J. R. Hoidal, and J. E. Repine.** 1982. Opsonic requirements for the uptake of *Cryptococcus neoformans* by human polymorphonuclear leukocytes and monocytes. *J. Infect. Dis.* **145:**870–874.

38. **Diamond, R. D., and J. E. Bennett.** 1973. Growth of *Cryptococcus neoformans* within human macrophages in vitro. *Infect. Immun.* **7:**231–236.

39. **Diamond, R. D., and N. F. Erickson III.** 1982. Chemotaxis of human neutrophils and monocytes induced by *Cryptococcus neoformans*. *Infect. Immun.* **38:**380–382.

40. **Diamond, R. D., J. E. May, M. Kane, M. M. Frank, and J. E. Bennett.** 1973. The role of late complement component and the alternate complement pathway in experimental cryptococcosis (37580). *Proc. Soc. Exp. Biol. Med.* **144:**312–315.

41. **Diamond, R. D., J. E. May, M. C. Kane, M. M. Frank, and J. E. Bennett.** 1974. The role of the classical and alternate complement pathways in host defenses against *Cryptococcus neoformans* infection. *J. Immunol.* **112:**2260–2270.

42. **Diamond, R. D., R. K. Root, and J. E. Bennett.** 1972. Factors influencing killing of *Cryptococcus neoformans* by human leukocytes *in vitro*. *J. Infect. Dis.* **125:**367–376.

43. **Dickson, D. W., L. A. Mattiace, K. Kure, K. Hitchings, W. D. Lyman, and C. F. Brosnan.** 1991. Biology of disease: microglia in human disease, with an emphasis on acquired immune deficiency syndrome. *Lab. Invest.* **64:**135–156.

44. **Dixon, D. M., and A. Polak.** 1986. *In vivo* and *in vitro* studies with an atypical, rhinotrophic isolate of *Cryptococcus neoformans*. *Mycopathologia* **96:**33–40.

45. **Ellis, D. H., and T. J. Pfeiffer.** 1990. Ecology, life cycle, and infectious propagule of *Cryptococcus neoformans*. *Lancet* **336:**923–925.

46. **Fazekas, G., and J. Schwarz.** 1958. Histology of experimental murine cryptococcosis. *Am. J. Pathol.* **34:**517–529.

47. **Feldmesser, M., and A. Casadevall.** 1997. Effect of serum IgG1 against murine pulmonary infection with *Cryptococcus neoformans. J. Immunol.* **158:**790–799.

48. **Feldmesser, M., A. Casadevall, Y. Kress, G. Spira, and A. Orlofski.** 1997. Eosinophil-*Cryptococcus neoformans* interactions in vivo and in vitro. *Infect. Immun.* **65:**1899–1907.

49. **Felsenfeld, O.** 1944. Yeast-like fungi in the intestinal tract of chronically institutionalized patients. *Am. J. Med.* **207:**60–62.

50. **Flesch, I. E. A., G. Schwamberger, and S. H. E. Kaufman.** 1989. Fungicidal activity of IFN-gamma activated macrophages. *J. Immunol.* **142:**3219–3224.

51. **Friedman, M., A. Brenski, and L. Taylor.** 1994. Treatment of aphthous ulcers in AIDS patients. *Laryngoscope* **104:**566–570.

52. **Fromtling, R. A., G. K. Abruzzo, and A. Ruiz.** 1988. *Cryptococcus neoformans*: a central nervous system isolate from an AIDS patient that is rhinotropic in a normal mouse model. *Mycopathologia* **102:**79–86.

53. **Gadebusch, H. H.** 1961. Natural host resistance to infection with *Cryptococcus neoformans*. The effect of the properdin system on the experimental disease. *J. Infect. Dis.* **109:**147–153.

54. **Gadebusch, H. H.** 1966. On the mechanism of cytotoxicity by cationic tissue proteins for *Cryptococcus neoformans. Z. Naturforsch. Sect. B* 21:1048–1051.

55. **Galli, S. J.** 1997. The mast cell: a versatile effector cell for a challenging world. *Int. Arch. Allergy Immunol.* **113:**14–22.

56. **Ganz, T., M. E. Selsted, D. Szklarek, S. S. L. Harwig, K. Daher, D. F. Bainton, and R. I. Lehrer.** 1985. Defensins. Natural peptide antibiotics of human neutrophils. *J. Clin. Invest.* **76:**1427–1435.

57. **Garrity, J. A., D. C. Herman, R. Imes, P. Fries, C. F. Hughes, and R. J. Campbell.** 1993. Optic nerve sheath decompression for visual loss in patients with acquired immunodeficiency syndrome and cryptococcal meningitis with papilledema. *Am. J. Ophthalmol.* **116:**472–478.

58. **Gentry, L. O., and J. S. Remington.** 1971. Resistance against *Cryptococcus* conferred by intracellular bacteria and protozoa. *J. Infect. Dis.* **123:**22–31.

59. **Glaser, J. B., and A. Garden.** 1985. Inoculation of cryptococcosis without transmission of the acquired immunodeficiency syndrome. *N. Engl. J. Med.* **313:**266.

60. **Goldman, D., S. C. Lee, and A. Casadevall.** 1994. Pathogenesis of pulmonary *Cryptococcus neoformans* infection in the rat. *Infect. Immun.* **62:**4755–4761.

61. **Goldman, D. L., S. C. Lee, and A. Casadevall.** 1995. Tissue localization of *Cryptococcus neoformans* glucuronoxylomannan in the presence and absence of specific antibody. *Infect. Immun.* **63:**3448–3453.

62. **Goren, M. B., and J. Warren.** 1968. Immunofluorescence studies of reactions at the cryptococcal capsule. *J. Infect. Dis.* **118:**215–229.

63. **Granger, D. L., J. B. Hibbs, J. R. Perfect, and D. T. Durack.** 1988. Specific amino acid (L-arginine) requirement for the microbiostatic activity of murine macrophages. *J. Clin. Invest.* **81:**1129–1136.

64. **Granger, D. L., J. R. Perfect, and D. T. Durack.** 1986. Macrophage-mediated fungistasis *in vitro*: requirements for intracellular and extracellular cytotoxicity. *J. Immunol.* **136:**672–680.

65. **Graybill, J. R., and J. Ahrens.** 1981. Immunization and complement interaction in host defense against murine cryptococcosis. RES *J. Reticuloendothel. Soc.* **30:**347–357.

66. **Graybill, J. R., R. Bocanegra, C. Lambros, and M. F. Luther.** 1997. Granulocyte colony stimulating factor therapy of experimental cryptococcal meningitis. *J. Med. Vet. Mycol.* **35:**243–247.

67. **Graybill, J. R., and L. Mitchell.** 1978. Cyclophosphamide effects on murine crypto-coccosis. *Infect. Immun.* **21:**674–677.

68. **Green, J. R., and G. S. Bulmer.** 1979. Gastrointestinal inoculation of *Cryptococcus neoformans* in mice. *Sabouraudia* **17:**233–240.

69. **Griffin, F. M.** 1980. Roles of macrophage Fc and C3b receptors in phagocytosis of im-munologically coated *Cryptococcus neoformans. Proc. Natl. Acad. Sci. USA* **78:**3853–3857.

70. **Gross, N. T., K. Nessa, P. Camner, M. Chinchilla, and C. Jarstrand.** 1997. Interaction between *Cryptococcus neoformans* and alveolar macrophages. *J. Med. Vet. Mycol.* **35:**263–269.

71. **Grover, D., E. Brummer, and D. A. Stevens.** 1996. Study of the role of iron in the anticryptococcal activity of human serum and fluconazole. *Mycopathologia* **133:**71–77.

72. **Harrison, T. S., H. Kornfeld, and S. M. Levitz.** 1995. The effect of infection with human immunodeficiency virus on the anticryptococcal activity of lymphocytes and monocytes. *J. Infect. Dis.* **172:**665–671.

73. **Harrison, T. S., and S. M. Levitz.** 1997. Priming with IFN-gamma restores deficient IL–12 production by peripheral blood mononuclear cells from HIV-seropositive do-nors. *J. Immunol.* **158:**459–463.

74. **Harrison, T. S., and S. M. Levitz.** 1997. Mechanisms of impaired anticryptococcal activity of monocytes from donors infected with human immunodeficiency virus. *J. Infect. Dis.* **176:**537–540.

75. **Harrison, T. S., S. Nong, and S. M. Levitz.** 1997. Induction of human immunodefi-ciency virus type 1 expression in monocytic cells by *Cryptococcus neoformans* and *Candida albicans. J. Infect. Dis.* **176:**485–491.

76. **Hendry, A. T., and A. Bakerspigel.** 1969. Factors affecting serum inhibited growth of *Candida albicans* and *Cryptococcus neoformans. Sabouraudia* **7:**219–229.

77. **Hidore, M. R., T. W. Mislan, and J. W. Murphy.** 1991. Response of murine natural killer cells to binding of the fungal target *Cryptococcus neoformans. Infect. Immun.* **59:**1489–1499.

78. **Hidore, M. R., and J. W. Murphy.** 1986. Correlation of natural killer cell activity and clearance of *Cryptococcus neoformans* from mice after adoptive transfer of splenic nylon wool-nonadherent cells. *Infect. Immun.* **51:**547–555.

79. **Hidore, M. R., and J. W. Murphy.** 1986. Natural cellular resistance of beige mice against *Cryptococcus neoformans. J. Immunol.* **137:**3624–3631.

80. **Hidore, M. R., and J. W.** Murphy. 1989. Murine natural killer cell interactions with a fungal target, *Cryptococcus neoformans. Infect. Immun.* **57:**1990–1997.

81. **Hidore, M. R., N. Nabavi, C. W. Reynolds, P. A. Henkart, and J. W. Murphy.** 1990. Cytoplasmic components of natural killer cells limit the growth of *Cryptococcus neo-formans. J. Leukocyte Biol.* **48:**15–26.

82. **Hidore, M. R., N. Nabavi, F. Sonleitner, and J. W. Murphy.** 1991. Murine natural killer cells are fungicidal to *Cryptococcus neoformans. Infect. Immun.* **59:**1747–1754.

83. **Hiemstra, P. S., P. B. Eisenhauer, L. S. Harwig, M. T. van den Barselaar, R. van Furth, and R. I. Lehrer.** 1993. Antimicrobial proteins of murine macrophages. *Infect. Immun.* **61:**3038–3046.

84. **Hiles, D. A., and R. L. Font.** 1968. Bilateral intraocular cryptococcosis with unilateral spontaneous regression. *Am. J. Ophthalmol.* **65:**98–108.

85. **Hill, J. O.** 1992. CD4+ T cells cause multinucleated giant cells to form around *Crypto-coccus neoformans* and confine the yeast within the primary site of infection in the respiratory tract. *J. Exp. Med.* **175:**1685–1695.

86. **Horn, C. A., and R. G. Washburn.** 1995. Anticryptococcal activity of NK cell-enriched peripheral blood lymphocytes from human immunodeficiency virus-infected sub-jects: responses to interleukin-2, interferon-gamma, and interleukin-12. *J. Infect. Dis.* **172:**1023–1027.

87. **Houpt, D. C., G. S. T. Pfrommer, B. J. Young, T. A. Larson, and T. R. Kozel.** 1994. Occurrences, immunoglobulin classes, and biological activities of antibodies in normal human serum that are reactive with *Cryptococcus neoformans* glucuronoxylomannan. *Infect. Immun.* **62:**3857–2864.

88. **Howard, D. H.** 1961. Some factors which affect the initiation of growth of *Cryptococcus neoformans*. *J. Bacteriol.* **82:**430–435.

89. **Huffnagle, G. B., N. E. Street, and M. Lipscomb.** 1992. In contrast to Balb/c mice, a *Cryptococcus neoformans* infection in C57BL/6 mice generates protective T-cell immunity in the periphery and non-protective T-cell mediated eosinophilia in the lungs. *FASEB J.* **6:**A1689.

90. **Hutto, J. O., C. S. Bryan, F. L. Greene, C. J. White, and J. I. Gallin.** 1988. Cryptococcosis of the colon resembling Crohn's disease in a patient with the hyperimmunoglobulinemia E-recurrent infection (Job's) syndrome. *Gastroenterology* **94:**808–812.

91. **Ibrahim, A. S., S. G. Filler, M. S. Alcouloumre, T. R. Kozel, J. E. Edwards, and M. A. Ghannoum.** 1995. Adherence to and damage of endothelial cells by *Cryptococcus neoformans* in vitro: role of the capsule. *Infect. Immun.* **63:**4368–4374.

92. **Igel, H., and R. P. Bolande.** 1966. Humoral defense mechanisms in cryptococcosis: substances in normal human serum, saliva, and cerebrospinal fluid affecting the growth of *Cryptococcus neoformans*. *J. Infect. Dis.* **116:**75–83.

93. **Ikeda, R., T. Shinoda, K. Kagaya, and Y. Fukazawa.** 1984. Role of serum factors in the phagocytosis of weakly or heavily encapsulated *Cryptococcus neoformans* strains by guinea pig blood leukocytes. *Microbiol. Immunol.* **28:**51–61.

94. **Jacobs, D. H., A. M. Macher, R. Handler, J. E. Bennett, M. J. Collen, and J. I. Gallin.** 1984. Esophageal cryptococcosis in a patient with the hyperimmunoglobulin E-recurrent infection (Job's) syndrome. *Gastroenterology* **87:**201–203.

95. **Jacobson, E. S., and S. B. Tinnell.** 1993. Antioxidant function of fungal melanin. *J. Bacteriol.* **175:**7102–7104.

96. **Jimenez-Lucho, V., V. Ginsburg, and H. Krivan.** 1990. *Cryptococcus neoformans, Candida albicans*, and other fungi bind specifically to the glycosphingolipid lactosylceramide (Galβ1–4Glcβ1–1Cer), a possible adhesion receptor for yeasts. *Infect. Immun.* **58:**2085–2090.

97. **Kagaya, K., T. Yamada, Y. Miyakawa, Y. Fukazawa, and S. Saito.** 1985. Characterization of pathogenic constituents of *Cryptococcus neoformans* strains. *Microbiol. Immunol.* **29:**517–532.

98. **Kalina, M., Y. Kletter, and M. Aronson.** 1974. The interaction of phagocytes and the large-sized parasite *Cryptococcus neoformans*: cytochemical and ultrastructural study. *Cell Tissue Res.* **152:**165–174.

99. **Karaoui, R. M., N. K. Hall, and H. W. Larsh.** 1977. Role of macrophages in immunity and pathogenesis of experimental cryptococcosis induced by the airborne route. II. Phagocytosis and intracellular fate of *Cryptococcus neoformans*. *Mykosen* **20:**409–422.

100. **Keller, R. G., G. S. Pfrommer, and T. R. Kozel.** 1994. Occurrences, specificities, and functions of ubiquitous antibodies in human serum that are reactive with the *Cryptococcus* neoformans cell wall. *Infect. Immun.* **62:**215–220.

101. **Kestelyn, P., H. Taelman, J. Bodaerts, A. Kagame, M. A. Aziz, J. Batungwanayo, A. M. Stevens, and P. Van de Perre.** 1993. Ophthalmic manifestations of infections with *Cryptococcus neoformans* in patients with the acquired immuodeficiency syndrome. *Am. J. Ophthalmol.* **116:**721–727.

102. **Klebanoff, S. J.** 1980. Oxygen metabolism and the toxic properties of phagocytes. *Ann. Intern. Med.* **93:**480–489.

103. **Kligman, A. M., A. P. Crane, and R. F. Norris.** 1951. Effect of temperature on survival of chick embryos infected intravenously with *Cryptococcus neoformans* (Torula histolytica). *Am. J. Med. Sci.* **221:**273–278.

104. **Kligman, A. M., and F. D. Weidman.** 1949. Experimental studies on treatment of human torulosis. *Arch. Dermatol. Syph.* **60:**726–741.

105. **Kozel, T. R.** 1993. Opsonization and phagocytosis of *Cryptococcus neoformans. Arch. Med. Res.* **24:**211–218.

106. **Kozel, T. R.** 1996. Activation of the complement system by pathogenic fungi. *Clin. Microbiol. Rev.* **9:**34–46.

107. **Kozel, T. R., R. R. Brown, and G. S. T. Pfrommer.** 1987. Activation and binding of C3 by *Candida albicans. Infect. Immun.* **55:**1890–1894.

108. **Kozel, T. R., B. Highison, and C. J. Stratton.** 1984. Localization on encapsulated *Cryptococcus neoformans* of serum components opsonic for phagocytosis by macrophages and neutrophils. *Infect. Immun.* **43:**574–579.

109. **Kozel, T. R., and G. S. T. Pfrommer.** 1986. Activation of the complement system by *Cryptococcus neoformans* leads to binding of iC3b to the yeast. *Infect. Immun.* **52:**1–5.

110. **Kozel, T. R., G. S. T. Pfrommer, A. S. Guerlain, B. A. Highison, and G. J. Highison.** 1988. Strain variation in phagocytosis of *Cryptococcus neoformans* dissociation of susceptibility to phagocytosis from activation and binding of opsonic fragments of C3. *Infect. Immun.* **56:**2794–2800.

111. **Kozel, T. R., G. S. T. Pfrommer, and D. Redelman.** 1987. Activated neutrophils exhibit enhanced phagocytosis of *Cryptococcus neoformans* opsonized with normal human serum. *Clin. Exp. Immunol.* **70:**238–246.

112. **Kozel, T. R., A. Tabuni, B. J. Young, and S. M. Levitz.** 1996. Influence of opsonization conditions on C3 deposition and phagocyte binding of large- and small-capsule *Cryptococcus neoformans* cells. *Infect. Immun.* **64:**2336–2338.

113. **Kozel, T. R., L. C. Weinhold, and D. M. Lupan.** 1996. Distinct characteristics of initiation of the classical and alternative complement pathways by *Candida albicans. Infect. Immun.* **64:**3360–3368.

114. **Kozel, T. R., M. A. Wilson, and J. W. Murphy.** 1991. Early events in initiation of alternative complement pathway activation by the capsule of *Cryptococcus neoformans. Infect. Immun.* **59:**3101–3110.

115. **Kozel, T. R., M. A. Wilson, G. S. Pfrommer, and A. M. Schlagetter.** 1989. Activation and binding of opsonic fragments of C3 on encapsulated *Cryptococcus neoformans* by using an alternative complement pathway reconstituted from six isolated patients. *Infect. Immun.* **57:**1922–1927.

116. **Kozel, T. R., M. A. Wilson, and W. H. Welch.** 1992. Kinetic analysis of the amplification phase for activation and binding of C3 to encapsulated and nonencapsulated *Cryptococcus neoformans. Infect. Immun.* **60:**3122–3127.

117. **Kuhn, L. R.** 1939. Growth and viability of *Cryptococcus hominis* at mouse and rabbit body temperatures. *Proc. Soc. Exp. Biol. Med.* **41:**573–574.

118. **Kuhn, L. R.** 1939. Experimental cryptococcic infection. *Arch. Pathol.* **27:**803.

119. **Kuttin, E. S., M. Feldman, A. Nyska, B. A. Weissman, J. Muller, and H. B. Levine.** 1988. Cryptococcosis of the nasopharynx in mice and rats. *Mycopathologia* **101:**99–104.

120. **Kwon-Chung, K. J., and J. E. Bennett.** 1978. Distribution of α and a mating types of *Cryptococcus neoformans* among natural and clinical isolates. *Am. J. Epidemiol.* **108:**337–340.

121. **Laxalt, K. A., and T. R. Kozel.** 1979. Chemotoxigenesis and activation of the alternative complement pathway by encapsulated and non-encapsulated *Cryptococcus neoformans. Infect. Immun.* **26:**435–440.

122. **Lee, S. C., A. Casadevall, and D. W. Dickson.** 1996. Immunohistochemical localization of capsular polysaccharide antigen in the central nervous system cells in cryptococcal meningoencephalitis. *Am. J. Pathol.* **148:**1267–1274.

123. **Lee, S. C., D. W. Dickson, C. F. Brosnan, and A. Casadevall.** 1994. Human astrocytes

inhibit *Cryptococcus neoformans* growth by a nitric oxide-mediated mechanism. *J. Exp. Med.* **180:**365–369.

124. **Lee, S. C., D. W. Dickson, and A. Casadevall.** 1996. Pathology of cryptococcal meningoencephalitis: analysis of 27 patients with pathogenetic implications. *Hum. Pathol.* **27:**839–847.

125. **Lee, S. C., D. W. Dickson, W. Liu, and C. F. Brosnan.** 1993. Induction of nitric oxide synthesase activity in human astrocytes by IL–1β and IFN–γ. *J. Neuroimmunol.* **41:**19–24.

126. **Lee, S. C., Y. Kress, D. W. Dickson, and A. Casadevall.** 1995. Human microglia mediate anti-*Cryptococcus neoformans* activity in the presence of specific antibody. *J. Neuroimmunol.* **16:**152–161.

127. **Lee, S. L., Y. Kress, M.-L. Zhao, D. W. Dickson, and A. Casadevall.** 1995. *Cryptococcus neoformans* survive and replicate in spacious phagosomes in human microglia. *Lab. Invest.* **73:**871–879.

128. **Lehrer, R. I., and T. Ganz.** 1990. Antimicrobial polypeptides of human neutrophils. *Blood* **76:**2176–2181.

129. **Lehrer, R. I., and K. M. Ladra.** 1977. Fungicidal components of mammalian granulocytes active against *Cryptococcus neoformans*. *J. Infect. Dis.* **136:**96–99.

130. **Lehrer, R. I., A. K. Lichenstein, and T. Ganz.** 1993. Defensins: antimicrobial and cytotoxic peptides of mammalian cells. *Annu. Rev. Immunol.* **11:**105–128.

131. **Levine, S., H. M. Zimmerman, and A. Scorza.** 1957. Experimental cryptococcosis (torulosis). *Am. J. Pathol.* **33:**385–409.

132. **Levitz, S. M.** 1991. Activation of human peripheral blood mononuclear cells by interleukin-2 and granulocyte-macrophage colony stimulating factor to inhibit *Cryptococcus neoformans*. *Infect. Immun.* **59:**3393–3397.

133. **Levitz, S. M., and D. J. DiBenedetto.** 1989. Paradoxical role of capsule in murine bronchoalveolar macrophage-mediated killing of *Cryptococcus neoformans*. *J. Immunol.* **142:**659–665.

134. **Levitz, S. M., D. J. DiBenedetto, and R. D. Diamond.** 1990. Inhibition and killing of fungi by the polyamine oxidase-polyamine system. *Antonie van Leeuwenhoek* **58:**107–114.

135. **Levitz, S. M., and T. P. Farrell.** 1990. Growth inhibition of *Cryptococcus neoformans* by cultured human monocytes: role of the capsule, opsonins, the culture surface and cytokines. *Infect. Immun.* **58:**1201–1209.

136. **Levitz, S. M., T. P. Farrell, and R. T. Maziarz.** 1991. Killing of *Cryptococcus neoformans* by human peripheral blood mononuclear cells stimulated in culture. *J. Infect. Dis.* **163:**1108–1113.

137. **Levitz, S. M., T. S. Harrison, A. Tabuni, and X. Liu.** 1997. Chloroquine induces human mononuclear phagocytes to inhibit and kill *Cryptococcus neoformans* by a mechanism independent of iron deprivation. *J. Clin. Invest.* **100:**1640–1646.

138. **Levitz, S. M., and A. Tabuni.** 1991. Binding of *Cryptococcus neoformans* by human cultured macrophages. Requirement for multiple complement receptors and actin. *J. Clin. Invest.* **87:**528–535.

139. **Levitz, S. M., A. Tabumi, H. Kornfeld, C. C. Reardon, and D. T. Golenbock.** 1994. Production of tumor necrosis factor alpha in human leukocytes stimulated by *Cryptococcus neoformans*. *Infect. Immun.* **62:**1975–1981.

140. **Levitz, S. M., A. Tabuni, T. R. Kozel, R. S. MacGill, R. R. Ingalls, and D. T. Golenbock.** 1997. Binding of *Cryptococcus neoformans* to heterologously expressed human complement receptors. *Infect. Immun.* **65:**931–935.

141. **Levitz, S. M., A. Tabuni, and C. Treseler.** 1993. Effect of mannose-binding protein on binding of *Cryptococcus neoformans* to human phagocytes. *Infect. Immun.* **61:**4891–4893.

142. **Lima, C., and J. P. Vital.** 1994. Olfactory pathways in three patients with cryptococcal meningitis and acquired immune deficiency syndrome. *J. Neurol. Sci.* **123:**195–199.

143. **Lima, C., and J. P. Vital.** 1994. Olfactory mucosa response in guinea pigs following intranasal instillation with *Cryptococcus neoformans*. *Mycopathologia* **126:**65–73.

144. **Lipscomb, M. F., T. Alvarellos, G. B. Tows, R. Tompkins, Z. Evans, G. Koo, and V. Kumar.** 1987. Role of natural killer cells in resistance to *Cryptococcus neoformans* infections in mice. *Am. J. Pathol.* **128:**354–361.

145. **Lipson, B. K., W. R. Freeman, J. Beniz, M. H. Goldbaum, J. R. Hesselink, R. N. Weinreb, and A. A. Sadun.** 1989. Optic neuropathy associated with cryptococcal arachnoiditis in AIDS patients. *Am. J. Ophthalmol.* **107:**523–527.

146. **Louie, A., A. L. Baltch, M. A. Franke, R. P. Smith, and M. A. Gordon.** 1994. Comparative efficacy of four antifungal agents to stimulate murine macrophages to produce tumour necrosis factor alpha: an effect that is attenuated by pentoxifylline, liposomal vesicles, and dexamethasone. *J. Antimicrob. Chemother.* **34:**975–987.

147. **Lowhagen, P., B.B. Johansson, and C. Nordborg.** 1993. The nasal route of cerebrospinal fluid drainage in man. A light-microscope study. *Neuropathol. Appl. Neurobiol.* **19:**480–488.

148. **Macher, A. M., J. E. Bennett, J. E. Gadek, and M. M. Frank.** 1978. Complement depletion in cryptococcal sepsis. *J. Immunol.* **120:**1686–1690.

149. **Malik, R., D. I. Wigney, D. B. Muir, and D. N. Love.** 1997. Asymptomatic carriage of *Cryptococcus* neoformans in the nasal cavity of dogs and cats. *J. Med. Vet. Mycol.* **35:**27–31.

150. **Mare, M., M. T. Sartori, M. Carretta, A. Bertaggia, and A. Girolami.** 1990. Thinophyma-like cryptococcal infection as an early manifestation of AIDS in a hemophilia B patient. *Acta Haematol.* **84:**101–103.

151. **Martin, E., A. Stuben, A. Gorz, U. Weller, and S. Bhakdi.** 1994. Novel aspect of amphotericin B action: accumulation in human monocytes potentiates killing of phagocytosed *Candida albicans*. *Antimicrob. Agents Chemother.* **38:**13–22.

152. **Merkel, G. J., and R. K. Cunningham.** 1992. The interaction of *Cryptococcus neoformans* with primary rat lung cell cultures. *J. Med. Vet. Mycol.* **30:**115–121.

153. **Merkel, G. J., and B. A. Scofield.** 1993. Conditions affecting the adherence of *Cryptococcus neoformans* to glial and lung cells *in vitro*. *J. Med. Vet. Mycol.* **31:**55–64.

154. **Merkel, G. J., and B. A. Scofield.** 1994. Comparisons between *in vitro* glial cell adherence and internalization of non-encapsulated and encapsulated strains of *Cryptococcus neoformans*. *J. Med. Vet. Mycol.* **32:**361–372.

155. **Merkel, G. J., B. A. Scofield, F. J. Rescorla, R. Yang, and J. L. Grosfeld.** 1995. Reduced recovery of a *Cryptococcus neoformans* adherence mutant from a rat model of cryptococcosis. *Can. J. Microbiol.* **41:**428–432.

156. **Miller, G. P. G., and S. Kohl.** 1983. Antibody-dependent leukocyte killing of *Cryptococcus neoformans*. *J. Immunol.* **131:**1455–1459.

157. **Miller, M. F., and T. G. Mitchell.** 1991. Killing of *Cryptococcus neoformans* strains by human neutrophils and monocytes. *Infect. Immun.* **59:**24–28.

158. **Miller, M. F., T. G. Mitchell, W. J. Storkus, and J. R. Dawson.** 1990. Human natural killer cells do not inhibit growth of *Cryptococcus neoformans* in the absence of antibody. *Infect. Immun.* **58:**639–645.

159. **Mitchell, T. G., and L. Friedman.** 1972. In vitro phagocytosis and intracellular fate of variously encapsulated strains of *Cryptococcus neoformans*. *Infect. Immun.* **5:**491–498.

160. **Mody, C. H., G.-H. Chen, C. Jackson, J. L. Curtis, and G. B. Toews.** 1994. In vivo depletion of murine CD8 positive T cells impairs survival during infection with a highly virulent strain of *Cryptococcus neoformans*. *Mycopathologia* **125:**7–17.

161. **Mody, C. H., C. L. Tyler, R. G. Sitrin, C. Jackson, and G. B. Toews.** 1991. Interferon-gamma activates rat alveolar macrophages for anticryptococcal activity. *Am. J. Respir. Cell. Mol. Biol.* **5:**19–26.

162. **Monga, D. P.** 1981. Role of macrophages in resistance of mice to experimental crypto-coccosis. *Infect. Immun.* **32:**975–978.

163. **Mozaffarian, N., J. W. Berman, and A. Casadevall.** 1995. Immune complexes increase nitric oxide production by interferon-gamma-stimulated murine macrophage-like J774.16 cells. *J. Leukocyte Biol.* **57:**657–662.

164. **Mucke, L., and M. Eddleston.** 1993. Astrocytes in infectious and immune-mediated diseases of the central nervous system. *FASEB J.* **7:**1226–1232.

165. **Mukherjee, S., M. Feldmesser, and A. Casadevall.** 1996. J774 murine macrophage-like cell interactions with *Cryptococcus neoformans* in the presence and absence of opsonins. *J. Infect. Dis.* **173:**1222–1231.

166. **Mukherjee, S., S. Lee, J. Mukherjee, M. D. Scharff, and A. Casadevall.** 1994. Mono-clonal antibodies to *Cryptococcus neoformans* capsular polysaccharide modify the course of intravenous infection in mice. *Infect. Immun.* **62:**1079–1088.

167. **Mukherjee, S., S. C. Lee, and A. Casadevall.** 1995. Antibodies to *Cryptococcus neofor-mans* glucuronoxylomannan enhance antifungal activity of murine macrophages. *Infect. Immun.* **63:**573–579.

168. **Murphy, J. W., M. R. Hidore, and N. Nabavi.** 1991. Binding interactions of murine natural killer cells with the fungal target *Cryptococcus neoformans*. *Infect. Immun.* **59:**1476–1488.

169. **Murphy, J. W., and D. O. McDaniel.** 1982. In vitro reactivity of natural killer (NK) cells against *Cryptococcus neoformans*. *J. Immunol.* **128:**1577–1583.

170. **Muth, S. M., and J. W. Murphy.** 1995. Direct anticryptococcal activity of lymphocytes from *Cryptococcus neoformans*-immunized mice. *Infect. Immun.* **63:**1637–1644.

171. **Nabavi, N., and J. W. Murphy.** 1985. In vitro binding of natural killer cells to *Crypto-coccus neoformans* targets. *Infect. Immun.* **50:**50–57.

172. **Nabavi, N., and J. W. Murphy.** 1986. Antibody-dependent natural killer cell-mediated growth inhibition of *Cryptococcus neoformans*. *Infect. Immun.* **51:**556–562.

173. **Nassar, F., E. Brummer, and D. A. Stevens.** 1995. Different components in human serum inhibit multiplication of *Cryptococcus neoformans* and enhance fluconazole ac-tivity. *Antimicrob. Agents Chemother.* **39:**2490–2493.

174. **Nathan, C.** 1992. Nitric oxide as a secretary product of mammalian cells. *FASEB J.* **6:**3051–3064.

175. **Nathan, C.F., and J. B. Hibbs, Jr.** 1991. Role of nitric oxide synthesis in macrophage antimicrobial activity. *Curr. Opin. Immunol.* **3:**65–70.

176. **Neilson, J. B., R. A. Fromtling, and G. S. Bulmer.** 1977. *Cryptococcus neoformans*: size range of infectious particles from aerosolized soil. *Infect. Immun.* **17:**634–638.

177. **Nikawa, H., L. P. Samaranayake, J. Tenovuo, K. M. Pang, and T. Hamada.** 1993. The fungicidal effect of human lactoferrin on *Candida albicans* and *Candida krusei*. *Arch. Oral Biol.* **38:**1057–1063.

178. **Okun, E., and W. T. Butler.** 1964. Ophthalmologic complications of cryptococcal meningitis. *Arch. Ophthalmol.* **71:**52–57.

179. **Papadimitriou, J. M., T. A. Robertson, Y. Kletter, M. Aronson, and M. N.-I. Walters.** 1978. An ultrastructural examination of the interaction between macrophages and *Cryptococcus neoformans*. *J. Pathol.* **124:**103–109.

180. **Perfect, J. R., and D. T. Durack.** 1985. Chemotactic activity of cerebrospinal fluid in experimental cryptococcal meningitis. *Sabouraudia* **23:**37–45.

181. **Perfect, J. R., D. L. Granger, and D. T. Durack.** 1987. Effects of antifungal agents and gamma-interferon on macrophage cytotoxicity for fungi and tumor cells. *J. Infect. Dis.* **156:**316–323.

182. **Perfect, J. R., M. M. Hobbs, D. L. Granger, and D. T. Durack.** 1988. Cerebrospinal fluid macrophage response to experimental cryptococcal meningitis: relationship be-tween in vivo and in vitro measurements of cytotoxicity. *Infect. Immun.* **56:**849–854.

183. **Perfect, J. R., S. D. R. Lang, and D. T. Durack.** 1980. Chronic cryptococcal meningitis. Am. J. Pathol. **101:**177–193.

184. **Pfrommer, G. S. T., S. M. Dickens, M. A. Wilson, B. J. Young, and T. R. Kozel.** 1993. Accelerated decay of C3b to iC3b when C3b is bound to the *Cryptococcus neoformans* capsule. *Infect. Immun.* **61:**4360–4366.

185. **Powell, K. E., B. A. Dahl, R. J. Weeks, and F. E. Tosh.** 1972. Airborne *Cryptococcus neoformans*: particles from pigeon excreta compatible with alveolar deposition. *J. Infect. Dis.* **125:**412–415.

186. **Randhawa, H. S., and P. Mahendra.** 1977. Occurrence and significance of *Cryptococcus neoformans* in the respiratory tract of patients with bronchopulmonary disorders. *J. Clin. Microbiol.* **5:**5–8.

187. **Reiss, F., and G. Szilagyi.** 1965. Ecology of yeast-like fungi in a hospital population. Detailed investigation of *Cryptococcus neoformans. Arch. Dermatol.* **91:**611–614.

188. **Reiss, F., and G. Szilagyi.** 1967. The effect of mammalian and avian sera on the growth of *Cryptococcus neoformans. J. Invest. Dermatol.* **48:**264–265.

189. **Rex, J. H., R. A. Larsen, W. E. Dismukes, G. A. Cloud, and J. E. Bennett.** 1993. Catastrophic visual loss due to *Cryptococcus neoformans* meningitis. *Medicine* **72:**207–224.

190. **Rhodes, J. C.** 1985. Contribution of complement component C5 to the pathogenesis of experimental murine cryptococcosis. *J. Med. Vet. Mycol.* **23:**225–234.

191. **Rhodes, J. C., L. S. Wicker, and W. Urba.** 1980. Genetic control of susceptibility to *Cryptococcus neoformans* in mice. *Infect. Immun.* **29:**494–499.

192. **Roseff, S. A., and S. M. Levitz.** 1993. Effect of endothelial cells on phagocyte-mediated anticryptococcal activity. *Infect. Immun.* **61:**3818–3824.

193. **Sahu, A., T. R. Kozel, and M. K. Pangburn.** 1994. Specificity of the thioester-containing site of human C3 and its significance to complement activation. *Biochem. J.* **302:**429–436.

194. **Sakaguchi, N.** 1993. Ultrastructural study of hepatic granulomas induced by *Cryptococcus neoformans* by quick-freezing and deep-etching method. *Virchows Arch. Cell. Pathol.* **64:**57–66.

195. **Salkowski, C. A., and E. Balish.** 1991. Cryptococcosis in beige mice: the effect of congenital effects in innate immunity on susceptibility. *Can. J. Microbiol.* **37:**128–135.

196. **Salkowski, C. A., and E. Balish.** 1991. Role of natural killer cells in resistance to systemic cryptococcosis. *J. Leukocyte Biol.* **50:**151–159.

197. **Salkowski, C. A., K. F. Bartizal, M. Balish, and E. Balish.** 1987. Colonization and pathogenesis of *Cryptococcus neoformans* in gnotobiotic mice. *Infect. Immun.* **55:**2000–2005.

198. **Sarosi, G. A., P. M. Silberfarb, and F. E. Tosh.** 1971. Cutaneous cryptococcosis: a sentinel of disseminated disease. *Arch. Dermatol.* **104:**1–3.

199. **Sawyer, D. W., G. R. Donowitz, and G. L. Mandell.** 1989. Polymorphonuclear neutrophils: an effective antimicrobial force. *Rev. Infect. Dis.* **11:**S1532–S1542.

200. **Schelenz, S., R. Malhotra, R. B. Sim, U. Holmskov, and G. J. Bancroft.** 1995. Binding of host collectins to the pathogenic yeast *Cryptococcus neoformans*: human surfactant D acts as an agglutinin for acapsular cells. *Infect. Immun.* **63:**3360–3366.

201. **Schneerson-Porat, S., A. Sharar, and M. Aronson.** 1965. Formation of histiocyte rings in response to *Cryptococcus neoformans* infection. *RES J. Reticuloendothel. Soc.* **2:**249–255.

202. **Sethi, K. K.** 1967. Attempts to produce experimental intestinal cryptococcosis and sporotrichosis. *Mycopathol. Mycol. Appl.* **31:**245–250.

203. **Shahar, A., Y. Kletter, and M. Aronson.** 1969. Granuloma formation in cryptococcosis. *Isr. J. Med. Sci.* **5:**1164–1172.

204. **Sohnle, P. G., C. Collins-Lech, and J. H. Weiessner.** 1991. Antimicrobial activity of an abundant calcium-binding protein in the cytoplasm of human neutrophils. *J. Infect. Dis.* **163:**187–192.

205. **Steinbakk, M., C.-F. Naess-Andresen, E. Lingaas, I. Dale, P. Brandtzaeg, and M. K. Fagerhol.** 1990. Antimicrobial actions of calcium binding leucocyte L1 protein, calprotectin. *Lancet* **336:**763–765.

206. **Szilagyi, G., F. Reiss, and J. C. Smith.** 1966. The anticryptococcal factor of blood serum. *J. Invest. Dermatol.* **46:**306–308.

207. **Tacker, J. B., F. Farhi, and G. S. Bulmer.** 1972. Intracellular fate of *Cryptococcus neoformans*. *Infect. Immun.* **6:**162–167.

208. **Takos, M. J.** 1956. Experimental cryptococcosis produced by the ingestion of virulent organisms. *N. Engl. J. Med.* **254:**598–601.

209. **Tesfa-Selase, F., and R. J. Hay.** 1995. Superoxide dismutase of *Cryptococcus neoformans*: purification and characterization. *J. Med. Vet. Mycol.* **33:**253–259.

210. **Tohyama, M., K. Kawakami, and A. Saito.** 1996. Anticryptococcal effect of amphotericin B is mediated through macrophage production of nitric oxide. *Antimicrob. Agents Chemother.* **40:**1919–1923.

212. **Trinchieri, G.** 1989. Biology of natural killer cells. *Adv. Immunol.* **47:**187–396.

213. **Truelsen, K., T. Young, and T. R. Kozel.** 1992. In vivo complement activation and binding of C3 to encapsulated *Cryptococcus neoformans*. *Infect. Immun.* **60:**3937–3939.

214. **Vartivarian, S. E., R. E. Cowart, E. J. Anaissie, T. Tashiro, and H. A. Sprigg.** 1995. Iron acquisition by *Cryptococcus neoformans*. *J. Med. Vet. Mycol.* **33:**151–156.

215. **Vazquez-Torres, A., J. Jones-Carson, and E. Balish.** 1996. Peroxynitrite contributes to the candidacidal activity of nitric oxide-producing macrophages. *Infect. Immun.* **64:**3127–3133.

216. **Vecchiarelli, A., M. Dottorini, D. Pietrella, C. Monari, C. Retini, T. Todisco, and F. Bistoni.** 1994. Role of human alveolar macrophages as antigen-presentation cells in *Cryptococcus neoformans* infection. *Am. J. Respir. Cell. Mol. Biol.* **11:**130–137.

217. **Vecchiarelli, A., D. Pietrella, M. Dottorini, C. Monari, C. Retini, and T. Todisco.** 1995. Encapsulation of *Cryptococcus neoformans* regulates fungicidal activity and the antigen presentation process in human alveolar macrophages. *Clin. Exp. Immunol.* **98:**217–223.

218. **Wagner, D. K., and P. G. Sohnle.** 1995. Cutaneous defenses against dermatophytes and yeasts. *Clin. Microbiol. Rev.* **8:**317–335.

219. **Wagner, R. P., S. M. Levitz, A. Tabuni, and H. Kornfeld.** 1992. HIV–1 envelope protein (gp120) inhibits the activity of human bronchoalveolar macrophages against *Cryptococcus neoformans*. *Am. Rev. Respir. Dis.* **146:**1434–1438.

220. **Wang, Y., P. Aisen, and A. Casadevall.** 1995. *Cryptococcus neoformans* melanin and virulence: mechanism of action. *Infect. Immun.* **63:**3131–3136.

221. **Wang, Y., and A. Casadevall.** 1994. Susceptibility of melanized and nonmelanized *Cryptococcus neoformans* to nitrogen- and oxygen-derived oxidants. *Infect. Immun.* **64:**3004–3007.

222. **Washburn, R. G., B. J. Bryant-Varela, N. C. Nulian, and J. E. Bennett.** 1991. Differences in *Cryptococcus neoformans* capsular polysaccharide structure influence assembly of alternative complement pathway C3 convertase on fungal surfaces. *Mol. Immunol.* **28:**465–470.

223. **Weinberg, P. B., S. Becker, D. L. Granger, and H. S. Koren.** 1987. Growth inhibition of *Cryptococcus neoformans* by human alveolar macrophages. *Am. Rev. Respir. Dis.* **136:**1242–1247.

224. **Weiss, C., I. H. Perry, and M. C. Shevky.** 1948. Infection of the human eye with *Cryptococcus neoformans* (*Torula histolytica*; *Cryptococcus hominis*). *Arch. Ophthalmol.* **39:**739–751.

225. **Wickes, B. L., M. E. Mayorga, U. Edman, and J. C. Edman.** 1996. Dimorphism and haploid fruiting in *Cryptococcus neoformans*: association with the alpha-mating type. *Proc. Natl. Acad. Sci. USA* **95:**7327–7331.

226. **Wilson, E., L. Thorson, and D. P. Speert.** 1991. Enhancement of macrophage super-oxide anion production by amphotericin B. *Antimicrob. Agents Chemother.* **35**:796–800.

227. **Wilson, M. A., and T. R. Kozel.** 1992. Contribution of antibody in normal human serum to early deposition of C3 onto encapsulated and nonencapsulated *Cryptococcus neoformans. Infect. Immun.* **60**:754–761.

228. **Wolf, J. E., and S. E. Massof.** 1990. In vivo activation of macrophage oxidative burst activity by cytokines and amphotericin B. *Infect. Immun.* **58**:1296–1200.

229. **Wright, G. W.** 1961. Structure and function of respiratory tract in relation to infection. *Bacteriol. Rev.* **25**:219–227.

230. **Yamoaka, H., N. Sakaguchi, K. Sano, and M. Ito.** 1996. Intravascular granuloma induced by intravenous inoculation of *Cryptococcus neoformans. Mycopathologia* **133**:149–158.

231. **Yeaman, M. R., A. S. Ibrahim, J. E. Edwards, A. S. Bayer, and M. A. Ghannoum.** 1993. Thrombin-induced rabbit platelet microbicidal protein is fungicidal in vitro. *Antimicrob. Agents Chemother.* **37**:546–553.

232. **Yoshida, K., T. Akaike, T. Doi, K. Sato, S. Ijiri, M. Suga, M. Ando, and H. Maeda.** 1993. Pronounced enhancement of NO-dependent antimicrobial activity by a NO oxidizing agent, imidazolineoxyl N-oxide. *Infect. Immun.* **61**:3552–3555.

233. **Young, B. J., and T. R. Kozel.** 1993. Effects of strain variation, serotype, and structural modification on kinetics for activation and binding of C3 to *Cryptococcus neoformans. Infect. Immun.* **61**:2966–2972.

234. **Yuan, R., A. Casadevall, J. Oh, and M. D. Scharff.** 1997. T cells cooperate with passive antibody to modify *Cryptococcus neoformans* infection in mice. *Proc. Natl. Acad. Sci. USA* **94**:2483–2488.

8 | Specific Immunity and Cytokines

Specific immune mechanisms are composed of humoral and cellular arms. Specific immunity develops in response to infection or immunization with *Cryptococcus neoformans* products and functions to contain and/or eradicate infection. The humoral arm involves the synthesis of antibodies to cryptococcal antigens. The cellular arm involves the generation of T lymphocytes that recognize cryptococcal antigens. The function and efficacy of specific immune mechanisms are often intrinsically linked to, and dependent on, nonspecific immune mechanisms such as professional phagocytic cells and complement. Historically, cell-mediated immune mechanisms have been regarded as the primary specific immune defense mechanism against *C. neoformans*. However, there is considerable evidence that both antibody- and cell-mediated immune mechanisms can have profound effects on the outcome of cryptococcal infection. Furthermore, there is some evidence that antibody- and cell-mediated immune mechanisms cooperate against *C. neoformans* and that each immune arm is more effective in the presence of the other.

HUMORAL IMMUNITY

Antibody Response to *C. neoformans* Antigens

Capsular polysaccharides and cryptococcal proteins can elicit antibody responses. Most studies of the antibody response to cryptococcal antigens have focused on the immunogenicity of the capsular polysaccharide and the protective efficacy of such antibodies. In contrast, relatively little work has been done on the antibody response to nonpolysaccharide antigens.

Antibody response to cryptococcal antigens in humans

The human antibody response to *C. neoformans* consists of antibodies to the polysaccharide capsule (14, 38, 42, 43, 46, 57, 70, 123, 123a, 125, 203, 224, 240, 247, 255) and to protein antigens (89, 90). The antibody response to the capsular polysaccharide antigen has been extensively studied. In contrast, relatively little work has been done to characterize the antibody response to protein antigens. A 115-kDa protein antigen of *C. neoformans* has been described and antibody to this antigen is found in sera from patients with cryptococcosis (90). Hamilton et al. (89) studied the serum antibody response to cytoplasmic *C. neoformans* proteins by Western blot analysis and found immunoglobulin G (IgG) to several protein antigens in human immunodeficiency virus (HIV)-positive individuals with cryptococcosis.

Numerous studies have investigated the antibody response to capsular polysaccharide antigens in humans. Many of the studies from the early 20th century reported no serum antibody response in patients with *C. neoformans* infection. This could have reflected the absence of sensitive serological techniques at the time. Since the 1960s, several studies have consistently shown antibodies to capsular polysaccharide antigen in patients with cryptococcosis. Kimball et al. (123a) developed a bentonite flocculation test for detection of capsule-binding antibody and demonstrated serum antibody to *C. neoformans* in 42% and 50% of patients with meningeal and nonmeningeal cryptococcosis, respectively. Gordon and Lapa (70) developed a charcoal particle agglutination assay for cryptococcal polysaccharide and used it to demonstrate serum antibody titer in 27 patients with *C. neoformans* infection. Charcoal particle agglutination was much more sensitive than the classical yeast agglutination technique (70). Vogel (240) developed an indirect immunofluorescent assay for serum antibody and detected strong reactions in 80% of patients with cryptococcosis and weak reactions in the majority of control sera. He observed that infection limited to the central nervous system was associated with lower antibody response. Furthermore, several patients with cryptococcosis in the setting of malignancies had no measurable serum antibody (240). Diamond and Bennett (42) used an indirect immunofluorescence assay to demonstrate serum antibody in 40% of patients with cryptococcosis at the beginning of therapy and correlated the absence of serum antibody with treatment failure. Hence, cryptococcal infections can elicit antibody responses to the polysaccharide in humans, but the likelihood of detecting such antibodies is highly dependent on the sensitivity and specificity of the serological method used in the study.

Several studies indicate that anticryptococcal antibodies to cryptococcal polysaccharide are common in the serum of individuals with or without cryptococcosis. Dromer et al. (43) used an enzyme-linked immunoabsorbent assay (ELISA) to study HIV-infected and control individuals without cryptococcosis and showed that the majority of individuals in both groups had serum antibody reactive with cryptococcal polysaccharide. AIDS patients had lower levels of IgG, raising the possibility that a defect in IgG synthesis resulting from HIV-associated B-cell dysregulation could predispose AIDS patients to *C. neoformans* infection. A follow-up study by Dromer and her collaborators (46) revealed a correlation between the level of antibody measured by ELISA and fluorescence-activated cell sorting in patients without cryptococcosis but not in patients with cryptococcosis. This observation was interpreted to indicate differences in the specificity of antibodies produced in response to clinical infection. Houpt et al. (105) studied 40 normal human subjects and found that the percentages of individuals with class-specific antibodies were 28% for IgG, 98% for IgM, and 3% for IgA. These naturally occurring antibodies were not opsonic and did not appear to significantly enhance complement activation (105). DeShaw and Pirofski (38) performed a detailed seroprevalence study of anticryptococcal antibodies in 70 HIV-positive and 53 HIV-negative individuals using a sensitive ELISA and demonstrated that IgM, IgG, and IgA to glucuronoxylomannan (GXM) polysaccharide were ubiquitous in both patient populations. The serum IgG to GXM was composed almost exclusively of IgG2, the IgG subclass associated with antibody responses to polysaccharides in humans (38). Higher serum IgA levels to GXM in HIV-infected individuals were attributed to the polyclonal B cell activation that accompanies HIV infection (38). Speed et al. (224) employed an enzyme immunoassay to demonstrate that the prevalence of IgA and IgG antibodies was 75% and 100%, respectively, in immunocompetent patients with *C. neoformans* var. *gattii* infection in Australia. Infection with *C. neoformans* var. *gattii* was more likely to be associated with a greater antibody response than was infection with *C. neoformans* var. *neoformans* (224). Furthermore, there were differences in the isotype composition of the antibody response to polysaccharide among patients infected with two varieties: those infected with var. *gattii* had higher IgA titer than those infected with var. *neoformans* (224).

The origin of serum antibody to GXM in individuals without a history of *C. neoformans* infection is uncertain. One explanation is that the antibody response is a consequence of a subclinical infection (144, 145). The observation that serum antibody to *C. neoformans* is significantly more common in pigeon fanciers and in patients with pigeon breeders' disease is consistent with chronic exposure to cryptococcal antigens and infection (57, 241). An alternative explanation is that this antibody is not specific for *C. neoformans* but rather is cross-reactive with other capsular polysaccharides. Antigens from other pathogens, including *Trichosporon* sp. (151, 222) and *Streptococcus pneumoniae* (206) have been shown to have cross-reactive antigens with *C. neoformans* polysaccharide (see chapter 4). The cross-reactivity of *C. neoformans* with *Trichosporon* sp. has made difficult the association of cryptococcal antibodies with hypersensitivity pneumonitis. For example, Miyagawa et al. (157) reported high titers of antibody against *C. neoformans* in sera from all patients with hypersensitivity pneumonitis. However, a subsequent study revealed that the anticryptococcal antibodies could be removed by absorption with *Trichosporon* antigens, and this finding raised questions concerning the validity of the association between *C. neoformans* and hypersensitivity pneumonitis (222).

There are several problems in the interpretation of serological data from patients with *C. neoformans* infections. Unlike other fungal infectious agents, *C. neoformans* can produce a large amount of capsular polysaccharide antigen that can theoretically bind and sequester specific antibody. Kimball et al. (123a) have demonstrated an inverse correlation between serum antigen and antibody levels, lending support to the notion that negative serology for antibody may reflect sequestration of antibody in antigen-antibody complexes. Antibody is known to reduce serum antigen by promoting its uptake by reticuloendothelial cells (67). Hence, the detection of antibody and the magnitude of antibody titer may reflect a dynamic equilibrium between the synthesis of antibody molecules, the synthesis of capsular polysaccharide, and the phagocytic capacity of the reticuloendothelial cell system.

In summary, the literature indicates that antibody reactive with *C. neoformans* polysaccharide is commonly found in individuals with or without cryptococcal infection. Discrepancies in seroprevalence between studies may reflect the various sensitivities of the assays used and the immune status of the patient populations studied. The functional efficacy of these antibody responses is uncertain, but at least one study indicates that absence of serum antibody implies a poorer prognosis for patients with cryptococcal infection (42). Patients with cryptococcosis secondary to malignancies may have lower antibody responses (240), a finding that may contribute to the particularly poor prognosis of cancer patients with *C. neoformans* infection (244). In patients with AIDS there is evidence that the antibody response is quantitatively and qualitatively different (38, 43, 46). The contribution of serum antibody deficits to susceptibility in patients with AIDS is uncertain, but the identification of serological differences in these patients suggests the need for additional studies.

Antibody response to cryptococcal antigens in laboratory animals
As in humans, the majority of serological studies of experimental infection in animals have focused on the response to polysaccharide antigens. In recent years, three studies have demonstrated antibody responses to several protein antigens during infection (28, 89, 117). A 77-kDa heat shock protein antigen appears to elicit strong antibody responses during infection in mice (117). Antibody responses to extracellular proteins like proteases (27) or phospholipases (29) could conceivably be beneficial to the host. The biological function of antibodies to *C. neoformans* proteins, if any, is unknown. In contrast, the antibody response to the capsular polysaccharide has received considerable attention since the early 20th century. The results from the early studies are highly variable, probably the result of the poor immunogenicity of capsular polysaccharides and differences in animal species, vaccination techniques, and serological assays. Some investigators were able to obtain high-titer serum, whereas others were unable to elicit antibodies in laboratory animals.

Stoddard and Cutler (227) studied the antibody response to *C. neoformans* infection in mice, rats, dogs, rabbits, and guinea pigs but were unable to measure agglutinins (agglutinating antibody) in the majority of infected animals. Benham (10) successfully elicited agglutinins by immunizing rabbits with HCl-treated *C. neoformans* cells and used the resulting sera to carry out the first systematic serological study of cryptococcus strains and related yeasts. Kligman (124) developed a method for the purification of polysaccharide but was unable to elicit agglutinins in rabbits and mice and concluded that the polysaccharide was nonantigenic. Hoff (103) was able to elicit high titer of agglutinins in a rabbit immunized with multiple injections of heat-killed organisms. Evans (52) studied the rabbit antibody response to formalin-killed cells from 12 strains of *C. neoformans* and found significant differences in immunogenicity among strains. By immunizing rabbits with multiple injections of *C. neoformans* cells killed with 1% formalin, Evans (52, 53) and Evans and Kessel (51) reproducibly elicited sera with high titer of agglutinins. The success of this immunization strategy may be due to the use of formalin-killed cells, the use of rabbits, and the administration of multiple injections (up to 36) to each animal. With this high-titer rabbit sera, it was possible to categorize *C. neoformans* strains into various isotypes by reciprocal agglutination assays.

Because of the difficulties associated with consistently eliciting high-titer antibody responses to GXM by either immunization or infection in other laboratory animals, rabbits have become the standard species for the generation of polyclonal sera. Rabbit sera have been used for serological studies (51–53), passive antibody protection experiments (62, 80), and human therapy (68). Among common laboratory animals it is interesting that rabbits are both resistant to infection and consistently mount strong antibody responses to cryptococcal polysaccharide. The difficulty in infecting rabbits cannot be attributed to higher body temperatures, since rabbits can be successfully infected if given corticosteroids (200). Hence, the ability of rabbits to mount a strong antibody response to the cryptococcal polysaccharide may contribute to the resistance of this species to *C. neoformans* infection.

In summary, laboratory animals can mount antibody responses to cryptococcal polysaccharide, but most do so poorly. Rabbits have historically provided the best source of antibody to cryptococcal polysaccharide. The antibody response to protein antigens remains largely unexplored.

The phenomenon of antibody unresponsiveness to capsular polysaccharide

C. neoformans capsular polysaccharides are poorly immunogenic. In 1950, Neill et al. (192) noted that large doses of capsular polysaccharide depressed the antibody response in rabbits. In 1958, Gadebusch (63) studied the protective efficacy of immunization with *C. neoformans* capsular polysaccharide and described little or no protection with most immunization protocols. Furthermore, he described an "immunological paralysis"-like phenomenon after the injection of *C. neoformans* polysaccharide in mice, which was characterized by shortened survival of mice immunized with large doses of polysaccharide (63). In 1960, Abrahams and Gilleran (2) studied the protective effect of immunization with formalin-killed vaccines and found that immunization with a small-capsule strain was more effective than immunization with a large-capsule strain. Vaccine efficacy was greatly dependent on the dose of killed yeast cells, and these effects were interpreted as being caused by different amounts of polysaccharide in the preparations. In 1972, Murphy and Cozad (184) studied the effect of polysaccharide immunization in mice by the hemolytic plaque technique and showed that immunological unresponsiveness was associated with very low numbers of antibody-producing cells in the spleen. Immunization with 0.5 µg of cryptococcal polysaccharide elicited the greatest number of antibody-producing cells, whereas immunization with 500 µg of cryptococcal polysaccharide had a paralyzing effect on the production of antibody-producing cells. These results were later confirmed by other investigators (15, 126). Administration of polysaccharide with Freund's incomplete adjuvant can augment the antibody response, but the effect is small and serum antibody levels are induced only transiently (126).

Kozel et al. (127) demonstrated that mice injected with 100 to 400 µg of polysaccharide had a reduced ability to produce an antibody response when challenged with polysaccharide emulsified in Freund's incomplete adjuvant. Interestingly, repeated small doses of polysaccharide injections did not produce an unresponsive state until the cumulative dose reached 100 to 400 µg. This suggested that antibody unresponsiveness did not occur until a threshold amount of polysaccharide was sequestered in vivo. Immunofluorescence studies revealed deposition of polysaccharide in kidney tubular epithelial cells (127).

The *C. neoformans* capsular polysaccharide behaves as a T-cell-independent antigen with regard to eliciting antibody responses (15, 49, 228) and has been classified as a type II T-cell-independent antigen (49, 228). However, the regulation of antibody response to cryptococcal polysaccharide appears to be under T-cell control; mice injected with either antilymphocyte or antithymocyte serum mount greater antibody responses when challenged with cryptococcal polysaccharide (15). This observation led to the suggestion that the antibody unresponsiveness observed with higher doses of cryptococcal polysaccharide resulted from induction of T-cell suppressor activity (15). Similar mechanisms have been suggested for the phenomenon of antibody unresponsiveness to bacterial polysaccharides (9). Sundstrom and Cherniak (229) studied the tolerogenic effects of cryptococcal poly-saccharide and produced evidence for both high-dose and low-dose tolerance to GXM. The low- and high-dose tolerance experiments used 5 and 50 μg of GXM for immunizing doses, respectively (229). The low-dose tolerance effect appeared to be T cell dependent, whereas the high-dose tolerance effect was T cell independent (229). Low-dose tolerance, but not high-dose tolerance, could be adoptively trans-ferred with CD4$^+$ T cells (229).

The antibody response to *C. neoformans* capsular polysaccharide in mice is under genetic control (49). Inbred mouse strains exhibit considerable variation in their ability to produce antibody responses after vaccination with cryptococcal polysaccharide (49). BALB/cJ and C3H/HeJ strains mount a strong antibody re-sponse when challenged with purified polysaccharide, whereas B10 and A/J mice mount weak responses (49). Interstrain variation in antibody responsiveness to cryptococcal polysaccharide may have contributed to the variable results reported in the literature for polysaccharide immunization (49). In BALB/c mice, systemic infection elicited high-titer responses in only 5% of the animals (23, 170).

In summary, cryptococcal polysaccharide is weakly immunogenic in labora-tory animals. The magnitude of the antibody response is highly dependent on the immunizing dose and the genetic background of the host. The mechanism respon-sible for the tolerogenic effects of cryptococcal polysaccharide are complex and poorly understood. Poor immunogenicity, combined with a propensity for induc-ing "immunological paralysis" or antibody unresponsiveness, undoubtedly con-tributes to the low antibody titers that often accompany human infection.

Antibody response to polysaccharide-protein conjugates

In contrast to purified capsular polysaccharide alone, polysaccharide-protein con-jugates are highly immunogenic and reproducibly elicit high-titer sera after immu-nization (71, 72). Polysaccharide-protein conjugates elicit antibody responses to GXM that are characteristic of T-cell-dependent antigens, with a predominance of IgG isotypes (22, 40). The mechanism by which polysaccharide-protein conjugates circumvent the tolerogenic effects of cryptococcal polysaccharide is not well under-stood. A GXM-tetanus toxoid (GXM-TT) conjugate vaccine has been synthesized that elicits high-titer sera in mice (22, 40). The specificity and molecular structure of antibodies elicited by the GXM-TT vaccine and by natural infection are very similar (170). Murine monoclonal antibodies (MAbs) generated from mice immu-nized with the GXM-TT vaccine have been shown to be protective in murine infection, indicating that this conjugate contains epitopes that can elicit protective antibody responses to *C. neoformans* (175). Administration of the GXM-TT vaccine was protective in mice: the mortality of GXM-TT-immunized mice was only 20 to 30% compared to 100% in a control group vaccinated with an *Escherichia coli* polysaccharide-TT conjugate after 4 months (39). GXM-TT vaccine has undergone initial clinical testing in humans and can elicit high-titer antibody responses (245, 246). For example, administration of 50 μg of GXM-TT vaccine to normal human subjects increases serum IgG 70-fold from preimmune levels (245). In humans, antibodies elicited by the vaccine are opsonic (255).

The conjugate vaccine strategy is based on the premise that high-titer antibody responses in patients at risk for *C. neoformans* infection can enhance the efficacy of nonspecific (e.g., macrophage, neutrophil, natural killer cell) and cellular immunity to prevent clinical infection. A potential problem with this approach is that conjugate vaccines are sometimes poorly immunogenic in patients at high risk for cryptococcosis, such those with late-stage HIV infection (226). Zhang et al. (253) studied the antibody response in GXM-TT-vaccinated HIV-positive and HIV-negative individuals using a peptide mimetope of GXM to inhibit binding of sera from vaccinated individuals to GXM. The peptide mimetope was a short, 10-amino-acid sequence that reacted with the binding site of a human IgM MAb to GXM generated in response to GXM-TT (253). This study revealed that the peptide was able to inhibit binding of serum antibody from HIV-positive individuals but not from HIV-negative individuals (253), implying differences in the fine specificity of anti-cryptococcal antibodies elicited in the presence or absence of HIV infection (253).

In summary, the GXM-TT vaccine represents the application of an established technology for generating vaccines against bacteria to a fungal pathogen. The vaccine is highly immunogenic and is intended to elicit highly protective antibody responses that could even prevent infection. Whether the GXM-TT vaccine can reduce infection in severely immunosuppressed individuals is not known; the answer to this question must await efficacy trials (see also chapter 15).

Molecular genetics of antibody responses to GXM
Variable-region gene usage in antibodies to cryptococcal polysaccharide has been studied in mice (21, 23, 170, 197) and humans (203). The antibody response to *C. neoformans* polysaccharide in mice is unusual in that the majority of antibodies are assembled from only a few variable-region genes (21, 23, 170, 197). This restriction in variable-region gene usage is exemplified by the fact that antibodies made in different laboratories have had a very similar molecular structure (21). Two human MAbs to cryptococcal polysaccharide have been found to utilize similar variable-region genes (203). These findings, together with serological studies of idiotype prevalence, strongly suggest that the human response is also restricted in variable-region gene usage (38, 203).

The biological consequences of mounting an antibody response to the polysaccharide antigen that is restricted in variable region are unknown. Analysis of hybridomas producing MAbs to GXM showed that the response in both infection (23) and conjugate vaccine immunization (170) is oligoclonal. This implies that few B cells respond to GXM polysaccharide antigen. The hybridomas generated from a mouse immunized with the GXM-TT vaccine were mostly derived from a single B cell, and the MAbs produced by these hybridomas had only minor differences in amino acid sequence resulting from somatic mutations (170). Protective and non-protective MAbs have been shown to be derived from a single B cell and to differ by only a few amino acid differences resulting from somatic mutations (173). This result indicates that small differences in antibody primary structure can produce differences in epitope specificity that translate into major differences in protective efficacy (173).

The structure of a protective MAb to GXM (MAb 2H1) has been solved by X-ray crystallography (249). The structure of the antibody-binding site and its complex with polysaccharide has been modeled by sophisticated computer-assisted programs (197). The antibody-binding pocket consists of a deep groove flanked by protein regions from both the antibody heavy and light chains (197).

Peptides that are mimetopes of GXM have been described, and the crystal structure of MAb 2H1 with a peptide mimetope is now available (234, 249). The existence of peptide mimetopes to GXM is interesting, because it raises the possibility that some GXM-reactive antibodies were, in fact, elicited by protein antigens. Furthermore, the identification of peptide mimetopes suggests that it may be possible to use peptides as vaccines to elicit antibody responses to GXM while bypassing all the problems associated with polysaccharide antigens (e.g., poor immunogenicity and antibody unresponsiveness).

In summary, the antibody response to *C. neoformans* polysaccharide is highly restricted in variable-region gene usage. The biological consequences of this phenomenon are not well understood. Protective and nonprotective MAbs can use the same variable-region gene elements but differ in specificity as a result of amino acid differences originating from somatic mutation during the generation of the antibody response.

Functional Efficacy of Antibody Responses

The usefulness of antibody immunity in protection against pathogens is usually inferred from one or more of the following criteria: (i) prevention or modification of infection by passive antibody administration, (ii) association of specific antibody (usually serum) with protection against infection, and/or (iii) correlation of antibody deficiency with susceptibility to infection. In vitro observation of antibody-mediated killing or antibody enhancement of cellular activity can provide supportive evidence for the usefulness of antibody immunity. For *C. neoformans* the role of natural antibody immunity has remained uncertain despite several decades of study. The literature contains multiple studies providing evidence for or against an important role for antibody in host defense (reviewed in references 20 and 202). In recent years, studies with MAbs have demonstrated both the existence of protective MAbs and unexpected complexity in the function and efficacy of antibody immunity against *C. neoformans* infection. Table 1 lists the criteria and supporting evidence for antibody-mediated protection.

C. neoformans has an antiphagocytic polysaccharide capsule composed primarily of GXM (31). Antibodies to GXM are potent opsonins that greatly enhance the phagocytosis of *C. neoformans*. Hence, antibody responses are theoretically useful for host defense. Administration of immune rabbit sera before experimental infection has been shown in several studies to prolong survival (1, 62, 74, 80). However, administration of immune sera 1 day after infection had no effect on survival of lethally infected mice (146). Inability to protect animals with antibody after infection is not surprising, since it is difficult to demonstrate antibody-mediated protection against established infection with most pathogens (including antibody-susceptible bacteria like *S. pneumoniae*) (24). In rabbits, administration of pre-formed antibody had no effect on the course of intracisternal infection (201). The strongest evidence that antibody can be protective against *C. neoformans* infections comes from passive transfer experiments with MAbs to GXM in multiple models of murine infection (Fig. 1). Three independent research groups have shown that administration of MAbs to GXM prolongs survival and/or reduces organ CFU (44, 45, 47, 170, 174–176, 180, 216). In addition, administration of MAbs to GXM after infection has been shown to enhance the therapeutic efficacy of the

Table 1 Evidence that antibody immunity contributes to protection against *C. neoformans* infection

Criterion	Supporting evidence	References
Association of specific antibody with protection against infection	Presence of specific antibody is associated with improved prognosis	42
	Mortality in mice is correlated to decreasing antibody titer	220
	Vaccination with a GXM-TT conjugate vaccine confers protection in mice	246
	Anecdotal reduction in cerebrospinal fluid yeast cells associated with intrathecal antibody production	130
Antibody administration is protective	Administration of polyclonal antibody prolongs survival in mice	1, 62, 80
	Administration of MAbs to GXM prolongs survival in mice	45, 47, 54, 174, 175
	Antibody-treated mice have stronger granulomatous inflammation	54
Antibody deficiency increases susceptibility to infection	Cryptococcosis in children with hyper-IgM syndrome	115, 129, 231
	Occasional cases of cryptococcosis in isolated hypoglobulinemia	87, 217
	Enhanced susceptibility of mice treated with cyclophosphamide	50, 81
	Reduced levels of antibodies to GXM in AIDS patients at risk for infection	38, 43
	xid mutation enhances susceptibility to *C. neoformans* in mice	148
	Reconstitution of immunity in SCID mice requires B and T cells	5
Studies in vitro demonstrating antibody-mediated killing or enhancement of cellular activity provide supportive evidence for the usefulness of antibody immunity	Antibody enhances leukocyte activity efficacy against *C. neoformans*	136
	NK cells require antibody for fungistatic activity	156, 189
	Antibodies are potent opsonins for a variety of effector cells	86, 225, 255
	Human microglia antifungal activity requires antibody	131
	GXM-MAb complexes enhance nitrogen- and oxygen-related oxidants by murine macrophages	167, 235
	MAbs to GXM promote killing of *C. neoformans* by murine macrophages	179, 181
	MAbs to GXM enhance expression of B7-1 costimulatory molecule on human monocytes	237
	MAbs to GXM promote murine eosinophil phagocytosis and degranulation	55
	Antibody stimulates classical complement activation	105, 247
Antibody additivity or synergy with antifungal drugs	Amphotericin B	44, 69, 178
	Fluconazole	172
	Flucytosine	56

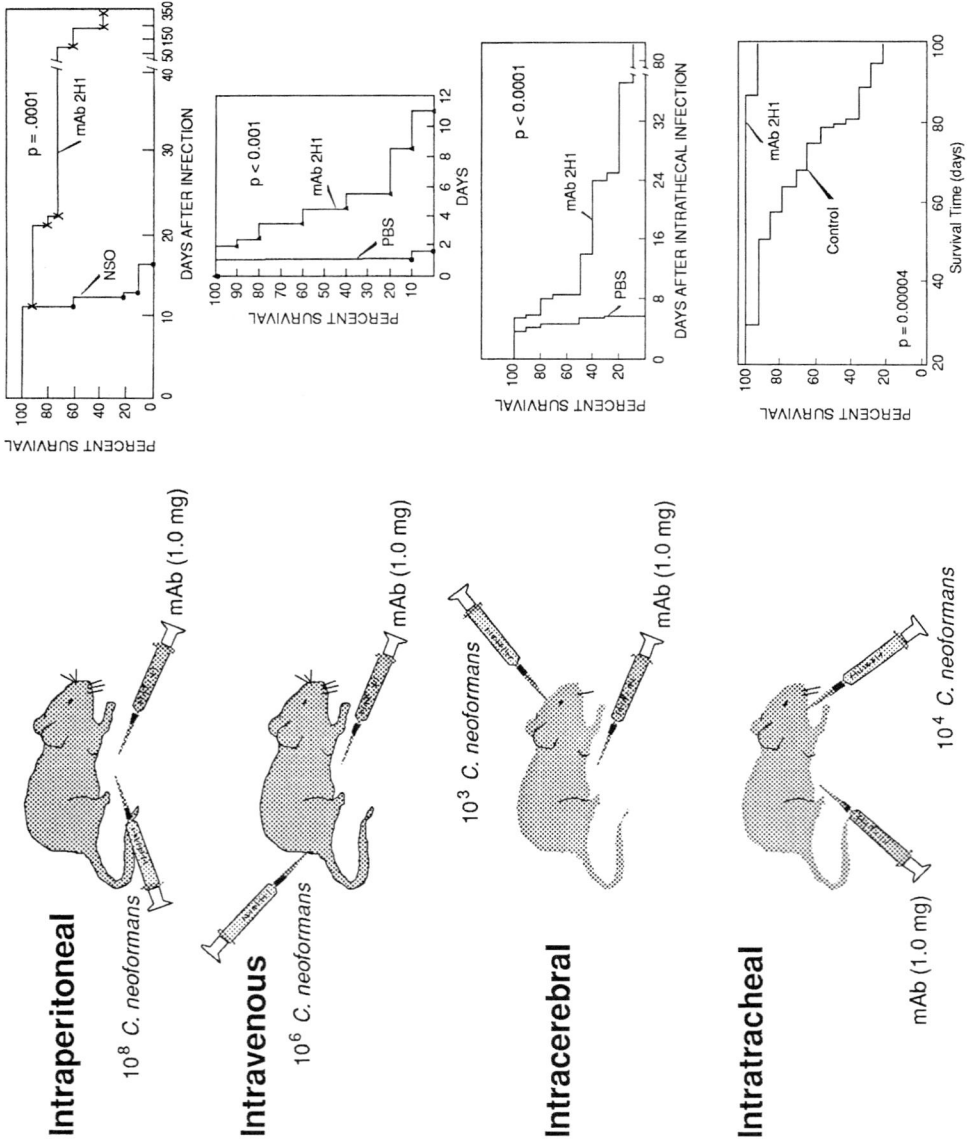

antifungal agents amphotericin B (44, 69, 178), flucytosine (56), and fluconazole (172) against *C. neoformans* in murine experimental infection. Hence, antibody administration can modify the course of *C. neoformans* infection to the benefit of the host.

A relationship between the presence of human serum antibody to *C. neoformans* and protection against infection has not been conclusively established. Several methodologies used to study the human antibody response to *C. neoformans* capsular polysaccharide (38, 43, 95, 105, 123) have provided complex and inconclusive results on the role of antibody immunity. Antibodies to GXM are ubiquitous in the sera of patients with or without HIV infection (38). However, the functional efficacy of these antibodies is unknown. Quantitative and qualitative differences in serum antibodies to GXM have been described in HIV-positive and HIV-negative individuals (38, 43). HIV-infected patients at risk for *C. neoformans* infection have lower levels of IgG to the capsular polysaccharide (38, 43). HIV-positive individuals have higher IgA titers to GXM than do HIV-negative individuals (38). Given the importance of isotype in determining antibody efficacy in murine antibodies, these results suggest that quantitative differences in isotype profile could contribute to the marked susceptibility of AIDS patients to *C. neoformans* infections.

A role for antibody in protection is suggested by reports of cryptococcosis in children with hyper-IgM syndrome (115, 129, 231) and occasional infections in patients with isolated hypogammaglobulinemia (87, 217). Other observations supporting a role for antibody in defense against *C. neoformans* are (i) a reduction in brain tissue CFU temporally correlated with the appearance of cerebrospinal fluid antibody in rabbit intracisternal infection (102), (ii) a shift to intracellular location for *C. neoformans* in phagocytic cells that parallels the appearance of serum antibody in rat pulmonary infection (67), (iii) death associated with a decline in serum antibody titers in mice (220), and (iv) the presence of serum antibody associated with recovery of infection in human cryptococcal meningitis (42). The appearance of specific antibody in cerebrospinal fluid has been anecdotally associated with recovery from infection (130).

It has been difficult to demonstrate a role for natural antibody immunity in protection against *C. neoformans* infection in mice. A major problem has been the lack of responsiveness of mice to *C. neoformans* polysaccharide antigen. Monga et al. (166) studied the outcome of *C. neoformans* infection in normal and B-cell-deficient mice and found no difference in mortality or organ CFU between B-cell-depleted mice and controls. However, the implications of this result are clouded by the fact that cryptococcal infection seldom elicits antibody responses in mice; hence, the two groups may not have been significantly different with regard to serum antibody responses. A role for B cells in resistance to *C. neoformans* has recently been demonstrated with the use of IgM-deficient mice (5). The importance of B cells was demon-

Figure 1 Effect of antibody administration (MAb 2H1, IgG1) on survival of mice given lethal cryptococcal infection by various routes. The degree to which antibody prolongs survival is dependent on the route of infection. For details of individual models, see the following references: intraperitoneal (175), intravenous (45, 177, 178, 180, 216), intracerebral (174), and intratracheal (54). Part of this figure was published previously (128) and is reproduced here with permission from Cold Spring Harbor Press.

strated in reconstitution experiments of severe combined immunodeficiency (SCID) mice, in which the presence of B cells resulted in enhanced resistance to infection as measured by lower tissue burden (5). The exact mechanism by which B cells contribute to protection is uncertain, but possible roles include antibody protection and/or interactions with T cells leading to more effective T-cell function (5). Additional evidence for a B-cell role in protection comes from experiments using cyclophosphamide-treated mice. Cyclophosphamide administration at doses that suppress B-cell function impaired resistance to intravenous infection (50). Furthermore, susceptibility to cryptococcal infection was associated with the *xid* mutation, which results in impaired antibody immunity (148).

An early conjugate vaccine composed of capsular polysaccharide and protein failed to protect mice despite eliciting high antibody titers (71). However, another conjugate vaccine has been shown to protect mice against lethal intravenous infection (39, 246). A potential explanation for these divergent observations is suggested by the finding that protective, nonprotective, and deleterious antibodies to *C. neoformans* exist (173, 175, 196, 251). Antibody efficacy in mice has been shown to depend on isotope and fine specificity (173, 175, 216, 251). Among the murine IgG subclasses, the relative efficacy is IgG2a ~ IgG1 > IgG2b (181, 216), and IgG3 is nonprotective (175, 216, 251). Murine MAbs of the IgG3 subclass have been shown consistently to be nonprotective, to enhance infection, and to function as blocking antibodies by reducing the efficacy of protective antibodies (196). Isotype switching from IgG3 to IgG1 has been demonstrated to convert a nonprotective MAb into a protective MAb (251). Some IgMs are protective and others are not, depending on their epitope specificity (173, 175). Analysis of two IgM MAbs derived from one B cell revealed that the IgM that produced a rim immunofluorescence pattern on binding to the *C. neoformans* capsule was protective, whereas the IgM that produced a punctate immunofluorescence pattern inside the capsule was nonprotective (173). This observation suggests that the efficacy of antibody-mediated protection depends on where it binds in the capsule. Affinity for antigen may also influence protective efficacy: higher-affinity MAbs have been demonstrated to be better opsonins than lower-affinity MAbs (179). Since polyclonal sera are complex mixtures of antibodies differing in isotype, epitope specificity, and affinity, the efficacy of the immune sera is likely to reflect the relative proportion of protective and nonprotective antibodies (20, 196).

Comparative studies of passive antibody-mediated protection in mice against *C. neoformans* strains indicate that antibody efficacy varies by strain (177). Administration of antibody prolonged the survival of mice infected with some strains but not others, despite binding to the polysaccharide capsule of all strains (177). The mechanisms by which *C. neoformans* strains vary in susceptibility to antibody are not understood (177). Differences in polysaccharide antigen production (177), melanin composition (242, 243), and proteinase production may affect the relative efficacy of antibody in promoting clearance of *C. neoformans* by host effector cells. The variation in strain susceptibility to antibody immunity may have contributed to the inconsistent results in previous studies of the efficacy of antibody immunity (20, 177).

In summary, the role of humoral immunity in defense against *C. neoformans* is complex and incompletely understood. There is convincing evidence that administration of GXM-binding MAbs can protect mice. However, efforts to establish a correlation between natural antibody immunity and protection have been inconclusive. Protective, nonprotective, and disease-enhancing antibodies have been described. The functional efficacy of individual antibodies represented by MAbs has been shown to be dependent on isotype, epitope specificity, and affinity. *C. neoformans* strain differences in susceptibility to antibody-mediated protection may have contributed to interexperimental variability. There is also strong evidence that antibody-mediated protection is dependent on T-cell function (250) (see below), a finding that complicates the assessment of antibody function in hosts with impaired immune function. Nevertheless, protective antibodies exist and may be particularly useful against heavily encapsulated strains that are opsonized poorly by complement alone. The functional efficacy of the antibody response is likely to be a function of the relative composition of protective and nonprotective antibodies. Hence, the development of an antibody response to *C. neoformans* is not synonymous with immunity. At the present time there are no serological assays to distinguish between useful, indifferent, and harmful immune responses.

Mechanisms of antibody-mediated protection

Antibodies are multifunctional molecules that are able to mediate a variety of biological effects, including opsonization, complement activation, toxin neutralization, and antibody-dependent cellular cytotoxicity. Some of these effects have been demonstrated for the interaction of antibodies to *C. neoformans*. Antibody binding to capsular polysaccharide produces structural changes in the capsule (171) and alters the cell charge (194). However, antibody has not been reported to have a direct toxic effect on *C. neoformans*. Instead, there is evidence that the mechanism of antibody-mediated protection involves enhanced antifungal activity by host effector cells. Capsule-binding antibodies act as opsonins that potentiate the anti-*C. neoformans* activity of neutrophils (153), natural killer (NK) cells (156, 189), macrophages (179, 181), eosinophils (55), and microglia (131). Antibody complexes with *C. neoformans* GXM can enhance the production of oxygen- and nitrogen-related oxidants by murine macrophages through an Fc receptor-mediated mechanism (167, 168, 235). Antibody facilitates the interaction between macrophages and *C. neoformans* cells, resulting in more cell-to-cell contacts (198).

Antibody to GXM enhances the elimination of polysaccharide antigen from serum and tissue (45, 67, 175, 177, 180), a phenomenon that could benefit host immunity given the multiple deleterious effects associated with polysaccharide on host immunity and host response. Antibody increases clearance of serum cryptococcal polysaccharide by promoting polysaccharide uptake by reticuloendothelial system cells in liver and spleen (67). Antibody-treated mice have been shown to have a more intense granulomatous response than control mice, suggesting that antibody may enhance host resistance by promoting granuloma formation. In this regard, administration of antibody to *C. neoformans* has been demonstrated to enhance giant cell formation in vivo after intraperitoneal infection in mice (8, 219, 221). The mechanism by which antibody promotes granuloma formation may involve alterations in cytokine production as a result of activating Fc receptors. Opsonic antibody to *C. neoformans* has been shown to enhance B7-1 (CD80) costimulatory molecules on human monocytes, suggesting another mechanism by which antibody could affect the inflammatory response (237). At this time it is uncertain which of the antibody-mediated effects is primarily responsible for the beneficial effects of some antibodies.

Characteristics of protective antibodies

Studies using murine MAbs to the capsular polysaccharide provide insight into the characteristics of protective antibodies. The protective efficacy of a MAb is dependent on its isotype and epitope specificity. The first protective MAb (E1) was described by Dromer and collaborators (45, 48) and is an IgG1 with strong affinity for serotype A polysaccharide. Sanford et al. (216) demonstrated that isotype is important in protective efficacy by showing that isotype variants of an IgG1 MAb varied in their ability to reduce tissue fungal burden. This observation was subsequently confirmed and extended to show that anticryptococcal antibodies of the murine IgG3 class were either nonprotective or disease-enhancing (196, 250, 251). Epitope location was demonstrated to be a crucial determinant of protective efficacy using MAbs (173, 195). Antibodies that bind to the cryptococcal capsule in a punctate pattern have consistently been shown to be nonprotective for several *C. neoformans* strains. Figure 2 provides a summary diagram of what is known about the requirements for antibody protective efficacy. (For reviews, see references 20 and 202.)

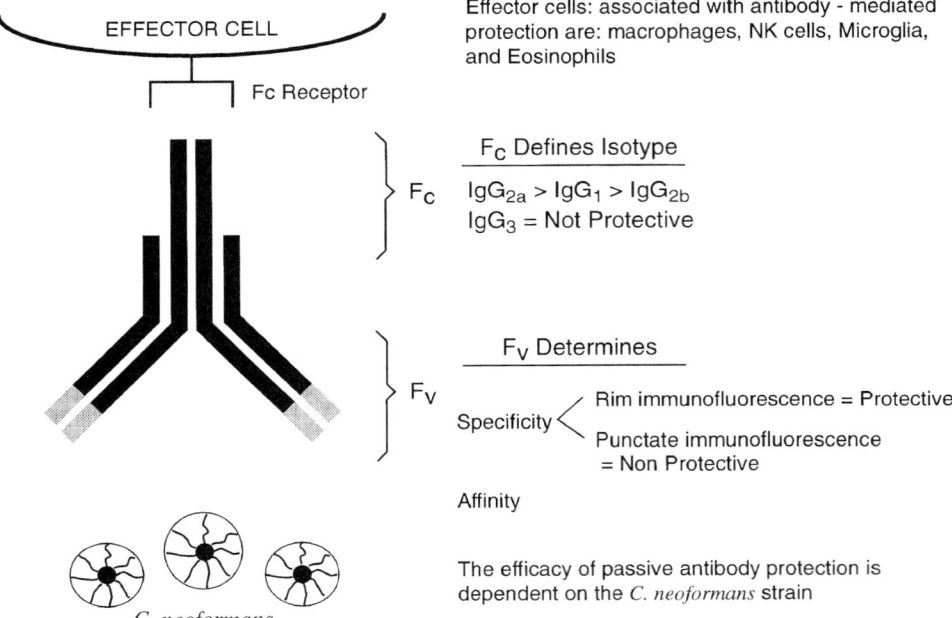

Figure 2 Schematic diagram highlighting the fact that antibody molecules serve as a bridge between host effector cells and yeast cells. The antibody molecule is multifunctional, with the Fc portion being responsible for several biological effects, including half-life, phagocytosis, and complement activation. The Fv portion is composed of the variable region and is responsible for binding to antigen. For details of interactions of effector cells with *C. neoformans*, see references 49, 54, 55, 131, 156, 179, 181, 189, and 252. For details of the relative efficacy of Fc region isotypes, see references 128, 175, 180, 181, 196, 216, 250, and 251. For details of the variable regions responsible for antigen-binding region Fv, see references 21, 170, and 173. For details about differences in passive antibody protection depending on the strain, see reference 177.

Evasion of antibody-mediated protection

Administration of antibody to mice before infection often prolongs survival but seldom eradicates the infection. The mechanism by which some yeast cells evade the effects of antibody and establish themselves in tissue is uncertain. Serial isolates from patients with cryptococcal infection have been shown to have GXM structure changes resulting in antigenic differences (30). Presumably, variants arise during the course of infection that may be selected and in time become the predominant population. Although this mechanism has not been proved to operate in allowing some cells to escape antibody-mediated immune mechanisms, the fact that it has been described suggests that it may occur. Melanization reduces the efficacy of antibody-mediated phagocytosis (242), and some strains of *C. neoformans* produce proteases that can degrade immunoglobulin (27). Antibody may promote phagocytosis by effectors that are unable to kill the yeast cells and thus provide *C. neoformans* with an intracellular niche outside the reach of serum antibody. In this regard, antibody-mediated phagocytosis has been shown to lead to intracellular replication in human microglia, resulting in eventual lysis of the host cell (132).

CELL-MEDIATED IMMUNITY

The importance of cell-mediated immunity in protection against *C. neoformans* has been firmly established. The strongest evidence that cell-mediated immunity, and in particular T cells, has a crucial role in protection against *C. neoformans* infection comes from animal studies showing that (i) depletion of T cells is associated with increased susceptibility, (ii) mice with T-cell defects are more susceptible to infection, and (iii) T cells transfer protection from immune to naive mice. In addition, the association of cryptococcosis with diseases that primarily affect cell-mediated immunity in humans provides strong supportive evidence for the importance of T-cell immunity. Although the exact mechanisms by which T cells function in host defense are not fully understood, there is compelling evidence that T cells are involved in all aspects of a successful immune response. T cells recruit and activate phagocytic effector cells (polymorphonuclear leukocytes, macrophages) against *C. neoformans* and may have direct effects against yeast cells. Furthermore, T-cell function is required for giant cell and granuloma formation and appears to make a decisive contribution to the generation and function of anticryptococcal antibody responses.

Animal Experiments Establish the Importance of T-Cell Immunity

Administration of antilymphocyte serum significantly shortens survival time in mice infected with *C. neoformans* (3, 82). Studies in congenitally T-cell-deficient mice provide strong evidence for the importance of T cells in host defense against *C. neoformans*. Congenitally athymic (*nu/nu*) mice lack a thymus, are deficient in cell-mediated immunity, fail to mount T-dependent antibody responses, and are incapable of delayed-type hypersensitivity (DTH) responses. BALB/c *nu/nu* mice are more susceptible to *C. neoformans* infection than are thymus-containing heterozygous (*nu/X*) mice given intraperitoneal infection (25, 79). Comparison of organ fungal burden between *nu/nu* and phenotypically normal mice revealed no difference in the first 2 weeks of infection (25). However, in heterozygous mice there was a reduction in organ CFU after 2 weeks that paralleled the development of DTH to cryptococcal antigens (25). Thymus transplantation to *nu/nu* mice followed by *C. neoformans* infection resulted in a marked prolongation of survival relative to *nu/nu* control mice that did not receive transplants (83). Histopathological studies of cryptococcal infection in *nu/nu* mice revealed the absence of granuloma formation and less intense inflammation, with a predominance of neutrophils (210). Systemic *C. neoformans* infection in *nu/nu* (but not heterozygous) mice results in cutaneous lesions, suggesting that this fungus may be dermatotropic in the absence of T-cell immunity (210). A role for T cells in protection against mucosal infection is suggested by the observation that oral infection in *nu/nu* mice results in colonization of mucosal surfaces and disseminated cryptococcosis with cerebral involvement (213). Interestingly, amphotericin B therapy for experimental infection was much less effective in *nu/nu* mice than in *nu/X* mice, strongly suggesting an important role for cell-mediated immunity in effective antifungal therapy with amphotericin B (78). The absence of functional T cells in *nu/nu* mice compromises anticryptococcal defenses to the extent that chronic infection can be induced in such mice with an avirulent nonencapsulated *C. neoformans* mutant (212). Normal rats mount effective immune responses to *C. neoformans* pulmonary infection that result in localization of infection to the lung with minimal dissemination (65, 75). Control of pulmonary infection in the rat is associated with an influx of lymphocytes into lung tissue and granuloma formation (65). In contrast, pulmonary infection in *nu/nu* rats is rapidly lethal (75). These studies with *nu/nu* mice and rats indicate the importance of thymic function in defense against *C. neoformans*.

Direct evidence for the role of T cells in defense against *C. neoformans* comes from adoptive transfer experiments with T-cell-enriched splenocyte preparations (82, 142). Graybill and Mitchell (82) showed that adoptive transfer of splenocytes from immune mice prolonged survival relative to splenocyte transfer from immunologically naive mice in mice given lethal infection and that this protection was associated with T cells. In similar experiments, Lim and Murphy (142) showed that transfer of T cells from mice infected with *C. neoformans* conferred protection against infection in recipient mice. Splenic T cells from mice 7 days after infection were not effective in transferring DTH or immunity to naive mice (142). However, transfer of splenic T cells from mice 35 days after infection transferred DTH and immunity to naive mice, as evidenced by reduced organ CFU after challenge with *C. neoformans* (142). In contrast, serum from infected mice had no protective value (142). Adoptive transfer experiments with sensitized T cells in SCID mice given pulmonary infection provide further evidence for the importance of T-cell immunity (113). SCID mice have nonfunctional B and T cells and are highly susceptible to pulmonary infection (113). Transfer of T cells from either the spleen or lymph nodes of immunocompetent mice increases pulmonary clearance (113). However, T cells from lung and hilar nodes are significantly more effective at enhancing pulmonary clearance than are splenic T cells (113).

In summary, the observations that (i) ablation of T cells increases susceptibility to infection, (ii) T-cell defects enhance susceptibility to infection, and (iii) immunity can be transferred by T cells provide a compelling case for the importance of T cells in host defense.

CD4+ and CD8+ T Cells Are Each Important in Host Defense

Experiments in mice depleted of T-cell subsets by the administration of antibody to CD4+ or CD8+ T cells indicate that both CD4+ and CD8+ T cells are important in protection against *C. neoformans* (98, 99, 110, 114, 158–160). CD4+ and CD8+ T-cell-deficient mice (knockout mice) are more susceptible to *C. neoformans* infection than are their parental strains (250). Clearance of pulmonary infection requires the presence of both CD4+ and CD8+ T cells (114). Depletion of CD4+ cells enhances dissemination and reduces survival of mice given intratracheal infection (160). Depletion of CD8+ T cells also reduces survival, but the effect on organ fungal burden appears to depend on the virulence of the strain (158, 159). Interestingly, depletion of CD8+ T cells reduces survival in experimental murine infection without increasing organ fungal burden for a highly virulent strain (159). Depletion of either CD4+ or CD8+ T cells reduced the number of macrophages and neutrophils in the mouse inflammatory response to pulmonary infection (110). The effect of CD4+ T-cell depletion on pulmonary inflammation is greater than that observed for CD8+ T-cell depletion, but only CD4+ T-cell depletion abrogated pulmonary eosinophilia (110). CD4+ and CD8+ T cells appear to have complementary roles in resistance to pulmonary infection (114). In the lung, CD4+ T cells have been proposed to recruit and activate effector cells, while CD8+ T cells function to lyse inactivated phagocytic cells with ingested cryptococci (114). CD4+ cells are also required for giant cell formation, which has been associated with containment of infection in the lung (98). In the brain, containment of *C. neoformans* is dependent on CD4+ T-cell immunity (99). Hence, there is strong experimental evidence that both CD4+ and CD8+ T cells contribute to defense against *C. neoformans*.

Direct Antifungal Effects of T Cells on *C. neoformans*

Mixed lymphocyte populations can inhibit the growth of *C. neoformans*. Among lymphocyte populations, CD4+ T cells, CD8+ T cells, and NK cells (but not B cells) have been demonstrated to have direct antifungal effects. T cells can directly inhibit the growth of *C. neoformans* in vitro (134, 186). Figure 3 shows a lymphocyte in direct contact with a *C. neoformans* cell. Purified preparations of T cells can mediate nearly complete fungistasis when activated with interleukin (IL)-2 (134). IL-2-responsive T cells that inhibit *C. neoformans* growth include CD4+, CD8+, and CD56+ cells (134). T-cell-mediated fungistasis occurs after close contact with the fungal target, but the interaction does not appear to require complement or specific antibody (134, 135). T-cell binding to *C. neoformans* occurs through broad areas of cell membrane attached to the fungal surface (186). Protease treatment of T cells reduces fungistatic activity, suggesting the involvement of surface protein structures in the interaction with fungi (135). The mechanism by which T cells mediate fungistasis is not known (139). Treating T cells with scavengers of oxygen-derived radicals and inhibitors of the cyclooxygenase system does not produce consistent reductions in lymphocyte-mediated fungistasis (139). It is not known whether the fungistatic activity of T cells against *C. neoformans* is mediated by a sensitized subset (as a result of previous exposure to this organism) or is the result of nonspecific interactions of T cells with a fungal target (186). The Jurkat human T-cell leukemia line is also able to inhibit the growth of *C. neoformans* (91). Infection

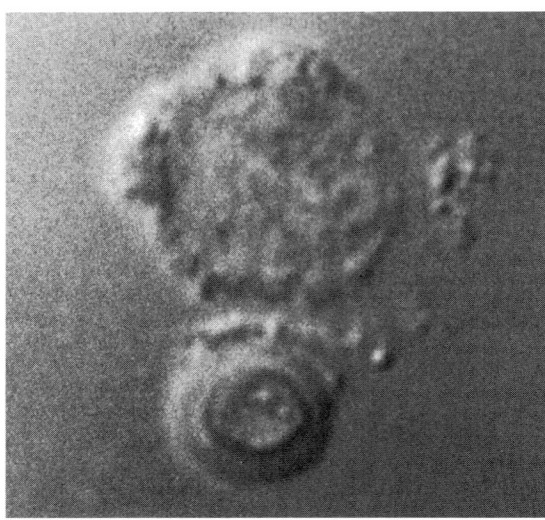

Figure 3 Direct interaction between a T lymphocyte and a *C. neoformans* cell. Photograph kindly provided by Stuart Levitz (Boston, Mass.). Details of this experiment can be found in reference 135. T lymphocytes have been shown to inhibit the replication of *C. neoformans* in vitro (134, 135, 186). For a review of this phenomenon, see reference 137.

of Jurkat cells with HIV resulted in a significant reduction of antifungal activity against *C. neoformans* (91). The physiological role of the phenomenon of T-cell-mediated fungistasis of *C. neoformans* in vivo is uncertain, but the in vitro results suggest that T cells may function during infection as direct antifungal effector cells.

T Cells and Granuloma Formation

Granulomatous responses are strongly associated with containment and resolution of infection in both humans and laboratory animals (see chapter 9). Granuloma formation is a classical manifestation of cell-mediated immune responses that are dependent on T-cell function. *C. neoformans* infection in *nu/nu* mice is characterized by the absence of granuloma formation (210). Multinucleated giant cells are found in cryptococcal granulomas and are probably important for the ingestion and containment of large encapsulated yeast forms found in tissue. Hill (98) has shown that giant cell formation in the respiratory tract is associated with containment of *C. neoformans* in the lung and is dependent on CD4+ T cells.

Studies of granulomatous inflammation in rat cryptococcal meningitis revealed that both CD4+ and CD8+ T cells were found in granulomas (66). Granuloma formation in the rat brain was associated with an influx of CD4+ and CD8+ T cells, strongly suggesting a requirement for both T-cell subsets in the control of cryptococcal infection.

Dependence of Antibody Immunity on T-Cell Function

There is evidence that T cells may be important for the generation of effective antibody responses to *C. neoformans* and for proper antibody function. Immunization of mice with the GXM-TT conjugate vaccine elicits a T-cell-dependent antibody response (22, 170). Analysis of MAbs obtained with the conjugate vaccine indicate a predominance of IgG isotypes, affinity maturation, and frequent somatic mutations, characteristics consistent with and indicative of T-cell-dependent antibody response (22, 170). Protective and nonprotective antibodies against *C. neoformans* differ by a few amino acid residues, which are presumably the result of somatic mutation of immunoglobulin genes (173), a process that may be influenced by T-cell regulation (59). CD4$^+$ T cells are necessary for protective antibodies to prolong survival, and this effect appears to be mediated through gamma interferon (IFN-γ) (250). Disease-enhancing antibodies of the IgG3 isotype require CD8$^+$ T cells in mice to mediate their effects. These observations indicate that antibody-mediated effects are dependent on both CD4$^+$ and CD8$^+$ cells.

Correlate Measures of Cellular Immunity

DTH

C. neoformans antigens can elicit strong DTH skin reactions. The DTH skin hypersensitivity reaction to *C. neoformans* antigens has been studied extensively since the 1950s. Initially, it was hoped that skin hypersensitivity testing for *C. neoformans* antigens would help elucidate the epidemiology of infection and be as useful as tuberculin testing has been for latent exposure to *Mycobacterium tuberculosis*. Specifically, it was hoped that an effective skin antigen preparation would also provide information on the prevalence of sensitization to *C. neoformans* in asymptomatic populations (214, 215). Cryptococcal antigen preparations used in DTH testing have sometimes been referred to as cryptococcin.

Salvin and Smith (215) prepared an alkaline extract from *C. neoformans* cells disrupted in a pressure cell, which elicited skin induration in infected guinea pigs. The cell extract preparation did not elicit DTH reactions in guinea pigs infected with *Candida albicans* or *Histoplasma capsulatum*, suggesting specificity for *C. neoformans* (215). A patient injected with this material produced a positive skin reaction (205). This material probably contained a complex mixture of protein and carbohydrate antigens and was used in human studies of the prevalence of DTH (169). Muchmore et al. (169) found that 32% of individuals in a rural Oklahoma community had typical DTH skin reactions after intradermal testing. The Salvin and Smith preparation was shown to inhibit the migration of leukocytes from infected animals and humans (164). This in vitro phenomenon was felt to be useful in detecting immunological responsiveness to *C. neoformans* antigens and could be analogous to a similar effect described with tuberculin in patients with tuberculosis (164). The Salvin and Smith preparation was cumbersome and had the inherent problem that it was difficult to achieve the same amount of cell breakage in different preparations (185). Furthermore, this preparation elicited DTH reactions in patients with other mycoses and additional research was undertaken to develop more sensitive and specific preparations.

Atkinson and Bennett (5a) prepared a urea extract from a poorly encapsulated strain of *C. neoformans*. This material elicited DTH reactions in laboratory animals and human patients. Reactivity among patients with active and inactive *C. neoformans* infection was 58% and 77%, respectively. Unfortunately, 50 to 70% of patients with other mycoses had positive skin reactions, indicating low specificity for *C. neoformans*. Interestingly, 81% of mycology laboratory workers had a positive skin reaction, consistent with either subclinical infection or chronic exposure to fungal antigens (5a). Skin testing with this preparation revealed that apparently normal patients who had had cryptococcal infections and had been cured often had impaired responsiveness to skin testing with cryptococcal and other antigens, suggesting subtle defects in cell-mediated immunity (218). Biochemical characterization of the urea extract revealed that the activity resided in a protein fraction with a molecular mass of approximately 10,000 Da. Polysaccharide was present in the preparation but had no DTH activity (11). Antigen prepared from live cells gave stronger DTH reactions than did antigen prepared from formalin-killed cells (11).

Newberry et al. (193) used the Atkinson and Bennett cryptococcin preparation (5a) to study several aspects of *C. neoformans* epidemiology. Pigeon fanciers were found to be more likely to have positive skin reactivity than were control subjects (32.1 versus 4.2%), suggesting that asymptomatic infections with *C. neoformans* had occurred and implicating pigeon habitats as potential sources for infection (193). Testing of five patients who were cured of *C. neoformans* infection revealed that only two had positive skin reactivity to cryptococcin, suggesting nonreactivity to cryptococcal antigens among some survivors of infection.

Using a modification of the cryptococcin protocol devised by Atkinson and Bennett (5a), Graybill and Alford (76) studied the skin test reactivity of patients with cryptococcosis in the pre-AIDS era. Consistent with the observations of Newberry et al. (193), Graybill and Alford found that four of eight patients with *C. neoformans* infection had a positive skin test to cryptococcin. In contrast, only 2 of 13 healthy control subjects manifested positive skin tests when challenged with cryptococcin. Interestingly, one patient was negative early in the course of the illness but gained skin test reactivity during antifungal therapy. A skin biopsy of a responder patient demonstrated monocytic inflammation consistent with a DTH skin reaction.

Murphy and coworkers (61, 107, 142, 143, 182, 183, 185, 187, 188) have performed extensive studies of the immune response to antigens in *C. neoformans* culture filtrates. A culture filtrate from formalin-killed cells elicited DTH reactions in guinea pigs infected with *C. neoformans* (185). No reactivity was elicited in guinea pigs sensitized to other medically important fungi (185). Histological studies of skin reactions by this antigen are similar to those elicited by tuberculin, suggesting that the antigen elicits a tuberculin type of hypersensitivity reaction (185). Using the culture filtrate preparations (known as CneF), Lim and Murphy (142) established that DTH could be transferred to naive mice by splenic T cells from sensitized mice. In their system, transfer of the splenic T cells conferred both DTH and immunity, as indicated by a reduction in organ fungal burden among recipient mice relative to control mice (142). Follow-up studies established a temporal correlation between the development of DTH reactivity to soluble antigens and induction of splenic lymphoid cells capable of inhibiting *C. neoformans* (61). In mice, CneF elicits DTH and protective immune responses when administered subcutaneously, but intravenous administration elicits antigen-specific immunological suppression (182). CneF contains capsular polysaccharides (GXM and galactoxylomannan) and mannoprotein (187). The mannoprotein fraction appears to be responsible for eliciting the DTH response.

Murphy (183) has also studied the cytokine responses associated with development of DTH. CneF immunization elicits two splenic populations of CD4$^+$ cells. One splenocyte population can transfer the DTH response from immune to naive mice and is known as T_{DH}. The other splenocyte population amplifies the DTH response if given to mice at the time of CneF immunization and is known as T_{AMP} (183). The T_{AMP} response, but not the T_{DH} response, can be ablated with cyclosporine. Supernatants from spleen cells immunized with CneF in complete Freund's adjuvant show that the T-helper-1 (TH1)-associated cytokines IFN-γ and IL-2 are produced, but the TH2-associated cytokines IL-4 and IL-5 are not (183). By analyzing the response to CneF in cyclosporine-treated mice, Murphy (183) further dissected the cytokines involved in the induction of T_{DH} and T_{AMP} cells. Conditions low in IFN-γ and IL-2 favored development of T_{DH}, whereas both T_{DH} and T_{AMP} cells developed when production of IFN-γ and IL-2 was high. Using gelatin implant sponges injected with CneF, Buchanan and Murphy (18) studied the cellular and cytokine responses associated with the development of DTH reaction in mouse footpads. The cellular response consists primarily of neutrophils and mononuclear cells (18). CD4$^+$ lymphocytes and NK cells are also present. As with spleen cells, the DTH reaction in CneF-injected sponges in either infected or immunized mice contains IFN-γ and IL-2-producing cells, supporting the association of DTH with TH1-associated cytokines (18). However, the TH2-associated cytokine IL-5 is also present in the DTH response to CneF-injected sponges (18). These results strongly implicate TH1-associated cytokines in the DTH response to cryptococcal antigen but also suggest the involvement of some TH2-associated cytokines.

Hay and Reiss (94) studied cytoplasmic fractions for their ability to elicit DTH reactions in infected mice. Two fractions, the postmitochondrial supernatant and the culture filtrate, elicit strong reactions in the footpad test (94). Cellular immune responses, as measured by DTH, are weaker in mice given intravenous infection than in mice given subcutaneous infection (94).

The ability of mice to develop DTH reactions to cryptococcal antigens may or may not correlate with protection against infection. Graybill and Taylor (85) demonstrated that immunization of BALB/c mice with cryptococcal extracts made from either culture filtrates or disrupted cells elicited a strong DTH reaction within 1 week, and mice lived longer when challenged with a lethal infection. However, DTH was also observed after nonimmunized mice were infected with C. neoformans, but in that case development of DTH was not associated with prolonged survival (85).

In summary, C. neoformans antigens can elicit strong DTH reactions in infected patients and in laboratory animals. Considerable effort has been devoted over the past four decades to develop a suitable skin test based on DTH reactivity to cryptococcal antigens for epidemiological studies. However, those efforts have been frustrated by a combination of cross-reactivity with other fungal pathogens and a relatively low sensitivity in detecting infection. Nevertheless, skin reactivity has provided some important insights into the possibility of asymptomatic infection in laboratory workers and into the development of cell-mediated immune responses. In laboratory animals, a positive DTH reaction may be a marker for effective cell-mediated immunity, but its occurrence does not necessarily imply effective immunity. Skin testing for reactivity to cryptococcal antigen is not widely used in clinical practice despite some evidence that it may be useful for identifying patients with severe immune defects (76).

Lymphocyte proliferation
Like DTH, the ability of lymphocytes to proliferate in vitro in response to microbial

antigens is a classical response associated with development of cell-mediated immunity. Diamond and Bennett (41) studied lymphocyte proliferation in response to cryptococcal antigens among individuals with positive and negative skin reactivity to cryptococcin and in patients who had recovered from *C. neoformans* infection. Lymphocytes from individuals with positive skin tests consistent with previous exposure to *C. neoformans* had higher proliferative responses than did lymphocytes from individuals with negative skin tests. Lymphocyte proliferation in patients with negative skin tests presumably reflects mitogenic activity in response to *C. neoformans* antigens and/or cross-reactivity with antigens to which these individuals were previously sensitized. Lymphocytes from patients who had recovered from cryptococcal infection with antifungal therapy had lower proliferative responses than did normal individuals with positive skin reactivity to cryptococcin. This result suggested a relative deficiency in cellular immunity among individuals who had contracted symptomatic *C. neoformans* infections and recovered with therapy. However, it is unclear whether this defect predates infection or results from the immunosuppressive effects associated with infection (see below).

Graybill and Alford (76) compared lymphocyte transformation and DTH skin reactions in patients who had recovered from cryptococcosis with those of healthy control subjects. Patients with previous or ongoing cryptococcal infection had strong lymphocyte transformation responses to cryptococcin (76). In contrast, healthy control subjects rarely had strong lymphocyte transformation responses to cryptococcin, except for laboratory workers exposed to *C. neoformans* (76). Lymphocyte transformation responses to whole *C. neoformans* cells were even stronger than those to soluble cryptococcin preparations, suggesting that some of the proliferative effect is, in part, nonspecific. The results of this study were also interpreted as suggesting baseline immune defects in patients with cryptococcosis. There was dissociation between the results obtained by DTH skin testing and lymphocyte transformation for some individuals in both the recovered patient and control groups. In a subsequent study, Graybill and collaborators (84) fractionated cellular and supernatant extracts to determine their immunological properties. There was good correlation between skin reactivity and cell migration, but there was poor correlation between the ability of a fraction to induce strong DTH reactions and to protect mice against lethal *C. neoformans* challenge (84).

Schimpff and Bennett (218) studied lymphocyte transformation in response to killed *C. neoformans* in normal individuals with or without positive cryptococcin skin reactivity and in individuals who had recovered from cryptococcal infection. Positive skin reactivity was associated with a stronger lymphocyte transformation reaction. The magnitude of lymphocyte proliferation in normal individuals with positive skin reactivity to cryptococcin was comparable to that found in recovered patients. As in the Graybill and Alford study (76), this study found considerable interpatient variation in skin reactivity and lymphocyte transformation assays. The Schimpff and Bennett study (218) described depressed skin reactivity and lymphocyte proliferation in patients successfully treated for cryptococcal meningitis. This finding was interpreted as suggestive of subtle immunological defects that predisposed to infection. Although this may be the case, the possibility of long-lasting immunosuppression resulting from a bout of cryptococcosis may also contribute to this effect.

Hall and colleagues (88) studied various antigen preparations for their efficacy in eliciting DTH reactions and lymphocyte proliferation. Comparison of the immunological reactivity of various preparations revealed that urea extracts elicited the strongest responses, followed by alkali extracts and soluble capsular polysaccharide preparations (88). The immunological reactivity of the preparations correlated with their protein content (88).

Miller and Puck (155) studied the lymphocyte response to *C. neoformans* in normal individuals and in a small number of patients recovering from cryptococcal meningitis. All normal individuals had lymphocyte proliferative responses to killed cryptococci, but recovering patients had more rapid and stronger responses. T cells from patients with cryptococcosis produced IL-2 after initial stimulation in vitro, whereas T cells from control patients required in vitro priming with cryptococci and antigen-presenting cells to produce IL-2. The finding of augmented proliferative responses in recovering patients is different from that of the Schimpff and Bennett (218) and Graybill and Alford (76) studies. This discrepancy may be explained by the fact that the Miller and Puck (155) findings were obtained with patients who had responded to therapy; the successful response to therapy may reflect the development of strong cellular immunity.

T cells from HIV-positive and HIV-negative individuals without cryptococcal infection show differences in their ability to proliferate in vitro in response to *C. neoformans* antigens (106). Lack of T-cell responsiveness in HIV-infected patients to *C. neoformans* antigens is strongly associated with progression to AIDS (106). Loss of the T-cell proliferative response is observed even before quantitative differences in CD4+ are evident, suggesting a subtle qualitative defect in T-cell function with HIV infection (106). These observations are consistent with and suggestive of a T-cell defect in HIV infection as a major predisposing factor to *C. neoformans* infection.

Mody et al. (161, 162) have recently investigated cellular fractions of *C. neoformans* to identify the yeast components that stimulate human lymphocytes to proliferate. Lymphocyte proliferation is stimulated by cell wall and membrane components (161). Treatment of cell wall and membrane components with proteinase eliminates the ability of the material to induce lymphocyte proliferation, indicating that the active ingredient is a protein. The lymphocyte proliferative response to cell wall and membrane-associated protein is similar to that induced by *Staphylococcus* enterotoxin B, suggesting that the protein has mitogenic properties. The *C. neoformans* mannoprotein component that has been implicated in eliciting DTH reactions does not induce lymphocyte proliferation. Furthermore, since culture filtrates do not stimulate lymphocytes to proliferate, this study suggests that different *C. neoformans* cellular components are responsible for eliciting DTH reactions and lymphocyte proliferation. Both CD4+ and CD8+ cells proliferate in response to *C. neoformans* antigens (230). CD4+ T cells can proliferate by themselves upon exposure to *C. neoformans* antigens, but CD8+ T-cell proliferation is dependent on IL-2 and other CD4+ T-cell signals (230).

Pitzurra and collaborators (204) have recently identified a 105-kDa mannoprotein antigen in *C. neoformans* extracts that stimulates strong lymphoproliferative responses in human peripheral blood mononuclear cells (PBMCs). This antigen appears to be a cell wall constituent that binds to concanavalin A and reacts with antibodies to mannan. The function of this antigen in the cryptococcal cell is not known.

In summary, lysates and extracts of cryptococcal cells can elicit DTH reactions in some sensitized hosts and lymphocyte proliferation in vitro. These preparations have proved useful in laboratory investigations of cryptococcal infections and in some epidemiological assays. There are significant differences in the biological activity of the cellular fractions that probably reflect differences in the extraction process. It is likely that future studies with defined antigen preparations prepared with recombinant DNA techniques will permit the identification of the proteins responsible for DTH and lymphoproliferative responses.

CYTOKINES

Cytokines constitute a diverse set of proteins that include lymphokines, inter-leukins, monokines, and colony-stimulating factors (CSF) (208). Cytokines are produced by a variety of cell types, including macrophages, T cells, fibroblasts, endothelial cells, hepatocytes, and stromal cells of the spleen and thymus (208), and they perform a myriad of activities in coordinating the immune response to pathogens, including recruitment and stimulation of immune effector cells (208). Cytokine stimulation can also result in undesirable inflammatory responses (208), but it is unclear whether this is a problem in cryptococcosis, for which histological studies often show little or no inflammation.

Many cytokines have been shown to have important roles in host defense against *C. neoformans*. The importance of individual cytokines in protection against *C. neoformans* has usually been established by animal experimentation. Most animal experiments fall into one of the following categories: (i) neutralizing the cytokine by administration of a specific antibody to the cytokine and observing the effect on infection, (ii) studying pathogenesis in cytokine-deficient mice generated by gene disruption (i.e., knockout mice), (iii) administering exogenous cytokine and observing its effect on the course of infection, and (iv) measuring cytokine responses and correlating the expression of specific cytokines with effective host responses against *C. neoformans*. The study of cytokines in mice is aided by the availability of many murine reagents and several mouse strains deficient in specific cytokines. However, extrapolation to humans of results in mice should be done cautiously, because most experiments have been done with genetically inbred strains, there are significant species differences, and the relevance of experimental infection to natural infection is uncertain. In contrast to mice, most studies of cytokine function in humans have been done in vitro. These studies usually seek to determine the effect of cytokines on effector cells and/or the cytokine response of various cell types to *C. neoformans* or fungal products.

The TH1-TH2 Paradigm

In recent years, studies of cytokines and tissue responses have been frequently interpreted within a theoretical construct based on the ability of various subsets of $CD4^+$ T cells to produce specific cytokines. This construct was derived from observations made with *Leishmania major* and has become a paradigm for evaluating cytokine responses. (For reviews, see references 6, 60, and 209.) On the basis of their pattern of cytokine production, $CD4^+$ T-helper (TH) cells have been divided into two subsets: TH1 cells secrete IFN-γ and IL-2, whereas TH2 cells secrete IL-4, IL-5, and IL-10. The TH1-associated cytokines IFN-γ and IL-2 have been associated with effective immune responses to *C. neoformans* infection. The current dogma is that TH1 responses are effective against intracellular pathogens (such as *C. neoformans*), whereas TH2 responses are effective against extracellular pathogens. This is called the TH1-TH2 paradigm (6, 60, 209).

In mice, both CD4$^+$ and CD8$^+$ T cells have been shown to be important in the development of effective cellular immune responses to *C. neoformans* infection (110, 158–160). Given that granulomatous responses are associated with control of *C. neoformans* infection, the T-cell TH1 and TH2 paradigm would predict that TH1-phenotype T cells secreting pro-inflammatory cytokines would be crucial for effective anticryptococcal responses. This prediction has generally been borne out by experimental studies; several studies have associated the TH1-associated cytokines IFN-γ and IL-2 with resistance to *C. neoformans* infection (see below). Despite its usefulness as a theoretical model, there are many caveats and exceptions to the TH1-TH2 paradigm, and it is likely that this view reflects an oversimplification of a very complex process (6).

The Cytokine Response to *C. neoformans* in Mice

Buchanan and Murphy (18, 19) have used a system of implantable sponges to detect the cytokines associated with DTH and the development of cell-mediated immune responses in mouse tissues. Gelatin sponges soaked with saline or CneF (an antigen preparation that can elicit DTH reactions) are implanted in infected and control mice, removed 4 days later, and analyzed for cytokine content (18). IFN-γ, IL-2, and IL-5 are found in greater concentration in CneF-impregnated sponges in both *C. neoformans*-infected and CneF-immunized mice, suggesting that these cytokines are important contributors to the DTH reaction (18). The sponge system implicates two TH1-associated cytokines (IFN-γ and IL-2) and one TH2-associated cytokine (IL-5) in the DTH reaction to cryptococcal antigen (18). The function of IL-5 in the murine response to CneF in sponges is unclear (18). IL-5 has been associated with eosinophilic pneumonia in C57BL/6 mice given intratracheal *C. neoformans* infection (101). These observations implicate TH1-associated cytokines in DTH, which in turn has been correlated with development of strong tissue inflammatory responses.

In lung tissue the pattern of cytokine expression in response to *C. neoformans* infection may differ (for review, see reference 108). Analysis of TH1- and TH2-associated cytokine expression in the lungs of (BALB/c × DNA/2)F1 mice given intratracheal infection revealed stronger and earlier expression of TH2-associated cytokines and transforming growth factor β in the early stages of infection, with little or no expression of the TH1-associated cytokines IL-2 and IFN-γ (120). Exogenous administration of IL-12, a cytokine shown to protect against infection in the mouse pulmonary model (122), enhanced expression of both TH1- and TH2-associated cytokines (120). For tumor necrosis factor (TNF)-α, experiments in mice have produced inconsistent results. In CBA/J mice, *C. neoformans* lung infection elicited early production of TNF-α, and neutralization of this cytokine with a MAb to TNF-α interfered with the inflammatory response (112). In contrast, no TNF-α expression was detected in lungs of (BALB/c × DNA/2)F1 after *C. neoformans* infection, and administration of a MAb to TNF-α had no effect on survival (119).

The discrepancies between these studies may reflect the use of different mouse and *C. neoformans* strains.

C.B-17 and C57BL/6 mouse strains differ in susceptibility to cryptococcal infection. Analysis of early cytokine production after intratracheal infection revealed the production of more TH1-associated cytokines in the C.B-17 strain (101). This result suggests that an early regional TH1 response to infection may contribute to localization and control of infection in mice.

C. neoformans infection can elicit TNF-α and IL-1α expression in brain tissue of C57BL/6 mice (13). Intracerebral immunization of mice with heat-killed *C. neoformans* resulted in the early expression of IL-6 and IL-1β mRNA (13). Intracerebral administration of IL-6 or IL-1β, but not TNF-α, to C57BL/6 mice before infection reduced the severity of infection, suggesting that these cytokines enhanced resistance to *C. neoformans* infection in brain tissue (13). Intracerebral chloroquine administration before infection reduces brain fungal burden; this phenomenon was associated with enhanced IL-6 and IL-1β expression in C57BL/6 mice (150).

In summary, cryptococcal infection in mice elicits expression of TH1-associated (IL-2, IFN-γ), TH2-associated (IL-4, IL-5, IL-10), and macrophage-derived (TNF-α, IL-1β, transforming growth factor α) cytokines. Expression of TH1-associated cytokines has been associated with the development of enhanced host resistance to infection. The role of individual cytokines may differ, depending on the infected tissue and the mode of infection. The interplay of the various cytokines in eliciting effective tissue responses is poorly understood.

Cytokine Deficiencies and Predisposition to Infection

Derangements in cytokine regulation may predispose to cryptococcal infection. There is one case report of a child with isolated cryptococcal osteomyelitis who had IL-2 deficiency (223). Deficient IFN-γ production in patients with AIDS as a consequence of CD4+ T-cell lymphopenia has been suggested to contribute to their marked susceptibility to invasive cryptococcal infections (163). PBMCs from patients with HIV-associated cryptococcosis and from healthy donors demonstrated similar lymphoproliferative kinetics but different cytokine profiles when challenged with *C. neoformans* (138). Neutralization of several cytokines in mice results in more-severe infections (Table 2). Many cytokines have been shown to enhance the antifungal efficacy of inflammatory cells in vitro (Table 3), and it is likely that genetic or acquired deficiencies of any of these cytokines would predispose to cryptococcal infection.

Selected Cytokines in *C. neoformans* Infection

IFN-γ

Mouse studies have established a crucial role for IFN-γ in protection against *C. neoformans* infection. Neutralization of IFN-γ results in more-severe infections (121, 147, 211) in both murine pulmonary (147) and intravenous infection models (4). Administration of antibody against IFN-γ can block the enhancement of NK cell activity that accompanies experimental infection in mice (211). IFN-γ knockout mice are more susceptible to *C. neoformans* infection (250). Conversely, administra-

Table 2 Effects of selected cytokine administration or neutralization against *C. neoformans* infection in mice

Cytokine	Action	Effect	References
INF-γ	Neutralization	↑ In organ CFU	4, 147
		↓ Pulmonary macrophage phagocytosis	118
		↓ NK cell activity against *C. neoformans*	211
	Administration	↑ Survival; ↓ organ CFU; synergy with amphotericin B	116
IL-12	Administration	↓ Organ CFU; synergy with fluconazole	32, 122
IL-6 or IL-1β	Administration	↓ Brain CFU	13
TNF-α	Neutralization	↑ Brain CFU	4

tion of IFN-γ to mice enhances resistance to infection, as manifested by reduced organ colony counts (116, 121), and potentiates the efficacy of amphotericin B in vitro (199) and in vivo (152). Differences in the susceptibility of C57BL/6 (suscep-tible) and C.B-17 (resistant) mice to infection have been associated with differences in IFN-t and IL-2 production by lung-associated lymph nodes (101). Both IFN-γ and IL-12 are necessary in C.B-17 mice for initiating an effective TH1-like response, which results in clearance of pulmonary infection (100).

IFN-γ is a pro-inflammatory cytokine that has multiple effects on host im-mune cells. IFN-γ enhances the activity of murine macrophages against *C. neofor-mans* in vitro (199) and may activate alveolar macrophages for anticryptococcal activity in vivo (163). IFN-γ primes murine macrophages for production of nitric oxide synthase, which can generate nitric oxide. In vitro, nitric oxide has been shown to be an important antimicrobial effector molecule against *C. neoformans* (73, 248). Some of the anticryptococcal effects of IFN-γ-primed macrophages are mediated through nitric oxide (7, 73). IFN-γ and TNF-α act synergistically to elicit nitric oxide production from murine macrophages. Blocking of IFN-γ activity in C.B-17 mice with specific antibody inhibits inducible NOS production and inter-feres with clearance of pulmonary *C. neoformans* infection (147). The enhanced anticryptococcal activity of macrophages treated with IFN-γ may result from in-creased production of reactive nitrogen intermediates (232). IFN-γ may also have an important role in mediating or enhancing the therapeutic efficacy of am-photericin B, which augments nitric oxide production by IFN-γ-activated murine macrophages (233).

IFN-γ appears to be necessary for antibody-mediated effects against *C. neofor-mans* infection (250). Protective antibodies do not mediate protection in T-cell-defi-cient mice, but administration of IFN-γ restores their efficacy (250). This observation indicates that IFN-γ is a critical cytokine in modulating the effects of both humoral and cellular immune responses against *C. neoformans*.

TNF-α

TNF-α is a pro-inflammatory cytokine produced by monocytes and macrophages

Table 3 In vitro cytokine effects on the efficacy of immune effector cells against C. neoformans

Cytokine	Effector cells	Effect	References
IFN-γ	Murine macrophages	↑ Fungicidal activity	26, 58, 199
	Rat alveolar macrophages	↑ Antifungal activity	163
	Murine macrophage cell lines	↑ Expression of proteins fungicidal to C. neoformans	97
	Murine macrophage-like J774 cells	↑ Antibody-mediated phagocytosis	181, 179
	Murine macrophages	↑ Antifungal efficacy; synergy with amphotericin B	96, 199
	Human monocytes	Restores IL-12 production	93
IL-2	T cells, NK cells	↑ Fungistatic activity	134
	Human PBMCs	↑ Fungicidal activity	133
IL-12	NK cells	Restores antifungal activity to cells from HIV-positive patients	104
TNF-α	Human PBMCs	↑ Fungistatic activity	136
	Murine peritoneal macrophages	↑ Complement-mediated phagocytosis	33
GM-CSF	Human PBMCs	↑ Fungistatic activity	133
	Rat alveolar macrophages	↑ Antifungal activity	26
	Murine peritoneal macrophages	↑↑ Complement-mediated phagocytosis	33
GM-CSF + IFN-γ	Murine alveolar and peritoneal macrophages	↑ Fungistatic activity	16, 17
	Rat alveolar macrophages	↑↑ Antifungal activity	26
IL-12 + IL-2	Mixed lymphocyte populations	↑ Antifungal activity	104
IL-12 + IL-18	Murine leukocytes	↑ Antifungal activity	254
G-CSF	Neutrophils	↑ Fungicidal activity	236
GM-CSF + TNF-α	Murine peritoneal macrophages	↑↑ Complement-mediated phagocytosis	33

that is believed to be important for activation and recruitment of inflammatory cells during microbial infection. TNF-α is made in response to *C. neoformans* infection and is likely to have an important role as a cellular signal for coordinating the inflammatory response and stimulating phagocytic cells for antifungal action. Neutralization of TNF-α by specific antibody has been shown to enhance infection in some mouse strains (33, 112) but not in others (121). Administration of TNF-α to mice with lethal cryptococcal infection can prolong survival time (121).

In conjunction with granulocyte-macrophage stimulating factor (GM-CSF), TNF-α may have a major role in activating cells for phagocytosis of *C. neoformans*. Acapsular *C. neoformans* cells are rapidly phagocytosed through mannose and β-glucan receptors on murine macrophages, and phagocytosis results in the production of the macrophage-derived pro-inflammatory cytokines TNF-α and GM-CSF (34). This observation may be relevant to cryptococcal pathogenesis because environmental strains are often poorly encapsulated (31, 34). It has been suggested that the initial fungal-host interaction involves poorly encapsulated cryptococcal cells and alveolar macrophages, with the production of highly encapsulated cells occurring after establishment of infection. Since TNF-α and GM-CSF can enhance complement-mediated phagocytosis of encapsulated *C. neoformans* cells (33, 35), ingestion of acapsular mutants can enhance phagocytosis of encapsulated strains (34). These results suggest an important role for TNF-α and GM-CSF in macrophage activation in response to *C. neoformans* infection.

Incubation of human PBMCs with heat-killed cryptococci stimulates these cells to kill subsequent inocula of live encapsulated yeasts (136). A subsequent study revealed that incubation of PBMCs or bronchoalveolar macrophages with heat-killed *C. neoformans* elicited the production of TNF-α, implicating TNF-α in this effect (140). However, the exact role of TNF-α in stimulating effector cell antifungal activity is unclear: incubation of human PBMCs with TNF-α does not enhance their fungistatic properties, but TNF-α production may be a more crucial event in inflammatory cell recruitment to infection. The observation that *C. neoformans* elicits TNF-α production by leukocytes may have an in vivo parallel: patients with HIV-associated cryptococcal meningitis have high levels of TNF-α in the cerebrospinal fluid (149). Since TNF-α induces HIV replication in latently infected cells, it is possible that concurrent infection of HIV and *C. neoformans* can accelerate HIV progression through the effect of TNF-α (140). At this time, however, the exact role of TNF-α in protecting against or contributing to the pathogenesis of *C. neoformans* in human infection remains to be established.

IL-1β

Inoculation of heat-killed cryptococci into the brains of mice produces local resistance to infection that is associated with increased production of IL-1β mRNA (13). Administration of IL-1β before intracerebral infection is associated with reduced fungal counts in tissue (13). These observations suggest an important role for IL-1β in brain defense against *C. neoformans*. Intracerebral administration of chloroquine increases IL-1β mRNA transcription in the brain and is associated with enhanced resistance to intracerebral *C. neoformans* infection (150).

IL-2

There is circumstantial evidence that the pro-inflammatory IL-2 is important in human protection against *C. neoformans* infection. As noted above, IL-2 deficiency was implicated as a potential predisposing factor in a child with *C. neoformans* infection (223). Lymphocytes from individuals who recover from *C. neoformans* infection produce large amounts of IL-2 when stimulated with cryptococci in vitro (154). Addition of IL-2 to PBMCs stimulates their fungicidal activity in vitro (133). The IL-2-responsive component of PBMCs was subsequently shown to be T cells and NK cells, which interact directly with *C. neoformans* to mediate antifungal effects (134). The mechanism by which T cells and NK cells mediate anticryptococcal effects is uncertain (139). IL-2 has been shown to consistently enhance the antifungal efficacy of human mixed lymphocyte populations in vitro against *C. neoformans* (104).

IL-5

IL-5 has been shown to be produced in response to *C. neoformans* infection in C57BL/6 mice (101). IL-5 was also detected in CneF-impregnated sponges in both *C. neoformans*-infected and CneF-immunized mice (18). IL-5 may contribute to the marked pulmonary eosinophilia that occurs in response to cryptococcal infection in some strains of mice (101).

IL-6

Increased production of IL-1β mRNA has been demonstrated in the brains of mice immunized with heat-killed *C. neoformans*; these mice are more resistant to intracerebral infection (13). Administration of IL-6 before intracerebral infection is associated with reduced fungal counts in brain tissue (13). Intracerebral administration of chloroquine has been shown to increase IL-6 mRNA transcription, which is also associated with enhanced resistance to infection (150).

IL-10

IL-10 is a TH2-associated cytokine that can down-regulate a variety of immunological processes, including antigen presentation. Addition of IL-10 to mixtures of human macrophages laden with cryptococci and T lymphocytes inhibited lymphoproliferation, IL-2 secretion by T lymphocytes, and expression of major histocompatibility complex class II molecules (165). IL-10 has also been shown to inhibit the release of TNF-α and IL-1β from human PBMCs (141). *C. neoformans* polysaccharide can induce IL-10 production by monocytes in vitro (238). These observations suggest a mechanism for the reduced inflammatory response in cryptococcosis: *C. neoformans* polysaccharide may enhance production of IL-10, which down-regulates the inflammatory response.

IL-12

Production of IL-12 by macrophages is induced by various pathogens. IL-12 stimulates IFN-γ production by NK cells and facilitates development of TH1 lymphocytes, which in turn produce more IFN-γ and IL-2 (12). Administration of IL-12 to mice with pulmonary infection enhances IFN-γ mRNA expression and is associated with stronger inflammatory responses that reduce lung fungal burden (122). In mice with intravenous infection, IL-12 has little effect on yeast counts in the spleen and lung but reduces the number of organisms in the brain (32). IL-12 also induces differentiation of T cells to a TH1 phenotype, resulting in increased IFN-γ production (12). Treatment of mice with IL-12 after intravenous infection reduces the number of C. neoformans cells in brain tissue but not in pulmonary or splenic tissues (32). Nevertheless, IL-12 appears to be essential for development of an effective inflammatory response in the lung (100). Early administration of IL-12 to mice with intratracheal infection enhanced fungal clearance from lung tissue and prevented dissemination to the brain, but late treatment with IL-12 was not effective in modifying the course of infection (122). Since IL-12 administration enhances lung IFN-γ production, this effect has been interpreted as suggesting that early IL-12 therapy enhances a TH1 response (122). The lack of efficacy associated with late IL-12 administration on the course of infection has been attributed to previous differentiation of the T-cell response such that it would not be affected by exogenous IL-12 administration (122). Human PBMCs produce IL-12 in response to challenge with C. neoformans (92). PBMCs from HIV-positive individuals produce significantly less IL-12 than do those from HIV-negative individuals, suggesting that HIV infection produces defects in IL-12 expression (92).

In human pulmonary C. neoformans infection, IL-12 production by alveolar macrophages may be an important early signal for an appropriate cell-mediated response. Incubation of human PBMCs with C. neoformans elicits IL-12 secretion (92). Comparison of PBMCs from HIV-positive and HIV-negative donors revealed that the cells from seropositive donors released significantly less IL-12 (92). This defect in IL-12 secretion by PBMCs from HIV-positive donors was not found at the level of mRNA expression, suggesting a posttranscriptional effect in NK cells as a consequence of HIV infection. Anticryptococcal activity of NK cells from patients with HIV infection can be restored in vitro by addition of IL-12 (104). A deficiency in IL-12 may be responsible for the diminution of NK cell activity against C. neoformans in patients with HIV infection (104).

IL-18
IL-18, which was originally known as IFN-γ-inducing factor, has been shown to work synergistically with IL-12 to enhance mouse leukocyte antifungal activity (254). The mechanism may be mediated by stimulating NK cells to produce IFN-γ, which in turn enhances nitric oxide production by mouse effector cells (254).

Colony-stimulating factors G-CSF and GM-CSF
Human recombinant granulocyte colony-stimulating factor (rhG-CSF) enhances the fungicidal activity of human neutrophils against C. neoformans (236). The effect appears to be due to enhanced superoxide production by neutrophils (236). rhG-CSF administration has been shown to prolong survival and reduce fungal burden in mice and to enhance the therapeutic effect of fluconazole in murine cryptococcal meningitis (77). GM-CSF has been shown to cooperate with TNF-α in promoting C3 receptor CR3 expression on mouse cells, a necessary event for efficient complement-mediated phagocytosis (33, 35). GM-CSF also enhances the fungistatic activity of murine alveolar macrophages against C. neoformans (191). Furthermore, the combination of GM-CSF-activated macrophages and fluconazole produces synergistic effects against C. neoformans in vitro (16).

MCP-1

Monocyte chemotactic protein-1 (MCP-1) is a chemokine that has been demonstrated to perform a crucial role in recruitment of pulmonary monocytes and CD4[+] T cells during murine lung infection (111).

Effects of *C. neoformans* Components on Cytokine Production

In many cryptococcal infections there is a striking absence of an effective inflammatory response. This suggests infection-induced cytokine dysregulation. Several studies have shown that *C. neoformans* components can affect cytokine production, supporting the hypothesis that fungus-induced alterations in cytokine regulation are responsible for suboptimal inflammatory responses.

Microbial products such as lipopolysaccharide (LPS) can have profound stimulatory effects on cytokine expression. In contrast to LPS, *C. neoformans* capsular polysaccharide reduces the production of pro-inflammatory cytokines and elicits the production of cytokines that are associated with inhibitory effects on the immune system. Incubation of human monocytes with acapsular strains induces more TNF-α and IL-1β than does incubation with an encapsulated strain (190, 239). Purified *C. neoformans* capsular polysaccharide can inhibit TNF-α secretion from LPS-stimulated human monocytes (239). Furthermore, purified capsular polysaccharide can induce production of IL-10 from human monocytes (238). IL-10 produces a dose-dependent inhibition of TNF-α production by human monocytes in response to either *C. neoformans* or LPS, an effect that appears to occur at the level of transcription for TNF-α and both transcriptionally and posttranslationally for IL-β (141). Since IL-10 is a potent down-regulator of TNF-α and IL-1β, this observation suggests that some of the immunosuppressive effects associated with polysaccharide may be a result of inhibitory cytokine secretion (238). However, *C. neoformans* capsular polysaccharide may also induce the release of pro-inflammatory cytokines from nonmonocyte phagocytic cells (207). Incubation of human neutrophils with encapsulated *C. neoformans* strains elicits the release of IL-1β, IL-6, IL-8, and TNF-α (207). This effect is related to the size of the capsule, with large capsular isolates releasing more cytokines than thinly encapsulated strains (207). Hence, the effect of capsular polysaccharide on cytokine expression appears to be complex and may be dependent on the type of effector cell that interacts with the polysaccharide antigen. Considering that the human cellular response to *C. neoformans* infection may include both monocytes and neutrophils, the net result of polysaccharide-mediated effects on cytokine release may be a function of the relative abundance of the various inflammatory cells at the site of infection.

Several components of the yeast cell can elicit release of TNF-α from human leukocytes, including GXM, galactoxylomannan, mannoprotein, and β(1→3) glucan (37). Of these, mannoprotein is the most potent inducer of TNF-α, and GXM is the weakest (37). Human leukocytes were observed to release TNF-α after incubation with small-capsule *C. neoformans* and serum, indicating that the availability of opsonins and the state of yeast cell encapsulation may be important variables for this effect (37, 140).

There is indirect evidence that *C. neoformans* melanin may affect the cytokine response to infection. Comparison of high- and low-melanin-producing strains of *C. neoformans* revealed differences in cytokine and inflammatory response in mice

(109). Incubation of murine alveolar macrophages with melanized *C. neoformans* resulted in inhibition of TNF-α production (109). The mechanism for this effect is not known.

Overview of Cytokine Responses to Cryptococcal Infection

Many cytokines have now been shown to have an important role in host defense against cryptococcal infection. The interplay between the various cytokines produced in response to *C. neoformans* infection remains poorly understood. Much of the information available on cytokine effects in vivo has been obtained in mice, and its exact relevance to human infection is uncertain. Further complicating the interpretation of animal studies are mouse strain differences; this raises the question of whether the information obtained in inbred mouse strains can be generalized. For humans, most of the information available has been obtained in in vitro studies that may or may not reflect in vivo conditions. Cytokine expression in response to *C. neoformans* infection appears to be influenced by fungal cell components, including shed capsular polysaccharide. Furthermore, different *C. neoformans* strains can elicit different tissue inflammatory responses (36), suggesting that the cytokine profile in response to infection may vary with the cryptococcal strain. Hence, the cytokine response to *C. neoformans* infection is likely to depend on the genetic background of both the host and the pathogen. The absence of inflammation in tissues infected with *C. neoformans* suggests cytokine dysregulation resulting in inappropriate cellular recruitment and activation. Cytokine studies remain an exciting area of investigation, because a better understanding of cytokine responses and their effect on host immunity is likely to enhance our understanding of pathogenesis and may translate into opportunities for immunotherapy.

REFERENCES

1. **Abrahams, I.** 1966. Further studies on acquired resistance to murine cryptococcosis: enhancing effect of *Bordetella pertussis*. *J. Immunol.* **96:**525–529.
2. **Abrahams, I., and T. G. Gilleran.** 1990. Studies on actively acquired resistance to experimental cryptococcosis in mice. *J. Immunol.* **86:**629–635.
3. **Adamson, D. M., and G. C. Cozad.** 1969. Effect of antilymphocyte serum on animals experimentally infected with *Histoplasma capsulatum* or *Cryptococcus neoformans*. *J. Bacteriol.* **100:**1271–1276.
4. **Aguirre, K., E. A. Havell, G. W. Gibson, and L. L. Johnson.** 1995. Role of tumor necrosis factor and gamma interferon in acquired resistance to *Cryptococcus neoformans* in the central nervous system of mice. *Infect. Immun.* **63:**1725–1731.
5. **Aguirre, K. M., and L. L. Johnson.** 1997. A role for B cells in resistance to *Cryptococcus neoformans* in mice. *Infect. Immun.* **65:**525–530.
5a. **Atkinson, A. J., and J. E. Bennett.** 1968. Experience with a new skin test antigen prepared from *Cryptococcus neoformans*. *Am. Rev. Respir. Dis.* **97:**637–643.
6. **Allen, J. E., and R. M. Maizels.** 1997. Th1-Th2: reliable paradigm or dangerous dogma. *Immunol. Today* **18:**387–392.
7. **Alspaugh, J. A., and D. L. Granger.** 1991. Inhibition of *Cryptococcus neoformans* replication by nitrogen oxide supports the role of these molecules as effectors of macrophage-mediated cystostasis. *Infect. Immun.* **59:**2291–2296.
8. **Aronson, M., and J. Kletter.** 1973. Aspects of the defense against a large-sized parasite, the yeast, *Cryptococcus neoformans*. *Isr. J. Med. Sci.* **1:**132–162.

9. **Baker, P. J.** 1990. Regulation of magnitude of antibody response to bacterial polysaccharide antigens by thymus-derived lymphocytes. *Infect. Immun.* **58:**3465–3468.

10. **Benham, R. W.** 1935. Cryptococci—their identification by morphology and by serology. *J. Infect. Dis.* **57:**255–274.

11. **Bennett, J. E.** 1981. Cryptococcal skin test antigen: preparation variables and characterization. *Infect. Immun.* **32:**373–380.

12. **Biron, C. A., and R. T. Gazzinelli.** 1995. Effects of IL-12 on immune responses to microbial infections: a key mediator in regulating disease outcome. **Curr. Opin. Immunol. 7:**485–496.

13. **Blasi, E., R. Barluzzi, R. Mazzola, L. Pitzurra, M. Puliti, S. Saleppico, and F. Bistoni.** 1995. Biomolecular events involved in anticryptococcal resistance in the brain. *Infect. Immun.* **63:**1218–1222.

14. **Blumer, S. O., and L. Kaufman.** 1977. Characterization of immunoglobulin classes of human antibodies to *Cryptococcus neoformans. Mycopathologia* **61:**55–60.

15. **Breen, J. F., I. C. Lee, F. R. Vogel, and H. Friedman.** 1982. Cryptococcal capsular polysaccharide-induced modulation of murine immune responses. *Infect. Immun.* **36:**47–51.

16. **Brummer, E., F. Nassar, and D. A. Stevens.** 1994. Effect of macrophage colony-stimulating factor on anticryptococcal activity of bronchoalveolar macrophages: synergy with fluconazole for killing. *Antimicrob. Agents Chemother.* **38:**2158–2161.

17. **Brummer, E., and D. A. Stevens.** 1994. Macrophage colony-stimulating factor induction of enhanced macrophage anticryptococcal activity: synergy with fluconazole for killing. *J. Infect. Dis.* **170:**173–179.

18. **Buchanan, K. L., and J. W. Murphy.** 1993. Characterization of cellular infiltrates and cytokine production during the expression of the anticryptococcal delayed-type hypersensitivity response. *Infect. Immun.* **61:**2854–2865.

19. **Buchanan, K. L., and J. W. Murphy.** 1994. Regulation of cytokine production during the expression phase of the anticryptococcal delayed-type hypersensitivity response. *Infect. Immun.* **62:**2930–2939.

20. **Casadevall, A.** 1995. Antibody immunity and invasive fungal infections. *Infect. Immun.* **63:**4211–4218.

21. **Casadevall, A., M. DeShaw, M. Fan, F. Dromer, T. R. Kozel, and L. Pirofski.** 1994. Molecular and idiotypic analysis of antibodies to *Cryptococcus neoformans* glucuronoxylomannan. *Infect. Immun.* **62:**3864–3872.

22. **Casadevall, A., J. Mukherjee, S. J. N. Devi, R. Schneerson, J. B. Robbins, and M. D. Scharff.** 1992. Antibodies elicited by a *Cryptococcus neoformans* glucuronoxylomannan-tetanus toxoid conjugate vaccine have the same specificity as those elicited in infection. *J. Infect. Dis.* **65:**1086–1093.

23. **Casadevall, A., and M. D. Scharff.** 1991. The mouse antibody response to infection with *Cryptococcus neoformans*: V_H and V_L usage in polysaccharide binding antibodies. J. Exp. Med. **174:**151–160.

24. **Casadevall, A., and M. D. Scharff.** 1994. "Serum therapy" revisited: animal models of infection and the development of passive antibody therapy. *Antimicrob. Agents Chemother.* **38:**1695–1702.

25. **Cauley, L. K., and J. W. Murphy.** 1979. Response of congenitally athymic (nude) and phenotypically normal mice to *Cryptococcus neoformans* infection. *Infect. Immun.* **23:**644–651.

26. **Chen, G.-H., J. L. Curtis, C. H. Mody, P. J. Christensen, L. R. Armstrong, and G. B. Toews.** 1994. Effect of granulocyte-macrophage colony stimulating factor on rat alveolar macrophage anticryptococcal activity in vitro. *J. Immunol.* **152:**724–734.

27. **Chen, L.-C., E. Blank, and A. Casadevall.** 1996. Extracellular proteinase activity of *Cryptococcus neoformans. Clin. Diagn. Lab. Immunol.* **3:**570–574.

28. **Chen, L.-C., L. Pirofski, and A. Casadevall.** 1997. Extracellular proteins of *Cryptococcus neoformans* and host antibody response. *Infect. Immun.* **65:**2599–2605.

29. **Chen, S. C. A., L. C. Wright, R. T. Santangelo, M. Muller, V. R. Moran, P. W. Kuchel, and T. C. Sorrell.** 1997. Identification of extracellular phospholipase B, lysophospholipase, and acyltransferase produced by *Cryptococcus neoformans*. *Infect. Immun.* **65:**405–411.

30. **Cherniak, R., L. C. Morris, T. Belay, E. D. Spitzer, and A. Casadevall.** 1995. Variation in the structure of glucuronoxylomannan in isolates from patients with recurrent cryptococcal meningitis. *Infect. Immun.* **63:**1899–1905.

31. **Cherniak, R., and J. B. Sundstrom.** 1994. Polysaccharide antigens of the capsule of *Cryptococcus neoformans*. *Infect. Immun.* **62:**1507–1512.

32. **Clemons, K. V., E. Brummer, and D. A. Stevens.** 1994. Cytokine treatment of central nervous system infection: efficacy of interleukin-12 alone and synergy with conventional antifungal therapy in experimental cryptococcosis. *Antimicrob. Agents Chemother.* **38:**460–464.

33. **Collins, H. L., and G. J. Bancroft.** 1992. Cytokine enhancement of complement-dependent phagocytosis by macrophages: synergy of tumor necrosis-alpha and granulocyte-macrophage colony stimulating factor for phagocytosis of *Cryptococcus neoformans*. *Eur. J. Immunol.* **22:**1447–1454.

34. **Cross, C. E., and G. J. Bancroft.** 1995. Ingestion of acapsular *Cryptococcus neoformans* occurs via mannose and beta-glucan receptors, resulting in cytokine production and increased phagocytosis of the encapsulated form. *Infect. Immun.* **63:**2604–2611.

35. **Cross, C. E., H. L. Collins, and G. J. Bancroft.** 1997. CR3-dependent phagocytosis by murine macrophages: different cytokines regulate ingestion of a defined CR3 ligand and complement-opsonized *Cryptococcus neoformans*. *Immunology* **91:**289–296.

36. **Curtis, J. L., G. B. Huffnagle, G. H. Chen, M. L. Warnock, M. Gyetko, R. McDonald, P. Scott, and G. B. Toews.** 1994. Experimental murine pulmonary cryptococcosis. *Lab. Invest.* **71:**113–126.

37. **Delfino, D., L. Cianci, M. Migliardo, G. Mancuso, V. Cusumano, C. Corradini, and G. Teti.** 1996. Tumor necrosis factor-inducing activities of *Cryptococcus neoformans* components. *Infect. Immun.* **64:**5199–5204.

38. **DeShaw, M., and L.-A. Pirofski.** 1995. Antibodies to the *Cryptococcus neoformans* capsular glucuronoxylomannan are ubiquitous in serum from HIV+ and HIV– individuals. *Clin. Exp. Immunol.* **99:**425–432.

39. **Devi, S. J. N.** 1996. Preclinical efficacy of a glucuronoxylomannan-tetanus toxoid conjugate vaccine of *Cryptococcus neoformans* in a murine model. *Vaccine* **14:**841–842.

40. **Devi, S. J. N., R. Schneerson, W. Egan, T. J. Ulrich, D. Bryla, J. B. Robbins, and J. E. Bennett.** 1991. *Cryptococcus neoformans* serotype A glucuronoxylomannan-protein conjugate vaccines: synthesis, characterization, and immunogenicity. *Infect. Immun.* **59:**3700–3707.

41. **Diamond, R. D., and J. E. Bennett.** 1973. Disseminated cryptococcosis in man: decreased lymphocyte transformation in response to *Cryptococcus neoformans*. *J. Infect. Dis.* **127:**694–697.

42. **Diamond, R. D., and J. E. Bennett.** 1974. Prognostic factors in cryptococcal meningitis. *Ann. Intern. Med.* **80:**176–181.

43. **Dromer, F., P. Aucouturier, J.-P. Clauvel, G. Saimot, and P. Yeni.** 1988. Cryptococcus neoformans antibody levels in patients with AIDS. *Scand. J. Infect. Dis.* **20:**283–285.

44. **Dromer, F., and J. Charreire.** 1991. Improved amphotericin B activity by a monoclonal anti-*Cryptococcus neoformans* antibody: study during murine cryptococcosis and mechanisms of action. *J. Infect. Dis.* **163:**1114–1120.

45. **Dromer, F., J. Charreire, A. Contrepois, C. Carbon, and P. Yeni.** 1987. Protection of

mice against experimental cryptococcosis by anti-*Cryptococcus neoformans* monoclonal antibody. *Infect. Immun.* **55:**749–752.

46. **Dromer, F., D. W. Denning, D. A. Stevens, A. Nobel, and J. R. Hamilton.** 1995. Anti-*Cryptococcus neoformans* antibodies during cryptococcosis in patients with the acquired immunodeficiency syndrome. *Serodiagn. Immunother. Infect. Dis.* **7:**181–188.

47. **Dromer, F., C. Perrone, J. Barge, J. L. Vilde, and P. Yeni.** 1989. Role of IgG and complement component C5 in the inital course of experimental cryptococcosis. *Clin. Exp. Immunol.* **78:**412–417.

48. **Dromer, F., J. Salamero, A. Contrepois, C. Carbon, and P. Yeni.** 1987. Production, characterization, and antibody specificity of a mouse monoclonal antibody reactive with *Cryptococcus neoformans* capsular polysaccharide. *Infect. Immun.* **55:**742–748.

49. **Dromer, F., P. Yeni, and J. Charreire.** 1988. Genetic control of the humoral response to cryptococcal capsular polysaccharide in mice. *Immunogenetics* **28:**417–424.

50. **Duke, S. S., and R. A. Fromtling.** 1984. Effects of diethylstilbestrol and cyclophosphamide on the pathogenesis of experimental *Cryptococcus neoformans* infections. *J. Med. Vet. Mycol.* **22:**125–135.

51. **Evans, E. D., and J. F. Kessel.** 1951. The antigenic composition of *Cryptococcus neoformans*. *J. Immunol.* **67:**109–114.

52. **Evans, E. E.** 1949. An immunologic comparison of twelve strains of *Cryptococcus neoformans* (*Torula histolytica*). *Proc. Soc. Exp. Biol. Med.* **71:**644–646.

53. **Evans, E. E.** 1950. The antigenic composition of *Cryptococcus neoformans*. I. A serologic classification by means of the capsular and agglutination reactions. *J. Immunol.* **64:**423–430.

54. **Feldmesser, M., and A. Casadevall.** 1997. Effect of serum IgG1 against murine pulmonary infection with *Cryptococcus neoformans*. *J. Immunol.* **158:**790–799.

55. **Feldmesser, M., A. Casadevall, Y. Kress, G. Spira, and A. Orlofski.** 1997. Eosinophil-*Cryptococcus neoformans* interactions in vivo and in vitro. *Infect. Immun.* **65:**1899–1907.

56. **Feldmesser, M., J. Mukherjee, and A. Casadevall.** 1996. Combination of 5-flucytosine and capsule binding monoclonal antibody in therapy of murine *Cryptococcus neoformans* infections and *in vitro*. *J. Antimicrob. Chemother.* **37:**617–622.

57. **Fink, J. N., J. J. Barboriak, and L. Kaufman.** 1968. Cryptococcal antibodies in pigeon breeder's disease. *J. Allergy* **41:**297–301.

58. **Flesch, I. E. A., G. Schwamberger, and S. H. E. Kaufman.** 1989. Fungicidal activity of IFN-gamma activated macrophages. *J. Immunol.* **142:**3219–3224.

59. **French, D. L., R. Laskov, and M. D. Scharff.** 1989. The role of somatic hypermutation in the generation of antibody diversity. *Science* **244:**1152–1157.

60. **Fresno, M., M. Kopf, and L. Rivas.** 1997. Cytokines and infectious diseases. *Immunol. Today* **18:**56–58.

61. **Fung, P. Y. S., and J. W. Murphy.** 1982. In vitro interactions of immune lymphocytes and *Cryptococcus neoformans*. *Infect. Immun.* **36:**1128–1138.

62. **Gadebusch, H. H.** 1958. Passive immunization against *Cryptococcus neoformans*. *Proc. Soc. Exp. Biol. Med.* **98:**611–614.

63. **Gadebusch, H. H.** 1958. Active immunization against *Cryptococcus neoformans*. *J. Infect. Dis.* **102:**219–226.

65. **Goldman, D., S. C. Lee, and A. Casadevall.** 1994. Pathogenesis of pulmonary *Cryptococcus neoformans* infection in the rat. *Infect. Immun.* **62:**4755–4761.

66. **Goldman, D. L., A. Casadevall, Y. Cho, and S. C. Lee.** 1996. *Cryptococcus neoformans* meningitis in the rat. *Lab. Invest.* **75:**759–770.

67. **Goldman, D. L., S. C. Lee, and A. Casadevall.** 1995. Tissue localization of *Cryptococcus neoformans* glucuronoxylomannan in the presence and absence of specific antibody. *Infect. Immun.* **63:**3448–3453.

68. **Gordon, M. A., and A. Casadevall.** 1995. Serum therapy of cryptococcal meningitis. *Clin. Infect. Dis.* **21:**1477–1479.

69. **Gordon, M. A., and E. Lapa.** 1964. Serum protein enhancement of antibiotic therapy in cryptococcosis. *J. Infect. Dis.* **114:**373–378.

70. **Gordon, M. A., and E. Lapa.** 1971. Charcoal particle agglutination test for detection of antibody to *Cryptococcus neoformans*. *Am. J. Clin. Pathol.* **56:**354–359.

71. **Goren, M. B.** 1967. Experimental murine cryptococcosis: effect of hyperimmunization to capsular polysaccharide. *J. Immunol.* **98:**914–922.

72. **Goren, M. B., and G. M. Middlebrook.** 1967. Protein conjugates of polysaccharide from *Cryptococcus neoformans*. *J. Immunol.* **98:**901–913.

73. **Granger, D. L., J. B. Hibbs, J. R. Perfect, and D. T. Durack.** 1988. Specific amino acid (L-arginine) requirement for the microbiostatic activity of murine macrophages. *J. Clin. Invest.* **81:**1129–1136.

74. **Graybill, J. R., and J. Ahrens.** 1981. Immunization and complement interaction in host defense against murine cryptococcosis. *RES J. Reticuloendothel. Soc.* **30:**347–357.

75. **Graybill, J. R., J. Ahrens, T. Nealon, and R. Paque.** 1983. Pulmonary cryptococcosis in the rat. *Am. Rev. Respir. Dis.* **127:**636–640.

76. **Graybill, J. R., and R. H. Alford.** 1974. Cell-mediated immunity in cryptococcosis. *Cell. Immunol.* **14:**12–21.

77. **Graybill, J. R., R. Bocanegra, C. Lambros, and M. F. Luther.** 1997. Granulocyte colony stimulating factor therapy of experimental cryptococcal meningitis. *J. Med. Vet. Mycol.* **35:**243–247.

78. **Graybill, J. R., P. C. Craven, L. F. Mitchell, and D. J. Drutz.** 1978. Interaction of chemotherapy and immune defences in experimental murine cryptococcosis. *Antimicrob. Agents Chemother.* **14:**659–667.

79. **Graybill, J. R., and D. J. Drutz.** 1978. Host defense in cryptococcosis. II. Cryptococcosis in the nude mouse. *Cell. Immunol.* **40:**263–274.

80. **Graybill, J. R., M. Hague, and D. J. Drutz.** 1981. Passive immunization in murine cryptococcosis. *Sabouraudia* **19:**237–244.

81. **Graybill, J. R., and L. Mitchell.** 1978. Cyclophosphamide effects on murine cryptococcosis. *Infect. Immun.* **21:**674–677.

82. **Graybill, J. R., and L. Mitchell.** 1979. Host defense in cryptococcosis. III. In vivo alteration of immunity. *Mycopathologia* **69:**171–178.

83. **Graybill, J. R., L. Mitchell, and D. J. Drutz.** 1979. Host defense in cryptococcosis. III. Protection of nude mice by thymus transplantation. *J. Infect. Dis.* **140:**546–552.

84. **Graybill, J. R., D. C. Straus, T. J. Nealon, M. Hague, and R. E. Paque.** 1982. Immunogenic fractions of *Cryptococcus neoformans*. *Mycopathologia* **78:**31–39.

85. **Graybill, J. R., and R. L. Taylor.** 1978. Host defense in cryptococcosis. I. An *in vivo* model for evaluating immune response. *Int. Arch. Allergy Appl. Immunol.* **57:**101–113.

86. **Griffin, F. M.** 1980. Roles of macrophage Fc and C3b receptors in phagocytosis of immunologically coated *Cryptococcus neoformans*. *Proc. Natl. Acad. Sci. USA* **78:**3853–3857.

87. **Gupta, S., M. Ellis, T. Cesario, M. Ruhling, and B. Vayuvegula.** 1987. Disseminated cryptococcal infection in a patient with hypogammaglobulinemia and normal T-cell function. *Am. J. Med.* **82:**129–131.

88. **Hall, N. K., K. C. Maluf, and R. Blackstock.** 1984. Functional testing and chemical composition of cryptococcal extracts. *Sabouraudia* **22:**439–442.

89. **Hamilton, A. J., J. I. Figueroa, L. Jeavons, and R. A. Seaton.** 1997. Recognition of cytoplasmic yeast antigens of *Cryptococcus neoformans* and *Cryptococcus neoformans* var. *gattii* by immune human sera. *FEMS Immunol. Med. Microbiol.* **17:**111–119.

90. **Hamilton, A. J., and J. Goodley.** 1993. Purification of the 115-kilodalton exoantigen of *Cryptococcus neoformans* and its recognition by immune sera. *J. Clin. Microbiol.* **31:**335–339.

91. **Harrison, T. S., H. Kornfeld, and S. M. Levitz.** 1995. The effect of infection with human immunodeficiency virus on the anticryptococcal activity of lymphocytes and monocytes. *J. Infect. Dis.* **172:**665–671.

92. **Harrison, T. S., and S. M. Levitz.** 1996. Role of IL-12 in peripheral blood mononuclear cells responses to fungi in persons with and without HIV infection. *J. Immunol.* **156:**4492–4497.

93. **Harrison, T. S., and S. M. Levitz.** 1997. Priming with IFN-gamma restores deficient IL-12 production by peripheral blood mononuclear cells from HIV-seropositive donors. *J. Immunol.* **158:**459–463.

94. **Hay, R. J., and E. Reiss.** 1978. Delayed-type hypersensitivity responses in infected mice elicited by cytoplasmic fractions of *Cryptococcus neoformans. Infect. Immun.* **22:**72–79.

95. **Henderson, D. K., J. E. Bennett, and M. A Huber.** 1982. Long-lasting, specific immunologic unresponsiveness associated with cryptococcal meningitis. *J. Clin. Invest.* **69:**1185–1190.

96. **Herrman, J. L., N. Dubois, M. Fourgeaud, D. Basset, and P. H. Lagrange.** 1994. Synergic inhibitory activity of amphotericin-B and gamma interferon against intracellular *Cryptococcus neoformans* in murine macrophages. *J. Antimicrob. Chemother.* **34:**1051–1058.

97. **Hiemstra, P. S., P. B. Eisenhauer, L. S. Harwig, M. T. van den Barselaar, R. van Furth, and R. I. Lehrer.** 1993. Antimicrobial proteins of murine macrophages. *Infect. Immun.* **61:**3038–3046.

98. **Hill, J. O.** 1992. CD4+ T cells cause multinucleated giant cells to form around *Cryptococcus neoformans* and confine the yeast within the primary site of infection in the respiratory tract. *J. Exp. Med.* **175:**1685–1695.

99. **Hill, J. O., and K. M. Aguirre.** 1994. CD4+ T cell-dependent acquired state of immunity that protects the brain against *Cryptococcus neoformans. J. Immunol.* **152:**2344–2350.

100. **Hoag, K. A., M. F. Lipscomb, A. A. Izzo, and N. E. Street.** 1997. IL-12 and IFN-gamma are required for initiating the protective Th1 response to pulmonary cryptococcosis in resistant C.B-17 mice. *Am. J. Respir. Cell. Mol. Biol.* **17:**733–739.

101. **Hoag, K. A., N. E. Street, G. B. Huffnagle, and M. F. Lipscomb.** 1995. Early cytokine production in pulmonary *Cryptococcus neoformans* infections distinguishes susceptible and resistant mice. *Am. J. Respir. Cell. Mol. Biol.* **13:**487–495.

102. **Hobbs, M. M., J. R. Perfect, D. L. Granger, and D. T. Durack.** 1990. Opsonic activity of cerebrospinal fluid in experimental cryptococcal meningitis. *Infect. Immun.* **58:**2115–2119.

103. **Hoff, C. L.** 1942. Immunity studies of *Cryptococcus hominis* (*Torula histolytica*) in mice. *J. Lab. Clin. Med.* **27:**751–754.

104. **Horn, C. A., and R. G. Washburn.** 1995. Anticryptococcal activity of NK cell-enriched peripheral blood lymphocytes from human immunodeficiency virus-infected subjects: responses to interleukin-2, interferon-gamma, and interleukin-12. *J. Infect. Dis.* **172:**1023–1027.

105. **Houpt, D. C., G. S. T. Pfrommer, B. J. Young, T. A. Larson, and T. R. Kozel.** 1994. Occurrences, immunoglobulin classes, and biological activities of antibodies in normal human serum that are reactive with *Cryptococcus neoformans* glucuronoxylomannan. *Infect. Immun.* **62:**3857–2864.

106. **Hoy, J. F., D. E. Lewis, and G. G. Miller.** 1988. Functional versus phenotypic analysis of T cells in subjects seropositive for the human immunodeficiency virus: a prospective study of in vitro responses to *Cryptococcus neoformans. J. Infect. Dis.* **158:**1071–1078.

107. **Hoy, J. F., J. W. Murphy, and G. G. Miller.** 1989. T cell response to soluble cryptococcal antigens after recovery from cryptococcal infection. *J. Infect. Dis.* **159:**116–119.

108. **Huffnagle, G. B.** 1996. Role of cytokines in T cell immunity to a pulmonary *Cryptococcus neoformans* infection. *Biol. Signals* **5**:215–222.

109. **Huffnagle, G. B., G.-H. Chen, J. L. Curtis, R. A. McDonald, R. M. Strieter, and G. B. Toews.** 1995. Down-regulation of the afferent phase of T cell-mediated pulmonary inflammation and immunity by a high melanin-producing strain of *Cryptococcus neoformans*. *J. Immunol.* **155**:3507–3516.

110. **Huffnagle, G. B., M. F. Lipscomb, J. A. Lovchik, K. A. Hoag, and N. E. Street.** 1994. The role of CD4+ and CD8+ T-cells in protective inflammatory response to a pulmonary cryptococcal infection. *J. Leukocyte Biol.* **55**:35–42.

111. **Huffnagle, G. B., R. M. Strieter, T. J. Standiford, R. A. McDonald, M. D. Burdick, S. L. Kunkel, and G. B. Toews.** 1995. The role of monocyte chemotactic protein-1 (MCP-1) in the recruitment of monocytes and CD4+ T cells during a pulmonary *Cryptococcus neoformans* infection. *J. Immunol.* **155**:4790–4797.

112. **Huffnagle, G. B., G. B. Toews, M. D. Burdick, M. B. Boyd, K. S. McAllister, R. A. McDonald, S. L. Kunkel, and R. M. Strieter.** 1996. Afferent phase production of TNF-alpha is required for the development of protective T cell immunity to *Cryptococcus neoformans*. *J. Immunol.* **157**:4529–4539.

113. **Huffnagle, G. B., J. L. Yates, and M. F. Lipscomb.** 1991. T cell-mediated immunity in the lung: a *Cryptococcus neoformans* pulmonary infection model using SCID and athymic nude mice. *Infect. Immun.* **59**:1423–1433.

114. **Huffnagle, G. B., J. L. Yates, and M. F. Lipscomb.** 1991. Immunity to pulmonary *Cryptococcus neoformans* infection requires both CD4+ and CD8+ T cells. *J. Exp. Med.* **173**:793–800.

115. **Iseki, M., M. Anzo, N. Yamashita, and N. Matsuo.** 1994. Hyper-IgM immunodeficiency with disseminated cryptococcosis. *Acta Paediatr.* **83**:780–782.

116. **Joly, V., L. Saint-Julien, C. Carbon, and P. Yeni.** 1994. In vivo activity of interferongamma in combination with amphotericin B in the treatment of experimental cryptococcosis. *J. Infect. Dis.* **170**:1331–1334.

117. **Kakeya, H., H. Udono, N. Ikuno, Y. Yamamoto, K. Mitsutake, T. Miyazaki, K. Tomono, H. Koga, T. Tashiro, E. Nakayama, and S. Kohno.** 1997. A 77-kilodalton protein of *Cryptococcus neoformans*, a member of the heat shock protein 70 family, is a major antigen detected in the sera of mice with pulmonary cryptococcosis. *Infect. Immun.* **65**:1653–1658.

118. **Kawakami, K., S. Kohno, J. -I. Kadota, M. Tohyama, K. Teruya, N. Kedeken, A. Saito, and K. Hara.** 1995. T cell-dependent activation of macrophages and enhancement of their phagocytic activity in the lungs of mice inoculated with heat-killed *Cryptococcus neoformans*: involvement of IFN-gamma and its protective effect against cryptococcal infection. *Microbiol. Immunol.* **39**:135–143.

119. **Kawakami, K., X. Qifeng, M. Tohyama, M. H. Quereshi, and A. Saito.** 1996. Contribution of tumour necrosis factor-alpha (TNF-α) in host defense mechanism against *Cryptococcus neoformans*. *Clin. Exp. Immunol.* **106**:468–474.

120. **Kawakami, K., M. Tohyama, X. Qifeng, and A. Saito.** 1997. Expression of cytokines and inducible nitric oxide synthase mRNA in the lungs of mice infected with *Cryptococcus neoformans*: effects of interleukin-12. *Infect. Immun.* **65**:1307–1312.

121. **Kawakami, K., M. Tohyama, K. Teruya, N. Kedeken, Q. Xie, and A. Saito.** 1996. Contribution of interferon-gamma in protecting mice during pulmonary and disseminated infection with *Cryptococcus neoformans*. *FEMS Immunol. Med. Microbiol.* **13**:133–140.

122. **Kawakami, K., M. Tohyama, Q. Xie, and A. Saito.** 1996. IL-12 protects mice against pulmonary and disseminated infection caused by *Cryptococcus neoformans*. *Clin. Exp. Immunol.* **104**:208–214.

123. **Keller, R. G., G. S. Pfrommer, and T. R. Kozel.** 1994. Occurrences, specificities, and

functions of ubiquitous antibodies in human serum that are reactive with the *Cryptococcus neoformans* cell wall. *Infect. Immun.* **62**:215–220.

123a. **Kimball, H. R., H. F. Hasenclever, and S. M. Wolff.** 1967. Detection of circulating antibody in human cryptococcosis by means of a bentonite flocculation technique. *Am. Rev. Respir. Dis.* **95**:631–637.

124. **Kligman, A. M.** 1947. Studies of the capsular substance of torula histolytica and the immunologic properties of torula cells. *J. Immunol.* **57**:395–401.

125. **Kong, Y., and H. B. Levine.** 1967. Experimentally induced immunity in the mycoses. *Bacteriol. Rev.* **31**:35–53.

126. **Kozel, T. R., and J. Cazin, Jr.** 1974. Induction of humoral antibody response by soluble polysaccharide of *Cryptococcus neoformans*. *Mycopathol. Mycol. Appl.* **54**:21–30.

127. **Kozel, T. R., W. F. Gulley, and J. Cazin, Jr.** 1977. Immune response to *Cryptococcus neoformans* soluble polysaccharide: immunological unresponsiveness. *Infect. Immun.* **18**:701–707.

128. **Kristensson, K., S. M. Glaser, S. L. Zebedee, W. D. Huse, A. Casadevall, and M. D. Scharff.** 1995. Humanization of a murine antibody against *Cryptococcus neoformans* polysaccharide using a novel approach, p. 39–43. *In Vaccines 95.* Cold Spring Harbor Laboratory Press, Cold Spring Harbor, N.Y.

129. **Kyong, C. U., G. Virella, H. H. Fudenberg, and C. P. Darby.** 1978. X-linked immunodeficiency with increased IgM: clinical, ethnic, and immunologic heterogeneity. *Pediatr. Res.* **12**:1024–1026.

130. **La Mantia, L., A. Salmaggi, L. Tajoli, D. Cerrato, E. Lamperti, A. Nespolo, and G. Bussone.** 1986. Cryptococcal meningoencephalitis: intrathecal immunological response. *J. Neurol.* **233**:362–366.

131. **Lee, S. C., Y. Kress, D. W. Dickson, and A. Casadevall.** 1995. Human microglia mediate anti-*Cryptococcus neoformans* activity in the presence of specific antibody. *J. Neuroimmunol.* **16**:152–161.

132. **Lee, S. L., Y. Kress, M.-L. Zhao, D. W. Dickson, and A. Casadevall.** 1995. *Cryptococcus neoformans* survive and replicate in spacious phagosomes in human microglia. *Lab. Invest.* **73**:871–879.

133. **Levitz, S. M.** 1991. Activation of human peripheral blood mononuclear cells by interleukin-2 and granulocyte-macrophage colony-stimulating factor to inhibit *Cryptococcus neoformans*. *Infect. Immun.* **59**:3393–3397.

134. **Levitz, S. M., and M. P. Dupont.** 1993. Phenotypic and functional characterization of human lymphocytes activated by interleukin-2 to directly inhibit growth of *Cryptococcus neoformans* in vitro. *J. Clin. Invest.* **91**:1490–1498.

135. **Levitz, S. M., M. P. Dupont, and E. H. Smail.** 1994. Direct activity of human T lymphocytes and natural killer cells against *Cryptococcus neoformans*. *Infect. Immun.* **62**:194–202.

136. **Levitz, S. M., T. P. Farrell, and R. T. Maziarz.** 1991. Killing of *Cryptococcus neoformans* by human peripheral blood mononuclear cells stimulated in culture. *J. Infect. Dis.* **163**:1108–1113.

137. **Levitz, S. M., H. L. Mathews, and J. W. Murphy.** 1995. Direct antimicrobial activity by T cells. *Immunol. Today* **16**:387–391.

138. **Levitz, S. M., and E. A. North.** 1997. Lymphopriliferation and cytokine profiles in human peripheral blood mononuclear cells stimulated by *Cryptococcus neoformans*. *J. Med. Vet. Mycol.* **35**:229–236.

139. **Levitz, S. M., E. A. North, M. Dupont, and T. S. Harrison.** 1995. Mechanisms of inhibition of *Cryptococcus neoformans* by human lymphocytes. *Infect. Immun.* **63**:3550–3554.

140. **Levitz, S. M., A. Tabumi, H. Kornfeld, C. C. Reardon, and D. T. Golenbock.** 1994. Production of tumor necrosis factor alpha in human leukocytes stimulated by *Cryptococcus neoformans*. *Infect. Immun.* **62**:1975–1981.

141. **Levitz, S. M., A. Tabuni, S.-H. Nong, and D. T. Golenbock.** 1996. Effects of interleukin-10 on human peripheral blood mononuclear cell responses to *Cryptococcus neoformans, Candida albicans,* and lipopolysaccharide. *Infect. Immun.* **64**:945–961.

142. **Lim, T. S., and J. W. Murphy.** 1980. Transfer of immunity to cryptococcosis by T-enriched splenic lymphocytes from *Cryptococcus neoformans*-sensitized mice. *Infect. Immun.* **30**:5–11.

143. **Lim, T. S., J. W. Murphy, and L. K Cauley,.** 1980. Host-etiological agent interactions in intranasally and intraperitoneally induced cryptococcosis in mice. *Infect. Immun.* **29**:633–641.

144. **Littman, M. L.** 1959. Cryptococcosis (torulosis). *Am. J. Med.* **27**:976–998.

145. **Littman, M. L.** 1968. Cryptococcosis: current status. *Am. J. Med.* **45**:922–932.

146. **Louria, D. B., and T. Kaminski.** 1965. Passively-acquired immunity in experimental cryptococcosis. *Sabouraudia* **4**:80–84.

147. **Lovchik, J. A., C. R. Lyons, and M. F. Lipscomb.** 1995. A role for gamma interferon-induced nitric oxide in pulmonary clearance of *Cryptococcus neoformans. Am. J. Respir. Cell. Mol. Biol.* **13**:116–124.

148. **Marquis, G., S. Montplaisir, M. Pelletier, S. Mousseau, and P. Auger.** 1985. Genetic resistance to murine cryptococcosis: increased susceptibility in the CBA/N XID mutant strain of mice. *Infect. Immun.* **47**:282–287.

149. **Mastroianni, C. M., M. Lichtner, F. Mengoni, P. Santopadre, V. Vullo, and S. Delia.** 1996. Marked activation of the tumour necrosis factor system in AIDS-associated cryptococcosis. *AIDS* **10**:1436–1438.

150. **Mazzola, R., R. Barluzzi, A. Brozzetti, J. R. Boelaert, T. Luna, S. Saleppico, F. Bistoni, and E. Blasi.** 1997. Enhanced resistance to *Cryptococcus neoformans* infection induced by chloroquine in a murine model of meningoencephalitis. *Antimicrob. Agents Chemother.* **41**:802–807.

151. **McManus, E. J., and J. M. Jones.** 1985. Detection of a *Trichosporon beigelii* antigen cross-reactive with *Cryptococcus neoformans* capsular polysaccharide in serum from a patient with disseminated *Trichosporon* infection. *J. Clin. Microbiol.* **21**:681–685.

152. **McNulty, A., J. M. Kaldor, A. M. McDonald, K. Baumgart, and D. A. Cooper.** 1994. Acquired immunodeficiency without evidence of HIV infection: national retrospective survey. *Br. Med. J.* **308**:825–826.

153. **Miller, G. P. G., and S. Kohl.** 1983. Antibody-dependent leukocyte killing of *Cryptococcus neoformans. J. Immunol.* **131**:1455–1459.

154. **Miller, G. P. G., and D. E. Lewis.** 1987. In vitro effect of cyclosporine on interleukin-2 receptor expression stimulated by *Cryptococcus neoformans. J. Infect. Dis.* **155**:799–802.

155. **Miller, G. P. G., and J. Puck.** 1984. *In vitro* human lymphocyte responses to *Cryptococcus neoformans.* Evidence for primary and secondary responses and normals and infected patients. *J. Immunol.* **133**:166–172.

156. **Miller, M. F., T. G. Mitchell, W. J. Storkus, and J. R. Dawson.** 1990. Human natural killer cells do not inhibit growth of *Cryptococcus neoformans* in the absence of antibody. *Infect. Immun.* **58**:639–645.

157. **Miyagawa, T., T. Ochi, and H. Takahashi.** 1978. Hypersensitivity pneumonitis with antibodies to *Cryptococcus neoformans. Clin. Allergy* **8**:501–509.

158. **Mody, C. H., G.-H. Chen, C. Jackson, J. L. Curtis, and G. B. Toews.** 1993. Depletion of murine CD8+ T-cells in vivo decreases pulmonary clearance of a moderately virulent strain of *Cryptococcus neoformans. J. Lab. Clin. Med.* **121**:765–773.

159. **Mody, C. H., G.-H. Chen, C. Jackson, J. L. Curtis, and G. B. Toews.** 1994. In vivo depletion of murine CD8 positive T cells impairs survival during infection with a highly virulent strain of *Cryptococcus neoformans. Mycopathologia* **125**:7–17.

160. **Mody, C. H., M. F. Lipscomb, N. E. Street, and G. B. Toews.** 1990. Depletion of CD4+

(L3T4$^+$) lymphocytes in vivo impairs murine host defense to *Cryptococcus neoformans.* *J. Immunol.* **144:**1472–1477.

161. **Mody, C. H., K. L. Sims, C. J. Wood, R. M. Syme, J. C. L. Spurrel, and M. M. Sexton.** 1996. Proteins in the cell wall and membrane of *Cryptococcus neoformans* stimulate lymphocytes from both adults and fetal cord blood to proliferate. *Infect. Immun.* **64:**4811–4819.

162. **Mody, C. H., and R. M. Syme.** 1993. Effect of polysaccharide capsule and methods of preparation on human lymphocyte proliferation in response to *Cryptococcus neoformans. Infect. Immun.* **61:**464–469.

163. **Mody, C. H., C. L. Tyler, R. G. Sitrin, C. Jackson, and G. B. Toews.** 1991. Interferon-gamma activates rat alveolar macrophages for anticryptococcal activity. *Am. J. Cell. Mol. Biol.* **5:**19–26.

164. **Mohr, J. A., H. G. Muchmore, F. G. Felton, E. R. Rhoades, and B. A. McKwon.** 1970. The effect of cryptococcin on leukocytic migration. *J. Infect. Dis.* **122:**454–458.

165. **Monari, C., C. Retini, B. Palazetti, F. Bistoni, and A. Vecchiarelli.** 1997. Regulatory role of exogenous IL-10 in the development of immune response versus *Cryptococcus neoformans. Clin. Exp. Immunol.* **109:**242–247.

166. **Monga, D. P., R. Kumar, L. N. Mahapatra, and A. N. Malaviya.** 1979. Experimental cryptococcosis in normal and B-cell-deficient mice. *Infect. Immun.* **26:**1–3.

167. **Mozaffarian, N., J. W. Berman, and A. Casadevall.** 1995. Immune complexes increase nitric oxide production by interferon-gamma-stimulated murine macrophage-like J774.16 cells. *J. Leukocyte Biol.* **57:**657–662.

168. **Mozaffarian, N., J. W. Berman, and A. Casadevall.** 1997. Enhancement of nitric oxide synthesis by macrophages represents an additional mechanism of action for amphotericin B. *Antimicrob. Agents Chemother.* **41:**1825–1829.

169. **Muchmore, H. G., F. G. Felton, S. B. Salvin, and E. R. Rhoades.** 1968. Delayed hypersensitivity to cryptococcin in man. *Sabouraudia* **6:**285–288.

170. **Mukherjee, J., A. Casadevall, and M. D. Scharff.** 1993. Molecular characterization of the antibody responses to *Cryptococcus neoformans* infection and glucuronoxylomannan-tetanus toxoid conjugate immunization. *J. Exp. Med.* **177:**1105–1106.

171. **Mukherjee, J., W. Cleare, and A. Casadevall.** 1995. Monoclonal antibody mediated capsular reactions (quellung) in *Cryptococcus neoformans. J. Immunol. Methods* **184:**139–143.

172. **Mukherjee, J., M. Feldmesser, M. D. Scharff, and A. Casadevall.** 1995. Monoclonal antibodies to *Cryptococcus neoformans* glucuronoxylomannan enhance fluconazole activity. *Antimicrob. Agents Chemother.* **39:**1398–1405.

173. **Mukherjee, J., G. Nussbaum, M. D. Scharff, and A. Casadevall.** 1995. Protective and non-protective monoclonal antibodies to *Cryptococcus neoformans* originating from one B-cell. *J. Exp. Med.* **181:**405–409.

174. **Mukherjee, J., L. Pirofski, M. D. Scharff, and A. Casadevall.** 1993. Antibody mediated protection in mice with lethal intracerebral *Cryptococcus neoformans* infection. *Proc. Natl. Acad. Sci. USA* **90:**3636–3640.

175. **Mukherjee, J., M. D. Scharff, and A. Casadevall.** 1992. Protective murine monoclonal antibodies to *Cryptococcus neoformans. Infect. Immun.* **60:**4534–4541.

176. **Mukherjee, J., M. D. Scharff, and A. Casadevall.** 1994. *Cryptococcus neoformans* infection can elicit protective antibodies in mice. *Can. J. Microbiol.* **40:**888–892.

177. **Mukherjee, J., M. D. Scharff, and A. Casadevall.** 1995. Variable efficacy of passive antibody administration against diverse *Cryptococcus neoformans* strains. *Infect. Immun.* **63:**3353–3359.

178. **Mukherjee, J., L. Zuckier, M. D. Scharff, and A. Casadevall.** 1994. Therapeutic efficacy of monoclonal antibodies to *Cryptococcus neoformans* glucuronoxylomannan alone and in combination with amphotericin B. *Antimicrob. Agents Chemother.* **38:**580–587.

179. **Mukherjee, S., M. Feldmesser, and A. Casadevall.** 1996. J774 murine macrophage-like cell interactions with *Cryptococcus neoformans* in the presence and absence of opsonins. *J. Infect. Dis.* **173**:1222–1231.

180. **Mukherjee, S., S. Lee, J. Mukherjee, M. D. Scharff, and A. Casadevall.** 1994. Monoclonal antibodies to *Cryptococcus neoformans* capsular polysaccharide modify the course of intravenous infection in mice. *Infect. Immun.* **62**:1079–1088.

181. **Mukherjee, S., S. C. Lee, and A. Casadevall.** 1995. Antibodies to *Cryptococcus neoformans* glucuronoxylomannan enhance antifungal activity of murine macrophages. *Infect. Immun.* **63**:573–579.

182. **Murphy, J. W.** 1988. Influence of cryptococcal antigens on cell-mediated immunity. *Rev. Infect. Dis.* **10**(Suppl. 2):S432–S435.

183. **Murphy, J. W.** 1993. Cytokine profiles associated with induction of the anticryptococcal cell-mediated immune response. *Infect. Immun.* **61**:4750–4759.

184. **Murphy, J. W., and G. C. Cozad.** 1972. Immunological unresponsiveness induced by cryptococcal polysaccharide assayed by the hemolytic plaque technique. *Infect. Immun.* **5**:896–901.

185. **Murphy, J. W., J. Gregory, and H. W. Larsh.** 1974. Skin testing of guinea pigs and footpad testing of mice with a new antigen for detecting delayed hypersensitivity to *Cryptococcus neoformans*. *Infect. Immun.* **9**:404–409.

186. **Murphy, J. W., M. R. Hidore, and S. C. Wong.** 1993. Direct interactions of human lymphocytes with the yeast-like organism *Cryptococcus neoformans*. *J. Clin. Invest.* **91**:1553–1566.

187. **Murphy, J. W., R. L. Mosley, R. Cherniak, G. H. Reyes, T. R. Kozel, and E. Reiss.** 1988. Serological, electrophoretic, and biological properties of *Cryptococcus neoformans* antigens. *Infect. Immun.* **56**:424–431.

188. **Muth, S. M., and J. W. Murphy.** 1995. Effects of immunization with *Cryptococcus neoformans* cells or cryptococcal culture filtrate antigen on direct anticryptococcal activities of murine T lymphocytes. *Infect. Immun.* **63**:1645–1651.

189. **Nabavi, N., and J. W. Murphy.** 1986. Antibody-dependent natural killer cell-mediated growth inhibition of *Cryptococcus neoformans*. *Infect. Immun.* **51**:556–562.

190. **Naslund, P. K., W. C. Miller, and D. L. Granger.** 1995. *Cryptococcus neoformans* fails to induce nitric oxide synthase in primed murine macrophage-like cells. *Infect. Immun.* **63**:1298–1304.

191. **Nassar, F., E. Brummer, and D. A. Stevens.** 1994. Effect of in vivo macrophage colony stimulating factor on fungistasis of bronchoalveolar and peritoneal macrophages against *Cryptococcus neoformans*. *Antimicrob. Agents Chemother.* **38**:2162–2164.

192. **Neill, J. M., I. Abrahams, and C. E. Kapros.** 1950. A comparison of the immunogenicity of weakly encapsulated and of strongly encapsulated strains of *Cryptococcus neoformans* (*Torula histolytica*). *J. Bacteriol.* **59**:263–275.

193. **Newberry, W. M., J. Walter, J. W. Chandler, and F. E. Tosh.** 1967. Epidemiologic study of *Cryptococcus neoformans*. *Ann. Intern. Med.* **67**:727–732.

194. **Nosanchuk, J. D., and A. Casadevall.** 1997. Cellular charge of *Cryptococcus neoformans*: contributions from the capsular polysaccharide, melanin, and monoclonal antibody binding. *Infect. Immun.* **65**:1836–1841.

195. **Nussbaum, G.** 1996. Protective determinants of antibodies to *Cryptococcus neoformans*. Ph.D. thesis. Albert Einstein College of Medicine, Bronx, N.Y.

196. **Nussbaum, G., R. Yuan, A. Casadevall, and M. D. Scharff.** 1996. Immunoglobulin G3 blocking antibodies to *Cryptococcus neoformans*. *J. Exp. Med.* **183**:1905–1909.

197. **Otteson, E. W., W. H. Welch, and T. R. Kozel.** 1994. Protein-polysaccharide interactions. A monoclonal antibody specific for the capsular polysaccharide of *Cryptococcus neoformans*. *J. Biol. Chem.* **269**:1858–1864.

198. **Papadimitriou, J. M., T. A. Robertson, Y. Kletter, M. Aronson, and M. N.-I. Walters.**

1978. An ultrastructural examination of the interaction between macrophages and *Cryptococcus neoformans*. *J. Pathol.* **124**:103–109.

199. **Perfect, J. R., D. L. Granger, and D. T. Durack.** 1987. Effects of antifungal agents and gamma-interferon on macrophage cytotoxicity for fungi and tumor cells. *J. Infect. Dis.* **156**:316–323.

200. **Perfect, J. R., S. D. R. Lang, and D. T. Durack.** 1980. Chronic cryptococcal meningitis. *Am. J. Pathol.* **101**:177–193.

201. **Perfect, J. R., S. D. R. Lang, and D. T. Durack.** 1981. Influence of agglutinating antibody in experimental cryptococcal meningitis. *Br. J. Exp. Pathol.* **62**:595–599.

202. **Pirofski, L., and A. Casadevall.** 1996. Antibody immunity to *Cryptococcus neoformans*: Paradigm for antibody immunity to the fungi? *Zentralbl. Bakteriol.* **284**:475–495.

203. **Pirofski, L., R. Lui, M. DeShaw, A. B. Kressel, and Z. Zhong.** 1995. Analysis of human monoclonal antibodies elicited by vaccination with a *Cryptococcus neoformans* glucuronoxylomannan capsular polysaccharide vaccine. *Infect. Immun.* **63**:3005–3014.

204. **Pitzurra, L., A. Vecchiarelli, R. Peducci, A. Cardinali, and F. Bistoni.** 1997. Identification of a 105 kilodalton *Cryptococcus neoformans* mannoprotein involved in human cell-mediated immune response. *J. Med. Vet. Mycol.* **35**:299–303.

205. **Procknow, J. J., J. R. Benfield, J. W. Rippon, C. F. Diener, and F. L. Archer.** 1965. Cryptococcal hepatitis presenting as a surgical emergency. First isolation of *Cryptococcus neoformans* from point source in Chicago. *JAMA* **191**:93–98.

206. **Rebers, P. A., S. A. Barker, M. Heidelberger, Z. Dische, and E. E. Evans.** 1958. Precipitation of the specific polysaccharide of *Cryptcococcus neoformans* A by types II and XIV antipneumococcal antisera. *J. Am. Chem. Soc.* **80**:1135–1137.

207. **Retini, C., A. Vecchiarelli, C. Monari, C. Tascini, F. Bistoni, and T. R. Kozel.** 1996. Capsular polysaccharide of *Cryptococcus neoformans* induces pro-inflammatory cytokine release by human neutrophils. *Infect. Immun.* **64**:2897–2903.

208. **Roilides, E., and P. A. Pizzo.** 1992. Modulation of host defenses by cytokines: evolving adjuncts in prevention and treatment of serious infections in immunocompromised hosts. *Clin. Infect. Dis.* **15**:508–524.

209. **Romagnani, S.** 1996. Understanding the role of the Th1/Th2 cells in infection. *Trends Microbiol.* **4**:470–473.

210. **Salkowski, C. A., and E. Balish.** 1991. Inflammatory responses to cryptococcosis in congenitally athymic mice. *J. Leukocyte Biol.* **49**:533–541.

211. **Salkowski, C. A., and E. Balish.** 1991. A monoclonal antibody to gamma interferon blocks augmentation of natural killer cell activity induced during systemic cryptococcosis. *Infect. Immun.* **59**:486–493.

212. **Salkowski, C. A., and E. Balish.** 1991. Susceptibility of congenitally immunodeficient mice to a nonencapsulated strain of *Cryptococcus neoformans*. *Can. J. Microbiol.* **37**:834–839.

213. **Salkowski, C. A., K. F. Bartizal, M. Balish, and E. Balish.** 1987. Colonization and pathogenesis of *Cryptococcus neoformans* in gnotobiotic mice. *Infect. Immun.* **55**:2000–2005.

214. **Salvin, S. B.** 1959. Current concepts of diagnostic serology and skin hypersensitivity in the mycoses. *Am. J. Med.* **27**:97–114.

215. **Salvin, S. B., and R. F. Smith.** 1961. An antigen for detection of hypersensitivity to *Cryptococcus neoformans* (26977). *Proc. Soc. Exp. Biol. Med.* **108**:498–501.

216. **Sanford, J. E., D. M. Lupan, A. M. Schlagetter, and T. R. Kozel.** 1990. Passive immunization against *Cryptococcus neoformans* with an isotype-switch family of monoclonal antibodies reactive with cryptococcal polysaccharide. *Infect. Immun.* **58**:1919–1923.

217. **Sarosi, G. A., J. D. Parker, I. L. Doto, and F. E. Tosh.** 1992. Amphotericin B in cryptococcal meningitis. *Ann. Intern. Med.* **71**:1079–1087.

218. **Schimpff, S. C., and J. E. Bennett.** 1975. Abnormalities in cell-mediated immunity in patients with *Cryptococcus neoformans* infection. *J. Allergy Clin. Immunol.* **55:**430–441.

219. **Schneerson-Porat, S., A. Sharar, and M. Aronson.** 1965. Formation of histiocyte rings in response to *Cryptococcus neoformans* infection. *RES J. Reticuloendothel. Soc.* **2:**249–255.

220. **Scott, E. N., H. G. Muchmore, and F. G. Felton.** 1981. Enzyme-linked immunosorbent assays in murine cryptococcosis. *Sabouraudia* **19:**257–265.

221. **Shahar, A., Y. Kletter, and M. Aronson.** 1969. Granuloma formation in cryptococcosis. *Isr. J. Med. Sci.* **5:**1164–1172.

222. **Shimazu, K., M. Ando, T. sakata, K. Yoshida, and S. Araki.** 1984. Hypersensitivity pneumonitis induced by *Trichosporon cutaneum. Am. Rev. Respir. Dis.* **130:**407–411.

223. **Sorensen, R. U., K. D. Boehm, D. Kaplan, and M. Berger.** 1992. Cryptococcal osteomyelitis and cellular immunodeficiency associated with interleukin-2 deficiency. *J. Pediatr.* **121:**873–879.

224. **Speed, B. R., J. Kaldor, B. Cairns, and M. Pegorer.** 1995. Serum antibody response to active infection with *Cryptococcus neoformans* and its varieties in immunocompetent hosts. *J. Med. Vet. Mycol.* **34:**187–193.

225. **Spira, G., M. Paizi, S. Mazar, G. Nussbaum, S. Mukherjee, and A. Casadevall.** 1996. Generation of biologically active anti-*Cryptococcus neoformans* IgG, IgE, and IgA isotype switch variant antibodies by acridine-orange mutagenesis. *Clin. Exp. Immunol.* **105:**436–442.

226. **Steinhoff, M. C., B. S. Auerbach, K. E. Nelson, D. Vlahov, R. L. Becker, N. M. Graham, D. H. Schwartz, A. H. Lucas, and R. E. Chaisson.** 1991. Antibody responses to *Haemophilus influenzae* type B vaccines in men with human immunodeficiency virus infection. *N. Engl. J. Med.* **325:**1837–1842.

227. **Stoddard, J. L., and E. C. Cutler.** 1916. *Torula Infection in Man.* Waverly Press, The Williams and Wilkins Co., Baltimore.

228. **Sundstrom, J. B., and R. Cherniak.** 1992. A glucuronoxylomannan of *Cryptococcus neoformans* serotype A is a type 2 T-cell-independent antigen. *Infect. Immun.* **60:**4080–4087.

229. **Sundstrom, J. B., and R. Cherniak.** 1993. T-cell dependent and T-cell-independent mechanisms of tolerance to glucuronoxylomannan of *Cryptococcus neoformans* serotype A. *Infect. Immun.* **61:**1340–1345.

230. **Syme, R. M., C. J. Wood, H. Wong, and C. H. Mody.** 1997. Both CD4+ and CD8+ human lymphocytes are activated and proliferate in response to *Cryptococcus neoformans. Immunology* **92:**194–200.

231. **Tabone, M.-C., G. Leverger, J. Landman, C. Aznar, L. Boccon-Gibod, and G. Lasfargues.** 1994. Disseminated lymphonodular cryptococcosis in a child with X-linked hyper-IgM immunodeficiency. *Pediatr. Infect. Dis. J.* **13:**77–79.

232. **Tohyama, M., K. Kawakami, M. Futenma, and A. Saito.** 1996. Enhancing effect of oxygen radical scavengers on murine macrophage anticryptococcal activity through production of nitric oxide. *Clin. Exp. Immunol.* **103:**436–441.

233. **Tohyama, M., K. Kawakami, and A. Saito.** 1996. Anticryptococcal effect of amphotericin B is mediated through macrophage production of nitric oxide. *Antimicrob. Agents Chemother.* **40:**1919–1923.

234. **Valadon, P., G. Nussbaum, L. F. Boyd, D. H. Margulies, and M. D. Scharff.** 1996. Peptide libraries define the fine specificity of anti-polysaccharide antibodies to *Cryptococcus neoformans. J. Mol. Biol.* **261:**11–22.

235. **Vazquez-Torres, A., J. Jones-Carson, and E. Balish.** 1996. Peroxynitrite contributes to the candidacidal activity of nitric oxide-producing macrophages. *Infect. Immun.* **64:**3127–3133.

236. **Vecchiarelli, A., C. Monari, F. Baldelli, D. Pietrella, C. Retini, C. Tascini, D. Francisci, and F. Bistoni.** 1995. Beneficial effects of recombinant human granulocyte col-

ony-stimulating factor on fungicidal activity of polymorphonuclear leukocytes from patients with AIDS. *J. Infect. Dis.* **171**:1448–1454.

237. **Vecchiarelli, A., C. Monari, C. Retini, D. Pietrella, B. Palazzetti, L. Pitzurra, and A. Casadevall.** 1997. *Cryptococcus neoformans* differently regulates B7–1 (CD80) and B7–2 (CD86) expression on human monocytes. *Eur. J. Immunol.* **28**:114–121.

238. **Vecchiarelli, A., C. Retini, C. Monari, C. Tascini, F. Bistoni, and T. R. Kozel.** 1996. Purified capsular polysaccharide of *Cryptococcus neoformans* induces interleukin-10 secretion by human monocytes. *Infect. Immun.* **64**:2846–2849.

239. **Vecchiarelli, A., C. Retini, D. Pietrella, C. Monari, C. Tascini, T. Beccari, and T. R. Kozel.** 1995. Downregulation by cryptococcal polysaccharide of tumor necrosis factor alpha and interleukin-1β secretion from human monocytes. *Infect. Immun.* **63**:2919–2923.

240. **Vogel, R. A.** 1966. The indirect fluorescent antibody test for the detection of antibody in human cryptococcal disease. *J. Infect. Dis.* **116**:573–580.

241. **Walter, J. E., and R. W. Atchison.** 1966. Epidemiological and immunological studies of *Cryptococcus neoformans*. *J. Bacteriol.* **92**:82–87.

242. **Wang, Y., P. Aisen, and A. Casadevall.** 1995. *Cryptococcus neoformans* melanin and virulence: mechanism of action. *Infect. Immun.* **63**:3131–3136.

243. **Wang, Y., P. Aisen, and A. Casadevall.** 1996. Melanin, melanin "ghosts," and melanin composition in *Cryptococcus neoformans*. *Infect. Immun.* **64**:2420–2424.

244. **White, M., C. Cirrincione, A. Blevins, and D. Armstrong.** 1992. Cryptococcal meningitis: outcome in patients with AIDS and patients with neoplastic disease. *J. Infect. Dis.* **165**:960–963.

245. **Williamson, P. R., J. E. Bennett, M. A. Polis, J. B. Robbins, and R. Schneerson.** 1993. Immunogenicity and safety of a conjugate glucuronoxylomannan-tetanus conjugate vaccine in volunteers. *Clin. Infect. Dis.* **17**:540. (Abstract 56.)

246. **Williamson, P. R., J. E. Bennett, J. B. Robbins, and R. Schneerson.** 1993. Vaccination for prevention of cryptococcosis, abstr. L22, p. 60. 2nd International Conference on Cryptococcus and Cryptococcosis, Milan.

247. **Wilson, M. A., and T. R. Kozel.** 1992. Contribution of antibody in normal human serum to early deposition of C3 onto encapsulated and nonencapsulated *Cryptococcus neoformans*. *Infect. Immun.* **60**:754–761.

248. **Yoshida, K., T. Akaike, T. Doi, K. Sato, S. Ijiri, M. Suga, M. Ando, and H. Maeda.** 1993. Pronounced enhancement of NO-dependent antimicrobial activity by a NO-oxidizing agent, imidazolineoxyl N-oxide. *Infect. Immun.* **61**:3552–3555.

249. **Young, A. C. M., P. Valadon, A. Casadevall, M. D. Scharff, and J. C. Sacchettini.** 1997. The three-dimensional structures of a polysaccharide binding antibody to *Cryptococcus neoformans* and its complex with a peptide from a phage display library: implication for the identification of peptide mimotopes. *J. Mol. Biol.* **274**:622–634.

250. **Yuan, R., A. Casadevall, J. Oh, and M. D. Scharff.** 1997. T cells cooperate with passive antibody to modify *Cryptococcus neoformans* infection in mice. *Proc. Natl. Acad. Sci. USA* **94**:2483–2488.

251. **Yuan, R., A. Casadevall, G. Spira, and M. D. Scharff.** 1995. Isotype switching from IgG3 to IgG1 converts a non-protective murine antibody to *C. neoformans* into a protective antibody. *J. Immunol.* **154**:1810–1816.

252. **Zebedee, S. L., R. K. Koduri, J. Mukherjee, S. Mukherjee, S. Lee, D. F. Sauer, M. D. Scharff, and A. Casadevall.** 1994. Mouse-human immunoglobulin G1 chimeric antibodies with activity against *Cryptococcus neoformans*. *Antimicrob. Agents Chemother.* **38**:1507–1514.

253. **Zhang, H., Z. Zhong, and L. Pirofski.** 1997. Peptide epitopes recognized by a human anti-cryptococcal glucuronoxylomannan antibody. *Infect. Immun.* **65**:1158–1164.

254. **Zhang, T., K. Kawakami, M. H. Qureshi, H. Okamura, M. Kurimoto, and A. Saito.**

1997. Interleukin-12 (IL-12) and IL-18 synergistically induce fungicidal activity of murine peritoneal exudate cells against *Cryptococcus neoformans* through production of gamma interferon by natural killer cells. *Infect. Immun.* **65:**3594–3599.

255. **Zhong, Z., and L. Pirofski.** 1996. Opsonization of *Cryptococcus neoformans* by human anticryptococcal glucuronoxylomannan antibodies. *Infect. Immun.* **64:**3446–3450.

9 | Tissue Responses and Special Topics in Immunity

The host response to *Cryptococcus neoformans* infection can best be described as protean. Review of the literature reveals tremendous diversity in the histopathological presentation, pathogenesis, clinical course, and outcome of *C. neoformans* infection in humans and animals. Generalizations are difficult and often require qualification when describing the characteristics of the tissue response to *C. neoformans* infection. This chapter explores the host response to infection, with emphasis on integrating clinical observations with those from animal studies and in vitro observations. The diversity in host response to infection appears to be a function of both the immunological status of the host and the virulence characteristics of the infecting *C. neoformans* strains.

The rarity of clinical cryptococcal infections in immunocompetent hosts indicates that the normal human host response to infection is highly effective. Since the fungus is prevalent in most environments, human exposure must be a frequent event. The high prevalence of serum antibodies reactive with the capsular polysaccharide in normal individuals (53) can be interpreted as indicative of frequent exposure to this fungus. If exposure to *C. neoformans* is frequent and symptomatic infection is rare, then those individuals who come to medical attention represent exceptions to the normal host response. This concept is important, because it highlights an essential uncertainty in our current knowledge base: we know very little about the normal human host response except that it is effective.

Most of the available knowledge on the host response to *C. neoformans* infection is derived from two sources, pathological studies of human specimens and animal experimentation. Each has significant limitations. The majority of clinical pathological studies involve the analysis of material from patients who come to medical attention because their immune response was inadequate to contain the initial infection. Hence, clinical pathological studies should not be considered to reflect the normal response to infection. Animal experimentation can provide insight into the host response and clinical course of infection. However, animal studies are limited in their implications for human infection because of species differences, the artificiality of experimental infection, and the inbred nature of most laboratory animal species. Further complicating our understanding of host response are uncertainties regarding basic issues of the pathogenesis of *C. neoformans*.

For example, it is not known if cryptococcal meningitis results from recent infection or reactivation of latent infection or the form of the infectious propagule (i.e., basidiospore or yeast cell).

In this chapter the host response to infection is described in detail for lung, brain, and skin, since these are the organs for which most information is available. For each organ, an attempt has been made to synthesize the available information into a coherent story that incorporates human clinical and pathological data with the results of animal experimentation.

OVERVIEW OF THE INFLAMMATORY RESPONSE

The inflammatory response to *C. neoformans* infection has four unusual features that are discussed in detail elsewhere in this chapter. First, the inflammatory response can range in intensity from florid, intense granulomatous inflammation to the virtual absence of inflammatory cells. The variables responsible for differences in the magnitude and intensity of the inflammatory response include the immune status of the host and poorly defined characteristics of *C. neoformans* strains. Second, *C. neoformans* can be found in tissue as both an intracellular pathogen and an extracellular pathogen. Intracellular parasitism is associated with granulomatous inflammation, whereas extracellular growth is associated with a weak or absent inflammatory response. Unchecked extracellular growth can result in the formation of gelatinous collections of organisms that appear cystic. Third, the type of inflammatory response is a function, in part, of the organ tissue where the infection occurs. In general, infection of organs rich in tissue macrophages tends to elicit stronger granulomatous responses. Hence, granulomatous reactions are more likely to be encountered in the macrophage lung and liver than in the brain and kidney (134, 283). Fourth, there is considerable variation in the type of effector cells found in inflammatory responses. Monocytes and lymphocytes predominate, but neutrophils and sometimes eosinophils have each been described in inflammatory responses. The relative frequency of effector cell types is highly variable. The combination of variable inflammatory response, differences in host immunity, intracellular and extracellular parasitism, and differences in organ tissue response result in a bewildering number of histopathological possibilities that defy a straightforward pathological classification system.

The most striking theme in reviewing the literature is the variability of the host inflammatory response to this fungus. One can find examples of granulomatous inflammation, lymphocytic inflammation, neutrophilic inflammation, no inflammation, calcified lesions, necrotic inflammatory responses, eosinophilic granulomas, gelatinous lesions consisting of well-encapsulated masses of yeast cells, and mixtures of these responses. In general, the more immunocompetent hosts tend to mount stronger inflammatory responses, whereas little inflammation may be found in hosts with severe immune deficits.

For humans, descriptions of strong inflammatory responses were more prevalent in the literature before the human immunodeficiency virus (HIV) epidemic was recognized in 1981. This may reflect the fact that most infections since 1981 have occurred in individuals with AIDS who have severely weakened immune systems. However, even among patients with HIV infection, there is a broad range of individual variation in the inflammatory response to *C. neoformans*, suggesting that unique host and/or fungal strain genetic factors can influence the host response in certain circumstances. One caveat in evaluating the immune response to cryptococcus is the possibility that antifungal therapy can modulate the host response. Amphotericin B, the mainstay of antifungal therapy since the 1950s, can have stimulatory effects on host immunity (29, 269). There is a report describing a change in inflammation from gelatinous to granulomatous lesions in a patient receiving amphotericin B therapy (213). Hence, the description of natural tissue reactions in this chapter often emphasizes the literature from the pre-amphotericin B era.

The range of tissue response to *C. neoformans* infection can be illustrated by several cases from the pre-amphotericin B era. Freeman (82) published an extensive monograph in 1931 in which he reviewed 19 cases of cryptococcal meningitis and the available literature. He noted a specific characteristic of cryptococcal meningitis, the "soapsuds" appearance of the brain. This pathological feature is a result of many small cysts that protrude slightly above the surface of the brain and are composed of gelatinous masses of yeast cells. Cysts filled with organisms may also be found in the gray matter. In general, the presence of cysts is associated with little inflammatory response. "Soapsuds" represent an extreme in the continuum of tissue responses in cryptococcosis, in which the absence of inflammation leads to unchecked replication of *C. neoformans* in the brain. At the other extreme are strong granulomatous responses that result in hard giant granulomas that behave like tumors and compress the cerebrum (52, 154, 251, 265). Microscopic examination of these granulomas reveals encapsulated lesions with dense infiltrates of giant cells, monocytes, lymphocytes, and plasma cells. Between the two extremes of gelatinous "soapsuds" lesions and tumorlike granulomas, practically every variation of tissue immune response has been described. The inflammatory response to *C. neoformans* in humans involves giant cells, lymphocytes, and plasma cells. There is considerable evidence from a variety of sources that *C. neoformans* infections in immunocompetent hosts elicit granulomatous inflammation and that granuloma formation is an effective host response for containing this fungal pathogen. Granulomatous inflammation in response to cryptococcal infection has been described in the brain and meninges (52, 154, 251, 265), spinal cord (168), lung (10), skin (170, 193, 284), peritoneum (98), liver (232), larynx (84), and kidney (234).

The Schwartz Classification: Reactive and Paucireactive Histological Patterns

Schwartz (250) grouped the histological patterns of cryptococcal infections into two general categories based on the host reaction: reactive and paucireactive. The reactive pattern is characterized by a granulomatous inflammatory response composed of macrophages, multinucleated giant cells, and lymphocytes, and the yeast forms are found primarily in an intracellular location. Neutrophilic infiltration is sometimes observed to be associated with necrotizing inflammatory responses. Schwartz considered fibrotic nodules or cryptococcomas to be a variant form of the reactive pattern. Cryptococcomas can calcify (but generally do not), are often fibrotic, and appear to represent healing cryptococcal lesions. The paucireactive pattern is characterized by minimal to absent inflammatory response, and tissue destruction results from compression necrosis by masses of cryptococci. In paucireactive patterns, yeast forms are often found extracellularly. Paucireactive inflammatory patterns may be indicative of a poor prognosis. To characterize the biological activity of *C. neoformans* in surgical specimens, Schwartz developed a scale that relied on determining the budding index and the carminophilic index. The budding index is calculated from the percentage of cryptococci in tissue that have one or more identifiable buds and presumably indicate replication in vivo. The carminophilic index is calculated by determining the percentage of cryptococci with capsules that stain positively for mucicarmine stain (250), which presumably indicates capsular synthesis in vivo. Paucireactive histological patterns have higher budding and carminophilic indexes than do reactive patterns. Hence, there is an inverse correlation between the presence of higher capsular production in tissue and less intense inflammatory responses. The Schwartz classification is potentially useful, because it provides an interpretation of the host response, prognosis, and biological activity and demonstrates that granulomatous responses are often associated with control of infection.

Intracellular and Extracellular Parasitism

C. neoformans can be found either as an extracellular or an intracellular pathogen, and the location of yeast cells can be a function of the type of immune response. In granulomatous responses, yeast cells are often observed within macrophages and giant cells (250) (Fig. 1). In infections in which there is little or no inflammatory response, *C. neoformans* cells are often observed as extracellular masses in tissues. There is evidence that *C. neoformans* can survive and replicate within phagocytic cells in vitro (56, 161). In vivo, intracellular replication probably occurs but has not been proved. Multiple yeast forms are often observed within single macrophages or giant cells, a finding consistent with intracellular replication. The polysaccharide capsule may contribute to extracellular survival and replication, since well-encapsulated forms found in tissues in vivo are often so large as to preclude efficient phagocytosis. Furthermore, the difficulties involved in phagocytosing encapsulated cells may be heightened by the absence of an antibody response to the capsular polysaccharide and/or acquired complement depletion, which can accompany cryptococcal infections (178).

In one case of overwhelming *C. neoformans* infection, yeast cells in the blood smear of the patient were noted inside and outside phagocytic leukocytes (290). There was also one report of yeast cells inside megakaryocytes in a patient with AIDS and disseminated *C. neoformans* infection (79). The yeast cells were consistent with cryptococci, but their identity was not unequivocally established by culture. The physiological or pathogenic relevance of this finding is uncertain, but it appears that cryptococci may occasionally be found in cells that are not usually associated with phagocytosis of microorganisms.

Apart from histological evidence placing *C. neoformans* inside phagocytic cells in many types of infections and hosts, this fungal pathogen has several characteristics in common with intracellular pathogens, such as *Mycobacterium tuberculosis* (105). First, most *C. neoformans* strains can cause chronic infections that kill the host only after a prolonged course of infection. Second, cell-mediated host defense mechanisms are critically important for containing and resolving *C. neoformans* infection. Third, effective host defense requires T-cell-dependent macrophage activation and granuloma formation. The similarities between the pathogenesis of

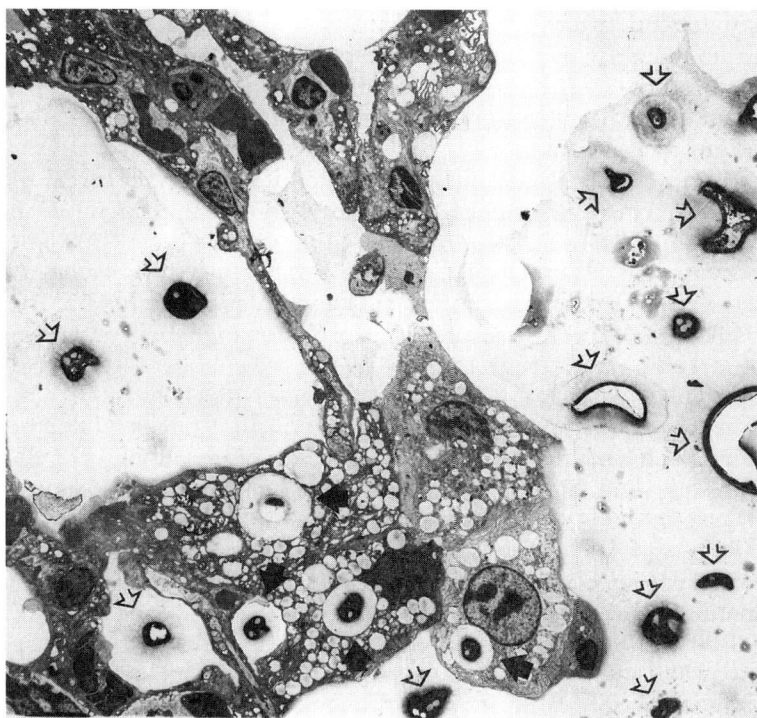

Figure 1 Electron micrograph of *C. neoformans*-infected mouse lung tissue at day 7 of infection. The photomicrograph shows an area of lung with cryptococci filling the airspace, whereas other fungal cells are found in the intracellular space. Open arrows indicate extracellular cryptococci. Closed arrows indicate intracellular cryptococci. Most intracellular cryptococci are inside alveolar macrophages. The tissue comes from C57BL/6 mice, and the cryptococcal strain is ATCC 24067. For details of experimental infection and microscopy, see references 39 and 77. Micrograph obtained at a magnification of ×3,000. Figure courtesy of Marta Feldmesser (Bronx, N.Y.).

tuberculosis and cryptococcosis have been noted by several investigators (9, 10, 107, 223, 245).

In summary, *C. neoformans* can be found in both intracellular and extracellular spaces. In hosts that mount intense and effective inflammatory response, most of the cells are found in intracellular spaces of phagocytic cells. In hosts that mount little or no inflammation, *C. neoformans* is found predominantly in extracellular spaces. For the majority of infections, *C. neoformans* probably functions as both an intracellular and an extracellular pathogen.

The Neutrophil Response

Neutrophils are highly effective microbicidal cells that often arrive first at sites of infection. Scattered neutrophils are commonly found in granulomatous responses to *C. neoformans*, but whether or not an initial neutrophilic response to infection occurs may depend on the specific circumstances of infection. Baker and Haugen (10) found no evidence of a neutrophil response in human cryptococcosis. However, the occurrence of a primary neutrophilic response to infections is best studied in experimental animal infection, because human specimens are unlikely to include those from individuals with primary asymptomatic early infection. Kuhn (155) found no neutrophilic response to intraperitoneal cryptococcal infection in mice or rabbits. However, other groups have shown that in mice and rabbits injected intraperitoneally with *C. neoformans*, there is an early influx of neutrophils that phagocytose the yeast cells (87, 88, 248, 254). Neutrophil phagocytosis can be enhanced in vivo by the administration of anticapsular antibody (87). The presence of anticapsular antibody also enhances the formation of histiocytic rings composed of neutrophils and macrophages, which appear to be a host response to the large well-encapsulated yeast forms (248, 254). Gadebusch (88) believed that neutrophils play a crucial role in the early response to infection and suggested that neutrophils are the primary phagocytic cells in the early stages of infection, when the majority of yeast cells are small and have relatively small capsules. Later, as surviving *C. neoformans* cells become well encapsulated, macrophages and giant cells become the primary phagocytic cell population (88). However, an early neutrophilic response has not been consistently observed in pulmonary murine infection. Intranasal instillation of *C. neoformans* produces an early increase in neutrophil count in lavage fluid, but this effect also occurs with saline and may reflect aspiration of upper airway bacteria (2). In mice given intratracheal infection, yeast cells are rapidly phagocytosed by alveolar macrophages without a prominent neutrophilic response (76).

In summary, it appears that in some conditions, *C. neoformans* infection can elicit a rapid and short-lived neutrophilic response. In most studies, neutrophils are components of the host inflammatory response but are seldom the predominant host effector cell. Neutrophils may be effective in eliminating organisms with small capsules and probably have an important role in protection. However, once infection is established, macrophages and giant cells appear to be the most prominent phagocytic cells in inflammatory infiltrates.

Granuloma Formation

Once *C. neoformans* infection is established in a human or animal host, the most effective host response for containing infection is granulomatous inflammation. Furthermore, resolution of infection, when it occurs, almost always follows granuloma formation. Granulomatous inflammation is more likely to be reported in non-HIV-associated cryptococcosis. The paucity of granulomatous inflammation in patients with HIV infection may reflect their profound impairment of cell-mediated immunity and CD4+ lymphopenia. In the central nervous system (CNS), granulomatous inflammation is observed in patients with non-HIV-associated cryptococcal meningitis but occurs less frequently in HIV-associated cryptococcosis (159). The absence of granulomatous inflammation may contribute to the fact that cryptococcal meningitis in the setting of HIV infection is difficult to eradicate with antifungal agents alone.

Granuloma formation may be a necessary response for containment of cryptococcal infection, given that well-encapsulated yeast cells are large and difficult to phagocytose by single cells alone. Encapsulated cryptococcal cells in tissue can be 10 to 20 μm in diameter, and variants up to 50 μm have been described (176). Giant cell formation in the lung is dependent on CD4$^+$ T cells (112), suggesting a mechanism by which abnormalities of cell-mediated immunity can translate into poor inflammatory responses.

Granuloma formation in experimental infection in animals

Despite strong evidence that granulomatous inflammation is the effective tissue response for containment and eradication of established infection, relatively few animal studies have focused on studying granuloma formation in response to *C. neoformans* infection. In the cryptococcal literature, there is considerable evidence that a strong granulomatous response is dependent on intact T-cell function. Nude mice (*nu/nu*) infected with *C. neoformans* develop cryptococcal cysts without granuloma formation or cellular infiltrate (218). Granulomas are also absent on severe combined immunodeficiency (SCID) mice that lack functional B or T-cell lymphocytes (46). Similarly, lymphocyte-deficient (nude) rats given pulmonary infection developed masses of cryptococci in the alveoli with little or no inflammation (99), whereas normal rats effectively contained the infection in the lung with a strong granulomatous response (93).

Schneerson-Porat et al. (248) demonstrated that intraperitoneal inoculation of mice with *C. neoformans* resulted in the formation of rings of histiocytes that surrounded the yeast cells in peritoneal exudates and the omentum. Histiocytic ring formation occurred only for large *C. neoformans* cells and was enhanced by the presence of specific antibody (rabbit antibody in their experimental system) (248). Histiocytic ring formation appears to be a mechanism by which phagocytic cells surround yeast cells that are too large to be phagocytosed (248). Subsequent studies suggested that the histiocytic rings were precursors of granuloma formation (254). Interestingly, there appear to be significant species differences in the requirements for granuloma formation. In mice, intraperitoneal inoculation does not elicit histiocytic ring formation unless the yeast cells were pretreated with anticapsular antibody (254). In rabbits, the inflammatory response to infection includes plasma cells, and there is no requirement for the addition of immune serum (254). Presumably, rabbits are able to mount an antibody response to cryptococcal polysaccharide and do not require exogenous specific antibody. Relatively little follow-up work has been done on the observation that antibody is required for histiocytic ring formation in laboratory animals. In this regard, it is noteworthy that *C. neoformans* var. *gattii* is more likely to cause granulomatous inflammation (260) and elicit antibody responses than is *C. neoformans* var. *neoformans* (261).

Detailed studies of granuloma formation in response to *C. neoformans* infection have been done in rats (92–94, 289). In contrast to mice, rats effectively contain *C. neoformans* infection, even when the fungus is inoculated directly into the CNS. Intratracheal infection of rats with *C. neoformans* elicits a strong granulomatous response that controls infection in the lung with minimal or no dissemination (92, 93). Granuloma formation is temporally associated with reduction in the tissue yeast burden in lung (92, 93). Rats also mount strong granulomatous inflammation in response to intraventricular infection, which is also temporally associated with a reduction of yeast burden (94).

Yamaoka et al. (289) studied granuloma formation in the intravascular space of rats given intravenous inoculation, a route of inoculation that results in disseminated infection. Three to five days after intravenous infection, granulomas begin to form in brain, lungs, liver, spleen, kidneys, and endothelium of large blood vessels. In the lung, liver, and spleen, the infection loci elicit strong granulomatous tissue responses. In brain and kidney, a weaker response results and gradually evolves to mucoid cystic loci. Analysis of endogenous peroxidase activity suggests that lung granulomas arise from alveolar macrophages, whereas intravascular granulomas arise from blood monocytes. Intravascular granulomas are covered by endothelial cells that express higher levels of adhesion molecules. Yamaoka et al. (289) suggested that the magnitude of the granulomatous response to *C. neoformans* in a given organ reflects the number of macrophages in the tissue.

In mice, the type and magnitude of the inflammatory response is also a function of the organ in which the infection occurs (283). Intravenous inoculation of *C. neoformans* in BALB/c mice results in disseminated infection that involves multiple organs, including the brain and the liver (283). A temporal analysis of the histological lesions in brain and liver indicates major differences in tissue response to infection (283). In the liver, granulomatous inflammation develops by the second week of infection and controls the infection, as shown by a gradual reduction in organ CFU (283). In the brain, a lack of granulomatous inflammation results in the formation of cysts filled with *C. neoformans* that continue to expand unchecked (283). Hence, granulomatous inflammation and cystic lesions are found at the same time in the same animal, depending on the organ sampled. Similar results were observed by Miyaji and Nishimura (190) in ddY mice, in which the probability of granuloma formation was significantly higher in organs rich in sessile macrophages (190). These studies indicate differences in the tissue response to infection and suggest that local organ factors may influence the strength of the inflammatory response and the likelihood of granuloma formation in response to *C. neoformans* infection.

In summary, animal studies indicate that the likelihood of granuloma formation is dependent on the immune status of the host, the site of infection, and possibly the route of infection.

C. neoformans strain factors

The inflammatory response to cryptococcal infection is dependent, in part, on *C. neoformans* strain characteristics. There appears to be an inverse relationship between the size of the capsule in tissue and the intensity of the granulomatous inflammatory response. Several case reports have associated intense inflammation and granuloma formation with poorly encapsulated strains (5, 75, 104, 134, 253). For several of these cases, a diagnosis of cryptococcal infection was delayed because the strong inflammatory response was assumed to be atypical of *C. neoformans* infections (5, 75, 104). This clinical observation suggests interference of capsular polysaccharide with the development of an inflammatory response, a hypothesis that is supported by several in vitro and in vivo experiments (see section below). A comparison of lightly and heavily encapsulated strains (not congenic) revealed stronger inflammatory responses associated with the lightly encapsulated strain (134). A role for capsular polysaccharide in inhibition of the inflammatory response was also suggested by the observation that infection with the perfect state of *C. neoformans* (nonencapsulated) elicited a strong cellular reaction relative to the encapsulated yeast form (190). Similarly, it has been reported that infection with a pseudohyphal strain of *C. neoformans* elicited a strong inflammatory response that was inhibited around encapsulated yeast forms of the same strain (217).

Apart from the capsule, other strain characteristics are also likely to contribute to the type and intensity of the inflammatory response. One clinical strain of *C. neoformans* has been reported to form tumorlike masses during experimental infection (273). The tumorlike formations can reach 19% of the mass of the mouse and appear to be composed of aggregates of yeast cells, capillaries, collagen fibrils, and reticular stroma. A comparison of the timing and cellular composition of inflammatory responses elicited by two encapsulated *C. neoformans* strains in mice given intratracheal infection revealed major differences in tissue histological response (50).

In summary, there is strong evidence that the tissue inflammatory response is influenced by the *C. neoformans* strain factors. The strain characteristics responsible for differences in the immune response are not understood, but there is circumstantial evidence that cells with small capsules elicit stronger granulomatous responses than do highly encapsulated strains.

Polysaccharide-mediated effects on the inflammatory response

Large deposits of capsular polysaccharide are found in infected tissues of patients with cryptococcosis (see color plate 2, following p. 456) (157, 158). Tissue polysaccharide deposits have also been demonstrated after experimental infection in mice (200) and rats (93, 94). Several lines of evidence indicate that the polysaccharide antigen of *C. neoformans* interferes with the development of strong inflammatory responses that could otherwise eliminate this fungus from infected tissues. There is anecdotal clinical data to suggest that poorly encapsulated *C. neoformans* strains elicit stronger inflammatory responses than do heavily encapsulated strains. Farmer and Komorowski (75) isolated a poorly encapsulated strain from a patient with a suppurative lung granuloma that resembled blastomycosis. The isolate produced nonmucoid colonies on Sabouraud dextrose agar and had a very small capsule on India ink examination (75). Comparison of the histological response to infection by the poorly encapsulated strain with that to infection by a well-encapsulated strain revealed that the capsule-deficient strain elicited significantly more intense inflammation (75). Gutierrez et al. (104) described a patient with a lung granuloma with nonencapsulated yeast cells that did not stain with mucicarmine staining. In this case there was intense inflammation, and the lesion was initially misdiagnosed as histoplasmosis (104). Attal et al. (5) described a case from India in which infection with a poorly encapsulated *C. neoformans* strain produced a histological picture that initially caused it to be misdiagnosed as histoplasmosis (5). In this patient, lymphatic and hepatic involvement was accompanied by granulomatous inflammation with a mixed cellular response that included eosinophils (5). Hence, case reports of human infection with poorly encapsulated *C. neoformans* strains suggest that strains with little or no capsule elicit strong inflammatory responses that can be confused with those produced by other infections.

Experiments in mice provide additional evidence for the hypothesis that the capsule inhibits the inflammatory response. Kagaya et al. (134) studied the organ inflammatory response to intravenous infection in mice with seven *C. neoformans* strains that varied in capsule size. A strain with a small capsule induced stronger granulomatous inflammatory responses than did several strains with large capsules. Cystic lesions composed of masses of cryptococci growing extracellularly were more prevalent in infection with well-encapsulated strains (134). Studies with a hypha-forming strain of *C. neoformans* in mice also implicated the polysaccharide capsule in inhibiting inflammatory responses. Some *C. neoformans* strains, like the Coward strain, produce aerial hypha-like structures under certain conditions (177, 253). Inoculation of mice with preparations of the hypha-form Coward strain produced lethal infection in mice from which the yeast form was recovered (177, 253). Histopathological analysis of inflammatory responses following intravenous infection with hyphal forms in mice revealed a strong granulomatous response to hyphal fragments in tissue but weak or absent inflammation around hyphal forms that had germinated to well-encapsulated yeast forms (253).

Indirect support for the concept that polysaccharide inhibits tissue inflammation comes from rats infected with *C. neoformans* and treated with hyaluronidase (233). Hyaluronidase treatment was associated with stronger granulomatous inflammatory responses than those seen in control rats (233). Intense phagocytic activity in hyaluronidase-treated rats has been interpreted as representing phagocytosis of *C. neoformans* following enzymatic disruption of the *C. neoformans* capsule (233). Hyaluronidase is most effective when administered early in the course of infection (233). Although this study did not show that hyaluronidase digested the capsule and did not exclude a direct effect of hyaluronidase on host immunity, the results are consistent with the hypothesis that the polysaccharide capsule inhibits inflammatory responses.

Various studies suggest several potential mechanisms by which the capsular polysaccharide can interfere with the development of strong inflammatory responses (Table 1). The polysaccharide capsule is antiphagocytic and, through this property, may interfere with the development of inflammatory responses by inhibiting phagocytosis and, consequently, antigen presentation (49). *C. neoformans* antigens (including polysaccharide) have been associated with the induction of T-suppressor cell responses in mice that may act to reduce the inflammatory response (see below). Capsular polysaccharide can directly inhibit leukocyte migration (57, 62–64, 70) and reduce lymphocyte proliferation (192). Capsular polysaccharide can interfere with the serum-dependent chemotactic response that results when cryptococci are incubated in serum (57). One potential mechanism for this effect is suggested by the observation that capsular polysaccharide can produce shedding of L-selectin from neutrophils (65). Support for this mechanism comes from unrelated studies that found that the inflammatory response in endothelial selectin-deficient mice with cytokine-induced meningitis is greatly reduced relative to normal mice (266). Another mechanism by which polysaccharide can inhibit the migration of leukocytes is through binding to the CD18 adhesion molecule and blocking their interaction with ligands in the endothelium (66). Presumably, a combination of L-selectin loss from neutrophils and masking of CD18 ligands by polysaccharide interferes with leukocyte migration in areas of *C. neoformans* infections where capsular polysaccharides are released.

The observation that polysaccharide can inhibit leukocyte migration contrasts with the observation that, under certain circumstances, the polysaccharide can be shown to have chemotactic properties. Leukocyte chemotaxis to a site of infection

Table 1 Immunological effects associated with *C. neoformans* capsule and polysaccharide

Effect	References[a]
Antibody unresponsiveness	45, 149, 151, 209, 264
Immunological tolerance	109, 110, 264
Inhibition of phagocytosis	147, 150, 153
Suppressor responses	15, 16, 207
Interference with antigen presentation	49, 278
Enhancement of L-selectin shedding from leukocytes	65
Inhibition of leukocyte migration	53, 62–64, 70
Binding to CD18 cell antigen	66
Enhancement of HIV infection in vitro	226–229
Cytokine release	236, 279, 280
Reduction of survival time in mice	11
Brain cell edema	117–120
Costimulatory molecule regulation	277
Complement activation and depletion	148, 152, 178

[a]Table references are not exhaustive. See text for more detailed discussion of effects. Additional information can be found in chapter 4 of this volume.

occurs either as a direct result of microbial products that serve as chemoattractants or indirectly through microbial activation of the complement system and generation of chemotactic factors. There is agreement that whole *C. neoformans* cells can activate complement and elicit leukocyte chemotaxis (57, 62, 156). However, others studying leukocyte migration in response to cryptococcal antigens have found no difference in migration relative to serum and have concluded that polysaccharide lacks chemotactic properties (57). Dong and Murphy (62) demonstrated that capsular polysaccharide antigens, in particular glucuronoxylomannan (GXM), have chemotactic properties for neutrophils compared to saline. However, the chemotactic activity of GXM was weak (62), and the extent to which this effect contributes to inflammation in vivo is unclear. Cryptococcal polysaccharide from the two varieties of *C. neoformans* can have very different effects on neutrophil migration (64); antigen from var. *neoformans* stimulates neutrophil chemotaxis, whereas antigen from var. *gattii* has no effect. Furthermore, antigen preparations from var. *gattii* strains inhibit neutrophil chemotactic responses to pooled human serum with a chemoattractant (64). These results highlight the complexity of the neutrophil chemotactic response to *C. neoformans* and the multitude of effects that cryptococcal polysaccharide can have on host cells. It appears that, under certain conditions, cryptococcal antigen and cells can stimulate neutrophil chemotaxis. Strain differences in the chemotactic activity of polysaccharide antigen may also contribute to some of the differences in host inflammatory responses. Whether the polysaccharide induces or inhibits leukocyte migration may depend on many factors, including polysaccharide structure, polysaccharide tissue concentration, presence of polysaccharide-binding antibodies, and the type of tissue in which the infection occurs.

Cryptococcal polysaccharide may also influence the inflammatory response through effects on cytokine production. The *C. neoformans* polysaccharide can stimulate the secretion of various cytokines and, in doing so, may disrupt the signal cascade necessary for the generation of protective inflammatory responses

(280). In this regard, capsule size may be a contributing variable. Cryptococci with small capsules induce stronger secretion of tumor necrosis factor (TNF)-α by human leukocytes than do cryptococci with large capsules (166). Capsular polysaccharide reduces the secretion of TNF-α by human macrophages in vitro after stimulation by lipopolysaccharide (280). Furthermore, capsular polysaccharide elicits the secretion of interleukin (IL)-10 by human monocytes (279). Since TNF-α is a pro-inflammatory cytokine and IL-10 is a potent down-regulator of the host inflammatory response, these effects, if reproduced in vivo, could have profound effects on inflammatory response.

In contrast to other fungi, such as *Candida albicans* and *Histoplasma capsulatum*, encapsulated *C. neoformans* cells do not induce production of nitric oxide synthase (NOS) from gamma interferon (IFN-γ)-primed J774 mouse macrophage-like cells (214). However, nonencapsulated mutants of *C. neoformans* are able to induce NOS. This result suggests that the capsular polysaccharide interferes with the cellular signaling necessary for TNF-α production and consequently NOS induction. The mechanism by which these effects occur is not known. The observations with NOS induction distinguish encapsulated *C. neoformans* from other intracellular microbes, such as *M. tuberculosis*, *Listeria monocytogenes*, and *Leishmania major*, which induce NOS synthesis in macrophages. The ability of *C. neoformans* to interfere with NOS induction in mouse macrophages may contribute to the survival of yeast cells in the host.

Capsular polysaccharide could also undermine immunity by potentiating other infections, such as HIV. In vitro capsular polysaccharide has been shown to increase HIV infection in human lymphocytes through a mechanism that may involve better attachment of the viral particle to the cell (227–229). This effect was observed with low polysaccharide concentrations (2 μg) comparable to those found in human infection. This phenomenon, if it occurs in vivo, could enhance HIV replication and further degrade the immune system in HIV-infected individuals. Synergistic interactions between HIV and *C. neoformans* that degrade the host immune response could contribute to the paucity of inflammatory responses observed for cryptococcal infections in most patients with AIDS.

In summary, there is strong evidence to link the polysaccharide capsule and the soluble polysaccharide antigen with the weak inflammatory responses that are commonly associated with cryptococcal infection. In vitro studies strongly suggest that there are multiple mechanisms by which polysaccharides can interfere with the generation of effective inflammatory responses. *C. neoformans* polysaccharide could theoretically inhibit host inflammatory responses directly by interfering with leukocyte migration, inducing suppressor responses, and producing cytokine dysregulation, and indirectly by reducing antigen presentation, enhancing HIV infection, failing to elicit an antibody response, and contributing to depletion of complement components. Polysaccharide antigen infiltrates tissue and can be found in the intracellular and extracellular spaces (158). Capsular polysaccharide is resistant to digestion in macrophages (241), and long-term persistence of polysaccharide in tissue is likely to contribute to the pathogenesis of infection. Considering the large variation in polysaccharide structure among cryptococcal strains, the protean manifestations in the tissue response to cryptococcal infection could reflect strain differences in the immunoregulatory properties of their capsular polysaccharides.

RESPONSE TO INFECTION IN THE LUNG

Human Infection

In 1955 Baker and Haugen (10) described tissue inflammation in lung tissue from 26 cases of pulmonary cryptococcosis obtained from autopsies, biopsies, and lung resections (10). As with Freeman's (81) pathological study of the brain three decades earlier, Baker and Haugen (10) noted considerable patient-to-patient variation in the lung tissue response to *C. neoformans* infection. Giant cells, lymphocytes, and macrophages were found consistently, whereas polymorphonuclear leukocytes (PMNs) were prominent in only a few cases. Baker and Haugen (10) argued that *C. neoformans* cells in tissue behaved more like a foreign body than an acute inflammatory irritant. In some cases, "caseous" necrosis was observed in pulmonary nodules. However, these investigators noted that caseation in cryptococcosis, unlike that in *M. tuberculosis* infection, was associated with large accumulations of organisms and hence was a histologically different process.

There has been considerable interest in the relationship of subpleural nodules to the pathogenesis of cryptococcosis. In 1948, Terplan (268) reported a case of cryptococcal meningitis in which a bean-sized, well-encapsulated subpleural nodule that contained encapsulated yeasts was found at autopsy. Terplan proposed that this tubercle-like lesion was the primary site of infection (268). Haugen and Baker (107) later identified subpleural nodules in autopsies of patients with or without *C. neoformans* infections and also proposed that these nodules represented the primary site of infection. The largest subpleural nodule was 1.5 cm in diameter. These subpleural nodules were considered to be an effective response in the human lung that contained *C. neoformans* infection (107). The tissue response in subpleural nodules was granulomatous inflammation and consisted of giant cells, lymphocytes, macrophages, and fibrous tissue (107, 268). Caseous necrosis was described on some nodules (107). In subsequent studies, Baker and collaborators (9, 245) established that subpleural nodules could be associated with *C. neoformans* in hilar lymph nodes, which led to the suggestion that the pathogenesis of cryptococcal infection was similar to that of tuberculosis. It was proposed that primary *C. neoformans* infection in an immunocompetent individual resulted in a primary "complex" that consisted of a pulmonary lesion and, possibly, the draining hilar lymph nodes. In this view, containment of cryptococcal infection at the site of primary infection resulted in resolution of the lesion and/or formation of a subpleural nodule (9). Extrapulmonary dissemination could occur in immunocompromised patients. Baker (9) pointed out that the primary complex of cryptococcal infection was much rarer than similar lesions in tuberculosis or histoplasmosis but convincingly argued for a similar pathogenic mechanism in each infection. Implicit in this scenario is the possibility that, like tuberculosis and histoplasmosis, *C. neoformans* yeast cells remain dormant in the primary complex and capable of reactivation and dissemination at later times, especially if immune defenses are weakened.

Intense granulomatous reactions in the lung can result in pulmonary masses that resemble carcinoma. Dormer et al. (68) described a huge tumorlike mass in the pleura of a 12-year-boy that, on histological examination, was a huge cryptococcal granuloma. Baker (8) described three patients with solid pulmonary lesions "varying in size from that of a plum to that of an orange." Meighan (187a) described three cases that were initially thought to be lung carcinomas but were in fact large tumorlike cryptococcal inflammatory lesions. One case was remarkable for the occurrence of hematemesis and sudden death resulting from the rupture of a *C. neoformans* mycotic aneurysm in the left subclavian artery (187a). Surgical resection of such lesions was curative in the pre-amphotericin B era (8, 13, 68). However, many patients developed cryptococcal meningoencephalitis subsequent to surgical resection (267). Some authorities considered disseminated cryptococcosis after surgical resection to be the result of intraoperative *C. neoformans* dissemination and recommended early ligation of the pulmonary veins, draining the lesion to avoid hematogenous spread (267). However, this was never proved, and it is possible that *C. neoformans* had already spread to the CNS before surgery. In addition to solid lesions that resemble tumors, it is noteworthy that cryptococcosis can, in occasional patients, produce cavitary pulmonary lesions (1, 6). Rarely, *C. neoformans* pulmonary lesions calcify, but calcification is not a common feature of healed pulmonary cryptococcosis.

A comprehensive review of 36 patients with pulmonary *C. neoformans* infection seen at Johns Hopkins Hospital in Baltimore from 1936 to 1983 was published by McDonnell and Hutchins (186). The majority of their patients had underlying malignancies or had received steroid therapy. The histological response to *C. neoformans* infection in the lung among individual patients was highly variable and included peripheral pulmonary granulomas (subpleural nodules), granulomatous pneumonia with intra-alveolar proliferating yeasts and varying degrees of inflammatory response, miliary granulomas, and interstitial pneumonia (186). Despite the diversity in histological response, McDonnell and Hutchins (186) were able to discern four distinct patterns in pulmonary cryptococcosis: (i) peripheral pulmonary granuloma (19% of cases); (ii) granulomatous pneumonia (53%); (iii) intracapillary and interstitial infection (19%); and (iv) massive pulmonary involvement (8%). However, the histological patterns could not be associated with specific conditions predisposing to *C. neoformans* infection. Peripheral pulmonary granulomas were either walled off (quiescent) or had disrupted granulomatous walls with spread to the surrounding lung parenchyma. Granulomatous pneumonia presented with a variable histological pattern, with inflammation ranging from nil, with masses of cryptococci in alveoli, to diffuse intra-alveolar granuloma formation with confinement of yeast forms to multinucleated cells. Some patients had intracapillary and interstitial infection, with organisms in the capillary lumens. Massive pulmonary involvement consisted of large numbers of cryptococci in both intra-alveolar and intravascular spaces with minimal inflammatory infiltrate. The different histological patterns may reflect differences in the pathogenesis of lung infection. The histological forms of pulmonary granuloma and granulomatous pneumonia were thought to be consistent with a primary cryptococcal infection acquired through inhalation of infectious particles. However, the intracapillary-interstitial and massive pulmonary involvement patterns were thought to reflect hematogenous dissemination to the lung from another primary site that was not identified in the study (186). Kent and Layton (139) had previously described three cases of massive pulmonary involvement and proposed that this condition reflected hematogenous dissemination from a primary lobar pneumonia, from a primary infection elsewhere, or from reactivation of a latent systemic infection.

Pulmonary *C. neoformans* infections in patients with AIDS usually elicit significantly less intense inflammatory responses. Gal et al. (89) studied pathological specimens from 11 patients with AIDS who had pulmonary cryptococcosis and found that the majority of patients had an interstitial pattern of dissemination with minimal cellular infiltrate. No granulomas, fibrosis, or calcification were noted in samples from patients with AIDS. *C. neoformans* organisms were found in lymphatics and capillaries, suggesting that extrapulmonary dissemination occurred by both lymphatic and hematogenous routes. Consistent with the pathological findings, a radiological study of cryptococcal pneumonia in 14 patients with AIDS revealed that the most common chest X-ray pattern was interstitial infiltrates, with pulmonary nodules being rare (83). Another study also described interstitial patterns of involvement in 11 of 12 AIDS patients with *C. neoformans* pneumonia (35). In a report of cryptococcal pneumonia in 27 patients with HIV infection and a review of the literature, Meyohas et al. (188) identified 92 patients for whom radiographic data was available. The relative frequency of radiographic findings in HIV-related cryptococcosis is as follows: normal, 10%; interstitial infiltrates, 65%; alveolar lesions, 13%; nodular densities, 11%; pleural effusions, 14%; and cavitary lesions, 11% (95, 188). The differences in the pulmonary pathology of *C. neoformans* infection in patients with or without AIDS are likely to reflect the severe immunodeficiency that accompanies HIV infection. Among patients with AIDS, localized pulmonary infection is more common in those individuals with higher CD4$^+$ T-cell counts (188). In patients with AIDS, pulmonary nodules and granulomatous inflammation are presumably less common because strong inflammatory responses are impaired. The inability of the host response to contain the fungus through granuloma formation may be responsible for the observation that *C. neoformans* pneumonia in the setting of HIV infection is more likely to involve diffuse areas of the lung and to disseminate than it is in relatively immunocompetent hosts (35, 83, 89, 140).

Pleural involvement is usually considered rare in *C. neoformans* infections, but several well-documented cases have been described (74, 201, 244, 255). In 1969 Sokolowski et al. (255) reviewed the literature and identified only nine cases of pleural cryptococcosis. Pleural involvement in most patients is unilateral and results in serosanguineous effusions that may or may not be positive for *C. neoformans* organisms (244, 255). The character of the pleural effusion has been described in the literature for pleural involvement in six patients with AIDS: four were exudative, one was transudative, and one was empyemic (201). The pathogenesis of pleural involvement appears to be contiguous, spread from a subpleural nodule (244). In one case, a subpleural nodule ulcerated, spilling its contents into the pleural space and resulting in extensive invasion of pleural lymphatics (268). In another case, perforation of a lung nodule led to a large pleural effusion with cavitation (6). Pneumothorax has been a rare presentation of pulmonary cryptococcosis in some patients with AIDS (270). Rupture of subpleural nodules into the pleura could conceivably result in a pneumothorax. Tissue reaction in the pleural space usually appears to be granulomatous (244). Pleural involvement may be more common in patients with AIDS. Pleural effusions were noted in radiographic studies in 13 of 92 patients with HIV-associated pulmonary cryptococcosis (188).

C. neoformans can sometimes be found with other pathogens in the lung. The coexistence of *C. neoformans* and *M. tuberculosis* in a solitary upper lobe pulmonary nodule has been described (135). The pathological analysis of this infection revealed both acid-fast bacteria and yeast forms in necrotizing granulomas (135). The pathogenesis of the mixed cryptococcal-tuberculous nodule was proposed to represent *C. neoformans* colonization of an *M. tuberculosis* granuloma (135). Mixed pulmonary infection of *C. neoformans* with other pathogens, such as *Pneumocystis carinii* (83), *Streptococcus anginosus* (6), and *Haemophilus influenzae* (83) has been described. Hence, the possibility of mixed infection should always be considered in the analysis of the host response to pulmonary *C. neoformans* infection.

Experimental Infection

Experimental *C. neoformans* pulmonary infection has been studied extensively in mice, rats, and guinea pigs (2, 34, 92, 93, 99, 112, 114, 125–127, 129, 130, 137, 138, 169, 171, 172). Rodent models provide insights into the pathogenesis of cryptococcal infection and the effective mammalian immune responses. In mice, pulmonary infection is usually induced by either intranasal or intratracheal infection. Intranasal infection is simpler, and possibly more physiological, but has the significant drawback that there is uncertainty about the size of the inoculum that reaches the lungs. Intratracheal infection in mice delivers a more consistent inoculum to the lungs but requires a surgical procedure in mice and guinea pigs to expose the trachea. In rats, the larger size of the animal permits cannulation of the trachea from the oral cavity without the need for surgery (93). Experimental pulmonary cryptococcal infection in mice and rats provides two useful animal models, because the infection closely mimics human pulmonary cryptococcosis.

Pulmonary infection with *C. neoformans* can elicit strong inflammatory responses and delayed-type hypersensitivity (DTH) in mice (126, 169). The initial defense against *C. neoformans* infection in the lung probably consists of alveolar macrophages. Electron microscopy of murine lung tissue infected with *C. neoformans* has shown that yeast cells are rapidly internalized by alveolar macrophages (76). A variety of cell types are recruited into the lung, including CD4$^+$ and CD8$^+$ T-cell lymphocytes, which appear to be crucial in generating an effective immune response to contain infection (126, 130). Depletion of either CD4$^+$ or CD8$^+$ T cells in mice can prevent pulmonary clearance and enhance extrapulmonary dissemination (126, 130). CD4$^+$ T cells have been proposed to recruit and activate phagocytic cells (macrophages and neutrophils), whereas CD8$^+$ T cells lyse inactivated phagocytic cells engorged with *C. neoformans* cells (126, 130, 191). Experiments involving intratracheal inoculation of heat-killed *C. neoformans* in mice indicate that macrophage activation occurs approximately 1 week after inoculation (138). Macrophage activation is dependent on the production of IFN-γ by T cells (137).

The inflammatory response to pulmonary *C. neoformans* infection in mice includes macrophages, neutrophils, lymphocytes, and eosinophils (126). The precise contribution of each cell type to host defense is uncertain. A transient neutrophilia has been observed in the pulmonary lavage fluid of mice given intranasal infection, but this may have been the result of an influx of bacteria into the lungs (2). In some mouse strains, pulmonary infection elicits an eosinophilic pneumonia, and eosinophil recruitment appears to be dependent on CD4$^+$ lymphocytes (126, 127). Electron microscopy studies have shown that pulmonary eosinophils can be found in close association with *C. neoformans* cells and may be important effector cells in the murine inflammatory response (77). The majority of studies of pulmonary infection in mice have shown that reduction of lung fungal burden does not begin until 7 to 14 days after infection, when granulomatous inflammation develops.

Urokinase (urokinase-like plasminogen activator) is important for the production of an effective pulmonary cellular response to *C. neoformans* infection (105). Urokinase is a protease that has been hypothesized to function in inflammation by degrading tissue matrix and permitting migration of effector cells across tissue planes. Genetically engineered urokinase-deficient mice have increased susceptibility for pulmonary *C. neoformans* infection that is characterized by markedly reduced pulmonary infiltrates. In urokinase-deficient mice, pulmonary inoculation of *C. neoformans* results in an uncontrolled and rapidly fatal infection. Histological examination of urokinase-deficient mice with cryptococcal infection reveals large numbers of organisms in alveoli with little or no inflammation. Such a histological pattern is similar to that observed in some cases of human infection. These results suggest an important role for urokinase in facilitating an adequate inflammatory response in mice. Furthermore, the results suggest that deficiency or dysregulation of urokinase expression could theoretically predispose to *C. neoformans* infection.

Another mediator of inflammation in the mouse lung is the chemokine monocyte chemotactic protein-1 (MCP-1) (128). MCP-1 is a small protein produced by a variety of cell types that is involved in monocyte and T-cell recruitment. Mice given antibody to MCP-1 have decreased recruitment of macrophages and CD4$^+$ cells into the lungs after pulmonary infection (128). Interestingly, neutrophil and B-cell recruitment is also reduced by MCP-1 neutralization, presumably as a secondary effect of decreased CD4$^+$ cell influx into the lungs (128). The net result of MCP-1 neutralization is diminished inflammation and alteration in pulmonary cytokine levels. This, in turn, reduces the clearance of *C. neoformans*. Hence, MCP-1 appears to have an important role in the formation of an effective cell-mediated response against *C. neoformans* infection in the lung (128).

The development of an effective inflammatory response in the murine lung is dependent on the production of IFN-γ and nitric oxide (175). Blocking of either IFN-γ or NOS by treatment with antibody to IFN-γ and N^G-monomethyl-L-arginine severely impairs the clearance of pulmonary *C. neoformans* in mice. However, the exact role of nitric oxide in host defense is uncertain. Neither inhibition of IFN-γ nor inhibition of NOS production affected the numbers or types of pulmonary inflammatory cells in the lung. Nitric oxide may be a direct antimicrobial effector molecule or may perhaps affect host defenses through its effects on vasodilation and vascular permeability (175).

CD4$^+$ T cells are necessary for the formation of multinucleated giant cells in the lung that surround and confine well-encapsulated yeast cells (112). Depletion of CD8$^+$ T cells abrogates DTH responses to cryptococcal antigens after pulmonary infection (191). Studies in SCID mice provide additional evidence supporting a critical role for T cells in orchestrating the immune response to cryptococcal infection in the lung (129). SCID mice are unable to contain cryptococcal infection in the lung, but adoptive transfer of T cells from immunocompetent donors significantly enhances pulmonary clearance (129).

A study comparing the inflammatory response to pulmonary infection in CBA/J mice using two strains of *C. neoformans* demonstrated significant strain differences (50). Strain ATCC 24067 (also known as 52D) induced dense perivascular and alveolar inflammation, whereas strain ATCC 62070 (also known as 145A) induced less intense lymphocytic infiltration and alveolitis. Analysis of lymphocyte subsets revealed that strain ATCC 24067 induced higher levels of CD4+ and CD8+ T-cell recruitment. The disparity between the strains was greatest in terms of CD4+ T-cell recruitment to lung tissues (50). This study strongly suggests that some of the differences observed in the human inflammatory response may be the result of unique *C. neoformans* strain differences in host inflammatory cell recruitment. The mechanism responsible for *C. neoformans* strain differences in inflammation remains unknown.

Intratracheal infection with *C. neoformans* in rats produces cryptococcal pneumonia (93, 99). Unlike mice, rats mount an extremely effective inflammatory response that contains the infection in the lungs with minimal or no dissemination (93, 99). Reduction of fungal burden in the rat lung is temporally associated with granuloma formation (93). CD4+ and CD8+ T cells are prominent in the rat inflammatory response to pulmonary infection (92). Athymic rats are unable to contain infection in the lung (99).

In summary, animal models support the view that granuloma formation represents an effective host response to *C. neoformans* infection in the lung. Granulomatous inflammation is associated with containment of infection and reduction in the number of yeast cells in lung tissues of mice and rats. Differences in the inflammatory response to infection can result both from *C. neoformans* strain factors and from the genetic background of the laboratory animals. The association between disseminated infection and disorders of cell-mediated immunity in human cryptococcosis is paralleled in rodent infection, where depletion of CD4+ and CD8+ cells invariably results in an inability to contain infection in the lung. The role of the antibody response in pulmonary protection is less certain. B cells do not appear to be necessary for the formation of multinucleated cells or the histiocytic rings that initially contain the infection (112). Pulmonary infection does not appear to elicit strong antibody responses in mice (76). However, administration of antibody prolongs survival, reduces fungal burden, and enhances granulomatous inflammation in mice (76). In would appear that the most effective pulmonary responses to *C. neoformans* infection involve cellular, humoral, and nonspecific immune responses. Granuloma formation and pulmonary containment are the hallmarks of a successful lung response.

RESPONSE TO INFECTION IN THE CNS

Human Infection

There are three general types of pathological reactions in the CNS: meningitis, meningoencephalitis, and granuloma formation in the brain parenchyma. Meningitis most commonly affects the base of the brain and the spinal subarachnoid space. The inflammatory response in meningitis is variable and can range from granulomatous inflammation to little or no inflammatory infiltrate. Granulomatous meningitis due to *C. neoformans* can resemble tuberculous meningitis. Meningitis is believed to evolve into meningoencephalitis as a result of invasion of the brain parenchyma by *C. neoformans* through Virchow-Robin spaces. It is likely that the perivascular Virchow-Robin spaces are infected by direct extension from the subarachnoid space. Virchow-Robin spaces that are swollen with organisms can give a "soapsuds" appearance on gross examination of the brain (82).

The pathology of cryptococcal infection in the CNS has been extensively described in the literature. Freeman (81) reported histopathological observations on 17 cases of cryptococcal meningoencephalitis in 1930 and categorized his cases as either diffuse or granulomatous meningitis. He noted enormous variation in the tissue response to infection in the brain and meninges, ranging from the complete absence of inflammatory cells to vigorous inflammation. Diffuse meningitis was characterized by proliferation of endothelial cells, with fibrosis and tissue infiltration by lymphocytes, plasma cells, and the occurrence of macrophage giant cells (Langerhans cells). Neutrophils were not observed. Granulomatous meningitis consisted of sharply defined nodules with whorls of fibrous tissue, giant cells, and epithelioid cells. Necrosis was rare, but there were instances of caseation in the larger inflammatory nodules. Organisms were found both enclosed in giant cells and free in tissue. Freeman (81) noted an inverse relation between the inflammatory response and the formation of cystic masses of *C. neoformans* organisms, with the latter being present in individuals with minimal tissue response. He also discounted the evidence for tissue digestion that had led to the naming of the organism as *Torula histolytica* and instead argued that tissue spaces were related to the capsule.

A recent autopsy study of cryptococcal meningitis in patients with or without HIV infection documented significant differences in the type of host inflammatory response found in these patient groups (159). Of 13 patients with HIV infection and cryptococcal meningitis, none had granulomatous inflammation. In contrast, most of the 14 patients with cryptococcal meningitis without HIV infection had granulomatous inflammation in the CNS. Lymphocytic and neutrophil infiltration was common in patients without HIV infection but rare among patients with AIDS. Considering the significantly better prognosis of cryptococcal meningitis in the non-HIV-infected host, this study suggests that granulomatous inflammation is a more effective response to *C. neoformans* infection in brain tissue.

There are reports in the literature of well-circumscribed *C. neoformans* granulomas in the brain (for review, see reference 37). *C. neoformans* can cause tumorlike masses in the brain (202) and spinal cord (37). *C. neoformans* granulomas are rarer in the CNS than in the lung, but they have been described in the cerebrum, cerebellum, and spinal cord (37, 52, 154, 168, 202, 251). The relative rarity of granulomas in the CNS compared to the lung may reflect the fact that extrapulmonary dissemination in itself could occur because of immune deficiencies and/or differences in the immune response in the brain and lung. Like pulmonary nodules, cryptococcal granulomas in the CNS are often well demarcated. In 1944 Swanson and Smith (265) described two patients with large cerebral granulomas that were originally diagnosed as tumors. Histopathological examination of the tumorlike tissues revealed cryptococcal granulomas in which the yeast cells were walled off by macrophages and lymphocytes (265). Liu (172a) described a patient who suffered a seizure and was found to have five round calcified masses in the frontal lobe, each measuring about 1.2 cm in diameter. The patient was treated by neurosurgical excision, but unfortunately no detailed histopathological analysis was provided. This case is unusual and noteworthy, because *C. neoformans* was

contained in the brain within granulomas and because calcification probably indicated a long-standing process (172a). This case emphasizes that calcification of
tissue can, in rare instances, occur in cryptococcal infection. In contrast, granulomas resulting from other fungal infections such as *H. capsulatum* frequently
calcify. Yeast cells can be so well contained within granulomas that cultures of the
cerebrospinal fluid (CSF) are negative for *C. neoformans* (202). Ley et al. (168) have
provided an excellent description of a cervical spinal cord granuloma in a young
girl. In this patient the granuloma produced symptoms of spinal cord compression,
and tumor was suspected. Removal was accompanied by symptomatic improvement, and histological examination of the lesion showed granulomatous inflammation with intracellular cryptococci in giant cells (168). These large *C. neoformans*
granulomas have been termed torulomas or cryptococcomas in the literature. In
general, cryptococcal granulomas in the CNS present as space-occupying masses
that are often initially assumed to be tumors (154, 251). Histologically, such lesions
are usually large granulomas.

C. *neoformans* infection can induce specific antibody in the CSF (231). Whether
such antibody is locally produced or reflects diffusion from serum is unknown.
Little work has been done on the physiological relevance or clinical usefulness of
antibody responses in the CSF.

Experimental Infection

The host response to experimental *C. neoformans* infection in the CNS has been
studied in a variety of animals, including mice (25, 26, 27, 61, 198), rats (94), and
rabbits (225). Analysis of animal models has provided important insights into the
effective immune response in the CNS. Animal infection can be induced by direct
inoculation of *C. neoformans* into the cranium. However, experimental animal infection is artificial by nature, and this fact, combined with species differences, pre-
cludes an exact correlation between animal models and human infection.

In mice, inoculation of *C. neoformans* into the cranium results in a lethal,
rapidly progressive meningoencephalitis (25, 26, 198, 274). Vanbreuseghem (274)
carried out some of the first systematic studies of intracerebral murine cryptococcosis and demonstrated significant strain differences in virulence. All strains examined produced a lethal infection, but the survival time after infection varied
significantly with the strain used. Interestingly, the size of the inoculum in intracerebral infection made relatively little difference in the survival time of infected
mice (274). For example, the average survival time of mice given either 100,000
or 5 cells was approximately 2 and 8 days, respectively. The rapid mortality that
often follows intracerebral infection of *C. neoformans* in mice is indicative of the
extreme susceptibility of this animal to direct CNS infection and can preclude
development of an effective immune response in the brain. The histological response to murine intracerebral infection has been described (163). Since mice die
rapidly after intracerebral infection, detailed histological studies have been limited
to the first few days of infection in mice that received relatively low infecting
inocula (163). Intracerebral inoculation results in both parenchymal and meningeal infection (163). In brain tissue, there is little or no inflammatory response,
but a cellular response consisting of PMNs and monocytes may occur in the
meninges (163).

Blasi and coworkers (25, 26) have studied the role of phagocytic defense mechanisms and the need for effective opsonization in the development of effective host responses against intracerebral infection in mice. Intracerebral inoculation of *C. neoformans* in CD-1 mice results in meningoencephalitis (25). The inflammatory cellular infiltrate is primarily monocytic and can vary from necrotizing inflammation to cystic parenchymal lesions with little or no inflammatory response (25). Treatment of mice with colchicine results in higher fungal burden and reduced survival, which has been interpreted as resulting from interference with brain phagocytic mechanisms (25). Mice given intracerebral infection with serum-opsonized *C. neoformans* cells have prolonged survival relative to those infected with the same number of unopsonized yeast cells (25). This finding is interpreted as suggesting that opsonization promotes phagocytosis and clearance of *C. neoformans* in the brain (25). Consistent with this observation is the finding that systemic administration of an opsonic monoclonal antibody to the *C. neoformans* capsule prolongs survival in murine intracerebral infection (198). The relative paucity of opsonins in the CNS is likely to be a major obstacle for effective host responses to *C. neoformans*. In this regard, *C. neoformans* yeast cells in brain tissue have been shown to be devoid of complement component C3 (271).

Once cryptococci disseminate to the CNS, there is evidence that systemic immune responses function to inhibit replication in the brain. In mice given intravenous infection, approximately 0.1% of the yeast cells colonize the brain, where they multiply rapidly (113). After 1 week, the increase in brain *C. neoformans* burden slows, consistent with the development of a protective immune response in the CNS (113). This response probably represents a systemic response to infection, since a previous pulmonary infection can prime the host for a more effective inflammatory response in the CNS (113). Histopathological studies revealed greater inflammation around *C. neoformans* lesions in mice with previous primary pulmonary infection. In contrast, there is no slowing of infection in the brain of SCID mice, but resistance can be transferred to SCID mice by CD4+ T cells from immune donors (113). Hence, the development of an effective systemic immune response confers some protection in CNS tissue, as measured by prolongation of survival and reduced fungal burden, and this response is dependent on T-cell immunity.

A study of brain inflammation in C57BL/6 mice infected intravenously revealed the presence of both L3T4 (CD4+) and Lyt2 (CD8+) cells around cystic collections of cryptococci in the brain (61). After intravenous cryptococcal infection, lymphocytic infiltration into the brain tissue was slower than into liver intravenously, suggesting slower kinetics for the leukocyte response in the brain relative to peripheral organs (61).

Local brain immune responses are also likely to be important in defense against *C. neoformans* in the CNS. Blasi et al. (28) have demonstrated that injection of heat-killed *C. neoformans* cells into mouse brains enhanced resistance to a subsequent local challenge with a lethal inoculum. Mice immunized with heat-killed cells intracerebrally have massive local inflammatory responses when challenged with lethal inocula of live *C. neoformans*. Intracerebral immunization elicits DTH responses in the hind feet in response to cryptococcal antigens, thus indicating that systemic immune responses arise from CNS inoculation of killed cryptococci (28). Furthermore, CNS inoculation of heat-killed cryptococcal cells delays brain coloni-

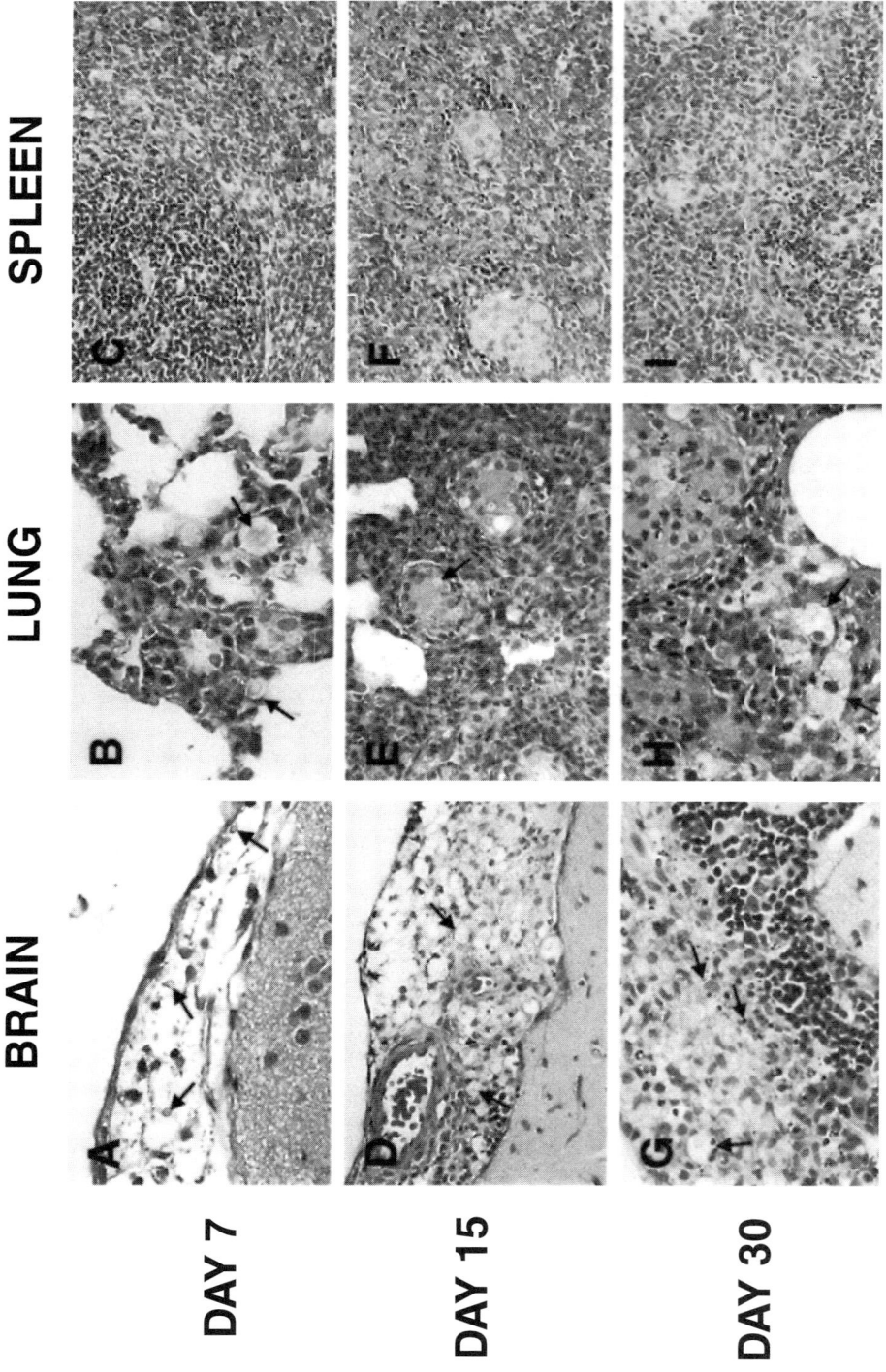

zation in mice given intravenous infection. This effect is associated with the production of IL-6 and IL-1β in brain tissue, and transient protection against brain infection can be demonstrated by intracerebral administration of these cytokines (28).

The course of intracerebral infection in mice can be altered by several compounds through their effects on host immune function. Chloroquine given to mice after intracerebral infection significantly enhances their survival time (185). Although the mechanism of chloroquine-induced protection is not fully understood, it appears to involve enhanced toxicity to intracellular organisms and changes in cytokine production (185). Pentoxifylline and dexamethasone have been demonstrated to enhance the efficacy of amphotericin B in mice given intracerebral infection, providing indirect evidence for the existence of neuroexcitatory pathogenic mechanisms (220).

In contrast to mice, inoculation of *C. neoformans* into the CSF of rats results in chronic infection that disseminates to extrameningeal organs (94) (Fig. 2). The rats show no obvious clinical signs of infection, and histopathological examination of the brain and meninges reveals a granulomatous inflammatory response to *C. neoformans* that effectively reduces fungal burden (94). In parallel with granuloma formation, there is increased blood-brain barrier permeability, with the appearance of complement-derived opsonins in the rat CSF that may promote phagocytosis and enhance the efficacy of phagocytic cells. The rat granulomas include both CD4$^+$ and CD8$^+$ T cells and macrophages that express inducible NOS (94). Interestingly, temporal analysis of histological changes in brain, lung, and spleen tissue reveals that the inflammatory reaction in the brain is delayed relative to lung and

Figure 2 Temporal inflammatory changes in three organs following intracisternal infection in rats. Rats given *C. neoformans* infection thorough the intracisternal route controlled the infection and seldom died (94). Histopathological sections of rat brain, lung, and spleen tissue on days 7, 15, and 30 after infection with *C. neoformans* are shown. (A) Brain tissue on day 7 reveals extracellular yeast forms (arrows) within the subarachnoid space. Original magnification, ×540. (B) Lung tissue on day 7 shows cryptococci surrounded by mononuclear cells. Original magnification, ×540. Cryptococci in lung tissue presumably are a result of hematogenous dissemination from brain tissue. (C) Spleen tissue on day 7 reveals no inflammation or cryptococci. Original magnification, ×335. (D) By day 15 the brain infection has changed histopathologically such that there are increased numbers of lymphocytes and macrophages in the subarachnoid space. Original magnification, ×335. (E) In the lung on day 15, inflammation has progressed to granuloma formation. Arrows point to cryptococci within Langerhans giant cells. Original magnification, ×335. (F) Spleen tissue on day 15 reveals discrete granulomas. Original magnification, ×335. (G) By day 30, inflammation has progressed in the subarachnoid space to include a large infiltration of lymphocytes, monocytes, and epithelioid cells. By this time, several layers of lymphocytes coat the pial surface. Original magnification, ×540. (H) In the lung on day 30, the inflammatory response continues with the formation of foamy macrophages. Original magnification, ×540. (I) In spleen tissue on day 30, inflammation is resolving, paralleling the resolution of infection in other organs. The figure, courtesy of David Goldman (Bronx, N.Y.), originally appeared in reference 94 and is reproduced here with permission.

spleen, suggesting that induction of an effective response in the CNS follows T-cell activation in peripheral tissues (94). The granulomatous inflammation observed in rat brains and meninges is similar to that observed in some non-AIDS patients with *C. neoformans* infection. The differences in the outcome and response between rat and mouse experimental systems indicate significant species differences and reinforce the concept that granuloma formation is an effective host response to *C. neoformans* infection in the brain (94).

Rabbits are resistant to intracisternal infection unless they are given cortisone to suppress their immune systems (225). The high level of resistance to infection in rabbits is unlikely to be related to the elevated body temperature of this animal, since cortisone administration does not eliminate fever and allows establishment of lethal infection in the brain. Histological studies reveal that, in the absence of cortisone treatment, rabbits mount an exuberant granulomatous-like mononuclear cellular reaction that controls the infection. In cortisone-treated rabbits, the inflammatory response is significantly inhibited, and this inhibition correlates with progression of infection. The ratio of mononuclear cells to PMNs is >10:1 at all times of infection. Analysis of CSF macrophages recovered from infected rabbits reveals that there is rapid macrophage activation, as measured by hydrogen peroxide production and cytotoxicity assays. However, CSF macrophages are unable to inhibit *C. neoformans* in vitro, suggesting that either these cells are not the primary effector cell in rabbit CSF or that the in vitro conditions lack the signals necessary for antifungal activity. Rabbits also respond to infection by producing opsonic IgG antibodies in the CSF, and there is a temporal association between the development of antibody opsonic activity in the CSF and a decrease in yeast burden (121). Rabbits also produce soluble chemotactic factors for PMNs and monocytes that contribute to recruitment of host inflammatory cells to the subarachnoid space (222). A florid inflammatory response in the subarachnoid space is a hallmark of the successful rabbit CNS immune response (222).

In summary, animal studies indicate that inoculation of *C. neoformans* into the CNS can elicit strong inflammatory responses in rabbits and rats that are effective in controlling infection. The effective inflammatory response is primarily monocytic and granulomatous. Humoral factors in the form of complement and antibody are likely to contribute to an effective response by providing opsonins for efficient phagocytosis by macrophages and chemotactic factors for cell recruitment. In mice, the infection is fulminant, and many animals die before an adequate response can develop. Nevertheless, measures that enhance the efficacy of phagocytic cells (i.e. complement opsonization or passive antibody administration) can prolong survival in mice. PMNs are commonly found in the inflammatory response but they do not appear to be a prominent feature of the immune response to infection in either species.

RESPONSE TO INFECTION IN THE SKIN

The first description of cutaneous cryptococcosis is often attributed to Rappaport and Kaplan (235), who reported an acneiform pustular lesion late in the clinical course of a patient with cryptococcal meningitis. At autopsy, the lesion contained fibrosis, granulomatous changes, and large numbers of yeastlike organisms (235). In the 1930s, cutaneous cryptococcosis was associated with systemic disease (193,

285); Weidman (284) provided an early review of this condition. Primary cutaneous cryptococcosis is actually an extremely rare condition, and the overwhelming majority of cases in which the skin is involved result from disseminated systemic infection (131, 219, 246). Hence, all cases of cutaneous cryptococcosis should be assumed to be due to disseminated infection until proved otherwise. The response to *C. neoformans* in the skin has been carefully described in humans and in experimental animal infection.

Human Infection

The inflammatory response in the skin to *C. neoformans* infection is dependent upon the immune status of the host. Wile (285) described the skin nodules histologically as consisting of large numbers of degenerating multinucleated giant cells engorged with yeast forms with minimal inflammatory reaction. Mook and Moore (193) described a case of cutaneous infection in the setting of disseminated infection that was characterized by masses of organisms with granuloma formation and fibrosis. Weidman (284) described the cutaneous lesion of *C. neoformans* as a granuloma. Linell et al. (170) described a case with large tumorlike masses and reviewed the literature. They noted that tissue reaction was often slight, with the inflammatory infiltrates consisting of sparse infiltrations of plasma cells, lymphocytes, and occasionally histiocytes and giant cells (170). Noble and Fajardo (219) described a patient with an inguinal ulcer that contained a heavy dermal inflammatory infiltrate composed of histiocytes and neutrophils, with yeasts found in compact masses within multinucleated cells or individually within macrophages. Many reports have described strong granulomatous inflammation in the skin and the presence of yeast cells within macrophages and giant cells (42, 193, 285). Hence, cutaneous cryptococcal infection has been associated with a diverse histological presentation, and the only specific characteristic is the presence of encapsulated fungi on skin specimens (170). Several reviews of the clinical literature of cutaneous cryptococcosis have been published (42, 170, 219, 221, 249, 284, 285). Six types of cutaneous lesions were recognized in the pre-AIDS literature: crusted granulomas, ulcers, subcutaneous swellings, cellulitis, dermal plaques or nodules, and pyoderma gangrenosum (221). Of these, crusted granulomas, ulcers, and subcutaneous swellings accounted for the majority of cases (221).

The ultrastructure of *C. neoformans* skin lesions has been studied in three patients without AIDS (115, 219). Electron microscopy of one lesion revealed that the majority of yeasts were intracellular, and frequent budding suggested intracellular replication (219). Giant cells contained multiple yeast forms, and up to 39 cryptococci were identified within one plane of a single giant cell (219). Another group studied lesions from two patients by electron microscopy; in one patient the majority of organisms were in the extracellular space, whereas in the other patient only intracellular fungi were observed (115). In both studies, macrophages and/or giant cells were intimately involved with *C. neoformans*. It is noteworthy that the histological appearance of cutaneous *C. neoformans* infection may change in response to antifungal therapy. Narisawa et al. (213) described a patient with acute lymphocytic leukemia complicated by cutaneous cryptococcosis in whom the histological appearance of the cryptococcal lesions changed from a gelatinous tissue reaction to a granulomatous tissue response. This case was unusual in that later lesions revealed giant cells with asteroid bodies.

Skin involvement is common in AIDS-related cryptococcosis, and it is almost always a consequence of systemic infection (179, 182, 203, 221, 246). Cutaneous lesions caused by *C. neoformans* have been reported in up to 27% of AIDS patients with cryptococcal meningoencephalitis (179). The shape of cutaneous lesions in patients with AIDS is highly variable, and the following types have been described: pustular, ulcerative, nodular, papulo-nodular, herpes-like, and molluscum contagiosum-like (51, 72, 179, 180). Cutaneous involvement in AIDS-related cryptococcosis has been associated with a rapidly progressive course of systemic infection (179). Histologically, the inflammatory response in AIDS-associated cutaneous cryptococcosis can range from granulomatous to gelatinous with a complete absence of inflammation (203). The presentation of cutaneous cryptococcal infection in patients with AIDS is often significantly different from that described in the pre-AIDS literature. Some cryptococcal skin lesions in patients with AIDS resemble Kaposi's sarcoma (221). Lesions resembling those of molluscum contagiosum have been described in several reports (72, 90, 180, 230). The skin lesions in patients with AIDS have commonly been described as raised, nodular, with a central umbilication that has been attributed to exuberant growth of *C. neoformans* in the setting of profound immunosuppression (221).

Experimental Infection

Cutaneous *C. neoformans* infection has been studied in a variety of animals, including mice (12, 73, 163, 195), rats (163), rabbits (163), guinea pigs (275), hamsters (252), and monkeys (141, 284). Sethi et al. (252) showed that inoculation of *C. neoformans* into the forelegs of hamsters produced chronic lesions in all animals and disseminated disease in occasional animals. Administration of cortisone to hamsters accelerated the course of infection such that all animals die within 4 weeks. The histopathological studies of the cutaneous lesions revealed histiocytic proliferation and granuloma formation. Interestingly, Schaumann bodies (calcified corpuscles) have been found in some of the hamster skin lesions (252). Weidman (284) serially inoculated a monkey intradermally and described the formation of nodules at the site of each intradermal injection. Histological analysis of cutaneous lesions in the monkey revealed granulomatous inflammation.

The most detailed studies of animal cutaneous infection have been performed in mice. In 1961, Bergman (12) published an extremely detailed, comprehensive study of subcutaneous infection in mice. He demonstrated that rapid lymphatic dissemination occurs within hours after cutaneous infection. Subcutaneous inoculation is followed by fungal proliferation at the infection site for several days, but, ultimately, the skin infection appears to resolve in most mice. In contrast, fungal replication continues in organs seeded by disseminating yeast and eventually results in lethal visceral infection. Other groups have described similar results. Inoculation of *C. neoformans* into the skin of immunologically intact mice produces cutaneous infection that can disseminate and kill the mice (73, 195, 256). However, a vigorous immune response occurs at the site of infection; this immune response contains the infection in the skin and leads to a reduction in yeast CFU within cutaneous lesions (73, 195). The histological description of skin lesions resulting from experimental infection in mice (272) is similar to that described for some cases of human cutaneous infection (40, 91). The murine inflammatory response 3 weeks after cutaneous inoculation consists of a granulomatous reaction dominated by macrophages and monocytes, with occasional neutrophils, lymphocytes, and multinucleated giant cells (195). Yeast forms are frequently found inside macrophages (195).

Some *C. neoformans* strains may exhibit a dermatotropic phenotype during experimental infection. Kessel and Holtzwart (141) injected *Macacus rhesus* monkeys intracardially and discovered that some developed nodular cutaneous lesions in the face and ears 4 weeks after inoculation. Dixon and Polak (60) described a rhinotrophic *C. neoformans* strain that caused prominent dermal lesions in the nose area after systemic infection. Fromtling et al. (85) described another rhinotrophic *C. neoformans* isolate that formed granulomas in periocular tissues, ears, feet, and tails. Both rhinotrophic isolates grew better at lower temperatures, suggesting that the thermotolerance of the strain contributed to the dermatotropic phenotype (108, 122).

Experiments in mice suggest that the immunological status of the host is a major factor in determining whether cutaneous lesions develop after systemic infection. Salkowski and Balish (242) discovered that inoculation of *C. neoformans* into the tail vein of mice with congenital T-cell defects (athymic and beige-athymic mice) resulted in dissemination to the skin, with the formation of small subcutaneous nodules without ulceration. Microscopic examination of these nodules revealed a minimal inflammatory response that consisted primarily of macrophages (242). Such lesions are not usually found in immunocompetent mice, and their occurrence in athymic and beige-athymic mice suggests that immunosuppression may contribute to cutaneous involvement in systemic cryptococcosis.

In summary, there is considerable diversity in the results of experimental cutaneous infection with *C. neoformans* in animals. In some reports, skin infection heals rapidly, whereas in others a chronic infection ensues. Interstudy differences may reflect differences in animal species, inoculum, immunological status of the host, and dermatotropism of the *C. neoformans* strain. The animal experiments suggest that the occurrence of cutaneous lesions in an individual cryptococcal infection is a function of both the immunological status of the host and the dermatotropism of the *C. neoformans* strain.

SPECIAL TOPICS IN IMMUNITY

Relative Efficacy of Anti-*C. neoformans* Defenses

In reviewing the literature on host defenses, it is striking that all humoral and cellular systems studied appear to contribute to defense against *C. neoformans*. Furthermore, the "importance" of any one system can often be shown by carefully selecting the experimental conditions. The question naturally arises as to which defense systems are more important and why. However, despite intensive studies over the last 3 decades, we do not know the answers to the following questions: Which is the primary effector cell against *C. neoformans*? What is the exact immune defect that predisposes to invasive cryptococcal infections? What is the relative contribution of complement and antibody to opsonization in vivo and to protection against infection? Is there a genetic predisposition to invasive cryptococcosis?

One approach to understanding the relative importance of the components of the immune system is to define the conditions that predispose to infection. For cryptococcal infections, there is a general consensus that cell-mediated immunity is the primary line of defense against infection (164). However, invasive cryptococcal infections tend to occur in individuals with profound derangements of immunity in which multiple defects of immune function occur. For example, conditions associated with *C. neoformans* infections include AIDS, lymphoproliferative disorders, and corticosteroid therapy. In each of these conditions, specific and nonspecific cellular defenses are impaired, and there is usually defective B-cell function. Furthermore, only a minority of patients with AIDS suffer from disseminated *C. neoformans* infections despite severely compromised T-cell, B-cell, and nonspecific immunity. Thus, for *C. neoformans*, inferences about the relative importance of humoral immune mechanisms based on the epidemiology of infection are likely to be equivocal and inconclusive.

A more controlled approach is to define the immune defects that predispose to infection or are associated with more severe infection in congenic animal strains. Reversal of increased susceptibility to infection by complementing or repairing the immune defect would provide strong evidence for the importance of the immune component in question. The overwhelming majority of animal studies that have explored immune function have been done in mice. The popularity of mice as an experimental system is the result of several factors: (i) mice are highly susceptible to experimental infection; (ii) many strains of mice are available, including dozens of strains with specific gene defects; (iii) murine immunological reagents are widely available from commercial sources; (iv) mice are well-established laboratory animals; and (v) the cost of experiments in mice is modest relative to those in other mammalian species. However, major differences in the immune systems of mice and humans necessitate caution in extrapolating to humans results obtained in mice. For example, human and murine immune systems differ in the ability of macrophages to generate nitric oxide (215, 216) and the defensin content of neutrophils (162). Despite these differences, the murine model remains a useful system for studying the pathogenesis of cryptococcal infection. However, the usefulness of animal models for determining the relative importance of the various immune components is less clear. For example, depending on experimental design, mice can be used to support an important role for many components of the immune system, including complement, antibody, T cells, natural killer (NK) cells, macrophages, and neutrophils, in protection against infection.

Inferences about immune cell function made from epidemiological observations and/or animal studies are often tested in vitro by studying the ability of a given cell type to mediate antifungal efficacy. One difficulty with such studies is that in vitro conditions may have little or no relevance to in vivo conditions. Furthermore, differences in the in vitro conditions can have profound effects on cell function as has been demonstrated for human monocytes (165, 167). Since the optimal conditions for anti-*C. neoformans* function by each cell type are unknown and may differ with the cell type, it is difficult to draw conclusions about the relative efficacy of immune effector cells from in vitro studies. Because of these uncertainties, negative studies on immune cell efficacy against *C. neoformans* in vitro have to be interpreted cautiously.

In summary, one cannot make definitive conclusions on the relative efficacy of the various components of the immune system in protection against *C. neoformans* at this time. *C. neoformans* is a low-virulence pathogen for individuals with intact immunity. There appears to be considerable redundancy in humoral and cellular defenses, and the fungus is susceptible to various microbicidal molecular systems present in immune effector cells. The redundancy in host defenses against cryptococcosis suggests that susceptibility to this pathogen may require multiple immune defects. If this is the case, an integrated approach to studying host defenses is likely to prove more fruitful in increasing our understanding of cryptococcal pathogenesis than are attempts to rank the relative efficacy of immune mechanisms.

Integrated Concepts of Immunity

Over the past quarter century, considerable effort has been made to dissect the mechanisms of host defense that contribute to defense against *C. neoformans* and to establish the basis for effective immunity. As discussed above, this goal has been elusive because of the complexity of the host response to *C. neoformans*, redundancy in host defense mechanisms, and the interrelationships between humoral and cellular immunity and specific and nonspecific defense mechanisms. Several investigators have argued that a successful host defense against *C. neoformans* is dependent on multiple mechanisms, resulting in integrated proposals for anticryptococcal immunity. In considering host defense mechanisms against *C. neoformans*, it is important to keep in mind that this fungus can exist in both intra- and extracellular compartments and that the effective host response must be able to deal with yeast cells in either venue. Furthermore, capsular synthesis after infection can result in very large cells in tissue that present a formidable problem for phagocytosis.

In the 1960s and early 1970s, Aronson and collaborators (3, 136, 248, 254) carried out a remarkable series of experiments to investigate early *C. neoformans* interactions with PMNs and monocytes. Their results clearly illustrate collaboration between humoral and cellular defense mechanisms. The question studied was the mechanism by which host inflammatory cells deal with a large microbial target such as well-encapsulated *C. neoformans* cells. These rarely cited studies show that antibody and complement promote the formation of PMN or histiocytic rings around *C. neoformans* yeast cells after peritoneal infection. Early in infection, the rings are composed of PMNs, whereas later in infection, the rings are composed of monocytes. Histiocytic ring formation needs exogenous antibody administration in mice but not in rabbits or guinea pigs. Inflammatory cells surrounding *C. neoformans* secrete lysosomal products that damage the organism and ultimately the fungal cell. Yeast replication is apparently inhibited by the natural fungistatic properties of serum that involve iron chelation. These early studies show the efficacy of collaboration between complement, antibody, serum iron-binding proteins, PMNs, and monocytes.

In 1980 Griffin (103) recognized the potential for collaboration between the antibody and complement systems in promoting phagocytosis of *C. neoformans* by macrophages and the requirement for macrophage activation for phagocytosis. He noted that a lymphokine produced from macrophages stimulated with immune complexes activated macrophages to ingest *C. neoformans* through C3 receptors. This report predated the explosion in knowledge of cytokine biology that occurred in the 1980s, and the lymphokine was never characterized. Griffin (103) proposed that the outcome of cryptococcal infection (i.e., containment versus dissemination) was determined by a complex interplay between the timing of fungal replication, antibody response, and macrophage activation. Considering that macrophage activation for killing of intracellular microorganisms is usually the result of T-cell immunity, the Griffin model implies that a successful host defense requires collaboration between humoral, cellular, and nonspecific immune mechanisms.

In 1982, Fromtling and Shadomy (86) reviewed the literature on the immunology of *C. neoformans* infection and concluded that "immunity to *C. neoformans* may depend upon a complex interaction of humoral and cellular immune factors." They noted the many apparent contradictions in the literature but discerned the crucial roles played by macrophages, opsonins, and T cells. They suggested that humoral immune mechanisms (i.e., complement and antibody) were responsible for opsonization of *C. neoformans* for macrophage phagocytosis, and cellular immunity (i.e., T cells) was responsible for macrophage activation for efficient fungicidal activity.

In 1986, Miller (189) reviewed the immunology of cryptococcal infection, noted the central role of T-cell function in effective host responses, and suggested the need for both humoral and cell-mediated immunity in providing protection against infection. She considered antibody immunity to play an "adjunctive role in recovery from infection by facilitating phagocytosis and intra- and extracellular killing of organisms" (189). In her view, the outcome of infection depends upon a race between the host's mobilization of humoral and cellular immunity and the organism's replication and production of immunosuppression through polysaccharide products. Macrophages and PMNs with complement-derived opsonins provide the first line of defense by phagocytosing and killing yeast cells. This interaction results in secretion of inflammatory cytokines, antigen presentation, and recruitment of T cells, which then secrete a variety of cytokines that activate effector cells and recruit additional inflammatory cells (e.g., NK cells, monocytes, eosinophils). B cells would respond to cryptococcal antigen by secreting antibody that enhances phagocytosis and killing of *C. neoformans* cells. In normal hosts this sequence of events results in containment and eradication of infection. However, in immunocompromised patients, the response is slower and less effective, allowing the fungal proliferation with release of polysaccharide antigens that suppress both the humoral and cellular immune responses (see chapter 9 for a description of the immunosuppressive effects of capsular polysaccharide). Miller (189) argued that the existence of both intact humoral and cellular immunity is responsible for the rarity of disease in immunocompetent individuals.

In summary, a common theme in immunological studies of *C. neoformans* infection is that an effective host response requires most, if not all, components of the immune system to be present and functional.

Host Defense and Susceptibility to Infection

The major conditions predisposing to *C. neoformans* infections are HIV infection, steroid use, and lymphoproliferative disorders. Each of these conditions weakens host defenses and permits dissemination of cryptococcal infection. At this time the exact immune defect(s) responsible for susceptibility to *C. neoformans* infection is unknown. Susceptibility to cryptococcal infection has traditionally been associated with conditions that weaken cell-mediated immunity and, in particular, T-cell immunity. It is not clear if cryptococcosis is a direct or an indirect result of defective T-cell immunity. For example, susceptibility resulting from a defect in T-cell immunity could involve the inability of T cells to mediate direct antifungal effects, produce pro-inflammatory cytokines, and/or coordinate granuloma formation. Alternatively, derangements in T-cell function could impair the ability of the host to mount effective antibody responses and predispose the host to infection by reducing opsonins. Considering the central role of T lymphocytes in the generation of effective immune responses, it is likely that susceptibility to cryptococcal infection is the result of global defects in immunity that follow defective T-cell immunity. However, it is worth remembering that only 1 in 20 patients with HIV infection and low CD4$^+$ T-cell counts actually develops symptomatic infection. Hence, CD4$^+$ T-cell lymphopenia alone may not be sufficient to predispose to cryptococcosis. Furthermore, some patients develop cryptococcosis despite normal lymphocyte counts.

C. neoformans infection has been described in patients with hypogammaglobulinemia and in children with the hyper-IgM syndrome. The relative paucity of *C. neoformans* infections among patients with antibody disorders has been used to argue against an important role for antibody immunity. However, cryptococcosis is a relatively unusual infection among most classes of immunocompromised patients, and, considering that severe disorders of antibody immunity are themselves rare, the lack of association may reflect the inherent difficulty in determining a link between two rare events. Hyper-IgM syndrome is an antibody disorder that has been associated with pediatric cryptococcosis. Quantitative and qualitative differences have been reported between antibody levels to *C. neoformans* polysaccharide among individuals with or without HIV infection, suggesting that subtle defects of antibody immunity may contribute to the marked predisposition in patients with AIDS. In mice, there are protective, nonprotective, and infection-enhancing antibodies to *C. neoformans* polysaccharide. If the same occurs in humans, one could anticipate that the relative composition of the serum antibody response could protect against or even predispose to *C. neoformans* infection. Furthermore, there is evidence that the T-cell status of the host may determine the protective efficacy of a given antibody. In the absence of CD4$^+$ cells, protective antibodies are not effective, whereas enhancing antibodies function to shorten survival of infected mice (291). Hence, the antibody composition of sera combined with the CD4$^+$/CD8$^+$ T-cell ratio may protect against or predispose to *C. neoformans*.

Neither congenital nor acquired defects in the complement system have been associated with human susceptibility to *C. neoformans* infection. In laboratory animals, there is convincing evidence that defects in the complement system or complement depletion can result in more severe cryptococcal infection (58, 59, 148, 237). Nevertheless, there is a consensus of opinion that the complement system is an important source of opsonins and mediators of inflammation in *C. neoformans* infection (148). It is not clear whether the lack of association between complement defects in humans reflects a true phenomenon or stems from the absence of careful studies that may detect subtle complement defects in these patients. However, such studies may be difficult, because cryptococcal infection is accompanied by a state of acquired complement deficiency as a result of complement pathway activation by the cryptococcal capsule (178).

The macrophage is a major effector cell against *C. neoformans* infection. Macrophages are probably the first line of defense in the alveoli, and monocyte- or macrophage-derived granulomatous inflammation appears to be the most effective tissue response in clearing and containing *C. neoformans*. No evidence for a global phagocytic defect in reticuloendothelial phagocytic function was found in patients with disseminated cryptococcosis, as measured by the uptake of radioiodinated albumin aggregates (14). However, in patients with HIV infection, there is evidence for defects in macrophage (36, 106) and NK cell function (123) against *C. neoformans*. Reduced cryptococcal activity in macrophages from HIV-infected patients could be a direct result of macrophage infection with macrophage-tropic HIV strains or an indirect result of cytokine dysregulation from reduced numbers of CD4+ T cells. Studies with rhesus monkey macrophages infected with simian immunodeficiency virus (SIV) also demonstrate killing defects against *C. neoformans* (30). However, the killing function of SIV-infected macrophages can be restored by culturing the cells in lymphocyte supernatants with elevated concentrations of IFN-γ and IL-10 (30). This result suggests that the functional defect is not intrinsic to the macrophages but rather is dependent on other SIV-induced effects, including CD4+ lymphopenia. In similar experiments, Horn and Washburn (123) demonstrated that the anticryptococcal activity of NK cells from HIV-infected individuals could be restored by addition of exogenous IL-12.

Genetic factors are likely to be important contributors to susceptibility to cryptococcal infection, but to date the human genetic traits associated with resistance and susceptibility are not known. In mice, there are clear genetic differences in susceptibility to experimental *C. neoformans* infection (238). Two loci associated with susceptibility to infection in mice are the *Hc* locus on chromosome 2, which determines the level of serum complement component 5 (238), and the *nu* locus on chromosome 11, which, when homozygous, results in hairless athymic mice (41, 102, 181). The influence of host genetics is also evident in differences in the host inflammatory response to pulmonary infection in mice (126, 174). Analysis of nine inbred mouse strains for susceptibility to *C. neoformans* infection showed that susceptibility to cryptococcosis was under multigenic control and associated the *xid* locus (X-linked immunodeficiency) with susceptibility to infection (181). The *xid* locus results in the absence of a B-cell subpopulation and a marked deficit in the ability to mount antibody responses (181). In humans, maleness is associated with enhanced susceptibility to *C. neoformans* infection (see chapter 11). The basis for the increased male susceptibility is not understood but could reflect either differences in the hormonal milieu of the host or male and female immune responses. Hence, there is compelling evidence from murine studies that genetic factors contribute to susceptibility to *C. neoformans* infection, and it is likely that the same applies to humans.

The most common defect associated with susceptibility to *C. neoformans* infection is abnormal cell-mediated immunity. In patients with AIDS, HIV infection induces CD4+ T-cell depletion, which has profound effects on specific and nonspecific host defense mechanisms. Corticosteroids also have global effects on immune function and result in decreased humoral and cellular immune responses. Similarly, lymphoproliferative disorders result in replacement of lymphatic tissue with malignant cells and are associated with decreased humoral and cellular immune responses. Several investigators have studied DTH skin reactivity and lymphocyte proliferation in vitro in response to cryptococcal antigens in non-AIDS patients who have recovered from *C. neoformans* infection (100, 247) and found subtle defects in cellular immunity. These defects were interpreted as predisposing to infection. However, *C. neoformans* infection can result in specific immunosuppression (see chapter 9), and it is unclear whether the immune deficits predated or were a consequence of infection.

In summary, the literature strongly supports the view that impairment of cell-mediated immunity (in particular, T-cell function) is associated with enhanced susceptibility to *C. neoformans* infection. However, the exact mechanism by which defective cell-mediated immunity permits *C. neoformans* to colonize and disseminate in mammalian hosts is uncertain. Defects in T-cell immune function may have secondary effects on the generation of antibody responses and/or the functional efficacy of nonspecific cellular immune mechanisms that predispose to fungal infection. The fact that only a small minority of individuals with CD4$^+$ T-cell lymphopenia develop cryptococcosis strongly suggests that other immune factors are important in both susceptibility and resistance to *C. neoformans* infection.

Fungus-Induced Immunosuppression

There is considerable evidence that cryptococcal infection produces detrimental effects on host immunity. Most human infections occur in patients with underlying immune disorders, and cryptococcal infection may further undermine the immune system. Many of the deleterious effects on host immunity have been associated with the capsular polysaccharide (see above). Capsular polysaccharide can induce antibody unresponsiveness (i.e., immune paralysis-like effect) (149, 151, 209, 263, 264), which can prevent the formation of an effective antibody response specific for the polysaccharide antigen. Capsular polysaccharide may also interfere with the development of effective inflammatory responses through its effects on leukocyte migration and stimulation of inhibitory cytokine secretion (see above). However, cryptococcal infection also has significant effects on cell-mediated immunity, and this section focuses on the phenomenon of suppressor responses that result from the induction of T-suppressor cell circuits.

Suppressor responses

Many individuals with cryptococcosis manifest anergy when challenged with cryptococcal antigen by skin testing. For example, Atkinson and Bennett (4a) demonstrated that only 58% of patients with active infection have positive skin reactivity to a cryptococcal antigen preparation (cryptococcin). Among individuals who recover from infection, only 70% have positive skin reactivity, suggesting prolonged anergy in a significant proportion of patients. Similar results were reported by Graybill and Alford (100), who demonstrated that only one third of patients who recover have positive skin reactivity to cryptococcin. There is also evidence for long-lasting antibody unresponsiveness to polysaccharide antigens (109, 110). The results of lymphocyte transformation studies in patients parallel the results of DTH skin testing (55). Lymphocytes from asymptomatic individuals with positive skin testing to cryptococcin exhibit significantly higher proliferation than do lymphocytes from controls with negative skin testing (55). However, patients who recover from cryptococcal infection have reduced lymphocyte proliferation in response to cryptococcal antigens, consistent with either tolerance to fungal antigens and/or persistence of immunological deficits (55, 247).

The findings of immunological deficits among patients who recover from infection have been interpreted as indicative of subtle immune deficits that originally predisposed the patient to infection (247). However, an alternative explanation is that the immunological deficits are the result of cryptococcal infection and reflect a long-lasting derangement of cellular immunity following exposure to cryptococcal antigens. Although this question is not resolvable with the available human data, the experience with animal experiments indicates that cryptococcal infection can induce suppression of cell-mediated immune responses.

Blackstock and coworkers (15, 18–24, 194, 239, 262) have studied a suppression phenomenon that occurs during experimental *C. neoformans* infection in mice. *C. neoformans* infection in immunocompetent C57BL/6J mice produces suppression both to specific responses to cryptococcus antigens and to unrelated antigens (18, 239). Suppression is transferred by cells but not by serum. Specific suppression results in reduced lymphocyte proliferation in response to cryptococcal antigens after about 2 weeks of infection (239). Nonspecific suppression results in a variety of defects in the immune response to noncryptococcal antigens, including reduced antibody responses, reduced lymphocyte proliferation, and diminished DTH reactions to sheep erythrocytes (18). The specific and nonspecific suppression phenomena may reflect one suppressor mechanism. The induction of suppressor responses has been associated with at least two types of cell: one found in the adherent cell fraction and the other found in the T-cell fraction (18). These cells begin to appear about 2 weeks after infection, but their phenotypes have not been characterized. The specific and nonspecific suppressor responses were also observed when mice were infected with an avirulent pseudohyphal strain of *C. neoformans* (17). Hence, infection-related immunosuppression may not directly relate to the virulence of the organism but rather appears to be a function of cryptococcal components that stimulate suppressor responses (17). The induction of suppressor responses even by avirulent strains may help to explain why it has been difficult to elicit protective responses by vaccination with cryptococcal preparations (17).

Both infection and soluble *C. neoformans* polysaccharide induce macrophage-regulatory suppressor T cells in the spleens of mice, which produce a suppressor factor that inhibits phagocytosis in a subset of macrophage cells (22, 194). The suppressor T cells are phenotypically CD8+ and I-J+ (20) and can be elicited by inoculation of as little as 25 μg of polysaccharide antigen in mice and persist for at least 2 months (22). These suppressor T cells inhibit macrophage phagocytosis of *C. neoformans* through the synthesis of a T-suppressor factor (TsF) (19). TsF has a mass of about 70 kDa, is I-J+, and can bind to soluble cryptococcal polysaccharide (19). Hence, TsF appears to be a bifunctional molecule capable of binding both to the I-J+ receptor and to cryptococcal polysaccharide. The macrophages that respond to TsF are a small proportion of the macrophage pool that express I-A antigen and have an I-J receptor (21). Because only a small proportion of macrophages are I-J receptor positive, it is unlikely that the effect on phagocytosis is responsible for the widespread immunological deficits observed during infection (15). Instead, the mechanism of suppression has been proposed to involve binding of TsF to macrophages and sensitizing (or arming) to respond to capsular polysaccharide stimulation with the release of a nonspecific macrophage suppressor factor (15, 24). TsF inhibits phagocytosis through Fc and mannan receptors but does not affect complement-mediated phagocytosis (21).

Analysis of the TsF effect associated with *C. neoformans* infection reveals functional equivalence to similar suppressor effects described for hapten-specific suppressor T cells, such as antipicryl TsF (4, 23, 24). Induction of suppressor T cells to produce TsF requires an antigen-presenting cell and cryptococcal polysaccharide (15). TsF factors have multiple effects on host immunity, including inhibition of contact sensitivity, DTH, and granulomatous formation. The functional similarities of the TsF induced by capsular polysaccharide and that induced by haptens suggests that TsF made in response to cryptococcal polysaccharide may contribute to the loss of DTH reactivity and poor granuloma responses observed during *C. neoformans* infection.

Another type of suppressor effect was described by Murphy and coworkers (33, 80, 196, 204–206, 208, 210–212). Using a culture filtrate preparation known as CneF, which contains both soluble capsular polysaccharide antigen and extracellular proteins, they elicited either DTH or suppression of DTH, depending on how CneF was administered. DTH phenomena have been divided into two phases: the afferent phase, which corresponds to induction of the immune response, and the efferent phase, which corresponds to expression of DTH after antigen challenge (205). Injection of mice with CneF subcutaneously in complete Freund's adjuvant induced a T-cell population that mediated DTH. In contrast, when CneF was given intravenously, it elicited specific T-cell-mediated suppression of the cryptococcal DTH response (205, 210). The suppressor lymphocytes were phenotypically $Lyt1^+2^-$ T cells and could suppress the cryptococcal DTH response if transferred before immunization with an antigen preparation (204, 210).

The regulation of the suppressor responses studied by Murphy and coworkers (33, 80, 142, 196, 204, 210–212) is complex. Two types of suppressor cells were identified. Ts1 and Ts2 are first- and second-order suppressor cells that inhibit the afferent and efferent limbs of the DTH response, respectively (204). Ts1 cells suppress the induction of DTH and induce the formation of Ts2 cells (212). Ts1 cells are $Lyt1^+$, $I-J^+$, and cyclophosphamide sensitive and are produced in lymph nodes in response to intravenous injection of CneF (210, 212). Ts2 cells are $Lyt1^-,2^+$, $I-J^+$ T cells produced from cyclophosphamide-resistant precursors in the spleen in response to Ts1 cells of TsF1 (80, 211). Ts1 and Ts2 cells are believed to produce suppressive factors, known as TsF1 and TsF2, respectively (196). TsF1, but not TsF2, binds to heat-killed cryptococci (196). Neither suppressor factor contains cryptococcal antigen, anticryptococcal antibody, or C1q immune complexes (196). However, it appears that Ts2 cells do not suppress the DTH response unless third-order suppressor T cells (Ts3) are also present (142). Furthermore, there appears to be a $CD4^+$ T-cell population in the spleen that up-regulates and amplifies the anticryptococcal DTH responses (33). The suppressor pathway described by Murphy and coworkers is similar to the antigen-specific suppressor T-cell pathway elicited in response to hapten immunization (210–212).

A physiological correlate for the results obtained with the CneF preparation is suggested by the observation that serum from infected mice can elicit similar suppressor cells that inhibit the DTH reaction (205). Furthermore, the degree of suppression correlates directly with the amount of polysaccharide in the serum (205). Subsequent experiments have shown that the suppressive factor in the sera of infected mice is a cryptococcal polysaccharide antigen (208). Hence, the available data strongly suggest that the cryptococcal antigen in CneF that elicits DTH is the mannoprotein (205). A biologically relevant consequence of the DTH-suppressive effects of intravenous CneF is suggested by the higher fungal burden in mice suppressed with CneF and subsequently challenged with *C. neoformans* (206). The suppressive effect of CneF appears to be specific for *C. neoformans*; no suppression has been noted for DTH to *L. monocytogenes* or for inhibition of clearance of *L. monocytogenes* bacteria from organ tissues in mice given intravenous CneF (206).

The relationship between the suppressor effects described by the Blackstock and Murphy groups can be inferred from what is known about classical suppressor responses to haptens (22). The suppressor phenomenon described by Blackstock and colleagues appears to be analogous to the picryl pathway (23). In contrast, the suppressor phenomenon described by Murphy and colleagues appears to be analogous to the 4-hydroxy-3-nitrophenyl acetyl (NP) pathway (67). Mechanistic studies of the picryl and NP haptenic pathways suggest convergence for some of the cellular and functional components of both pathways. The Ts3 cell of the NP pathway (analogous to the Ts3 cell described in the Murphy suppressor pathway) and the suppressor T cell of the antipicryl TsF pathway (analogous to the suppressor T cell that secretes TsF in the Blackstock suppressor pathway) both function to arm T-acceptor (T_{ACC}) cells and macrophages. Therefore, these cells are functionally equivalent but differ in their binding-site antigen specificity. The suppressor pathway described by Murphy and colleagues appears to be specific for mannoprotein antigens, whereas the suppressor pathway described by Blackstock and colleagues is responsive to GXM. Furthermore, these suppressor pathways may require different induction stimuli. The suppressor pathway described by Murphy may require the induction of Ts1, Ts2, and Ts3 cells, whereas that described by Blackstock may be induced by GXM directly. Both suppressor systems share important similarities. For example, both are induced by similar amounts of capsular antigen, and their suppressor T cells have similar phenotypes and produce suppressor factors that bind to polysaccharide antigen. The role of cytokines, prostaglandins, and nitric oxide in the induction and production of the immunological suppressive effects associated with *C. neoformans* infection have not been fully defined. Since cryptococcal polysaccharide has been shown to induce cytokine production in vitro (236, 279), some of the immunosuppressive effects associated with *C. neoformans* infection may be cytokine related.

C. neoformans-induced suppressor responses have also been described in rats. Rats infected with *C. neoformans* produce suppressor cells that inhibit a cell-mediated response (DTH) to the unrelated antigen human serum albumin (183, 184, 257–259). The suppressor cell population includes both nylon wool adherent and nonadherent splenic mononuclear cells (258). The nonadherent suppressor cells induced by *C. neoformans* are cyclophosphamide sensitive and have been characterized as Ts1 cells (184, 258). In *C. neoformans*-infected rats, there is a significant reduction in peritoneal cells expressing the I-A antigen (major histocompatibility complex class II) (183). However, the phenomenon is more complex: a reduction in macrophage cells bearing the I-A antigen is associated with T-helper responses, and an increase in cells bearing the I-E antigen is associated with T-suppressor responses (240). Incubation of peritoneal macrophages with suppressor T cells reduces macrophage I-A expression in vitro. Analysis of lymphocyte subsets in infected rats has revealed changes in the ratio of the number of molecules expressing the I-A and I-E histocompatibility molecules in spleen, thymus, and lymph nodes (259). Furthermore, splenic CD4$^+$ and CD8$^+$ cells increase, and B cells decrease (259). These results indicate that *C. neoformans* infection in rats can induce changes in the cellular composition of the lymphoid compartment (259). This suppressor phenomenon described in rats has certain similarities to suppressor pathways described by the Blackstock and Murphy groups (see above).

In summary, three research groups have described the induction of immunosuppressive phenomena as a consequence of *C. neoformans* infection. The suppressor effects in mice are similar to those described for classical haptenic T-suppressor cell responses (4, 67). Suppressor effects have been observed in two rodent species. In humans, clinical experience indicates anergy and poor inflammatory responses; this provides circumstantial evidence consistent with immunosuppression in association with *C. neoformans* infection. It is noteworthy that the amount of capsular polysaccharide required to elicit suppressor T cells in C57BL/6 mice (22) is similar to that required to produce antibody unresponsiveness (209). Antibody unresponsiveness to *C. neoformans* polysaccharide may be analogous to the immunological paralysis effect described for vaccination with pneumococcal polysaccharide, which appears to be mediated by induction of T-suppressor cell responses (7). The relationship between the hapten-like T-suppressor cell phenomena and antibody unresponsiveness is not known, but it is possible that all the effects are different manifestations of a common pathway.

In summary, there is strong evidence for the occurrence of immunosuppression during cryptococcal infection that is associated with the induction of T-suppressor cell responses. Fungus-induced immunosuppression could, in turn, produce global deficits in the host immune response involving B and T lymphocytes and macrophages, which presumably contribute to virulence by weakening the ability of the host to fight infection. The mechanisms by which suppressor responses contribute to pathogenesis, the relative importance of the various suppressive phenomena, and the correlation of in vitro and in vivo effects are important areas for continued investigation.

Other fungal products

In addition to the soluble *C. neoformans* antigens that have been classically associated with immunosuppression, fungal cells produce a variety of compounds that may interfere with the induction of effective immune responses (276). (For a more complete description of extracellular products and their effects on the immune system, see chapter 4.) An "endotoxic substance" in extracts from cryptococcal cells reduces survival when administered to mice 24 h before intravenous infection (146). The biochemical nature of this endotoxic substance is not known (146). *C. neoformans* strains have associated extracellular proteinase activity that could degrade immunoglobulin, complement, and other immunologically important proteins (31, 44). Melanin synthesis by *C. neoformans* cells may also influence the immune response; comparison of high- and low-melanin-producing strains suggests that melanin interferes with induction of T-cell-mediated immunity in the lung (124). Melanin is an efficient free radical scavenger that may quench reactive oxygen- and nitrogen-derived oxidants produced by phagocytic cells (132, 281, 282). It has been suggested that melanin may interfere with macrophage TNF-α production (124). *C. neoformans* cells have been shown to produce the hexitol mannitol during human cryptococcal meningitis (187) and experimental infection in rabbits (288). Mannitol may contribute to tissue edema and could function as a scavenger of reactive oxygen intermediates produced by neutrophils and macrophages (43).

Pharmacologic Enhancement of Host Defense Mechanisms

Since *C. neoformans* infections occur primarily in patients with immunosuppression, there is considerable interest in the development of agents that enhance host defense against this fungal pathogen. Agents that enhance host defense against *C. neoformans* include drugs, chemicals, and natural products of the immune system. Furthermore, the fact that *C. neoformans* infection results in immunosuppression, possibly through the induction of suppressor responses (see chapter 9), raises the possibility that agents that interfere with the development of host suppressor mechanisms enhance host defenses.

Amphotericin B is one of the most important anticryptococcal drugs. The classical mechanism by which amphotericin B kills fungi is by binding fungal membrane sterols and disrupting their cell membranes (29). However, there is considerable evidence that amphotericin B is a potent immunostimulatory agent that has powerful effects on host immune function. Administration of amphotericin B to mice enhances the ability of peritoneal macrophages to make superoxide (287). The combination of amphotericin B and IFN-γ produced synergistic effects on macrophage superoxide production (287) and activated macrophages for fungistatic and tumoricidal activity (224). Amphotericin B has also been shown to activate human monocyte-derived macrophages to produce more superoxide anion and express more I-A antigen (286). Amphotericin B has been shown to stimulate the production of TNF-α by mouse macrophages (173). This phenomenon may be responsible for the fact that amphotericin B enhances IFN-γ fungicidal activity by mouse macrophages through a nitric oxide-related mechanism (269).

Administration of the aranucleoside Ara-A can modify the course of *C. neoformans* infection in mice (116). Ara-A has antiviral activity and is used for the treatment of herpes virus infection in humans. When Ara-A was given to mice infected with *C. neoformans,* it prolonged survival of C57BL/6 mice, had no effect on DBA/2 mice, and reduced the survival of AKR mice. Ara-A appears to have no effect on *C. neoformans* but can augment both humoral and cell-mediated immunity. The widely divergent results of Ara-A on the course of *C. neoformans* infection, depending on the strain of inbred mouse, highlight genetic differences between strains and unpredictable effects on host defenses. Hinrichs et al. (116) proposed no mechanism for the effects observed but speculated that Ara-A could be acting by interfering with *C. neoformans*-induced immune suppressor responses.

Kitchen (143) has suggested that diethylcarbamazine (DEC) may be a useful immunomodulator to enhance the host response to *C. neoformans.* DEC is a compound used in the treatment of lymphatic filariasis. Administration of DEC to mice given intraperitoneal *C. neoformans* infection resulted in a diminution of CFU in the peritoneal cavity but not in the brain (144). An interesting effect of DEC was a reduction in the weight of infected mouse brains consistent with reduced brain edema (144). DEC has also been demonstrated to enhance blood microbicidal activity against *C. neoformans* in vitro after overnight incubation (145). The usefulness of DEC as a therapeutic agent in human cryptococcosis is not known. If DEC is active against *C. neoformans* in humans, then the widespread use of DEC in areas where both filariasis and cryptococcosis are endemic provides a confounding variable for evaluation of anticryptococcal vaccines or the application of other interventions to reduce the prevalence of cryptococcosis.

Cytokine administration is a potential therapeutic strategy for enhancing immune function in immunosuppressed patients with *C. neoformans* infections. Several cytokines have been shown to enhance the efficacy of host inflammatory cells in vitro and to prolong survival in vivo in experimental animal models of cryptococcosis. Cytokines can enhance the efficacy of various antifungal drugs in vitro and in vivo (32, 47, 101, 133, 224). Blockage of certain cytokines may also benefit the host. Pentoxifylline is a compound that inhibits TNF-α, IL-1, and IL-6. Administration of pentoxifylline to mice in combination with amphotericin B was superior to administration of amphotericin B alone in prolonging survival and reducing brain CFU in experimental infection with *C. neoformans* var. *gattii* (220).

Administration of antibody to the capsular polysaccharide has been shown by three research groups to enhance amphotericin B efficacy in mice by (69, 97, 199). In humans, administration of rabbit immune serum leads to rapid clearance of serum cryptococcal antigen without significant side effects (96). Monoclonal antibodies to the capsule have been shown to potentiate the efficacy of amphotericin B (69, 199), fluconazole (197), and flucytosine (78) in mice and in in vitro assays with macrophages. Human-mouse chimeric antibodies have been constructed that are active against *C. neoformans* in mice, but these compounds have not been tested in humans.

Administration of transfer factor prepared from lymphocytes of guinea pigs that survived infection prolonged the survival of naive guinea pigs after *C. neoformans* infection (54). In contrast, levaminasole had no effect on the survival of guinea pigs with cryptococcal infection and may be detrimental at high doses (54). Diethylstilbestrol administration was also ineffective at enhancing host resistance to *C. neoformans* in mice (71).

In summary, many biological agents and chemical compounds have provided encouraging results either in vitro or in animal models. The difficulty of treating cryptococcosis has led to the proposal that immunotherapies be developed and used as adjuncts of antifungal chemotherapy (38). Although no immune-enhancing therapies are currently available, ongoing research may produce useful adjunctive therapies that target host immunity and are synergistic with antifungal drugs.

REFERENCES

1. **Amundson, D. E.** 1992. Cavitary pulmonary cryptococcosis complicating Churg-Strauss vasculitis. *South. Med. J.* **85:**700–702.
2. **Anderson, D. M., and M. A. Dykstra.** 1984. Pulmonary cell response in mice following intranasal instillation with *Cryptococcus neoformans. Mycopathologia* **86:**179–184.
3. **Aronson, M., and J. Kletter.** 1973. Aspects of the defense against a large-sized parasite, the yeast, *Cryptococcus neoformans. Isr. J. Med. Sci.* **1:**132–162.

4. **Asherton, G. L., V. Colizzi, and M. Zembala.** 1986. An overview of T-suppressor cell circuits. *Annu. Rev. Immunol.* **4:**37–68.

4a. **Atkinson, A. J., and J. E. B. Bennett.** 1968. Experience with a new skin test antigen prepared from *Cryptococcus neoformans. Am. Rev. Respir. Dis.* **97:**637–643.

5. **Attal, H. C., S. Grover, M. P. Bansai, B. S. Chaubey, and V. K. Joglekar.** 1983. Capsule deficient *Cryptococcus neoformans* an unusual clinical presentation. *J. Assoc. Physicians India* **31:**49–51.

6. **Baird, R. W., and A. S. Garfield.** 1994. Cavitation and rupture of a pulmonary cryptococcoma in an immunocompetent man. *Am. J. Med.* **97:**309–311.

7. **Baker, P. J.** 1990. Regulation of magnitude of antibody response to bacterial polysaccharide antigens by thymus-derived lymphocytes. *Infect. Immun.* **58:**3465–3468.

8. **Baker, R. D.** 1952. Resectable mycotic lesions and acutely fatal mycoses. *JAMA* **159:**1579–1581.

9. **Baker, R. D.** 1976. The primary pulmonary lymph node complex of cryptococcosis. *Am. J. Clin. Pathol.* **65:**83–92.

10. **Baker, R. D., and R. K. Haugen.** 1955. Tissue changes and tissue diagnosis in cryptococcosis. A study of 26 cases. *Am. J. Clin. Pathol.* **25:**14–24.

11. **Bennett, J. E., and H. F. Hasenclever.** 1965. *Cryptococcus neoformans* polysaccharide: studies of serologic properties and role in infection. *J. Immunol.* **94:**916–920.

12. **Bergman, F.** 1961. Pathology of experimental cryptococcosis. A study of course and tissue response in subcutaneously induced infection in mice. *Acta Pathol. Microbiol. Scand.* **147**(Suppl.):1–163.

13. **Berk, M., and B. Gerstl.** 1952. Torulosis (Cryptococcosis) producing a solitary pulmonary lesion. *JAMA* **149:**1310–1312.

14. **Bird, D. C., J. N. Sheagren, and J. E. Bennett.** 1969. Reticuloendothelial phagocytic function during systemic mycotic infections in man. *J. Lab. Clin. Med.* **74:**340–345.

15. **Blackstock, R.** 1996. Cryptococcal capsular polysaccharide utilizes an antigen-presenting cell to induce a T-suppressor cell to secrete TsF. *J. Med. Vet. Mycol.* **34:**19–30.

16. **Blackstock, R., and A. Casadevall.** 1997. Presentation of cryptococcal capsular polysaccharide (GXM) on activated antigen presenting cells inhibits the T-suppressor response and enhances delayed-type hypersensitivity and survival. *Immunology* **92:**334–339.

17. **Blackstock, R., and N. K. Hall.** 1982. Immunosuppression by avirulent, pseudohyphal forms of *Cryptococcus neoformans. Mycopathologia* **80:**95–99.

18. **Blackstock, R., and N. K. Hall.** 1984. Non-specific immunosuppression by *Cryptococcus neoformans* infection. *Mycopathologia* **86:**35–43.

19. **Blackstock, R., N. K. Hall, and N. C. Hernandez.** 1989. Characterization of a suppressor factor that regulates phagocytosis by macrophages in murine cryptococcosis. *Infect. Immun.* **57:**1773–1779.

20. **Blackstock, R., and N. C. Hernandez.** 1988. Inhibition of macrophage phagocytosis in cryptococcosis: phenotypic analysis of the suppressor cell. *Cell. Immunol.* **114:**174–187.

21. **Blackstock, R., and N. C. Hernandez.** 1989. Characterization of the macrophage subset affected and its response to a T suppressor factor (TsFmp) found in cryptococcosis. *Infect. Immun.* **57:**2931–2937.

22. **Blackstock, R., J. M. McCormack, and N. K. Hall.** 1987. Induction of a macrophage-suppressive lymphokine by soluble cryptococcal antigens and its association with models of immunologic tolerance. *Infect. Immun.* **55:**233–239.

23. **Blackstock, R., M. Zembala, and G. L. Asherson.** 1991. Functional equivalence of cryptococcal and haptene-specific T suppressor factor (TsF). I. Picryl and oxazolone-specific TsF, which inhibit transfer of contact sensitivity, also inhibit phagocytosis by a subset of macrophages. *Cell. Immunol.* **136:**435–447.

24. **Blackstock, R., M. Zembala, and G. L. Asherson.** 1991. Functional equivalence of cryptococcal and haptene-specific T suppressor factor (TsF). II. Monoclonal anti-cryp-tococcal TsF inhibits both phagocytosis by a subset of macrophages and transfer of contact sensitivity. *Cell. Immunol.* **136:**448–461.

25. **Blasi, E., R. Barluzzi, R. Mazzola, P. Mosci, and F. Bistoni.** 1992. Experimental model of intracerebral infection with *Cryptococcus neoformans*: roles of phagocytes and op-sonization. *Infect. Immun.* **60:**3682–3688.

26. **Blasi, E., R. Barluzzi, R. Mazzola, L. Pitzurra, M. Puliti, S. Saleppico, and F. Bistoni.** 1995. Biomolecular events involved in anticryptococcal resistance in the brain. *Infect. Immun.* **63:**1218–1222.

27. **Blasi, E., R. Mazzola, R. Barluzzi, P. Mosci, A. Bartoli, and F. Bistoni.** 1991. Intracere-bral transfer of an in vitro established microglial cell line: local induction of a protective state against lethal challenge with *Candida albicans. J. Neuroimmunol.* **32:**249–257.

28. **Blasi, E., R. Mazzola, R. Barluzzi, P. Mosci, and F. Bistoni.** 1994. Anticryptococcal resistance in the mouse brain: beneficial effects of local administration of heat-inacti-vated yeast cells. *Infect. Immun.* **62:**3189–3196.

29. **Brajtburg, J., W. G. Powderly, G. Kobayashi, and G. Medoff.** 1993. Amphotericin B: current understanding of mechanisms of action. *Antimicrob. Agents Chemother.* **34:**183–188.

30. **Brodie, S. J., V. G. Sasseville, K. A. Reinmann, M. A. Simon, P. K. Sehgal, and D. J. Ringler.** 1994. Macrophage function in simian AIDS. Killing defects in vivo are inde-pendent of macrophage function, associated with alterations in Th phenotype, and reversible with IFN-gamma. *J. Immunol.* **153:**5790.

31. **Brueske, C. H.** 1986. Proteolytic activity of a clinical isolate of *Cryptococcus neoformans. J. Clin. Microbiol.* **23:**631–633.

32. **Brummer, E., and D. A. Stevens.** 1994. Macrophage colony-stimulating factor induc-tion of enhanced macrophage anticryptococcal activity: synergy with fluconazole for killing. *J. Infect. Dis.* **170:**173–179.

33. **Buchanan, K. L., P. L. Fidel, and J. W. Murphy.** 1991. Effects of *Cryptococcus neofor-mans*-specific suppressor T cells on the amplified anticryptococcal delayed-type hy-persensitivity response. *Infect. Immun.* **59:**29–35.

34. **Bulmer, G. S., and J. R. Tacker.** 1975. Phagocytosis of *Cryptococcus neoformans* by alveolar macrophages. *Infect. Immun.* **11:**73–79.

35. **Cameron, M. L., J. A. Bartlett, H. A. Gallis, and H. A. Waskin.** 1991. Manifestations of pulmonary cryptococcosis in patients with acquired immunodeficiency syndrome. *Rev. Infect. Dis.* **13:**64–67.

36. **Cameron, M. L., D. L. Granger, T. J. Matthews, and J. B. Weinberg.** 1994. Human immunodeficiency virus (HIV)-infected human blood monocytes and peritoneal macrophages have reduced anticryptococcal activity whereas HIV-infected alveolar macraphages retain normal activity. *J. Infect. Dis.* **170:**60–70.

37. **Carton, C. A., and L. A. Mount.** 1951. Neurosurgical aspects of cryptococcosis. *J. Neurosurg.* **8:**143–156.

38. **Casadevall, A.** 1993. Cryptococcosis: the case for immunotherapy. *Cliniguide to Fungal Infections* **4:**1–5.

39. **Casadevall, A., D. L. Goldman, and M. Feldmesser.** 1997. Antibody-based therapies for infectious diseases: renaissance for an abandoned arsenal? *Bull. Inst. Pasteur* **95:**247–257.

40. **Casadevall, A., J. Mukherjee, Y. RuiRong, and J. Perfect.** 1994. Management of *Cryptococcus neoformans* contaminated needle injuries. *Clin. Infect. Dis.* **19:**951–953.

41. **Cauley, L. K., and J. W. Murphy.** 1979. Response of congenitally athymic (nude) and phenotypically normal mice to *Cryptococcus neoformans* infection. *Infect. Immun.* **23:**644–651.

42. **Cawley, E. P., R. H. Grekin, and A. C. Curtis.** 1950. Torulosis. A review of the cutaneous and adjoining mucous membrane manifestations. *J. Invest. Dermatol.* **14:**327–344.

43. **Chatuverdi, V., B. Wong, and S. L. Newman.** 1996. Oxidative killing of *Cryptococcus neoformans* by human leukocytes. Evidence that fungal mannitol protects by scavenging reactive oxygen intermediates. *J. Immunol.* **156:**3836–3840.

44. **Chen, L.-C., E. Blank, and A. Casadevall.** 1996. Extracellular proteinase activity of *Cryptococcus neoformans. Clin. Diagn. Lab. Immunol.* **3:**570–574.

45. **Cherniak, R., and J. B. Sundstrom.** 1994. Polysaccharide antigens of the capsule of *Cryptococcus neoformans. Infect. Immun.* **62:**1507–1512.

46. **Clemons, K. V., R. Azzi, and D. A. Stevens.** 1996. Experimental systemic cryptococcosis in SCID mice. *J. Med. Vet. Mycol.* **34:**331–335.

47. **Clemons, K. V., E. Brummer, and D. A. Stevens.** 1994. Cytokine treatment of central nervous system infection: efficacy of interleukin-12 alone and synergy with conventional antifungal therapy in experimental cryptococcosis. *Antimicrob. Agents Chemother.* **38:**460–464.

49. **Collins, H. L., and G. J. Bancroft.** 1991. Encapsulation of *Cryptococcus neoformans* impairs antigen-specific T-cell responses. *Infect. Immun.* **59:**3883–3888.

50. **Curtis, J. L., G. B. Huffnagle, G. H. Chen, M. L. Warnock, M. Gyetko, R. McDonald, P. Scott, and G. B. Toews.** 1994. Experimental murine pulmonary cryptococcosis. *Lab. Invest.* **71:**113–126.

51. **Cusini, M., P. Cagliani, R. Grimalt, G. Tadini, E. Alessi, and M. Fasan.** 1991. Primary cutaneous cryptococcosis in a patient with the Acquired Immunodeficiency Syndrome. *Arch. Dermatol.* **127:**1848–1849.

52. **Daniel, P. M., F. Shiller, and R. L. Vollum.** 1949. Torulosis of the central nervous system. Report of two cases. *Lancet* **i:**53–56.

53. **DeShaw, M., and L.-A. Pirofski.** 1995. Antibodies to the *Cryptococcus neoformans* capsular glucuronoxylomannan are ubiquitous in serum from HIV+ and HIV− individuals. *Clin. Exp. Immunol.* **99:**425–432.

54. **Diamond, R. D.** 1977. Effects of stimulation and suppression of cell-mediated immunity on experimental cryptococcosis. *Infect. Immun.* **17:**187–194.

55. **Diamond, R. D., and J. E. Bennett.** 1973. Disseminated cryptococcosis in man: decreased lymphocyte transformation in response to *Cryptococcus neoformans. J. Infect. Dis.* **127:**694–697.

56. **Diamond, R. D., and J. E. Bennett.** 1973. Growth of *Cryptococcus neoformans* within human macrophages in vitro. *Infect. Immun.* **7:**231–236.

57. **Diamond, R. D., and N. F. Erickson III.** 1982. Chemotaxis of human neutrophils and monocytes induced by *Cryptococcus neoformans. Infect. Immun.* **38:**380–382.

58. **Diamond, R. D., J. E. May, M. Kane, M. M. Frank, and J. E. Bennett.** 1973. The role of late complement component and the alternate complement pathway in experimental cryptococcosis (37580). *Proc. Soc. Exp. Biol. Med.* **144:**312–315.

59. **Diamond, R. D., J. E. May, M. C. Kane, M. M. Frank, and J. E. Bennett.** 1974. The role of the classical and alternate complement pathways in host defenses against *Cryptococcus neoformans* infection. *J. Immunol.* **112:**2260–2270.

60. **Dixon, D. M., and A. Polak.** 1986. *In vivo* and *in vitro* studies with an atypical, rhinotrophic isolate of *Cryptococcus neoformans. Mycopathologia* **96:**33–40.

61. **Dobrick, P., K. Miksits, and H. Hahn.** 1995. L3T4(CD4)-, Lyt-2(CD8)- and Mac-1(CD11b)-phenotypic leukocytes in murine cryptococcal meningoencephalitis. *Mycopathologia* **131:**159–166.

62. **Dong, Z. M., and J. W. Murphy.** 1993. Mobility of human neutrophils in response to *Cryptococcus neoformans* cells, culture filtrate antigen, and individual components of the antigen. *Infect. Immun.* **61:**5067–5077.

63. **Dong, Z. M., and J. W. Murphy.** 1995. Intravascular cryptococcal culture filtrate (CneF) and its major component, glucuronoxylomannan, are potent inhibitors of leukocyte accumulation. *Infect. Immun.* **63:**770–778.

64. **Dong, Z. M., and J. W. Murphy.** 1995. Effects of the two varieties of *Cryptococcus neoformans* cells and culture filtrate antigens on neutrophil locomotion. *Infect. Immun.* **63:**2632–2644.

65. **Dong, Z. M., and J. W. Murphy.** 1996. Cryptococcal polysaccharides induce L-selectin shedding and tumor necrosis receptor loss from the surface of human neutrophils. *J. Clin. Invest.* **97:**689–698.

66. **Dong, Z. M., and J. W. Murphy.** 1997. Cryptococcal polysaccharide bind to CD18 on human neutrophils. *Infect. Immun.* **65:**557–563.

67. **Dorf, M., and B. Benacerraf.** 1984. Suppressor cells and immunoregulation. *Annu. Rev. Immunol.* **2:**127–158.

68. **Dormer, B. A., J. Friedlander, F. J. Wiles, and F. W. Simson.** 1945. Tumor of the lung due to Cryptococcus histolyticus (blastomycosis). *J. Thoracic Surg.* **14:**322–329.

69. **Dromer, F., and J. Charreire.** 1991. Improved amphotericin B activity by a monoclonal anti-*Cryptococcus neoformans* antibody: study during murine cryptococcosis and mechanisms of action. *J. Infect. Dis.* **163:**1114–1120.

70. **Drouhet, E., and G. Segretain.** 1951. Inhibition de la migration leucocytaire *in vitro* par un polyoside capsulaire de *Torulopsis (Cryptococcus) neoformans*. *Ann. Inst. Pasteur* **81:**674.

71. **Duke, S. S., and R. A. Fromtling.** 1984. Effects of diethylstilbestrol and cyclophosphamide on the pathogenesis of experimental *Cryptococcus neoformans* infections. *J. Med. Vet. Mycol.* **22:**125–135.

72. **Durden, F. M., and B. Elewski.** 1994. Cutaneous involvement with *Cryptococcus neoformans* in AIDS. *J. Am. Acad. Dermatol.* **30:**844–848.

73. **Dystra, M. A., and L. Friedman.** 1978. Pathogenesis, lethality, and immunizing effect of experimental cutaneous cryptococcosis. *Infect. Immun.* **20:**446–455.

74. **Epstein, R., R. Cole, and K. K. Hunt.** 1972. Pleural effusion secondary to pulmonary cryptococcosis. *Chest* **61:**296–298.

75. **Farmer, S. G., and R. A. Komorowski.** 1973. Histologic response to capsule-deficient *Cryptococcus neoformans*. *Arch. Pathol.* **96:**383–387.

76. **Feldmesser, M., and A. Casadevall.** 1997. Effect of serum IgG1 against murine pulmonary infection with *Cryptococcus neoformans*. *J. Immunol.* **158:**790–799.

77. **Feldmesser, M., A. Casadevall, Y. Kress, G. Spira, and A. Orlofski.** 1997. Eosinophil-*Cryptococcus neoformans* interactions in vivo and in vitro. *Infect. Immun.* **65:**1899–1907.

78. **Feldmesser, M., J. Mukherjee, and A. Casadevall.** 1996. Combination of 5-flucytosine and capsule binding monoclonal antibody in therapy of murine *Cryptococcus neoformans* infections and *in vitro*. *J. Antimicrob Chemother.* **37:**617–622.

79. **Ferry, J. A., C. K. Pettit, A. E. Rosenberg, and N. L. Harris.** 1991. Fungi in megakaryocytes. An unusual manifestation of fungal infection in the bone marrow. *Am. J. Clin. Pathol.* **96:**577–581.

80. **Fidel, P. L., and J. W. Murphy.** 1988. Characterization of an in vitro-stimulated, *Cryptococcus neoformans*-specific second-order suppressor T cell and its precursor. *Infect. Immun.* **56:**1267–1272.

81. **Freeman, W.** 1930. Torula meningo-encephalitis. Comparative histopathology in seventeen cases. *Trans. Am. Neurol. Assoc.* **56:**203–217.

82. **Freeman, W.** 1931. Torula infection of the central nervous system. *J. Psychol. Neurol.* **43:**236–245.

83. **Friedman, E. P., R. F. Miller, A. Severn, I. G. Williams, and P. J. Shaw.** 1995. Cryptococcal pneumonia in patients with the acquired immunodeficiency syndrome. *Clin. Radiol.* **50:**756–760.

84. **Frisch, M., and D. R. Gnepp.** 1995. Primary cryptococcal infection of the larynx: report of a case. *Otolaryngol. Head Neck Surg.* **113:**477–480.

85. **Fromtling, R. A., G. K. Abruzzo, and A. Ruiz.** 1988. *Cryptococcus neoformans*: a central nervous system isolate from an AIDS patient that is rhinotropic in a normal mouse model. *Mycopathologia* **102:**79–86.

86. **Fromtling, R. A., and H. J. Shadomy.** 1982. Immunity in cryptococcosis: an overview. *Mycopathologia* **77:**183–190.

87. **Gadebusch, H. H.** 1961. Natural host resistance to infection with *Cryptococcus neoformans*. The effect of the properdin system on the experimental disease. *J. Infect. Dis.* **109:**147–153.

88. **Gadebusch, H. H.** 1972. Mechanisms of native and acquired resistance to infection with *Cryptococcus neoformans*. *Crit. Rev. Microbiol.* **1:**311–320.

89. **Gal, A. A., M. N. Koss, J. Hawkins, S. Evans, and H. Einstein.** 1986. The pathology of pulmonary cryptococcal infections in the acquired immunodeficiency syndrome. *Arch. Pathol. Lab. Med.* **110:**502–507.

90. **Ghigliotti, G., G. Carrega, A. Farris, A. Burroni, A. Nigro, G. Pagano, and R. De Marchi.** 1992. Cutaneous cryptococcosis resembling molluscum contagiosum in a homosexual man with AIDS. *Acta Dermatol. Venereol.* **72:**182–184.

91. **Glaser, J. B., and A. Garden.** 1985. Inoculation of cryptococcosis without transmission of the acquired immunodeficiency syndrome. *N. Engl. J. Med.* **313:**266.

92. **Goldman, D., Y. Cho, M.-L. Zhao, A. Casadevall, and S. C. Lee.** 1996. Expression of inducible nitric oxide synthase in rat pulmonary *Cryptococcus neoformans* granulomas. *Am. J. Pathol.* **148:**1275–1282.

93. **Goldman, D., S. C. Lee, and A. Casadevall.** 1994. Pathogenesis of pulmonary *Cryptococcus neoformans* infection in the rat. *Infect. Immun.* **62:**4755–4761.

94. **Goldman, D. L., A. Casadevall, Y. Cho, and S. C. Lee.** 1996. *Cryptococcus neoformans* meningitis in the rat. *Lab. Invest.* **75:**759–770.

95. **Goodman, J. S., L. Kaufman, and G. Koenig.** 1971. Diagnosis of cryptococcal meningitis. *N. Engl. J. Med.* **285:**434–436.

96. **Gordon, M. A., and A. Casadevall.** 1995. Serum therapy of cryptococcal meningitis. *Clin. Infect. Dis.* **21:**1477–1479.

97. **Gordon, M. A., and E. Lapa.** 1964. Serum protein enhancement of antibiotic therapy in cryptococcosis. *J. Infect. Dis.* **114:**373–378.

98. **Gordon, S. M., A. A. Gal, and J. R. Amerson.** 1994. Granulomatous peritoneal cryptococcomas. An unusual sequela of disseminated cryptococcosis. *Arch. Pathol. Lab. Med.* **118:**194–195.

99. **Graybill, J. R., J. Ahrens, T. Nealon, and R. Paque.** 1983. Pulmonary cryptococcosis in the rat. *Am. Rev. Respir. Dis.* **127:**636–640.

100. **Graybill, J. R., and R. H. Alford.** 1974. Cell-mediated immunity in cryptococcosis. *Cell. Immunol.* **14:**12–21.

101. **Graybill, J. R., R. Bocanegra, C. Lambros, and M. F. Luther.** 1997. Granulocyte colony stimulating factor therapy of experimental cryptococcal meningitis. *J. Med. Vet. Mycol.* **35:**243–247.

102. **Graybill, J. R., and D. J. Drutz.** 1978. Host defense in cryptococcosis. II. Cryptococcosis in the nude mouse. *Cell. Immunol.* **40:**263–274.

103. **Griffin, F. M.** 1980. Roles of macrophage Fc and C3b receptors in phagocytosis of immunologically coated *Cryptococcus neoformans*. *Proc. Natl. Acad. Sci. USA* **78:**3853–3857.

104. **Gutierrez, F., Y. S. Fu, and H. I. Lurie.** 1975. Cryptococcosis histologically resembling histoplasmosis. *Arch. Pathol.* **99:**347–353.

105. **Gyetko, M. R., G.-H. Chen, R. A. McDonald, R. Goodman, G. B. Huffnagle, C. C. Wilkinson, J. A. Fuller, and G. B. Toews.** 1996. Urokinase is required for the pulmonary inflammatory response to *Cryptococcus neoformans*. *J. Clin. Invest.* **97:**1818–1826.

106. **Harrison, T. S., H. Kornfeld, and S. M. Levitz.** 1995. The effect of infection with human immunodeficiency virus on the anticryptococcal activity of lymphocytes and monocytes. *J. Infect. Dis.* **172:**665–671.

107. **Haugen, R. K., and R. D. Baker.** 1954. The pulmonary lesions in cryptococcosis with special reference to subpleural nodules. *Am. J. Clin. Pathol.* **24:**1381–1390.

108. **Heere, L. J., T. A. Mahvi, and M. M. Annable.** 1975. Effect of temperature on growth and macromolecular biosynthesis in Cryptococcus species. *Mycopathologia* **55:**105–113.

109. **Henderson, D. K., J. E. Bennett, and M. A. Huber.** 1982. Long-lasting, specific immunologic unresponsiveness associated with cryptococcal meningitis. *J. Clin. Invest.* **69:**1185–1190.

110. **Henderson, D. K., V. L. Kan, and J. E. Bennett.** 1986. Tolerance to cryptococcal polysaccharide in cured cryptococcosis patients: failure of antibody secretion *in vitro*. *Clin. Exp. Immunol.* **65:**639–646.

112. **Hill, J. O.** 1992. CD4$^+$ T cells cause multinucleated giant cells to form around *Cryptococcus neoformans* and confine the yeast within the primary site of infection in the respiratory tract. *J. Exp. Med.* **175:**1685–1695.

113. **Hill, J. O., and K. M. Aguirre.** 1994. CD4+ T cell-dependent acquired state of immunity that protects the brain against *Cryptococcus neoformans*. *J. Immunol.* **152:**2344–2350.

114. **Hill, J. O., and P. L. Dunn.** 1993. A T cell-independent protective response against *Cryptococcus neoformans* expressed at the primary site of infection in the lung. *Infect. Immun.* **61:**5302–5308.

115. **Hino, H., K. Takizawa, and G. Asboe-Hansen.** 1982. Ultrastructure of *Cryptococcus neoformans*. *Acta Dermatol. Venereol.* **62:**113–117.

116. **Hinrichs, J., D. Kitz, G. Kobayashi, and J. R. Little.** 1983. Immune enhancement in mice by Ara-A. *J. Immunol.* **130:**829–833.

117. **Hirano, A., H. M. Zimmerman, and S. Levine.** 1964. Fine structure of cerebral fluid accumulation. V. Transfer of fluid from extracellular compartments in acute phase of cryptococcal polysaccharide lesions. *Arch. Neurol.* **11:**632–641.

118. **Hirano, A., H. M. Zimmerman, and S. Levine.** 1964. Fine structure of cerebral fluid accumulation. III. Extracellular spread of cryptococcal polysaccharides in the acute stage. *Am. J. Pathol.* **46:**1–11.

119. **Hirano, A., H. M. Zimmerman, and S. Levine.** 1965. Fine structure of cerebral fluid accumulation. VI. Intracellular accumulation of fluid and cryptococcal polysaccharide in oligodendria. *Arch. Neurol.* **12:**189–196.

120. **Hirano, A., H. M. Zimmerman, and S. Levine.** 1965. Fine structure of cerebral fluid accumulation. VII. Reactions of astrocytes to cryptococcal polysaccharide implantation. *J. Neuropathol. Exp. Neurol.* **24:**386–396.

121. **Hobbs, M. M., J. R. Perfect, D. L. Granger, and D. T. Durack.** 1990. Opsonic activity of cerebrospinal fluid in experimental cryptococcal meningitis. *Infect. Immun.* **58:**2115–2119.

122. **Hoff, C. L.** 1942. Immunity studies of *Cryptococcus hominis (Torula histolytica)* in mice. *J. Lab. Clin. Med.* **27:**751–754.

123. **Horn, C. A., and R. G. Washburn.** 1995. Anticryptococcal activity of NK cell-enriched peripheral blood lymphocytes from human immunodeficiency virus-infected subjects: responses to interleukin-2, interferon-gamma, and interleukin-12. *J. Infect. Dis.* **172:**1023–1027.

124. **Huffnagle, G. B., G.-H. Chen, J. L. Curtis, R. A. McDonald, R. M. Strieter, and G. B. Toews.** 1995. Down-regulation of the afferent phase of T cell-mediated pulmonary inflammation and immunity by a high melanin-producing strain of *Cryptococcus neoformans*. *J. Immunol.* **155:**3507–3516.

125. **Huffnagle, G. B., and M. F. Lipscomb.** 1992. Animal model of human disease: pulmonary cryptococcosis. *Am. J. Pathol.* **141:**1517–1520.

126. **Huffnagle, G. B., M. F. Lipscomb, J. A. Lovchik, K. A. Hoag, and N. E. Street.** 1994. The role of CD4+ and CD8+ T-cells in protective inflammatory response to a pulmonary cryptococcal infection. *J. Leukocyte Biol.* **55:**35–42.

127. **Huffnagle, G. B., N. E. Street, and M. Lipscomb.** 1992. In contrast to Balb/c mice, a *Cryptococcus neoformans* infection in C57BL/6 mice generates protective T cell immunity in the periphery and non-protective T-cell mediated eosinophilia in the lungs. *FASEB J.* **6:**A1689. (Abstract.)

128. **Huffnagle, G. B., R. M. Strieter, T. J. Standiford, R. A. McDonald, M. D. Burdick, S. L. Kunkel, and G. B. Toews.** 1995. The role of monocyte chemotactic protein-1 (MCP-1) in the recruitment of monocytes and CD4+ T cells during a pulmonary *Cryptococcus neoformans* infection. *J. Immunol.* **155:**4790–4797.

129. **Huffnagle, G. B., J. L. Yates, and M. F. Lipscomb.** 1991. T cell-mediated immunity in the lung: a *Cryptococcus neoformans* pulmonary infection model using SCID and athymic nude mice. *Infect. Immun.* **59:**1423–1433.

130. **Huffnagle, G. B., J. L. Yates, and M. F. Lipscomb.** 1991. Immunity to pulmonary *Cryptococcus neoformans* infection requires both CD4+ and CD8+ T cells. *J. Exp. Med.* **173:**793–800.

131. **Iarobellis, F. W., M. I. Jacobs, and R. P. Cohen.** 1984. Primary cutaneous cryptococcosis. *Arch. Dermatol.* **23:**673–675.

132. **Jacobson, E. S., and S. B. Tinnell.** 1993. Antioxidant function of fungal melanin. *J. Bacteriol.* **175:**7102–7104.

133. **Joly, V., L. Saint-Julien, C. Carbon, and P. Yeni.** 1994. In vivo activity of interferon-gamma in combination with amphotericin B in the treatment of experimental cryptococcosis. *J. Infect. Dis.* **170:**1331–1334.

134. **Kagaya, K., T. Yamada, Y. Miyakawa, Y. Fukazawa, and S. Saito.** 1985. Characterization of pathogenic constituents of *Cryptococcus neoformans* strains. *Microbiol. Immunol.* **29:**517–532.

135. **Kahn, F. W., D. N. England, and J. M. Jones.** 1985. Solitary pulmonary nodule due to *Cryptococcus neoformans* and *Mycobacterium tuberculosis*. *Am. J. Med.* **78:**677–681.

136. **Kalina, M., Y. Kletter, and M. Aronson.** 1974. The interaction of phagocytes and the large-sized parasite *Cryptococcus neoformans*: cytochemical and ultrastructural study. *Cell Tissue Res.* **152:**165–174.

137. **Kawakami, K., S. Kohno, J. -I. Kadota, M. Tohyama, K. Teruya, N. Kedeken, A. Saito, and K. Hara.** 1995. T cell-dependent activation of macrophages and enhancement of their phagocytic activity in the lungs of mice inoculated with heat-killed *Cryptococcus neoformans*: involvement of IFN-gamma and its protective effect against cryptococcal infection. *Microbiol. Immunol.* **39:**135–143.

138. **Kawakami, K., S. Kohno, N. Morikawa, J. Kadota, A. Saito, and K. Hara.** 1994. Activation of macrophages and expansion of specific T lymphocytes in the lungs of mice intratracheally inoculated with *Cryptococcus neoformans*. *Clin. Exp. Immunol.* **96:**230–237.

139. **Kent, T. H., and J. M. Layton.** 1962. Massive pulmonary cryptococcosis. *Am. J. Clin. Pathol.* **38:**596–604.

140. **Kerkering, T. M., R. J. Duma, and S. Shadomy.** 1981. The evolution of pulmonary cryptococcosis. *Ann. Intern. Med.* **94:**611–616.

141. **Kessel, J. F., and F. Holtzwart.** 1935. Experimental studies with torula from a knee infection in man. *Am. J. Trop. Med.* **15:**467–481.

142. **Khakpour, F. R., and J. W. Murphy.** 1987. Characterization of a third-order suppressor T cell (Ts3) induced by cryptococcal antigen. *Infect. Immun.* **55:**1657–1662.

143. **Kitchen, L. W.** 1996. Adjunctive immunologic therapy for *Cryptococcus neoformans* infections. *Clin. Infect. Dis.* **23:**209–210.

144. **Kitchen, L. W., J. A. Ross, J. E. Hernandez, A. L. Zarraga, and F. J. Mather.** 1992. Effect

of administration of diethylcarbamazine on experimental bacterial and fungal infections in mice. *Int. J. Antimicrob. Agents* **1**:259–268.

145. **Kitchen, L. W., J. A. Ross, B. S. Turner, J. E. Hernandez, and F. J. Mather.** 1995. Diethylcarbamazine enhances blood microbicidal activity. *Adv. Therapy* **12**:22–29.

146. **Kobayashi, T., I. Nakashima, and N. Kato.** 1975. Factors affecting experimental infection with *Cryptococcus neoformans* in mice with special reference to an endotoxic substance of *C. neoformans*. *Mycopathologia* **55**:17–22.

147. **Kozel, T. R.** 1993. Opsonization and phagocytosis of *Cryptococcus neoformans*. *Arch. Med. Res.* **24**:211–218.

148. **Kozel, T. R.** 1996. Activation of the complement system by pathogenic fungi. *Clin. Microbiol. Rev.* **9**:34–46.

149. **Kozel, T. R., and J. Cazin, Jr.** 1972. Immune response to *Cryptococcus neoformans* soluble polysaccharide. *Infect. Immun.* **5**:35–41.

150. **Kozel, T. R., and E. Gotschlich.** 1982. The capsule of *Cryptococcus neoformans* passively inhibits phagocytosis of the yeast by macrophages. *J. Immunol.* **129**:1675–1680.

151. **Kozel, T. R., W. F. Gulley, and J. Cazin, Jr.** 1977. Immune response to *Cryptococcus neoformans* soluble polysaccharide: immunological unresponsiveness. *Infect. Immun.* **18**:701–707.

152. **Kozel, T. R., and G. S. T. Pfrommer.** 1986. Activation of the complement system by *Cryptococcus neoformans* leads to binding of iC3b to the yeast. *Infect. Immun.* **52**:1–5.

153. **Kozel, T. R., E. Reiss, and R. Cherniak.** 1980. Concomitant but not causal association between surface charge and inhibition of phagocytosis by cryptococcal polysaccharide. *Infect. Immun.* **29**:295–300.

154. **Krainer, L., J. M. Small, A. B. Hewlitt, and T. Deness.** 1946. A case of systemic torula infection with tumour formation in the meninges. *J. Neurol. Neurosurg. Psychiatr.* **9**:158–162.

155. **Kuhn, L. R.** 1939. Experimental cryptococcic infection. *Arch. Pathol.* **27**:803.

156. **Laxalt, K. A., and T. R. Kozel.** 1979. Chemotoxigenesis and activation of the alternative complement pathway by encapsulated and non-encapsulated *Cryptococcus neoformans*. *Infect. Immun.* **26**:435–440.

157. **Lee, S. C., and A. Casadevall.** 1996. Polysaccharide antigen in brain tissue of AIDS patients with cryptococcal meningitis. *Clin. Infect. Dis.* **23**:194–195.

158. **Lee, S. C., A. Casadevall, and D. W. Dickson.** 1996. Immunohistochemical localization of capsular polysaccharide antigen in the central nervous system cells in cryptococcal meningoencephalitis. *Am. J. Pathol.* **148**:1267–1274.

159. **Lee, S. C., D. W. Dickson, and A. Casadevall.** 1996. Pathology of cryptococcal meningoencephalitis: analysis of 27 patients with pathogenetic implications. *Hum. Pathol.* **27**:839–847.

161. **Lee, S. L., Y. Kress, M.-L. Zhao, D. W. Dickson, and A. Casadevall.** 1995. *Cryptococcus neoformans* survive and replicate in spacious phagosomes in human microglia. *Lab. Invest.* **73**:871–879.

162. **Lehrer, R. I., A. K. Lichenstein, and T. Ganz.** 1993. Defensins: antimicrobial and cytotoxic peptides of mammalian cells. *Annu. Rev. Immunol.* **11**:105–128.

163. **Levine, S., H. M. Zimmerman, and A. Scorza.** 1957. Experimental cryptococcosis (torulosis). *Am. J. Pathol.* **33**:385–409.

164. **Levitz, S. M.** 1992. Overview of host defenses in fungal infections. *Clin. Infect. Dis.* **14**(Suppl. 1):S37BS42.

165. **Levitz, S. M., and T. P. Farrell.** 1990. Growth inhibition of *Cryptococcus neoformans* by cultured human monocytes: role of the capsule, opsonins, the culture surface and cytokines. *Infect. Immun.* **58**:1201–1209.

166. **Levitz, S. M., A. Tabumi, H. Kornfeld, C. C. Reardon, and D. T. Golenbock.** 1994.

Production of tumor necrosis factor alpha in human leukocytes stimulated by *Cryptococcus neoformans*. *Infect. Immun.* **62:**1975–1981.

167. **Levitz, S. M., A. Tabuni, and C. Treseler.** 1993. Effect of mannose-binding protein on binding of *Cryptococcus neoformans* to human phagocytes. *Infect. Immun.* **61:**4891–4893.

168. **Ley, A., R. Jacas, and C. Oliveras.** 1951. Torula granuloma of the cervical spinal cord. *J. Neurosurg.* **8:**327–335.

169. **Lim, T. S., J. W. Murphy, and L. K. Cauley.** 1980. Host-etiological agent interactions in intranasally and intraperitoneally induced cryptococcosis in mice. *Infect. Immun.* **29:**633–641.

170. **Linell, F., B. Magnusson, and A. Norden.** 1953. Cryptococcosis. Review and report of a case. *Acta Dermatol. Venereol.* **33:**103–122.

171. **Lipscomb, M. F.** 1989. Lung defenses against opportunistic infections. *Chest* **96:**1393–1399.

172. **Lipscomb, M. F., G. B. Huffnagle, J. A. Lovchik, C. R. Lyons, A. M. Pollard, and J. L. Yates.** 1993. The role of T lymphocytes in pulmonary microbial defense mechanisms. *Arch. Pathol. Lab. Med.* **117:**1225–1232.

172a. **Liu, C. T.** 1953. Intracerebral cryptococcic granuloma. *J. Neurosurg.* **10:**686–689.

173. **Louie, A., A. L. Baltch, M. A. Franke, R. P. Smith, and M. A. Gordon.** 1994. Comparative efficacy of four antifungal agents to stimulate murine macrophages to produce tumour necrosis factor alpha: an effect that is attenuated by pentoxifylline, liposomal vesicles, and dexamethasone. *J. Antimicrob. Chemother.* **34:**975–987.

174. **Lovchik, J. A., and M. F. Lipscomb.** 1993. Role for C5 and neutrophils in the pulmonary intravascular clearance of circulating *Cryptococcus neoformans*. *Am. J. Respir. Cell. Mol. Biol.* **9:**617–627.

175. **Lovchik, J. A., C. R. Lyons, and M. F. Lipscomb.** 1995. A role for gamma interferon-induced nitric oxide in pulmonary clearance of *Cryptococcus neoformans*. *Am. J. Respir. Cell. Mol. Biol.* **13:**116–124.

176. **Love, G. L., G. D. Boyd, and D. L. Greer.** 1985. Large *Cryptococcus neoformans* isolated from brain abscess. *J. Clin. Microbiol.* **22:**1068–1070.

177. **Lurie, H. I., and H. J. Shadomy.** 1971. Morphological variations of a hypha-forming strain of *Cryptococcus neoformans* (Coward strain) in tissues of mice. *Sabouraudia* **9:**10–14.

178. **Macher, A. M., J. E. Bennett, J. E. Gadek, and M. M. Frank.** 1978. Complement depletion in cryptococcal sepsis. *J. Immunol.* **120:**1686–1690.

179. **Manfredi, R., A. Mazzoni, A. Nanetti, A. Mastroianni, O. Coronado, and F. Chiodo.** 1996. Morphologic features and clinical significance of skin involvement in patients with AIDS-related cryptococcosis. *Acta Dermatol. Venereol.* **76:**72–76.

180. **Manrique, P., J. Mayo, J. A. Alvarez, X. Ganchegui, I. Zabalza, and M. Flores.** 1992. Polymorphous cutaneous cryptococcosis: nodular, herpes-like, and molluscum-like lesions in a patient with the acquired immunodeficiency syndrome. *J. Am. Acad. Dermatol.* **26:**122–124.

181. **Marquis, G., S. Montplaisir, M. Pelletier, S. Mousseau, and P. Auger.** 1985. Genetic resistance to murine cryptococcosis: increased susceptibility in the CBA/N XID mutant strain of mice. *Infect. Immun.* **47:**282–287.

182. **Martinez, A. J., M. Sell, T. Mitrovics, G. Stoltenburg-Didinger, J. R. Iglesias-Rozas, M. A. Giraldo-Velasquez, G. Gosztonyi, V. Schneider, and J. Cervos-Navarro.** 1995. The neuropathology and epidemiology of AIDS. A Berlin experience. A review of 200 cases. *Pathol. Res. Pract.* **191:**427–443.

183. **Masih, D. T., C. E. Sotomayor, L. A. Cervi, C. M. Riera, and H. R. Rubinstein.** 1991. Inhibition of I-A expression in rat peritoneal macrophages due to T-suppressor cells induced by *Cryptococcus neoformans*. *J. Med. Vet. Mycol.* **29:**125–128.

184. **Masih, D. T., C. E. Sotomayor, H. R. Rubinstein, and C. M. Riera.** 1991. Immunosup-

pression in experimental cryptococcosis in rats. Induction of efferent T suppressor cells to a non-related antigen. *Mycopathologia* **114:**179–186.

185. **Mazzola, R., R. Barluzzi, A. Brozzetti, J. R. Boelaert, T. Luna, S. Saleppico, F. Bistoni, and E. Blasi.** 1997. Enhanced resistance to *Cryptococcus neoformans* infection induced by chloroquine in a murine model of meningoencephalitis. *Antimicrob. Agents Chemother.* **41:**802–807.

186. **McDonnell, J. M., and G. M. Hutchins.** 1985. Pulmonary cryptococcosis. *Hum. Pathol.* **16:**121–128.

187. **Megson, G. M., D. A. Stevens, J. R. Hamilton, and D. W. Denning.** 1996. D-Mannitol in cerebrospinal fluid of patients with AIDS and cryptococcal meningitis. *J. Clin. Microbiol.* **34:**218–221.

187a. **Meighan, J. W.** 1972. Pulmonary cryptococcosis mimicking carcinoma of the lung. *Radiology* **103:**61–62.

188. **Meyohas, M.-C., P. Roux, D. Bollens, C. Chouaid, W. Rozenbaum, J.-L. Meynard, J.-L. Poirot, J. Frottier, and C. Mayaud.** 1995. Pulmonary cryptococcosis: localized and disseminated infections in 27 patients with AIDS. *Clin. Infect. Dis.* **21:**628–633.

189. **Miller, G. P. G.** 1986. The immunology of cryptococcal disease. *Semin. Respir. Infect.* **1:**45–52.

190. **Miyaji, M., and K. Nishimura.** 1981. Studies on organ specificity in experimental murine cryptococcosis. *Mycopathologia* **76:**145–154.

191. **Mody, C. H., G.-H. Chen, C. Jackson, J. L. Curtis, and G. B. Toews.** 1993. Depletion of murine CD8+ T-cells in vivo decreases pulmonary clearance of a moderately virulent strain of *Cryptococcus neoformans*. *J. Lab. Clin. Med.* **121:**765–773.

192. **Mody, C. H., and R. M. Syme.** 1993. Effect of polysaccharide capsule and methods of preparation on human lymphocyte proliferation in response to *Cryptococcus neoformans*. *Infect. Immun.* **61:**464–469.

193. **Mook, W. H., and M. Moore.** 1936. Cutaneous torulosis. *Arch. Dermatol. Syph.* **33:**951–962.

194. **Morgan, M. A., R. A. Blackstock, G. S. Bulmer, and N. K. Hall.** 1983. Modification of macrophage phagocytosis in murine cryptococcosis. *Infect. Immun.* **40:**493–500.

195. **Moser, S. A., F. L. Lyon, J. E. Domer, and J. E. Williams.** 1982. Immunization of mice by intracutaneous inoculation with viable virulent *Cryptococcus neoformans*: immunological and histopathological parameters. *Infect. Immun.* **35:**685–696.

196. **Mosley, R. L., J. W. Murphy, and R. A. Cox.** 1986. Immunoadsorption of *Cryptococcus*-specific suppressor T-cell factors. *Infect. Immun.* **51:**844–850.

197. **Mukherjee, J., M. Feldmesser, M. D. Scharff, and A. Casadevall.** 1995. Monoclonal antibodies to *Cryptococcus neoformans* glucuronoxylomannan enhance fluconazole activity. *Antimicrob. Agents Chemother.* **39:**1398–1405.

198. **Mukherjee, J., L. Pirofski, M. D. Scharff, and A. Casadevall.** 1993. Antibody mediated protection in mice with lethal intracerebral *Cryptococcus neoformans* infection. *Proc. Natl. Acad. Sci. USA* **90:**3636–3640.

199. **Mukherjee, J., L. Zuckier, M. D. Scharff, and A. Casadevall.** 1994. Therapeutic efficacy of monoclonal antibodies to *Cryptococcus neoformans* glucuronoxylomannan alone and in combination with amphotericin B. *Antimicrob. Agents Chemother.* **38:**580–587.

200. **Mukherjee, S., S. Lee, J. Mukherjee, M. D. Scharff, and A. Casadevall.** 1994. Monoclonal antibodies to *Cryptococcus neoformans* capsular polysaccharide modify the course of intravenous infection in mice. *Infect. Immun.* **62:**1079–1088.

201. **Mulanovich, V. E., W. E. Dismukes, and N. Markowitz.** 1995. Cryptococcal empyema: case report and review. *Clin. Infect. Dis.* **20:**1396–1398.

202. **Muller, W., W. Schorre, R. Suchenwirth, H. M. Zitz, and G. Konorza.** 1978. A case of fatal cryptococcus meningitis with intracerebral granuloma. *Acta Neurochirur.* **44:**223–235.

203. **Murakawa, G. J., R. Kerschmann, and T. Berger.** 1996. Cutaneous *Cryptococcus* infection and AIDS. *Arch. Dermatol.* **132:**545–548.

204. **Murphy, J. W.** 1985. Effects of first-order *Cryptococcus*-specific T-suppressor cells on induction of cells responsible for delayed-type hypersensitivity. *Infect. Immun.* **48:**439–445.

205. **Murphy, J. W.** 1988. Influence of cryptococcal antigens on cell-mediated immunity. *Rev. Infect. Dis.* **10**(Suppl. 2):S432BS435.

206. **Murphy, J. W.** 1989. Clearance of *Cryptococcus neoformans* from immunologically suppressed mice. *Infect. Immun.* **57:**1946–1952.

207. **Murphy, J. W.** 1992. Cryptococcal immunity and immunostimulation. *Adv. Exp. Med. Biol.* **319:**225–230.

208. **Murphy, J. W., and R. A. Cox.** 1988. Induction of antigen-specific suppression by circulating *Cryptococcus neoformans* antigen. *Clin. Exp. Immunol.* **73:**174–180.

209. **Murphy, J. W., and G. C. Cozad.** 1972. Immunological unresponsiveness induced by cryptococcal polysaccharide assayed by the hemolytic plaque technique. *Infect. Immun.* **5:**896–901.

210. **Murphy, J. W., and J. W. Moorhead.** 1982. Regulation of cell-mediated immunity in cryptococcosis. I. Induction of specific afferent T suppressor cells by cryptococcal antigen. *J. Immunol.* **126:**276–283.

211. **Murphy, J. W., and R. L. Mosley.** 1985. Regulation of cell-mediated immunity in cryptococcosis. III. Characterization of second-order T suppressor cells (Ts2). *J. Immunol.* **134:**577–583.

212. **Murphy, J. W., R. L. Mosley, and J. W. Moorhead.** 1983. Regulation of cell- immunity in cryptococcosis. II. Characterization of first-order T suppressor cells (Ts1) and induction of second-order suppressor cells. *J. Immunol.* **130:**2876–2880.

213. **Narisawa, Y., T. Kojima, A. Iriki, J. Masaki, and H. Kohda.** 1986. Tissue changes in cryptococcosis: histologic alteration from gelatinous to suppurative granulomatous tissue response with asteroid body. *Mycopathologia* **126:**65–73.

214. **Naslund, P. K., W. C. Miller, and D. L. Granger.** 1995. *Cryptococcus neoformans* fails to induce nitric oxide synthase in primed murine macrophage-like cells. *Infect. Immun.* **63:**1298–1304.

215. **Nathan, C.** 1992. Nitric oxide as a secretary product of mammalian cells. *FASEB J.* **6:**3051–3064.

216. **Nathan, C. F., and J. B. Hibbs, Jr.** 1991. Role of nitric oxide synthesis in macrophage antimicrobial activity. *Curr. Opin. Immunol.* **3:**65–70.

217. **Neilson, J. B., R. A. Fromtling, and G. S. Bulmer.** 1981. Pseudohyphal forms of *Cryptococcus neoformans*: decreased survival in vivo. *Mycopathologia* **73:**57–59.

218. **Nishimura, K., and M. Miyaji.** 1979. Histopathological studies on experimental cryptococcosis in nude mice. *Mycopathologia* **68:**145–153.

219. **Noble, R. C., and L. F. Fajardo.** 1971. Primary cutaneous cryptococcosis: review and morphologic study. *Am. J. Clin. Pathol.* **57:**13–22.

220. **Ostrosky-Zeichner, L., J. L. Soto-Hernandez, A. Angeles-Morales, F. Teixeira, C. Nava-Ruiz, C. Rios, F. Solis, and J. Sotelo.** 1996. Effects of pentoxifylline or dexamethasone in combination with amphotericin B in experimental murine cerebral cryptococcosis: evidence of neuroexcitatory pathogenic mechanisms. *Antimicrob. Agents Chemother.* **40:**1194–1197.

221. **Pema, K., J. Diaz, L. G. Guerra, D. Nabhan, and A. Verghese.** 1994. Disseminated cutaneous cryptococcosis. Comparison of clinical manifestations in the pre-AIDS and AIDS eras. *Arch. Intern. Med.* **154:**1032–1034.

222. **Perfect, J. R., and D. T. Durack.** 1985. Chemotactic activity of cerebrospinal fluid in experimental cryptococcal meningitis. *Sabouraudia* **23:**37–45.

223. **Perfect, J. R., D. T. Durack, and H. A. Gallis.** 1983. Cryptococcemia. *Medicine* **62:**98–109.

224. **Perfect, J. R., D. L. Granger, and D. T. Durack.** 1987. Effects of antifungal agents and gamma-interferon on macrophage cytotoxicity for fungi and tumor cells. *J. Infect. Dis.* **156:**316–323.

225. **Perfect, J. R., S. D. R. Lang, and D. T. Durack.** 1980. Chronic cryptococcal meningitis. *Am. J. Pathol.* **101:**177–193.

226. **Pettoello-Mantovani, M., A. Casadevall, and H. Goldstein.** 1992. The presence of cryptococcal capsular polysaccharide increases the sensitivity of HIV-1 coculture in children. *In* Proc. Pediatric AIDS: Clinical, Pathologic and Basic Science Perspectives, Washington, D.C. (Abstract.)

227. **Pettoello-Mantovani, M., A. Casadevall, and H. Goldstein.** 1993. The presence of cryptococcal capsular polysaccharide increases the sensitivity of HIV-1 coculture in children. *Ann. N.Y. Acad. Sci.* **693:**281–283.

228. **Pettoello-Mantovani, M., A. Casadevall, T. R. Kollman, A. Rubinstein, and H. Goldstein.** 1992. Enhancement of HIV-1 infection by the capsular polysaccharide of *Cryptococcus neoformans. Lancet* **339:**21–23.

229. **Pettoello-Mantovani, M., A. Casadevall, P. Smarnworawong, and H. Goldstein.** 1994. HIV-1 infectivity is increased in vitro by the presence of the capsular polysaccharide of *Cryptococcus neoformans* and *Haemophilus influenzae. AIDS Res. Hum. Retroviruses* **10:**1079–1087.

230. **Picon, L., L. Vaillant, T. Duong, G. Lorette, Y. Bacq, J. M. Besnier, and P. Choutet.** 1989. Cutaneous cryptococcosis resembling molluscum contagiosum: a first manifestation of AIDS. *Acta Dermatol. Venereol.* **69:**365–367.

231. **Porter, K. G., D. G. Sinnamon, and R. R. Gillies.** 1977. *Cryptococcus neoformans*-specific oligoclonal immunoglobulins in the cerebrospinal fluid in cryptococcal meningitis. *Lancet* **i:**1262.

232. **Procknow, J. J., J. R. Benfield, J. W. Rippon, C. F. Diener, and F. L. Archer.** 1965. Cryptococcal hepatitis presenting as a surgical emergency. First isolation of *Cryptococcus neoformans* from point source in Chicago. *JAMA* **191:**93–98.

233. **Radhakrishnan, V. V., A. Mathai, J. Shanmugham, and G. J. Mathews.** 1982. The role of hyaluronidase in experimental cryptococcal infections. *Surg. Neurol.* **17:**239–244.

234. **Randall, R. E., W. K. Stacy, E. C. Toone, G. R. Prout, G. E. Madge, H. J. Shadomy, S. Shadomy, and J. P. Utz.** 1968. Cryptococcal pyelonephritis. *N. Engl. J. Med.* **27:**60–65.

235. **Rappaport, B. Z., and B. Kaplan.** 1926. Generalized torula mycosis. *Arch. Pathol.* **1:**720–741.

236. **Retini, C., A. Vecchiarelli, C. Monari, C. Tascini, F. Bistoni, and T. R. Kozel.** 1996. Capsular polysaccharide of *Cryptococcus neoformans* induces proinflammatory cytokine release by human neutrophils. *Infect. Immun.* **64:**2897–2903.

237. **Rhodes, J. C.** 1985. Contribution of complement component C5 to the pathogenesis of experimental murine cryptococcosis. *J. Med. Vet. Mycol.* **23:**225–234.

238. **Rhodes, J. C., L. S. Wicker, and W. Urba.** 1980. Genetic control of susceptibility to *Cryptococcus neoformans* in mice. *Infect. Immun.* **29:**494–499.

239. **Robinson, B. E., N. K. Hall, G. S. Bulmer, and R. Blackstock.** 1982. Suppression of responses to cryptococcal antigen in murine cryptococcosis. *Mycopathologia* **80:**157–163.

240. **Rubinstein, H. R., C. E. Sotomayor, L. A. Cervi, C. M. Riera, and D. T. Masih.** 1993. Modulation of I-A and I-E expression in macrophages by T-suppressor cells induced in *Cryptococcus neoformans* infected rats. *Mycopathologia* **123:**141–148.

241. **Sakaguchi, N., T. Baba, M. Fukuzawa, and S. Ohno.** 1993. Ultrastructural study of *Cryptococcus neoformans* by quick-freezing and deep-etching method. *Mycopathologia* **121:**133–141.

242. **Salkowski, C. A., and E. Balish.** 1991. Cutaneous cryptococcosis in athymic and beige-athymic mice. *Infect. Immun.* **59:**1785–1789.

244. **Salyer, W. R., and D. C. Salyer.** 1974. Pleural involvement in cryptococcosis. *Chest* **66:**139–140.

245. **Salyer, W. R., D. C. Salyer, and R. D. Baker.** 1974. Primary complex of *Cryptococcus* and pulmonary lymph nodes. *J. Infect. Dis.* **130:**74–77.

246. **Sarosi, G. A., P. M. Silberfarb, and F. E. Tosh.** 1971. Cutaneous cryptococcosis: a sentinel of disseminated disease. *Arch. Dermatol.* **104:**1–3.

247. **Schimpff, S. C., and J. E. Bennett.** 1975. Abnormalities in cell-mediated immunity in patients with *Cryptococcus neoformans* infection. *J. Allergy Clin. Immunol.* **55:**430–441.

248. **Schneerson-Porat, S., A. Sharar, and M. Aronson.** 1965. Formation of histiocyte rings in response to *Cryptococcus neoformans* infection. *RES J. Reticuloendothel. Soc.* **2:**249–255.

249. **Schupbach, C. W., C. E. Wheeler, R. A. Briggaman, N. A. Warner, and E. P. Kanof.** 1976. Cutaneous manifestations of disseminated cryptococcosis. *Arch. Dermatol.* **112:**1734–1740.

250. **Schwartz, D. A.** 1988. Characterization of the biological activity of *Cryptococcus* infections in surgical pathology. The budding index and the carminophilic index. *Ann. Clin. Lab. Sci.* **18:**388–397.

251. **Selby, R. C., and N. M. Lopes.** 1973. Torulomas (cryptococcal granulomata) of the central nervous system. *J. Neurosurg.* **38:**40–46.

252. **Sethi, K. K., K. Salfelder, and J. Schwarz.** 1965. Experimental cutaneous primary infection with *Cryptococcus neoformans* (Sanfelice) Vuillemin. *Mycopathol. Mycol. Appl.* **27:**357–368.

253. **Shadomy, H. J., and H. I. Lurie.** 1971. Histopathological observations in experimental cryptococcosis caused by hypha-producing strain of *Cryptococcus neoformans* in mice. *Sabouraudia* **9:**6–9.

254. **Shahar, A., Y. Kletter, and M. Aronson.** 1969. Granuloma formation in cryptococcosis. *Isr. J. Med. Sci.* **5:**1164–1172.

255. **Sokolowski, J. W., R. F. Schillaci, and T. E. Motley.** 1969. Disseminated cryptococcosis complicating sarcoidosis. *Am. Rev. Respir. Dis.* **100:**717–722.

256. **Song, M. M.** 1971. Experimental cryptococcosis of the skin. *Sabouraudia* **12:**133–137.

257. **Sotomayor, C. E., H. R. Rubinstein, L. Cervi, C. M. Riera, and D. T. Masih.** 1989. Immunosuppression in experimental cryptococcosis in rats. Induction of thymic suppressor cells. *Mycopathologia* **108:**5–10.

258. **Sotomayor, C. E., H. R. Rubinstein, C. M. Riera, and D. T. Masih.** 1987. Immunosuppression in experimental cryptococcosis in rats. Induction of afferent T suppressor cells to a non-related antigen. *J. Med. Vet. Mycol.* **25:**67–75.

259. **Sotomayor, C. E., H. R. Rubinstein, C. M. Riera, and D. T. Masih.** 1995. Immunosuppression in experimental cryptococcosis: variation of splenic and thymic populations and expression of class II major histocompatibility complex gene products. *Clin. Immunol. Immunopathol.* **77:**19–26.

260. **Speed, B., and D. Dunt.** 1995. Clinical and host differences between infections with the two varieties of *Cryptococcus neoformans*. *Clin. Infect. Dis.* **21:**28–34.

261. **Speed, B. R., J. Kaldor, B. Cairns, and M. Pegorer.** 1995. Serum antibody response to active infection with *Cryptococcus neoformans* and its varieties in immunocompetent hosts. *J. Med. Vet. Mycol.* **34:**187–193.

262. **Stevens, D. A., J. E. Domer, R. B. Ashman, R. Blackstock, and E. Brummer.** 1994. Immunomodulation in the mycoses. *J. Med. Vet. Mycol.* **32**(Suppl. 1):253–265.

263. **Sundstrom, J. B., and R. Cherniak.** 1992. A glucuronoxylomannan of *Cryptococcus neoformans* serotype A is a type 2 T-cell-independent antigen. *Infect. Immun.* **60:**4080–4087.

264. **Sundstrom, J. B., and R. Cherniak.** 1993. T-cell dependent and T-cell-independent mechanisms of tolerance to glucuronoxylomannan of *Cryptococcus neoformans* serotype A. *Infect. Immun.* **61:**1340–1345.

265. **Swanson, H., and W. A. Smith.** 1944. Torular granulomas simulating cerebral tumor. Report of two cases. *Arch. Neurol. Psychiatr.* **51:**426–431.

266. **Tang, T., P. S. Frenette, R. O. Hynes, D. D. Wagner, and T. N. Mayadas.** 1996. Cytokine-induced meningitis is dramatically attenuated in mice deficient in enothelial selectins. *J. Clin. Invest.* **97:**2485–2490.

267. **Taylor, E. R.** 1970. Pulmonary cryptococcosis. Analysis of 15 cases from the Columbia area. *Ann. Thorac. Surg.* **10:**309–316.

268. **Terplan, K.** 1948. Pathogenesis of cryptococcic (Torula) meningitis. *Am. J. Pathol.* **24:**711–712.

269. **Tohyama, M., K. Kawakami, and A. Saito.** 1996. Anticryptococcal effect of amphotericin B is mediated through macrophage production of nitric oxide. *Antimicrob. Agents Chemother.* **40:**1919–1923.

270. **Torre, D., R. Martegani, F. Speranza, C. Zeroli, and G. P. Fiori.** 1995. Pulmonary cryptococcosis presenting as pneumothorax in a patient with AIDS. *Clin. Infect. Dis.* **21:**1524–1525.

271. **Truelsen, K., T. Young, and T. R. Kozel.** 1992. In vivo complement activation and binding of C3 to encapsulated *Cryptococcus neoformans*. *Infect. Immun.* **60:**3937–3939.

272. **Turner, S. H., and R. Cherniak.** 1990. Multiplicity in the structure of the glucuronoxylomannan of *Cryptococcus neoformans*, p. 123–142. *In* J. P. Latge and D. Boucias (ed.), *Fungal Cell Wall and Immune Response*. Springer-Verlag, Berlin.

273. **Twonley-Price, J., and G. S. Bulmer.** 1972. Tumor induction by *Cryptococcus neoformans*. *Infect. Immun.* **6:**199–205.

274. **Vanbreuseghem, R.** 1967. White mice sensitivity to *Cryptococcus neoformans*. Production of resistance against a cerebral infection. *Ann. Soc. Belg. Med. Trop.* **47:**281–294.

275. **Van Cutsem, J., J. Fransen, F. Van Gerven, and P. A. J. Janssen.** 1986. Experimental cryptococcosis: dissemination of *Cryptococcus neoformans* and dermatropism in guinea pigs. *Mykosen* **29:**561–575.

276. **Vartivarian, S. E.** 1992. Virulence properties and nonimmune pathogenetic mechansims of fungi. *Clin. Infect. Dis.* **14**(Suppl. 1):S30–S36.

277. **Vecchiarelli, A., C. Monari, C. Retini, D. Pietrella, B. Palazzetti, L. Pitzurra, and A. Casadevall.** 1997. *Cryptococcus neoformans* differently regulates B7-1 (CD80) and B7-2 (CD86) expression on human monocytes. *Eur. J. Immunol.* **28:**114–121.

278. **Vecchiarelli, A., D. Pietrella, M. Dottorini, C. Monari, C. Retini, and T. Todisco.** 1995. Encapsulation of *Cryptococcus neoformans* regulates fungicidal activity and the antigen presentation process in human alveolar macrophages. *Clin. Exp. Immunol.* **98:**217–223.

279. **Vecchiarelli, A., C. Retini, C. Monari, C. Tascini, F. Bistoni, and T. R. Kozel.** 1996. Purified capsular polysaccharide of *Cryptococcus neoformans* induces interleukin-10 secretion by human monocytes. *Infect. Immun.* **64:**2846–2849.

280. **Vecchiarelli, A., C. Retini, D. Pietrella, C. Monari, C. Tascini, T. Beccari, and T. R. Kozel.** 1995. Downregulation by cryptococcal polysaccharide of tumor necrosis factor alpha and interleukin-1beta secretion from human monocytes. *Infect. Immun.* **63:**2919–2923.

281. **Wang, Y., P. Aisen, and A. Casadevall.** 1995. *Cryptococcus neoformans* melanin and virulence: mechanism of action. *Infect. Immun.* **63:**3131–3136.

282. **Wang, Y., and A. Casadevall.** 1994. Susceptibility of melanized and nonmelanized *Cryptococcus neoformans* to nitrogen- and oxygen-derived oxidants. *Infect. Immun.* **64:**3004–3007.

283. **Watabe, T., M. Miyaji, and K. Nishimura.** 1984. Studies on the relationship between cysts and granulomas in murine cryptococcosis. *Mycopathologia* **86:**113–120.

284. **Weidman, F. D.** 1933. Cutaneous torulosis. The identification of yeast cells in general in histologic sections. *South. Med. J.* **26:**851–863.

285. **Wile, U. J.** 1935. Cutaneous torulosis. *Arch. Dermatol. Syph.* **31**:58–66.

286. **Wilson, E., L. Thorson, and D. P. Speert.** 1991. Enhancement of macrophage super-oxide anion production by amphotericin B. *Antimicrob. Agents Chemother.* **35**:796–800.

287. **Wolf, J. E., and S. E. Massof.** 1990. In vivo activation of macrophage oxidative burst activity by cytokines and amphotericin B. *Infect. Immun.* **58**:1296–1200.

288. **Wong, B., J. R. Perfect, S. Beggs, and K. A. Wright.** 1990. Production of the hexitol D-mannitol by *Cryptococcus neoformans* in vitro and in rabbits with experimental meningitis. *Infect. Immun.* **58**:1664–1670.

289. **Yamaoka, H., N. Sakaguchi, K. Sano, and M. Ito.** 1996. Intravascular granuloma induced by intravenous inoculation of *Cryptococcus neoformans. Mycopathologia* **133**:149–158.

290. **Yao, J. D. C., C. F. Arkin, J. P. Doweiko, and S. M. Hammer.** 1990. Disseminated cryptococcosis diagnosed on peripheral blood smear in a patient with acquired im-munodeficiency syndrome. *Am. J. Med.* **89**:100–102.

291. **Yuan, R., A. Casadevall, J. Oh, and M. D. Scharff.** 1997. T cells cooperate with passive antibody to modify *Cryptococcus neoformans* infection in mice. *Proc. Natl. Acad. Sci. USA* **94**:2483–2488.

10 | Animal Models and Veterinary Aspects of Cryptococcosis

INTRODUCTION

Cryptococcus neoformans has been reported to cause infection in a wide variety of mammalian hosts (4, 77, 90, 129) (Table 1). Of the mammals reported to be naturally infected with *C. neoformans*, the most common species are cats (17, 26, 42, 49, 62, 67, 84, 85, 92, 93, 106, 166), dogs (73, 82, 83, 85, 143), cows (18, 94, 122, 140, 146), horses (5, 14, 44, 82, 96, 121, 142, 170), and primates (3, 98, 107, 131, 149, 159). Pathologically, these infections range from local infection—cats, for example, commonly present with nasal and/or ocular discharges and ulcerated masses in the oral cavity—to widely disseminated fungal disease, which involves most organs of the body. Infected animals may have typical symptoms of cryptococcosis, including cryptococcal meningoencephalitis, depending on the infected body site. For instance, cryptococcal meningitis syndrome has been described naturally in dogs, cats, horses, and cows, and yeasts have been isolated from cerebrospinal fluid (CSF) in some of these cases. *C. neoformans* has caused infections or colonizations in such a broad spectrum of animals that even the saltwater dolphin (47, 97) has been found to develop cryptococcal pneumonia, presumably through fungal aerosols tracking down its blowhole. Furthermore, like other cryptococcal species, *C. neoformans* has been isolated from insects such as cockroaches (156) and sow bugs and from mites. Most strains survive ingestion by free-living amebae in the environment, and this exposure to amebae may have also selected for survival in mammalian host cells (132).

C. neoformans can also cause natural outbreaks of infection in a geographically restricted area. For instance, 18 elephant, large tree, or lesser tree shrews developed cryptococcosis over a 30-month period in a national zoological park (161). Goats have been known to develop cryptococcosis (21); in Spain, from 1990 to 1994, cryptococcosis outbreaks occurred in goat herds in five different geographical areas (2). From 2 to 12% of grazing goats developed cryptococcosis with symptoms of pneumonia and cachexia. The strain that produced infection was *C. neoformans* var. *gattii*, and at least one herd was grazing in a eucalyptus grove (2).

The adaptability of *C. neoformans* to such a large number of animal and insect species suggests that it has the survival properties for a myriad of environmental

Table 1 Cryptococcosis in mammals (natural infection)

Species	Reference
Bat	25
Camel	123
Cat	17, 26, 49, 67, 85, 92, 93, 106
Cheetah	6, 167
Civet	77
Cow	94
Dog	73, 82, 83, 143
Dolphin	47, 97
Ferret	54, 148
Fox/fennec	134, 153
Goat	21
Guinea pig	9
Horse	5, 14, 57, 142, 170
Koala	12, 151
Mangabey	98
Monkey	3, 131, 149, 159, 160
Mouse	72, 162, 169
Pig	165
Rat	72, 94
Shrew	160, 161

or host niches and supports the notion that this species is one of the premier animal pathogens and/or colonizers in the kingdom Fungi.

HISTORY OF ANIMAL INFECTIONS AND MODELS

The reports of cryptococcosis in animals have followed a pattern similar to those in human infections. The clinical history of both natural and experimental animal cryptococcosis began over 100 years ago with the initial description of the pathogenic yeast. Around the same time that Sanfelice discovered an environmental yeast isolate from peach juice, he discovered a similar yeast from the lymph node of an ox (140). Sanfelice initially considered the yeast from the ox to be different from the *C. neoformans* isolate and named it *Saccharomyces lithogenes* because of its tendency to produce calcifications in tissue (141), a trait that cryptococcus rarely displays. However, it is likely that this report is the first described case of animal cryptococcosis; interestingly, the animal described actually died of liver carcinoma, so this may have been the first case of cryptococcosis reported in a host with an underlying disease. It was also at this time in the late 19th century that experimental animal models of cryptococcosis were first used. Sanfelice injected guinea pigs with *C. neoformans* to produce pseudotumor or sarcomalike lesions with lymphadenopathy after intraperitoneal inoculation (140, 141). In fact, these animal experiments misled Sanfelice into believing that this yeast caused cancerous tumors.

The pathologist Curtis similarly focused on experimental animal infections. He was able to produce lethal infections in rats with yeast cells inoculated subcu-

taneously. These inoculations produced enormous tumorlike, granulomatous lesions in the lungs, spleen, and kidney. He also extended the range of hosts by
infecting mice, guinea pigs, and dogs (27, 28). In the early experiments with these
infections, the granulomatous inflammation seen in human infections was also
present in animal tissue, but at the time there was more appreciation for subcutaneous infections than for the later-recognized ability of this yeast to infect the brain.
In 1901, Vuillemin (165) reported the first case of pulmonary cryptococcosis in a
pig, and a year later Frothingham (44) described a fungal, myxoma-like mass lesion
in the lung of a horse after a year of persistent nasal discharge. In 1934, Weidman
and Ratcliffe (167) described a case of widely disseminated natural infection of
cryptococcus, including involvement of the central nervous system (CNS) in a
cheetah.

Over the last 70 years, the description of natural animal infections with C.
neoformans has allowed us to further understand the pathogenesis of this encapsulated yeast. It is remarkable that many clinical conditions found in humans can be
similarly described in animals. For instance, clinical presentations in both animals
and humans include widely disseminated disease, localized pneumonia, and meningoencephalitis. Similarly, simple colonization of the nasal cavity or alimentary
canal can be found in both animals and humans (72, 85, 126). There is strong
evidence in some species, such as dogs, cats, and horses, that direct inoculation into
mucous membranes is a portal of entry for this yeast in some infections. These
animals generally present with facial infection, which may be produced by the
close proximity of their faces to soil containing the yeast. This clinical feature of
facial involvement is uncommonly found in humans and may simply represent the
more intense exposure of animals to the higher inocula found in contaminated soil.
There is also evidence of gastrointestinal disease produced by C. neoformans in
animals (53, 158). This clinical feature is rarely seen in humans (1, 13, 29, 65, 66),
but it is important to emphasize the potential for this site as a portal of entry in
some cases of infection.

BIRDS

Birds are a very common animal group associated with C. neoformans. From the
ecological observations of Emmons (37) on pigeons in 1955 to those of Staib (152)
on canary droppings a few years later, studies on the epidemiological factors
associated with birds and cryptococcosis have been emphasized. In particular, the
pigeon has been frequently linked with this encapsulated yeast (see chapters 3 and
11). It appears that the pigeon, although occasionally infected by this yeast, generally is only colonized. It either carries the yeast as a colonizer within its gastrointestinal tract and excretes it in its guano, or alternatively, the guano enriches the
soil for growth of C. neoformans already present in the soil. The lack of consistent
infection in pigeons has been attributed to their extremely high body temperature
(43 to 45°C) (75). Consistent with this hypothesis of physical protection is that C.
neoformans would not survive such extreme temperatures in vitro. In fact, C. neoformans var. gattii, which is even more sensitive to environmental temperatures than
is var. neoformans, is generally not found in pigeon droppings. These findings do,
however, suggest that yeasts in the alimentary tracts of birds are within a protected
sanctuary, and it is clear that C. neoformans can withstand the biological rigors of

the digestive system in some animals. For instance, in pigeons that were force-fed large amounts of *C. neoformans,* the yeast could still be cultured from their crops up to 2.5 months later (157). However, it should be emphasized that, although infections in pigeons and other birds are not common and high body temperatures may be partially protective, these birds are not completely immune to fungal infection. There are reported cases of invasive cryptococcosis in several avian species, including pigeons (38, 95), parrots (130), cockatoos (41), macaws (23), columbiformes (55), and kiwis (58). Even a cold-blooded reptile such as a snake, which may have variable temperatures, has been known to be infected with *C. neoformans* (91). These findings of infection in some birds and reptiles likely represent the adaptability of certain *C. neoformans* strains in concert with persistent exposure of the host to an environment that is heavily contaminated (large inocula) with this fungus. Furthermore, at least one bird species, the kiwi, has been shown to be susceptible to *C. neoformans* infection with the more temperature-intolerant var. *gattii* (58). These birds, like emus, have body temperatures that are similar to those in mammals and thus much lower than those in the typical avian species, which may make them even more susceptible to certain strains of *C. neoformans.*

The common finding of an association between isolation of *C. neoformans* in pigeons and in their guano may have epidemiological ramifications. For instance, pigeons feed on a variety of grasses and grains, which could actually be the primary ecological niche of *C. neoformans* var. *neoformans* as eucalyptus trees are for *C. neoformans* var. *gattii.* If this link is real, birds could be a major vector for disseminating the yeast to animals and humans. Through their droppings, they may transplant the fungus from its normal environment directly into the proximity of mammalian hosts. It has also been shown that *C. neoformans* var. *gattii* may colonize or infect birds and koalas (151), both of which inhabit eucalyptus trees. This finding is consistent with the known ecological niche for this variety. Thus, these observations suggest that birds may represent a significant part of the infectious disease life cycle of the yeast *C. neoformans* var. *neoformans* and var. *gattii.*

DOGS AND CATS

Of all natural infections of *C. neoformans* in animals, the two species with which we have had the most clinical experience are dogs and cats. Enough clinical information is available about these animals to examine certain features of infection to help in understanding pathogenesis and comparability to human disease. For example, Malik et al. (85) examined nasal washings from 56 dogs and 45 cats that were asymptomatic and randomly chosen. Several pathological findings in this study were of further interest. First, more than half of the culture-positive animals had a relatively high fungal burden in the nasopharynx, with more than 100 colonies of *C. neoformans* per plate found on swabs. Therefore, it appears that *C. neoformans* can probably grow in the nasopharynx without consistently invading the host tissue. Second, this yeast colonization was not associated with evidence of clinical disease, and there was no detectable cryptococcal polysaccharide antigen in the sera of these animals. This finding supports our clinical notion that detection of polysaccharide antigen in sera reflects tissue invasion with dissemination of infection and not colonization. Third, despite the fact that the study was conducted in an area where var. *gattii* is frequently isolated, 14% of the dogs and 7% of the cats were colonized exclusively with *C. neoformans* var. *neoformans*. The majority of cases of cryptococcosis, whether in AIDS patients or in animals, have var. *neoformans* as the primary pathogen or colonizer, even in areas endemic for var. *gattii*. This study also emphasizes the colonizing feature of *C. neoformans* in mammals and suggests that some mammals may have more intense exposure to these yeasts than others. For instance, dogs have higher rates of colonization than cats, possibly because of their propensity to sniff their environment. This colonization state as part of the pathogenic life cycle of *C. neoformans* may be a feature of many mammals. For example, the nasopharynx and upper airway of humans have occasionally been found to contain *C. neoformans* without active cryptococcal disease (36).

In cats, cryptococcosis frequently presents with a characteristic ulcerative, crusty nasofacial lesion. However, *C. neoformans* can also widely disseminate to other organs, including the brain, in which case behavioral changes may be the primary clinical presentation. It has also been suggested that presentation of skin infection versus disseminated disease in cats may have geographical influences, such as the prevalence of different strains, but this speculation remains unproved. If dissemination of this infection occurs, the most common organs infected are the lung, brain, and eyes. A common theme in these natural infections that has also been seen in experimental models with other species is that male cats are more likely than female cats to develop cryptococcosis (49). This is similar to the findings in humans, if immunosuppressive disease states are discounted. Also, there have been reports of seasonal variation in clinical cases within cat populations. For instance, in one study, cryptococcosis was primarily diagnosed in outdoor cats during the warm season and, in contrast, it was diagnosed only in strictly indoor cats during the cold season (84). If these findings continue to be observed, the relationship of environmental temperature to fungal growth may prove to be important in the contagion of disease. However, in humans, there is no known seasonal predisposition to infection. Seasonal infectivity in humans could be obscured by subclinical disease, which cannot be detected without a reproducible skin test, and/or by cases of reactivated disease associated with immunodepressive events.

The influence of immunosuppressive viral infections on cryptococcosis in cats has been an interesting pathological interaction to study. However, clear-cut answers to questions regarding the predisposition for cryptococcosis in cats have not been revealed, as they have for humans with human immunodeficiency virus (HIV) infection. Studies with both feline leukemia virus (FeLV) and feline immunodeficiency virus (FIV) infections in cats with cryptococcosis have given conflicting results in regard to the importance of these concomitant viral infections for the yeast infection (17, 42, 84, 86, 106, 124). In one study, the frequency of *C. neoformans* isolated from oropharyngeal swabs was significantly higher in FIV-seropositive cats (42). However, in some studies, FIV infection did not appear to impart an unfavorable prognosis (84, 89). On the other hand, FIV-seropositive cats can require longer courses of antifungal treatment, and with a concomitant retroviral infection, cats may present with more disseminated cryptococcal disease (84). Similarly, there was a study in which both FeLV- and FIV-seropositive cats did, in fact, have a poorer prognosis with treatment (67). One of the reasons for the uncertainty about outcome measures of cryptococcosis in FeLV- or FIV-infected cats compared to those in HIV-infected humans is that these virally infected animals have only minor alterations in the numbers of lymphocyte subsets (166). The fact that the predisposition to crypto-coccosis in cats cannot be explained by deficiencies in lymphocyte subsets is in direct contrast to findings of HIV infection and cryptococcosis in which low CD4[+] counts ($\leq 100/mm^3$) predict the risk of cryptococcosis (97a). Therefore, because FeLV or FIV infection may have some marginal effect on treatment outcomes for cryptococcosis, the fungal infection might require prolonged therapy in animals that have these underlying retroviral infections. However, these retroviral infections in animals do not approach the immunosuppressive effects of HIV and its profound influence on immunity against cryptococcosis in humans.

Cats have probably given us the most experience in treating natural crypto-coccal infections in animals. Itraconazole and fluconazole have become standard antifungal regimens for successful treatment of this infection in cats (67, 84, 92, 93). Cryptococcal polysaccharide antigen titers during treatment have been followed in cats, and there is a significant tendency for treatment to be successful in those animals who have a drop in their serum antigen (67). However, as in humans, antigen titers can persist for long periods (months to years) in these animals without relapses (42). Similarly, it is uncertain whether specific recommendations can be made to stop or start antifungal agents based on a specific antigen titer result. On the other hand, it is always encouraging when the serum antigen becomes undetectable during treatment and therapy. At least one group of clini-cians for animals has suggested that antifungal therapy be continued until titers decline to less than 1:1 or after a drop in titer of 1:32 or greater, with periodic monitoring of the serum antigen titer after therapy is discontinued (87).

In dogs, at least half of the cases of cryptococcosis develop within, adjacent to, or contiguous with the nasal cavity (83). This finding is consistent with dogs' direct exposure to the fungus in soil through sniffing, which can result in either coloniza-tion or infection in this anatomical area. Therefore, cryptococcal rhinosinusitis is a prominent clinical feature for dogs, and eye (48), orbit (125), and ear (108) involve-ment have also been described. However, disseminated disease, including menin-goencephalitis, can occur in this species (10). It is even possible that neurological disturbances produced by this fungus may mimic other CNS diseases. For in-stance, there has been a report of a fox that was considered to have rabies because of its neurological behavior, but instead had developed cryptococcal menin-goencephalitis (153). This disease has been reported to primarily infect both the lungs and stomach of dogs (144, 164). It has also been noted in dogs that iatrogenic immune suppression may lead to a risk of disseminated cryptococcal infection. For instance, there have been several case reports of dogs developing cryptococcosis while receiving corticosteroids for other medical conditions (82). There are also cases of canine cryptococcosis associated with Hodgkin's disease (87), other can-cers (70), and concomitant infections (24). This clinical scenario is reminiscent of that for human disease. Finally, treatment with amphotericin B and fluconazole has been successful in dogs with cryptococcosis, but relapses occur with flucytosine treatment (15, 39).

COWS AND MAMMARY GLANDS

Although birds may be a vector for the spread of *C. neoformans* in the environment, and dogs and cats are the best-studied animals in regard to clinical disease, many mammals develop invasive disease in nature (Table 1). While the facial cryptococcosis of cats is very suggestive of primary skin cryptococcosis through inoculation, the outbreaks of cryptococcal mastitis reported in several dairy herds even more firmly establish the fact that *C. neoformans* can produce infection by direct inoculation into the skin of susceptible mammals. In the first large outbreak of cryptococcal mastitis, with 106 cases in a dairy herd over a period of a year, the infection was apparently inoculated into the teats by contaminated milking machines (122). In a second outbreak, 50 animals were probably infected by the use of a contaminated glucose solution for injecting penicillin into the udder (146). The extensive granulomatous reaction within the mammary gland and regional lymph node enlargement generally limited infection to this site, and only rarely was a metastatic focus of infection found in these outbreaks. However, a significant consequence of these infections was a severe reduction in milk production by the animals infected. Follow-up work using experimental mastitis in goats demonstrated the ability to infect this gland with a direct inoculation of 2×10^6 CFU of yeasts into the mammary gland (147). This goat model produced both acute and chronic inflammation, and the severe mastitis in the goats also led to a sharp decrease in milk production. Possibly, as with the testicles of normal rabbits, in which cryptococcal infection can be easily established (8), the temperature of the mammary glands in cows might be particularly suited to the establishment of a cryptococcal infection. There may also be other factors conducive to development of infection at this site, since cryptococcal mastitis has been occasionally described in humans (see chapter 13). Although *C. neoformans* has been isolated from the milk of infected animals (18), routine pasteurization should readily kill *C. neoformans*. Milk properly treated for human consumption would not be a source of infection, even if the gastrointestinal tract is proved to be an important primary or secondary site for entry of *C. neoformans* into the body.

ANIMAL MODELS: GENERAL

Although animals in nature develop sporadic and uncontrolled cases of cryptococcosis, these infections are difficult to use for the study of pathogenesis. However, since *C. neoformans* was originally described in the late 19th century, a variety of animal species have been used to study its pathogenesis. In fact, excellent animal model systems now exist to carefully study the pathogenesis and treatment of this infection. Mice, guinea pigs, rats, and rabbits have been extensively used as models. All these species can be made susceptible to *C. neoformans* infection, either by the use of a large infectious inoculum of yeasts or by the induction of an immune defect(s). A summary of these models and their characteristics is found in Table 2. There are both advantages and disadvantages to each major animal model (Table 3). However, all four models can be used very effectively to study the pathogenesis and/or treatment of *C. neoformans* infections. Of these species, mice are generally the most susceptible to *C. neoformans*. In fact, some inbred strains are particularly susceptible to infection, and spontaneous natural infections do occur in outbred mice (162, 169). Conversely, rabbits are probably the most resistant to infection; the induction of an immune defect and/or a large inoculum is generally required to cause disease.

Infections in these models can be made subclinical or lethal depending on inoculum size, site of inoculation, strain of yeast, and/or species of animal. Inocu-

Table 2 Animal model characteristics for cryptococcosis of the four major species[a]

Species	Strain or treatment	Function	Outcome	Inoculation route	Site of infection	References
Mouse	BALB/c, CBA/J, CD-1, CB-17, C57BL/6	Normal	Variable	Intravenous, intraperitoneal, intracerebral, intratracheal	Brain, lung, spleen, liver, kidney, skin	—[b]
	C5-deficient DBA/2	↓ Opsonization and recruitment of cells	↓↓ Survival	Intravenous, intraperitoneal	Brain, lung, liver, spleen	34, 127, 128
	Athymic (nu/nu)	↓ T-cell function	↓↓ Yeast counts, then ↑↑ yeast counts	Intraperitoneal	Spleen, liver, lung, brain	19, 51, 64, 103, 136
	Beige (bg/bg)	↓ PMNs, MNCs, NK cells	↓ Survival; ↓↑ organ tissue counts	Intravenous	Brain, liver, spleen	56, 88, 138
	Beige/athymic (bg/bg, nu/nu)	↓ T-cell function; ↓↓ PMNs, MNCs, NK cells	↓↓ Survival; ↑ organ counts, same CNS counts	Intravenous	Kidney, liver, lung, spleen, brain, skin	135, 137
	CBA/N (xid)	↓ Antibody-forming ability; ↓ T- and B-cell function	↑↑ Organ counts (liver, spleen)	Intravenous	Liver, spleen, brain	89
	SCID	↓ T- and B-cell function	↑↑ Organ counts	Intravenous	Liver, spleen, brain	22, 59, 61, 64

Animal	Host condition	Response	Outcome	Route	Site	Reference
Mouse	Athymic (*nu/nu*), gnotobiotic	↓ T cells; ↓ bacterial colonization	↑↑Dissemination	Intragastrointestinal	Gastrointestinal tract, brain, liver, spleen, kidney	139
	Steroids		↓↓ Survival	Intravenous	Brain, lung, liver, spleen	79
Guinea pig	Complement-deficient (cobra venom factor)	↓ Late complement components; ↓ alternative pathway	↓↓ Survival	Intravenous	Brain, spleen	32
	Steroids	↓ Inflammatory cells	↓↓ Survival	Aerosol, intravenous	Lung	31, 46
Rat	Steroids	↓ Inflammatory cells	↓↓ Survival	Aerosol	Lung	46
	Normal	Intense granulomatous reaction	Survival	Intratracheal	Lung	16, 50
Rabbit	Steroids	↓ Inflammatory cells	↓↓ Survival; ↑↑ organ counts	Intracisternal	CSF	115
	Normal	Intense inflammation	Keratitis, iritis, endophthalmitis	Intraocular, intravenous	Eye	45, 69, 168
	Normal	↑ Inflammation	Self-limited	Intrapulmonary	Lung	40
	Normal	Active inflammation	Site-specific	Intratesticular	Testes	8
	Normal	Inflammation; inoculum dependent	Skin with some dissemination	Intradermal	Skin	155

[a]Abbreviations: MNCs, mononuclear cells; NK, natural killer cells; PMNs, polymorphonuclear cells.
[b]There are numerous references in the literature in which a variety of mouse strains were used, from outbred Swiss mice to a series of inbred strains.

Table 3 Advantages and disadvantages of the primary animal models for cryptococosis

Species	Advantages	Disadvantages
Mouse	Inexpensive; genetics: multiple inbred strains, gene knockouts; immunological regents available; ideal for early treatment studies	Highly susceptible to infection and thus may differ from human infection; small size precludes many procedures
Rat	Immunological reagents; moderately susceptible to infection; good granulomatous responses; larger size allows manipulations	Not as well studied as mice and rabbits; genetic backgrounds are less well defined; no CNS model; few treatment data
Rabbit	Large size allows manipulations; resistant to infection unless immunosuppressed (similar to humans); ability to analyze host cells and yeast sequentially at infection site; treatment data available	High body temperatures; few immunological reagents; primarily CNS infection
Guinea pig	Chronic infection; reactivation possible; sex differences present	No CNS model; few treatment data

lation of large numbers ($\leq 10^6$ CFU/ml) of yeasts into the bloodstream of these animals does not generally cause septic shock; cryptococcemia also does not generally produce septic shock clinically in humans (23).

MOUSE MODELS

Mice are extremely attractive for the study of cryptococcosis for two very important reasons. First, they provide both a very uniform set of immune responses and a wide variety of these responses in a series of inbred strains. This combination allows investigators to carefully examine the contributions of both the yeasts and the specific host factors to this infection. For example, strains of mice such as Swiss, BALB/c, C57BL/6, DBA/2, CD-1, and CBA/J vary in their susceptibility to *C. neoformans*. Furthermore, strains of *C. neoformans* can have different patterns of virulence within a single inbred strain (see chapter 6). Thus, in complex virulence studies it is helpful when the host response is relatively uniform within each set of experiments, and mouse models generally allow this to happen. The second reason that mice are attractive for the study of cryptococcosis is the ability to immunologically alter them, either genetically or through immunological reagents. This feature has been extremely helpful in understanding specific host-parasite interactions.

Most in vivo studies of *C. neoformans* with animal models have used mice, and a variety of methods have been used to establish infections. Intraperitoneal inoculations are the most convenient. Intracerebral inoculations allow direct study of the proliferation of these yeasts in the CNS, which is the most important clinical site of infection with this yeast. The technique of intracerebral inoculation can be performed by inserting a 26- to 27-gauge needle approximately 1 to 3 mm into the cerebral cortex in the posterior half of the skull and about one-quarter inch laterally from the midline to avoid the superior sagittal sinus. With expert technical experience and anesthesia, the mortality with this procedure is less than 1%. Intravenous inoculations are the most standardized and quantitative technique. Finally, intratracheal or intranasal inoculation (74) may best simulate natural infection, but infection rates can be inconsistent with the intranasal route. Gastrointestinal inoculations have also been used to study the possibility of a gastrointestinal mode of entry into the host.

Mouse models have great flexibility in the study of cryptococcal pathogenesis at all infection sites. The route of administration, the particular strain, and the inoculum, which may vary from 10^2 to 10^8 CFU/ml will influence the duration of infection prior to symptoms or death. It is also important to note that although intracerebral inoculation causes severe CNS disease, most inoculation routes that allow disseminated infection in mice will also eventually produce involvement of the CNS.

Besides immune factors, the mouse model can be used to examine the yeast's contribution to disease. Different strains and null mutants for certain genes of wild-type strains can produce a variety of outcomes measured in mouse models (20). This feature allows these models to be used to dissect a variety of specific factors associated with the intrinsic virulence of the yeast in a relatively small number of animals (see chapter 6). The virulence of a strain can be studied in mice both by survival and also by tissue counts of yeasts, which generally correlate with survival rates after infection. However, there is occasionally a discrepancy between quantitative tissue counts and survival. Mice can also be used to identify *C. neoformans* strains that have specific tissue tropisms, such as the skin or facial area (33, 43, 163).

Similar to investigations into important yeast factors, the mouse models also have been extremely useful in the investigation and dissection of the important host factors that are so relevant to the control of cryptococcosis (see chapters 7 and 8). The following mouse models represent examples of immune defects and their effect on cryptococcosis. For instance, in a challenge with *C. neoformans*, C5-deficient mice (DBA/2) will die quickly from acute cryptococcal pneumonia, but when smaller inocula are used, they will succumb to late progressive meningoencephalitis. The importance of inocula on the manifestation of infection can be seen in this host. In contrast, BALB/c mice will survive much longer than DBA/2 mice when similar inocula and strains are used; these results emphasize the contribution of host factors to infection outcome (34, 127, 128). Histopathologically, the C5-deficient mice show a significant reduction in the inflammatory response to *C. neoformans* that is probably related to the lack of the chemotactic peptide C5a. As in human infections with HIV or in animals treated with corticosteroids, the paucity of the quantitative cellular response at the site of infection appears to be a crucial host factor in the outcome of *C. neoformans* infection. Similarly, the beige mouse, which has defects in chemotaxis, impairment of phagosome-lysosome fusion, natural killer cell activity, and reduced microbicidal function, also has reduced cellular reaction at the site of cryptococcal infections and a delay in the transition from acute to chronic inflammation (56, 88, 138). This reduced inflammatory response leads to a more aggressive infection with *C. neoformans* and explains the importance of professional phagocytes and natural killer cells in the defense against cryptococcal infections.

Mouse models have also been used to dissect the importance of specific cell-mediated immunity in cryptococcosis. Nude mice, which are homozygous for the recessive gene that renders them athymic and hairless, were the first animal model in which the essential importance of the T-lymphocyte populations in *C. neoformans* infections was specifically confirmed (19, 51, 103, 136). It was demonstrated unequivocally that nude mice were more susceptible to infection than were their heterozygous littermates. Use of this model underscored the fundamental importance of the T lymphocyte and paved the way for more specific research on subpopulations of T lymphocytes. Recent work with mice has centered on a combination of cell-depletion experiments by monoclonal antibodies (63, 99, 100) and the use of cell replacement experiments in severe combined immunodeficiency (SCID) mice (59–61, 64). Mice depleted of CD4+ cells appear to have a poor inflammatory response to *C. neoformans* in the lung, and this cellular response is apparently important to protect the host from dissemination of infection to other organs. CD8+-depleted mice have decreased pulmonary clearance of *C. neoformans* and abrogate delayed-type hypersensitivity without effects on antigen recognition or lymphocyte proliferation in vitro. It is likely that these lymphocytes, along with natural killer cells and professional phagocytes, are responsible for the specific fungicidal activity at the site of initial infection. Because SCID mice are more susceptible to *C. neoformans* infection, additional experiments could also be performed. Instead of specific cell-depletion experiments, the effect of lymphocyte subpopulation replacements on cryptococcosis can now be observed in vivo. Findings similar to the depletion experiments confirmed the respective importance of the CD4+, CD8+, and other primed lymphocytes in these additive experiments with SCID mice. These elegant studies in understanding the host immunological circuitry for cryptococcosis make mice the animal model most used today for the study of pathogenesis.

It is clear that basic mouse strains vary in their individual susceptibility to cryptococcal infections, and these differences can be exploited to help understand pathogenesis. For instance, CBA/J and BALB/c mice are more resistant to pulmonary infection than are C57BL/6 mice. These differences in strains may then be used to find more subtle defects in cytokine networks and specific effector functions for the control of cryptococcosis. Differences in mouse strain susceptibility can even localize the important immune factors to specific areas of the host chromosome. CB-17 mice have better clearance of *C. neoformans* from the lung than do BALB/c mice. Genetic crosses of these mice suggest that the differences in yeast clearance can be mapped to sites on chromosome 12 in these mice, that differ only in the distal region of the chromosome. This chromosomal region includes sites of the immunoglobulin heavy-chain locus and its neighboring genes (80). These findings not only encourage further examinations into differences in antibody formation that may be crucial for the effective murine immune response, but also vividly demonstrate that host genetics play an important role in susceptibility to *C. neoformans* infection. It is interesting that both the immunoglobulin locus on chromosome 12 and the *xid* allele (89), which is also related to the humoral system and antibody production, are important to host resistance. On the other hand, these murine studies have also determined that the basic resistance to infection with *C. neoformans* is probably complex and multigenic. The mouse model vividly illustrates the breadth of the protective immune response to *C. neoformans*. It is multifactorial, including both cellular and humoral immunity, and both can be under genetic control. In humans, the genetic susceptibility to *C. neoformans* is less well understood, but even in this less controlled environment there have been several reports of familial cryptococcosis (71, 171).

Mouse models have even challenged a basic concept in pathophysiology that states that in human infections the lung is the primary organ of yeast entry into the body. In gnotobiotic nude mice, or even with bacterial flora, intragastric ingestion of *C. neoformans* can lead to colonization of gastrointestinal tissue with eventual dissemination of infection from this site (139). In contrast, normal mice rarely become colonized with oral inocula of *C. neoformans* and generally do not develop disseminated infection (53). However, even primates may become colonized with *C. neoformans* through the gastrointestinal tract and disseminate infection. For example, in one study with the use of high inocula in monkeys fed *C. neoformans*-coated bananas, yeasts were able to colonize and disseminate from the gastrointestinal tract (158). These findings continue to stimulate discussions about the hypothesis that oral ingestion of these yeasts could be an entry site for *C. neoformans* into the body for some severely immunocompromised patients or when large amounts of yeasts are ingested in food.

The mouse models of cryptococcosis also have great utility in the early comparative trials of antifungal agents and immune modulators (see chapter 14). The advantages of low cost, availability of inbred populations, reproducible infections with several endpoint evaluations, including both survival and quantitative colony counts from tissue, and the availability of immune modulators for this species make the mouse models one of the primary screening systems for new therapeutic regimens. Also, the models tend to be relatively uniform in their outcome so that small numbers (approximately 10 per group) can be used. However, as with all animal models, it is essential that particular attention be paid to drug pharmacokinetics within the species when specific antifungal drugs are studied.

GUINEA PIG MODELS

The guinea pig has been used in both intravenous (31, 32, 163) and pulmonary inoculation models (46). Female guinea pigs are more resistant to *C. neoformans* infection than are males, and this finding appears to mimic human findings if immunosuppressive events are discounted (31). This animal species is also considered to be "cortisone resistant," and there has been direct comparison between guinea pigs and rats, which are "cortisone susceptible" in their response to a *C. neoformans* challenge. The terms cortisone susceptible versus cortisone resistant refer to the relative response of lymphocyte depletion when steroids are given to animals. For example, guinea pigs and mice are cortisone-resistant species compared to rats, rabbits, and humans. In a series of creative and well-designed studies, Gadebusch and Gikas (46) examined the pathogenesis of cryptococcosis in guinea pigs and rats. They found that with intravenous inoculation of the two species, both were equally susceptible to infection with a certain *C. neoformans* strain (31, 32). However, when the same *C. neoformans* strain was introduced through the respiratory tract, the guinea pigs were more resistant to infection than were the rats. With corticosteroid therapy, there was increased mortality in both species, but the increase was more dramatic in the cortisone-susceptible rats (46).

These findings emphasize the importance of both the lymphocyte populations, with their impact on the total host immune response to *C. neoformans,* and the site of infection. It was further shown in guinea pigs and rats that a sublethal respiratory challenge can be performed on these animals when there is no clinical or roentgenographic evidence of infection. However, if corticosteroids are then administered within a month of the infectious challenge, clinical infection can be reactivated (46). This reactivation model suggests that these animals, like humans, can produce a dormant infection that can be reactivated by an immunosuppressive event. These models could be used for important studies into our understanding of host and yeast mechanisms for reactivation. However, present data suggest that this dormancy condition is transient in these models and may not last for more than several months. It is also important to note that many pulmonary infection models still use intratracheal instillation rather than aerosols, as these early guinea pig studies did. For the study of initial respiratory infections with basidiospores and/or small, weakly encapsulated yeasts, it may be important to use aerosol chambers to simulate possible natural infection through production of actual aerosols rather than direct intratracheal inoculation. On the other hand, the intratracheal model allows a consistent and reproducible inoculum to be administered for the study of pulmonary immune mechanisms.

RAT MODELS

Rats have been used in multiple immunological and pathological studies to define certain aspects of cryptococcosis (16, 46, 50, 154). After intravenous inoculation, rats will develop a more chronic pulmonary infection than mice will when the same *C. neoformans* strain and inoculum are used. Although some rats will develop chronic, progressive pulmonary disease and at times meningoencephalitis, it appears that rats are intrinsically more resistant to *C. neoformans* infection than are mice. Another finding in rats suggests that the relative virulence of a given *C. neoformans* strain is also not necessarily the same in mice. It is important to emphasize in virulence studies that certain host-range determinant genes are differentially expressed in different strains that are more or less important in one animal over another. It is also essential to recognize that some results in pathogenesis studies with these animal models will have no relevance to human disease but are specific to the host species studied. It is important for investigators to determine intrinsic limitations of each animal model in their quest to apply results to our understanding of human infections.

The rat has been further refined as a model to study the pathogenesis of pulmonary cryptococcosis (16, 46, 50). *C. neoformans* can be a natural pathogen for rats and has been described to cause pulmonary infection in rats in nature (94). Because rats are larger than mice, intratracheal inoculation can reliably produce infection without surgical techniques and without potential damage to tissue planes or pulmonary structures. In the rat model, extrapulmonary dissemination can occur early, but infection is generally contained within the lung; thus, the study of a progressive, effective subacute immune response and infection can progress without the need for specific immunosuppression. *C. neoformans* infection in rats shows an array of complex immunocellular reactions for study. Polysaccharide antigen moves into the systemic circulation from a peritoneal cavity infection but rarely is detected in sera when given in large amounts intratracheally. This observation may have direct relevance to the clinical use of serum antigen titers for determination of extrapulmonary spread of infection. In fact, findings in the rat model by Goldman et al. (50) suggest that when cryptococcal polysaccharide antigen is detected in serum, there is a distinct possibility that *C. neoformans* has already disseminated from the lung site of infection. Although the rat model lacks the genetic and immunological foundation of the mouse model, the greater size of rats gives investigators the technical ability to perform more site-specific and dynamic studies, such as performance of bronchoalveolar lavage and sampling of the CSF.

RABBIT MODELS

The rabbit is an interesting model for the study of cryptococcosis and has both positive and negative features for the study of cryptococcal pathogenesis. First, the body temperature of the rabbit is relatively high for mammals. The average body temperature of laboratory rabbits is between 39.3 and 39.5°C. These are stress temperatures for the growth of *C. neoformans*; in the early studies with this model, investigators were able to produce consistent *C. neoformans* infection only when yeasts were given intratesticularly because of the slightly lower temperatures in that site (8). Attempts to produce pulmonary disease in this species were successful when a large inoculum was given directly into the lung, but there was no consistent detection of dissemination to extrapulmonary sites (40). With high inocula introduced into the lung directly or intratracheally and the use of high doses of corticosteroid, *C. neoformans* can disseminate from the lung to the CNS in rabbits (108a). There have also been several ocular infection models with cryptococcus in rabbits (45, 69, 168). Finally, direct intradermal inoculations of *C. neoformans* in rabbits can produce skin lesions whose sizes correlate with the inocula sizes, and approximately 30% of the animals will show dissemination from the skin to another body site (155).

This species is attractive for the study of *C. neoformans* infection because the major site of infection for this fungus is the brain, and the rabbit model allows ready access to this site for study on a dynamic basis. Rabbits are considered a cortisone-susceptible species, and there is a direct clinical link between corticosteroid treatments and cryptococcal infection. When corticosteroids are given to rabbits, even at a body temperature of 40°C, *C. neoformans* can survive and replicate in the CNS if introduced into the subarachnoid space. The established infection can result in neurological symptoms and death of the host. There is also easy access to yeast during infection by repeated cisternal taps. Thus, the rabbit model of cryptococcal meningitis with its intrathecal inoculation has become a tool to study this disease at one of its most important sites of infection (115).

The rabbit model has several features that make it relevant to human disease. First, the progression of infection is not only contingent on the presence of cortisone immunosuppression but also is affected by the dose of this drug. The rabbit model also demonstrates that other classes of immunosuppressive agents can influence progression of disease, such as the immunophilins cyclosporine and tacrolimus (110). Another feature of this model, as in mice, is that infection is inoculum and strain dependent. A certain high inoculum is required to produce disease ($>10^6$ CFU), and there are *C. neoformans* strains with differences in virulence potential within the model.

In the early studies of this model, the natural history of infection was studied with multiple *C. neoformans* strains, including the original Busse-Buschke strain from 1895. With one *C. neoformans* strain (H99) a chronic, fatal meningitis developed over 1 to 2 months with accompanying neurological findings in the animals. However, over the last decade, the model has been used only during the first 1 to 3 weeks of infection to study host immune responses, yeast viability to molecular manipulation (i.e., mutants), or drug treatments. These measurements can be made by quantitative assessment of yeast burdens in the subarachnoid space and generally abrogate the use of clinical symptoms or death as an endpoint in these studies. Animals only rarely demonstrate neurological symptoms or die from infection during this observation period.

Through continuous sampling of CSF during infection, the rabbit model of cryptococcal meningitis allows a unique window of examination into an important site of infection for studies on pathogenesis and treatment in individual animals that is not generally available in smaller animals. This model has many features that are similar to those of human infections: (i) immunosuppression as a feature of infection, (ii) infection involving an immunologically sequestered site for host proteins, (iii) CSF leukopenia, (iv) chronic infection that is fatal if untreated, (v) dissemination of infection to other organs, and (vi) responses to the treatment of infection. Several of these points should be amplified. It is clear from these rabbit studies that *C. neoformans* can reseed the bloodstream from a CNS site. *C. neoformans* can migrate from a subarachnoid source of infection back into the systemic circulation. It is also impressive in this model that the entire quantitative CSF cellular response of the host is essential. This animal model, as in human infections with a concomitant HIV infection, demonstrates that CSF leukopenia is a major factor in susceptibility to infection at this site. The model has effectively been used to dynamically study host responses at this site, including humoral immunity, chemotactic factors, and CSF cellular types and their activation (111, 113, 114). It is now being used to understand the yeast's response to this hostile environment through the study of *C. neoformans* gene expressions at this site (78, 105, 116, 118). Finally, the model has been used to predict treatment outcomes (109, 111, 112, 117, 119, 120), and its results have been validated in human trials. For instance, the rabbit model predicted the more rapid sterilization of CSF with amphotericin B compared to currently available azoles in the treatment of cryptococcal meningitis (133). Furthermore, despite extremely low CSF levels of amphotericin B (112) and itraconazole (117) in this model, as in humans (7, 30), both drugs are still effective in the treatment of cryptococcal meningitis in both rabbits and humans.

The negative aspects of the rabbit model include the cost of the animals, a lack of studies with inbred strains, reduced immunological reagents, the need for immunosuppression, the use of an unnatural route of infection such as intracisternal inoculations, and at present only one well-characterized strain. As with all animal models, the drawbacks of this model emphasize the need for each investigator to create the host model to simulate disease and test hypotheses but to realize its limitations. No animal model will exactly simulate a human infection, but each model can yield extremely important information if there are proper controls and carefully defined clinical or pathological questions.

OTHER CRYPTOCOCCAL ANIMAL MODELS

Although much of this discussion on animal model systems has emphasized the four major models, *C. neoformans* has also been inoculated into other animals. It has been used to produce infections in both dogs and cats (11, 81, 150). Primates such as monkeys have been fed cryptococcus-laden bananas to produce infection through gastrointestinal dissemination (158). Two of three marmoset monkeys developed disseminated disease when fed a paste of cryptococci sandwiched between slices of banana. Although there was no evidence of gastrointestinal involvement pathologically, mesenteric lymph nodes were infected. In contrast, a calf given a broth of *C. neoformans* orally did not develop local or disseminated infection (122). Goats have been used to study experimental cryptococcal mastitis (147). Chick embryos, which can be inoculated intravenously, have been ingeniously used to study the effects of high temperature on *C. neoformans* infection (68). Finally, even the very resistant bird species have been used to produce experimental infections with *C. neoformans* (76, 145). Certain specific questions about pathology may be better answered in these less-developed model systems.

In the study of *C. neoformans,* there are a variety of animal models that have potential relevance to human infection. Some models are more sophisticated than others, but all provide an excellent foundation for studies of *C. neoformans* pathobiology and for assessment of treatment with new antifungal agents and immunomodulators.

CONCEPTS REGARDING MODEL SYSTEMS

One approach to the study of cryptococcosis and immune suppression in animal models is to use immunosuppressive agents such as corticosteroids, immunophilins, and cytotoxic agents. Corticosteroids have been shown to enhance cryptococcosis in all species tested, such as mice (79), guinea pigs (31, 46), rats (46), and rabbits (115). In rats and guinea pigs, corticosteroids can actually reactivate a dormant or latent infection (46). In both guinea pigs and rabbits, corticosteroid treatment dramatically reduces the inflammatory response. This reduction in inflammation directly correlates with increased numbers of viable yeasts at the site of infection, and this pancytopenia is comparable to histopathological findings in cryptococcosis with concomitant HIV infection. Cytotoxic agents such as cyclophosphamide have a minimal or no effect on the course of cryptococcosis in mice (35, 52). The immunophilins' effects on animal models of cryptococcosis are more complex. In a rabbit model, treatment with both cyclosporine and tacrolimus appeared to immunosuppress the animal, and infection was encouraged by these drugs (104, 110). In contrast, cyclosporine in mice manifested direct antifungal activity over immunosuppression and effectively treated cryptococcal infections (101, 102). These results led to the further investigation of cyclosporine's effects on *C. neoformans,* and eventually calcineurin A was identified as a temperature-sensitive molecular target for *C. neoformans* growth (105).

The findings with cyclosporine illustrate an important concept concerning animal models for cryptococcosis, which is that experimental designs in one model may not give the same results in another model. This concept emphasizes the importance of choosing the correct model for the hypothesis to be tested, but it also shows how the animal model can create and help answer new questions. Another example of animal model differences has been found in the importance of urease production by *C. neoformans* in the rabbit meningitis and mouse pulmonary models. Urease production was apparently not important to the outcome of infection in the rabbit meningitis model, but in the mouse pulmonary model the production of urease was found to be essential for the establishment of a fatal infection (24a). As molecular pathogenesis studies continue, careful assessment of virulence factors will require different inocula, different sites of infection, and different endpoints to assess the effect of various composite features of the fungus on the host. To assign the importance of various alleles for virulence, it is also likely that the molecularly manipulated strains may need to be tested in more than one animal model. *C. neoformans* is now well placed for these pathogenesis and therapeutic studies, since a variety of clinically relevant and well-characterized animal models have been created for cryptococcosis.

REFERENCES

1. **Allen, V. K., and L. Lowbeer.** 1945. Rectal ulcers with perirectal fistula as a port of entrance for *Torula encephalitis*. *South. Med. J.* **38**:564–569.
2. **Baro, T., J. M. Torres-Rodriquez, M. Hermosa de Mendoza, Y. Morera, and C. Alia.** 1998. First identification of autochthonous *Cryptococcus neoformans* var. *gattii* isolated from goats with predominantly severe pulmonary disease in Spain. *J. Clin. Microbiol.* **36**:458–461.
3. **Barrie, M. T., and C. K. Stadler.** 1995. Successful treatment of *Cryptococcus neoformans* infection in an Allen's swamp monkey (*Allenopithecus nigroviridis*) using fluconazole and flucytosine. *J. Zool. Wildlife Med.* **26**:109–114.
4. **Barron, C. N.** 1955. Cryptococcosis in animals. *J. Am. Vet. Med. Assoc.* **127**:125–132.
5. **Barton, M. D., and I. Knight.** 1972. Cryptococcal meningitis of a horse. *Aust. Vet. J.* **48**:534.
6. **Beehler, B. A.** 1982. Oral therapy for nasal cryptococcosis in a cheetah. *J. Am. Vet. Med. Assoc.* **181**:1400–1401.
7. **Bennett, J. E., W. Dismukes, R. J. Duma, G. Medoff, M. A. Sande, H. Gallis, J. Leonard, B. T. Fields, M. Bradshaw, H. Haywood, Z. A. McGee, T. R. Cate, C. G. Cobbs, J. F. Warner, and D. W. Alling.** 1979. A comparison of amphotericin B alone and combined with flucytosine in the treatment of cryptococcal meningitis. *N. Engl. J. Med.* **301**:126–131.
8. **Bergman, F.** 1966. Effect of temperature on intratesticular cryptococcal infection in rabbits. *Sabouraudia* **5**:54–58.
9. **Betty, M. J.** 1977. Spontaneous cryptococcal meningitis in a group of guinea pigs caused by a hyphae-producing strain. *J. Comp. Pathol.* **87**:377–382.
10. **Bistner, S., A. DeLahunta, and M. Lorenz.** 1971. Generalized cryptococcosis in a dog. *Cornell Vet.* **61**:440–457.
11. **Blovin, P., and R. M. Cello.** 1980. Experimental ocular cryptococcosis. *Invest. Ophthalmol. Visual Sci.* **19**:21–30.
12. **Bollinger, A., and E. S. Finckh.** The prevalence of cryptococcosis in the koala (*Phascolarctos cinereus*). *Med. J. Aust.* **14**:545–547.
13. **Bonacini, M., J. Nussbaum, and C. Ahluwalia.** 1990. Gastrointestinal, hepatic and pancreatic involvement with *Cryptococcus neoformans* in AIDS. *J. Clin. Gastroenterol.* **12**:296–297.
14. **Boulton, C. H., and L. Williamson.** 1984. Cryptococcal granuloma associated with jejunal intussusception in a horse. *J. Equine Vet. Sci.* **16**:548–551.

15. **Bovee, K. C., J. D. Conroy, and S. A. Rosenthal.** 1966. Cryptococcosis in a dog and treatment with amphotericin B. *J. Am. Vet. Med. Assoc.* **149:**49–55.

16. **Buchanan, K. L., and J. W. Murphy.** 1994. Regulation of cytokine production during the expression phase of the anticryptococcal delayed-type hypersensitivity response. *Infect. Immun.* **62:**2930–2939.

17. **Cabanes, F. J., M. L. Abarca, R. Bonavia, M. R. Bragulat, G. Castella, and L. Ferrer.** 1995. Cryptococcosis in a cat seropositive for feline immunodeficiency virus. *Mycoses* **38:**131–133.

18. **Carter, H. S., and J. L. Young.** 1950. Note on the isolation of *Cryptococcus neoformans* from a sample of milk. *J. Pathol. Bacteriol.* **62:**271–273.

19. **Cauley, L. K., and J. W. Murphy.** 1979. Response of congenital athymic (nude) and phenotypically normal mice to *Cryptococcus neoformans* infection. *Infect. Immun.* **23:**644–651.

20. **Chang, Y. C., and K. J. Kwon-Chung.** 1994. Complementation of a capsule-deficiency mutation of *Cryptococcus neoformans* restores its virulence. *Mol. Cell. Biol.* **14:**4912–4919.

21. **Chapman, H. M., W. F. Robinson, J. R. Bolton, and J. P. Robertson.** 1990. *Cryptococcus neoformans* infection in goats. *Aust. Vet. J.* **67:**263–265.

22. **Clemons, K. V., R. Azzi, and D. A. Stevens.** 1996. Experimental systemic cryptococcosis in SCID mice. *J. Med. Vet. Mycol.* **34:**331–335.

23. **Clipsham, R. C., and J. O. Britt, Jr.** 1983. Disseminated cryptococcosis in a macaw. *J. Am. Vet. Med. Assoc.* **183:**1303–1305.

24. **Collett, M. G., A. S. Doyle, F. Reyers, T. Kruse, and B. Fabian.** 1987. Fatal disseminated cryptococcosis and concurrent ehrlichiosis in a dog. *J. South Afr. Vet. Assoc.* **58:**197–202.

24a. **Cox, G. M., and J. R. Perfect.** Unpublished data.

25. **Crose, E., C. J. Marin-Kelle, and C. Striegel.** 1968. The use of tissue cultures in the identification of *Cryptococcus neoformans* from Columbia bats. *Sabouraudia* **6:**127–132.

26. **Curtis, A. J.** 1951. A case of torulosis in a domestic cat. *Aust. J. Med. Tech.* **1:**71.

27. **Curtis, F.** 1895. Note sur un nouveau parasite humain, megalococcus myxoides, trouvé dans un néoplasme de la région inguino-crurole. *C.R. Soc. Biol.* (Paris) **1895:**715–718.

28. **Curtis, F.** 1896. Contribution a l'étude de la saccharomycose humaine. *Ann. Inst. Pasteur* **10:**449–468.

29. **Daly, J. S., K. A. Porter, F. K. Chong, and R. J. Robillard.** 1990. Disseminated non-meningeal gastrointestinal cryptococcal infection in an HIV-negative patient. *Am. J. Gastroenterol.* **85:**1421–1424.

30. **Denning, D. W., R. M. Tucker, J. S. Hostetler, S. Gill, and D. A. Stevens.** 1990. Oral itraconazole therapy of cryptococcal meningitis and cryptococcosis in patients with AIDS, p. 305–324. *In* H. Vanden Bossche (ed.), *Mycoses in AIDS Patients.* Plenum Press, New York.

31. **Diamond, R. D.** 1977. Effects of stimulation and suppression of cell-mediated immunity on experimental cryptococcosis. *Infect. Immun.* **17:**187–194.

32. **Diamond, R. D., J. E. May, and M. A. Kane.** 1974. The role of the classical and alternative complement pathways in host defense against *Cryptococcus neoformans* infection. *J. Immunol.* **112:**2260–2270.

33. **Dixon, D. M., and A. Polak.** 1986. *In vivo* and *in vitro* studies with an atypical, rhinotrophic isolate of *Cryptococcus neoformans*. *Mycopathologia* **96:**33–40.

34. **Dromer, F., C. Perronne, J. Borge, J. L. Vilde, and P. Yeni.** 1989. Role of IgG and complement component C5 in the initial course of experimental cryptococcosis. *Clin. Exp. Immunol.* **78:**412–417.

35. **Duke, S. S., and R. A. Fromtling.** 1984. Effects of diethyl-stilbestrol and cyclophos-

phamide on the pathogenesis of experimental *Cryptococcus neoformans* infections. *J. Med. Vet. Mycol.* **22:**125–135.

36. **Duperval, R., P. E. Hermans, N. S. Brewer, and G. D. Roberts.** 1977. Cryptococcosis, with emphasis on the significance of isolation of *Cryptococcus neoformans* from the respiratory tract. *Chest* **72:**13–19.

37. **Emmons, C. W.** 1955. Saprophytic sources of *Cryptococcus neoformans* associated with the pigeon. *Am. J. Hyg.* **62:**227–232.

38. **Ensley, P. K., C. E. Davis, M. P. Anderson, and K. C. Fletcher.** 1979. Cryptococcosis in a male Beccari's crowned pidgeon. *J. Am. Vet. Med. Assoc.* **175:**992–994.

39. **Faggi, E., G. Gargani, C. Pizzirani, S. Pizzirani, and N. Saponetto.** 1993. Cryptococcosis in domestic mammals. *Mycoses* **36:**165–170.

40. **Felton, F. G., W. E. Maldonado, H. G. Muchmore, and E. R. Rhoades.** 1966. Experimental cryptococcal infection in rabbits. *Am. Rev. Respir. Dis.* **94:**589–594.

41. **Fenwick, B., K. Takushita, and A. Wong.** 1985. A moluccan cockatoo with disseminated cryptococcosis. *J. Am. Vet. Med. Assoc.* **187:**1218–1219.

42. **Flatland, B., R. T. Greene, and M. R. Lappin.** 1996. Clinical and serologic evaluation of cats with cryptococcosis. *J. Am. Vet. Med. Assoc.* **209:**1110–1113.

43. **Fromtling, R. A., G. K. Abruzzo, and A. Ruiz.** 1988. *Cryptococcus neoformans*: a central nervous system isolate from an AIDS patient that is rhinotropic in a normal mouse model. *Mycopathologia* **102:**72–86.

44. **Frothingham, L.** 1902. A tumor-like lesion in the lung of a horse caused by a blastomyces (torula). *J. Med. Res.* **3:**31–43.

45. **Fujita, N.** 1983. Experimental hematogenous endophthalmitis due to *Cryptococcus neoformans*. *Invest. Ophthalmol. Visual Sci.* **24:**368–375.

46. **Gadebusch, H. H., and P. W. Gikas.** 1965. The effect of cortisone upon experimental pulmonary cryptococcosis. *Am. Rev. Respir. Dis.* **92:**64–74.

47. **Gales, N., G. Wallace, and J. Dickson.** 1985. Pulmonary cryptococcosis in a striped dolphin (*Stenella coeruleoalba*). *J. Wildlife Dis.* **21:**443–446.

48. **Gelatt, K. N., L. D. McGill, and V. Perman.** 1973. Ocular and systemic cryptococcosis in a dog. *J. Am. Vet. Med. Assoc.* **162:**370–375.

49. **Gerds-Grogan, S., and B. Dayrell-Hart.** 1997. Feline cryptococcosis: a retrospective evaluation. *J. Am. Anim. Hosp. Assoc.* **33:**118–122.

50. **Goldman, D., S. C. Lee, and A. Casadevall.** 1994. Pathogenesis of pulmonary *Cryptococcus neoformans* infection in the rat. *Infect. Immun.* **62:**4755–4761.

51. **Graybill, J. R., and D. J. Drutz.** 1978. Host defense in cryptococcosis. II. Cryptococcosis in the nude mouse. *Cell. Immunol.* **40:**263–274.

52. **Graybill, J. R., and L. Mitchell.** 1978. Cyclophosphamide effects on murine cryptococcosis. *Infect. Immun.* **21:**674–677.

53. **Green, J. R., and G. S. Bulmer.** 1979. Gastrointestinal inoculation of *Cryptococcus neoformans* in mice. *Sabouraudia* **17:**233–240.

54. **Greenlee, P. G., and E. Stephena.** 1984. Meningeal cryptococcosis and ingestive cardiomyopathy in a ferret. *J. Am. Vet. Med. Assoc.* **184:**840–841.

55. **Griner, L. A., and H. A. Walch.** 1978. Cryptococcosis in Columbiformes at the San Diego Zoo. *J. Wildlife Dis.* **14:**389–394.

56. **Hidore, M. R., and J. W. Murphy.** 1986. Natural cellular resistance of beige mice against *Cryptococcus neoformans*. *J. Immunol.* **137:**3624–3631.

57. **Hilbert, B. J., C. R. Huxtable, and S. E. Pawley.** 1980. Cryptococcal pneumonia in a horse. *Aust. Vet. J.* **56:**391–392.

58. **Hill, F. I., A. J. Woodgyer, and M. A. Lintott.** 1995. Cryptococcosis in a North Island brown kiwi (*Apteryx australis mantelli*) in New Zealand. *J. Med. Vet. Mycol.* **33:**305–309.

59. **Hill, J. O.** 1992. CD4+ T cells cause multinucleated giant cells to form around

Cryptococcus neoformans and confine the yeast within the primary site of infection in the respiratory tract. *J. Exp. Med.* **175:**1685–1695.

60. **Hill, J. O., and K. M. Aquirre.** 1994. CD4+ T cell-dependent acquired state of immunity that protects the brain against *Cryptococcus neoformans. J. Immunol.* **152:**2344–2350.

61. **Hill, J. O., and P. C. Dunn.** 1993. A T-cell independent protective response against *Cryptococcus neoformans* expressed at the primary site of infection in the lung. *Infect. Immun.* **61:**5302–5308.

62. **Holzworth, J.** 1952. Cryptococcosis in a cat. *Cornell Vet.* **42:**12–15.

63. **Huffnagle, G. B., M. F. Lipscomb, J. A. Lovchik, K. A. Hoag, and N. E. Street.** 1994. The role of CD4+ and CD8+ T cells in the protective inflammatory response to a pulmonary cryptococcal infection. *J. Leukocyte Biol.* **55:**35–42.

64. **Huffnagle, G. B., J. L. Yates, and M. F. Lipscomb.** 1991. T cell-mediated immunity in the lung: a *Cryptococcus neoformans* pulmonary infection model using SCID and athymic nude mice. *Infect. Immun.* **59:**1423–1433.

65. **Hutto, J. O., C. S. Bryan, F. L. Greene, C. J. White, and J. I. Gallin.** 1988. Cryptococcosis of the colon resembling Crohn's disease in a patient with hyperimmunoglobulin E-recurrent infection (Job's) syndrome. *Gastroenterology* **94:**808–812.

66. **Jacobs, D. H., A. M. Macher, R. Handler, J. E. Bennett, M. J. Collern, and J. I. Gallin.** 1984. Esophageal cryptococcosis in a patient with the hyperimmunoglobulin E-recurrent infection (Job's) syndrome. *Gastroenterology* **87:**201–203.

67. **Jacobs, G. J., L. Medleau, C. Calvert, and J. Brown.** 1997. Cryptococcal infection in cats: factors influencing treatment outcome, and results of sequential serum antigen titers in 35 cats. *J. Vet. Intern. Med.* **11:**1–4.

68. **Kligman, A. M., A. P. Crane, and R. F. Norris.** 1951. Effect of temperature on survival of chick embryos infected intravenously with *Cryptococcus neoformans* (*Torula histolytica*). *Am. J. Med. Sci.* **221:**273–278.

69. **Kligman, A. M., and F. D. Weidman.** 1949. Experimental studies on treatment of human torulosis. *Arch. Dermatol. Syph.* **60:**726–741.

70. **Kock, N. D., E. P. Lane, F. Rowbotham, A. Pawandiwa, and F. W. Hill.** 1991. Concurrent systemic cryptococcosis and haemangiosarcoma in a dog. *J. Comp. Pathol.* **104:**117–120.

71. **Krick, J. A.** 1981. Familial cryptococcal meningitis. *J. Infect. Dis.* **143:**133.

72. **Kuttin, E. S., M. Feldman, A. Nyska, B. A. Weissman, J. Muller, and H. B. Levine.** 1988. Cryptococcosis of the nasophauynx in mice and rats. *Mycopathologia* **101:**99–104.

73. **Lichtensteiger, C. A., and L. E. Hilf.** 1994. Atypical cryptococcal lymphadenitis in a dog. *Vet. Pathol.* **31:**493–496.

74. **Lim, T. S., J. W. Murphy, and L. K. Cauley.** 1980. Host etiological agent interactions in intranasally and intraperitoneally induced cryptococcosis in mice. *Infect. Immun.* **29:**633–641.

75. **Littman, M. L., and R. Borok.** 1968. Relation of the pigeon to cryptococcosis: natural carrier state, heat resistance and survival of *Cryptococcus neoformans. Mycopathologia* **35:**329–345.

76. **Littman, M. L., R. Borok, and T. J. Dalton.** 1965. Experimental avian cryptococcosis. *Am. J. Epidemiol.* **82:**197–207.

77. **Littman, M. L., and L. E. Zimmerman.** 1956. *Cryptococcosis, Torulosis or European Blastomycosis*, p. 38–46. Grune and Stratton, New York.

78. **Lodge, J. K., E. Jackson-Machelski, D. L. Toffaletti, J. R. Perfect, and J. I. Gordon.** 1994. Targeted gene replacement demonstrates that myristoyl CoA: protein N-myristoyl-transferase is essential for the viability of *Cryptococcus neoformans. Proc. Natl. Acad. Sci. USA* **91:**12008–12012.

79. **Louria, D. B., N. Fallon, and H. G. Browne.** 1960. The influence of cortisone on experimental fungus infections in mice. *J. Clin. Invest.* **39:**1435–1449.

80. **Lovchik, J., J. Wilder, M. F. Lipscomb, R. Riblet, and R. C. Lyons.** 1996. The clearance rate of a pulmonary cryptococcal infection is determined by a gene on chromosome 12, abstr. 2:15. 3rd International Conference on Cryptococcus and Cryptococcosis, Paris.

81. **Lutsky, I., and J. Brodish.** 1964. Experimental canine cryptococcosis. *J. Infect. Dis.* **114:**273–276.

82. **MacDonald, D. W., and H. C. Stretch.** 1982. Canine cryptococcosis associated with prolonged corticosteriod therapy. *Can. J. Vet. Res.* **23:**200–202.

83. **Malik, R., E. Dill-Mackey, P. Martin, D. I. Wigney, D. B. Muir, and D. N. Love.** 1995. Cryptococcis in dogs: a retrospective study of 20 consecutive cases. *J. Med. Vet. Mycol.* **33:**291–297.

84. **Malik, R., D. I. Wigney, D. B. Muir, D. J. Gregory, and D. N. Love.** 1992. Cryptococcosis in cats: clinical and mycological assessment of 29 cases and evaluation of treatment using orally administered fluconazole. *J. Med. Vet. Mycol.* **30:**133–144.

85. **Malik, R., D. I. Wigney, D. B. Muir, and D. N. Love.** 1997. Asymptomatic carriage of *Cryptococcus neoformans* in the nasal cavity of dogs and cats. *J. Med. Vet. Mycol.* **35:**27–31.

86. **Mancianti, F., C. Giannelli, M. Bendinelli, and A. Poli.** 1992. Mycological findings in feline immunodeficiency virus-infected cats. *J. Med. Vet. Mycol.* **30:**257–259.

87. **Marcato, P. S.** 1966. A case of Hodgkin's disease associated with cryptococcosis in a dog. *J. Small Anim. Pract.* **7:**649–650.

88. **Marquis, G., S. Montplaisir, and M. Pelletier.** 1985. Genetic resistance to murine cryptococcosis: the beige mutation (Chediak-Higashi syndrome) in mice. *Infect. Immun.* **47:**288–293.

89. **Marquis, G., S. Montplaisir, M. Pelletier, S. Mousseau, and P. Auger.** 1985. Genetic resistance to murine cryptococcosis: increased susceptibility in the CBA/N XID mutant strain of mice. *Infect. Immun.* **47:**282–287.

90. **McGrath, J. T.** 1954. Cryptococcosis of the central nervous system in domestic animals. *Am. J. Pathol.* **30:**651.

91. **McNamara, T. S., R. A. Cook, J. L. Behler, L. Ajello, and A. A. Padhye.** 1994. Cryptococcosis in a common anaconder (Eunectes murinus). *J. Zool. Wildlife Med.* **25:**128–132.

92. **Medleau, L., C. E. Greene, and P. M. Rakich.** 1990. Evaluation of ketoconazole and itraconazole for treatment of disseminated cryptococcosis in cats. *Am. J. Vet. Res.* **51:**1454–1458.

93. **Medleau, L., G. J. Jacobs, and M. A. Marks.** 1995. Itraconazole for the treatment of cryptococcosis in cats. *J. Vet. Intern. Med.* **9:**39–42.

94. **Mehrotra, B. S.** 1983. Cryptococcal mastitis in dairy animals. *Mykosen* **26:**615–616.

95. **Mendoza, M. H., A. M. Garcia, and L. Vizcaino.** 1984. Spontaneous cryptococcosis in a pigeon. *Arch. Zootech.* **33:**27–41.

96. **Meyer, K. F.** 1914. A pathogenic blastomyces from the horse. *Proc. Pathol. Soc. Phil.* **16:**28.

97. **Migaki, G., R. D. Gunnels, and H. W. Casey.** 1978. Pulmonary cryptococcosis in an Atlantic bottlenosed dolphin (*Tursiops truncatus*). *Lab. Anim. Sci.* **28:**603–606.

97a. **Mitchell, T. G., and J. R. Perfect.** 1995. Cryptococcosis in the era of AIDS—100 years after the discovery of *Cryptococcus neoformans*. *Clin. Microbiol. Rev.* **8:**515–548.

98. **Miyake, T., T. Veda, and Y. Matsuura.** 1978. A case of cryptococcosis in a sooty mangabey, *Cercocebus torquatus atys*. *J. Jpn. Assoc. Zool. Gardens Aquariums* **20:**4–6.

99. **Mody, C. H., G. H. Chen, C. Jackson, J. L. Curtis, and G. B. Toews.** 1994. In vivo

depletion of murine CD8 positive T-cells impairs survival during infection with a highly virulent strain of *Cryptococcus neoformans*. *Mycopathologia* **125**:7–17.

100. **Mody, C. H., M. F. Lipscomb, N. E. Street, and G. B. Toews.** 1990. Depletion of CD4+ (L3T4+) lymphocytes *in vivo* impairs murine host defenses to *Cryptococcus neoformans*. *J. Immunol.* **144**:1472–1477.

101. **Mody, C. H., G. B. Toews, and M. F. Lipscomb.** 1988. Cyclosporin A inhibits the growth of *Cryptococcus neoformans* in a murine model. *Infect. Immun.* **56**:7–12.

102. **Mody, C. H., G. B. Toews, and M. F. Lipscomb.** 1989. Treatment of murine cryptococcosis with cyclosporin-A in normal and athymic mice. *Am. Rev. Respir. Dis.* **139**:8–13.

103. **Nishimura, K., and M. Miyaji.** 1979. Histopathological studies on experimental cryptococcosis in nude mice. *Mycopathologia* **68**:143–153.

104. **Odom, A., M. Del Poeta, J. Perfect, and J. Heitman.** 1997. The immunosuppressant FK506 and its non-immunosuppressive analog L-685,818 are toxic to *Cryptococcus neoformans* by inhibition of a common target protein. *Antimicrob. Agents Chemother.* **41**:156–161.

105. **Odom, A., S. Muir, E. Lim, D. L. Toffaletti, J. R. Perfect, and J. Heitman.** 1997. Calcineurin is required for virulence of *Cryptococcus neoformans*. *EMBO J.* **16**:2576–2589.

106. **Pal, M.** 1991. Feline meningitis due to *Cryptococcus neoformans* var. *neoformans* and review of feline cryptococcosis. *Mycoses* **34**:313–316.

107. **Pal, M., G. D. Dube, and B. S. Mehrota.** 1984. Pulmonary cryptococcosis in a rhesus monkey (macaca mulatta). *Mykosen* **27**:309–312.

108. **Pal, M., K. Ono, R. Goitsuka, and A. Hasegawa.** 1990. Isolation of *Cryptococcus neoformans* var. *neoformans* from canine otitis. *Mycoses* **33**:465–467.

108a. **Perfect, J. R.** Unpublished data.

109. **Perfect, J. R.** 1990. Fluconazole therapy for experimental cryptococcosis and candidiasis in the rabbit. *Rev. Infect. Dis.* **12**(Suppl. 3):299–302.

110. **Perfect, J. R., and D. T. Durack.** 1985. Effects of cyclosporine in experimental cryptococcal meningitis. *Infect. Immun.* **50**:22–26.

111. **Perfect, J. R., and D. T. Durack.** 1985. Chemotactic activity of cerebrospinal fluid in experimental cryptococcal meningitis. *J. Med. Vet. Mycol.* **23**:37–45.

112. **Perfect, J. R., and D. T. Durack.** 1985. Comparison of amphotericin B and N-D-ornithyl amphotericin B methyl ester in experimental crytococcal meningitis and *Candida albicans* endocarditis with pyelonephritis. *Antimicrob. Agents Chemother.* **28**:751–755.

113. **Perfect, J. R., M. M. Hobbs, D. L. Granger, and D. T. Durack.** 1988. Cerebrospinal fluid macrophage cytotoxicity: in vitro and in vivo correlation. *Infect. Immun.* **56**:849–854.

114. **Perfect, J. R., S. D. R. Lang, and D. T. Durack.** 1981. Influence of agglutinating antibody in experimental cryptococcal meningitis. *Br. J. Exp. Pathol.* **62**:595–599.

115. **Perfect, J. R., S. D. R. Lang, and D. T. Durack.** 1980. Chronic cryptococcal meningitis: a new experimental model in rabbits. *Am. J. Pathol.* **101**:177–194.

116. **Perfect, J. R., and T. H. Rude.** 1993. Identifying *Cryptococcus neoformans* by differential expression at the site of infection, abstr. L3, p. 24. 2nd International Conference on Cryptococcus and Cryptococcosis, Milan.

117. **Perfect, J. R., D. V. Savani, and D. T. Durack.** 1986. Comparison of itraconazole and fluconazole in treatment of cryptococcal meningitis and *Candida* pyelonephritis in rabbits. *Antimicrob. Agents Chemother.* **29**:579–583.

118. **Perfect, J. R., D. L. Toffaletti, and T. H. Rude.** 1993. The gene encoding for phosphoribosylaminoimidazole carboxylase (ADE2) is essential for growth of *Cryptococcus neoformans* in cerebrospinal fluid. *Infect. Immun.* **61**:4446–4451.

119. **Perfect, J. R., and K. A. Wright.** 1994. Amphotericin B lipid complex in the treatment

of experimental cryptococcal meningitis and disseminated candidiasis. *J. Antimicrob. Chemother.* **33:**73–81.

120. **Perfect, J. R., K. A. Wright, M. M. Hobbs, and D. T. Durack.** 1989. Treatment of experimental cryptococcal meningitis and disseminated candidiasis with SCH 39304. *Antimicrob. Agents Chemother.* **33:**1735–1740.

121. **Petrites-Murphy, M. B., L. A. Robbins, J. M. Donahue, and B. Smith.** 1996. Equine cryptococcal endometritis and placentitis with neonatal cryptococcal pneumonia. *J. Vet. Diagn. Invest.* **8:**383–386.

122. **Pounden, W. D., J. M. Amberson, and R. F. Jaeger.** 1952. A severe mastitis problem associated with *Cryptococcus neoformans* in a large dairy herd. *Am. J. Vet. Res.* **13:**121–128.

123. **Ramadan, R. D., A. A. Faged, and A. M. El-Hassan.** 1989. Cryptococcosis in a camel (*Camelus dromedarius*). *Vet. Med. J. Giza* **37:**77–82.

124. **Ramos-Vara, J. A., L. Ferrer, and J. Visa.** 1994. Pathological findings in a cat with cryptococcosis and feline immunodeficiency virus infection. *Histol. Histopathol.* **9:**305–308.

125. **Rebhun, W. C., and N. J. Edwards.** 1977. Cryptococcosis involving the orbit of a dog. *Vet. Med. Small Anim. Clin.* **72:**1447–1450.

126. **Reiss, F., and G. Szilagyi.** 1965. Ecology of yeast-like fungi in a hospital population. *Arch. Dermatol.* **91:**611–614.

127. **Rhodes, J. C.** 1985. Contribution of complement component C5 to the pathogenesis of experimental murine cryptococcosis. *Sabouraudia* 23:225–234.

128. **Rhodes, J. C., L. S. Wicker, and W. J. Urba.** 1980. Genetic control of susceptibility to *Cryptococcus neoformans* in mice. *Infect. Immun.* **29:**494–499.

129. **Rippon, J. W.** 1988. *Medical Mycology. The Pathogenic Fungi and the Pathogenic Actinomycetes.* W.B. Saunders, Philadelphia.

130. **Rosskopf, W. J., Jr., and R. W. Woespel.** 1984. Cryptococcosis in a thick-billed parrot. *Avian/Exotic Pract.* **1:**14–8.

131. **Roussilhon, C., J. . Postal, and P. Rauisse.** 1987. Spontaneous cryptococcosis of a squirrel monkey (Saimiri sciureus) in French Guyana. *J. Med. Primatol.* **16:**39–47.

132. **Ruiz, A., J. B. Neilson, and G. S. Bulmer.** 1982. Control of cryptococcus in nature by biotic factors. *Sabouraudia* 20:21–29.

133. **Saag, M. S., W. G. Powderly, G. A. Cloud, P. Robinson, M. H. Grieco, P. K. Sharkey, S. E. Thompson, A. Sugar, C. U. Tuazon, J. F. Fisher, N. Hyslop, J. M. Jacobson, R. Hafner, and W. F. Dismukes.** 1992. Comparison of amphotericin B with fluconazole in the treatment of acute AIDS-associated cryptococcal meningitis. *N. Engl. J. Med.* **326:**83–89.

134. **Saez, H., J. Rinjard, and M. R. Battesti.** 1978. Cryptococcosis in a fennec, *Fennecus zerda.* Characteristics and differentiation of *Cryptococcus neoformans. Bull. Soc. Fr. Mycol. Med.* **7:**69–72.

135. **Salkowski, C. A., and E. Balish.** 1990. Pathogenesis of *Cryptococcus neoformans* in congenitally immunodeficient beige athymic mice. *Infect. Immun.* **58:**3300–3306.

136. **Salkowski, C. A., and E. Balish.** 1991. Inflammatory responses to cryptococcosis in congenitally athymic mice. *J. Leukocyte Biol.* **49:**533–541.

137. **Salkowski, C. A., and E. Balish.** 1991. Cutaneous cryptococcosis in athymic and beige-athymic mice. *Infect. Immun.* **59:**1785–1789.

138. **Salkowski, C. A., and E. Balish.** 1991. Cryptococcosis in beige mice: the effect of congenital defects in innate immunity on susceptibility. *Can. J. Microbiol.* **37:**128–135.

139. **Salkowski, C. A., K. F. Bartizal, Jr., M. J. Balish, and E. Balish.** 1987. Colonization and pathogenesis of *Cryptococcus neoformans* in gnotobiotic mice. *Infect. Immun.* **55:**2000–2005.

140. **Sanfelice, F.** 1894. Uber eine neuen Pathogenen *Blastomyceten welcher* innerhalb der

Gewebe unter Bildung, kalkartig aussehender massendegeneriert. *Zentralbl. Bakteriol. Parasitol. Infekt. Hyg.* **18**:521–526.

141. Sanfelice, F. 1895. Sull'azione patogena dei blastomiceti. *Ann. Ig.* **5**:239–262.

142. Scott, E. A., J. R. Duncan, and J. E. McCormack. 1974. Cryptococcosis involving the postorbital area and frontal sinus in a horse. *J. Am. Vet. Med. Assoc.* **165**:626–627.

143. Seibold, H. R., C. S. Roberts, and E. M. Jordan. 1953. Cryptococcosis in a dog. *J. Am. Vet. Med. Assoc.* **122**:213–215.

144. Sethi, K. K., H. S. Randhawa, S. Abraham, and R. Viswanathan. 1967. Primary pulmonary canine cryptococcosis. *Indian J. Chest Dis.* **9**:222–224.

145. Sethi, K. K., and J. Schwartz. 1966. Experimental ocular cryptococcosis in pigeons. *Am. J. Ophthalmol.* **62**:95–98.

146. Siman, J., R. E. Nichols, and E. V. Morse. 1953. An outbreak of bovine cryptococcosis. *J. Am. Vet. Med. Assoc.* **122**:31–35.

147. Singh, M., P. P. Gupta, J. S. Rana, and S. K. Jand. 1994. Clinico-pathological studies on experimental cryptococcal mastitis in goats. *Mycopathologia* **126**:147–155.

148. Skulski, G., and W. S. C. Symmers. 1954. Actiomycesis and torulosis in the ferret (*Mustela furo*). *J. Comp. Pathol. Ther.* **64**:306–311.

149. Sly, D. L., W. T. London, A. E. Palmer, and J. M. Rice. 1977. Disseminated cryptococcosis in a Patas monkey (Erythrocebus patas). *Lab. Anim. Sci.* **27**:694–699.

150. Smith, G. W., W. H. Mosberg, L. O. J. Manganiello, and J. A. Alvarez de Choudens. 1953. Torulosis of the central nervous system in the laboratory animal. *Bull. Sch. Med.* **38**:32–41.

151. Spencer, A., C. Ley, P. Canfield, P. Martin, and R. Perry. 1993. Meningeocephalitis in a koala (*Phascolarctos cinereus*) due to *Cryptococcus neoformans* var *gattii* infection. *J. Zool. Wildlife Med.* **24**:519–22.

152. Staib, F. 1962. *Cryptococcus neoformans* beim Kanarienvogel. *Zentralbl. Bakteriol.* **185**:129–134.

153. Staib, F., W. Weller, S. Brem, R. Schindlmayr, and E. Schmittdial. 1985. A *Cryptococcus neoformans* strain from the brain of a wildlife fox (*Vulpes vulpes*) suspected of rabies: mycological observations and comments. *Zentralbl. Bakteriol. Mikrobiol. Hyg.* **260**:566–571.

154. Stoddard, J. L., and E. C. Cutler. 1916. *Torula Infection in Man*, p. 1–98. Monograph no. 6. Rockefeller Institute for Medical Research, New York.

155. Strippoli, V., N. Simonetti, and A. Cassone. 1978. Effect of a tetracycline antibiotic on experimental pathogenicity of *Cryptococcus neoformans*. *Chemotherapy* **24**:290–296.

156. Swinne, D., K. Kayembe, and M. Niyimi. 1986. Isolation of saprophytic *Cryptococcus neoformans* var. *neoformans* in Kinshasa, Zaire. *Ann. Soc. Belg. Med. Trop.* **66**:57–61.

157. Swinne-Desgain, D. 1976. *Cryptococcus neoformans* in the crops of pigeons following its experimental administration. *Sabouraudia* **14**:313–317.

158. Takos, M. J. 1956. Experimental cryptococcosis produced by the ingestion of virulent organisms. *N. Engl. J. Med.* **254**:598–601.

159. Takos, M. J., and N. W. Elton. 1953. Spontaneous cryptococcosis of marmoset monkeys in Panama. *Arch. Pathol. Lab. Med.* **55**:403–407.

160. Tell, L., and M. Bush. 1993. Cryptococcosis in tree shrews (tupaia tana and tupaia minor) and elephant shrews (macroscelides proboscides), p. 49–50. Proc. Am. Assoc. Zoo Vet.

161. Tell, L. A., D. K. Nichols, W. P. Fleming, and M. Bush. 1997. Cryptococcus in tree (*Tupaia tana* and *Tupaia minor*) and elephant shrews (*Macroscelides proboscides*). *J. Zool. Wildlife Med.* **28**:175–181.

162. Tortelly, R., A. Grune, G. M. D. Paiva, and M. D. Rossi. 1979. Spontaneous cryptococcosis in the laboratory mouse. *Cientifica Brazil* **6**:125–127.

163. Van Cutsem, J., J. Fransen, F. Van Gerven, and P. A. J. Janssen. 1986. Experimental

cryptococcosis: dissemination of *Cryptococcus neoformans* and dermatropism in guinea pigs. *Mykosen* **29:**561–575.

164. **Van der Gaag, I., M. H. van Niel, B. E. Belshaw, and W. T. Wolvekamp.** 1991. Gastric granulomatous cryptococcosis mimicking gastric carcinoma in a dog. *Vet. Q.* **13:**185–190.

165. **Vuillemin, P.** 1901. Les blastomycetes pathogenes. *Rev. Gen. Sci. Pures Appl.* **12:**732–751.

166. **Walker, C., R. Malik, and P. J. Canfield.** 1995. Analysis of leucocytes and lymphocytes in cats with naturally-occurring cryptococcosis, but differing feline immunodeficiency virus status. *Aust. Vet. J.* **72:**93–97.

167. **Weidman, F. D., and H. L. Ratcliffe.** 1934. Extensive generalized torulosis in a cheetah or hunting leopard (*Crynaelurus jabatus*). *Arch. Pathol.* **18:**362–369.

168. **Weiss, C., I. H. Perry, and M. C. Sheury.** 1948. Infections of the human eye with *Cryptococcus neoformans*: a clinical and experimental study with a new diagnostic method. *Arch. Ophthalmol.* **39:**739–751.

169. **Weitzman, I., P. Bonaparte, V. Guevin, and M. Crist.** 1973. Cryptotoccosis in a field mouse. *Sabouraudia* **11:**77–79.

170. **Welsh, R. D., and E. L. Stair.** 1995. Cryptococcal meningitis in horse. *J. Equine Vet. Sci.* **15:**80–82.

171. **Yinnon, A. M., B. Rudensky, E. Sagi, G. Brever, C. Brautbar, I. Polacheck, and J. Halevy.** 1997. Invasive cryptococcosis in a family with epidermodysplasia verruciformis and idiopathic CD4 cell depletion. *Clin. Infect. Dis.* **25:**1252–1253.

11 | Epidemiology

INTRODUCTION

The overwhelming majority of cases of symptomatic *Cryptococcus neoformans* infection now occur in patients with impaired immunity. This represents a major change in the epidemiology of cryptococcal infections since before 1950, when the majority of cases were described in apparently normal individuals. The prevalence of *C. neoformans* infections in a population appears to be a function of the number of immunocompromised individuals and their exposure to this environmental fungal pathogen. *C. neoformans* has a worldwide distribution, and human exposure to this fungus is presumably a common event (see chapter 3). Human *C. neoformans* infections are sporadic, and, unlike other mycoses, cases of cryptococcosis do not occur in clusters. In healthy human populations cryptococcosis is not a common infection. Hence, the prevalence of cryptococcosis in a population can be considered a sentinel marker for the overall percentage of individuals with defective immunity. The epidemiology of cryptococcal infection is an important undertaking because it provides clues as to the pathogenesis of infections and permits an assessment of infection risk in a given population.

THE ENVIRONMENT AND HUMAN INFECTION

The evidence for the environment as the source of most human *C. neoformans* infections is very strong but circumstantial. The case is strengthened by the following observations: (i) *C. neoformans* is commonly found in urban areas, (ii) cases due to human-to-human transmission are very rare and appear to be unique iatrogenic events, (iii) cases of animal-to-human transmission have never been documented, and (iv) *C. neoformans* is very rarely a human commensal (see chapter 3 for an in-depth discussion of commensalism, environmental sources, and acquisition of infection). The only documented cases of human-to-human transmission have involved organ transplantation with infected tissues (15, 151) and needlestick accidents with contaminated blood (80) (see below). Nosocomial infection has not been convincingly described. Several molecular epidemiological studies have shown that clinical and environmental *C. neoformans* isolates are indistinguishable. This observation is consistent with and supports the notion that cryptococcal infections are acquired from environmental reservoirs (see below). At the present time there is consensus that the overwhelming majority of human *C. neoformans* infections are acquired from the environment.

ROUTE OF HUMAN INFECTION

Human infection with *C. neoformans* is widely believed to result from inhalation of either desiccated cells, poorly encapsulated yeast forms, or basidiospores. Initial infection is presumed to occur in the lungs, with extrapulmonary dissemination occurring primarily in immunocompromised individuals who cannot contain the infection at the primary site. Despite the almost universal acceptance of the lung as the primary route of infection, there is little or no direct evidence for this. It is interesting that most late-20th century authoritative sources (52, 121) are likely to accept a pulmonary route of infection whereas earlier authorities (78) were more cautious on this issue. The consensus for a pulmonary route of infection appears to have emerged in the 1950s and 1960s after (i) the identification of avian excreta as a significant environmental reservoir with its potential for aerosol dissemination, (ii) the recognition of cryptococcal pneumonia as a distinct clinical syndrome, and (iii) the description of subpleural nodules and the primary pulmonary lymph node complex by Baker and collaborators (10, 87, 176). Knowledge of how and where *C. neoformans* infects the human host is crucial for understanding the epidemiology and pathogenesis of infection.

The Case for a Pulmonary Route of Infection

Several lines of evidence strongly suggest that human cryptococcosis results from inhalation of infectious particles. First, infectious particles of a size compatible with alveolar deposition can be isolated from the air above contaminated sites (162, 172). Second, pulmonary lesions, often in the form of healed granulomas, have been identified in patients with cryptococcal meningitis (10, 87, 176). These lesions are often found in subpleural nodules of asymptomatic patients at autopsy and resemble the primary lesions of pulmonary tuberculosis and histoplasmosis (10, 87, 176). Third, *C. neoformans* is occasionally isolated from human sputum (see chapter 3). Fourth, cryptococcal pneumonia is a well-recognized clinical entity that may precede cryptococcal meningitis (52). Fifth, experimental primary pulmonary *C. neoformans* infection in mice and rats can result in extrapulmonary dissemination (68, 81, 92). Taken together, the evidence for a pulmonary route of infection is compelling but still circumstantial. The finding of infectious particles in the ambient air does not necessarily imply that the particles reach the alveolar space and/or establish infection in the lung. Similarly, the existence of pulmonary lesions in patients with disseminated cryptococcosis does not necessarily imply a respiratory route of infection, since hematogenous spread could conceivably result in lung infection. In summary, the case for a pulmonary route for initial human infection is strong and widely accepted, but unequivocal evidence for the lung as the initial site of infection is lacking.

Other Routes of Initial Infection

If the case for pulmonary infection is not proved, then the evidence to suggest a non-pulmonary route of infection is even weaker. The possibility of a gastrointestinal route of infection is suggested by the isolation of *C. neoformans* from various foods, including fruits and milk. Human infection of the nasopharynx has been described (101). A localized rhinophyma-like cryptococcal infection has been described in a hemophilic patient with AIDS (130). *C. neoformans* has been isolated from some aphthous ulcers in patients with AIDS (73). Primary cryptococcal infection of the larynx has also been reported (75). Several cases of gastrointestinal cryptococcal infection in humans have been described (2, 19, 44, 94, 99). Animal studies provide some support for the feasibility of infection through the gastrointestinal tract. Monkeys given food heavily contaminated with *C. neoformans* developed disseminated infection (210). Similarly, disseminated infection can follow oral administration of yeast cells in mice, and encapsulated *C. neoformans* organisms can adhere to epithelial cells throughout the alimentary tract (174). Rectal ulcers have been suggested as a portal of entry in one patient with disseminated disease (2). These findings suggest that the gastrointestinal tract can serve as a portal of entry for *C. neoformans*.

C. neoformans has also been isolated from an inguinal groin abscess in a patient with a mucopurulent vaginal discharge (133). Skin has been considered by some a site for initial infection, and there are several reports in the literature of possible primary cutaneous cryptococcosis (13, 84, 95, 147, 204). However, the overwhelming majority of cases of cutaneous cryptococcosis involve disseminated disease in which skin involvement reflects dissemination from another source (178). In fact, an unequivocal diagnosis of primary cutaneous cryptococcosis is so rare that skin cryptococcal infection should always be considered a sign of disseminated infection until proved otherwise. However, rare cases of true primary cutaneous cryptococcosis have occurred following injuries with contaminated needles (28, 80).

SYMPTOMATIC AND ASYMPTOMATIC INFECTION

Cryptococcosis is a term that refers to invasive *C. neoformans* infections. Invasive or systemic infection is associated with a variety of immune defects, is usually symptomatic, and commonly involves the central nervous system (166). Untreated *C. neoformans* meningoencephalitis is always fatal. Campbell et al. (26) reviewed all literature cases of long-term survivors with cryptococcal meningitis who were not treated with amphotericin B and concluded that, although occasional remissions occurred, there was only one cure in a woman treated with sulfadiazine followed by potassium iodide. On the other hand, for every symptomatic infection, it is likely that there are many clinically mild and possibly asymptomatic infections in immunocompetent individuals. Mild and asymptomatic infections are often referred to as subclinical infections because these conditions are seldom diagnosed.

For many infectious diseases the prevalence of subclinical infections can be estimated by skin sensitivity testing using microbial antigens or the presence of antibody in sera. Delayed hypersensitivity reactions to cryptococcal antigens can be elicited in patients with active cryptococcosis (see chapter 9 for a full discussion of delayed hypersensitivity reactions to *C. neoformans* antigens). For example, Atkinson and Bennett (8) described a skin test antigen, cryptococcin, made by urea extract from *C. neoformans* cells. Cryptococcin elicited reactions in a high percentage of patients with cryptococcosis, in pigeon breeders, and in laboratory workers exposed to cryptococcus but seldom in control subjects (8). Newberry et al. (143) used cryptococcin skin testing to demonstrate asymptomatic infections in pigeon fanciers. Some early cryptococcin preparations were not specific and frequently elicited reactions in patients with other mycotic infections (175). However, Schimpff and Bennett (179) used a cryptococcin preparation that produced no response in 10 unexposed individuals but elicited a positive skin response in 11 of 13 laboratory workers exposed to *C. neoformans*. This suggests that everyday life exposure to *C. neoformans* is not sufficient to elicit a positive skin reaction to cryptococcin but that work in a laboratory where cryptococcus is cultured can result in a significantly higher level of exposure (179).

The use of specific antibody testing to determine the prevalence of infection has proved disappointing. Serological studies have shown a high prevalence of antibodies reactive with either cell wall or polysaccharide capsule components in healthy individuals (46, 90, 107). Unfortunately, there is considerable antigenic cross-reactivity among *C. neoformans* and other microorganisms (49, 134, 163), and it is not clear whether these serum antibodies are specific for *C. neoformans* or simply cross-reactive.

Hence, the study of *C. neoformans* epidemiology, unlike the case for infections with mycobacteria and other mycoses, has been hampered by the lack of sensitive and specific immunological tests to evaluate the true prevalence of infection in human populations. In the absence of conclusive data on the prevalence of delayed skin hypersensitivity or specific antibodies to *C. neoformans*, the ratio of clinical to subclinical *C. neoformans* infections is not known. However, it is likely that there are many more subclinical infections for every case of cryptococcosis. For example, Littman (120) estimated that there were 5,000 to 15,000 subclinical *C. neoformans* infections in New York City per year by extrapolating from the ratio of clinical to subclinical infections from other mycoses for which the epidemiology of infection was better understood. Although the accuracy of this calculation is questionable, considering the differences in pathogenesis between the endemic mycoses, it makes the point that asymptomatic infection is likely to be a common event.

Mention should be made of the possible association between *C. neoformans* exposure and summer-type hypersensitivity pneumonitis in Japan (86, 138, 139, 182). Several studies associated *C. neoformans* with hypersensitivity pneumonitis on the basis of antibody titers and strong lymphocyte proliferative responses to *C. neoformans* antigens (86, 138, 139, 182). Inhalation-provocation challenge studies of hypersensitivity pneumonitis patients with cryptococcal antigens produced pulmonary symptoms consistent with *C. neoformans* being the causative agent of this condition (138). However, the association between summer-type hypersensitivity and *C. neoformans* has been questioned because of cross-reactions between *C. neoformans* and another antigenically related fungus, *Trichosporon cutaneum* (182). At the present time, *T. cutaneum* appears to be a more likely causative agent for hypersensitivity pneumonitis than *C. neoformans*.

PREVALENCE OF HUMAN INFECTION

Cryptococcosis before 1981 (Pre-AIDS Era)

In the first half of the 20th century cryptococcosis was rarely reported, as indicated by the paucity of literature reports. In 1919, only 13 cases of cryptococcal meningitis (or torula meningitis in the old nomenclature) had been reported in the literature (118). In 1931, Freeman (71a) carried out an extensive review and summarized the 43 cases in the literature. In 1937, the number of cases in the literature was 60 (118). In 1941, Bindford (16) reported a case and counted 75 cases in the literature (16). By the mid-20th century, about 300 cases had been described, and appreciation of this infection was on the rise (123). In the late 1950s, Littman (120) estimated a prevalence of *C. neoformans* infection of 2 cases per million people per year. A similar prevalence was obtained by Friedman (72), who reviewed the incidence of cryptococcosis among Kaiser-Permanente Medical Care subscribers for the years 1970 to 1980 in Northern California and calculated an incidence of 0.8 cases per million people per year.

However, by the late 1970s it was clear that cryptococcosis was becoming more common. In 1977, Kaufman and Blumer (106) reviewed the rising incidence of cryptococcal infection and called cryptococcosis "the awakening giant" among the systemic mycoses. Epidemiological data collected by the Centers for Disease Control and Prevention (CDC, Atlanta, Ga.) reveal that between 1965 and 1975 the average number of reported cases in the United States per year was 104 (106). Since cryptococcosis is not a reportable disease in the United States, this figure is undoubtedly an underestimate of the true incidence of infection (106). Another measure of the rising incidence of *C. neoformans* infection in the 1960s and 1970s was a significant increase in the collection of cryptococcal isolates by the Fungus Immunology Branch of the CDC (106).

The steady rise in the number of reports of *C. neoformans* infections prior to 1981 is probably the result of a combination of factors. Improvements in microbiological diagnostics combined with greater awareness of this infection probably contributed to the increased reporting and more precise diagnosis. The availability of serological tests for the detection of cryptococcal polysaccharide in the 1960s made diagnosis easier (17, 83). However, the most important factor contributing to the rise in cryptococcal infections may have been advances in medical progress that resulted in prolonged survival at the price of weakened immunity. For example, steroid therapy introduced in the 1940s increased the risk for cryptococcal infections (50, 59, 82). Similarly, antineoplastic therapy, renal dialysis, and organ transplant therapies introduced in the second half of the 20th century are each associated with a higher risk for cryptococcal infections (see below).

In summary, accurate figures for the incidence and prevalence of *C. neoformans* infection prior to the AIDS epidemic are not available. Nevertheless, there is evidence that the incidence of cryptococcosis was rising rapidly before the AIDS epidemic was recognized in 1981. The increase in prevalence of *C. neoformans* infections has paralleled a general increase in all fungal infections and the emergence of new fungal pathogens (51).

Cryptococcosis after 1981 (AIDS Era)

Since 1981 the incidence of cryptococcal infections in some countries has increased dramatically as a result of the AIDS epidemic. AIDS (the acquired immunodeficiency syndrome) is caused by infection with human immunodeficiency virus (HIV), which results in profound immunosuppression. Patients with HIV infection are at risk for cryptococcosis late in the course of infection when their CD4$^+$ lymphocyte counts are less than 100 cells per mm^3 (40). The association between HIV infection and increased susceptibility to disseminated *C. neoformans* infection was noted early in the AIDS epidemic. AIDS patients represent the majority of cases reported in the literature since 1981, and during the late 20th century symptomatic *C. neoformans* infections are overwhelmingly found in HIV-infected populations. Today, the diagnosis of disseminated *C. neoformans* infection suggests an evaluation for infection with HIV.

In the United States, the majority of studies report a prevalence of *C. neoformans* infection among HIV-infected patients in the 5 to 10% range (Table 1). In 1991 Currie and Casadevall (42) surveyed all hospitals in New York City and identified over 1,200 cases of cryptococcosis. Using the available figures for the number of HIV patients at risk for infection, this study reported a yearly prevalence of 6 to 8% (42). Estimates of the prevalence of infection based on culture data underestimated the prevalence of infection measured by serum antigen determination by 100% (42). A review of neuropathological findings in AIDS patients at autopsy revealed that the percentage of individuals with cryptococcal brain lesions ranged from 3 to 14% among 9 studies (132). Differences in the prevalence of cryptococcal infection reported in the various studies of opportunistic infections associated with AIDS may reflect differences in the incidence of infection, susceptibility of populations, exposure depending on geography, and methodology. There is evidence that the prevalence of cryptococcal infection in patients with AIDS is declining as a result of widespread use of fluconazole in this population (9, 135, 144) (see chapter 15).

Cryptococcosis associated with HIV has been reported worldwide, with the incidence of cryptococcosis rising in direct proportion to the spread of the HIV epidemic. In most countries the prevalence of cryptococcosis among HIV-infected individuals has been consistently 5 to 10% (Table 2). Reports from Africa suggest a higher prevalence of HIV-associated cryptococcosis in that continent (37, 205, 209). The factors responsible for this apparent geographic variation in the prevalence of

Table 1 Prevalence of cryptococcal infection in HIV-infected patients in the United States

Geographic area	Study	Year[a]	Prevalence[b] (%)
United States	Kovacs et al. (111)	1985	7.5
New York City	Zuger et al. (226)	1986	8.5
New Jersey	Eng et al. (67)	1986	13.3
San Francisco	Chuck and Sande (35)	1989	5.1
New Orleans	Clark et al. (36)	1990	11.6
New York City	Currie and Casadevall (42)	1994	6–8
Los Angeles	Sorvillo et al. (191)	1997	2.9[c]

[a]Year the study was published. All studies were retrospective.
[b]Prevalence of infection is the number of cases of cryptococcosis divided by the number of HIV-infected patients.
[c]The authors of this study suggested that the drier climate of southern California may contribute to a lower prevalence of infection relative to other regions of the United States.

Table 2 Prevalence of cryptococcal infection in HIV-infected patients in several countries

Geographic area	Study	Year	Prevalence (%)
Africa	Clumeck et al. (37)	1984	30
	Desmet et al. (47)	1989	12.2
	Perriens et al. (158)	1992	5.6
	Maher and Mwandumba (127)	1994	6–12
Australia	Speed and Dunt (192)	1995	5.5
France	Dromer et al. (53)	1996	4.8–7.8
United Kingdom	Knight et al. (109)	1993	4

infection are not understood but may reflect the history of recent or past exposure to sources of infection.

Seasonal Incidence

At this time it is uncertain whether the incidence of cryptococcosis varies with the seasons. A study of 550 cases of *C. neoformans* infection in Thailand revealed no apparent seasonal variation in the incidence of infection (31). However, a study of 112 patients with cryptococcosis in Los Angeles County, Calif., suggested a higher incidence of infection in the fall and winter (191).

SEX DIFFERENCES IN SUSCEPTIBILITY TO *C. NEOFORMANS*

Numerous studies have reported a predominance of male patients among individuals with cryptococcosis, both before the HIV epidemic (78, 119, 121) and afterward (Table 3). Analysis of the incidence of *C. neoformans* infections among male and female HIV-infected patients in Italy and New York revealed a highly significant association between *C. neoformans* infection and maleness (128). Hence, the imbalance in the male-to-female ratio could not be attributed to a higher prevalence of HIV infection in men (128). There is no satisfying biological explanation for this phenomenon. It is difficult to attribute the difference in the sex ratio of cryptococcosis cases to variation in exposure, given that the exact route and timing of infection are unknown. A more likely explanation is that subtle differences in male and female host immune systems and physiology are responsible for the predominance of men among patients with *C. neoformans* infections. Human sex hormones can stimulate the growth and maturation of another fungal pathogen, *Coccidioides immitis* (57). Estrogens can inhibit the mycelium-to-yeast transformation in *Paracoccidioides brasiliensis*, a finding that suggests an explanation for the apparent resistance of women to paracoccidioidomycosis (164). It is conceivable that hormonal differences between men and women contribute to the higher prevalence of *C. neoformans* infection in men. Alternatively, women may simply be more resistant on a genetic basis. In mice, susceptibility to cryptococ-

Table 3 Predominance of male patients in studies of clinical cryptococcosis

Study	Year	No. of patients	Male patients (%)
Zimmerman and Rappaport (224)[a]	1954	59	91
Spickard et al. (195)	1963	30	77
Lewis and Rabinovich (119)[b]	1972	161	71
De Wytt et al. (45)	1982	30	63
Zuger et al. (226)[c]	1986	26	92
Stamm et al. (199)	1987	194	68
Chuck and Sande (35)[c]	1989	106	99
Clark et al. (36)[c]	1990	68	98
Seknon et al. (180)	1990	91	62
Rozenbaum and Goncalves (168)[c]	1994	129	75
Speed and Dunt (192)	1995	133	72
Dromer et al. (53)	1996	1,013	85
Chariyalertsak et al. (31)	1996	550	78

[a]Many of the patients in this study had lymphoproliferative neoplasms.
[b]This study reviewed several previous studies, and the total number of patients reported is shown here.
[c]All or most of the patients in these studies had been infected with the HIV virus.

cosis is under multigenic control and is influenced by the *xid* locus on the X chromosome (131).

AGE AND RACIAL SUSCEPTIBILITY

Cryptococcosis can occur in all age groups. However, several studies report that the majority of cases occur between the ages of 20 and 50 (45, 121). Cryptococcosis is uncommon in children (see below). Some populations may be at higher risk for cryptococcosis on the basis of either genetic susceptibility or increased exposure to infectious particles. A suggestion that human genetic factors could confer susceptibility to *C. neoformans* comes from the description of cryptococcosis in two siblings who had no obvious predisposing conditions for infection (112). Littman (121) suggested that *C. neoformans* infection was more common among whites. On the other hand, in Australia, several studies have noted a high frequency of infection among Aborigines (124). In Los Angeles County, Calif., the incidence of cryptococcosis in Hispanics was twice that in whites (191). However, data regarding differences in racial susceptibilities are scant, and at the present time there are no controlled studies to indicate true differences in racial susceptibility.

CRYPTOCOCCOSIS IN SELECTED POPULATIONS

Risk Factors for Cryptococcosis

Illnesses and therapies that impair host immune mechanisms predispose to disseminated *C. neoformans* infections. Cryptococcosis is primarily an infection of immunocompromised individuals. Numerous conditions have been associated with an increased risk for *C. neoformans* infections (Table 4). In the pre-AIDS era, major risk factors for cryptococcosis were lymphoproliferative disorders, corticosteroid therapy, sarcoidosis, organ transplantation, and possibly diabetes mellitus. Since 1981, the most common risk factor for cryptococcosis has been HIV infection.

In recent years, the increase in the use of organ transplantation to treat a variety of conditions has produced a larger population. Cryptococcal infections have been associated with idiopathic CD4+ T lymphocytopenia without HIV infection, a poorly characterized disorder that may predispose to infection with opportunistic pathogens, including *C. neoformans* (48, 58, 136, 152, 186, 196). Apart from HIV, other infections may predispose to cryptococcal infection by damaging the immune system, the lung, or both. There are two anecdotal reports of cryptococcosis following paracoccidioidomycosis (12). Both infections were localized, and it was suggested that paracoccidioidomycosis predisposed to *C. neoformans* infection (12). Arasteh et al. (7) have noted an association between *Pneumocystis carinii* pneumonia in patients with AIDS and concomitant or subsequent infection with *C. neoformans*. For both paracoccidioidomycosis and pneumocystis, however, larger studies with control groups will be necessary before making definitive conclusions.

Cryptococcosis in Normal Hosts

Does disseminated *C. neoformans* infection occur in normal hosts? The answer to this question may hinge on the definition of the term "normal" host. The overall incidence of cryptococcosis in immunocompetent individuals has been estimated at 0.2 per million people per year (72). Occasional cases are reported in patients without obvious underlying disease or immune defects, and it is generally accepted that cryptococcosis can occur in otherwise immunocompetent hosts (66, 70). However, careful immunological study of such patients can sometimes reveal subtle defects in immunity that may have predisposed to disseminated infection.

Table 4 Conditions associated with predisposition to *C. neoformans* infections

Condition	Reference
HIV infection	36, 67, 111, 137
Lymphoproliferative disorders	38, 78, 108, 119, 224
Sarcoid	148, 189, 195
Corticosteroid therapy	50, 59, 82
Hypogammaglobulinemia	85, 177
Hyper-IgM syndrome	114, 187, 208
Hyper-IgE syndrome	94, 99
Systemic lupus erythematosus	225
HIV-negative CD4+ T lymphopenia	48, 58, 136, 152, 186, 196
Diabetes mellitus[a]	195
Organ transplantation	77, 98, 100
Smoking habit	150
Peritoneal dialysis	129, 188, 223
Cirrhosis	125

[a]Diabetes mellitus has historically been considered a risk factor for cryptococcal infection. However, diabetes is a common disease, and it is unclear whether this condition is truly a risk factor for cryptococcosis.

Schimpff and Bennett (179) demonstrated decreased skin testing responses and minimal lymphocyte migration inhibition effects in cured patients subsequently challenged with cryptococcin and mumps antigen. This was interpreted as suggesting the existence of subtle immunological defects in apparently normal individuals who develop cryptococcosis (179). However, the possibility that the measured immunological deficits required an acquired defect resulting from cryptococcal infection was not ruled out. At this time the consensus is that cryptococcosis can occur in a small subset of normal individuals (51).

The field of immunology is advancing rapidly, and it is probable that as knowledge increases many of the "normal" patients will be found to have subtle defects in host defense that predispose to infection. Given the strong association between disseminated cryptococcal infection and defective immunity, one should be prudent and consider an immunological evaluation in apparently normal individuals who develop cryptococcosis. A minimal immunological evaluation may include HIV serology, CD4$^+$ lymphocyte count, immunoglobulin levels, and skin testing for common antigens.

Cryptococcosis in Children

Risk factors for pediatric cryptococcosis are similar to those for adults and include HIV infection, lymphoproliferative disorders, and immunosuppressive therapy (116). Several cases of C. neoformans infections have been described in children with hyper-immunoglobulin M (IgM) and hyper-IgE syndromes (114, 187, 201, 208). Nevertheless, C. neoformans infections are considerably rarer in children than in adults. In 1961, Emanuel et al. (65) identified only 23 cases of pediatric cryptococcosis in the literature. During the AIDS epidemic, cryptococcosis has continued to remain primarily an adult infection. From 1982 to 1985 cryptococcosis was diagnosed in only 4 of 307 children with AIDS (167). Another study reported cryptococcosis in only 31 (1%) of 2,786 HIV-infected children younger than 13 years (117). Hence, the prevalence of C. neoformans infection in HIV-infected children appears to be about 1%, whereas the prevalence in adults ranges from 5 to 10% (Table 1). Cases of disseminated C. neoformans infection among neonates and very young children have also been described (184, 187). Clinical manifestations of pediatric cryptococcosis are similar to those of adults (117, 212). In children with sarcomas, pulmonary cryptococcosis can present as lung nodules that can be confused with metastatic lesions (3). The relative paucity of cryptococcal infections in HIV-infected children is not well understood. Subtle differences in the immunological defects caused by HIV infection and/or differences in environmental exposure may account for the relatively low frequency of cryptococcosis in children. Evidence for a lack of exposure comes from serological studies that showed that the prevalence of antibody to polysaccharide capsule was 5% and 69% in children and adults, respectively (193).

Cryptococcosis in Patients with Neoplastic Disease

Among neoplastic disorders, disseminated cryptococcal infection has been historically associated with lymphoproliferative malignancies. This association was noted in reviews from the late 1940s and early 1950s (38, 78, 224). Hodgkin's disease was particularly associated with predisposition to disseminated cryptococcal infection. The most comprehensive early study was that of Zimmerman and Rappaport (224), who identified lymphoproliferative disease in 30% of 60 cases of *C. neoformans* infection collected at the Armed Forces Institute of Pathology over 20 years. They estimated that 5 to 10% of patients with cryptococcosis had concurrent malignant disease of the reticuloendothelial system. Kaplan et al. (104) reviewed the experience with cryptococcosis at Memorial Sloan-Kettering Cancer Center from 1952 to 1972 and also found a striking association between cryptococcosis and patients with leukemias and lymphomas. The incidence of cryptococcal infection among patients suffering from chronic lymphatic leukemia, Hodgkin's disease, chronic myelogenous leukemia, and multiple myeloma was 24.3, 13.3, 10.9, and 6.9, respectively, per 1,000 patients. In contrast, cryptococcosis was extremely rare among patients with solid tumors (104).

Since the association between cryptococcosis and lymphoproliferative malignancies predates the use of immunosuppressive chemotherapy, it is likely that immune derangements accompanying lymphoproliferative disorders increased susceptibility to *C. neoformans* (78, 224). The precise immune defect in patients with Hodgkin's disease and other lymphoproliferative disorders that is responsible for increased susceptibility to infection remains unknown. For example, Hodgkin's disease is associated with impairment of both cellular and humoral immunity, as evidenced by anergy to tuberculin and weak antibody responses to vaccines (78, 224). When *C. neoformans* infections complicate Hodgkin's disease, yeast forms can be found in lymph nodes containing neoplastic cells (78). The presence of neoplastic cells in lymphoid tissue could compromise local lymphoid tissue architecture and reduce their ability to clear *C. neoformans* organisms from lymph nodes.

White et al. (220) compared the outcome of cryptococcal meningitis in 14 patients with lymphoproliferative neoplasia to that in 41 patients with AIDS after antifungal therapy. The median overall survival of patients with lymphoproliferative neoplasia and AIDS was 2 months and 9 months, respectively. These findings suggest that the prognosis of cryptococcal meningitis in patients with lymphoproliferative neoplastic disorders is worse than that in patients with AIDS. Factors that may contribute to a worse prognosis in cancer patients relative to AIDS patients include older age, immunosuppressive therapy, the underlying immune deficit from neoplastic disease, and the clinical stage of the malignancy (220).

Cryptococcosis in Transplant Patients

Organ transplantation is associated with an increased risk for disseminated cryptococcal infection. When cryptococcosis occurs in organ transplant patients, it tends to occur 4 to 6 months after transplantation (89). Cryptococcosis in transplant patients often presents with skin lesions (89). Renal transplant recipients appear to be at particular risk during the early days of immunosuppressive therapy. Several studies have reported a (lifetime) prevalence of cryptococcosis in renal transplant patients of 3 to 4% after transplantation (77, 100, 142, 181). In 1975, Gallis et al. (77) described 10 cases of *C. neoformans* infection among 171 patients following renal transplantation. The median time to the onset of cryptococcal infection was 13.5 months after renal transplantation (range 0 to 61 months). The usual course was subacute meningitis, and the mortality was 60% (77). Shaariah et al. (181) described

16 cases of cryptococcosis among 448 renal transplant patients over 17 years in Kuala Lumpur, Malaysia. In this study the mean time that patients were on immunosuppressive therapy before the onset of cryptococcal infection was 48 months, and mortality was 30%. Cryptococcosis is a relatively common infection among renal transplant patients despite the fact that renal transplantation has the lowest incidence of fungal infection among solid organ transplantation procedures (153). This suggests that predisposition of renal transplant patients to cryptococcosis may be a function of both the therapeutic immunosuppressive regimens and the underlying illness. In this regard, uremia can decrease lymphocyte transformation in *C. neoformans*-infected mice (76). Patients receiving renal transplants are also likely to have been treated with dialysis for kidney failure. Whether dialysis contributes to the risk of cryptococcosis is not known, but several cases of cryptococcal peritonitis have been reported in peritoneal dialysis patients (129, 188, 223).

In contrast to renal transplantation, recipients of other organ transplants appear to be at a lower risk for cryptococcosis despite comparable immunosuppressive therapies. Most studies of liver transplantation have shown a lower risk for cryptococcosis than that reported in renal transplant recipients. In 1996 Jabbour et al. (98), in an extensive review of liver transplantation, and described 10 cases of cryptococcal meningitis occurring after 3,767 transplants, an incidence of 0.26%. The diagnosis of cryptococcal infection was made at a median time of 3.5 months after liver transplantation. The usual course was subacute meningitis, and the mortality with therapy was 50% (98). However, Singh (185) described six cases of *C. neoformans* infection (6%) in 102 of the consecutive patients who received liver transplants at a hospital in Pittsburgh, Pa. (185). The most common presentation of cryptococcosis in these patients was cutaneous or osseous lesions (185). Dummer et al. (57a) studied infections in heart-lung transplant recipients and described one case among 42 patients (2.3%). In summary, organ transplantation is a risk factor for cryptococcosis, but there are significant differences in the magnitude of the risk depending on the type of transplantation. Unfortunately, there is little or no specific information on the risk factors that predispose an individual transplant patient to *C. neoformans* infection, and thus guidelines for prophylaxis in this patient group will likely be focused on prevention of all fungal infections.

In rare cases, cryptococcosis in transplant patients can result from transplantation of an infected organ. There are several cases in the literature documenting human-to-human transmission of cryptococcal infection through organ transplants (15, 151). In lung transplantation there is the possibility of introducing *C. neoformans* infection as a consequence of dormancy or colonization states in the donor lung. In one series of fungal infections in lung- and lung-heart transplant patients, there were three cases of cryptococcosis among 59 patients with fungal infections (102). In at least one case it was highly likely that the *C. neoformans* infection came from the transplanted lung (102). In 1971, Ooi et al. (151) described a woman who received a kidney from a donor with undiagnosed cryptococcosis. The patient subsequently excreted *C. neoformans* in the urine and was successfully treated with amphotericin B and flucytosine (151). In 1978, Beyt and Waltman (15) described a woman who received a cornea from a donor with undiagnosed disseminated cryptococcosis. The patient developed an anterior-chamber mass and was successfully treated with amphotericin B and flucytosine (15). Transplanted infections can result from either donor infection or tissue contamination during organ processing and harvesting. In some cases cryptococcal infection is a consequence of using donor organ tissues that harbor unsuspected *C. neoformans* infection (15, 151). These cases highlight the fact that difficulties inherent in the premortem diagnosis of some cases of cryptococcosis can pose a risk for transplant recipients.

The primary factor contributing to cryptococcosis in transplant patients is probably the immunosuppressive drugs used to prevent transplant rejection. Corticosteroids are associated with an increased risk for cryptococcal infection (50, 59). However, not all immunosuppressive regimens may predispose to *C. neoformans* infection. Cyclosporine has been shown to protect mice (140, 141), but not rabbits (155), against *C. neoformans* infection. Cyclosporine-mediated protection appears to be the result of a direct toxic effect on the fungus (140). However, transplant patients treated with cyclosporine are unlikely to be protected from cryptococcosis despite the antifungal activity of this compound because of its potent immunosuppressive properties (141, 149).

Cryptococcosis in Patients with AIDS

HIV infection produces severe cell-mediated and humoral immune defects that are associated with a high risk for disseminated *C. neoformans* infection (36, 67, 111). *C. neoformans* var. *neoformans* is the variety overwhelmingly recovered from patients with AIDS (20, 165, 170, 183, 192). *C. neoformans* var. *gattii* infections are extremely rare in patients with AIDS, even in geographic areas where this variety is found in the environment (55, 103, 169, 194, 199). Cryptococcal infections almost always occur in advanced HIV infection when the CD4$^+$ lymphocyte count is less than 100 cells per mm^3 (40). Cryptococcal meningoencephalitis is the most common disseminated fungal infection in patients with AIDS. Cryptococcosis in patients with AIDS is usually incurable, and individuals who survive the initial illness require lifelong therapy to prevent recurrence of the disease (226). The frequency of cryptococcosis can be significantly reduced in patients with AIDS by fluconazole prophylaxis (4, 135, 144, 145). Antifungal prophylaxis is not routinely recommended for the prevention of *C. neoformans* infection because of concerns about cost and the selection of resistant fungi (154) (see chapter 15 for a full discussion of prevention-related issues). However, there is some evidence that the frequency of *C. neoformans* infections may be declining in the United States with the increased utilization of fluconazole in patients with AIDS (9, 135, 144).

CRYPTOCOCCOSIS IN SELECTED REGIONS

Cryptococcosis occurs worldwide, but the overwhelming majority of literature reports are from countries in the Northern Hemisphere, such as the United States. Nevertheless, cryptococcosis is a significant problem in many countries. Insights on the epidemiology and pathogenesis of infection have resulted from comparative studies of cryptococcosis in various geographic regions. The salient features of cryptococcosis in several world areas will be briefly summarized.

Cryptococcosis in Africa

Early reports of the AIDS epidemic in Africa suggested that the prevalence of cryptococcal infection was much higher than that in the United States (37, 73, 105, 159, 214). Of 18 African patients treated in Belgium, 6 (30%) had cryptococcal meningitis (37). A study of 14 African patients treated in Paris described 5 (35%) cases of *C. neoformans* infection (105). At a hospital in Kigali, Rwanda, the incidence of cryptococcosis rose from 1 case per year in 1983 to 2 cases per week in 1990 (209). However, other studies have reported a lower prevalence of infection among African HIV-infected patients. Among some sub-Saharan HIV-infected patient populations, the prevalence of cryptococcosis was 6 to 12% (127), which is the same range reported in the United States (Table 1). Another study from Zaire reported the prevalence of cryptococcal meningitis among HIV-infected patients with neurological complications as 5.6% (158). The higher prevalence of cryptococcal infection in Africa, if true, could reflect significant rates of HIV infection combined with increased exposure to *C. neoformans*. In fact, *C. neoformans* has been recovered from a high percentage of homes in one African city (205). *C. neoformans* was recovered from 40% of homes of patients with cryptococcosis and in 20% of control homes (205). In contrast, a survey of Canadian homes revealed no isolates of *C. neoformans* (203).

A major problem in Africa is a lack of resources for adequate care. Maher and Mwandumba (127) reported that for the patients in Malawi who cannot afford anticryptococcal therapy, the median survival time after diagnosis is 4 days. Effective anticryptococcal drugs such as amphotericin B and fluconazole are not available in government hospitals in Malawi and other sub-Saharan countries, because cost-benefit considerations mitigate against provision of expensive antifungal drugs (127). African *C. neoformans* isolates have been predominantly var. *neoformans*. In Zaire, six of seven isolates recovered before 1969 were *C. neoformans* var. *gattii*, but all subsequent clinical and environmental isolates from that country have been classified as *C. neoformans* var. *neoformans* (206, 207). In Burundi all clinical and environmental isolates were classified as var. *neoformans* (205). In Rwanda two clinical isolates of *C. neoformans* var. *gattii* have been described (18).

Cryptococcosis in Asia

Information about cryptococcosis in Asia is available from several studies carried out in various countries. Imwidthaya et al. (96) reported 13 cases of *C. neoformans* infection in Thailand. All isolates were *C. neoformans* var. *neoformans* but, interestingly, women constituted a majority of patients in that study (97). Another study from Thailand classified 4 of 14 clinical isolates as belonging to var. *gattii* (211). Before the AIDS epidemic the most common predisposing cause of cryptococcosis in Thailand was systemic lupus erythematosus, occurring in 7 of 32 patients (96).

Recent studies from India, Taiwan, and China reveal a picture of cryptococcosis similar to that described in other countries. A study from India analyzed 18 clinical *C. neoformans* isolates and found that 3 were var. *gattii*, documenting the presence of that variety in the Indian subcontinent. A case of *C. neoformans* var. *gattii* (serotype B) infection has been described in an Indian patient with AIDS (1). In Taiwan, varietal analysis of 21 strains revealed all but one to be var. *neoformans*

(91). In China, analysis of 19 clinical isolates revealed 2 that were classified as var. *gattii*, with the majority of isolates belonging to serotype A var. *neoformans* (6).

Cryptococcosis in Australia

Cryptococcosis in Australia has been the subject of extensive studies since the 1940s, when Cox and Tolhurst (39) published a monograph describing the experience with cryptococcosis in Australian patients. Several publications over the past half-century provide a detailed picture of cryptococcosis in Australia (45, 60, 62, 69, 124, 192). Furthermore, studies of the epidemiology and ecology of *C. neoformans* in Australia yielded two important insights into the biology of this fungus: (i) the association of *C. neoformans* var. *gattii* with eucalyptus trees (61, 63, 64) and (ii) the association between immune status and susceptibility to one of the two *C. neoformans* varieties (192).

The incidence of *C. neoformans* infections in Australia may be higher than that in other areas of the world (60). Within Australia the incidence of cryptococcosis in the pre-AIDS era was significantly higher in the north (4.7 per 100,000) than in the southern parts of the country (1.8 per 100,000) (60). Cryptococcosis in Australia is caused by both varieties of *C. neoformans*, var. *neoformans* and var. *gattii*. In the study of Speed and Dunt (192), 90% of *C. neoformans* var. *neoformans* infections occurred in immunocompromised hosts, whereas all *C. neoformans* var. *gattii* infections occurred in immunocompetent hosts. Cryptococcosis due to var. *neoformans* had a high mortality, whereas none of the patients with var. *gattii* infections died (192). However, var. *gattii* infections had more neurological sequelae and often required surgery and prolonged therapy (192).

Cryptococcosis in Central and South America and the Caribbean

The most comprehensive information about cryptococcosis in South America is available from Brazil, where Rozenbaum and Goncalves (168) conducted a detailed study of 171 patients. Among Brazilian AIDS patients, most isolates were *C. neoformans* var. *neoformans*, whereas *C. neoformans* var. *gattii* was common among immunocompetent patients (168, 170). Screening of 233 isolates from Brazil for variety classification revealed that 194 (83%) were var. *neoformans* and 39 (17%) were var. *gattii* (146). Studies from other countries reveal a picture of *C. neoformans* epidemiology similar to that in other tropical and subtropical countries. In Venezuela the percentage of isolates belonging to var. *neoformans* and var. *gattii* were 67% and 29%, respectively (218). In Puerto Rico, analysis of 129 environmental isolates revealed that all were var. *neoformans* (173). In Havana, Cuba, all 22 strains from clinical sources were also var. *neoformans* (5).

Some reports from South America provide tantalizing hints about the epidemiology and pathogenesis of cryptococcosis. Bava et al. (11) reported the second known urease-negative isolate of *C. neoformans* from a clinical source (for the first report, see reference 171). The South American urease-negative strain was isolated from a patient with AIDS who died of cryptococcal meningitis and was classified as having var. *neoformans*. In other reports, *C. neoformans* var. *gattii* strains have been isolated from bat guano (115) and a wasp nest (79).

Cryptococcosis in France

Dromer and collaborators (53, 54) have carefully studied the epidemiology of *C. neoformans* infection in France and have provided the most detailed picture of cryptococcosis in a European country. HIV infection was the predisposing condition for 86% of cases of cryptococcosis in France (53). Among patients without HIV infection, the most common predisposing conditions were malignancies, organ transplantation, and corticosteroid therapy. The frequency of *C. neoformans* infection among HIV-infected individuals was 4.8 to 7.8% (53). The overwhelming majority of *C. neoformans* infections in France were due to var. *neoformans*, but three cases of var. *gattii* occurred among 1,057 infections (53). Of the *C. neoformans* var. *neoformans* isolates, 21% were serotype D and the remainder were serotype A or AD (53, 54). This finding is consistent with the previous observation that serotype D infections were more frequent in European countries than in other parts of the world (14, 113). Serotype D infections were more frequent in patients that were either born in Europe, received corticosteroid therapy, or had skin lesions (54).

Cryptococcosis in New York City

In the United States, New York City is an epicenter for the AIDS epidemic. There has been an epidemic of HIV-associated *C. neoformans* infection in New York City since the early 1980s (42). A survey of New York City hospitals found more than 1,200 cases of cryptococcosis in 1991 (42). The prevalence of *C. neoformans* infection among HIV patients at risk for cryptococcal infection (i.e., CD4+ T-cell count of <200 cells per mm^3) was 6 to 8% (42). New York City has a large population of pigeons, and two studies three decades apart have documented that sites contaminated with pigeon excreta are often positive for *C. neoformans* (43, 122). DNA typing of *C. neoformans* clinical isolates in New York City has shown extensive genetic variation in clinical strains (27, 43). Some environmental and clinical isolates from New York City were indistinguishable by DNA restriction fragment length polymorphism (RFLP) analysis, supporting an association between *C. neoformans* in pigeon excreta and clinical infection (43).

MOLECULAR EPIDEMIOLOGY

Genetic Diversity among Clinical *C. neoformans* Strains

In the last decade, DNA typing techniques have been applied to *C. neoformans* isolates to investigate the epidemiology of infection (157, 160, 215). In recent years, clinical and environmental *C. neoformans* strains have been studied by RFLPs (27, 43, 110, 157, 160, 161, 197, 198, 215–217), electrophoretic karyotype (32, 56, 74, 156, 157, 160), randomly amplified polymorphic DNA (RAPD) analysis (23, 33, 88, 219), multilocus enzyme typing (21, 23), and sequence analysis (27, 32). All DNA typing techniques applied to *C. neoformans* isolates have shown significant genetic variation between individual isolates. Extensive genetic variation provides the means to discriminate among isolates, and DNA typing methods can be applied to the investigation of certain questions of *C. neoformans* pathogenesis. Hunter and Fraser (93) have interpreted the genetic diversity of *C. neoformans* strains as consistent with the theory of adaptive polymorphism, which predicts that the genetic diversity of a species is proportional to the diversity of ecological sites occupied. Since *C. neoformans* can occupy diverse ecological niches (see chapter 3), this theory would explain the finding of significant genetic variation relative to that of other fungi (93).

The mechanisms responsible for generating and maintaining a high degree of interstrain genetic diversity are not understood (32, 93). DNA sequence analysis of clinical isolates from New York City suggested that some strains of *C. neoformans* have clonal lineages and may be evolving independently, despite the potential for sexual reproduction (32). RAPD analysis of isolates from the United States also provided findings consistent with a clonal population structure (22).

Despite extensive genetic diversity among clinical isolates, there is evidence that some strains are more common than others in particular groups of patients. Currie et al. (43) examined 17 clinical strains from AIDS patients in New York City by Southern blot analysis with the use of the highly discriminatory repetitive CNRE-1 DNA probe and found that half of the isolates could be grouped into three strains. Strains within CNRE-1 groups were clonally related (32). In Australia, Chen et al. (33) studied 58 serotype A isolates by RAPD analysis and found that 42 could be assigned to a particular RAPD profile. Detailed analysis of these 42 isolates revealed that those from AIDS patients were predominantly assigned to a particular RAPD subgroup (33). These studies suggest that subsets of *C. neoformans* strains exist that could have enhanced virulence or high affinity for particular populations at risk for infection.

Infection with Only One Strain

The majority of DNA typing studies of multiple serial isolates from individuals with *C. neoformans* infections reveal the existence of only one strain (24, 156, 198, 217). Haynes et al. (88) studied multiple isolates from several patients by RAPD analysis, found differences in DNA typing, and concluded that some infections were due to more than one strain. The interpretation of these findings was questioned on the basis of the high sensitivity of RAPD typing and the possibility of in vivo DNA changes (29). Evidence for the possibility of in vivo DNA changes is provided by the demonstration of chromosome rearrangements for some strains during murine infection (74). Brandt et al. (24) also described three patients with serial isolates demonstrating karyotype differences despite identical RFLPs, RAPD profiles, and multilocus enzyme electrophoresis patterns. The ability of strains to undergo DNA changes during infection complicates the interpretation of the occasional cases where differences have been shown between multiple isolates from one individual. Two other large studies using RAPD analysis have identified no cases of mixed infections. Brandt et al. (22) conducted a large study of serial isolates from patients with chronic infection and identified one strain in each patient by RAPD techniques. Chen et al. (33) found that isolates from the same person taken at different times and from different body sites produced identical RAPD profiles and concluded that the population of cryptococci for individual patients was homogeneous. However, a detailed follow-up study of two patients with cryptococcal meningitis who relapsed provides strong evidence for infection with a second strain and suggests that some cases of recurrent infection are probably new infections (202). Hence, at this time, most studies indicate that *C. neoformans* infections are caused by one strain, but it is likely that some cases of human cryptococcosis are caused by more than one strain.

Recurrences and the Persistence of Infection

Cryptococcal infections often recur after initial therapy with antifungal agents (226). Given that most patients with cryptococcosis are immunosuppressed, recurrences could, in theory, be due to new infection or persistence of the initial infection. This question was investigated by Spitzer et al. (198), who analyzed initial and recurrent isolates from four patients with *C. neoformans* meningitis by RFLP and karyotype analysis. In each case, the initial and relapse isolates were genetically related, and these investigators concluded that recurrent infections were due to persistence of the initial strain despite therapy (198). Brandt et al. (24) confirmed this result in a detailed study of serial isolates from 31 patients. Using the strain typing techniques of multilocus enzyme electrophoresis, electrophoretic karyotyping, RAPD analysis, and RFLP, they demonstrated that in every case the same strain persisted with time (24). Hence, several studies using diverse strain typing techniques have confirmed that recurrence of *C. neoformans* infection is generally caused by persistence of the initial strain (22, 24, 126, 156, 217). Most of the molecular studies of relapse isolates have involved patients with HIV infection. The finding that recurrences are almost always due to persistence of infection implies that existing antifungal therapy is unable to eradicate most infections in the setting of defective immunity.

Genetic Analysis of Environmental and Clinical Strains

Five studies have analyzed *C. neoformans* strains from patients and their environment by DNA typing (43, 71, 190, 217, 222). Three studies have analyzed *C. neoformans* var. *neoformans* strains (43, 217, 222), and one study has analyzed *C. neoformans* var. *gattii* strains (190). Currie et al. (43) compared RFLPs of *C. neoformans* var. *neoformans* strains from patients in New York City and their environment and found extensive genetic diversity among clinical and environmental isolates. Despite the extensive genetic diversity, two strains were found to be represented among clinical and environmental isolates (43). Varma et al. (217) studied several *C. neoformans* var. *neoformans* strains from African patients and their immediate environment but found them to be different. Yamamoto et al. (222) studied clinical and environmental *C. neoformans* var. *neoformans* isolates from Nagasaki, Japan, and identified the same strains in patients and in the environment. Sorrell et al. (190) studied *C. neoformans* var. *gattii* isolates recovered from patients and eucalyptus trees in Australia and concluded that they were the same strains. Franzot et al. (71) demonstrated that some clinical and environmental isolates from Belo Horizonte, Brazil, were indistinguishable. The identification of the same strains in clinical and environmental *C. neoformans* isolates is consistent with, and supportive of, acquisition of human infection from environmental reservoirs.

Karyotype Changes for *C. neoformans* during Human Infection

Electrophoretic karyotyping is an extremely sensitive epidemiological tool, because most *C. neoformans* strains have different karyotypes (32, 56, 156, 160, 221).

The mechanism responsible for the extreme chromosomal variability among *C. neoformans* strains is unexplained, but some investigators have suggested the possibility of chromosomal rearrangements during infection (74). Two independent groups have described electrophoretic karyotype variants among serial isolates from some patients with chronic infection (25, 74, 198). These variants exhibit minor karyotype differences that involve one or two changes in chromosome band migration. This observation suggested the possibility of chromosomal changes occurring during human infection. Fries et al. (74) tested this hypothesis by passaging *C. neoformans* strains in mice and comparing the electrophoretic karyotype of the initial and passaged strains. They discovered several examples of apparent karyotype changes during experimental infection. The demonstration of karyotype changes during murine infection, combined with the finding of karyotype variants in 25 to 40% of serial isolates from human infection, strongly suggests the possibility of genetic changes during the course of human cryptococcosis. Genetic changes may occur as a result of in vivo selection and may be responsible for the observation that serial isolates can differ in antifungal susceptibility (30), sterol composition (41), and capsular polysaccharide structure (34).

REFERENCES

1. **Abrahan, M., V. Mathews, A. Ganesh, J. John, and M. S. Mathews.** 1997. Infection caused by *Cryptococcus neoformans* var. *gattii* serotype B in an AIDS patient in India. *J. Med. Vet. Mycol.* **35:**283–284.

2. **Allen, V. K., and L. Lowbeer.** 1945. Rectal ulcers with perirectal fistula as a port of entrance for torula encephalitis. *South. Med. J.* **38:**564–569.

3. **Allende, M., P. A. Pizzo, M. Horowitz, H. I. Pass, and T. J. Walsh.** 1993. Pulmonary cryptococcosis presenting as metastases in children with sarcomas. *Pediatr. Infect. Dis. J.* **12:**240–243.

4. **Ammassari, A., A. Linzalone, R. Murri, G. Marasca, G. Morace, and A. Antinori.** 1995. Fluconazole for primary prophylaxis of AIDS-associated cryptococcosis: a case-control study. *Scand. J. Infect. Dis.* **27:**235–237.

5. **Andreu, C. F., G. M. Machin, L. P. A. Bernal, R. Morales, and C. A. Herrera.** 1990. *Cryptococcus neoformans* var. *neoformans* isolated in Havana City. *Mem. Inst. Oswaldo Cruz* **85:**245.

6. **Ansheng, L., K. Nishimura, H. Taguchi, R. Tanaka, W. Shaoxi, and M. Miyaji.** 1993. The isolation of *Cryptococcus neoformans* from pigeon droppings and serotyping of naturally and clinically sourced isolates in China. *Mycopathologia* **124:**1–5.

7. **Arasteh, K., F. Staib, G. Crosse, U. Futh, and M. L'age.** 1996. Cryptococcosis in HIV infection in man: an epidimiological and immunological indicator? *Zentralbl. Bakteriol.* **284:**1563–163.

8. **Atkinson, A. J., and J. E. Bennett.** 1968. Experience with a new skin test antigen prepared from *Cryptococcus neoformans*. *Am. Rev. Respir. Dis.* **97:**637–643.

9. **Bacellar, H., A. Munoz, E. N. Miller, B. A. Cohen, D. Besley, O. A. Selnes, J. T. Becker, and J. C. McArthur.** 1994. Temporal trends in the incidence of HIV-1-related neurologic diseases. *Neurology* **44:**1892–1900.

10. **Baker, R. D.** 1976. The primary pulmonary lymph node complex of cryptococcosis. *Am. J. Clin. Pathol.* **65:**83–92.

11. **Bava, A. J., R. Negroni, and M. Bianchi.** 1993. Cryptococcosis produced by a urease negative strain of *Cryptococcus neoformans*. *J. Med. Vet. Mycol.* **31:**87–89.

12. **Benard, G., R. C. B. Gryscheck, A. J. S. Duarte, and M. A. Shikanai-Yasuda.** 1996.

Cryptococcosis as an opportunistic infection in immunodeficiency secondary to para-coccidioidomycosis. *Mycopathologia* **133**:65–69.

13. **Beng Bee, O., T. Tan, and R. Pang.** 1981. A case of primary cutaneous cryptococcosis successfully treated with miconazole. *Arch. Dermatol.* **117**:290–291.

14. **Bennett, J. E., K. J. Kwon-Chung, and D. H. Howard.** 1977. Epidemiologic differences among serotypes of *Cryptococcus neoformans*. *Am. J. Epidemiol.* **105**:582–586.

15. **Beyt, B. E., and S. R. Waltman.** 1978. Cryptococcal endoththalmitis after corneal transplantation. *N. Engl. J. Med.* **298**:825–826.

16. **Bindford, C. H.** 1941. Torulosis of the central nervous system. Review of recent literature and report of a case. *Am. J. Clin. Pathol.* **11**:242–251.

17. **Bloomfield, N., M. A. Gordon, and D. F. Elmendorf, Jr.** 1963. Detection of *Cryptococcus neoformans* antigen in body fluids by latex particle agglutination. *Proc. Soc. Exp. Biol. Med.* **114**:64–67.

18. **Bogaerts, J., H. Taelman, J. Batungwanayo, P. Van de Perre, and D. Swinne.** 1993. Two cases of HIV-associated cryptococcosis due to variety *gattii* in Rwanda. *Trans. R. Soc. Trop. Med. Hyg.* **63**:64.

19. **Bonacini, M., J. Nussbaum, and C. Ahluwalia.** 1990. Gastrointestinal, hepatic, and pancreatic involvement with *Cryptococcus neoformans* in AIDS. *J. Clin. Gastroenterol.* **12**:296–297.

20. **Bottone, E. J., I. F. Salkin, N. J. Hurd, and G. P. Wormser.** 1987. Serogroup distribution of *Cryptococcus neoformans* in patients with AIDS. *J. Infect. Dis.* **156**:242.

21. **Brandt, M., S. L. Bragg, and W. R. Pinner.** 1993. Multilocus enzyme typing of *Cryptococcus neoformans*. *J. Clin. Microbiol.* **31**:2819–2823.

22. **Brandt, M., L. C. Hutwagner, L. A. Klug, W. S. Baughman, D. Rimland, E. A. Graviss, R. J. Hamill, C. Thomas, P. G. Pappas, A. L. Reingold, and R. W. Pinner.** 1996. Molecular subtype distribution of *Cryptococcus neoformans* in four areas of the United States. *J. Clin. Microbiol.* **34**:912–917.

23. **Brandt, M. E., L. C. Hutwagner, R. J. Kuykendall, R. W. Pinner, and The Cryptococcal Disease Active Surveillance Group.** 1995. Comparison of multilocus enzyme electrophoresis and random amplifed polymorphic DNA analysis for molecular subtyping of *Cryptococcus neoformans*. *J. Clin. Microbiol.* **33**:1890–1895.

24. **Brandt, M. E., M. A. Pfaller, R. A. Hajjeh, E. A. Graviss, J. Rees, E. D. Spitzer, R. W. Pinner, and L. W. Mayer.** 1996. Molecular subtypes and antifungal susceptibilities of serial *Cryptococcus neoformans* isolates in human immunodeficiency virus-associated cryptococcosis. *J. Infect. Dis.* **174**:812–820.

25. **Brandt, M. E., M. A. Pfaller, R. A. Hajjeh, and R. W. Pinner.** 1995. DNA subtypes and antifungal susceptibilities in recurrent cryptococcosis, abstr. E-77. Program Abstr. 35th Intersci. Conf. Antimicrob. Agents Chemother., 1995.

26. **Campbell, G. D., R. D. Currier, and J. F. Busey.** 1981. Survival in untreated cryptococcal meningitis. *Neurology* **31**:1154–1157.

27. **Casadevall, A., L. Freundlich, L. Marsh, and M. D. Scharff.** 1992. Extensive allelic variation in *Cryptococcus neoformans*. *J. Clin. Microbiol.* **30**:1080–1084.

28. **Casadevall, A., J. Mukherjee, Y. RuiRong, and J. Perfect.** 1994. Management of *Cryptococcus neoformans* contaminated needle injuries. *Clin. Infect. Dis.* **19**:951–953.

29. **Casadevall, A., and A. Spitzer.** 1995. Involvement of multiple *Cryptococcus neoformans* strains a single episode of cryptococcosis and reinfection with novel strains in recurrent infection demonstrated by random amplification of polymorphic DNA and DNA fingerprinting. *J. Clin. Microbiol.* **33**:1682–1683.

30. **Casadevall, A., E. D. Spitzer, D. Webb, and M. G. Rinaldi.** 1993. Susceptibilities of serial *Cryptococcus neoformans* isolates from patients with recurrent cryptococcal meningitis to amphotericin B and fluconazole. *Antimicrob. Agents Chemother.* **37**:1383–1386.

31. **Chariyalertsak, S., T. Sirisanthana, K. Supparatpinyo, and K. E. Nelson.** 1996. Sea-

sonal variation of disseminated *Penicillium marneffei* infections in Northern Thailand: a clue to the reservoir? *J. Infect. Dis.* **173:**1490–1493.

32. **Chen, F., B. P. Currie, L.-C. Chen, S. G. Spitzer, E. D. Spitzer, and A. Casadevall.** 1995. Genetic relatedness of *Cryptococcus neoformans* clinical isolates grouped with the repetitive DNA probe CNRE-1. *J. Clin. Microbiol.* **33:**2818–1822.

33. **Chen, S. C. A., A. G. Brownlee, T. C. Sorrell, P. Ruma, D. H. Ellis, T. Pfeiffer, B. R. Speed, and G. Nimmo.** 1996. Identification by random amplification of polymorphic DNA of a common molecular type of *Cryptococcus neoformans* var. *neoformans* in patients with AIDS or other immunosuppressive conditions. *J. Infect. Dis.* **172:**754–758.

34. **Cherniak, R., L. C. Morris, T. Belay, E. D. Spitzer, and A. Casadevall.** 1995. Variation in the structure of glucuronoxylomannan in isolates from patients with recurrent cryptococcal meningitis. *Infect. Immun.* **63:**1899–1905.

35. **Chuck, S. L., and M. A. Sande.** 1989. Infections with *Cryptococcus neoformans* in the acquired immunodeficiency syndrome. *N. Engl. J. Med.* **321:**794–799.

36. **Clark, R. A., D. Greer, W. Atkinson, G. T. Valianis, and N. Hyslop.** 1990. Spectrum of *Cryptococcus neoformans* infection on 68 patients infected with human immunodeficiency virus. *Rev. Infect. Dis.* **12:**768–777.

37. **Clumeck, N., J. Sonnet, H. Taelman, F. Mascart-Lemont, M. De Bruyere, P. Vandeperre, J. Dasnoy, L. Marcelis, M. Lamy, C. Jonas, L. Eyckmans, H. Noel, M. Vanhaeverbeek, and J.-P. Butzler.** 1984. Acquired immunodeficiency syndrome in African patients. *N. Engl. J. Med.* **320:**492–497.

38. **Collins, V. P., A. Gellhorn, and J. R. Trimble.** 1995. The coincidence of cryptococcosis and disease of the reticulo-endothelial and lymphatic systems. Cancer **4:**883–889.

39. **Cox, L. B., and J. C. Tolhurst.** 1946. *Human Torulosis.* Melbourne University Press, Melbourne.

40. **Crowe, S. M., J. B. Carlin, K. I. Stewart, C. R. Lucas, and J. F. Hoy.** 1991. Predictive value of CD4 lymphocyte numbers for the development of opportunistic infections and malignancies in HIV-infected persons. *J. Acquired Immune Defic. Syndr.* **4:**770–776.

41. **Currie, B., H. Sanati, A. S. Ibrahim, J. E. Edwards, A. Casadevall, and M. A. Ghannoum.** 1995. Sterol compositions and susceptibilities to amphotericin B of environmental *Cryptococcus neoformans* isolates are changed by murine passage. *Antimicrob. Agents Chemother.* **39:**1934–1937.

42. **Currie, B. P., and A. Casadevall.** 1994. Estimation of the prevalence of cryptococcal infection among HIV infected individuals in New York City. *Clin. Infect. Dis.* **19:**1029–1033.

43. **Currie, B. P., L. F. Freundlich, and A. Casadevall.** 1994. Restriction fragment length polymorphism analysis of *Cryptococcus neoformans* isolates from environmental (pigeon excreta) and clinical isolates in New York City. *J. Clin. Microbiol.* **32:**1188–1192.

44. **Daly, J. S., K. A. Porter, F. K. Chong, and R. J. Robillard.** 1990. Disseminated, nonmeningeal gastrointestinal cryptococcal infection in an HIV-negative patient. *Am. J. Gastroenterol.* **85:**1421–1424.

45. **De Wytt, C., P. L. Dikson, and G. Holt.** 1982. Cryptococcal meningitis. A review of 32 years experience. *J. Neurol. Sci.* **53:**283–292.

46. **DeShaw, M., and L.-A. Pirofski.** 1995. Antibodies to the *Cryptococcus neoformans* capsular glucuronoxylomannan are ubiquitous in serum from HIV+ and HIV− individuals. *Clin. Exp. Immunol.* **99:**425–432.

47. **Desmet, P., K. D. Kayembe, and C. De Vroey.** 1989. The value of cryptococcal antigen screening among HIV-positive/AIDS patients in Kinshasa, Zaire. *AIDS* **3:**77–78.

48. **Dev, D., G. S. Basran, D. Slater, P. Taylor, and M. Wood.** 1994. Consider HIV negative immunodeficiency in cryptococcosis. *Br. Med. J.* **308:**1436.

49. **Devi, S. J. N., P. G. Reddy, C. A. Lyman, T. J. Walsh, C. E. Frasch, and A. C. Bush.** 1996. Immunohistochemical properties of a polysaccharide antigen of *Trichosporon*

beigelii that cross-reacts with the capsular polysaccharide of *Cryptococcus neoformans*. *Immunol. Infect. Dis.* **6:**87–92.

50. **Diamond, R. D., and J. E. Bennett.** 1974. Prognostic factors in cryptococcal meningitis. *Ann. Intern. Med.* **80:**176–181.

51. **Dixon, D. M., M. M. McNeil, M. L. Cohen, B. G. Gellin, and J. R. LaMontagne.** 1996. Fungal infections. A growing threat. *Public Health Rep.* **111:**226–235.

52. **Driver, J. A., C. A. Saunders, B. Heinze-Lacey, and A. M. Sugar.** 1995. Cryptococcal pneumonia in AIDS: is cryptococcal meningitis preceded by clinically recognizable pneumonia. *J. Acquired Immune Defic. Syndr.* **9:**168–171.

53. **Dromer, F., S. Mathoulin, B. Dupont, and A. Laporte.** 1996. Epidemiology of cryptococcosis in France: a 9-year survey (1985–1993). *Clin. Infect. Dis.* **23:**82–90.

54. **Dromer, F., S. Mathoulin, B. Dupont, L. Letenneur, and O. Ronin.** 1996. Individual and environmental factors associated with infection due to *Cryptococcus neoformans* serotype D. *Clin. Infect. Dis.* **23:**91–96.

55. **Dromer, F., O. Ronin, and B. Dupont.** 1992. Isolation of *Cryptococcus neoformans* var. *gattii* from an Asian patient in France: evidence for dormant infection in healthy subjects. *J. Med. Vet. Mycol.* **30:**395–397.

56. **Dromer, F., A. Varma, O. Ronin, S. Mathoulin, and B. Dupont.** 1994. Molecular typing of *Cryptococcus neoformans* serotype D clinical isolates. *J. Clin. Microbiol.* **32:**2364–2371.

57. **Drutz, D. J., M. Huppert, S. H. Sun, and W. L. McGuire.** 1981. Human sex hormones stimulate the growth and maturation of *Coccidioides immitis. Infect. Immun.* **32:**897–907.

57a. **Dummer, J. M., C. G. Montero, B. P. Griffith, R. L. Hardesty, I. L. Paradis, and M. Ho.** 1986. Infections in heart-lung transplant recipients. *Transplantation* **41:**725–728.

58. **Duncan, R. A., C. Fordhan von Reyn, G. M. Alliegro, Z. Toosi, A. M. Sugar, and S. M. Levitz.** 1993. Idiopathic CD4+ T-lymphocytopenia—four patients with opportunistic infections and no evidence of HIV infection. *N. Engl. J. Med.* **328:**393–398.

59. **Duperval, R., P. E. Hermans, N. S. Brewer, and G. S. Roberts.** 1977. Cryptococcosis, with emphasis on the significance of isolation of *Cryptococcus neoformans* from the respiratory tract. *Chest* **72:**13–19.

60. **Edwards, V. E., J. M. Sutherland, and J. H. Tyrer.** 1970. Cryptococcosis of the central nervous system. Epidemiological, clinical, and therapeutic features. *J. Neurol. Neurosurg. Psychiatr.* **33:**415–425.

61. **Ellis, D., and T. J. Pfeiffer.** 1992. The ecology of *Cryptococcus neoformans. Eur. J. Epidemiol.* **8:**321–325.

62. **Ellis, D. H.** 1987. *Cryptococcus neoformans* var. *gattii* in Australia. *J. Clin. Microbiol.* **25:**430–431.

63. **Ellis, D. H., and T. J. Pfeiffer.** 1990. Ecology, life cycle, and infectious propagule of *Cryptococcus neoformans. Lancet* **336:**923–925.

64. **Ellis, D. H., and T. J. Pfeiffer.** 1990. Natural habitat of *Cryptococcus neoformans* var. *gattii. J. Clin. Microbiol.* **28:**1642–1644.

65. **Emanuel, B., E. Ching, A. D. Leiberman, and M. Goldin.** 1961. Cryptococcus meningitis in a child successfully treated with amphotericin B, with a review of the pediatric literature. *J. Pediatr.* **59:**577–591.

66. **Emmons, W. W., III, S. Luchsinger, and L. Miller.** 1995. Progressive pulmonary cryptococcosis in a patient who is immunocompetent. *South. Med. J.* **88:**657–660.

67. **Eng, R. H. K., E. Bishburg, S. M. Smith, and R. Kapila.** 1986. Cryptococcal infections in patients with acquired immune deficiency syndrome. *Am. J. Med.* **81:**19–23.

68. **Feldmesser, M., and A. Casadevall.** 1997. Effect of serum IgG1 against murine pulmonary infection with *Cryptococcus neoformans. J. Immunol.* **158:**790–799.

69. **Fisher, D., J. Burrow, D. Lo, and B. Currie.** 1993. *Cryptococcus neoformans* in tropical northern Australia: predominantly variant *gattii* with good outcomes. *Aust. N.Z. J. Med.* **23:**678–682.

70. **Flickinger, F. W., M. D. Sathyanarayana, J. E. Whote, E. J. Stinger, and R.-M. Fincher.** 1993. Cryptococcocal pneumonia presenting as an infiltrative mass simulating carcinoma in an immunocompetent host: plain film, CT, and MRI. *South. Med. J.* **86:**450–452.

71. **Franzot, S. P., J. S. Hamdan, B. P. Currie, and A. Casadevall.** 1997. Molecular epidemiology of *Cryptococcus neoformans* in Brazil and the United States: evidence for both local genetic differences and a global clonal population structure. *J. Clin. Microbiol.* **35:**2243–2251.

71a. **Freeman, W.** 1931. Torula infection of the central nervous system. *J. Psychol. Neurol.* **43:**236–345.

72. **Friedman, G. D.** 1983. The rarity of cryptococcosis in Northern California: the 10-year experience of a large defined population. *Am. J. Epidemiol.* **117:**230–234.

73. **Friedman, M., A. Brenski, and L. Taylor.** 1994. Treatment of aphthous ulcers in AIDS patients. *Laryngoscope* **104:**566–570.

74. **Fries, B. C., F. Chen, B. P. Currie, and A. Casadevall.** 1996. Karyotype instability in *Cryptococcus neoformans* infection. *J. Clin. Microbiol.* **34:**1531–1534.

75. **Frisch, M., and D. R. Gnepp.** 1995. Primary cryptococcal infection of the larynx: report of a case. *Otolaryngol. Head Neck Surg.* **113:**477–480.

76. **Fromtling, R. A., A. M. Fromtling, S. Staib, and S. Muller.** 1981. Effect of uremia on lymphocyte transformation and chemiluminescence by spleen cells of normal and *Cryptococcus neoformans*-infected mice. *Infect. Immun.* **32:**1073–1078.

77. **Gallis, H. A., R. A. Berman, T. R. Cate, J. D. Hamilton, J. C. Gunnells, and D. L. Stickel.** 1975. Fungal infection following renal transplantation. *Arch. Intern. Med.* **135:**1163–1171.

78. **Gendel, B. R., M. Ende, and S. L. Norman.** 1950. Cryptococcosis. A review with special reference to apparent association with Hodgkin's disease. *Am. J. Med.* **9:**343–355.

79. **Gezuele, E., L. Calegari, D. Sanabria, G. Davel, and E. Civila.** 1993. Isolation in Uruguay of *Cryptococcus neoformans* var. *gattii* from a nest of the wasp *Polybia occidentalis. Rev. Iber. Micol.* **10:**5–6.

80. **Glaser, J. B., and A. Garden.** 1985. Inoculation of cryptococcosis without transmission of the acquired immunodeficiency syndrome. *N. Engl. J. Med.* **313:**266.

81. **Goldman, D., S. C. Lee, and A. Casadevall.** 1994. Pathogenesis of pulmonary *Cryptococcus neoformans* infection in the rat. *Infect. Immun.* **62:**4755–4761.

82. **Goldstein, E., and O. N. Rambo.** 1962. Cryptococcal infection following steroid therapy. *Ann. Intern. Med.* **56:**114–120.

83. **Gordon, M. A., and D. K. Vedder.** 1966. Serologic tests in diagnosis and prognosis of cryptococcosis. *JAMA* **197:**131–137.

84. **Granier, F., J. Kanitakis, C. Hermier, Y. Y. Zhu, and J. Thivolet.** 1987. Localized cutaneous cryptococcosis successfully treated with ketoconazole. *J. Am. Acad. Dermatol.* **16:**243–249.

85. **Gupta, S., M. Ellis, T. Cesario, M. Ruhling, and B. Vayuvegula.** 1987. Disseminated cryptococcal infection in a patient with hypogammaglobulinemia and normal T-cell function. *Am. J. Med.* **82:**129–131.

86. **Hamagami, S., T. Miyagawa, T. Ochi, I. Tsuyuguchi, and S. Kishimoto.** 1992. A raised level of soluble CD8 in bronchoalveolar lavage fluid in summer-type hypersensitivity pneumonitis in Japan. *Chest* **101:**1044–1049.

87. **Haugen, R. K., and R. D. Baker.** 1954. The pulmonary lesions in cryptococcosis with special reference to subpleural nodules. *Am. J Clin. Pathol.* **24:**1381–1390.

88. **Haynes, K. A., D. J. Sullivan, D. C. Coleman, J. C. K. Clarke, R. Emilianus, and K. J. Cann.** 1995. Involvement of multiple *Cryptococcus neoformans* strains in a single episode of cryptococcosis and reinfection with novel strains in recurrent infection

demonstrated by random amplification of polymorphic DNA and DNA fingerprinting. *J. Clin. Microbiol.* **33**:99–102.

89. **Hibberd, P. L., and R. H. Rubin.** 1994. Clinical aspects of fungal infections in organ transplant patients. *Clin. Infect. Dis.* **19**(Suppl. 1):S33–S44.

90. **Houpt, D. C., G. S. T. Pfrommer, B. J. Young, T. A. Larson, and T. R. Kozel.** 1994. Occurrences, immunoglobulin classes, and biological activities of antibodies in normal human serum that are reactive with *Cryptococcus neoformans* glucuronoxylomannan. *Infect. Immun.* **62**:3857–2864.

91. **Hsu, M. M., J.-C. Chang, K. Yokoyama, K. Nishimura, and M. Miyaji.** 1994. Serotypes and mating types of clinical strains of *Cryptococcus neoformans* in Taiwan. *Mycopathologia* **125**:77–81.

92. **Huffnagle, G. B., J. L. Yates, and M. F. Lipscomb.** 1991. Immunity to pulmonary *Cryptococcus neoformans* infection requires both CD4+ and CD8+ T cells. *J. Exp. Med.* **173**:793–800.

93. **Hunter, P. R., and C. A. M. Fraser.** 1990. Application of the theory of adaptive polymorphism to the ecology and epidemiology of pathogenic yeasts. *Appl. Environ. Microbiol.* **56**:2219–2222.

94. **Hutto, J. O., C. S. Bryan, F. L. Greene, C. J. White, and J. I. Gallin.** 1988. Cryptococcosis of the colon resembling Crohn's disease in a patient with the hyperimmunoglobulinemia E-recurrent infection (Job's) syndrome. *Gastroenterology* **94**:808–812.

95. **Iarobellis, F. W., M. I. Jacobs, and R. P. Cohen.** 1984. Primary cutaneous cryptococcosis. *Arch. Dermatol.* **23**:673–675.

96. **Imwidthaya, P.** 1994. One year's experience with *Cryptococcus neoformans* in Thailand. *Trans. R. Soc. Trop. Med. Hyg.* **88**:208.

97. **Imwidthaya, P., P. Dithaprasop, and C. Egtasaeng.** 1989. Clinical and environmental isolates of *Cryptococcus neoformans* in Bangkok (Thailand). *Mycopathologia* **108**:65–67.

98. **Jabbour, N., J. Reyes, S. Kusne, M. Martin, and J. Fung.** 1996. Cryptococcal meningitis after liver transplantation. *Transplantation* **61**:156–167.

99. **Jacobs, D. H., A. M. Macher, R. Handler, J. E. Bennett, M. J. Collen, and J. I. Gallin.** 1984. Esophageal cryptococcosis in a patient with the hyperimmunoglobulin E-recurrent infection (Job's) syndrome. *Gastroenterology* **87**:201–203.

100. **John, G. T., M. Mathew, E. Snehalatha, V. Anandi, A. Date, C. K. Jacob, and J. C. M. Shastry.** 1994. Cryptococcosis in renal allograft recipients. *Transplantation* **58**:855–856.

101. **Jones, E. L.** 1927. Torula infection of the nasopharynx. *South. Med. J.* **20**:120–126.

102. **Kanj, S. S., K. Welty-Wolf, J. Madden, V. Tapson, M. A. Baz, R. D. Davis, and J. R. Perfect.** 1996. Fungal infections in lung and heart-lung transplant recipients. *Medicine* **75**:142–156.

103. **Kapend'a, K., K. Komichelo, D. Swinne, and J. Vandepitte.** 1987. Meningitis due to *Cryptococcus neoformans* biovar *gattii* in a Zairean AIDS patient. *Eur. J. Clin. Microbiol.* **6**:320–321.

104. **Kaplan, M. H., P. P. Roses, and D. Armstrong.** 1977. Cryptococcosis in a cancer hospital. Clinical and pathological correlates in forty-six patients. *Cancer* **39**:2265–2274.

105. **Katlama, C., C. Leport, F. Brun-Vezinet, C. Rouzioux, D. Vittecoq, T. Lambolez, P. Lebras, P. Petitprez, G. Offendstadt, F. Vachon, J. L. Vilde, J. P. Coulaud, and A. G. Saimot.** 1984. Acquired Immunodeficiency syndrome (AIDS) in Africans. *Ann. Soc. Belg. Med. Trop.* **64**:379–389.

106. **Kaufman, L., and S. Blumer.** 1977. Cryptococcosis: the awakening giant, abstr. 176–182. Proc. Fourth International Conference on the Mycoses. PAHO Scientific Publication no. 356.

107. **Keller, R. G., G. S. Pfrommer, and T. R. Kozel.** 1994. Occurrences, specificities, and functions of ubiquitous antibodies in human serum that are reactive with the *Cryptococcus neoformans* cell wall. *Infect. Immun.* **62**:215–220.

108. **Keye, J. D., and W. E. Magee.** 1956. Fungal diseases in a general hospital. *Am. J. Clin. Pathol.* **26:**1235–1253.

109. **Knight, F. R., D. W. Mackenzie, B. G. Evans, K. Porter, N. J. Barrett, and G. C. White.** 1993. Increasing incidence of cryptococcosis in the United Kingdom. *J. Infect.* **27:**185–191.

110. **Kohno, S., A. Varma, K. J. Kwon-Chung, and K. Hara.** 1994. Epidemiology studies of clinical isolates of *Cryptococcus neoformans* of Japan by restriction fragment length polymorphism. *J. Jpn. Assoc. Infect. Dis.* **68:**1512–1517.

111. **Kovacs, J. A., A. A. Kovacs, M. Polis, W. C. Wright, V. J. Gill, C. U. Tuazon, E. P. Gelmann, H. C. Lane, R. Longfield, G. Overturf, A. M. Macher, A. S. Fauci, J. E. Parrillo, J. E. Bennett, and H. Masur.** 1985. Cryptococcosis in the acquired immunodeficiency syndrome. *Ann. Intern. Med.* **103:**533–538.

112. **Krick, J. A.** 1981. Familial cryptococcal meningitis. *J. Infect. Dis.* **143:**133.

113. **Kwon-Chung, K. J., and J. E. Bennett.** 1984. Epidemiologic differences between the two varieties of *Cryptococcus neoformans. Am. J. Epidemiol.* **120:**123–130.

114. **Kyong, C. U., G. Virella, H. H. Fudenberg, and C. P. Darby.** 1978. X-linked immunodeficiency with increased IgM: clinical, ethnic, and immunologic heterogeneity. *Pediatr. Res.* **12:**1024–1026.

115. **Lazer, M. S., B. Wanke, and M. M. Nishikawa.** 1993. Isolation of both varieties of *Cryptococcus neoformans* from saprophytic sources in the city of Rio de Janeiro, Brazil. *J. Med. Vet. Mycol.* **31:**449–454.

116. **Leggiadro, R. J., F. F. Barrett, and W. T. Hughes.** 1992. Extrapulmonary cryptococcosis in immunocompromised infants and children. *Pediatr. Infect. Dis. J.* **11:**43–47.

117. **Leggiadro, R. J., M. W. Kline, and W. T. Hughes.** 1991. Extrapulmonary cryptococcosis in children with acquired immunodeficiency syndrome. *Pediatr. Infect. Dis. J.* **10:**658–662.

118. **Levin, E. A.** 1937. Torula infection of the central nervous system. *Arch. Intern. Med.* **59:**667–684.

119. **Lewis, J. L., and S. Rabinovich.** 1972. The wide spectrum of cryptococcal infections. *Am. J. Med.* **53:**315–322.

120. **Littman, M. L.** 1959. Cryptococcosis (torulosis). *Am. J. Med.* **27:**976–998.

121. **Littman, M. L.** 1968. Cryptococcosis: current status. *Am. J. Med.* **45:**922–932.

122. **Littman, M. L., and S. S. Schneierson.** 1959. *Cryptococcus neoformans* in pigeon excreta in New York City. *Am. J. Hyg.* **69:**49–59.

123. **Littman, M. L., and L. E. Zimmerman.** 1956. *Cryptococcosis, Torulosis or European Blastomycosis.* Grune & Stratton, New York.

124. **Lo, D.** 1976. Cryptococcosis in the northern territory. *Med. J. Aust.* **27:**825–826.

125. **Mabee, C. L., S. W. Mabee, R. B. Kirkpatrick, and S. L. Koletar.** 1995. Cirrhosis: a risk factor for cryptococcal peritonitis. *Am. J. Gastroenterol.* **90:**2042–2045.

126. **Magee, J. T., C. Philpot, J. Yang, and I. K. Hosein.** 1994. Pyrolysis typing of isolates from a recurrence of systemic cryptococcosis. *J. Med. Microbiol.* **40:**165–169.

127. **Maher, D., and H. Mwandumba.** 1994. Cryptococcal meningitis in Lilongwe and Blantyre, Malawi. *J. Infect.* **28:**59–64.

128. **Manfredi, R., G. Rezza, V. G. Coronado, A. C. Lepri, O. V. Coronado, A. Mastroianni, and F. Chiodo.** 1995. Is AIDS-related cryptococcosis more frequent among men? *AIDS* **9:**397–398.

129. **Mansoor, G. A., and D. B. Ornt.** 1997. Cryptococcal peritonitis in peritoneal dialysis patients: a case report. *Clin. Nephrol.* **41:**230–232.

130. **Mare, M., M. T. Sartori, M. Carretta, A. Bertaggia, and A. Girolami.** 1990. Thinophyma-like cryptococcal infection as an early manifestation of AIDS in a hemophilia B patient. *Acta Haematol.* **84:**101–103.

131. **Marquis, G., S. Montplaisir, M. Pelletier, S. Mousseau, and P. Auger.** 1985. Genetic

resistance to murine cryptococcosis: increased susceptibility in the CBA/N XID mutant strain of mice. *Infect. Immun.* **47**:282–287.

132. **Martinez, A. J., M. Sell, T. Mitrovics, G. Stoltenburg-Didinger, J. R. Iglesias-Rozas, M. A. Giraldo-Velasquez, G. Gosztonyi, V. Schneider, and J. Cervos-Navarro.** 1995. The neuropathology and epidimiology of AIDS. A Berlin experience. A review of 200 cases. *Pathol. Res. Pract.* **191**:427–443.

133. **McGehee, J. L., and I. D. Michelson.** 1926. Torula infection in man, report of a case. *Surg. Gynecol. Obstet.* **42**:803–808.

134. **McManus, E. J., and J. M. Jones.** 1985. Detection of a *Trichosporon beigelii* antigen cross-reactive with *Cryptococcus neoformans* capsular polysaccharide in serum from a patient with disseminated *Trichosporon* infection. *J. Clin. Microbiol.* **21**:681–685.

135. **McNeill, J. I., and V. L. Kan.** 1995. Decline in the incidence of cryptococcosis among HIV-infected patients. *J. Acquired Immune Defic. Syndr. Hum. Retrovirol.* **9**:206–208.

136. **McNulty, A., J. M. Kaldor, A. M. McDonald, K. Baumgart, and D. A. Cooper.** 1994. Acquired immunodeficiency without evidence of HIV infection: national retrospective survey. *Br. Med. J.* **308**:825–826.

138. **Miyagawa, T., S. Hamagami, T. Ochi, T. Osugi, M. Kikui, and H. Takahashi.** 1982. Japanese summer-type hypersensitivity pneumonitis: studies using *Cryptococcus* antigen. *Clin. Allergy* **12**:343–354.

139. **Miyagawa, T., T. Ochi, and H. Takahashi.** 1978. Hypersensitivity pneumonitis with antibodies to *Cryptococcus neoformans*. *Clin. Allergy* **8**:501–509.

140. **Mody, C. H., G. B. Toews, and M. F. Lipscomb.** 1988. Cyclosporin A inhibits the growth of *Cryptococcus neoformans* on a murine model. *Infect. Immun.* **56**:7–12.

141. **Mody, C. H., G. B. Toews, and M. F. Lipscomb.** 1989. Treatment of murine cryptococcosis with cyclosporin-A in normal and athymic mice. *Am. Rev. Respir. Dis.* **139**:8–13.

142. **Murphy, J. J., F. D. McDonald, M. Dawson, A. Reite, J. Turcotte, and R. Fekety.** 1976. Factors affecting the frequency of infection in renal transplant recipients. *Arch. Intern. Med.* **136**:670–677.

143. **Newberry, W. M., J. Walter, J. W. Chandler, and F. E. Tosh.** 1967. Epidemiologic study of *Cryptococcus neoformans*. *Ann. Intern. Med.* **67**:727–732.

144. **Newton, J. A., S. A. Tasker, W. D. Bone, E. C. Oldfield III, P. E. Olson, and M. T. Nguyen.** 1995. Weekly fluconazole for the suppression of recurrent thrush in HIV-seropositive patients: impact on the incidence of disseminated cryptoccal infection. *AIDS* **9**:1286–1287.

145. **Nightingale, S. D., S. X. Cal, D. M. Peterson, S. D. Loss, B. A. Gamble, D. A. Watson, C. P. Manzone, J. E. Baker, and J. D. Jockusch.** 1992. Primary prophylaxis with fluconazole against systemic fungal infections in HIV-positive patients. *AIDS* **6**:191–194.

146. **Nishikawa, M. M., O. D. Sant'anna, M. S. Lazera, and B. Wanke.** 1996. Use of D-proline assimilation and CGB medium for screening Brazilian *Cryptococcus neoformans* isolates. *J. Med. Vet. Mycol.* **34**:365–366.

147. **Noble, R. C., and L. F. Fajardo.** 1972. Primary cutaneous cryptococcosis. *Am. J. Clin. Pathol.* **57**:13–22.

148. **Nottenbart, H. C., R. F. McGehee, and J. P. Utz.** 1973. Cryptococcosis complicating sarcoidosis. *Am. Rev. Respir. Dis.* **107**:1060–1063.

149. **Odom, A., M. Del Poeta, J. Perfect, and J. Heitman.** 1997. The immunosuppressant FK506 and its nonimmunosuppressive analog L-685,818 are toxic to *Cryptococcus neoformans* by inhibition of a common target protein. *Antimicrob. Agents Chemother.* **41**:156–161.

150. **Olson, P. E., K. C. Earhart, R. J. Rossetti, J. A. Newton, and M. R. Wallace.** 1997. Smoking and risk of cryptococcosis in patients with AIDS. *JAMA* **277**:629.

151. **Ooi, B., B. T. M. Chen, C. H. Lim, O. T. Khoo, and K. T. Chan.** 1971. Survival of a

patient transplanted with a kidney infected with *Cryptococcus neoformans. Transplantation* **11**:428–429.

152. **Ostrowski, M., I. E. Salit, W. L. Gold, D. Sutton, M. L. Montpetit, D. Lepine, and T. Salas.** 1993. Idiopathic CD4+ T-lymphocytopenia in two patients. *Can. Med. Assoc. J.* **149**:1679–1683.

153. **Paya, C. V.** 1993. Fungal infections in solid-organ transplantation. *Clin. Infect. Dis.* **16**:677–688.

154. **Perfect, J. R.** 1993. Antifungal prophylaxis: to prevent or not. *Am. J. Med.* **94**:233–234.

155. **Perfect, J. R., and D. T. Durack.** 1985. Effects of cyclosporine in experimental cryptococcal meningitis. *Infect. Immun.* **50**:22–26.

156. **Perfect, J. R., N. Ketabchi, G. M. Cox, C. W. Ingram, and C. L. Beiser.** 1993. Karyotyping of *Cryptococcus neoformans* as an epidemiological tool. *J. Clin. Microbiol.* **31**:3305–3309.

157. **Perfect, J. R., B. B. Magee, and P. T. Magee.** 1989. Separation of chromosomes of *Cryptococcus neoformans* by pulsed field gel electrophoresis. *Infect. Immun.* **57**:2624–2627.

158. **Perriens, J. H., M. Mussa, M. K. Luabeya, K. Kayembe, B. Kapita, C. Brown, P. Piot, and R. Janssen.** 1992. Neurological complications of HIV-1-seropositive internal medicine in patients in Kinshasa, Zaire. *J. Acquired Immune Defic. Syndr.* **5**:333–340.

159. **Piot, P., T. C. Quinn, H. Telman, F. M. Feinsod, K. B. Minlangu, O. Wobin, N. Mbendi, P. Mazebo, K. Ndandi, W. Stevens, K. Kalambayi, S. Mitchell, C. Bridts, and J. B. McCormick.** 1984. Acquired immunodeficiency syndrome in a heterosexual population in Zaire. *Lancet* **ii**:65–68.

160. **Polacheck, I., and G. Lebens.** 1989. Electrophoretic karyotype of the pathogenic yeast *Cryptococcus neoformans. J. Gen. Microbiol.* **135**:65–71.

161. **Polacheck, I., G. Lebens, and J. B. Hicks.** 1992. Development of DNA probes for early diagnosis and epidemiological study of cryptococcosis in AIDS patients. *J. Clin. Microbiol.* **30**:925–930.

162. **Powell, K. E., B. A. Dahl, R. J. Weeks, and F. E. Tosh.** 1972. Airborne *Cryptococcus neoformans*: particles from pigeon excreta compatible with alveolar deposition. *J. Infect. Dis.* **125**:412–415.

163. **Rebers, P. A., S. A. Barker, M. Heidelberger, Z. Dische, and E. E. Evans.** 1958. Precipitation of the specific polysaccharide of *Cryptococcus neoformans* A by types II and XIV antipneumococcal antisera. *J. Am. Chem. Soc.* **80**:1135–1137.

164. **Restrepo, A., M. E. Salazar, L. E. Cano, E. P. Stover, D. Feldman, and D. A. Stevens.** 1984. Estrogens inhibit mycelium-to-yeast transformation in the fungus *Paracoccidioides brasiliensis*: implications for resistance of females to paracoccidioidomycosis. *Infect. Immun.* **46**:346–353.

165. **Rinaldi, M. G., D. J. Drutz, A. Howell, M. A. Sande, C. B. Wofsy, and W. K. Hadley.** 1986. Serotypes of *Cryptococcus neoformans* in patients with AIDS. *J. Infect. Dis.* **153**:642.

166. **Rippon, J. W.** 1988. *Medical Mycology. The Pathogenic Fungi and the Pathogenic Actinomycetes*, p. 582. W.B. Saunders, Philadelphia.

167. **Rogers, M. F., P. Thomas, E. T. Starcher, M. C. Noa, T. J. Bush, and H. W. Jaffe.** 1987. Acquired immunodeficiency syndrome in children: report of the Centers for Disease Control national surveillance, 1982 to 1985. *Pediatrics* **79**:1008–1014.

168. **Rozenbaum, R., and A. J. Goncalves.** 1994. Clinical epidemiological study of 171 cases of cryptococcosis. *Clin. Infect. Dis.* **18**:369–380.

169. **Rozenbaum, R., A. J. Goncalves, B. Wanke, and W. Viera.** 1990. *Cryptococcus neoformans* var. *gattii* in a Brazilian AIDS patient. *Mycopathologia* **112**:33–34.

170. **Rozenbaum, R., A. J. Goncalves, B. Wanke, M. J. Caiuby, H. Clemente, M. D. S. Lazera, P. C. F. Monteiro, and A. T. Londero.** 1992. *Cryptococcus neoformans* varieties as agents of cryptococcosis in Brazil. *Mycopathologia* **119**:133–136.

171. **Ruane, P. J., L. J. Walker, and W. L. George.** 1988. Disseminated infection caused by urease-negative *Cryptococcus neoformans. J. Clin. Microbiol.* **26:**2224–2225.

172. **Ruiz, A., and G. S. Bulmer.** 1981. Particle size of airborne *Cryptococcus neoformans* in a tower. *Appl. Environ. Microbiol.* **41:**1225–1229.

173. **Ruiz, A., D. Velez, and R. A. Fromtling.** 1989. Isolation of saprophytic *Cryptococcus neoformans* from Puerto Rico: distribution and variety. *Mycopathologia* **106:**167–170.

174. **Salkowski, C. A., K. F. Bartizal, M. Balish, and E. Balish.** 1987. Colonization and pathogenesis of *Cryptococcus neoformans* in gnotobiotic mice. *Infect. Immun.* **55:**2000–2005.

175. **Salvin, S. B.** 1959. Current concepts of diagnostic serology and skin hypersensitivity in the mycoses. *Am. J. Med.* **27:**97–114.

176. **Salyer, W. R., D. C. Salyer, and R. D. Baker.** 1974. Primary complex of *Cryptococcus* and pulmonary lymph nodes. *J. Infect. Dis.* **130:**74–77.

177. **Sarosi, G. A., J. D. Parker, I. L. Doto, and F. E. Tosh.** 1992. Amphotericin B in cryptococcal meningitis. *Ann. Intern. Med.* **71:**1079–1087.

178. **Sarosi, G. A., P. M. Silberfarb, and F. E. Tosh.** 1971. Cutaneous cryptococcosis: a sentinel of disseminated disease. *Arch. Dermatol.* **104:**1–3.

179. **Schimpff, S. C., and J. E. Bennett.** 1975. Abnormalities in cell-mediated immunity in patients with *Cryptococcus neoformans* infection. *J. Allergy Clin. Immunol.* **55:**430–441.

180. **Sekhon, A. S., S. N. Bannerjee, B. M. Mielke, H. Idikio, G. Wood, and J. M. S. Dixon.** 1990. Current status of cryptococcosis in Canada. *Mycoses* **33:**73–80.

181. **Shaariah, W., Z. Morad, and A. B. Suleiman.** 1992. Cryptococcosis in renal transplant recipients. *Transplant. Proc.* **24:**1898–1899.

182. **Shimazu, K., M. Ando, T. Sakata, K. Yoshida, and S. Araki.** 1984. Hypersensitivity pneumonitis induced by *Trichosporon cutaneum. Am. Rev. Respir. Dis.* **130:**407–411.

183. **Shimizu, R. Y., D. H. Howard, and M. N. Clancy.** 1986. The variety of *Cryptococcus neoformans* in patients with AIDS. *J. Infect. Dis.* **154:**1042.

184. **Siewers, C. M. F., and H. G. Cramblett.** 1964. Cryptococcosis (torulosis) in children. A report of four cases. *Pediatrics* **34:**393–400.

185. **Singh, N.** 1997. Clinical and environmental factors associated with cryptococcosis in liver transplant patients. *Clin. Infect. Dis.* **24:**744–745.

186. **Smith, D. K., J. J. Neal, and S. D. Holmberg.** 1993. Unexplained opportunistic infections and CD4+ T-lymphocytopenia without HIV infection. *N. Engl. J. Med.* **328:**373–379.

187. **Smith, J. H., M. M. Nichols, A. Goldman, F. C. Schmalstieg, and R. M. Goldblum.** 1982. Disseminated cryptococcosis in an infant with severe combined immunodeficiency. *Hum. Pathol.* **13:**500–503.

188. **Smith, J. W., and W. C. Arnold.** 1988. Cryptococcal peritonitis in patients with peritoneal dialysis. *Am. J. Kidney Dis.* **11:**430–433.

189. **Sokolowski, J. W., R. F. Schillaci, and T. E. Motley.** 1969. Disseminated cryptococcosis complicating sarcoidosis. *Am. Rev. Respir. Dis.* **100:**717–722.

190. **Sorrell, T. C., S. C. Chen, P. Ruma, W. Meyer, T. J. Pfeiffer, D. H. Ellis, and A. G. Brownlee.** 1996. Concordance of clinical and environmental isolates of *Cryptococcus neoformans* var. *gattii* by random amplification of polymorphic DNA analysis and PCR fingerprinting. *J. Clin. Microbiol.* **34:**1253–1260.

191. **Sorvillo, F., G. Beall, P. A. Turner, V. L. Beer, A. A. Kovacs, and P. R. Kerndt.** 1997. Incidence and factors associated with extrapulmonary cryptococcosis among persons with HIV infection in Los Angeles County. *AIDS* **11:**673–679.

192. **Speed, B., and D. Dunt.** 1995. Clinical and host differences between infections with the two varieties of *Cryptococcus neoformans. Clin. Infect. Dis.* **21:**28–34.

193. **Speed, B. R., and J. Kaldor.** 1997. Rarity of cryptococcal infection in children. *Pediatr. Infect. Dis. J.* **16:**536–537.

194. **Speed, B. R., L. Strawbridge, and D. H. Ellis.** 1993. *Cryptococcus neoformans* var. *gattii* meningitis in an Australian patient with AIDS. *J. Med. Vet. Mycol.* **31:**395–399.

195. **Spickard, A., W. T. Butler, V. Andriole, and J. P. Utz.** 1963. The improved prognosis of cryptococcal meningitis with amphotericin B therapy. *Ann. Intern. Med.* **58:**66–83.

196. **Spira, T. J., B. M. Jones, J. K. A. Nicholson, R. B. Lala, T. Rowe, A. C. Mawle, C. B. Lauter, J. A. Shulman, and R. A. Monson.** 1993. Idiopathic CD4+ T-lymphocytopenia—an analysis of five patients with unexplained opportunistic infections. *N. Engl. J. Med.* **328:**386–392.

197. **Spitzer, E. D., and S. G. Spitzer.** 1992. Use of a dispersed repetitive DNA element to distinguish clinical isolates of *Cryptococcus neoformans*. *J. Clin. Microbiol.* **30:**1094–1097.

198. **Spitzer, E. D., S. G. Spitzer, L. F. Freundlich, and A. Casadevall.** 1993. Persistence of the initial infection in recurrent cryptococcal meningitis. *Lancet* **341:**595–596.

199. **Stamm, A. M., R. B. Diasio, W. E. Dismukes, S. Shadomy, G. A. Cloud, C. A. Bowles, and G. H. Karam.** 1987. Toxicity of Amphotericin B plus flucytosine in 194 patients with cryptococcal meningitis. *Am. J. Med.* **83:**236–242.

200. **St.-Germain, G., G. Noel, and K. J. Kwon-Chung.** 1988. Disseminated cryptococcosis due to *Cryptococcus neoformans* variety *gattii* in a Canadian patient with AIDS. *Eur. J. Clin. Microbiol. Infect. Dis.* **7:**587–588.

201. **Stone, B. D., and J. G. Wheeler.** 1990. Disseminated cryptococcal infection in a patient with hyperimmunoglobulinemia E syndrome. *J. Pediatr.* **117:**92–95.

202. **Sullivan, D., K. Haynes, G. Moran, D. Shanley, and D. Coleman.** 1996. Persistence, replacement, and microevolution of *Cryptococcus neoformans* strains in recurrent meningitis in AIDS patients. *J. Clin. Microbiol.* **34:**1739–1744.

203. **Summerbell, R. C., F. Staib, R. Dales, N. Nolard, J. Kane, H. Zwanenburg, R. Burnett, S. Kradjen, D. Fung, and D. Leong.** 1992. Ecology of fungi in human dwellings. *J. Med. Vet. Mycol.* **30**(Suppl. 1):279–285.

204. **Sussman, E. J., F. MacMahon, D. Wright, and H. M. Friedman.** 1984. Cutaneous cryptococcosis without evidence of systemic involvement. *J. Am. Acad. Dermatol.* **11:**371–374.

205. **Swinne, D., M. Deppner, S. Maniratunga, R. Laroche, J.-J. Floch, and P. Kadende.** 1991. AIDS-associated cryptococcosis in Bujumbura, Burundi: an epidimiological study. *J. Med. Vet. Mycol.* **29:**25–30.

206. **Swinne, D., K. Kayembe, and M. Niyimi.** 1986. Isolation of saprophytic *Cryptococcus neoformans* var. *neoformans* in Kinshasa, Zaire. *Ann. Soc. Belg. Med. Trop.* **66:**57–61.

207. **Swinne, D., J. B. Nkurikiyinfura, and T. L. Muyembe.** 1986. Clinical isolates of *Cryptococcus neoformans* from Zaire. *Eur. J. Clin. Microbiol.* **5:**50–51.

208. **Tabone, M.-C., G. Leverger, J. Landman, C. Aznar, L. Boccon-Gibod, and G. Lasfargues.** 1994. Disseminated lymphonodular cryptococcosis in a child with X-linked hyper-IgM immunodeficiency. *Pediatr. Infect. Dis. J.* **13:**77–79.

209. **Taelman, H., J. Clerinx, A. Kagame, J. Batungwanayo, A. Nyirabareja, and J. Bogaerts.** 1991. Cryptococcosis, another growing burden in central Africa. *Lancet* **338:**761.

210. **Takos, M. J.** 1956. Experimental cryptococcosis produced by the ingestion of virulent organisms. *N. Engl. J. Med.* **254:**598–601.

211. **Tanphaichitra, D., S. Sahaphongs, and S. Srimuang.** 1988. Cryptococcal antigen survey among racing pigeon workers and patients with cryptococcosis, pythiosis, histoplasmosis, and penicilliosis. *Int. J. Clin. Pharm. Res.* **8:**433–439.

212. **Ting, S. F., B. E. Glader, and C. G. Prober.** 1991. *Cryptococcus* infection in a nine-year-old child with hemophilia and the acquired immunodeficiency syndrome. *Pediatr. Infect. Dis. J.* **10:**76–77.

214. **Van de Perre, P., P. Lepage, P. Kestelyn, A. Hekker, D. Rouvroy, J. Bogaerts, J.**

Kayihigi, and J.-P. Butzler. 1984. Acquired immunodeficiency syndrome in Rwanda. *Lancet* **ii:**62–65.

215. **Varma, A., and K. J. Kwon-Chung.** 1989. Restriction fragment polymorphism in mitochondrial DNA of *Cryptococcus neoformans. J. Gen. Microbiol.* **135:**3353–3362.

216. **Varma, A., and K. J. Kwon-Chung.** 1992. DNA probe for typing of *Cryptococcus neoformans. J. Clin. Microbiol.* **30:**2960–2967.

217. **Varma, A., D. Swinne, F. Staib, J. E. Bennett, and K. J. Kwon-Chung.** 1995. Diversity of DNA fingerprints in *Cryptococcus neoformans. J. Clin. Microbiol.* **33:**1807–1814.

218. **Villanueva, E., M. Mendoza, E. Torres, M. B. Albornoz, M. E. Cavazza, and G. Urbina.** 1989. Serotipificacion de 27 cepas de *Cryptococcus neoformans* aisladas en Venezuela. *Acta Cientifica Venezolana* **40:**151–154.

219. **Wang, F., Y. So, E. Vittinghoff, H. Malani, A. Reingold, E. Lewis, J. Giordano, and R. Janssen.** 1995. Incidence proportion of and risk factors for AIDS patients diagnosed with HIV dementia, central nervous system toxoplasmosis, and cryptococcal meningitis. *J. Acquired Immune Defic. Syndr.* **8:**75–82.

220. **White, M., C. Cirrincione, A. Blevins, and D. Armstrong.** 1992. Cryptococcal meningitis: outcome in patients with AIDS and patients with neoplastic disease. *J. Infect. Dis.* **165:**960–963.

221. **Wickes, B. L., D. E. Moore, and K. J. Kwon-Chung.** 1994. Comparison of the electrophoretic karyotypes and chromosomal location of ten genes in the two varieties of *Cryptococcus neoformans. Microbiology* **140:**543–555.

222. **Yamamoto, Y., S. Kohno, H. Koga, H. Kakeya, K. Tomono, M. Kaku, T. Yamazaki, M. Arisawa, and K. Hara.** 1995. Random amplified polymorphic DNA analysis of clinically and environmentally isolated *Cryptococcus neoformans* in Nagasaki. *J. Clin. Microbiol.* **33:**3328–3332.

223. **Yinnon, A. M., A. Solages, and J. J. Treanor.** 1993. Cryptococcal peritonitis: report of a case developing during continuous peritoneal dialysis and review of the literature. *Clin. Infect. Dis.* **17:**736–741.

224. **Zimmerman, L. E., and H. Rappaport.** 1954. Occurrence of cryptococcosis in patients with malignant disease of reticuloendothelial system. *Am. J. Clin. Pathol.* **24:**1050–1072.

225. **Zimmermann, B., III, M. Spiegel, and E. V. Lally.** 1992. Cryptococcal meningitis in systemic lupus erythematosus. *Semin. Arthritis Rheum.* **22:**18–24.

226. **Zuger, A., E. Louie, R. S. Holzman, M. S. Simberkoff, and J. J. Rahal.** 1986. Cryptococcal disease in patients with the acquired immunodeficiency syndrome: diagnostic features and outcome of treatment. *Ann. Intern. Med.* **104:**234–240.

12 | Diagnosis and Laboratory Techniques

DIRECT EXAMINATION OF SPECIMENS

The India ink examination is a particularly useful and rapid diagnostic test for cryptococcal meningitis. Although *Cryptococcus neoformans* can become fragile, collapse, or be crescentic when dried or fixed, the encapsulated cells are rapidly distinguished in a colored colloidal medium such as India ink when mixed with fluids such as cerebrospinal fluid (CSF). A modified India ink preparation for *C. neoformans* in CSF employs 2% chromium mercury and India ink, which allows for identification of both external and internal features in the yeast (120). This simple test has been found to be positive in more than 80% of patients with AIDS and cryptococcal meningitis and in 30 to 50% of patients with non-AIDS cryptococcal meningitis. In our experience the sensitivity of an India ink examination generally allows detection of yeast cells in a CSF specimen with careful observation when there are between 10^3 and 10^4 CFU of yeasts per ml of CSF or greater concentrations. Before the AIDS epidemic, the quantity of yeast in CSF during meningitis ranged between 10^3 and 10^7 CFU/ml (93); it is likely that the heavily positive results of India ink examinations in many AIDS patients with meningitis represent concentrations between 10^5 and 10^7 CFU of yeasts per ml in CSF. Furthermore, the sensitivity of the examination may even be improved by centrifuging the CSF specimen (i.e., 500 rpm for 10 min) and using the pellet for staining. The specificity of the examination occasionally can be reduced by false positives associated with leukocytes, myelin globules, fat droplets, and tissue cells. Although not as sensitive or specific as the serological tests for cryptococcal meningitis, India ink examination is a rapid test that can often deliver an immediate diagnosis within minutes of a lumbar puncture and give the clinician an appreciation for the burden of yeasts. For instance, a heavily positive examination may identify a patient with a higher risk for sudden increase in intracranial pressure during early administration of antifungal drugs and thus alert the clinician to the potential complication. Therefore, this examination should be performed as a screening test on all initial CSF specimens from patients at high risk for this infection.

Histopathologically, the 5- to 20-μm encapsulated yeasts in tissues have been relatively easy to identify because of their prominent capsules (see color plates following p. 456). Yeasts have been identified from various body sites and tissues with histological stains ranging from the nonspecific Papanicolaou, hematoxylin and eosin, and acridine orange preparations to more specific fungal stains such as Calcofluor, which stains fungal chitin, or a Gomori methenamine silver stain. Even a Gram stain may show *C. neoformans* as a poorly stained gram-positive yeast. However, because of the polysaccharide capsule possessed by *C. neoformans* and in most cases the enlargement of this capsule during infection, several specific stains have been developed, such as the mucicarmine, periodic acid-Schiff, and alcian blue stains, that identify the capsular material surrounding the yeasts (see color plates). Although *Histoplasma* capsules will not stain with mucicarmine, *Blastomyces* strains may occasionally stain with it. Therefore, other morphological characteristics, such as the wide-based budding of *Blastomyces,* should be helpful in distinguishing it from *Cryptococcus*. However, it should be emphasized that there are cases in which *C. neoformans* in tissue is poorly encapsulated (12, 36, 50, 52, 72, 73, 85, 97), and the yeast may be identified only by a Gomori methenamine silver or Fontana-Masson stain. On the other hand, the size and narrow budding characteristics of this yeast may still be helpful in predicting *C. neoformans* within a lesion, even without a positive capsular stain. The Fontana-Masson stain, which appears to identify melanin, has been used occasionally to specifically identify *C. neoformans* in tissue when the capsule is not apparent (70, 97). This stain can occasionally give false-positives with other fungi, such as *Cryptococcus laurentii,* and even young *Coccidioides immitis* endospores can be stained in tissue with the Fontana-Masson stain.

In clinical specimens, yeast cells of *C. neoformans* are mostly globose in shape, although some may be oval to lemon-shaped, and var. *gattii* cells may actually be elliptical. At times when specimens are dried, heat-fixed, and stained, *C. neoformans* cells will collapse or become crescent-shaped, making them more difficult to identify. *C. neoformans* generally does not produce hyphae in tissue, but occasionally—under certain conditions, such as severe temperature elevations (38 to 42°C) and in some mammalian hosts such as the rabbit—the budding and separation of dividing yeast cells will be impaired. Chains of cells will then form with unusual shapes that can mimic pseudohyphae or actually appear to possess true hypha formation on direct examination (35). Furthermore, with sputum or pus, it may be reasonable to digest the specimens with 10% potassium hydroxide to remove tissue artifacts, making yeast cells easier to identify.

Ease and accuracy have made cytopathological techniques extremely useful in the diagnosis of cryptococcus infection (24, 57, 65, 80). For instance, ultrasound guidance of percutaneous aspirates of lung nodules and bronchoalveolar lavage fluid for cytological examination may provide the initial findings in the diagnosis of a cryptococcal infection.

GROWTH CHARACTERISTICS AND CULTURE METHODS

C. neoformans is a mammalian pathogen in large part because, of all the cryptococcal species, it is the only species that can routinely grow at 37°C. Some strains of *Cryptococcus albidus* and *C. laurentii* can grow at 37°C, but this characteristic is not common for these species. On the other hand, most strains of *C. neoformans* optimally grow at 30 to 37°C, and their growth is tightly regulated at these temperatures. In fact, the growth rates of many *C. neoformans* strains are significantly reduced at temperatures between 39 and 40°C. *C. neoformans* var. *gattii* appears to be even more sensitive to these high temperatures than *C. neoformans* var. *neoformans*, and strains will die rapidly when environmentally exposed to 40°C. Furthermore, *C. neoformans* has other carefully controlled growth and survival features that respond to certain environmental cues. The yeast grows well in acidic media (pH 5–7) but does not tolerate alkaline pH conditions above 7.6 and will die. The yeast also grows well aerobically, but growth stops in anaerobic conditions. Furthermore, as the pCO_2 tension rises in the environment to host levels (i.e., from 0.04 to 5%), capsular synthesis is induced in most strains (48), along with a concomitant slight decrease in cell cycle replications. Despite the lack of production of known siderophores (62), *C. neoformans* growth crucially requires the presence of at least some iron in the medium, and studies show that assimilation of iron occurs through its conversion by the yeast cells to the ferrous state (63, 111). Also, fluxes in iron concentrations can affect the structure of the yeast. For instance, in low-iron states, capsule synthesis is actually stimulated (110). Under optimal growth conditions in the laboratory, the generation time of a typical *C. neoformans* strain is between 2.7 and 3.4 h. It will not generally grow as fast as *C. albicans* or *Saccharomyces cerevisiae* strains. The growth of *C. neoformans* will be detected on most standard fungal and bacterial media; it is not a particularly fastidious yeast, and colonies on agar plates can usually be observed by visual inspection in 48 to 72 h. Although *C. neoformans* grows well at 24°C, it has been recommended that plates be incubated at 30 to 35°C for clinical specimens. This range of temperatures may actually speed up isolation of *C. neoformans*, since some strains of var. *gattii* may require 5 to 7 days to grow and may not be easily identified at higher temperatures (35 to 37°C). Standard fungal media such as Sabouraud glucose agar work well for the isolation of *C. neoformans*, but when antibiotics are used in the medium to reduce bacterial contamination, chloramphenicol is a better choice than cycloheximide. Cycloheximide has been tried as a treatment for cryptococcosis (see chapter 14). Most isolates are inhibited by drug concentrations under 25 μg/ml of cycloheximide. The yeast will grow on artificial media as opaque, creamy colonies that may turn orange-tan or brown after prolonged incubation (see color plates, following p. 456).

There are two very noticeable features of *C. neoformans* colonies. First, there is variability among strains in their ability to appear mucoid. The mucoid appearance of the colony is directly related to the capsule size around the yeast. As the colony matures, the increase in capsule size can be significant in some strains, and the colony appears very mucoid. Capsule size or thickness is determined by the genetics of the strain. The size of the capsule for a given strain can vary considerably. In fact, capsule production in most strains can be optimized by in vitro growth on solid or liquid medium in the presence of 1% glucose, $\geq$1 mg/ml of glutamate, neutral pH, temperature at 37°C, elevated pCO_2, or decreased iron concentrations (32, 48, 76, 110). However, the in vitro size or thickness of the capsule in various strains is not directly correlated with their virulence potential (32). Conversely, capsule production by many strains can be decreased in vitro by growth at lower temperatures, under conditions of high osmolarity (16% glucose or 2.9% NaCl) or acid pH, or after storage in soil (32, 48, 61, 76, 110). Although the receptors and specific pathways for the response to these environmental cues for capsule synthesis remain to be elucidated, it is clear that part of the signal transduction of these nutritional responses passes through the signaling heterotrimeric protein GPA1 and its mediation of cyclic AMP levels (1). The observation that the effects of these growth conditions on capsule size are strain dependent provides evidence of phenotypic and probably genotypic variation among individual isolates of *C. neoformans*. However, no specific phenotype or genotype of *C. neoformans* has yet been correlated with specific host risk factors.

The second interesting feature of cryptococcal colonies is the development of sectoring within some colonies after prolonged incubation on agar plates (see color plates following p. 456). Benham (5) first described variations in the morphological appearance of *C. neoformans*. Drouhet and Couteau (29) categorized the sectoring of colonies into three types, mucoid, rough, and smooth. Littman (77) emphasized that this sectoring is related to cells within the population that acquire thick capsules or tend to form pseudomycelium. These variations in morphological appearance are reminiscent of the switching phenomenon characterized in *Candida albicans* and other fungi. Recently, Goldman et al. (42) further characterized this high-frequency event (~1 $\times$ 10^{-3}); it is reversible, influenced by environmental conditions, and associated with differences in the capsular and cellular characteristics of the yeast and its virulence. For instance, Granger et al. (48) were able to isolate a thinly encapsulated, avirulent clone (C3D) from a wild-type sectoring colony under elevated pCO$_2$ conditions. This clone did not respond to elevated CO$_2$ concentrations by increasing capsular production and was isolated by repeated selection for a single nonmucoid and nonswitching phenotype, which then lost its ability to produce disease in an animal model. The genetics of the switching mechanisms for certain phenotypes in the morphological changes of colonies and their relevance to the pathobiology of the yeast await further studies. However, it is apparent that a single strain may represent a heterogeneous population of cells in which specific phenotypic clones can be selected. This phenotype switching property is not unique to *C. neoformans* among the fungal kingdom, but because of the genetics and molecular biology of *C. neoformans*, this fungus should be considered in any attempts to understand the biological importance of the phenotypic switching mechanism(s).

Despite the ease of cultivating *C. neoformans* in vitro on a variety of media and even in minimal media without vitamins such as thiamine (14), there have been several reports that the growth of some clinical strains of *C. neoformans* may require particular attention. First, there has been a report of a nutritionally aberrant strain of *C. neoformans* that was incapable of growing on brain-heart infusion agar but grew well on Sabouraud dextrose agar (31). The reason for this isolate's specific nutritional requirements remains unclear, but it emphasizes that for some strains, different media may be required for primary isolation of *C. neoformans*. For instance, a histologically positive tissue but culture-negative specimen may need to be placed on several different media before yeast viability or nonviability can be definitely determined. In a difficult case of culture-negative chronic meningitis, several fungal media should be used, although most of these culture-negative cases are probably due to too few organisms at the site of infection to be detected.

Second, there have been reports of labile forms of the yeast (i.e., L-forms) that required clinical material to be recovered first in a hypertonic medium of salt and/or 0.3 M sucrose before culture onto standard media (78). Third, blood cultures may need careful attention. While positive blood cultures were uncommon

(98) and a relatively poor prognostic factor in the era before human immunodeficiency virus (HIV) infection (27, 93), cryptococcemia has become a common occurrence with disseminated cryptococcosis and AIDS. In several series of disseminated cases in AIDS, the rates of positive blood cultures ranged from 35 to 68% of cases (21, 22, 33, 69). There has been some concern that radiometric methods such as the BACTEC system may not identify positive cultures when the numbers of yeasts in the blood are very low (99). In fact, colony counts of <1 CFU/ml are characteristic of cryptococcemia in patients with AIDS, as determined by the lysis-centrifugation system (118). Therefore, there has been a recommendation to subculture blood culture bottles from high-risk patients despite the low radiometric readings. Also, lysis-centrifugation methods may be more sensitive when there are small numbers of yeasts in the bloodstream (13, 118). However, it should be emphasized that the radiometric methods do frequently identify cryptococcemia in AIDS patients, and this probably reflects the high number of yeasts in the blood. Also, it has been shown that continuous agitation of BACTEC blood culture bottles for the full incubation time significantly improves the detection of CO_2 and recovery of *C. neoformans* from blood cultures (58, 96).

Fourth, although most *C. neoformans* strains grow on artificial media within several days of incubation, there are strains that may require longer periods to produce visible colonies, particularly if the patient has received previous antifungal treatment. Therefore, specific fungal cultures for *C. neoformans* should be well vented and agitated (if broth) and observed for 3 to 4 weeks for detection of yeast growth. It has been shown that oxygen may be a growth-limiting nutrient for *C. neoformans,* and this fact should be taken into account when cultures and in vitro susceptibility testing for this yeast are performed (92). Finally, as noted by Mess and Daar (84), many times the diagnosis of cryptococcosis in AIDS is already made by other methods before the blood cultures containing *C. neoformans* are identified.

Quantitative cultures of the CSF are not routinely determined in clinical practice, but when performed, the yeast concentrations can range from 10^3 to 10^7 CFU/ml of CSF (93). As previously noted, it generally takes more than 10^3 CFU/ml in a specimen to obtain a positive India ink examination, and heavily positive India ink examinations can reflect more than 10^6 CFU/ml of yeasts in the CSF. Blood contains less than 10 CFU of yeasts per ml of blood, and we suspect that urine cultures may frequently be negative because this site contains less than 1,000 yeasts per ml, which would not be routinely detected by current screening methods.

There are two clinical situations in which cultures of the CSF for identification of viable *C. neoformans* may be difficult. First, in some cases of chronic cryptococcal meningitis, the lumbar fluid is found to be culture negative. If treatment has not been started, this finding is likely to be due to a low burden of organisms that have not reached the lumbar sac. Therefore, there are two possibilities for improving recovery of yeast: (i) withdraw more CSF (10 to 20 ml) and then culture the entire CSF-centrifuged pellet of this volume, or (ii) try another site for CSF removal, such as the cisternal space, where organisms might be causing an active basilar meningitis but are not yet present in the lumbar sac (43). On the other hand, the ventricular CSF fluid is not commonly culture positive for this yeast and is not a high-yield area for sampling, but it will occasionally contain yeast.

Second, some patients who receive an antifungal treatment course will have positive India ink examinations at the end of therapy despite clinical improvement, and these yeasts will not be viable on routine culture media when the CSF is sampled. If electron microscopy is performed on these cells seen in the India ink examination, organelles are in severe disarray and it is likely that these yeasts are dead. However, the finding does suggest that there may still be a viable focus of yeast cells in the central nervous system contributing to these nonviable progeny. For example, if normal rabbits are inoculated intracisternally with killed cryptococci, they will rapidly eliminate the yeast forms from the subarachnoid space. Therefore, the normal central nervous system and its subarachnoid space generally have the potential to eliminate yeasts rapidly from this infection site. Why these nonviable yeast forms occasionally persist for months after successful treatment in humans remains a mystery. Although the presence of nonviable yeast cells in CSF does not necessarily require institution of treatment, it is always clinically worrisome that clinical relapse with positive cultures may occur in the future. Other clinical areas where nonviable yeasts are occasionally found are in lung biopsies of normal hosts with cryptococcal nodules and in renal transplant patients in whom a return to dialysis and thus elimination of immunosuppressive regimens will produce cellulitis with nonviable *C. neoformans*.

For the isolation of *C. neoformans* from heavily contaminated specimens such as sputum in clinical practice or environmental samples, an excellent selective medium has been developed. Staib's work (105, 106) showed that niger seed (*Guizotia abyssinica*, birdseed) could be used to exploit the ability of *C. neoformans* to break down caffeic acid to melanin; *C. neoformans* can be identified as the brown colonies on the plates containing this substrate (see color plates, following p. 456). Using the birdseed agar as a primary culture medium for sputum and urine in patients with AIDS will increase sensitivity for detection of *C. neoformans,* since these patients are frequently colonized with other yeasts (25, 105, 106). Birdseed agar can be supplemented with several antibiotics and biphenyl to minimize the growth of bacterial contaminants. The medium should contain no more than 0.1% glucose to ensure induction of the phenoloxidase enzyme by the yeast, which is up-regulated under low glucose conditions. Other diphenolic compounds can also be used in the medium, such as L-dopa, dopamine, epinephrine, and norepinephrine, but their shelf life is considerably shorter than that of niger seed, which has become the standard substrate for selective media. Although formal sensitivity studies for identification of *C. neoformans* in sputum of patients with pulmonary infection have not been done, Campbell in his retrospective review (17) found only 19 of 101 patients with positive sputum cultures in confirmed cases of pulmonary cryptococcosis.

Canavanine-glycine-bromothymol blue (CGB) agar can be used to separate *C. neoformans* var. *neoformans* from var. *gattii* (71). This medium exploits the biochemical differences between the two varieties. All isolates of var. *gattii* can utilize glycine and are resistant to canavanine (94). The medium turns blue as a result of alkalinization by the ammonia released during the degradation of glycine. Var. *neoformans* cannot utilize glycine as either a carbon or a nitrogen source and be simultaneously resistant to canavanine in the same medium. There is also a D-proline medium that separates the two varieties since only var. *gattii* can use D-proline as a sole nitrogen source (30). This medium compares favorably to CGB medium, though with an occasional false negative (91).

For the study of the sexual stages of *C. neoformans* (*Filobasidiella neoformans*) and its hyphal growth by haploid fruiting, several media have been used to identify and isolate the hyphal structures and basidiospores in genetic crosses. The characteristics of these media are similar in that they generally represent low nutrient conditions. Nutrient starvation is an important signal for pheromones (88) and mating pathways (1) in this fungus. In the mating of haploid isolates, opposite mating types of MAT**a** and MATα are mixed together in a streak on either V-8 juice agar or hay infusion agar and incubated at 23 to 25°C without sealing of the petri dishes. For instance, V-8 juice agar contains 50 ml of V-8 juice, 0.5 g of KH_2PO_4, and 40 g of agar in 1,000 ml of distilled water and is adjusted to pH 7.2. Hyphae and basidiospores can be seen under a microscope from 4 to 14 days after the crosses are made. Timing of this filamentous appearance on the edges of the streak depends on the mating pairs used. Filament agar to identify haploid fruiting with *C. neoformans* contains yeast nitrogen base without amino acids and ammonium sulfate, 0.5% glucose, and 4% Bacto Agar (Difco) at pH 5.0 (116). The three important conditions that need to be controlled for optimal mating and haploid fruiting efficiencies are nutrient starvation, moisture, and temperature. For instance, mating agar plates need to be dry, with no condensation on the lids, and incubation must occur at room temperature (23 to 25°C).

IDENTIFICATION

The general characteristics of species within the genus *Cryptococcus* include the absence of true pseudohyphae or hyphae, ballistospores, the inability to ferment sugars, and the ability to assimilate inositol and produce urease. There are at least 38 different species of *Cryptococcus* excluding *C. neoformans*, and except for the few clinical cases of *C. albidus*, *C. curvatus*, and *C. laurentii*, these other species have not been shown to cause cryptococcosis. Occasionally, some *Cryptococcus* species other than *C. neoformans* will be isolated from clinical specimens as colonizers or contaminants. The specific identification of these *Cryptococcus* species will be left to standard microbiology texts.

C. neoformans is similar in many features to other *Cryptococcus* species. In its identification, certain features, such as the presence of encapsulation, growth at 37°C, and production of a brown colony on niger seed agar, will presumptively identify the yeast as *C. neoformans*. *C. neoformans* also produces a significant amount of extracellular urease. Zimmer and Roberts (121) used this observation to develop a rapid test for detection of urease production from yeast isolates that could then be used to distinguish *Cryptococcus* and other urease-positive yeasts from *Candida* species that are urease negative. For example, over 99% of 286 *C. neoformans* isolates showed urease activity in colonies after 15 min in this assay. In contrast, other urease-positive yeasts required more than 3 h for detection of urease activity (121). This rapid enzymatic test on yeast colonies can be performed prior to specific biochemical studies (18). There are a few reported clinical cases of cryptococcosis caused by urease-negative *C. neoformans* strains, but they are not common (3, 74). A formal biochemical profile of *C. neoformans* shows that it does not assimilate nitrate. In its carbon assimilation profile, it will utilize galactose, maltose, galactitol, and sucrose. However, it will not assimilate lactose or melibiose, and its growth is strain variable with erythritol.

In the clinical laboratory, there are a series of commercially available micromethod systems employing modified conventional biochemical tests, such as the API 20C system, the Flow Laboratories Uni-yeast-Tek System, the BBL Minitek System, and the Vitek Yeast Biochemical Card. These systems use carbohydrate assimilation and generally require 24-h incubations. There are commercial mul-titest identification systems based on detection of pre-formed enzymes that can identify yeasts with chromogenic substrates within 4 h of inoculation (API Yeast Identification System and Microscan Rapid Yeast Identification). Both systems can reliably identify *C. neoformans* (15, 107a).

Most *C. neoformans* strains can be identified by the previously described struc-tural and biochemical characteristics. However, there is a commercially available DNA probe to rRNA (AccuProbe) that can confirm or identify a yeast isolate as *C. neoformans* with 100% sensitivity and specificity (59). In the last few years, there have been a series of studies that used rapid PCR-based identification schemes to identify yeast strains (56, 87, 112). The diagnostic methods can now be adapted directly to clinical specimens. The strategies generally utilize specific primers from the gene encoding the 18S rRNA. This gene has both universally conserved fungal sequences and specific fungal species sequences. With the use of nested specific PCR, these molecular techniques can rapidly distinguish *C. neoformans* from other yeasts within a mixed sample or in tissue. However, at this time it is uncertain when these molecular strategies will prove to supplement, complement, or com-pletely replace culture methods on a clinical basis for *C. neoformans* and other fungi.

Serotyping is not specifically relevant to diagnosis but has been used in com-bination with biological tests for *C. neoformans* identification (64). However, sero-typing may have relevance to characteristics of epidemiology, pathogenesis, and clinical presentations. Isolates can be placed into four serotypes: A, B, C, and D. There is now a commercially available serological typing method that will distin-guish the four serotypes based on the capsular structure of *C. neoformans* isolates (60). A commercial kit that uses monoclonal antibodies to identify isolates is avail-able (Iatron, Tokyo, Japan). Studies have also used fluorescent-antibody reagents for typing *C. neoformans* in tissue (66), and serotyping based on direct analysis of culture supernatants for the major type-specific capsular antigen with a dot en-zyme assay has been described (4).

SEROLOGY

Serological tests for the diagnosis of cryptococcosis are one of the true success stories for rapid and accurate diagnosis of fungal infections in the last 30 to 40 years. Cryptococcal antibodies have been detected in humans during and after infection, and there has even been some suggestion that their detection represents a favorable prognostic sign (27). However, detection of antibody responses in many of these immunocompromised hosts is complex (8, 45, 47, 113), and in general, measurements of cryptococcal antibodies in body fluids have not been of consistent clinical benefit for diagnosis. With a class-specific enzyme immunoassay (EIA) for immunoglobulin G (IgG) and IgA, Speed et al. (104) found that antibodies are generally not readily detected during acute infection in immunocompetent patients. However, antibodies in sera may be detected during antigenemia, and although IgA antibodies fall over 1 to 2 years in the recovery phase, IgG levels persist. This assay may be particularly useful for seroepidemiology studies of infection with *C. neoformans*.

In contrast to antibody assays, detection of cryptococcal polysaccharide antigen in body fluids has become the specific focus for the development of many commercial tests for rapid diagnosis. In 1963, Bloomfield et al. (10) published a sentinel paper describing the use of antibody-coated latex particles that detected soluble antigens of *C. neoformans* in the sera or CSF of seven of nine patients with the diagnosis of disseminated cryptococcosis. These investigators found few nonspecific reactions and reported that the titers of these specific antigens declined during successful treatment of the infections. The conclusions of this important study supported the latex agglutination test as a potential rapid diagnostic test for cryptococcal infection, even when the India ink examination is negative. These investigators suggested that the test might have prognostic as well as diagnostic value.

From its humble beginnings as a test reported in less than a dozen patients (10), there have now been over three decades of development and clinical experience with latex agglutination testing in the detection of cryptococcal polysaccharide, in various body fluids but with a particular emphasis on sera and CSF (6, 8, 44, 47, 67, 114). The latex agglutination test is now approximately 95% sensitive and specific for identification of invasive cryptococcosis (28, 103). In fact, this test is routinely used in clinical practice in developed countries for testing CSF specimens from immunocompromised patients and from those with subacute or chronic meningitis.

During the maturation of this test, various specimen treatment methods and reagents have been reported to eliminate false positives caused by rheumatoid factor (53, 68) and other unknown factors (46, 49). Most laboratories use commercial kits and can reduce the false positives with the pretreatment of specimens such as serum with proteolytic enzymes (pronase) (51) and, in some cases, reducing agents such as 2-β-mercaptoethanol (115) or dithiothreitol. The specimens can also be boiled for 5 min with EDTA. All these pretreatments can reduce the number of false positives and even the false negatives that arise because of interfering substances, and thus they will improve the sensitivity and specificity of this extremely useful diagnostic test. It is more likely that serum specimens will need this pretreatment procedure for interfering substances than will CSF.

False-positive reactions for cryptococcal antigen tests have been recorded in a variety of clinical circumstances. For example, false-positive tests have been caused by contamination of specimens with minute amounts of agar, agarose, or its syneresis fluid during pipetting in the laboratory (11, 54). It has been shown that EIAs do not interact with syneresis fluid, unlike latex agglutination assays (34). Cleaning of the slides with certain disinfectants and soaps during repeated washing has the potential to cause false reactions (9). Slides that are reused should be carefully washed with 10% hypochlorite. Another iatrogenic cause of positive antigen tests is the use of low-molecular-weight hydroxyethyl starch for vascular support treatment in patients, since it can cross-react with antibodies against the cryptococcal polysaccharide (86). Similarly, false-positive tests may occur when cross-reactive antigens produced from another infectious microorganism(s) are present in the specimen (53, 68, 79). This cross-reaction of antigens has been specifically seen with the polysaccharide of a related pathogen, *Trichosporon beigelii* (16, 82, 83), but has also been reported with other infections (19). A positive cryptococcal antigen in a patient with risk factors for trichosporonosis should be viewed as possibly indicating either a disseminated infection with *Trichosporon* or *Cryptococcus*.

False-negative tests can occur in cryptococcal meningitis (7, 23, 51, 55, 107, 108), but they are unusual. However, one study reported eight false-negative CSF latex agglutination tests for 88 AIDS patients with culture-confirmed cryptococcal meningitis (21). Generally, these tests are falsely negative because of (i) prozone effects, or immune complexes, (ii) low burden of organisms, or possibly (iii) a poorly encapsulated strain. False-negative tests can occur because of prozone-like reactions both in sera and in CSF. On the other hand, dilution of the specimen for detection of this effect will be useful on a practical basis only for patients at very high risk for cryptococcal infection or for unusual cases of meningitis without an apparent diagnosis. The tests can also be falsely negative in the very early stages of infection when the yeast is present in CSF but no other CSF abnormalities are found; there can be a lack of antigen detection when the yeast has not yet shed polysaccharide into the CSF (101). Although commercial kits may differ, most latex agglutination test kits can detect at least 10 ng of polysaccharide per ml of biological fluid. Although it is theoretically possible that poorly encapsulated or acapsular strains could cause false-negative tests, this hypothesis remains unproved. Such false negatives would be an uncommon event, since hypo- or acapsular strains rarely produce cryptococcosis. Even though pronase is generally not recommended for CSF specimens, it has been shown to help resolve a false-negative CSF titer (54); thus, this fluid may also contain interfering substances or antibodies. Finally, occasionally one kit is negative for a specimen but other kits are positive, and the difference may simply be related to less potent reagents in certain kits (117).

The popularity of the diagnostic test for cryptococcal polysaccharide antigen is readily apparent when it is noted that there are at least five commercial latex agglutination kits for cryptococcal polysaccharide. In the United States, there are four such kits: Crypto-LA (International Biological Laboratories, Cranbury, N.J.); CALAS (Meridian Diagnostics, Cincinnati, Ohio); Myco-Immune (American Micro Scan, Rahway, N.J.); and IMMY (Immuno-mycologies, Norman, Okla.). In Europe a kit is made by Sanofi Diagnostics Pasteur, Marnes la Cogvetti, France. In all these kits, the sensitivity and specificity for CSF samples range from 93 to 100% (38, 51, 109). The serum sensitivity varies from 83 to 97%, depending on the choice of pretreatment protocols. Despite these excellent results for diagnosis, the use of the latex agglutination test for determining changes in polysaccharide titer during therapy is more problematic. When investigators attempted to obtain concordant titer results with various kits, the range of success was 50 to 70% with serum and 85 to 95% with CSF (51). Therefore, if changing antigen titers in an individual case of cryptococcosis are to be interpreted accurately, it will probably be necessary to employ one kit from a single manufacturer. Then the same kit can be used for serial specimen measurements at the same time. The variability between kits may have partially contributed to our difficulties in using antigen titers to make therapeutic decisions in individual cases. In fact, Powderly et al. (95) reviewed serial serum antigen titers during acute infections in AIDS patients and found the titers unhelpful in detecting failure or relapse. However, when the antigen titer was ≥1:8 in CSF and remained unchanged or when CSF titer increased during therapy, it did correlate with a clinical and/or microbiological failure to respond to therapy; also, a rise in CSF antigen titer during suppressive therapy was associated with relapse of cryptococcal meningitis. In our opinion, precision in the use of sera and CSF antigen titers for making individual therapeutic decisions is still lacking for human cryptococcosis, and they should probably not be used as the sole clinical determinant on which to base changes in therapy. On the other hand, Malik et al. (80a) have suggested that cryptococcal antigen titers may be followed in animals to make treatment decisions. These investigators found a two- to fourfold titer drop per month during successful therapy in cats and dogs, and they recommended that therapy be stopped when serum titers decline to >1:1 or when there is a 32-fold drop in titer with periodic monitoring (80a).

EIAs have also been developed to detect either antigen or antibody. While most latex agglutination tests detect cryptococcal polysaccharide as the antigen, the currently available EIA measures the major component of the polysaccharide, which is the glucuronoxylomannan. In comparison with latex agglutination tests, the readings of EIAs are less subjective; they are unaffected by prozone reactions and do not react with rheumatoid factor or require pronase treatment. Finally, the EIAs may detect antigen earlier in infection and at lower antigen concentrations. There may also be fewer false-positive tests with the EIAs. Finally, there is no correlation between high titers by EIA or by latex agglutination. However, both tests will identify all *C. neoformans* serotype infections. A comparison between a commercial EIA (Premier; Meridian Diagnostics, Cincinnati, Ohio), which employs a monoclonal antibody that recognizes all serotypes, and standard latex agglutination tests showed 92 to 97.8% agreement between them (38, 39, 100, 109).

Despite the occasional false positive, all positive CSF/serum cryptococcal polysaccharide antigen titers should be taken seriously, and in the correct clinical setting, such as an abnormal CSF or a high-risk patient, they may be diagnostic. On the other hand, the quantitative aspects of antigen testing are less certain to be useful in management of human disease, as previously noted. For example, titers of >1:1,024 generally yield little additional prognostic information except to emphasize that the burden of organisms in the host is high. This assumption is of practical importance since the present use of commercial diluents can substantially raise the cost of a single test if titers are taken into the 10,000 to 1,000,000 range. Until further studies can prove the therapeutic utility of quantifying the actual titers for each patient, it may be reasonable to stop antigen titer determinations at 1:1,024 and simply assume the burden of organisms is high if titers are greater.

Primarily, CSF and serum specimens are used for clinical diagnosis of cryptococcosis, but other body fluids can be screened. In fact, urine and pulmonary fluids may become more sensitive body sites for detection of antigen than sera in some AIDS patients with lung or disseminated infections. Patients with HIV infection and primary pulmonary cryptococcosis without dissemination may have false-negative serum tests, since yeasts have not yet disseminated from the lung. It is our belief, with support from animal studies (41), that once antigen is detected in the serum of a patient with cryptococcosis, this detection is a sign that infection has now disseminated from its pulmonary location. In a prospective study by Baughman et al. (2) of bronchoalveolar lavage fluids from patients with HIV infection, all eight patients who developed cryptococcal pneumonia had positive antigen titers of ≥8 in their bronchoalveolar lavage specimens. However, the specificity of this test was poor in that four false-positive reactions at titers up to 1:8 were also found. Thus, titers of ≥8 in bronchoalveolar lavage fluids from high-risk symptomatic patients may be diagnostic of pulmonary cryptococcosis. Antigen testing can also detect infection in transthoracic aspirates. In fact, in eight patients with pulmonary cryptococcosis, percutaneous lung aspirates with titers of $\geq$1:8 were found, and there was only one false positive in a total of 41 patients (75). Also, in 92 AIDS patients, Chapin-Robertson et al. (20) compared antigen titers in urine with those in serum and CSF. In all cases, patients with positive titers in CSF and/or serum also had positive titers in urine. Quantification of titers showed a ranking of serum > CSF > urine. Pronase treatment of specimens could increase some urine titers, which suggests that there are interfering substances in urine. Thus, urine could be used as a sampling site for disseminated disease, although it is unlikely to replace serum and CSF in most clinical situations.

The latex agglutination test for serum cryptococcal antigen has developed in certain clinics for use with high-risk, HIV-infected patients as a screening test for fever of unknown etiology, for new neurological symptoms, or even when the patients are asymptomatic. Because of the prevalence of cryptococcal disease in certain geographic locations, despite its worldwide distribution, various conclusions have been drawn regarding the value of this screening method. In some areas of the world, such screening has been useful for early diagnosis (26), and in several other areas the incidence of infection in high-risk patients is simply too low to justify routine screening for antigen in serum (55, 90). Furthermore, some investigators have found that in asymptomatic patients with HIV infection, a periodic antigen test did not identify or predict infection that occurred only 4 to 8 months after a negative antigen screening test (89).

Despite the controversy regarding the value of screening tests, presumptive use of polysaccharide antigen has created a series of cases identified as "isolated cryptococcal polysaccharidemia." The management of these laboratory findings, in which there is a positive serum antigen titer and negative fluid or tissue cultures for *C. neoformans* and without other clinical findings in HIV-infected patients, is unclear. However, several groups have now reported that some patients in this classification rapidly progress to a documented cryptococcal infection. Feldmesser et al. (37) found 10 patients with positive isolated cryptococcal polysaccharide antigen tests ($\geq$1:4). Six patients were treated with fluconazole, one patient developed disseminated disease later, and three patients were found to have disseminated disease during their initial evaluation. Yuen et al. (119) reported 10 of 13 patients with isolated positive antigen titers who received therapy, and none of these patients developed disseminated cryptococcal disease. In contrast, two of three patients who did not receive therapy developed cryptococcal meningitis 2 to 4 months after the positive cryptococcal antigen test. Similarly, Manfredi et al. (81) found that 5 of 27 consecutive HIV-infected patients, who had positive CSF antigen titers and negative cultures and tissue microscopic examinations, eventually developed evidence of cryptococcosis clinically or at necropsy. These findings support an aggressive therapeutic approach to the discovery of isolated serum and CSF cryptococcal antigen titers ($\geq$1:4) in HIV-infected patients and probably in other severely immunosuppressed individuals, because the tests may actually identify early cryptococcal disease. Finally, we have also seen an occasional case in an immunocompromised high-risk patient in whom the antigen and all microscopic evaluations were negative in the CSF, but the culture grew *C. neoformans* (101). These conditions likely represent early disease in high-risk patients for infection and may occasionally occur when diagnostic tests are performed preemptively in these high-risk patients.

MOLECULAR TECHNIQUES

The techniques for the molecular biology of *C. neoformans* have advanced over the last decade. Although further development of technical expertise and improvement in molecular methods with this fungus will be welcomed by investigators in the field, the following protocols for DNA and RNA isolation, karyotyping, and DNA transformation have been used successfully in our laboratories. These techniques are discussed because they are more *C. neoformans*-specific; most other molecular techniques for working with *C. neoformans* are standard molecular biology procedures.

In the molecular manipulation of *C. neoformans* there are two difficult features. First, there is a need to eliminate the large amount of contaminating polysaccharide from DNA and RNA that generally elutes with these nucleic acids. Large amounts of polysaccharide can occasionally inhibit a ligation reaction and potentially alter PCR. The second problem is the low frequency of homologous recombination, which can be as low as 0.01% and as high as 10%. There are also frequent ectopic integrative events during transformation events, and at present there is a lack of stable extrachromosomal plasmids for *C. neoformans*.

DNA Extraction and Purification from *C. neoformans* by Spheroplasting

1. Take 1 or 2 colonies from Sabouraud dextrose agar or yeast extract-peptone-dextrose (YEPD) broth culture and prepare a starter culture of *C. neoformans* in 10 ml of YEPD broth. Incubate at 30°C, 250 rpm, overnight.
2. Make a dilution of 1:400 (100 μl in 40 ml) in yeast nitrogen base broth and incubate at 30°C, 250 rpm, for about 16 h (a mid-logarithmic phase).
3. Pellet cells at 2,800 rpm and wash 3 times in 50 ml of 0.5 M NaCl–50 mM EDTA.
4. Resuspend in 9.5 ml of sterile double-distilled water. Add 0.5 ml of β-mercaptoethanol and incubate at 37°C for 1 h, mixing occasionally.
5. Spin at 2,800 rpm, resuspend in 4 ml of SCE solution (1 M sorbitol, 0.1 M sodium citrate [pH 5.8], 0.01 M EDTA), and place on ice.
6. Add 1 ml of SCE solution containing 5 mg of Lysing enzyme (Sigma; *Trichoderma harzianum*; final concentration, 1 mg/ml) and incubate at 37°C for 30 to 45 min, mixing occasionally (no more than 1 h). (Several other enzyme preparations such as Mureinase and Novozyme 234 can also be used.)
7. Pellet at 1,000 to 1,200 rpm for 10 min. Gently wash the pellet twice with 5 ml of SCE solution and gently resuspend without vortexing.
8. Pellet at 1,000 to 1,200 rpm, resuspend in 2 to 5 ml of lysing solution (4% sodium dodecyl sulfate, 50 mM EDTA), and leave at room temperature for 5 min. Do not vortex, but occasionally mix gently.
9. Transfer to a 15-ml polypropylene tube and extract twice with an equal volume of phenol-chloroform (1:1).
10. Transfer the aqueous phase to a Corex glass tube and precipitate, adding 1/10 volume of 3 M sodium acetate (NaAc) (pH 5.2) and 2 volumes of 100% ethanol, and incubate at −20°C for 30 min or overnight.
11. Spin at 10,000 rpm for 30 min at 4°C and resuspend in 2 to 5 ml of Tris-EDTA buffer (TE) (pH 8.0).

12. Treat with 100 μg of RNase (stock solution: 10 mg/ml) per ml of DNA solution and incubate at 37°C for 30 min.
13. Treat with 100 μg of proteinase K (stock solution: 20 mg/ml) per ml of DNA solution and incubate at 37°C for 30 min.
14. Transfer to a 15-ml polypropylene tube and extract with an equal volume of phenol-chloroform (1:1).
15. Extract twice with an equal volume of chloroform–isoamyl alcohol (24:1) and transfer the aqueous phase to a glass tube.
16. Precipitate DNA by adding 1/10 volume of 3 M NaAc (pH 5.2) and 2 volumes of 100% ethanol; incubate at −20°C for 2 h or overnight.
17. Pellet DNA at 10,000 rpm for 30 min at 4°C.
18. Wash once with 500 μl of 70% ethanol, centrifuge at 10,000 rpm for 30 min at 4°C, and resuspend in TE (pH 8.0).

Simplified Isolation of Cryptococcal DNA by Glass Beads

1. (A) *From broth culture.* Pellet cells from YEPD broth culture (10 ml) in a table-top centrifuge for 5 to 10 min.
    Suspend the cells in 1.5 ml of sterile distilled water, transfer to a 2-ml screw-capped microcentrifuge tube, and pellet in a microcentrifuge for 30 s.
    Pour off the supernatant, add 0.5 ml of TENTS (10 mM Tris [pH 7.5], 1 mM EDTA [pH 8.0], 100 mM NaAc, 2% Triton X-100, 1% sodium dodecyl sulfate), and resuspend the cells with a sterile toothpick.
   (B) *From agar culture.* Transfer 2 or 3 loopfuls of colonies from YEPD agar plate to a 2-ml screw-capped tube containing 0.5 ml of TENTS and suspend the cells with a sterile toothpick.
2. Add 0.6 g of 0.45-mm glass beads and 0.5 ml of phenol-chloroform and cap the tubes, making sure that the tubes are sealed completely.
3. Vortex for 2 min, spin in a microcentrifuge for 10 min, and remove the aqueous phase to a fresh tube.
4. Add 1 ml of 100% ethanol. (You should see a nucleic acid precipitate immediately. If not, leave at −20°C for at least 2 h.)
5. Spin in a microcentrifuge at 4°C for 30 min, remove all supernatant, and dissolve in 500 μl of TE containing 100 μg of RNase A per ml.
6. Incubate at 37°C for 20 min and then extract with phenol-chloroform.
7. Transfer the aqueous phase to a fresh tube, precipitate by adding 50 μl of 3 M NaAc and 1 ml of 100% ethanol, and incubate at −20°C for at least 2 h.
8. Spin in a microcentrifuge at 4°C for 30 min, wash with 500 μl of 70% ethanol, and dry.
9. Dissolve in 100 to 200 μl of TE and store frozen. From 10 ml of culture you should get about 100 μg of DNA.

Chromosomal DNA and Karyotypes

1. Inoculate 5 ml of YEPD broth with 1 or 2 colonies of *C. neoformans* and incubate for 24 h (until saturated), shaking vigorously at 30°C.

2. Add 100 μl of this suspension to 40 ml of yeast nitrogen base broth (1:400 dilution) and incubate for approximately 16 h at 30°C.
3. Pellet cells and wash 3 times in 0.5 M NaCl–50 mM EDTA.
4. Resuspend cells in 9.5 ml of distilled water and add 0.5 ml of β-mercaptoethanol. Incubate in a 37°C water bath for 1 h, mixing occasionally.
5. Pellet cells and resuspend in 2 ml of SCE solution. Place cells on ice.
6. Add 4 mg of Lysing enzyme (Sigma) to 2 ml of SCE solution and add this mixture to the cell suspension.
7. Inoculate for 30 to 60 min at 37°C, mixing occasionally until spheroplasting is complete.
8. Pellet cells for 10 min at 1,000 to 1,200 rpm. Wash cells twice in 5 ml of SCE solution. Resuspend cells in 3 ml of SCE.
9. Add 5 ml of sterile low-melting-point agarose (1% agarose in 0.125 M EDTA) that has been melted and equilibrated to 55°C. Pour agarose/cell suspension into a 6-cm-diameter petri dish and allow to solidify at room temperature.
10. Add a 5-ml overlay solution (0.5 M EDTA, 0.01 M Tris [pH 8], 1% sarcosyl, 1 mg proteinase K) over the gel and incubate for 18 h at 50°C.
11. Remove the overlay and make plugs. Store plugs in 0.5 M EDTA at 4°C.

Pulsed-field gel electrophoresis is performed by using gel plugs containing *C. neoformans* DNA inserted into a 1% agarose gel (4 by 4 in. [ca. 10 by 10 cm]). The DNA bands are separated in the gel by a CHEF DRII apparatus (Bio-Rad, Richmond, Calif.) in 0.5× Trisborate-EDTA at 14°C. Electrophoretic conditions are as follows: for 20 h, ramped from 50 s to 130 s at 125 V; for 41 h, ramped from 170 s to 300 s at 125 V. Gels are stained with ethidium bromide and photographed.

RNA Isolation from *C. neoformans*

1. Harvest the cells in mid-logarithmic growth (approximately 5×10^7 to 7×10^7 cells/ml).
   (A) Pellet the cells at 3,000 to 5,000 rpm at 4°C for 10 min.
   (B) Wash the cells 3 times in cold sterile 500 mM NaCl–50 mM EDTA (for each wash, 1 culture volume).
   (C) Resuspend the cells to approximately 10^9/ml in 500 mM NaCl–50 mM EDTA.
   (D) Transfer 1-ml aliquots to sterile 2-ml screw-capped conical microcentrifuge tubes.
   (E) Pellet cells at $10,000 \times g$ for 30 s.
   (F) Aspirate supernatant, freeze cell pellet in dry ice, and store at −70°C.
2. RNA isolation
   (A) Thaw the cell pellet on ice.
   (B) Resuspend the cells in 200 μl to 500 μl of a denaturing solution (solution D) containing 4 M guanidium thiocyanate, 25 mM sodium citrate (pH 7.0), and 0.1 M 2-β-mercaptoethanol. Prepare a stock solution as follows: dissolve 25 g of guanidium thiocyanate in 32 ml of distilled water and 1.7 ml of 0.75 M sodium citrate (pH 7.0) (stable for 3 months at room temperature). Prepare solution D by adding 0.036 ml of 2-β-mercaptoethanol to 5.0 ml of stock solution (stable for 1 month at room temperature).

(C) Add RNase-free glass beads (0.45 μm in diameter) to the resuspended cells so that the volume of glass beads is one-half to two-thirds the volume of cells.

(D) Add an equal volume of phenol (saturated in Tris [pH 7.8])–chloroform–isoamyl alcohol (50:49:1).

(E) Vortex in a Mini Bead Beater (Biospec) or a Vortex-Genie 2 (Fisher), beat vigorously for 20 s, then cool on ice for 20 s; repeat 5 times.

(F) Spin at 10,000 $\times$ g for 5 min and transfer the aqueous phase to a sterile 1.5-ml microcentrifuge tube.

(G) Repeat phenol–chloroform–isoamyl alcohol extraction of the aqueous phase (2 to 4 times).

(H) Spin at 10,000 $\times$ g for 5 min and transfer the aqueous phase to a sterile 1.5-ml microcentrifuge tube.

(I) Treat the aqueous phase with an equal volume of chloroform–isoamyl alcohol (49:1).

(J) Spin at 10,000 $\times$ g for 5 min and transfer the aqueous phase to a sterile microcentrifuge tube.

(K) Precipitate RNA with 0.1 volume of 3 M NaAc (pH 5.2) and 3 volumes of cold 100% ethanol and place at $-20°$C or $-70°$C overnight.

(L) Pellet RNA at 10,000 $\times$ g for 30 min at 4°C and wash pellet with 200 μl of cold 80% ethanol.

(M) Dissolve pellet in 200 μl of RNase-free distilled water, dilute a small volume, and read at A_{260} and A_{280}. Store RNA ($\geq$1 μg/μl) at $-70°$C.

To remove polysaccharide and purify the RNA, either apply the sample over a Nucleobond AX 100 column (Nest Group, Southboro, Mass.) and precipitate with 0.8 volume of isopropanol for 5 to 15 min at room temperature, or treat with LiCl as follows:

1. To the RNA sample add 0.1 volume of 8 M LiCl (RNase-free) and incubate on ice for 2 h.
2. Centrifuge at 10,000 $\times$ g for 30 min at 40°C.
3. Carefully remove supernatant and wash pellet in 80% ethanol. Dissolve pellet in 200 μl of diethyl pyrocarbonate (DEPC)-treated water.
4. Poly(A)$^+$ RNA can be further purified by affinity chromatography on oligo(dT)-cellulose.

Alternatively, use the FAST RNA Red Optimization Kit (Bio 101) that utilizes the Savant Fast Prep Bead Beater (Bio 101), which also works well.

1. To a fast-prep tube containing the glass beads, add 500 μl of chaotropic RNA stabilizing reagent, 500 μl of phenol acid reagent, 100 μl of chloroform–isoamyl alcohol, and 200 μl (10^9 cells) of a *C. neoformans* cell suspension.
2. Place tube in the Fast Prep Bead Beater and process for 20 s at a speed rating of 6. Remove tube and chill on ice for 1 min. Repeat 5 times. Remove tube and place on ice for at least 5 min.
3. Spin in a microcentrifuge at 10,000 rpm for 15 min to separate phases. Transfer the top phase (avoid interphase) to a clean 1.5-ml microcentrifuge tube.

4. Add 500 μl of chloroform–isoamyl alcohol. Vortex for 10 s. Centrifuge at 10,000 rpm for 2 min. Transfer top phase to a clean microcentrifuge tube.

5. Add 500 μl of DEPC-treated isopropanol precipitation solution, vortex for 10 s, and incubate for 2 min at room temperature. Centrifuge for 5 min to pellet RNA.

6. Wash pellet twice with 250 μl of RNase-free salt ethanol wash, remove supernatant, and let pellet air dry for 10 min.

7. Dissolve pellet in 50 to 100 μl of DEPC-treated distilled water.

8. To remove polysaccharide, add 20 μl of 12 M LiCl to 100 μl of RNA solution. Incubate for 15 min on ice. Centrifuge for 10 min and wash pellet twice in 250 μl of RNase-free salt ethanol wash.

9. Remove supernatant and let air dry for 10 min. Dissolve RNA in 50 to 100 μl of DEPC-treated distilled water and store RNA ($\geq$1 μg/μl) at -70°C.

Transformation of *C. neoformans* by Biolistic Delivery of DNA

Material

1. *C. neoformans* auxotrophic strains (serotype A and D) and wild-type H99 (serotype A) are available. Circular or linearized plasmids carrying the compatible gene marker (i.e., *URA5*, *ADE2*, hygromycin B) can be constructed.

2. *Media and buffers.* YEPD: 1% yeast extract, 1% peptone, 2% dextrose. Selective medium: 0.67% yeast nitrogen base without amino acids, 2% glucose (maybe supplemented with a drop-out mix). Hygromycin B medium: YEPD supplemented with hygromycin B (200 U/ml). Solid medium contains 2% agar. Add 1 M sorbitol to solid medium for biolistic transformation. Regeneration medium: 1 M sorbitol, 1 M mannitol, 0.9% yeast nitrogen base without amino acids, 2.6% glucose, 2.67% yeast extract–peptone, 0.13% gelatin. Filter-sterilize regeneration medium and store at room temperature. 2.5 M CaCl$_2$: autoclave and store at 4°C. 1 M spermidine (free base): filter-sterilize and store at -20°C. Absolute ethanol.

3. Biolistics PDS-1000 He/Particle Delivery system (Bio-Rad), 1,100 or 1,350 lb/in^2 rupture disks, and 0.6-μm gold beads.

Methods

1. *Cell preparation.* Inoculate 1 or 2 colonies into 10 to 30 ml of YEPD broth and shake vigorously at 30°C for 24 to 36 h (stationary phase). Collect cells by centrifugation (3,000 to 5,000 rpm, 5 min, 4°C) and resuspend in regeneration medium (0.2 to 0.3 of the original culture volume) to give a final cell density of 5×10^8 cells/ml. With a hockey stick, spread 200 μl (10^8) of the cell suspension on selective medium containing 1 M sorbitol, and allow the lawn of cells to dry. If transforming with the hygromycin B selectable marker, spread the cells on YEPD supplemented with 1 M sorbitol.

2. *Gold beads/DNA/macrocarrier preparation.* Add 60 mg of gold beads and 1 ml of absolute ethanol to a 1.5-ml microcentrifuge tube and vortex for 2 to 3 min. Pellet the gold beads (10,000 $\times$ g, 1 min), wash twice in sterile distilled water, resuspend in 1 ml of sterile distilled water, and store at 4°C. To a 1.5-ml microcentrifuge tube, add the following reagents in order, with continuous vortexing (5 bombardments): 50 μl of gold bead suspension, 5 μl of DNA (1

μg/μl), 50 μl of 2.5 M CaCl$_2$, 10 μl of 1 M spermidine. Allow the DNA to precipitate onto the beads without agitation for 10 min. Pellet the DNA-bead mixture, wash in 0.25 ml of absolute ethanol, and resuspend in 60 μl of absolute ethanol. Spread 10 μl of the DNA-bead suspension on the macrocarrier and allow this slurry to dry in a desiccator box.

3. *Bombardment with DNA-coated gold beads.* Place the rupture disk, macrocarrier, and plate containing the recipient cells in the biolistic chamber. Generate a full vacuum (≥28.5 in Hg) within the chamber and fire the device. Incubate the plates at 30°C for 3 to 5 days (transformants will appear within 2 to 3 days). If transforming with the hygromycin B marker, incubate the plates for 2 h at 30°C. Add 1 ml of YEPD to the plate and resuspend the cells with a hockey stick. Spread 200 μl of the cell suspension onto hygromycin B selective medium and incubate at 30°C for 3 to 5 days. Store remaining cells at 4°C for later use.

Transformation of *C. neoformans* by Electroporation

Materials

1. *C. neoformans* auxotrophic strains (serotype A and D) and wild-type H99 (serotype A) are available. Linearized telomerized plasmids (pCnTel) carrying the compatible gene marker (i.e., *URA5*, *ADE2*, hygromycin B) can be used. Serotype D strains are preferred for high-frequency transformations with electroporation.

2. *Media and buffers.* YEPD: 1% yeast extract, 1% peptone, 2% dextrose. Selective medium: 0.67% yeast nitrogen base without amino acids, 2% glucose. Hygromycin B medium: YEPD supplemented with hygromycin B (200 U/ml). Solid medium contains 2% agar. Electroporation buffer: 10 mM Tris-HCl (pH 7.5), 1 mM MgCl$_2$, 270 mM sucrose. Filter sterilize and store at 4°C. Dithiothreitol solution: 1 M dithiothreitol prepared in sterile distilled water and stored at −20°C.

3. BTX ECM electroporation system or Bio-Rad Gene Pulser and 2-mm gap cuvettes.

Methods

1. *Cell preparation.* Inoculate 1 or 2 colonies into YEPD broth and shake vigorously overnight at 30°C. In a final volume of 50 ml, dilute the overnight culture 1:50 with prewarmed (30°C) YEPD broth and grow until the cells reach a density of ~2.5 × 10^7 cells/ml (early-log phase, 3 to 4 h). Collect the cells by centrifugation (5,000 rpm, 5 min, 4°C), wash twice in cold sterile water, resuspend in 50 ml of cold electroporation buffer (determine cell density), and add 200 μl of 1 M dithiothreitol. Place cells on ice for 5 to 30 min, pellet cells, and resuspend in cold electroporation buffer (~1 ml) without dithiothreitol at a cell density no less than 10^9/ml (higher cell densities are even better).

2. *Plasmid preparation.* Digest pCnTel plasmids with *Sce*I, precipitate the linearized plasmids with ethanol to remove enzyme and salt, and resuspend in TE or distilled water (~0.1 μg/ml).

3. *Electroporation and plating.* Add 2 μl of linearized DNA (~0.2 μg) and 50 μl of the cell suspension to a sterile 0.5-ml microcentrifuge tube and place on ice for 1 min. Transfer the DNA-cell mixture to iced 2-mm gap cuvettes and electropo-

rate using the following parameters: (for BTX) voltage = 500, resistance = 360 Ω, capacitance = 50 μF, charging voltage = 475; (for Bio-Rad) voltage = 470, resistance = infinity, capacitance = 25 μF. These settings should generate time constants of 10 to 22 ms (BTX) or 20 to 50 ms (Bio-Rad). Immediately remove cells and plate directly on selective medium. If transforming with the hygromycin B marker, add 1 ml of YEPD to the cuvette, transfer cells to a 14-ml Falcon tube, and shake at 30°C for 1 to 2 h. Spread 200 μl of the cell suspension onto hygromycin B selective plates. Incubate plates at 30°C for 3 to 5 days (transformants will appear in 2 to 3 days).

REFERENCES

1. **Alspaugh, J. A., J. R. Perfect, and J. Heitman.** 1997. *Cryptococcus neoformans* mating and virulence are regulated by the G-protein gamma subunit GPA1 and cAMP. *Genes Dev.* **11:**3206–3217.

2. **Baughman, R. P., J. C. Rhodes, M. N. Dohn, H. Henderson, and P. T. Frame.** 1992. Detection of cryptococcal antigen in bronchoalveolar lavage fluid: a prospective study of diagnostic utility. *Am. Rev. Respir. Dis.* **145:**1226–1229.

3. **Bava, A. J., R. Negroni, and M. Bianchi.** 1993. Cryptococcosis produced by a urease negative strain of *Cryptococcus neoformans*. *J. Med. Vet. Mycol.* **31:**87–89.

4. **Belay, T., R. Cherniak, E. B. O'Neill, and T. R. Kozel.** 1996. Serotyping of *Cryptococcus neoformans* by dot enzyme assay. *J. Clin. Microbiol.* **34:**466–470.

5. **Benham, R. W.** 1950. Cryptococcosis and blastomycosis. *Ann. N.Y. Acad. Sci.* **50:**1299–1314.

6. **Bennett, J. E., H. F. Hasenclever, and B. S. Tynes.** 1964. Detection of cryptococcal polysaccharide in serum and spinal fluid: value in diagnosis and prognosis. *Trans. Assoc. Am. Physicians* **77:**145–150.

7. **Berlin, L., and J. H. Pincus.** 1989. Cryptococcal meningitis: false negative antigen test results and cultures in nonimmunosuppressed patients. *Arch. Neurol.* **46:**1312–1316.

8. **Bindschadler, D. D., and J. E. Bennett.** 1968. Serology of human cryptococcosis. *Ann. Intern. Med.* **69:**45–52.

9. **Blevins, L. B., J. Fenn, H. Segal, P. Newcomb-Gayman, and K. C. Carroll.** 1995. False-positive cryptococcal antigen latex agglutination caused by disinfectants and soaps. *J. Clin. Microbiol.* **33:**1674–1675.

10. **Bloomfield, N., M. A. Gordon, and D. F. Elmendorf, Jr.** 1963. Detection of *Cryptococcus neoformans* antigen in body fluids by latex particle agglutination. *Proc. Soc. Exp. Biol. Med.* **114:**64–67.

11. **Boom, W. H., D. J. Piper, K. L. Ruoff, and M. J. Ferraro.** 1985. New cause for false-positive results with cryptococcal antigen test by latex agglutination. *J. Clin. Microbiol.* **22:**856–857.

12. **Bottone, E. J., B. E. Johnasson, M. Toma, and G. P. Wormser.** 1986. Poorly encapsulated *Cryptococcus neoformans* from patients with AIDS. I. Preliminary observations. *AIDS Res. Hum. Retroviruses* **2:**221.

13. **Brannon, P., and T. E. Kiehn.** 1985. Large-scale clinical comparison of the lysis-centrifugation and radiometric systems for blood culture. *J. Clin. Microbiol.* **22:**951–954.

14. **Bruatto, M., V. Vidotto, and A. M. Maine.** 1992. Growth of *Cryptococcus neoformans* in thiamine-free medium. *Mycopathologia* **119:**129–132.

15. **Buesching, W. J., K. Kurek, and G. D. Roberts.** 1979. Evaluation of the modified API 20C system for identification of clinically important yeasts. *J. Clin. Microbiol.* **9:**565–569.

16. **Campbell, C. K., A. L. Payne, A. J. Teall, A. Brownell, and D. W. R. Mackenzie.** 1985.

Cryptococcal latex antigen test positive in patient with *Trichosporon beigelii* infection. *Lancet* **ii:**43–44.

17. **Campbell, G. D.** 1966. Primary pulmonary cryptococcosis. *Am. Rev. Respir. Dis.* **94:**236–243.

18. **Canteros, C. E., L. Rodero, M. C. Rivas, and G. Davel.** 1996. A rapid urease test for presumptive identification of *Cryptococcus neoformans*. *Mycopathologia* **136:**21–23.

19. **Chanock, S. J., P. Toltzis, and C. Wilson.** 1993. Cross-reactivity between *Stomatococcus mucilaginosus* and latex agglutination for cryptococcal antigen. *Lancet* **342:**1119–1120.

20. **Chapin-Robertson, K., C. Bechtel, S. Waycott, C. Kdontnick, and S. C. Edberg.** 1993. Cryptococcal antigen detection from urine of AIDS patients. *Diagn. Microbiol. Infect. Dis.* **17:**197–201.

21. **Chuck, S. L., and M. A. Sande.** 1989. Infections with *Cryptococcus neoformans* in the acquired immunodeficiency syndrome. *N. Engl. J. Med.* **321:**794–799.

22. **Clark, R. A., D. Greer, W. Atkinson, G. T. Valainis, and N. Hyslop.** 1990. Spectrum of *Cryptococcus neoformans* infection in 68 patients infected with acquired immunodeficiency virus. *Rev. Infect. Dis.* **12:**768–777.

23. **Currie, B. P., L. F. Freundlich, M. A. Soto, and A. Casadevall.** 1993. False-negative cerebrospinal fluid cryptococcal latex agglutination tests for patients with culture-positive cryptococcal meningitis. *J. Clin. Microbiol.* **31:**2519–2522.

24. **Davies, S. F., B. J. Gormus, and R. Yarchoan.** 1978. Cryptococcal meningitis with false-positive cytology in the CSF. *JAMA* **239:**2369.

25. **Denning, D. W., D. A. Stevens, and J. R. Hamilton.** 1990. Comparison of *Guizotia abyssinica* seed extract (birdseed) agar with conventional media for selective identification of *Cryptococcus neoformans* in patients with acquired immunodeficiency syndrome. *J. Clin. Microbiol.* **28:**2565–2567.

26. **Desmet, P., K. D. Kayembe, and C. DeVroey.** 1989. The value of cryptococcal serum antigen screening among HIV positive AIDS patients in Kinshasa, Zaire. *AIDS* **3:**77–78.

27. **Diamond, R. D., and J. E. Bennett.** 1974. Prognostic factors in cryptococcal meningitis. A study of 111 cases. *Ann. Intern. Med.* **80:**176–181.

28. **Dolan, C. T.** 1972. Specificity of the latex cryptococcal antigen test. *Am. J. Clin. Pathol.* **58:**358.

29. **Drouhet, E., and M. Couteau.** 1951. Sur les variations sectorielles des colonies de Torulopsis neoformans. *Ann Inst. Pasteur* **80:**456–457.

30. **Dufait, R., R. Velho, and C. de Vroey.** 1987. Rapid identification of the two varieties of *Cryptococcus neoformans* by D-proline assimilation. *Mykosen* **30:**483.

31. **Dunn, W. M., Jr., G. Bhandarkar, and D. Nafziger.** 1995. Isolation of a nutritionally aberrant strain of *Cryptococcus neoformans* from a patient with AIDS. *Clin. Infect. Dis.* **21:**1512–1513.

32. **Dykstra, M. A., L. Friedman, and J. W. Murphy.** 1977. Capsule size of *Cryptococcus neoformans*: control and relationship to virulence. *Infect. Immun.* **16:**129–135.

33. **Eng, R. H. K., E. Bishburg, S. M. Smith, and R. Kapila.** 1986. Cryptococcal infections in patients with acquired immune deficiency syndrome. *Am. J. Med.* **81:**19–23.

34. **Engler, H. D., and Y. R. Shen.** 1994. Effect of potential interference factors on performance of enzyme immunoassay and latex agglutination assay for cryptococcal antigen. *J. Clin. Microbiol.* **32:**2307–2308.

35. **Erke, K. H., and J. D. Schneidau.** 1973. Relationship of some *Cryptococcus neoformans* hypha-forming strains to standard strains and to other species of yeasts as determined by deoxyribonucleic acid base ratios and homologies. *Infect. Immun.* **7:**941–948.

36. **Farmer, S. G., and R. A. Komorowski.** 1973. Histologic response to capsule-deficient *Cryptococcus neoformans*. *Arch. Pathol.* **96:**383–387.

37. **Feldmesser, M., C. Harris, S. Reichberg, S. Khan, and A. Casadevall.** 1996. Serum cryptococcal antigen in patients with AIDS. *Clin. Infect. Dis.* **23:**827–830.
38. **Frank, U. K., S. L. Nishimura, N. C. Li, K. Sugai, D. M. Yajko, W. K. Hadley, and V. L. Ng.** 1993. Evaluation of an enzyme immunoassay for detection of cryptococcal capsular polysaccharide antigen in serum and cerebrospinal fluid. *J. Clin. Microbiol.* **31:**97–101.
39. **Gade, W., S. W. Hinnefeld, L. S. Babcock, P. Gilligan, W. Kelly, K. Wait, D. L. Greer, M. Pinilla, and R. L. Kaplan.** 1991. Comparison of the premier cryptococcal antigen enzyme immunoassay and the latex agglutination assay for detection of cryptococcal antigens. *J. Clin. Microbiol.* **29:**1616–1619.
41. **Goldman, D., S. C. Lee, and A. Casadevall.** 1994. Pathogenesis of pulmonary *Cryptococcus neoformans* infection in the rat. *Infect. Immun.* **62:**4755–4761.
42. **Goldman, D. L., S. P. Franzot, B. Fries, and A. Casadevall.** 1997. Phenotypic switching in the human pathogenic fungus *Cryptococcus neoformans*, abstr. F-44. 97th Annu. Meet. Am. Soc. Microbiol. 1997. American Society for Microbiology, Washington, D.C.
43. **Gonyea, E. F.** 1973. Cisternal puncture and cryptococcal meningitis. *Arch. Neurol.* **28:**200–201.
44. **Goodman, J. S., L. Kaufman, and M. G. Loening.** 1971. Diagnosis of cryptococcal meningitis: detection of cryptococcal antigen. *N. Engl. J. Med.* **285:**434–436.
45. **Gordon, M. A., and E. Lapa.** 1971. Charcoal particle agglutination test for detection of antibody to *Cryptococcus neoformans*: a preliminary report. *Am. J. Clin. Pathol.* **56:**354–359.
46. **Gordon, M. A., and E. W. Lapa.** 1974. Elimination of rheumatoid factor in the latex test for cryptococcosis. *Am. J. Clin. Pathol.* **60:**488–494.
47. **Gordon, M. A., and D. K. Vedder.** 1966. Serologic tests in the diagnosis and prognosis of cryptococcosis. *JAMA* **197:**961–967.
48. **Granger, D. L., J. R. Perfect, and D. T. Durack.** 1985. Virulence of *Cryptococcus neoformans*: regulation of capsule synthesis by carbon dioxide. *J. Clin. Invest.* **76:**508–516.
49. **Gray, L. D., and G. D. Roberts.** 1988. Experience with the use of pronase to eliminate interference factors in the latex agglutination test for cryptococcal antigen. *J. Clin. Microbiol.* **26:**2450–2451.
50. **Gutierrez, F., Y. S. Fu, and H. Lurie.** 1975. Cryptococcosis histologically resembling histoplasmosis: a light and electron microscopic study. *Arch. Pathol. Lab. Med.* **99:**347.
51. **Hamilton, J. R., A. Noble, D. W. Denning, and D. A. Stevens.** 1991. Performance of cryptococcus antigen latex agglutination kits on serum and cerobrospinal fluid specimens of AIDS patients before and after pronase treatment. *Clin. Microbiol. Rev.* **29:**333–339.
52. **Harding, S. A., W. M. Scheld, P. S. Feldman, and M. A. Sande.** 1979. Pulmonary infection with capsule-deficient *Cryptococcus neoformans*. *Virchows Arch. (A)* **182:**113–118.
53. **Hay, R. J., and D. W. R. Mackenzie.** 1982. False-positive latex tests for cryptococcal antigen in cerebrospinal fluid. *J. Clin. Pathol.* **35:**244–245.
54. **Heelan, J. S., L. Corpus, and N. Kessimian.** 1991. False-positive reactions in the latex agglutination test for *Cryptococcus neoformans* antigen. *J. Clin. Microbiol.* **29:**1260–1261.
55. **Hoffmann, S., J. Stenderup, and L. R. Mathiesen.** 1991. Low yield of screening for cryptococcal antigen by latex agglutination assay on serum and cerebrospinal fluid from Danish patients with AIDS on ARC. *Scand. J. Infect. Dis.* **23:**697–702.
56. **Hopfer, R. L., P. Walden, S. Setterquist, and W. E. Highsmith.** 1993. Detection and differentiation of fungi in clinical specimens using polymerase chain reaction (PCR) amplification and restriction enzyme analysis. *J. Med. Vet. Mycol.* **31:**65–75.
57. **Hsu, C. Y.** 1993. Cytologic diagnosis of pulmonary cryptococcosis in immunocompetent hosts. *Acta Cytol.* **37:**667–672.

58. **Huahua, T., J. Rudy, and C. M. Kunin.** 1991. Effect of hydrogen peroxide on growth of *Candida, Cryptococcus,* and other yeasts in simulated blood culture bottles. *J. Clin. Microbiol.* **29:**328–332.

59. **Huffnagle, K. E., and R. M. Gander.** 1993. Evaluation of Gen-Probe's *Histoplasma capsulatum* and *Cryptococcus neoformans* AccuProbes. *J. Clin. Microbiol.* **31:**419–421.

60. **Ikeda, R., R. Shinoda, Y. Fukazawa, and L. Kaufman.** 1982. Antigenic characterization of *Cryptococcus neoformans* serotypes and its application of serotyping of clinical isolates. *J. Clin. Microbiol.* **16:**22–29.

62. **Jacobson, E. S., and M. J. Petro.** 1987. Extracellular iron chelation in *Cryptococcus neoformans.* *J. Med. Vet. Mycol.* **25:**415–418.

63. **Jacobson, E. S., and S. E. Vartivarian.** 1992. Iron assimilation in *Cryptococcus neoformans.* *J. Med. Vet. Mycol.* **30:**443–450.

64. **Kabasawa, K., H. Itagaki, R. Ikeda, T. Shinoda, K. Kagaya, and Y. Fukazawa.** 1991. Evaluation of a new method for identification of *Cryptococcus neoformans* which uses serologic tests aided by selected biological tests. *J. Clin. Microbiol.* **29:**2873–2876.

65. **Kanjanavirojkul, N., C. Sripa, and A. Puapairoj.** 1997. Cytologic diagnosis of *Cryptococcus neoformans* in HIV-positive patients. *Acta Cytol.* **41:**493–496.

66. **Kaplan, W., S. L. Bragg, S. Crane, and D. G. Ahearn.** 1981. Serotyping *Cryptococcus neoformans* by immunofluorescence. *J. Clin. Microbiol.* **14:**313–317.

67. **Kauffman, L., and S. Blumer.** 1968. Value and interpretation of serological tests for the diagnosis of cryptococcosis. *Appl. Microbiol.* **16:**1907–1912.

68. **Kornfeld, S. L., and M. Worthington.** 1980. False-positive CSF cryptococcal antigen tests. *Arch. Neurol.* **37:**603–605.

69. **Kovacs, J. A., A. A. Kovacs, M. Polis, W. C. Wright, V. J. Gill, C. U. Tuazon, E. P. Gelmann, H. C. Lane, R. Longfield, G. Overturf, A. M. Macher, A. S. Fauci, J. E. Parrillo, J. E. Bennett, and H. Masur.** 1985. Cryptococcosis in the acquired immunodeficiency syndrome. *Ann. Intern. Med.* **103:**533–538.

70. **Kwon-Chung, K. J., W. B. Hill, and J. E. Bennett.** 1981. New special stain for histopathological diagnosis of cryptococcosis. *J. Clin. Microbiol.* **13:**383–387.

71. **Kwon-Chung, K. J., I. Polacheck, and J. E. Bennett.** 1982. Improved diagnostic medium for separation of *Cryptococcus neoformans* var. *neoformans* (serotype A and D) and *Cryptococcus neoformans* var. *gattii* (serotype B and C). *J. Clin. Microbiol.* **15:**535–537.

72. **Lacaz, C. S., E. M. Heins-Vaccari, N. T. Melo, O. A. Moreno-Carvalho, M. L. Sampaio, L. S. Nogueira, R. J. Badaro, and J. A. Livramento.** 1993. Neurocryptococcosis caused by nonencapsulated *Cryptococcus neoformans.* *Arq. Neuropsiquiatr.* **51:**395–398.

73. **Levinson, D. J., D. C. Silcox, J. W. Rippon, and S. Thomsen.** 1974. Septic arthritis due to nonencapsulated *Cryptococcus neoformans* with co-existing sarcoidosis. *Arthritis Rheum.* **17:**1037–1047.

74. **Li, A., N. Guo, and S. Wu.** 1993. A strain of urease negative *Cryptococcus neoformans* isolated from the environment in China. *Chin. Med. Sci. J.* **8:**52–54.

75. **Liaw, Y.-S., P.-C. Yang, C.-J. Yu, D.-B. Chang, H.-J. Wang, L.-N. Lee, S.-H. Kuo, and K.-T. Luh.** 1995. Direct determination of cryptococcal antigen in transthoracic needle aspirate for diagnosis of pulmonary cryptococcosis. *J. Clin. Microbiol.* **33:**1588–1591.

76. **Littman, M. L.** 1958. Capsule synthesis by *Cryptococcus neoformans. Trans. N.Y. Acad. Sci.* **20:**623–648.

77. **Littman, M. L.** 1959. Cryptococcosis (torulosis). *Am. J. Med.* **27:**976–998.

78. **Louria, D. B., T. Kaminski, M. Grieco, and J. Singer.** 1969. Aberrant forms of bacteria and fungi found in blood or cerebrospinal fluid. *Arch. Intern. Med.* **124:**39–48.

79. **MacKinnon, S., J. G. Kane, and R. H. Parker.** 1978. False positive cryptococcal antigen test and cervical prevertebral abscess. *JAMA* **240:**1982–1983.

80. **Maesaki, S., S. Kohno, H. Mashinoto, J. Araki, S. Asai, and K. Hara.** 1995. Detection

of *Cryptococcus neoformans* in bronchial lavage cytology: report of four cases. *Intern. Med.* **34**:54–57.

80a. **Malik, R., R. McPetrie, D. I. Wigney, A. J. Craig, and D. N. Love.** 1996. A latex cryptococcal antigen agglutination test for diagnosis and monitoring of therapy for cryptococcosis. *Aust. Vet. J.* **74**:358–364.

81. **Manfredi, R., A. Moroni, A. Mazzoni, A. Nanetti, M. Donati, A. Mastroiannis, O. V. Coronado, and F. Chiodo.** 1996. Isolated detection of Cryptococcal polysaccloride antigen in cerebrospinal fluid samples from patients with AIDS. *Clin. Infect. Dis.* **23**:849–850.

82. **McManus, E. J., M. J. Bozdeck, and J. M. Jones.** 1985. Role of the latex agglutination test for cryptococcal antigen in diagnosing disseminated infections with *Trichosporon beigelii. J. Infect. Dis.* **151**:1167–1169.

83. **McManus, E. J., and J. M. Jones.** 1985. Detection of a *Trichosporon beigelii* antigen cross-reactive with *Cryptococcus neoformans* capsular polysaccharide in serum from a patient with disseminated *Trichosporon* infection. *J. Clin. Microbiol.* **21**:681–685.

84. **Mess, T., and E. S. Daar.** 1997. Utility of fungal blood cultures for patients with AIDS. *Clin. Infect. Dis.* **25**:1350–1353.

85. **Milchgrub, S., E. Visconti, and J. Avellini.** 1990. Granulomatous prostatitis induced by capsule-deficient cryptococcal infection. *J. Urol.* **143**:365–366.

86. **Millon, L., T. Barale, M. C. Julliot, J. Martinez, and G. Manton.** 1995. Interference by hydroxyl starch used for vascular filling in latex agglutination test for cryptococcal antigen. *J. Clin. Microbiol.* **33**:1917–1919.

87. **Mitchell, T. G., E. Z. Freedman, T. J. White, and J. W. Taylor.** 1994. Unique oligonucleotide primers in PCR for identification of *Cryptococcus neoformans. J. Clin. Microbiol.* **32**:253–255.

88. **Moore, T. D., and J. C. Edman.** 1993. The alpha-mating type locus of *Cryptococcus neoformans* contains a peptide pheromone gene. *Mol. Cell. Biol.* **13**:1962–1970.

89. **Negroni, R., C. Cendoya, A. I. Arechravala, A. M. Robles, M. Bianchi, A. J. Bava, and S. Helou.** 1995. Detection of *Cryptococcus neoformans* capsular polysaccharide antigen in asymptomatic HIV-infected patients. *Rev. Inst. Med. Trop. Sao Paulo* **37**:385–389.

90. **Nelson, M. R., M. Bower, D. Smith, C. Reed, D. Shanson, and B. Gazzard.** 1990. The value of serum cryptococcal antigen in the diagnosis of cryptococcal infection in patients infected with the human immunodeficiency virus. *J. Infect. Dis.* **21**:175–181.

91. **Nishikawa, M. M., O. D. Sant'Anna, M. S. Lazera, and B. Wanke.** 1996. Use of D-proline assimilation and CGB medium for screening Brazilian *Cryptococcus neoformans* isolates. *J. Med. Vet. Mycol.* **34**:365–366.

92. **Odds, F. C., T. DeBacker, G. Dams, X. Vranck, and F. Woestenboroghs.** 1995. Oxygen as a limiting nutrient for growth of *Cryptococcal neoformans. J. Clin. Microbiol.* **33**:995–997.

93. **Perfect, J. R., D. T. Durack, and H. A. Gallis.** 1983. Cryptococcemia. *Medicine* **62**:98–109.

94. **Polacheck, I., and K. J. Kwon-Chung.** 1986. Canavanine resistance in *Cryptococcus neoformans. Antimicrob. Agents Chemother.* **29**:468–473.

95. **Powderly, W. G., G. A. Cloud, W. E. Dismukes, and M. S. Saag.** 1994. Measurement of cryptococcal antigen in serum and cerebrospinal fluid: value in the management of AIDS-associated cryptococcal meningitis. *Clin. Infect. Dis.* **18**:789–792.

96. **Prevost-Smith, E., and N. Hutton.** 1994. Improved detection of *Cryptococcus neoformans* in the BACTEC NR 660 blood culture system. *Microbiol. Infect. Dis.* **102**:741–745.

97. **Ro, J. Y., S. S. Lee, and A. G. Ayala.** 1987. Advantage of Fontana-Masson stain in capsule-deficient cryptococcal infection. *Arch. Pathol. Lab. Med.* **111**:53–57.

98. **Roberts, G. D., and J. A. Washington II.** 1975. Detection of fungi in blood cultures. *J. Clin. Microbiol.* **1**:309–310.

99. **Robinson, P. G., M. J. Sulita, E. K. Matthews, and J. R. Warren.** 1987. Failure of the BACTEC 460 radiometer to detect *Cryptococcus neoformans* in an AIDS patient. *Am. J. Clin. Pathol.* **87:**783–786.

100. **Sekhon, A. S., A. K. Garg, L. Kaufman, G. S. Kobayashi, Z. Hamir, M. Jalbert, and N. Moledina.** 1993. Evaluation of a commercial enzyme immunoassay for the detection of a cryptococcal antigen. *Mycoses* **36:**31–34.

101. **Shaunak, S., W. A. Schell, and J. R. Perfect.** 1989. Cryptococcal meningitis with normal cerebrospinal fluid. *J. Infect. Dis.* **160:**912. (Letter.)

103. **Snow, R. M., and W. E. Dismukes.** 1975. Cryptococcal meningitis: diagnostic value of cryptococcal antigen in cerebrospinal fluid. *Arch. Intern. Med.* **135:**1155–1157.

104. **Speed, B. R., J. Kaldor, B. Cairns, and M. Pegorer.** 1997. Serum antibody response to active infection with *Cryptococcus neoformans* and its varieties in immunocompetent subjects. *J. Med. Vet. Mycol.* **34:**187–193.

105. **Staib, F., M. Seibold, E. Antweiler, and B. Frohlich.** 1989. Staib agar supplemented with a triple antibiotic combination for the detection of *Cryptococcus neoformans* in clinical specimens. *Mycoses* **32:**448–454.

106. **Staib, F., M. Seibold, E. Antweiler, B. Frohlich, S. Weber, and A. Blisse.** 1987. The brown color effect (BCE) of *Cryptococcus neoformans* in the diagnosis, control and epidemiology of *C. neoformans* infections in AIDS patients. *Zentralbl. Bakteriol. Hyg. A* **266:**167–177.

107. **Stamm, A. M., and S. S. Polt.** 1980. False-negative cryptococcal antigen test. *JAMA* **244:**1359.

107a. **St.-Germain, G., and D. Beauchesne.** 1991. Evaluation of the MicroScan Rapid Yeast Identification panel. *J. Clin. Microbiol.* **29:**2296–2299.

108. **Taelman, H., J. Bogaerts, J. Batungwanayo, P. Van de Perre, S. Lucas, and S. Allen.** 1994. Failure of the cryptococcal serum antigen test to detect primary pulmonary cryptococcosis in patients infected with human immunodeficiency virus. *Clin. Infect. Dis.* **18:**119–120.

109. **Tanner, D. C., M. P. Weinstein, B. Fedorciw, K. L. Joho, J. J. Thorpe, and L. B. Reller.** 1994. Comparison of commercial kits for detection of cryptococcal antigen. *J. Clin. Microbiol.* **32:**1680–1684.

110. **Vartivarian, S. E., E. J. Anaissie, R. E. Cowart, H. A. Sprigg, M. J. Tingler, and E. S. Jacobson.** 1993. Regulation of cryptococcal capsular polysaccharide by iron. *J. Infect. Dis.* **167:**186–190.

111. **Vartivarian, S. E., R. E. Cowart, E. J. Anaissie, T. Tashiro, and H. A. Sprigg.** 1995. Iron acquisition by *Cryptococcus neoformans*. *J. Med. Vet. Mycol.* **33:**151–156.

112. **Vilgalys, R., and M. Hester.** 1990. Rapid genetic identification and mapping of enzymatically amplified ribosomal DNA from several *Cryptococcus* species. *J. Bacteriol.* **172:**4238–4246.

113. **Vogel, R. A.** 1966. The indirect fluorescent antibody test for the detection of antibody in human cryptococcal disease. *J. Infect. Dis.* **116:**573–580.

114. **Walter, J. E., and R. D. Jones.** 1968. Serodiagnosis of clinical cryptococcosis. *Am. Rev. Respir. Dis.* **97:**275–282.

115. **Whittier, S., R. L. Hopfer, and P. H. Gilligan.** 1994. Elimination of false-positive serum reactivity in latex agglutination test for cryptococcal antigen in human immunodeficiency virus-infected population. *J. Clin. Microbiol.* **32:**2158–2161.

116. **Wickes, B. L., M. E. Mayorga, U. Edman, and J. C. Edman.** 1996. Dimorphism and haploid fruiting in *Cryptococcus neoformans* association with the alpha-mating type. *Proc. Natl. Acad. Sci. USA* **93:**7327–7331.

117. **Wu, T., and S. Y. Koo.** 1983. Comparison of three commercial cryptococcal latex kits for detection of cryptococcal antigen. *J. Clin. Microbiol.* **18:**1127–1130.

118. **Yagupsky, P., and M. A. Menegus.** 1990. Cumulative positivity rates of multiple blood

cultures for *Mycobacterium avium intracellulare* and *Cryptococcus neoformans* in patients with the acquired immunodeficiency syndrome. *Arch. Pathol. Lab. Med.* **114:**923–925.

119. **Yuen, C., A. Graziani, N. Pietroski, R. MacGregor, and M. Schuster.** 1994. Cryptococcal antigenemia in HIV-infected patients. *Clin. Infect. Dis.* **19:**579.

120. **Zerpa, R., L. Huicho, and A. Guillen.** 1996. Modified India ink preparation for *Cryptococcus neoformans* in cerebrospinal fluid specimens. *J. Clin. Microbiol.* **34:**2290–2291.

121. **Zimmer, B. L., and G. D. Roberts.** 1979. Rapid selective urease test for presumptive identification of *Cryptococcus neoformans*. *J. Clin. Microbiol.* **10:**380–381.

13 | Human Cryptococcosis

INTRODUCTION

In the first 100 years of clinical cases of *Cryptococcus neoformans*, the infection was described in most organs of the human and animal host. It produced either few inflammatory changes in tissues or a granulomatous inflammation as the tissue expanded in response to the yeast. Unlike histoplasma infection, *C. neoformans* rarely produced calcifications in tissue. This fact, which made lesions radiographically silent, combined with the lack of a clinically useful skin test, made the clinical pathogenesis and presentation of this fungal infection more difficult to define.

In 1976, Baker (15) made an important pathological discovery that has formed the basis for our present understanding of the clinical manifestations of cryptococcosis. In 1954–1955 Baker and Haugen (16) had described the histopathology of pulmonary lesions in 12 cases of cryptococcosis, but the lesions did not demonstrate the lymph node complex so typical of tuberculosis. Baker persisted in his pathological evaluations and in 1976 reported nine cases of a primary pulmonary lymph node complex with cryptococcosis (15). This lymph node complex was found in approximately 1% of autopsies or thoracotomies in patients with cryptococcosis.

In a normal host, the pulmonary lymph node complex of cryptococcosis is represented by a small pulmonary granulomatous cryptococcoma and/or a small granulomatous cryptococcal lymphadenitis. In contrast, in a severely immunocompromised host, the complex can become primarily or secondarily a diffuse cryptococcal pneumonia and lymphadenitis. In some patients who are either apparently immunocompetent or immunocompromised, the pathology is mixed or crosses over to the opposite histopathology. This complex is clinically diagnosed far less often in *C. neoformans* since it is generally smaller than a pulmonary complex in tuberculosis and does not readily calcify as in histoplasmosis, which makes it radiographically silent. However, this pathological discovery has formed the basis for our understanding of the clinical manifestations of cryptococcal disease.

The present hypothesis for clinical disease is the following scenario. The host is likely to be infected by the inhalation of propagules such as small, poorly encapsulated yeasts or 1- to 2-μm-sized basidiospores created from sexual crosses or haploid fruiting in the environment. The host develops a primary pulmonary lymph node complex. In many cases, the infection is asymptomatic and yeasts remain dormant within this complex either to die or to reactivate and cause a secondary infection with a later immunosuppressive event. This primary initial infection may also cause the host to have pulmonary symptoms if the host is immunosuppressed or receives a particularly large inoculum of yeast. Spreading outside the lung to other organs can potentially occur as a result of either a primary or secondary infection. In the stage of dissemination outside the pulmonary location, the brain becomes the organ with a unique and still unexplained propensity for becoming a body site for clinical disease. Therefore, most of our data describe either lung or central nervous system (CNS) manifestations of disease.

CLINICAL MANIFESTATIONS OF CRYPTOCOCCAL INFECTION

The clinical presentations of human cryptococcosis may be varied, and because of the variety of signs and symptoms, clinical diagnosis without laboratory studies can be difficult at times in both high- and low-risk patients. For instance, cryptococcosis ranges from the asymptomatic nodule in the lung that is surgically removed because of concern that it represents a malignancy to a severely immunocompromised patient who has widely disseminated disease and dies in septic shock with acute respiratory distress syndrome within a few days of the onset of symptoms despite treatment. Although there are few distinguishing clinical features of cryptococcosis, Table 1 lists the clinical manifestations by organ system and demonstrates the breadth of clinical signs attributed to *C. neoformans*. In Table 2 the relative differences in frequency of clinical symptoms and findings are compared in patients with or without human immunodeficiency virus (HIV) infection. There is significant overlap between patients with AIDS and other severely immunocompromised patients, such as recipients of organ transplants and patients treated with high doses of corticosteroids for other conditions. In severely immunocompromised hosts, the body sites of cryptococcal infection contain a high burden of yeasts and a concomitant paucity of host inflammatory cells. These features explain the common characteristics of cryptococcosis in AIDS and organ transplant patients: a high frequency of positive blood and urine cultures, frequent simultaneous extraneural disease with meningitis, heavily positive India ink examinations, high polysaccharide antigen titers with a sparse inflammatory cell response in the subarachnoid space or other host tissue, and common clinical relapses after initial antifungal treatment without suppressive therapy (66, 111, 189, 301, 342, 392). Table 3 lists the major risk groups for cryptococcosis and a ranking of their clinical importance. Even when AIDS patients are not included, the frequency of patients with identifiable risk factors and disseminated cryptococcosis still approaches 75 to 80%. Thus, although apparently normal individuals develop cryptococcosis, including disseminated infections such as meningitis, they are in a minority compared to those with defined risk factors.

Table 1. Clinical manifestations of cryptococcosis

Central nervous system	Abscesses	Endocarditis (native and
Acute, subacute, chronic	Cellulitis	prosthetic)
meningitis	Purpura	Mycotic aneurysm
Cryptococcomas of brain	Acneiform	Myocarditis
(abscess)	Draining sinuses	Pericarditis
Subdural effusion	Ulcers	Vascular foreign body
Spiral and granuloma	Bullae	
Dementia	Herpetiformis-like	**Gastrointestinal tract**
	Molluscum contagiosum-	Esophagitis
Lung	like	Biliary tract
Nodules (single or multi-	Concomitant tumor or	Duodenum and colon
ple)	infection	Hepatitis
Lobar infiltrates		Peritonitis
Interstitial infiltrates	**Eye**	Pancreatitis
Cavities	Extraocular muscle paresis	
Endobronchial masses	Keratitis	**Breast**
Tumor-like projections	Choroiditis	Mastitis
Allergy	Endophthalmitis	
Colonization	Optic nerve atrophy	**Lymph nodes**
Acute respiratory distress		Lymphadenopathy
syndrome	**Genitourinary tract**	
Concomitant opportunistic	Prostatitis	**Thyroid**
infection	Pyelonephritis	Thyroiditis
Bronchiolitis obliterans-	Genital lesions	Thyroid mass
organizing pneumonia		
Mediastinal masses	**Bone and joints**	**Adrenal gland**
Hilar adenopathy	Osteomyelitis (chronic)	Adrenal insufficiency
Pneumothorax	(single or multiple	Cushings disease
Pleural effusions/empyema	sites)	Adrenal mass
Miliary pattern	Arthritis (acute/chronic)	
		Head and neck
Skin	**Muscle**	Gingivitis
Papules	Myositis	Sinusitis
Tumorlike projections		Salivary gland involvement
Vesicles	**Heart**	Larynx
Plaques	Cryptococcemia	Neck mass

CLINICAL EPIDEMIOLOGY AND RISK GROUP CHARACTERISTICS

Sex

Cryptococcal infections prior to the AIDS epidemic were reported more commonly in men than in women (204, 208). This clinical observation could be related to certain occupational exposures or possibly to the influence of estrogenic hormones (85, 238). For instance, diethylstilbestrol can inhibit the growth of *C. neoformans* in vitro. Furthermore, male guinea pigs appear to be more susceptible to cryptococcal infection than are females (93). The sex differences have continued during the AIDS epidemic. The greater frequency of cryptococcal infections in men may simply represent more HIV infections in men in developed countries (221), or there could

Table 2 Differences in clinical symptoms and findings between AIDS and non-AIDS patients with cryptococcal meningitis[a]

Symptoms/findings	AIDS	Non-AIDS
Length of symptoms (>2 weeks)	++	+++
Headaches	++++	++++
Fever	+++	+++
Sensorium deficits	+	+
Meningismus	++	++
Vision changes	++	++
Positive India ink examination	++++	++
Abnormal CSF chemistries (glucose <40; protein >45)	++	+++
CSF antigen titer (>1:1,024)	++	+
CSF pleocytosis (<20/μl)	++++	+
CD4$^+$ counts (<100/μl)	++++	+
Serum antigen positive	+++	++
Cryptococcemia	+++	+
Extraneural site	++	+
Brain lesions[b]	++	+
Increased intracranial pressure	++	+
C. neoformans var. neoformans	++++	+++
Relapse[c]	++	+

[a]The following scale represents an approximate incidence of symptoms/findings: +, 0–25%; ++, 25–50%; +++, 50–75%; ++++, 75–100%.
[b]Reflects concomitant infectious diseases or tumors.
[c]Without antifungal suppression; with antifungal suppressive therapy, AIDS relapse is reduced to +.

Table 3 Risk groups for development of cryptococcosis

Risk group condition	Frequency (importance)[a]
HIV infection	++++
Corticosteroids (≥20 mg of prednisone)	++++
Transplants (solid organ)	+++
Diabetes	++
Chronic obstructive pulmonary disease or lung cancer[b]	++
Lymphoma[b]	++
Chronic leukemias[b]	++
Sarcoidosis	+
Cirrhosis	+
Connective tissue disease[b] (systemic lupus erythematosus, rheumatoid arthritis)	+
Pregnancy	+

[a]++++, must always consider in differential diagnosis of infection or fever; +++, a complicating infectious disease factor; ++, major risk groups without severe immunosuppression; +, more than a dozen cases reported.
[b]Steroids add to the risk.

be other factors such as localization of infection in the prostate. On the other hand, in our clinical experience with a cohort of immunosuppressed patients with cryptococcemia prior to the HIV epidemic, 7 of the 15 patients were female (274). Despite some apparent differences in the frequency of presentation for clinical infection, there have been no apparent differences between the sexes in the clinical manifestations or prognosis of this infection.

Pediatrics

An interesting and still unexplained observation is the relatively few cases of cryptococcosis reported in children despite HIV infection, corticosteroid use, and the occurrence of significant congenital immune deficiencies. There have now been a series of case reports of pediatric cases (72, 158), and in many respects the risk factors and clinical presentations are the same as those seen in adults. For instance, cryptococcosis has been described in children with sarcoidosis, systemic lupus erythematosus, solid organ transplantation, and Hodgkin's disease (8, 108, 163, 201, 324, 325, 376).

Cryptococcosis is observed during childhood malignancies but can be confused with the appearance of metastatic disease in radiographs (6). In children, cryptococcosis also tends to occur more commonly in acute leukemias, such as acute lymphocytic leukemia; almost a dozen cases have been reported (201, 241). In contrast, adult infections are primarily found in the chronic leukemias, such as chronic lymphocytic leukemia. Corticosteroids are suspected to be the common theme in these diseases when they are used in treatment of the underlying disease. In most respects, the clinical sites of infection and presentations in children and adults are also similar. However, as might be expected, congenital immune defects in both cell-mediated and humoral immunity have been uniquely recognized in pediatric cryptococcosis. These primary immunodeficiencies include chronic mucocutaneous candidiasis (134, 201, 377), Bruton's agammaglobulinemia (370), hyper-immunoglobulin M (IgM) immunodeficiency (167, 353), severe combined immunodeficiency (333), and hyperimmunoglobulinemia E syndrome (345).

Pediatric patients with HIV infection have also not escaped infection with *C. neoformans* (144, 202, 220, 283, 359). Gonzalez et al. (144) retrospectively reviewed 473 HIV-infected children and found 4 patients (0.85%) who developed cryptococcosis during an 8-year observation period. All patients had severe CD4+ lymphocytopenia, fever, localized pulmonary or disseminated infection, and onset of cryptococcosis in the second decade of life (144). In that study and in a large review of immunocompromised patients without AIDS (201), pediatric patients were generally treated similarly to adults, and there were no deaths directly attributable to *C. neoformans*. Most complications, including death, during treatment of cryptococcosis occurred secondary to other infections or to the patient's underlying disease. Similar observations regarding the natural history of treatment outcome have also been made in adults with cryptococcemia (274).

The second area that is unique to the pediatric population is neonatal cryptococcosis. Approximately half a dozen cases of neonatal disease have been reported (132, 233, 257). It is not certain how these neonates were infected, but no congenital cases of cryptococcosis have been reported despite over two dozen known cases of cryptococcosis during pregnancy in the mothers. However, both animals (278) and humans (240) have been shown to have the placenta infected with *C. neoformans*. In the human case, the placenta had focally abundant intervillous and perivillous cryptococcal yeast cells, but there was no evidence of chorioamnionitis or villitis. We suspect that these neonatal cases most likely represent transmission of infection during parturition and that the placenta is probably a barrier to spread of infection to the fetus, but because of the infrequent nature of the infection in pregnancy, the exact pathophysiology still remains uncertain. Finally, even in a premature infant, amphotericin B treatment has been successfully used in management of this infection (132). Except for neonatal cryptococcosis and disease associated with congenital immune defects, cryptococcosis in children mimics that in adults in most features.

Despite the unique risk groups in children, cryptococcosis still remains an uncommon infection in this age group. For instance, 0.85% of 473 HIV-infected children prospectively studied were found to have cryptococcosis (144), and in a population-based survey from 1980 to 1990, only 1 of 133 cryptococcosis cases occurred in a child younger than 16 years (335). Speed and Kaldor (336) support the hypothesis that the low incidence of cryptococcosis in children simply reflects their lower exposure to sources of *C. neoformans*. With the use of antibody serological tests, they found a seroprevalence of 4.1% in children, in contrast to 69% in an adult blood donor population from the same area (336). Despite a lower incidence of infection, in our opinion there is nothing except reduced time of exposure to this infectious agent that protects this large population group.

Pregnancy

More than two dozen cases of cryptococcosis during pregnancy have been described (2, 58, 63, 64, 81, 83, 88, 107, 116, 129, 173, 186, 240, 271, 280, 326, 340). Pregnancy is not a severe immunodepressive event, but there are alterations in immune function that might benefit the establishment or dissemination of a cryptococcal infection (279, 339). One of the most dramatic long-term cases of untreated cryptococcosis, described by Beeson (20), suggested that pregnancy may have had an effect on the patient's cryptococcosis. Beeson described an 18-year-old female patient who developed cryptococcal meningitis in 1935 before the discovery of specific antifungal therapy. Over the next 11 years she intermittently had symptoms but performed secretarial duties, married, and in 1946 had a normal pregnancy and child. In 1950, she had a second child, but neurological symptoms returned during the pregnancy and she never regained her health. She died 6 months later with extensive cryptococcal meningoencephalitis. There is some suggestion from this case report that the immunosuppression of the second pregnancy may have tipped the balance in favor of the yeast. On the other hand, it is likely that cryptococcosis is not as severe a fungal infection as coccidioidomycosis may be during pregnancy. There are also pregnant patients reported who have comorbid risk factors such as systemic lupus erythematosus (240) and HIV infection (186), but most cases occur in apparently normal hosts.

Ely et al. (107) presented five cases of pulmonary cryptococcosis during pregnancy and an excellent review of 24 cases of cryptococcosis in pregnancy. In their five cases, infection appeared to be limited to the lung, although in the literature most of the cases during pregnancy were meningitis. It is interesting that in this series, three of four pregnant patients with pulmonary disease tested positive for serum cryptococcal polysaccharide antigen. In our opinion, this finding suggests that they were probably at high risk for development of disseminated infection. In all cases studied so far, there has been no evidence that in utero infection occurred, despite cases of maternal disseminated cryptococcosis. There is evidence that the placenta can be infected in both animals (278) and humans (240, 283), but it probably acts as a barrier to in utero infection in most cases. On the other hand, neonatal infection has occurred (132), and horizontal transmission, possibly by aspiration of yeasts, may occur during childbirth. Seven of 29 mothers died of cryptococcosis, but in most cases the infant was well; only a few infant deaths were reported (107). Cryptococcosis is not limited to the period of gestation, since postpartum cryptococcemia has also been described (147). In pregnancy, CNS symptoms can range from overwhelming cryptococcal meningitis to a subacute complex of headaches, nausea, and neurological symptoms. In this group of patients, pulmonary disease may present with chest pain, cough, or fever or may simply be seen as an asymptomatic nodule on a chest radiograph.

The principles of management for cryptococcosis during pregnancy are similar to those for other mildly immunosuppressed patients, except for several features. First, asymptomatic lung disease, like some cases of pulmonary colonization with C. neoformans (103, 150, 371), has been managed with no treatment (107). This lack of treatment might be reasonable if both blood and cerebrospinal fluid (CSF) have been checked to ensure that there is no evidence of dissemination and the patient has no symptoms. Furthermore, the patient will need to be closely monitored both during and after delivery. Any patient with symptoms or evidence of invasive infection should be treated. Clinical experience suggests that amphotericin B is the drug of choice for cryptococcosis during pregnancy (88, 116, 129, 186, 199, 280). This drug is relatively safe and effective and can be found in amniotic fluid and fetal blood. At this point in our understanding of this infection, it is assumed that successful treatment of the mother will make the fetus or neonate unlikely to be at risk for invasive infection. Flucytosine has also been used successfully during pregnancy, with high levels found in cord blood and amniotic fluid (58, 63, 173, 271, 340). Both amphotericin B and flucytosine have been used successfully during pregnancy without development of any apparent fetal abnormalities, but there has been some concern about the use of triazoles during pregnancy. There have been reports of congenital malformations in bone and heart associated with the use of fluconazole during pregnancy (199, 290). The use of triazoles should probably be evaluated individually on a cost-benefit basis, taking into account the trimester of pregnancy in which they are to be administered. It should also be realized that these antifungal drugs can appear in milk secretions of nursing mothers, although no significant toxicity has ever been found in neonates from this mode of drug acquisition.

Solid Organ Transplantation

Organ transplantation remains one of the major risk factors for cryptococcosis in non-HIV-infected patients. Between 10 and 20% of cases of disseminated cryptococcosis in HIV-negative patients come from this risk group. Occasionally, a bone marrow transplant patient may present with cryptococcosis, but the overwhelming number of cases occur in solid organ transplants (265).

Renal transplants

Recipients of renal transplants were the first group of patients with severe, iatrogenic immunosuppression to develop invasive cryptococcosis. In the first two decades of renal transplantation, several series of disseminated cryptococcosis were reported (127, 315, 350). Gallis et al. (127) reviewed the early experience in a renal transplant program in the southeastern United States, an area with a significant number of cryptococcal cases in all at-risk patient populations (127). Of the first 171 renal transplants in that program, 22 patients (5.8%) developed cryptococcosis, a much higher rate than the 17 reported cases in 950 renal transplants (1.8%) reported in the late 1960s and early 1970s. From the western United States, Schroter et al. (315) then reported 10 cases of cryptococcosis in 650 renal transplant patients (1.5%). These findings suggest that certain geographic areas may have a higher incidence of subclinical cryptococcal infection that can be reactivated by the use of transplant immunosuppression. They also emphasize that cryptococcosis must be considered a definite complicating infection for renal transplantation.

The experience with renal transplants and cryptococcosis illustrates several clinical principles that extend to all solid organ transplants with this infection. First, many of these patients did not have successful outcomes because of other viral or bacterial infections. Second, the success rate of treatment was 40 to 50% (127, 315), and short courses of therapy for disseminated disease (≤4 weeks) were not sufficient for cure. It is likely that prolonged therapy of at least 3 to 6 months should be the initial goal of therapy. Third, transplant recipients should be carefully screened prior to transplantation, with chest radiographs and possibly serum cryptococcal antigen titers, for evidence of cryptococcosis. Fourth, patients with clinical evidence of infection at the time of transplantation will experience a rapid worsening of their infection after the transplantation and the beginning of immunosuppressive regimens (127). Rapid dissemination of *C. neoformans* can occur over several days when sudden high doses of corticosteroids are administered for rejection episodes (274). Fifth, it is possible to perform a solid organ transplant on a patient who previously had a successfully treated invasive cryptococcal infection (163). Sixth, in the treatment of cryptococcosis, immunosuppressive regimens should be scaled down with the goal of reducing the corticosteroid dosage to less than 20 mg/day; any renal transplant patient with uncontrolled cryptococcosis faces the prospect that the transplanted kidney might be lost. Since there is an alternative dialysis treatment in renal transplant patients with loss of graft, there should not be a clinical focus to save the kidney at all costs. On the other hand, disseminated cryptococcosis can be successfully managed without loss of graft function (374, 375). Seventh, with the maturity of immunosuppressive regimens for renal transplant, the incidence of renal transplantation and cryptococcosis has been reduced in the last 20 years in centers where it was previously common (272a). However, cryptococcosis still remains an occasional complicating infectious disease in renal transplant patients (256, 262). Unfortunately, there are no guidelines for identifying which transplant patients are at highest risk for cryptococcal disease. Finally, the three most common body sites for cryptococcal disease in the renal transplant group are brain, lung, and skin. All patients in this transplant group in whom skin and/or lung involvement with *C. neoformans* is detected should be checked for CNS involvement with a lumbar puncture.

Cryptococcosis in renal transplant patients generally occurs in the later post-transplant period (>3 months). Although some symptoms are acute, most patients have subacute symptoms such as fever and headaches for several weeks to months before diagnosis. As seen in AIDS patients, some cryptococcal infections in transplant patients will occur with severe limitation of inflammation at the sites of infection when immunosuppressive agents are administered in high doses. Since some cases occur months to years after transplantation, it is probable that some infections do represent primary rather than reactivation disease, but there are no guidelines for the restriction of transplant patients' environmental contacts.

Liver transplants

Cryptococcosis is a serious opportunistic infection in liver transplant recipients. Its incidence ranges from 0.3 to 1%, with a wide range of mortality (20 to 100%) in various reports (52, 268, 314, 360). There are geographic areas in which cryptococcosis has not been reported in liver transplants (360, 369). A new factor in liver transplants has been the use of tacrolimus and its possible effect on cryptococcosis. It is known that all the immunophilins (cyclosporine, tacrolimus, and sirolimus) have anticryptococcal activity in vitro (157), but the effect of these medications on human cryptococcosis in this group remains uncertain. A recent review by Singh et al. (329) of invasive cryptococcosis in liver transplant recipients receiving tacrolimus describes some important features. Invasive cryptococcosis was found to develop in 6 of 102 liver transplant patients (6%) (329). Interestingly, previous studies had found chronic or subacute meningitis as the most common presenting focus of cryptococcal infection. However, in this study, cutaneous and/or osteoarticular cryptococcosis was found in 67% of patients, whereas CNS cryptococcosis was diagnosed in only one patient. There was also some potential bias for infection to occur in patients living in the eastern part of the United States and in older patients. However, there were no other apparent risk factors for this infection, including severity of the underlying liver disease. In one liver transplant patient, cryptococcosis presented with refractory shock, disseminated intravascular coagulation, and multiorgan failure attributed solely to *C. neoformans*. This clinical manifestation is not a widely appreciated syndrome for this yeast, but *C. neoformans* can occasionally present with a syndrome of septic shock or acute respiratory distress syndrome. In fact, patients such as those with HIV infection will frequently have *C. neoformans* circulating in their blood with only minimal symptoms, such as fever. However, the septic shock syndrome, with or without the acute respiratory distress syndrome, can be an occasional presentation for cryptococcosis in severely immunosuppressed patients (e.g., organ transplant recipients, AIDS patients, cancer patients) with a higher tissue burden of yeasts.

Lung and heart-lung transplants

Cryptococcal infection is the third most common invasive fungal infection in solid organ transplants (156, 267), including heart transplants (42), but the infection has been notably absent during the first 4 to 6 months after organ transplantation. In one series of fungal infections in lung and lung-heart transplants from a single institution, there were three cases of cryptococcosis of the lung in a 3-year period (175).

Since cryptococcal infection is primarily acquired by inhalation, occurrence of an infection during the early posttransplantation period would suggest either an intense environmental exposure or the donor organs as a potential source of the fungus. Cases of cryptococcal pneumonia reported after lung transplantation have been considered to be related to inhalation of yeast (267). In fact, this entry mechanism might be common in this group. However, cryptococcus is known to form a silent lymph node complex in the lung (15, 16, 307), and it is likely that as more lung transplantations occur, more organisms such as mycobacteria, histoplasma, and cryptococcus will be transplanted into the recipient with the donor lung. At least one case has been reported in which the donor lung probably contained *C. neoformans* and the patient developed pulmonary cryptococcosis (175). *C. neoformans* was isolated from the transplanted lung of this patient 1 day after bilateral surgical transplantation of the lungs. This case is probably the second identified case of cryptococcosis acquired from some type of transplanted tissue. The first case was from a corneal transplant (26). Another concept in treatment of fungal infections acquired by transplantation was illustrated in the case of the donor lung infection. The patient, who had no symptoms, was treated with fluconazole for 5 months after isolation of *C. neoformans* from bronchoalveolar lavage fluid. Six months after therapy was discontinued, the patient relapsed, with pulmonary nodules containing an apparently azole-susceptible *C. neoformans* in vitro (MIC, 4 μg/ml). The patient required a second, longer course of azole antifungal agents. This case should impress upon the clinician the fact that in most transplant cases, therapies need to be extended for long periods of time. Although fluconazole has been successful in treatment of transplant patients with fungal infections (78), it is not particularly fungicidal in this patient population and thus requires prolonged therapy.

Infections with *C. neoformans* var. *neoformans* versus *C. neoformans* var. *gattii*

The possibility that certain cryptococcal strains can influence specific disease manifestations should not be entirely dismissed. The most developed understanding of the effect of strains on disease is probably in the observations of clinical cryptococcosis with *C. neoformans* var. *neoformans* and var. *gattii*. It is now known that certain environmental links to eucalyptus trees are associated with *C. neoformans* var. *gattii*, and this knowledge helps to distinguish the clinical epidemiology of the two varieties of infections (105, 337). However, Henderson et al. (154) proposed that there may be inherent differences in pathobiology between the two cryptococcal varieties. Although both varieties infect the CNS, var. *gattii* appears to invade the brain parenchyma more commonly. This hypothesis, along with the infrequent finding of var. *gattii* in AIDS patients, even in endemic areas (53, 299, 301, 337), stimulated two clinical groups to compare the clinical manifestations of 251 cases of cryptococcosis with both varieties. Both Mitchell et al. (236) and Speed and Dunt (335) came to similar conclusions. First, *C. neoformans* var. *gattii* generally occurs in apparently healthy hosts. In this respect, it is truly a primary fungal pathogen that rarely needs an identifiable immune defect for clinical disease to occur. Second, this variety frequently produces infections that have focal brain parenchymal lesions and are associated with CNS complications of hydrocephalus, cranial nerve palsies, and frequent seizures. Patients with *C. neoformans* var. *gattii* infection do appear to have their symptoms longer, and the radiographic appearance of subclinical brain involvement on computed tomography (CT) and nuclear magnetic resonance (NMR) scans is more frequently apparent. Third, many patients with CNS involvement with var. *gattii* will also have concomitant mass-like lesions in the lung. Focal pulmonary lesions are common in var. *gattii* infections, whereas more diffuse infiltrates are seen with var. *neoformans* infections. Fourth, despite a series of CNS or pulmonary complications such as hydrocephalus or mass lesions requiring surgical intervention, patients with *C. neoformans* var. *gattii* infection do appear to have a better survival rate than those with var. *neoformans* infection (121, 236, 335). Finally, there is some suggestion that var. *gattii* infection requires longer antifungal therapy and occasional surgical removal of cryptococcomas more frequently than does var. *neoformans* infection (335). However, even if there is a difference in the length of therapy, it is not apparent that the varieties differ in direct in vitro susceptibility to available antifungal agents (123). The longer duration of therapy for healthy hosts with var. *gattii* infections appears to be related to a high rate of complications, slow response, and persistent clinical abnormalities.

The most likely explanation for the apparent clinical differences in presentation between the two varieties of *C. neoformans* remains the host factors. It should be emphasized that the vast majority of patients with var. *gattii* infection of the CNS are apparently immunocompetent. The differences in clinical presentation will likely become less clear when the two varieties are compared specifically within a given high-risk or low-risk group. Table 4 summarizes some of the potential differences found in clinical manifestations between the two varieties.

Dromer et al. (101) have examined the potential clinical influence of serotypes within a variety. In a review of 452 cases of cryptococcosis diagnosed in France, approximately 21% of the cases were serotype D, and the risk of infection with serotype D strains was significantly higher for patients with skin lesions, those receiving corticosteroids, and those living in certain locations in France. In serotype D strains infecting AIDS patients, the risk was higher for those over 60 years of age, those with intravenous drug use, and those with skin lesions. Dromer et al. (101) suggested that possible individual and environmental factors may be associated with the particular serotype of the infecting strain even within the same variety.

Table 4 Differences in clinical manifestations of *C. neoformans* var. *neoformans* versus *C. neoformans* var. *gattii*[a]

Manifestation	var. *neoformans*	var. *gattii*
Immunosuppressed host	+ + + +	+
Lung involvement	+ +	+ + +
Cryptococcemia	+ +	+
Focal CNS involvement	+	+ +
Neurological sequelae	+	+ +
Mortality	+ +	+

[a]Scale represents percentage of patients with these manifestations: +, 0–25%; + +, 25–50%;+ + +, 50–75%; + + + +, 75–100%.

Therefore, there may be undefined differences in the genotype and phenotype for virulence between the varieties. For instance, it is possible that there is a more brain-invasive ability within strains of var. *gattii*. However, at this stage of our understanding, the CNS and pulmonary manifestations of human disease are more likely to be dominated by the host status than by the specific yeast strain's serotype or varietal capabilities.

Corticosteroid-Mediated Risk Groups

There is some small predisposition for cryptococcosis in several disease categories, but among the entire infectious disease complications of these clinical groups, cryptococcosis is relatively insignificant. Two of the categories in which patients occasionally present with cryptococcal infection are sarcoidosis and diabetes. These two risk groups are not necessarily related to corticosteroid use. On the other hand, the risk for cryptococcosis in patients with chronic obstructive pulmonary disease, connective tissue diseases such as systemic lupus erythematosus and rheumatoid arthritis, and lymphoma/leukemias is frequently increased because of their concomitant steroid use (74, 183, 272). For instance, among 249 deaths related to infection among patients with systemic lupus erythematosus reported by Dubois (102), two cases of cryptococcosis were found. Corticosteroids, if given for long periods or with high daily doses, can contribute to the development of cryptococcosis in connective tissue disease or even in a vasculitis such as temporal arteritis or other conditions requiring corticosteroids (23, 142, 153, 366).

Corticosteroids can also affect cirrhosis and the appearance of cryptococcosis when they are used to treat chronic active hepatitis (274). However, end-stage cirrhosis in itself can be a risk factor for invasive cryptococcosis. As cases of cirrhosis with cryptococcal meningitis and peritonitis are reviewed, it is impressive how poor the prognosis is for these infections in this comorbid process. Similar to cancers such as lymphoma, Hodgkin's disease, and leukemia that are associated with cryptococcosis and in which prognosis for long-term recovery is poor (75, 133, 390), the occurrence of disseminated cryptococcosis in a cirrhotic patient is probably a marker for very advanced liver disease, even without steroid use, and thus predicts a poor outcome. Since humans are a steroid-sensitive species (67), it takes only small daily doses of prednisone to cause some detectable immunosuppression. It is not certain what daily dose of prednisone places patients at increased risk for cryptococcosis, but it is probably between 10 and 20 mg/day; with higher daily doses there is probably even more risk for dissemination of infection. Besides HIV infection, corticosteroid use is the single biggest risk factor for cryptococcosis in humans. The importance of steroids for this infection has also been emphasized in animal models (93, 124, 275).

Miscellaneous Risk Factors

Several other risk factors for cryptococcosis need to be emphasized. When non-life-threatening cases of cryptococcosis in HIV-negative patients are examined, diabetes and chronic lung diseases will occur frequently as underlying diseases. Sarcoidosis has been linked to patients with cryptococcal bone disease (21). High-risk patients include those with malignancies (176, 378), specifically those in the classical tumor risk groups such as Hodgkin's disease, lymphoma, and chronic leukemias (59, 75, 133, 229). Cryptococcosis frequently manifests itself during the terminal phases of these neoplasms (176, 378). It still remains one of the most common infectious causes of meningitis in patients with malignancies (59).

In patients with the diagnosis of cryptococcosis, it has been common clinical practice to obtain both CD4+ lymphocyte counts and HIV serologies. This strategy has allowed clinicians to detect patients with CD4+ lymphocytopenia without HIV infection but with cryptococcosis (192, 194, 386). The numbers of these CD4+-depleted patients remain small. Therefore, general recommendations for therapeutic management are not certain, but it seems reasonable to prolong their treatment when this risk factor is identified. However, it should be emphasized that most non-HIV-infected patients with cryptococcosis actually have normal CD4+ cell counts when they are measured at the time of their diagnosis of cryptococcosis.

Another concept regarding risk factors and cryptococcosis is that patients with cryptococcosis may also have simultaneous infection with another deep-seated mycosis. Although dual fungal infections are not common, *C. neoformans* infection has been described in conjunction with *Coccidioides immitis* (161), *Blastomyces dermatitidis* (245), *Pneumocystis carinii* (308, 385), *Histoplasma capsulatum* (254, 291), *Paracoccidioides braziliensis* (25), and *Sporothrix schenckii* (an unpublished case in our clinic with CD4+ lymphocytopenia). This observation requires that clinicians, pathologists, and microbiologists carefully examine all specimens from immunocompromised hosts with the possibility that more fungi than *C. neoformans* may be contained within a specimen.

SITES OF CRYPTOCOCCAL INFECTION AND THEIR MANIFESTATIONS

There are a series of excellent reviews on cryptococcosis, both in the AIDS era (1, 39, 66, 96, 100, 111, 112, 188, 189, 243, 248, 272, 287, 301, 305, 379, 392) and in the pre-AIDS era (104, 114, 205, 209, 237, 274, 332, 373). Each review can supply a perspective on the clinical manifestations and pathology of this infection. In over 100 years of clinical experience with this fungus, *C. neoformans* has been found to infect essentially every major organ of the body (205, 208, 237, 272). Major sites of infection include the brain, lungs, skin, eyes, and prostate. However, adrenals, bone, heart, liver, lymph nodes, joints, muscles, and kidneys have also been found to be infected by this yeast (Table 1). For a discussion of the variety of clinical presentations, infections will be discussed by body site and ranked in order of both clinical importance and frequency of their occurrence. However, it should be emphasized that more than one site can be infected simultaneously, with signs and symptoms in multiple organs. In fact, host immunity probably plays a more significant part in the rapidity of signs and symptoms and the patient's clinical presentations than does the particular site of infection.

CNS

Cryptococcal meningoencephalitis is the most frequently encountered life-threatening manifestation of cryptococcosis. Although it is frequently called cryptococcal meningitis and almost always has a subarachnoid space component to the infection, close pathological inspection of human cases confirms that the underlying brain parenchyma is generally involved to some extent, and therefore the term meningoencephalitis is appropriate.

Clinical manifestations

C. neoformans has a unique and still unexplained predisposition for establishing clinical infection in the CNS, and it is one of the most common causes of chronic/subacute meningoencephalitis worldwide. It should always be included in the differential diagnosis of all cases of subacute meningoencephalitis. However, the clinical presentation of cryptococcal meningoencephalitis may be quite variable. Most patients (70 to 90%) present with signs and symptoms of subacute meningoencephalitis such as headache, fever, lethargy, nausea/vomiting, personality changes, memory loss, stupor, and/or coma. Unlike bacterial meningitis, the majority of these symptoms occur over 2 to 4 weeks, and although fever with nuchal rigidity does occur, it is not necessarily a common finding. The time of onset of these symptoms is variable; patients may present with severe headaches for only a few days, intermittent headaches for months, or even no headaches despite meningeal involvement. Many of these symptoms, such as persistent headaches with fever, and physical signs such as papilledema or a cranial nerve paresis will likely lead to an investigation of the CNS, but occasionally the diagnosis will be more subtle, with subacute dementia as the only finding (282). In these cases the clinician must be alert to the possibility that cryptococcal infection is in the differential diagnosis of subacute dementia (344), since a lumbar puncture will be needed for diagnosis and cryptococcal infection is a potentially reversible cause of dementia. In some areas of the world, such as Africa, nuchal rigidity and cranial nerve findings are more commonly reported in cryptococcal meningoencephalitis. For instance, Moosa and Coovadia (243) found neurological signs in 50% of HIV-infected patients and in 29% of non-HIV-infected patients.

Although the majority of patients have a subacute course of symptoms over several weeks, patients with sudden large increases in their immunosuppressive therapies or with a severe underlying immunosuppressive illness such as AIDS may present with neurological symptoms of less than a week's duration. In contrast, it has also been well documented that prior to the use of antifungal agents for effective treatment, some reported cases of chronic cryptococcal meningoencephalitis lasted for 9 to 22 years (20, 48). Thus, *C. neoformans* is included as an etiologic agent for true chronic meningitis (106). In fact, a silent CNS infection is occasionally diagnosed after the isolation of *C. neoformans* from another body site or months after a ventricular shunt has been placed for idiopathic hydrocephalus produced by an undiagnosed cryptococcal infection (164).

The lumbar puncture findings in cryptococcal meningoencephalitis generally reflect the parameters found in a case of chronic meningitis. The inflammatory cells are predominantly mononuclear, and numbers generally range from 50 to 500 cells/μl of CSF. It should be noted that if a Coulter counter is used to determine host cell counts, it will count both yeast cells and host cells. A case of pseudopleocytosis has been described in which cryptococcal meningoencephalitis was initially reported as leukocytosis in the CSF as a result of a cell counter identifying yeast cells as inflammatory cells (24). AIDS patients with their profound CD4$^+$ lymphocytopenia routinely have a severely depressed host inflammatory response, and CSF host cell counts may actually fall below 20 cells/μl of CSF. CSF proteins, which can include oligoclonal anticryptococcal antibodies (195), may be elevated in approximately one-half of the patients, but it is unusual to find CSF protein concentrations of >500 to 1,000 mg/dl. If these high protein levels are found, the clinician should consider a subarachnoid block of CSF circulation. Cryptococcal meningoencephalitis can also be included in the differential diagnosis of chronic or subacute meningitis with hypoglycorrhachia. For instance, approximately one-fourth of patients will have CSF glucose levels of ≤40 mg/dl. In fact, a persistently low CSF glucose level at the end of therapy in non-AIDS patients has been associated with higher rates of relapse (94). Persistent hypoglycorrhachia should be viewed with concern. If therapy is stopped when there is profound hypoglycorrhachia, close follow-ups of these patients are needed to ensure that relapse does not occur over the next year.

The India ink examination remains a simple and rapid test for the diagnosis of cryptococcal meningitis. While it may be positive in only half of the non-AIDS patients with meningitis, in severely immunocompromised hosts such as those with HIV infection this test will be positive in over 80% of cases (189, 392). Furthermore, there are patients in whom the India ink examination will remain positive during and after a treatment course despite negative CSF cultures. Although a positive CSF India ink examination and a negative culture at the end of therapy do not necessarily translate into a treatment failure, these patients are at a higher risk for relapse of their infection. A routine India ink examination with careful study of the entire slide will generally detect $\geq 10^4$ CFU of yeasts per ml of CSF, but lower concentrations of yeasts (10^3 CFU/ml) will generally not be detected by this test. Therefore, it is obvious that the fungal culture will always be a more sensitive test for diagnosis than is the India ink examination.

Routine cultures of the CSF will identify at least 75 to 90% of cases of meningitis. However, there are difficult-to-diagnose cases of chronic meningitis in which the yield of obtaining a positive culture for *C. neoformans* is improved by withdrawal of large amounts of CSF (~20 ml) for culture. Since at times cryptococcal meningitis will present as a basilar meningitis with active inflammation but few yeasts, culture-negative cases by lumbar fluid analysis for cryptococcal infection may need to be diagnosed by a cisternal tap and withdrawal of fluid (143, 274). A significant improvement in the rapid diagnosis of cryptococcal meningitis is the serological test for polysaccharide antigen. This test is discussed further in chapter 12. It is over 95% sensitive and specific and must be used in the evaluation of all cases of subacute and chronic meningitis. Finally, it should also be noted that in high-risk patients, a diagnosis of cryptococcal meningitis is rarely made by CSF culture alone when the CSF profile is otherwise completely normal (320). In a study of AIDS patients in Africa with cryptococcal meningitis, 7 of 41 (17%) had normal CSF parameters (243). In these cases the diagnosis was probably made in the very earliest stages of infection in a host with poor inflammatory responses.

Besides its ability to produce meningoencephalitis, *C. neoformans* has also been found to cause other CNS events, including subdural effusions (200), cryptococcomas, and disease within the spinal canal (346).

Radiographs in CNS infections

CT and NMR scans have become standard diagnostic and management procedures for cryptococcal meningitis in health care systems where they are routinely available. The radiographic appearance of CNS cryptococcosis can vary among groups with different types of immunosuppression and/or different underlying diseases. CT findings among non-AIDS patients with meningitis can be normal or can show hydrocephalus, gyral enhancement, and/or multiple nodules that may be enhancing and/or nonenhancing (79). One study attempted to quantify the relative number of different CT scan presentations. That review found that scans were normal (50%) or revealed hydrocephalus (25%), gyral enhancement (15%), or focal nodules (15%) (356). The cryptococcomas (nodules) can be either single or multiple (311) and have been found in up to 25% of patients (212). Cryptococcoma lesions with septa can also be observed in the intraventricular spaces around the choroid plexus of the trigone. In the spinal cord and its roots, cryptococcus can also show invasion by CT/NMR scans (260, 270). On the other hand, in patients with AIDS, approximately half the patients with meningitis had a normal CT scan, but in contrast with non-AIDS patients, 34% had diffuse cortical atrophy and only 9% had hydrocephalus (285). These findings likely reflect the effects of the underlying retroviral infection on brain parenchyma and also the rapidity of infection with the host's poor inflammatory response. However, similar to non-AIDS patients, only 11% of AIDS patients had focal mass lesions in the brain parenchyma. These lesions ranged from single to multiple enhancing nodules, either solid or ring-like, to multiple nonenhancing, low-density lesions in the basal ganglia and thalamus, which may represent gelatinous pseudocysts. On an NMR scan, these hydrodense lesions may be shown to be infection rather than infarcts (130).

NMR scans appear to be even more sensitive than CT scans for detecting abnormalities in this infection. For example, when CT scans are found to be normal in cryptococcal CNS infection, NMR scans may identify significant structural abnormalities (9). The NMR scan may show multiple miliary enhancing parenchymal and leptomeningeal nodules with gadopentetate dimeglumine (358). Other NMR findings in cryptococcal meningoencephalitis include numerous clustered tiny foci that are hyperintense on T2-weighted images and nonenhancing on postcontrast T1-weighted images within the basal ganglia and mid-brain, representing Virchow-Robin spaces, which do not enhance with gadopentetate dimeglumine. Although dilated Virchow-Robin spaces can be observed among all age groups and the perivascular spaces do appear larger with age, the NMR findings of dilated Virchow-Robin spaces in a high-risk patient may suggest cryptococcal meningoencephalitis (377). These findings probably represent advanced infection with large numbers of yeasts packed into these spaces, thus partially blocking the outflow of CSF; in these cases a lumbar puncture would also readily make the diagnosis. NMR scans have identified involvement of most brain structures with *C. neoformans*. This yeast can prominently infect the dentate nuclei (303) and the choroid plexus (266) within the brain.

CNS radiographic examinations during cryptococcal meningoencephalitis have certain suggestive features for this infection, but none are pathognomonic. However, it is reasonable to obtain one of these radiographs when available, for the following reasons. First, in follow-up, both AIDS and (rarely) non-AIDS patients may have concomitant CNS infections or tumors. In AIDS patients, *Toxoplasma gondii* infection or CNS lymphoma may occur concomitantly with cryptococcal meningitis (13), and in non-AIDS patients we have seen cases of dual CNS infections with nocardia and dematiaceous fungi. Therefore, these radiographs may help to further document the possibility of a second etiologic agent for a CNS disturbance. Second, like other mycoses, cryptococcal meningoencephalitis may either present with an apparent idiopathic hydrocephalus (164) or develop it while receiving appropriate therapy. This obstructive condition will require corrective surgery and needs to be defined, and it is best identified by these radiographs. Third, these scans can ensure that mass lesions are not the cause of the clinical development of increased intracranial pressure in some cases of cryptococcal meningitis (91) and thus help clinicians in their management of elevated intracranial pressures. Finally, as more of these sensitive CNS radiographs are being used to follow patients with meningitis, it has been noted that these radiographs may change. For instance, among non-AIDS patients, a focal parenchymal lesion(s) may actually increase in size and/or number over the first few months of adequate therapy. This enlargement does not necessarily indicate a failure of treatment but may represent increased enhancement by inflammatory cells within these granulomas as the foci are being cleared. Generally, patients are clinically well and this fact will help in making therapeutic decisions. On the other hand, in an AIDS patient or other severely immunosuppressed patient, the change may be a sign of infection with a drug-resistant yeast or an untreated second pathogen or tumor. Thus, follow-up of clinical progress of treated infection with these radiographs will need to be carefully correlated with clinical responses.

AIDS versus non-AIDS cryptococcal meningitis

Overall, the signs and symptoms of meningoencephalitis in patients with and without AIDS are remarkably similar. On the other hand, there are some differences in the severity of immunosuppression and burden of organisms that make the incidence of certain clinical features slightly different. In Table 2 is a comparison of clinical symptoms and findings between AIDS and non-AIDS patients with an approximate quantification of their frequency of occurrence. It is important to determine HIV status in all patients in whom *C. neoformans* is isolated from sterile body sites and not simply to rely on clinical history or predictions of HIV status by the clinical findings. It is also essential to emphasize that successful treatment of the underlying diseases, including HIV infection, is so important to the final outcome of cryptococcal meningitis that basic screens for these risk factors are necessary.

Several differences in clinical features between AIDS and non-AIDS patients with cryptococcal meningitis should be emphasized to help clinicians in diagnosis and management. First, the duration of symptoms and signs may be shorter in AIDS patients than in non-AIDS patients (66, 111, 189, 392). In areas with a high incidence of *C. neoformans* infection, a screening serum antigen titer may be helpful in the evaluation of recent headaches and fever (92). In a non-AIDS patient, symptoms commonly occur for 2 to 4 weeks before diagnosis, but neurological symptoms of chronic meningitis for longer than 4 to 6 weeks can occur. Prolonged symptoms for over a month would actually make cryptococcus more likely than tuberculosis to be the cause of a CNS infection, since symptomatic tuberculous meningitis left untreated for over a month would likely progress to a fatal infection, whereas cryptococcal meningoencephalitis would not necessarily do so. Second, AIDS patients with cryptococcal meningitis are more likely to have a second (extraneural) site of infection, such as the lung, skin, prostate, or blood, either before or at the time of diagnosis for CNS cryptococcosis. Also, AIDS patients are more likely to have or will develop a second opportunistic infection or neoplasm on follow-up, although the incidence of these complications may have been reduced or attenuated with the new, highly active antiretroviral agents. Finally, focal neurological symptoms occur in only approximately 18% of patients on initial presentation with cryptococcal meningoencephalitis. In fact, most AIDS and non-AIDS patients do not have evidence of cryptococcomas on initial presentation. Therefore, CNS mass lesions should be carefully followed in AIDS patients because of their propensity for a second neurological event, and patients may require biopsy of parenchymal lesions if symptoms and signs persist despite therapy.

There are few data that can relate the severity of CNS disease to the infecting strain of *C. neoformans*. Clearly, in animals there are well-documented strain-dependent differences in virulence and even tropism for certain tissues, such as the skin and nose (see chapter 6). However, despite approximately two decades of the HIV pandemic, a specific or unique phenotype or genotype of *C. neoformans* associated with this infection has not yet been conclusively found, except for the observation that over 98% of AIDS patients are infected with var. *neoformans*. Also, var. *gattii* infections in AIDS patients seem to behave similarly to var. *neoformans* infections in this high-risk group (37, 76). The observation that AIDS patients may be infected with less well-encapsulated strains (38, 125) is not conclusive and will require more rigorous study.

Lung

Pulmonary cryptococcosis is the second most relevant site of infection for *C. neoformans*. Although epidemiological and travel histories for pulmonary cryptococcosis are less relevant than those for certain endemic mycoses, pigeon breeders may be at risk (28, 118, 293), and even a bagpipe player's lung infection was possibly related to yeast contamination of his instrument (70). These findings suggest that lung infection occurs through the inhalation of yeasts. It is likely that yeasts or basidiospores enter the lung, and under certain circumstances the infection can later disseminate to most other sites of the body. Between those two stages there are a variety of clinical presentations and manifestations within the lung. Although the categorization is somewhat arbitrary, with overlap between the groups, we will discuss pulmonary cryptococcosis in three patient groups, depending on whether the host is immunocompetent or is immunosuppressed with or without HIV infection.

Pulmonary cryptococcosis in the immunocompetent host
Hundreds of cases of primary cryptococcosis are described in the literature, and each adds to our understanding of the clinical manifestations of pulmonary cryptococcosis (95, 103, 150, 155, 159, 179, 231, 258, 263, 298, 300, 363, 365, 371, 388). However, one of the most comprehensive reviews of lung infection remains the description in 1966 by Campbell (47) of 101 patients with pulmonary disease caused by cryptococcus. In that review approximately 10% of the patients had a known underlying disease, such as malignancy, diabetes, tuberculosis, alveolar proteinosis, or pemphigus vulgaris. In fact, in this retrospective review it was found that at least 32% of the patients with evidence of pulmonary infection were

asymptomatic, and the infection was discovered only by an abnormal finding on a chest radiograph taken for other clinical reasons. For instance, a frequent clinical presentation in such patients is the solitary chest nodule, which is removed or biopsied because of a concern about a malignancy. Today, this clinical scenario is common. However, the majority of patients in Campbell's review did indeed have clinical symptoms. *C. neoformans* can cause pneumonia in normal hosts, and the extent of pneumonitis might be related to the size of the yeast inoculum. Symptoms included cough (54%), chest pain (46%), increased sputum production (32%), weight loss and fever (26%), and hemoptysis (18%). Other pulmonary symptoms included in the review were dyspnea and night sweats.

In case reports, there have been a plethora of interesting clinical manifestations and presentations of pulmonary cryptococcosis. For instance, there has been the rare clinical presentation of "allergic" cryptococcal pneumonia characterized by urticaria, hypotension, and dyspnea (135). Cryptococcal pneumonia can mimic a Pancoast's tumor (235) or cause superior vena cava obstruction (203, 228). Pulmonary cryptococcus can present as endobronchial lesions, and sometimes with this high burden of organisms even normal hosts appear to have progressive pulmonary disease despite azole therapy (109, 217). This complication should be considered and investigated with bronchoscopy in patients who initially respond poorly to therapy. Frequently, pulmonary cryptococcosis is found in underlying lung diseases such as chronic obstructive pulmonary disease, interstitial lung diseases, and pulmonary malignancies, but it has also been shown to occasionally produce its own unusual histopathological processes within the lung. For instance, pulmonary cryptococcosis has been reported to cause chronic eosinophilic pneumonia (343) and development of bronchiolitis obliterans-organizing pneumonia (50) and even to confuse the clinical diagnosis of sarcoidosis (321). It may also present with a complication such as cavitation and rupture of a pulmonary cryptococcoma (14), or mimic other pulmonary problems, such as drug toxicities (198).

Although coinfections with other pathogens are found more commonly in the severely immunocompromised host, pulmonary cryptococcosis in the apparently immunocompetent host can occasionally be associated with other infections. Thus, the clinician must always be wary of the patient who does not respond well to appropriate therapy. For example, both tuberculosis (80, 174, 298) and echinococcosis (86) have been discovered in patients with pulmonary cryptococcosis. Cavitary pulmonary cryptococcosis has also been complicated by the development of an aspergilloma (300).

Although the majority of reported cases of pulmonary disease suggest a bias toward clinical manifestations, it remains likely that the vast majority of human pulmonary infections are asymptomatic. These silent clinical presentations are similar to cases of asymptomatic tuberculosis or histoplasmosis, but they are less clinically observable. *C. neoformans* of the lung generally does not produce as much scar tissue or encapsulation as tuberculosis and does not add as much calcium to infected necrotic pulmonary foci as histoplasmosis; therefore, infections are radiographically silent. Furthermore, unlike the other two microorganisms, there is no commercially available skin test to demonstrate the magnitude of asymptomatically infected individuals. However, studies based on a laboratory standardized cryptococcal skin test in workers with direct contact in a research laboratory suggest that apparent asymptomatic infections with *C. neoformans* are common (12, 22).

Radiographically, cryptococcal pneumonia in the normal host may present with well-defined, noncalcified single or multiple lung nodules, indistinct to mass-like infiltrates, lobar infiltrates, hilar and mediastinal lymphadenopathy, pleural effusions, and more rarely cavitation (115, 159, 185, 226, 252, 263, 388, 391). In our experience, single or multiple peripheral nodules are probably the most common radiographic findings, and many of these infections are discovered when nodules are aspirated or removed to rule out a malignancy (185). Cryptococcal infection in the hilum or mediastinum (331) can produce endobronchial masses (213) and can even present with lung collapse (49). This clinical picture could mimic lung cancer. A series of chest radiographs of pulmonary cryptococcosis (color plates 20 and 21, following p. 456) demonstrate the breadth of radiographic manifestations of this infection. It is our opinion that the breadth of pulmonary manifestations and radiographic presentations makes it impossible to be certain of a cryptococcal etiology for a radiographic lesion without supportive histopathology and/or cultures.

Pleural effusions do occur during pulmonary cryptococcosis. They may occur with or without evidence of pulmonary disease and can occur in both immunocompetent and immunosuppressed patients, including those with HIV infection (89, 113, 306, 388). There are no unique features of the fluid, which is generally exudative, except that it will likely contain polysaccharide antigen, and a latex agglutination test on the fluid will probably be positive. In most cases of pleural effusions containing *C. neoformans*, the infection can be treated successfully with antifungal agents alone without surgical drainage, but there are the occasional complex cases with gross empyema features in which tube drainage is considered beneficial (247, 357). In fact, prior to the development of amphotericin B, surgical resection of large pulmonary lesions was a main therapeutic modality. Surgery is rarely needed today except to aid in diagnosis by excisional biopsy or in cases of large pseudotumorlike masses that are unresponsive to aggressive antifungal chemotherapy (150, 165, 338). However, since our understanding of the pathophysiology of *C. neoformans* in the lung suggests a hilar lymph node complex (15, 307), it is unlikely that surgery alone on a mass or nodule can completely eliminate the yeast from the lung (273). This concept should be taken into consideration when deciding whether to institute antifungal therapy after surgery. Present data are also unconvincing that surgery, by itself, on the lung of a patient with pulmonary cryptococcosis predisposes to dissemination of this infection (218).

Diagnosis of pulmonary infection in the immunocompetent host has generally been made by lung biopsy, either excisional or incisional, for histopathology and/or culture, cytopathology, sputum culture, or antigen testing in response to a chest X-ray abnormality. Several reports on cryptococcal disease of the lung have shown asymptomatic colonization of the respiratory airways by *C. neoformans*. This condition of culture positivity for the yeast in the sputum without symptoms or radiographic evidence of infection has commonly occurred in patients with an underlying lung disease such as chronic obstructive pulmonary disease (103, 150, 347, 365, 371).

In the normal host, *C. neoformans* in the lung generally does not disseminate; therefore, blood, urine, and CSF cultures are negative. Occasionally, the serum cryptococcal antigen titer is positive; this test should probably be used when *C. neoformans* is isolated from the lung (87, 169, 178, 363). A positive serum cryptococcal antigen titer would suggest either an extremely high burden of organisms within the lung or, more likely, dissemination of infection outside the lung (273). Since *C. neoformans* has such a unique predilection for the CNS, there is always a concern that when it is isolated from the lung, it may have already traveled to the CNS. In the early stages of CNS infection, the patient may have minimal symptoms. Therefore, a major management question in pulmonary cryptococcosis is whether patients with *C. neoformans* isolated from the lung should be examined for meningitis with a lumbar puncture. If the patient is symptomatic or possesses a high-risk factor for invasive infection, the answer is probably yes. However, the asymptomatic normal host is more problematic, and until there are further data, the decision for a lumbar puncture in this group should probably be made on an individual basis by the attending clinician, who will take into account the potential risk factors, follow-up potential of the patient, and treatment strategy to be used. Other sites of dissemination found in patients who were originally classified as having pulmonary cryptococcosis include blood, bone, skin, and eyes (28, 205, 272, 274, 338).

Management of pulmonary cryptococcosis is discussed in chapter 14. The prognosis for isolated cryptococcal pneumonia in the normal host is excellent. Morbidity has occurred predominantly in patients with extensive disease and after dissemination of infection in patients with a documented relapse.

Immunocompromised host without HIV infection

Immunocompromised patients with cryptococcal pneumonia generally have a more rapid clinical course and a greater tendency to disseminate their infection. Hence, immunosuppressed patients normally present with a focus on the meningitis syndrome rather than on a pulmonary syndrome. However, an overwhelming pneumonia with development of adult respiratory distress syndrome can occur on initial presentation with this yeast infection (155, 179, 253) and be the primary clinical focus. Acute respiratory distress syndrome has also been described in an apparently normal host (389). Immunocompromised hosts at high risk for both pulmonary disease and disseminated cryptococcosis include patients with HIV infection, cirrhosis, diabetes, Cushing's syndrome, sickle cell disease, sarcoidosis, leukemia, and lymphoma, as well as patients treated with glucocorticoids and patients receiving organ transplants (30, 46, 65, 119, 152, 180, 190, 380). A new clinical development has been the use of lung transplantation, and it might be anticipated that organs from patients asymptomatically carrying *C. neoformans* could now be unknowingly transplanted into a newly created immunocompromised host. This clinical situation has already occurred (175), and the frequency of pulmonary cryptococcosis is likely to increase in lung transplants from highly endemic areas such as the southeastern United States without the use of prophylactic antifungal agents.

The classical review of pulmonary infections in immunocompromised patients prior to the AIDS epidemic was by Kerkering et al. in 1981 (180). This review described 34 of 41 patients with pulmonary cryptococcosis who had an underlying immunocompromised condition other than HIV. Twenty-eight of these 34 patients (83%) developed disseminated disease, compared with 1 of 7 patients (14%) who were apparently normal. This observation emphasizes the risk of spread from the lung to other sites when the host has an immunocompromised condition. Also, unlike immunocompetent hosts, who generally have an inapparent pneumonitis, the 28 patients with disseminated disease had constitutional symptoms. The most common presenting symptoms were fever (63%), malaise (61%), chest pain (44%), weight loss (37%), dyspnea (27%), night sweats (24%), cough (17%), and hemoptysis and/or headache (7%). Subsequent to the diagnosis of pulmonary cryptococcosis in the entire series of patients, dissemination to the meninges occurred in 25 patients (61%) within 2 to 20 weeks after clinical presentation. All of these patients had abnormal chest radiographs initially. The most common radiographic findings (in order of frequency of occurrence) were alveolar or interstitial infiltrates, single- or multiple-coin lesions, masses, cavitary lesions, and pleural effusions. A clinical principle emphasized in this study is that immunocompromised patients with pulmonary cryptococcosis frequently have disseminated infection and always require antifungal therapy. Confusion may arise about the definition of the immunocompromised state, but with the risk of dissemination, investigations into infection dissemination and aggressive treatment should probably be liberally used in patients considered to be immunosuppressed.

Immunocompromised HIV-infected host

Pulmonary cryptococcosis is less common than meningitis as a clinical entity in patients with HIV infection, but it will still be detected in at least one-third of HIV-infected patients with cryptococcal infection. In fact, it can occur in up to 10% of all patients within a select group of AIDS patients (349). Pulmonary cryptococcosis in patients with HIV infection has a slightly different frequency of manifestations than it has in other types of immunocompromised hosts, but generally correlates more with severity of immunosuppression (46, 56, 69, 184, 372). Most patients have symptoms, including fever (81 to 94%), cough (63 to 71%), dyspnea (5 to 50%), weight loss (47%), and headache (41%) (46, 69, 230, 372). In some series, severe pleuritic chest pain (372) and hemoptysis (46) have been found to occur in cases of pulmonary infection. Cryptococcal pneumonia can also be associated with development of a pneumothorax (7, 362). Physical examination may find lymphadenopathy, chest rales, tachypnea, splenomegaly, and concurrent oral candidiasis (46). The range of hypoxia is wide, with a spread from none to severe with acute respiratory distress syndrome and its sustained hypoxia and difficulties with mechanical ventilation (46, 69).

Driver et al. (99) in a retrospective study proposed that early identification of cryptococcal infection while it is still in its pulmonary phase might improve the prognosis in HIV-infected patients. This group reviewed medical records of 18 cases of cryptococcal meningitis in patients with HIV infection. Fourteen of the 18 patients (78%) did have respiratory symptoms during the 4-month period before meningitis appeared. Unfortunately, the specificity of this finding is not great, since other causes of pulmonary symptoms may occur in this high-risk group, which generally includes patients with very low CD4+ counts and common respiratory symptoms. However, it does suggest that careful diagnostic work-ups for respiratory symptoms may diagnose cryptococcal infection early in its course when it is limited to the lungs. In fact, Miller et al. (232) have reported three cases of isolated pulmonary nodules due to *C. neoformans* in HIV-infected patients who were asymptomatic. These nodules were found during routine surveillance radiographs and occurred in patients with CD4+ counts between 100 and 200/μl. These cases were easily treated without sequelae and thus suggest that with surveillance and higher CD4$^+$ counts, some cases of pulmonary cryptococcosis can have clinical manifestations similar to those in normal or mildly immunosuppressed patients. On the other hand, many of the symptomatic and clinically aggressive cases of pulmonary cryptococcosis, with or without dissemination, occur in patients with CD4$^+$ counts less than 100/μl.

Dissemination of infection to the meninges occurs in 65 to 94% of documented cases of pulmonary cryptococcosis during HIV infection (46, 69, 230, 372), and it is likely that untreated pulmonary cryptococcosis will eventually spread to the CNS in almost all patients with an underlying HIV infection. Concomitant pulmonary infections with other opportunistic pathogens can also be an important feature of HIV and cryptococcus. Pulmonary cryptococcosis has been shown to occur with frequent coinfection pathogens such as *P. carinii, Mycobacterium avium* complex, *Mycobacterium tuberculosis,* cytomegalovirus, and *H. capsulatum* (46, 56, 69). In addition, *C. neoformans* pneumonia may occur as a consequence of steroid therapy for *P. carinii* pneumonia (196).

Chest radiographs most often reveal interstitial infiltrates that are either focal or diffuse and can even mimic *P. carinii* pneumonia (46, 69, 210, 234). Lymphadenopathy and nodular and alveolar infiltrates are actually less common in HIV-infected patients than in normal hosts, but cavitary cryptococcomas can occur and be completely cleared with antifungal agents (316). Also, large masses and pleural effusions are distinctly unusual (46). This may reflect the lack of a controlling pulmonary inflammatory response in those patients with very low CD4$^+$ counts (126). In severely immunosuppressed HIV-infected patients, cryptococcal infection can even present with a miliary pattern on chest radiographs (98). Although standard chest radiographs will identify most cases of pulmonary cryptococcosis, CT scans can occasionally detect nodules in the lung before they are detected by a chest radiograph (323).

In HIV-infected patients, with their propensity for frequent dissemination, blood and CSF specimens are high-yield culture sites. A serum cryptococcal antigen test is positive in more than 90% of patients with pulmonary cryptococcosis and HIV infection (230), and this test may be used to make a rapid presumptive diagnosis. Sputum for cultures, cytopathology, or antigen detection can also be helpful in diagnosis, since they are frequently positive (46, 56, 219). However, it is important to distinguish *C. neoformans* from strains of *Candida,* which frequently colonize or infect the oral cavity. In fact, cultures may be more specific than cytopathology in this group of patients. Of course, pulmonary infections may first be diagnosed by culture from another site, such as blood, skin, bone marrow, or urine with or without prostatic massage (111, 189, 381, 392).

Further details of management of cryptococcosis are discussed in chapter 14, but a general principle is that pulmonary cryptococcosis in AIDS patients with low CD4 counts should be treated as already-disseminated disease, and the potential for relapse will be high without suppressive antifungal therapy.

Skin

The skin is the third most common organ, after the brain and the lung, to be a site for cryptococcal infection. In animals, *C. neoformans* strains have been identified that appear to either localize or produce infection directly in skin tissue (334). It has been suggested that serotype D strains may have a greater propensity for infecting the skin (255), but *C. neoformans* var. *gattii* can also produce skin disease (149). Over the last century, and particularly in the last several decades with an enlarging immunocompromised population, the spectrum of cutaneous cryptococcosis has substantially widened. In fact, almost every type of skin lesion has been described to be produced by *C. neoformans*; this was first reported by Littman and Zimmerman in their 1956 monograph on cryptococcosis (208). For instance, skin lesions may include papules, tumorlike projections, vesicles, plaques, abscesses, cellulitis, purpura, acneiform lesions, draining sinuses, ulcers, bullae, and subcutaneous swellings (55, 208, 309, 317). There have been cryptococcal cases that mimicked other skin diseases, such as basal cell carcinoma (251), pyoderma gangrenosum (224), or even viral diseases such as varicella (223, 384). With the advent of so many immunocompromised patients in the last half of this century, there have been reports of cellulitis or vasculitis (322) on an extremity or at an indwelling catheter site caused by *C. neoformans*. These cases can have the clinical appearance of a bacterial cellulitis (138, 330, 348). We have even seen cases in which the host produces a cellulitis after a change or elimination of an immunosuppressive regimen and a return to dialysis, in which the yeast infection presents as cellulitis and yeast can be seen in the lesion but cannot be grown from it. Cutaneous lesions have also been commonly seen in cryptococcosis in liver transplant recipients receiving tacrolimus, but the cause for this propensity is unknown (329). Skin lesions can occur anywhere on the body, from the top of the head to the genitalia. This array of cutaneous presentations emphasizes the need to biopsy all new skin lesions for culture and histopathological study in high-risk, immunologically compromised patients.

During the AIDS epidemic, the variety and frequency of cutaneous manifestations of cryptococcosis have expanded. It has been found that at least 6% of patients with HIV infection and cryptococcosis can present with cutaneous cryptococcosis (250). In many respects, the cutaneous lesions of cryptococcosis during HIV infection are not unique but may simply represent more burden of organisms and less inflammatory responses. For example, HIV infection and cryptococcosis have been diagnosed by examining ulcerated plaques (215) and herpetiformis lesions (35). However, there do appear to be some unique features of HIV infection and cutaneous cryptococcosis. First, there have been multiple reports of cutaneous cryptococcosis presenting as molluscum contagiosum-like lesions (60, 77, 249, 269), and the clinician should be aware of this presentation. There has even been a report of a lesion containing both *C. neoformans* and molluscum contagiosum in the same lesion (348). Second, there have been several other reports of combined infections in the same lesion, such as Kaposi's sarcoma and *C. neoformans* (137). Third, in a small number of patients, either HIV infected or severely immunocompromised by other events, secondary cryptococcal skin lesions may occur simultaneously with cutaneous lesions due to another systemic mycosis or dermatophyte infection (281). Therefore, during the examination of high-risk patients, skin lesions with different characteristics should be individually biopsied and processed. Finally, in all HIV-infected patients and most other severely immunocompromised patients, a skin lesion with *C. neoformans* should be considered a sentinel sign of disseminated disease. Thus, clinical work-ups and management strategies should take this pathophysiological consideration into account.

Primary (direct inoculation) cutaneous cryptococcosis is likely to be an uncommon event, but it has been reported (10, 145, 259). For example, primary cutaneous cryptococcosis has definitely occurred during clinical and laboratory accidents. Direct inoculation of yeast in tissue of a healthy host produces a papule at the site of entry, with or without a local immune reaction such as regional lymphadenopathy (51, 136). In one case, *C. neoformans* was transmitted to a health care worker with a needlestick from an AIDS patient with cryptococcosis without transmission of HIV infection (136). In contrast to this probable low inoculum of yeasts from a clinical needlestick, a laboratory accident resulted in inoculation of large numbers of yeasts into the skin (51). Both accidents produced skin lesions during the week after injury. There are other cases of direct skin inoculation that are associated with trauma and direct implantation of this yeast in the skin (33, 368). However, unless there is evidence of direct traumatic introduction of the yeast in the skin, most cases of cutaneous cryptococcosis should be considered a sign of disseminated infection and thus a secondary infection site. There have also been reported cases of spontaneous clearing of cryptococcal skin lesions (54, 310), but in most clinical situations, whether in immunocompromised or immunocompetent hosts, it is probably reasonable to treat cases of isolated cutaneous cryptococcosis with systemic antifungal agents.

Eyes

In the early reports of disseminated cryptococcosis, ocular involvement was common. In one study prior to the AIDS epidemic, ocular signs and symptoms, including blindness and extraocular muscle paralysis, were found in approximately 45% of all patients with meningitis (261). During the AIDS era, the involvement of the eye with *C. neoformans* has become even more common and complex, with the distinct possibility of simultaneous infections of the eye with this fungus and HIV or cytomegalovirus (97, 182).

Choroiditis caused by *C. neoformans* has been described in half a dozen cases (128, 207, 227, 246). Signs and symptoms may include decreased visual acuity, presence of blind spots, and photophobia. Findings suggestive of cryptococcal choroiditis in high-risk patients include deterioration of vision, papilledema, an infiltrative process consistent with chorioretinitis, and neural atrophy. This infection can progress into cases of full endophthalmitis.

Cryptococcal endophthalmitis has occasionally been reported (90), and even a case of *Cryptococcus laurentii* endophthalmitis has been identified (84). There is a remarkable degree of heterogeneity in the fundal appearance with endophthalmitis. Observations in the retinal examination range from small, glistening preretinal spheres to flat, white lesions typical of retinal infarcts with hemorrhage, and even to larger, white, elevated lesions within the retina. Retinal detachment and loss of vision can occur with progression of infection, and only occasionally is infection successfully managed (90). Early diagnosis by aspiration of the vitreous body for culture or the identification of *C. neoformans* at another site and rapid treatment are likely to be essential for preserving the patient's vision. Although concomitant infection of the CNS is likely, in approximately one fourth of cases endophthalmitis may actually present before the diagnosis of meningoencephalitis is made (82). Cryptococcal endophthalmitis is generally associated with either disseminated

disease or local inoculation and has occurred equally in both immunocompromised and immunocompetent patients.

Reports have recently described catastrophic loss of vision in patients without
evidence of endophthalmitis (171, 295). There has been the discovery of fulminant
necrosis of both optic nerves (71). In a review of 82 patients from Papua, New
Guinea with *C. neoformans* var. *gattii* infection, loss of vision associated with optic
atrophy occurred in more than one-half of the patients and progressed despite
therapy in one-fifth of those patients (318). Fundoscopic examinations in the patients either were normal or revealed evidence of papilledema. Vision loss in
patients with cryptococcal meningitis has been reported to be caused by direct
invasion of the optic pathways by the yeast (29, 193, 261, 355), by the effects of
increased pressure and papilledema, and possibly by adhesive arachnoiditis (207).
Rex et al. (295) cogently reviewed cases of blindness in cryptococcal meningitis.
They suggested that two pathological processes may be taking place and that, in
fact, distinguishing between these processes may have therapeutic relevance (295).
First, one group of patients will have rapid loss of vision in a period as short as 12
to 24 h. This clinical syndrome of rapid blindness is suggestive of optic neuritis and
may result from infiltration of yeast cells into the optic nerve and the blood supply
to this nerve. Therapeutic measures have not generally been successful for this
form of vision loss during cryptococcal infection. On the other hand, a second
group of patients will present with slow vision loss, which typically begins in the
later stages of therapy for cryptococcal meningoencephalitis, and symptoms
gradually progress over weeks to months. In this group, the symptoms may be
primarily related to increasing intracranial pressure from the development of
hydrocephalus. This type of vision loss may be more amenable to therapy. A CNS
shunt or optic nerve fenestration (131) may halt progression of vision loss.

The eye has even been found to be a portal of entry for *C. neoformans* infection,
and the cornea has little protection against *C. neoformans* if its structural integrity is
breached. For example, a corneal transplant carried the yeast from a donor to a
recipient, who later developed cryptococcal meningitis (26). Cryptococcal keratitis
has been introduced following a keratoplasty procedure (277). There was a case of
ocular cryptococcal infection following trauma of the eye in a barnyard accident;
despite enucleation, *C. neoformans* was able to travel into the CNS and cause infection there 9 months later (27). These reports illustrate that the eye can be a direct
entry point for *C. neoformans* into the CNS. This likely requires both some type of
trauma to the eye to circumvent its natural structural defenses and possibly a foreign
body contaminated with a significant inoculum of *C. neoformans*. However, it should
be noted that most cases of fungal keratitis are caused by molds, and *C. neoformans*
infection of the anterior chamber of the eye is actually quite rare.

Genitourinary Tract

The prostate is known as a sanctuary for *C. neoformans* after dissemination, and this
fact is not dissimilar to findings with other systemic fungi such as *B. dermatitidis*. The
infection does not generally cause symptoms of prostatitis, but the yeast can occasionally be isolated from prostatic tissue and even from blood cultures after urologic
surgery (284). However, the prostate's importance as a site potentially protected
from antifungal drug therapy for the yeast has been emphasized during the AIDS

era. Infection persists despite good penetration of fluconazole into the prostate (120). In careful follow-up of patients with HIV infection, fungal cultures of routinely voided urine, urine from postprostatic massage, and seminal fluid were frequently positive at the end of therapy (197, 341). Urine cultures in AIDS patients with disseminated infection can be positive in 2 to 8% of patients, but the percentage is probably higher when culture plates are kept long enough to detect *C. neoformans* growth and when the 10^3 CFU cut-off for positivity is not used. It has been hypothesized that this site may be the organ that harbors relapse isolates of infection after treatment, although it still is likely that most relapse isolates come directly from a CNS sanctuary. Furthermore, with the use of long-term azole suppressive therapy in HIV-infected patients, checking this site for fungal relapse has become less clinically important. However, if discontinuation of therapy is ever considered, it will be important to check this site for the possibility of persistent infection.

In non-AIDS patients the frequency of positive urine cultures is low, and these cultures are rarely useful for the diagnosis of disseminated disease. However, cases of cryptococcal pyelonephritis and prostatitis have been described in non-AIDS patients (40, 292). Genital lesions caused by *C. neoformans* have been reported in both men and women. A penile ulcer was described that was possibly a relapse from a cryptococcal infection of the lip 17 years before (276). There have been several reported cases of vulvar cryptococcosis with or without HIV infection (31, 361). There has also been a case report of vaginal cryptococcosis (57) that completely cleared with the use of fluconazole therapy. Despite *C. neoformans'* prominent localization in the prostate during dissemination, and the fact that it can also be expressed in seminal fluid, there have been no convincing reports of conjugal spread of this fungus in humans.

Bone and Joints

Although bone involvement with *C. neoformans* is not the most common clinical condition associated with cryptococcosis, it has historical interest in that the first described case of cryptococcosis, by Busse and Buschke in 1894–1895, demonstrated that the yeast produced osteomyelitis of the tibia (44, 45). Bone has become an occasionally reported site of extraneural infection (11, 21, 146, 151, 191, 244, 264). It has also been suggested that prior to the AIDS epidemic up to 5% of cases of disseminated disease had clinical evidence of bone or joint involvement (21). This site of infection more commonly presents in an apparently normal host. In fact, in a review of 39 cases of skeletal cryptococcosis by Behrman et al. (21), only 38% of the patients had an apparent underlying disease. If there is one disease most commonly linked to cryptococcal osteomyelitis, it is sarcoidosis. In the comprehensive review of Behrman et al., 10 of 39 cases had sarcoidosis listed as a risk factor (21). On the other hand, in AIDS patients, a review suggested that there were remarkably few bone and joint infections (296). However, bone marrow involvement is not uncommon in AIDS patients (73, 382); in one series of cryptococcosis cases with AIDS, three of seven patients had positive bone marrow cultures (189). It is likely that the rapidity of infection in severely immunosuppressed patients allows other sites to show clinical manifestations before the bones or joints do.

The initial symptoms of bone infection generally include tenderness and swelling of soft tissue for weeks to months. The infection may present with or without fever. In three-fourths of patients there is a single skeletal site, and when multiple sites are observed, a large proportion of these are related to local extension from contiguous areas. *C. neoformans* can infect the skull, particularly the temporal bone (11), and the extremities from the hands (383) to the hips (36). A variety of bones have been involved, but the vertebrae are the single most likely site for bony involvement (21, 148). The vertebral involvement can mimic tuberculosis of the spine (139), and untreated cases can have serious complications (225). Thus, radiographic lesions of the spine must be biopsied, and *C. neoformans* can be included in the differential diagnosis. Finally, cryptococcal osteomyelitis is generally a result of hematogenous spread, but it can also be produced by a direct inoculation into bone (244).

Radiographic and pathological manifestations of bone involvement are not specific for *C. neoformans* (21). Patients' radiographs are uniformly found to have osteolytic or eroded lesions with infrequent periosteal reaction or osteoblastic activity. The radiographs will generally improve or resolve with appropriate therapy. Pathologically, gross examination of the bone reveals circumscribed areas of destruction that may be associated with abscesses containing mucoid, gelatinous pus. Histologically, there is acute and chronic inflammation, often with granulomas in the bone. Fungal stains or cultures will sometimes identify yeasts associated with these abscesses containing mucoid, gelatinous pus.

Diagnosis of osteomyelitis will generally require biopsy unless *C. neoformans* is isolated from another body site, and less than half of the patients will have a positive serum cryptococcal antigen test. Cryptococcal osteomyelitis should be considered a manifestation of disseminated disease and treated as a systemic infection. However, there have been cases cured with surgery alone (4, 61), and the removal of bony sequestrum, soft tissue infection, and surrounding affected bone may help decrease the infectious burden and in some cases speed up recovery. On the other hand, all patients should also receive at least a standard course of systemic antifungal therapy.

Cryptococcal arthritis is an uncommon manifestation of a hematogenously disseminated infection, with approximately a dozen cases reported. The onset of infection is generally subacute and not functionally incapacitating. This mild synovitis may reflect the fact that most of the reported cases occurred in patients who did not have a known immunosuppressive event (19). However, in severely immunosuppressed patients such as those with AIDS or solid transplants a joint can be rapidly seeded with *C. neoformans*, and presentation appears to be an acute monoarticular arthritis (274, 296, 330). The knee joint has been the most commonly reported location for cryptococcal joint infection. The pathogenesis scheme suggests that the initial hematogenous spread is into the parasynovial bone; the infection then develops into a mature osteomyelitis and invades the adjacent synovial space. Joint fluids may have a range of abnormal parameters, but in some patients, host cell counts can be above $15,000/\mu l$. There are no data to support the value of drainage of these joints, but it is clear that systemic antifungal therapy can be curative, and arthritis should generally be considered part of a disseminated infection.

Muscle

Involvement of the muscle with *C. neoformans* is not common, but there have been several case reports of patients with a renal transplant (160), chronic lymphocytic leukemia (170), or AIDS who have clinically apparent myositis (18). It can be found in single or multiple muscle groups with swelling and tenderness and may be identified by NMR scan and confirmed by tissue histopathology and culture. In most cases muscle involvement probably represents disseminated infection, although direct traumatic inoculation is also possible. Cryptococcal myositis responds to antifungal therapy.

Cryptococcemia and the Heart

Cryptococcemia was not a common finding prior to the AIDS epidemic; it was detected during disseminated disease in less than 10% of cases (94, 205, 274). Early reports with amphotericin B treatment suggested that this finding was a very poor prognostic sign, with most patients dying if they had positive blood cultures (94). As treatments improved and blood cultures were more sensitive in their detection of fungemia, the prognosis for cryptococcemia improved (274). It is likely that most patients with cryptococcemia have a high tissue burden of organisms in the lung, CNS, or other body sites. This hypothesis has been supported by the observation that between 10 and 70% of AIDS patients with disseminated cryptococcosis have positive blood cultures (66, 68, 111, 189). This wide range probably reflects the frequency of blood cultures drawn, but it is also clear that in these severely immunosuppressed patients, circulating cryptococci in blood can frequently be found. Most of these patients have no worse symptoms than fever or complications from other infected body sites. Today, cryptococcemia does not commonly appear to be a major poor prognostic sign, and unlike yeasts such as *Candida* spp., it does not frequently cause hypotensive/shock states. However, there are occasionally patients who develop septic shock with disseminated intravascular coagulopathy (117) , hypotension, and acidosis when the only organism in the blood and tissue is *C. neoformans*. Therefore, although cryptococcemia rarely causes vascular instability, it has been known to do so.

Cryptococcemia also does not imply cardiac valvular involvement, even in the presence of cardiac abnormalities. Cryptococcal endocarditis is a rarely reported infection with approximately half a dozen cases, but it has occurred on both native (76, 211) and prosthetic (17, 32) valves. In patients with cryptococcemia and signs of endocarditis, a transthoracic and transesophageal echocardiogram may be helpful for diagnosis, although there are too few cases to determine whether cryptococcal endocarditis is associated with large vegetations. *C. neoformans* has also been implicated in the formation of a mycotic aneurysm (297) and has even infected a prosthetic dialysis fistula, for which the cure was removal (41). Finally, in severely immunosuppressed individuals such as AIDS patients, both cryptococcal myocarditis (172, 206) and pericarditis (43) have been reported. It is possible that myocardial involvement could lead to arrhythmias and sudden death.

Gastrointestinal Tract

Although *Cryptococcus* species have been isolated from human stool (294), and in an animal model they can cause disseminated infection by oral inoculation (354), the gastrointestinal tract has not yet been proved to be a significant portal of entry into the human body. There is also no proof that *C. neoformans* causes any specific gastrointestinal disturbances, such as diarrhea. On the other hand, cryptococcal hepatitis has been described and this infection site has even presented as an acute abdominal emergency (289). Cryptococcal infection is also a significant complication of primary liver disease, such as chronic active hepatitis and cirrhosis, and it is possible that it may actually contribute to liver damage in some patients (140, 304). For instance, some patients with disseminated cryptococcosis and severe liver disease seem to deteriorate functionally and die despite treatment (216); in at least one child, cryptococcal infection was suggested to cause cirrhosis (140). In our opinion, the presence of severe liver disease such as cirrhosis can be considered a high risk for treatment failure in cryptococcosis.

Although cryptococcal infections of the alimentary tract are rarely reported, C. neoformans has been described to cause infection from the esophagus (168) to the rectum (5). However, the rectal infection was poorly documented. C. neoformans can infect the biliary tract (187) and can directly infect the bowel, including both the duodenum (62) and the colon (162). In fact, in one immunosuppressed patient, it had the appearance of Crohn's disease (162). Finally, C. neoformans has also been found to invade the pancreas, in a case with widespread disease (34).

This yeast infection can spread to the peritoneal surface of the abdominal cavity; there have been almost two dozen reported cases of cryptococcal peritonitis (274, 376, 387). It has been the cause of small bowel obstruction with its adhesions (222). It is also possible that yeasts enter the peritoneum either from the catheter in patients with a kidney dialysis catheter, through the bowel wall, or in most cases by hematogenous spread. Clinical experience with cryptococcal peritonitis allowed Yinnon et al. (387) to divide it into two presentation groups. Group 1 includes patients who develop infection during chronic ambulatory peritoneal dialysis. These patients have symptoms consistent with peritonitis, such as abdominal pain, fever, nausea, and/or cloudy fluid. Accordingly, the patients' catheters are removed, they receive antifungal agents, and in a very high percentage of patients the infection is cured. In contrast, in group 2 are patients with severe underlying illnesses such as cirrhosis; the diagnosis is generally made late in the illness, and the mortality rate is very high. It is likely that the underlying disease and delayed diagnosis contribute to the poor prognosis in this group. Finally, clinically relevant cryptococcal peritonitis in AIDS patients is reported in less than 2% of cases. This is not a major organ for infection in this risk group, but even in this group a patient with abdominal pain presented with an incarcerated inguinal hernia because of a cryptococcoma in the omentum (189).

Breast

Although *Candida albicans* is the most common cause of fungal mastitis and usually occurs after childbirth, cryptococcal mastitis is a well-reported type of cryptococcosis. Retrospectively, Symmers (351) reported a case of probable cryptococcal mastitis in a 50-year-old woman after examining histopathological slides from 1894 that were labeled "colloid cancer" (351). This apocrine gland has also been the site for major outbreaks of animal cryptococcosis and appears to be a host site favorable to growth of C. neoformans. Herds of dairy cows have become infected with cryptococcal mastitis (286, 327), and an experimental model of mastitis in goats has been developed (328). The implications of this infection are that milk production decreases and *C. neoformans* can be isolated from milk but will be killed by proper pasteurization. In humans there have been less than a dozen reported cases of cryptococcal mastitis (141). The lesion of cryptococcal mastitis generally presents with breast masses that may or may not be painful and may or may not involve the overlying skin. The diagnosis of cryptococcal mastitis can be made before or after the finding of invasive disease elsewhere in the body. In fact, in one review, six of nine patients with mastitis had additional sites of infection (141). Occasionally, there is concomitant axillary lymphadenopathy, and contralateral breast involvement has also been reported. Results of clinical examination and mammography may not distinguish cryptococcal mastitis from malignancy; thus, biopsy of tissue is necessary for definitive diagnosis. The facts that several patients had disseminated disease at the time cryptococcal mastitis was discovered, and that several patients developed recurrent infection after initial surgery alone, argue for systemic antifungal treatment in all cases of cryptococcal mastitis.

Lymph Nodes

The diagnosis of cryptococcosis is not commonly made through clinical involvement of lymph nodes. However, cryptococcosis can involve the lymph nodes, particularly in those patients in whom lymphadenopathy might be multifactorial, such as in HIV infection. Most commonly, cervical or supraclavicular nodes are involved. Fine-needle aspiration for cytological examination of enlarged lymph nodes is a rapid method for diagnosing cryptococcal lymphadenitis in high-risk groups because the yeasts can be easily detected. This percutaneous procedure may obviate the need for formal biopsy and histopathology (3, 239, 364).

Thyroid

C. neoformans has been found in the thyroid. It is not as common a cause of fungal thyroiditis as *Aspergillus* spp., but it generally occurs during widely disseminated disease. It has been found to present as thyroiditis (352) or simply as a thyroid mass (177), and most cases were diagnosed with fine-needle aspiration for cytology.

Adrenal Gland

Cryptococcal invasion of the adrenals has been shown to cause adrenal insufficiency. However, considering all the manifestations of disseminated disease, it is a rare complication (319). Therefore, routine testing of adrenal function is not recommended for cryptococcosis, but as in all disseminated mycoses, adrenal insufficiency should be considered if signs and symptoms develop in patients with cryptococcosis. On the other hand, Cushing's disease, with endogenous overproduction of corticosteroids, has been shown to allow development of cryptococcosis (190). These reports of endogenous corticosteroids and frequent use of exogenous steroids emphasize the major influence of corticosteroids on this infection. Involvement of the adrenal gland may be identified by an enlargement or mass in the adrenal gland, and diagnosis can be made by fine-needle aspiration for cytology (288).

Head and Neck

It is interesting that there are only infrequently reported cases of cryptococcosis of the oral cavity and neck, since the infectious propagule must pass through this area and it is a common presentation in dogs and cats. There have been cases of gingival ulceration (313), salivary gland involvement (242), and sinusitis (214) produced by *C. neoformans*, but cryptococcal infection is not as common as histoplasma infection for these oral mucosa surfaces and does not approach molds in frequency as a fungal cause of sinusitis. Probably the most common location for *C. neoformans* to infect in the head and neck, excluding the CNS, is the larynx (122, 166, 181). It may present with hoarseness and can occur in both immunocompromised and immunocompetent patients. Laryngeal cryptococcosis may be a limited primary infection, but it also can occur during disseminated disease, so other sites of infection, such as the CNS, should be checked. It is best diagnosed by biopsy or scrapings that contain the yeast. *C. neoformans* infection can also present as a neck mass (312). In fact, the yeast infection in the neck may not be coming directly from an oral or respiratory site; it may be spreading from a cervical osteomyelitis.

REFERENCES

1. **Aberg, J. A., and W. G. Powderly.** 1997. Cryptococcosis. *Adv. Pharmacol.* **37:**215–251.
2. **Aitken, G. W., and E. M. Symonds.** 1962. Cryptococcal meningitis in pregnancy treated with amphotericin B. *Br. J. Obstet. Gynaecol.* **69:**677–679.
3. **Alfonso, F., L. Gallo, B. Winkler, and M. J. Suhrhand.** 1994. Fine needle aspiration cytology of peripheral lymph node cryptococcosis: a report of three cases. *Acta Cytol.* **38:**459–462.
4. **Allcock, E. A.** 1961. Torulosis. *J. Bone Joint Surg.* **43:**71–76.
5. **Allen, V. K., and L. Lowbeer.** 1945. Rectal ulcers with perirectal fistula as a port of entrance for *Torula encephalitis. South. Med. J.* **38:**564–569.
6. **Allende, M., P. A. Pizzo, M. Horowitz, H. I. Pass, and T. J. Walsh.** 1993. Pulmonary cryptococcosis presenting as metastases in children with sarcomas. *Pediatr. Infect. Dis. J.* **12:**240–243.
7. **Alonso-Villaverde, C., S. Hernandez-Flix, R. Tomas, and L. Masana.** 1992. Occurrence of pneumothorax and pneumomediastinum in a patient with AIDS and pulmonary infection caused by *Pneumocystis carinii* and *Cryptococcus neoformans. Rev. Clin. Esp.* **191:**397–398.
8. **Al-Rasheed, S. A., and I. M. Al-Fawaz.** 1990. Cryptococcal meningitis in a child with systemic lupus erythematosus. *Ann. Trop. Paediatr.* **10:**323–326.
9. **Andreula, C. F., N. Burdi, and A. Carella.** 1993. CNS cryptococcosis in AIDS: spectrum of MR findings. *J. Comput. Assist. Tomogr.* **17:**438–441.
10. **Antony, S. A., and S. J. Antony.** 1995. Primary cutaneous cryptococcus in nonimmunocompromised patients. *Cutis* **56:**96–98.
11. **Armonda, R. A., J. M. Fleckenstein, B. Brandvold, and S. L. Dndra.** 1993. Cryptococcal skull infection: a case report with review of the literature. *J. Neurosurg.* **32:**1034–1036.
12. **Atkinson, A. J., and J. E. Bennett.** 1968. Experience with a new skin test antigen prepared from *Cryptococcus neoformans. Am. Rev. Respir. Dis.* **97:**637–643.
13. **Bahls, F., and S. M. Sumi.** 1986. Cryptococcal meningitis and cerebral toxoplasmosis in a patient with acquired immune deficiency syndrome. *J. Neurol. Neurosurg. Psychiatr.* **49:**328–330.
14. **Baird, R. W., and A. S. Garfield.** 1994. Cavitation in an immunocompetent man. *Am. J. Med.* **97:**309–311.

15. **Baker, R. D.** 1976. The primary pulmonary lymph node complex of cryptococcosis. *Am. J. Clin. Pathol.* **65:**83–92.

16. **Baker, R. D., and R. K. Haugen.** 1955. Tissue changes and tissue diagnosis of cryptococcosis: a study of 26 cases. *Am. J. Pathol.* **25:**14.

17. **Banerjee, U., K. Gupta, and P. Venugopal.** 1997. A case of prosthetic valve endocarditis caused by *Cryptococcus neoformans* var. *neoformans. J. Med. Vet. Mycol.* **35:**139–141.

18. **Barber, B. A., J. M. Crotty, R. G. Washburn, and P. S. Pegram.** 1995. Cryptococcus neoformans myositis in a patient with AIDS. *Clin. Infect. Dis.* **21:**1510–1511.

19. **Bayer, A. S., C. Choi, and D. B. Tillman.** 1980. Fungal arthritis. V. Cryptococcal and histoplasmal arthritis. *Semin. Arthritis Rheum.* **9:**218.

20. **Beeson, P. B.** 1952. Cryptococcic meningitis of nearly sixteen years' duration. *Arch. Intern. Med.* **89:**797–801.

21. **Behrman, R. E., J. R. Masci, and P. Nicholas.** 1990. Cryptococcal skeletal infections: case report and review. *Rev. Infect. Dis.* **12:**181–190.

22. **Bennett, J. E.** 1981. Cryptococcal skin test antigen: preparation variables and characterization. *Infect. Immun.* **32:**373–380.

23. **Bennington, J. L., S. L. Haber, and N. L. Morgenstern.** 1964. Increased susceptibility to cryptococcosis following steroid therapy. *Dis. Chest* **45:**262.

24. **Berger, S. A., D. A. Waitman, and R. Baruch.** 1992. Coulter counter identifies *Cryptococcus neoformans* as leukocytes. A case of pseudopleocytosis. *Am. J. Clin. Pathol.* **97:**663–664.

25. **Bernard, G., R. C. Gryschek, A. J. Duarte, and M. A. Shikanai-Yasuda.** 1996. Cryptococcosis as an opportunistic immunodeficiency secondary to paracoccidioidomycosis. *Mycopathologia* **133:**65–69.

26. **Beyt, B. E., and S. R. Waltman.** 1978. Cryptococcal endophthalmitis after corneal transplantation. *N. Engl. J. Med.* **298:**825–826.

27. **Birkmann, L. W., and D. R. Bennett.** 1988. Meningoencephalitis following evaluation for cryptococcal endophthalmitis. *Ann. Neurol.* **4:**476–477.

28. **Bisseru, B., A. Bajaj, R. H. Carruthers, and H. N. Chhabra.** 1983. Pulmonary and bilateral retinochoroidal cryptococcosis. *Br. J. Opthalmol.* **67:**157–161.

29. **Blackie, J. D., G. Danta, T. Sorrell, and P. Collignon.** 1985. Ophthalmological complications of cryptococcal meningitis. *Clin. Exp. Neurol.* **21:**263.

30. **Blanc, M., P. Leuenberger, and G. Favez.** 1982. Cryptococcose pulmonaire invasive isolé. Presentation d'un cas. *Schweiz. Med. Wochenschr.* **112:**421–424.

31. **Blocker, K. S., J. A. Weeks, and R. C. Noble.** 1987. Cutaneous cryptococcal infection presenting as valvular lesion. *Genitourin. Med.* **63:**341–343.

32. **Boden, W. E., A. Fisher, A. Medeiros, I. Benham, and M. T. McEnany.** 1983. Bioprosthetic endocarditis due to *Cryptococcus neoformans. J. Cardiovasc. Surg.* **24:**164–166.

33. **Bohne, T., A. Sander, A. Pfister-Wartha, and E. Schopf.** 1996. Primary cutaneous cryptococcosis following trauma of the right forearm. *Mycoses* **39:**457–459.

34. **Bonacini, M., J. Nussbaum, and C. Ahluwalia.** 1990. Gastrointestinal, hepatic and pancreatic involvement with *Cryptococcus neoformans* in AIDS. *J. Clin. Gastroenterol.* **12:**296–297.

35. **Borton, L.K., and B. U. Wintroub.** 1984. Disseminated cryptococcosis presenting as herpetiform lesions in a homosexual man with acquired immunodeficiency syndrome. *J. Am. Acad. Dermatol.* **10:**387–390.

36. **Bosch, X., R. Ramon, J. Font, X. Alemary, and A. Coca.** 1994. Bilateral cryptococcosis of the hip. A case report. *J. Bone Joint Surg.* **76:**1234–1238.

37. **Bottone, E. J., I. F. Salkin, N. J. Hurd, and G. P. Wormser.** 1987. Serogroup distribution of *Cryptococcus neoformans* in patients with AIDS. *J. Infect. Dis.* **156:**242.

38. **Bottone, E. J., M. Toma, B. E. Johansson, and G. P. Wormser.** 1985. Capsule-deficient *Cryptococcus neoformans* in AIDS patients. *Lancet* **ii:**553.

39. **Bozzette, S. A.** 1993. The management of cryptococcal disease in patients with AIDS. *Curr. Clin. Top. Infect. Dis.* **13:**250–268.

40. **Braman, R. T.** 1981. Cryptococcosis (torulopsis) of prostate. *Urology* **17:**284–286.

41. **Braun, D. K., D. A. Janssen, J. R. Marcus, and C. A. Kauffman.** 1994. Cryptococcal infection of a prosthetic dialysis fistula. *Am. J. Kidney Dis.* **24:**864–867.

42. **Britt, R. H., D. R. Enzmann, and J. S. Remington.** 1981. Intracranial infection in cardiac transplant recipients. *Ann. Neurol.* **9:**107.

43. **Brivet, F., J. Livartowski, P. Herve, B. Rain, and J. Dormont.** 1987. Pericardial cryptococcal disease in acquired immunodeficiency syndrome. *Am. J. Med.* **82:**1273.

44. **Buschke, A.** 1895. Ueber eine durch Coccidien hemorgerufene Krankheit des Menschen. *Dtsch. Med. Wochenschr.* **21:**14.

45. **Busse, O.** 1894. Ueber parasitäre Zelleinschulusse and ihre Zuchtung. *Int. J. Med. Microbiol.* 16:175–180.

46. **Cameron, M. L., J. A. Bartlett, H. A. Gallis, and H. A. Waskin.** 1991. Manifestations of pulmonary cryptococcosis in patients with acquired immunodeficiency syndrome. *Rev. Infect. Dis.* **13:**64–67.

47. **Campbell, G. D.** 1966. Primary pulmonary cryptococcosis. *Am. Rev. Respir. Dis.* **94:**236–243.

48. **Campbell, G. D., R. D. Currier, and J. F. Busey.** 1981. Survival in untreated cryptococcal meningitis. *Neurology* **31:**1154–1157.

49. **Carter, E. A., D. W. Henderson, J. McBride, and M. R. Sage.** 1992. Case report: complete lung collapse: an unusual presentation of cryptococcosis. *Clin. Radiol.* **46:**292–294.

50. **Cary, C. F., L. Mueller, C. L. Fotopoulos, and L. Dall.** 1991. Bronchitis obliterans-organizing pneumonia associated with *Cryptococcus neoformans* infection. *Rev. Infect. Dis.* **13:**1253–1254.

51. **Casadevall, A. J., J. Mukherjee, R. Ruong, and J. R. Perfect.** 1994. Management of *Cryptococcus neoformans* contaminated needle injuries. *Clin. Infect. Dis.* **19:**951–953.

52. **Castaldo, P.** 1991. Clinical spectrum of fungal infections after orthotopic liver transplantation. *Arch. Surg.* **126:**149.

53. **Castavon-Olivares, L. R., R. Lopez-Martinez, G. Barriga-Angulo, and C. Rios-Rosas.** 1997. *Cryptococcus neoformans* var. *gattii* in an AIDS patient: first observation in Mexico. *J. Med. Vet. Mycol.* **35:**57–59.

54. **Castillo, J. L. M., G. Del Negro, and E. Heins-Vaccari.** 1986. Primary cutaneous cryptococcosis. *Mycopathologia* **96:**25–28.

55. **Cawley, E. P., R. H. Grekin, and A. C. Curtis.** 1950. Torulosis, a review of the cutaneous and adjoining mucous membrane manifestations. *J. Invest. Dermatol.* **14:**327–341.

56. **Chechani, V., and S. L. Kamholz.** 1990. Pulmonary manifestations of disseminated cryptococcosis in patients with AIDS. *Chest* **98:**1060–1066.

57. **Chen, C. K., D. Y. Chang, S. C. Chang, E. F. Lee, S. C. Huang, and S. N. Chow.** 1993. Cryptococcal infection of the vagina. *Obstet. Gynecol.* **81:**867–869.

58. **Chen, C. P., and K. G. Wang.** 1996. Cryptococcal meningitis in pregnancy. *Am. J. Perinatol.* **13:**35–36.

59. **Chernik, N. L., D. Armstrong, and J. B. Posner.** 1973. Central nervous system infections in patients with cancer. *Medicine* **52:**563–581.

60. **Chiewchanvit, S., B. Chuaychoo, and P. Mahanupab.** 1994. Disseminated cryptococosis presenting as molluscum-like lesions in three male patients with acquired immunodeficiency syndrome. *J. Med. Assoc. Thai.* **77:**322–326.

61. **Chleboun, J., and S. Nade.** 1977. Skeletal cryptococcosis. *J. Bone Joint Surg.* **59:**509–514.

62. **Cholasani, N., C. M. Wilcox, H. T. Hunter, and D. A. Schwartz.** 1997. Endoscopic features of gastroduodenal cryptococcosis in AIDS. *Gastrointest. Endosc.* **45:**315–317.

63. **Chotmongkol, V., and S. Siricharoensang.** 1991. Cryptococcal meningitis in pregnancy: a case report. *J. Med. Assoc. Thai.* **74:**421–422.

64. **Chotmongkol, V., and A. Sookprasert.** 1992. Itraconazole in cryptococcal meningitis in pregnancy: a case report. *J. Med. Assoc. Thai.* **75:**606–608.

65. **Christoph, I.** 1990. Pulmonary *Cryptococcus neoformans* and disseminated *Nocardia brasiliensis* in an immunocompromised host. *N.C. Med. J.* **51:**219–220.

66. **Chuck, S. L., and M. A. Sande.** 1989. Infections with *Cryptococcus neoformans* in the acquired immunodeficiency syndrome. *N. Engl. J. Med.* **321:**794–799.

67. **Claman, H. N.** 1972. Corticosteroids and lymphoid cells. *N. Engl. J. Med.* **287:**388.

68. **Clark, R. A., D. Greer, W. Atkinson, G. T. Valainis, and N. Hyslop.** 1990. Spectrum of *Cryptococcus neoformans* infection in 68 patients infected with acquired immunodeficiency virus. *Rev. Infect. Dis.* **12:**768–777.

69. **Clark, R. A., D. L. Greer, G. T. Valainis, and N. E. Hyslop.** 1990. *Cryptococcus neoformans* pulmonary infection in HIV-1-infected patients. *J. Acquired Immune Defic. Syndr.* **3:**480–485.

70. **Cobcroft, R., H. Kronenberg, and I. Wilkinson.** 1978. Cryptococcus in bagpipes. *Lancet* **i:**1368–1369.

71. **Cohen, D. B., and B. J. Glasgow.** 1993. Bilateral optic nerve cryptococcosis in sudden blindness in patients with acquired immune deficiency syndrome. *Ophthalmology* **100:**1689–1694.

72. **Cohen, I.** 1985. Isolated pulmonary cryptococcosis in a young adolescent. *Pediatr. Infect. Dis. J.* **4:**416–419.

73. **Colavecchio, A., G. Cannatelli, F. Dubini, and L. Riviera.** 1993. Bone marrow cryptococcosis as the first manifestation of AIDS. *Pathologica* **85:**411–415.

74. **Collins, J. V., D. Tong, and R. G. Bucknel.** 1972. Cryptococcal meningitis as a complication of systemic lupus erythematosus treated with systemic corticosteroids. *Postgrad. Med. J.* **48:**52.

75. **Collins, P., A. Gelhorn, and J. R. Trimble.** 1951. The coincidence of cryptococcus and diseases of the reticuloendothelial and lymphatic systems. *Cancer* **4:**883.

76. **Colmers, R. A., W. Irniger, and D. H. Steinberg.** 1967. *Cryptococcus neoformans* endocarditis cured by amphotericin B. *JAMA* **199:**762.

77. **Concus, A. P., R. F. Helfand, M. J. Imber, S. A. Lerner, and R. J. Sharpe.** 1988. Cutaneous cryptococcosis mimicking *Molluscum contagiosum* in a patient with AIDS. *J. Infect. Dis.* **158:**897–898.

78. **Conti, D. J., N. E. Tolkoff-Rubin, and G. P. Baker, Jr.** 1989. Successful treatment of invasive fungal infections with fluconazole in organ transplant recipients. *Transplantation* **48:**692–695.

79. **Cornell, S. H., and C. G. Jacoby.** 1982. The varied computed tomographic appearance of intracranial cryptococcosis. *Radiology* **143:**703–707.

80. **Corpe, R. F., and L. H. Parr.** 1953. Pulmonary torulosis complicating pulmonary tuberculosis treated by resection. *J. Thorac. Cardiovasc. Surg.* **27:**392–398.

81. **Crotty, J. M.** 1965. Systemic mycotic infections in Northern Territory Aborigines. *Med. J. Aust.* **1:**184–186.

82. **Crump, J. R., S. G. Elner, V. M. Elner, and C. A. Kauffman.** 1992. Cryptococcal endophthalmitis: case report and review. *Clin. Infect. Dis.* **14:**1069–1073.

83. **Curole, D. N.** 1981. Cryptococcal meningitis in pregnancy. *J. Reprod. Med.* **26:**317–319.

84. **Curtis, P. H., J. A. Haller, and E. de Juan.** 1995. An unusual case of cryptococcal endophthalmitis. *Retina* **15:**300–304.

85. **Dabe, S. S., and R. A. Fromtling.** 1984. Effects of diethylstibestrol and cyclophos-

phamide on pathogenesis of experimental *Cryptococcus neoformans* infections. *J. Med. Vet. Mycol.* **22**:125.

86. **Dalgleish, A. G.** 1981. Concurrent hydatid disease and cryptococcosis in a 16-year-old girl. *Med. J. Aust.* **2**:144–145.

87. **Davies, S. F., and G. A. Sarosi.** 1987. Role of serodiagnostic tests and skin tests in the diagnosis of fungal disease. *Clin. Chest Med.* **8**:135–146.

88. **Dean, J. L., J. E. Wolfe, A. C. Ranzini, and M. A. Laughlin.** 1994. Use of amphotericin B during pregnancy: case report and review. *Clin. Infect. Dis.* **18**:364–368.

89. **de Lalla, F., A. Vaglia, M. Franzetti, V. Manfrin, G. P. Pellizer, and P. Fabris.** 1993. Cryptococcal pleural effusion as first indicator of AIDS: a case report. *Infection* **21**:192.

90. **Denning, D. W., R. W. Armstrong, M. Fishman, and D. A. Stevens.** 1991. Endophthalmitis in a patient with disseminated cryptococcosis and AIDS who was treated with itraconazole. *Rev. Infect. Dis.* **13**:1126–1130.

91. **Denning, D. W., R. W. Armstrong, B. H. Lewis, and D. A. Stevens.** 1991. Elevated cerebrospinal fluid pressures in patients with cryptococcal meningitis and acquired immunodeficiency syndrome. *Am. J. Med.* **91**:267–272.

92. **Desmet, P., K. D. Kayembe, and C. DeVroey.** 1989. The value of cryptococcal serum antigen screening among HIV positive AIDS patients in Kinshasa, Zaire. *AIDS* **3**:77–78.

93. **Diamond, R. D.** 1977. Effects of stimulation and suppression of cell-mediated immunity on experimental cryptococcosis. *Infect. Immun.* **17**:187–194.

94. **Diamond, R. D., and J. E. Bennett.** 1974. Prognostic factors in cryptococcal meningitis. a study of 111 cases. *Ann. Intern. Med.* **80**:176–181.

95. **Diamond, R. D., and S. M. Levitz.** 1988. *Cryptococcus neoformans* pneumonia, p. 457–471. *In* J. E. Pennington (ed.), *Respiratory Infections: Diagnosis and Management.* Raven Press, New York.

96. **Dismukes, W. E.** 1988. Cryptococcal meningitis in patients with AIDS. *J. Infect. Dis.* **157**:624–628.

97. **Doft, B. H., and V. T. Curtin.** 1982. Combined ocular infection with cytomegalovirus and cryptococcosis. *Arch. Ophthalmol.* **100**:1800–1803.

98. **Douketis, J. O., and S. Kesten.** 1993. Miliary pulmonary cryptococcosis in a patient with the acquired immunodeficiency syndrome. *Thorax* **48**:402–403.

99. **Driver, J. A., C. A. Saunders, B. Heinze-Lacey, and A. M. Sugar.** 1995. Cryptococcal pneumonia in AIDS: is cryptococcal meningitis preceded by clinically recognizable pneumonia? *J. Acquired Immune Defic. Syndr. Hum. Retrovirol.* **9**:168–171.

100. **Dromer, F., S. Mathoulin, B. Dupont, and A. Laporte.** 1996. Epidemiology of cryptococcus in France: a 9-year survey. *Clin. Infect. Dis.* **23**:82–90.

101. **Dromer, F., S. Mathoulin, B. Dupont, L. Letenneur, and O. Ronin.** 1996. Individual and environmental factors associated with infection due to *Cryptococcus neoformans* serotype D. *Clin. Infect. Dis.* **23**:91–96.

102. **Dubois, E. L.** 1976. *Lupus Erythematosus.* University of Southern California Press, Los Angeles.

103. **Duperval, R., P. E. Hermans, N. S. Brewer, and G. D. Roberts.** 1977. Cryptococcosis, with emphasis on the significance of isolation of *Cryptococcus neoformans* from the respiratory tract. *Chest* **72**:13–19.

104. **Edwards, V. E., J. M. Sutherland, and J. H. Tyrer.** 1970. Cryptococcosis of the central nervous system. *J. Neurol. Neurosurg. Psychiatr.* **33**:415–425.

105. **Ellis, D. H., and T. J. Pfeiffer.** 1990. Natural habitat of *Cryptococcus neoformans* var. *gattii. J. Clin. Microbiol.* **28**:1642–1644.

106. **Ellner, J. J., and J. E. Bennett.** 1976. Chronic meningitis. *Medicine* **55**:341–369.

107. **Ely, E. W., J. E. Peacock, E. F. Haponik, and R. G. Washburn.** 1998. Cryptococcal

pneumonia complicating pregnancy: case series and review of 24 published cases of cryptococcosis during pregnancy. *Medicine*, in press.

108. **Emmanuel, B., E. Ching, A. D. Lieberman, and M. Goldin.** 1961. Cryptococcus meningitis in a child successfully treated with amphotericin B with a review of the pediatric literature. *J. Pediatr.* **59:**577–591.

109. **Emmons, W. W., S. Luchsinger, and L. Miller.** 1995. Progressive pulmonary cryptococcosis in a patient who is immunocompetent. *South. Med. J.* **88:**657–660.

111. **Eng, R. H. K., E. Bishburg, S. M. Smith, and R. Kapila.** 1986. Cryptococcal infections in patients with acquired immune deficiency syndrome. *Am. J. Med.* **81:**19–23.

112. **Ennis, D. M., and M. S. Saag.** 1993. Cryptococcal meningitis in AIDS. *Hosp. Pract.* **28:**99–112.

113. **Epstein, R., R. Cole, and K. K. Hunt.** 1972. Pleural effusion secondary to pulmonary cryptococcosis. *Chest* **61:**296.

114. **Falconer, H., S. I. Terry, and H. Spencer.** 1980. Cryptococcosis in the West Indies. *West Indian Med. J.* **29:**142.

115. **Feigin, D. S.** 1983. Pulmonary cryptococcosis: radiologic-pathologic correlates of its three forms. *Am. J. Roentgenol.* **141:**1263–1272.

116. **Feldman, R.** 1959. Cryptococcosis (torulosis) of the central nervous system treated with amphotericin B during pregnancy. *South. Med. J.* **52:**1415–1417.

117. **Fera, G., N. Semeraro, V. De Mitrio, and O. Schiraldi.** 1993. Disseminated intravascular coagulation associated with disseminated cryptococcosis in a patient with acquired immunodeficiency syndrome. *Infection* **21:**171–173.

118. **Fink, J. N., J. J. Barboriak, and L. Kaufman.** 1968. Cryptococcal antibodies in pigeon Breeder's disease. *J. Allergy Clin. Immunol.* **41:**297–301.

119. **Finke, R., E. S. Strobel, T. Kroepelin, J. Guzman, P. J. Klein, H. E. Schaefer, R. Kappe, and J. Muller.** 1988. Disseminated cryptococcosis in a patient with malignant lymphoma. *Mycoses* **31:**102–108.

120. **Finley, R. W., J. D. Cleary, J. Goolsby, and S. W. Chapman.** 1995. Fluconazole penetration into the human prostate. *Antimicrob. Agents Chemother.* **39:**553–555.

121. **Fisher, D., J. Burrow, D. Lo, and B. Currie.** 1993. *Cryptococcus neoformans* in tropical northern Australia: predominantly variant gattii with good outcome. *Aust. N.Z. J. Med.* **23:**678–682.

122. **Frisch, M., and D. R. Gnepp.** 1995. Primary cryptococcal infection of the larynx: report of a case. *Otolaryngol. Head Neck Surg.* **113:**477–480.

123. **Fromtling, R. A., G. K. Abruzzo, and G. S. Bulmer.** 1986. *Cryptococcus neoformans*: comparisons of *in vitro* antifungal susceptibilities of serotypes AD and BC. *Mycopathologia* **94:**27–30.

124. **Gadebusch, H. H., and P. W. Gikas.** 1965. The effect of cortisone upon experimental pulmonary cryptococcosis. *Am. Rev. Respir. Dis.* **92:**64–74.

125. **Gal, A. A., S. Evans, and P. R. Meyer.** 1987. The clinical laboratory evaluation of cryptococcal infections in the Acquired Immunodeficiency Syndrome. *Diagn. Microbiol. Infect. Dis.* **7:**249–254.

126. **Gal, A. A., M. N. Koss, J. Hawkins, S. Evans, and H. Einstein.** 1986. The pathology of pulmonary cryptococcal infections in the acquired immunodeficiency syndrome. *Arch. Pathol. Lab. Med.* **110:**502–507.

127. **Gallis, H. A., R. A. Berman, and T. R. Cate.** 1975. Fungal infection following renal transplantation. *Arch. Intern. Med.* **135:**1163–1172.

128. **Gandhi, S. A., A A. McMeeking, D. Friedberg, and R. S. Holzman.** 1996. Cryptococcal choroiditis in a patient with AIDS: case report and review. *Clin. Infect. Dis.* **23:**1193–1194.

129. **Gantz, J. A., J. A. Nuetzel, and J. B. Keller.** 1958. Cryptococcal meningitis treated with amphotericin B. *Arch. Intern. Med.* **102:**795–800.

130. **Garcia, C. A., L. A. Weisberg, and W. S. J. LaCorte.** 1985. Cryptococcal intracerebral mass lesions: CT pathologic considerations. *Neurology* **35:**731–734.

131. **Garrity, J. A., D. C. Herman, R. Imes, P. Fries, C. F. Hughes, and R. J. Campbell.** 1993. Optic nerve sheath decompression for visual loss in patients with acquired immunodeficiency syndrome and cryptococcal meningitis with papilledema. *Am. J. Ophthalmol.* **116:**472–478.

132. **Gavai, M., S. Gaur, and L. D. Frenkel.** 1995. Successful treatment of cryptococcosis in a premature neonate. *Pediatr. Infect. Dis. J.* **14:**1009–1010.

133. **Gendel, B. R., M. Ende, and S. L. Norman.** 1950. Cryptococcosis: a review with special reference to apparent association with Hodgkin's disease. *Am. J. Med.* **9:**343.

134. **Gerbeaux, J., A. Baculard, and G. Tournier.** 1975. Partial cell-mediated immune deficiency in a child with chronic mucocutaneous candidiasis: intercurrent meningeal and pulmonary cryptococcosis. *Ann. Intern. Med.* **126:**615–625.

135. **Gerstenhaber, B. J., B. Weiner, R. Morecki, R. Bernstein, and S. Luftschein.** 1977. "Allergic" cryptococcal pneumonia. *Lung* **154:**195–199.

136. **Glaser, J. B., and A. Garden.** 1985. Inoculation of cryptococcosis without transmission of the acquired immunodeficiency syndrome. *N. Engl. J. Med.* **313:**264.

137. **Glassman, S. J., and M. J. Hale.** 1995. Cutaneous cryptococcosis and Kaposi's sarcoma occurring in the same lesions in a patient with the acquired immunodeficiency syndrome. *Clin. Exp. Dermatol.* **20:**480–486.

138. **Gloster, H. M., R. A. Swerlick, and A. R. Solomon.** 1994. Cryptococcal cellulitis in a diabetic kidney transplant patient. *J. Am. Acad. Dermatol.* **30:**1025–1026.

139. **Glynn, M. J., G. Duckworth, J. A. Ridge, W. J. Grange, and D. D. Gibbs.** 1994. Cryptococcal spondylitis: solitary infective bone lesions are not always tuberculous. *Br. J. Rheumatol.* **33:**1085–1086.

140. **Goenka, M. K., S. Mehta, S. K. Yachha, B. Nagi, A. Chakraborty, and A. K. Malik.** 1995. Hepatic involvement culminating in cirrhosis in a child with disseminated cryptococcosis. *J. Clin. Gastroenterol.* **20:**57–60.

141. **Goldman, M., and J. C. Pottage, Jr.** 1995. Cryptococcal infection of the breast. *Clin. Infect. Dis.* **21:**1166–1169.

142. **Goldstein, E., and O. N. Rombo.** 1962. Cryptococcal infection following steroid therapy. *Ann. Intern. Med.* **56:**114.

143. **Gonyea, E. F.** 1973. Cisternal puncture and cryptococcal meningitis. *Arch. Neurol.* **28:**200–201.

144. **Gonzalez, C., D. Shetty, L. L. Lewis, B. U. Mueller, P. A. Pizzo, and T. J. Walsh.** 1996. Cryptococcosis in human immunodeficiency virus-infected children. *Pediatr. Infect. Dis. J.* **15:**796–800.

145. **Goonetilleke, A. K., K. Kraus, D. N. Slater, D. Deu, M. L. Wood, and G. S. Basran.** 1995. Primary cutaneous cryptococcosis in an immunocompromised pigeon keeper. *Br. J. Dermatol.* **133:**650–652.

146. **Gosling, H. R., and W. S. Gilmer.** 1956. Skeletal cryptococcosis: report of a case and review of the literature. *J. Bone Joint Surg.* **38A:**660–668.

147. **Gupta, S., and U. U. Deshmukh.** 1996. Cryptococcemia and meningococcemia in a post partum woman. *J. Assoc. Physicians India* **44:**339–341.

148. **Gurevitz, O., A. Goldschmied-Reuven, C. Block, J. Kopolovic, Z. Farfel, and D. Hassin.** 1994. *Cryptococcus neoformans* vertebral osteomyelitis. *J. Med. Vet. Mycol.* **32:**315–318.

149. **Hamann, I. D., R. J. Gillespie, and J. K. Ferguson.** 1997. Primary cryptococcal cellulitis caused by *Cryptococcus neoformans* var. *gattii* in an immunocompetent host. *Australas. J. Dermatol.* **38:**29–32.

150. **Hammerman, K. J., K. E. Powell, C. S. Christianson, P. M. Huggin, H. W. Larsh, J. R. Vivas, and F. E. Tosh.** 1973. Pulmonary cryptococcosis: clinical forms and treat-

ment. A Center for Disease Control Cooperative Mycoses Study. *Am. Rev. Respir. Dis.* **108:**1116–1123.

151. **Hammerschlag, M. R., J. Domingo, J. O. Haller, and D. Papayanopulos.** 1982. Cryptococcal osteomyelitis. Report of a case and review of the literature. *Clin. Pediatr.* **21:**109–112.

152. **Hardy, R. E., C. Cummings, F. Thomas, and D. Harrison.** 1986. Cryptococcal pneumonia in a patient with Sickle Cell disease. *Chest* **89:**892–894.

153. **Hedderwick, S. A., H. F. Bonilla, S. F. Bradley, and C. A. Kauffman.** 1997. Opportunistic infections in patients with temporal arteritis treated with corticosteroids. *J. Am. Geriatr. Soc.* **45:**334–337.

154. **Henderson, D. K., J. E. Edwards, W. E. Dismukes, and J. E. Bennett.** 1981. Meningitis produced by different serotypes of *Cryptococcus neoformans,* abstr. 11. Abstr. 81st Annu. Meet. Am. Soc. Microbiol. 1981. American Society for Microbiology, Washington, D.C.

155. **Henson, D. J., and A. R. Hill.** 1984. Cryptococcal pneumonia: a fulminant presentation. *Am. J. Med.* **228:**221.

156. **Hibberd, P. L., and R. H. Rubin.** 1994. Clinical aspects of fungal infection in organ transplant recipients. *Clin. Infect. Dis.* **19**(Suppl. 1):S33–S40.

157. **High, K. P.** 1994. The antimicrobial activities of cyclosporine, FK506, and rapamycin. *Transplantation* **57:**1689.

158. **Hung, P. C., H. S. Wang, M. L. Chou, P. C. Sun, and S. C. Huang.** 1995. Cerebral cryptococcosis in a child. *Acta Paediatr. Sin.* **36:**131–135.

159. **Hunt, K. K., Jr., R. W. Enquist, and T. E. Bowen.** 1976. Multiple pulmonary nodules with central cavitation. *Chest* **69:**529–530.

160. **Hurd, D. D., D. B. Staub, R. L. Roelofs, and L. P. Dehner.** 1989. Profound muscle weakness as the presenting feature of disseminated cryptococcal infection. *Rev. Infect. Dis.* **11:**970–974.

161. **Huth, R. G.** 1994. Concomitant systemic cryptococcosis and coccidioidomycosis in a patient with AIDS. *Clin. Infect. Dis.* **18:**262–263.

162. **Hutto, J. O., C. S. Bryan, F. L. Greene, C. J. White, and J. I. Gallin.** 1988. Cryptococcosis of the colon resembling Crohn's disease in a patient with hyperimmunoglobulin E-recurrent infection (Job's) syndrome. *Gastroenterology* **94:**808–812.

163. **Iitaka, K., P. T. McEnery, and C. D. West.** 1978. Successful renal transplantation after generalized cryptococcosis. *J. Pediatr.* **92:**422–423.

164. **Ingram, C. W., H. B. Haywood, V. M. Morris, R. L. Allen, and J. R. Perfect.** 1993. Cryptococcal ventricular peritoneal shunt infection: clinical and epidemiological evaluation of two closely associated cases. *Infect. Immun.* **4:**719–722.

165. **Inoue, N., T. Uozumi, T. Yamamoto, T. Ohishi, Y. Murai, M. Murakami, and H. Yoshimatsu.** 1984. Treatment of cryptococcal meningitis with pulmonary granuloma. *J. Neurol.* **231:**109–111.

166. **Isaacson, J. E., and M. A. Frable.** 1996. Cryptococcosis of the larynx. *Otolaryngol. Head Neck Surg.* **114:**106–109.

167. **Iseki, M., M. Anzo, N. Yamashita, and N. Matsuo.** 1994. Hyper-IgM immunodeficiency with disseminated cryptococcosis. *Acta Paediatr.* **83:**780–782.

168. **Jacobs, D. H., A. M. Macher, R. Handler, J. E. Bennett, M. J. Collern, and J. I. Gallin.** 1984. Esophageal cryptococcosis in a patient with the hyperimmunoglobulin E-recurrent infection (Job's) syndrome. *Gastroenterology* **87:**201–203.

169. **Jensen, W. A., R. M. Rose, S. M. Hammer, and A. W. Karchmer.** 1985. Serologic diagnosis of focal pneumonia caused by *Cryptococcus neoformans*. *Am. Rev. Respir. Dis.* **132:**189–191.

170. **Jimenez-Lucho, V., R. del Busto, and E. J. Fisher.** 1985. Disseminated cryptococcosis with muscular involvement. *J. Infect. Dis.* **151:**567.

171. **Johnston, S. R., E. L. Corbett, O. Foster, S. Ash, and J. Cohen.** 1992. Raised intracranial pressure and visual complications in AIDS patients with cryptococcal meningitis. *J. Infect.* **24:**185–189.

172. **Jones, I., E. Nassua, and P. Smith.** 1965. Cryptococcosis of the heart. *Br. Heart J.* **27:**462–464.

173. **Jones, J. M., and W. A. Craig.** 1983. Cryptococcal meningitis: resulution eight months after antifungal therapy. *South. Med. J.* **76:**1567–1569.

174. **Kahn, F. W., D. M. England, and J. M. Jones.** 1985. Solitary pulmonary nodule due to *Cryptococcus neoformans* and *Mycobacterium tuberculosis. Am. J. Med.* **78:**677–681.

175. **Kanj, S. S., K. Welty-Wolf, J. Madden, V. Tapson, M. A. Baz, R. D. Davis, and J. R. Perfect.** 1996. Fungal infections in lung and heart-lung transplant recipients, report of 9 cases and review of the literature. *Medicine* **75:**142–156.

176. **Kaplan, M. H., P. P. Rosen, and D. Armstrong.** 1977. Cryptococcus in a cancer hospital. *Cancer* **39:**2265–2274.

177. **Karo, Y. T., and C. Brunnemer.** 1994. Initial diagnosis of disseminated cryptococcosis and acquired immunodeficiency syndrome by fine needle aspiration of the thyroid. A case report. *Acta Cytol.* **38:**427–430.

178. **Kauffman, C. A., A. G. Bergman, P. J. Severance, and K. D. McClatchey.** 1981. Detection of cryptococcal antigen. Comparison of two latex agglutination tests. *Am. J. Clin. Pathol.* **75:**106–109.

179. **Kent, T. H., and J. M. Layton.** 1962. Massive pulmonary cryptococcosis. *Am. J. Clin. Pathol.* **38:**596–604.

180. **Kerkering, T. M., R. J. Duma, and S. Shadomy.** 1981. The evolution of pulmonary cryptococcosis. clinical implications from a study of 41 patients with and without compromising host factors. *Ann. Intern. Med.* **94:**611–616.

181. **Kershnev, J. E., M. B. Ridley, and J. N. Greene.** 1995. Laryngeal cryptococcus treatment with oral fluconazole. *Arch. Otolaryngol. Head Neck Surg.* **121:**1193–1195.

182. **Kestelyn, P., H. Taelman, J. Bogaerts, A. Kagme, M. Abdel Aziz, J. Batungwanayo, A. M. Stevens, and P. Van de Perre.** 1993. Ophthalmic manifestations of infections with *Cryptococcus neoformans* in patients with the acquired immunodeficiency syndrome. *Am. J. Ophthalmol.* **116:**721–727.

183. **Khan, M. A., and S. Sbar.** 1975. Cryptococcal meningitis in steroid-treated systemic lupus erythematosus. *Postgrad. Med. J.* **5:**660–662.

184. **Khardori, N., F. Butt, and K. V. I. Rolston.** 1988. Pulmonary cryptococcosis in AIDS. *Chest* **93:**1319–1320.

185. **Khoury, M. B., J. D. Godwin, C. E. Ravin, H. A. Gallis, R. A. Halvorsen, and C. E. Putnam.** 1984. Thoracic cryptococcosis: Immunologic competence and radiologic appearance. *Am. J. Roentgenol.* **141:**893–896.

186. **Kida, M., C. R. Abramowsky, and C. Santoscoy.** 1989. Cryptococcosis of the placenta in a woman with acquired immunodeficiency. *Hum. Pathol.* **20:**920–921.

187. **Kim, J. S., B. I. Choi, and H. C. Han.** 1994. Cryptococcal cholangiohepatitis with intraductal cryptococcoma. *Am. J. Roentgenol.* **163:**995–996.

188. **Knudsen, J. D., L. Jensen, T. L. Sorensen, T. Jensen, H. Kjersem, J. Stenderup, and C. Pederson.** 1997. Cryptococcosis in Denmark: an analysis of 28 cases in 1988–1993. *Scand. J. Infect. Dis.* **29:**51–55.

189. **Kovacs, J. A., A. A. Kovacs, M. Polis, W. C. Wright, V. J. Gill, C. U. Tuazon, E. P. Gelmann, H. C. Lane, R. Longfield, G. Overturf, A. M. Macher, A. S. Fauci, J. E. Parrillo, J. E. Bennett, and H. Masur.** 1985. Cryptococcosis in the acquired immunodeficiency syndrome. *Ann. Intern. Med.* **103:**533–538.

190. **Kramer, M., M. L. Corrado, V. Bacci, A. C. Carter, and S. H. Landesman.** 1983. Pulmonary cryptococcosis and Cushing's syndrome. *Ann. Intern. Med.* **143:**2179–2180.

191. **Kromminger, R., F. Sthib, V. Thalmann, G. Jautzke, M. Seibold, H. Hochrein, H.**

Witt, and H. Burger. 1990. Osteomyelitis due to *Cryptococcus neoformans* in advanced age. Case report and review of the literature. *Mycoses* **33**:157–166.

192. **Kumlin, U., L. G. Elmquist, M. Granlund, B. Olsen, and A. Tarnvik.** 1997. CD4 lymphopenia in a patient with cryptococcal osteomyelitis. *Scand. J. Infect. Dis.* **29**:205–206.

193. **Kupfer, C., and E. McCrane.** 1974. A possible cause of decreased vision in cryptococcal meningitis. Invest. Ophthalmol. *Visual Sci.* **13**:801.

194. **Lalonde, R. G., P. Rene, and M. A. Wainberg.** 1993. Opportunistic infections and CD4+ T-lymphocytopenia without HIV infection: report of two cases. *Can. Med. Assoc. J.* **149**:179–182.

195. **Lamantia, L., A. Salmaggi, L. Tajoli, D. Cerrato, E. Lamperti, A. Nespolo, and E. Bussone.** 1986. Cryptococcal meningoencephalitis: intrathecal immunological response. *J. Neurol.* **233**:362–366.

196. **Lambertus, M. W., M. B. Goetz, A. R. Murthy, and G. E. Mathisen.** 1990. Complications of corticosteroid therapy in patients with the acquired immunodeficiency syndrome and *Pneumocystis carinii* pneumonia. *Chest* **98**:38–43.

197. **Larsen, R. A., S. Bozzette, A. McCutchan, J. Chiu, M. A. Leal, and D. D. Richman.** 1989. Persistent *Cryptococcus neoformans* infection of the prostate after successful treatment of meningitis. *Ann. Intern. Med.* **111**:125–128.

198. **Law, K. F., C. P. Aranda, R. L. Smith, K. A. Berkowitz, M. M. Ittman, and M. L. Lewis.** 1993. Pulmonary cryptococcosis mimicking methotrexate pneumonitis. *J. Rheumatol.* **20**:872–873.

199. **Lee, B. E., M. Feinberg, J. J. Abraham, and A. R. Murthy.** 1992. Congenital malformation in an infant born to a woman treated with fluconazole. *Pediatr. Infect. Dis. J.* **11**:1062–1064.

200. **Lee, Y. T., N. W. Leung, R. Kay, and H. K. Ng.** 1997. Subdural effusion in chronic cryptococcal meningitis in a cirrhotic patient. *Int. J. Clin. Pract.* **51**:254–255.

201. **Leggiadro, R. J., F. F. Barrett, and W. T. Hughes.** 1992. Extrapulmonary cryptococcosis in immunocompromised infants and children. *Pediatr. Infect. Dis. J.* **11**:43–47.

202. **Leggiadro, R. J., M. W. Kline, and W. T. Hughes.** 1991. Extrapulmonary cryptococcosis in children with AIDS. *Pediatr. Infect. Dis. J.* **10**:658–662.

203. **Lehmann, P. F., R. J. Morgan, and E. H. Freimer.** 1984. Infection with *Cryptococcus neoformans* var. *gattii* leading to a pulmonary cryptococcoma and meningitis. *J. Infect.* **9**:301–306.

204. **Levitz, S. M.** 1982. The ecology of *Cryptococcus neoformans* and the epidemiology of cryptococcosis. *Rev. Infect. Dis.* **53**:283–292.

205. **Lewis, J. L., and S. Rabinovich.** 1972. The wide spectrum of cryptococcal infections. *Am. J. Med.* **53**:315–322.

206. **Lewis, N., J. Lipstick, and C. Cammarosano.** 1985. Cryptococcal myocarditis in acquired immune deficiency syndrome. *Am. J. Cardiol.* **55**:1240.

207. **Lipson, B. K., W. R. Freeman, and J. Beniz.** 1989. Optic neuropathy associated with cryptococcal arachnoiditis in AIDS patients. *Am. J. Ophthalmol.* **107**:523–527.

208. **Littman, M. L., and L. E. Zimmerman.** 1956. *Cryptococcosis, Torulosis or European Blastomycosis*, p. 38–46. Grune and Stratton, New York.

209. **Lo, D.** 1976. Cryptococcosis in the Northern Territory. *Med. J. Aust.* **2**:825–828.

210. **Loerinc, A. M., E. J. Bottone, L. J. Finkel, and A. S. Teirstein.** 1988. Primary cryptococcal pneumonia mimicking *Pneumocystis carinii* pneumonia in a patient with AIDS. *Mt. Sinai J. Med.* **55**:181–186.

211. **Lombardo, T. A., A. S. Rabson, and H. T. Dodge.** 1957. Mycotic endocarditis. *Am. J. Med.* **22**:664.

212. **Long, J. A., J. R. Herdt, G. DiChiro, and H. R. Cramer.** 1980. Cerebral mass lesions in torulosis demonstrated by computer tomography. *J. Comput. Assist. Tomogr.* **4**:766–769.

213. **Long, R. F., S. V. Berens, and G. R. Shambhag.** 1972. Case reports. An unusual manifestation of pulmonary cryptococcosis. *Br. J. Radiol.* **45:**757–759.

214. **Lucatorto, F., and L. R. Eversole.** 1993. Deep mycoses and palatal perforation with granulomatous pansinusitis in acquired immunodeficiency syndrome: a case report. *Quintessence Int.* **24:**743–748.

215. **Lui, H., W. A. McLeod, and W. S. Wood.** 1993. An ulcerated plaque in a gay man. Cutaneous cryptococcosis. *Arch. Dermatol.* **129:**499–502.

216. **Mabee, C. L., S. W. Mabee, R. B. Kirkpatrick, and S. L. Koletar.** 1995. Cirrhosis: a risk factor for cryptococcal peritonitis. *Am. J. Gastroenterol.* **90:**2042–2045.

217. **Mahida, P., R. Morar, M. A. Goolam, E. Song, J. P. Tissandie, and C. Feldman.** 1996. Cryptococcosis: an unusual cause of endobronchial obstruction. *Eur. Respir. J.* **9:**837–839.

218. **Majid, A. A.** 1989. Clinical notes. Surgical resection of pulmonary cryptococcomas in the presence of cryptococcal meningitis. *J. R. Coll. Surg. Edinb.* **34:**332–333.

219. **Malabonga, V. M., and S. L. Kamholz.** 1991. Utility of bronchoscopic sampling techniques for cryptococcal disease in AIDS. *Chest* **99:**370–373.

220. **Manfredi, R., O. V. Coronado, A. Mastroianni, and F. Chiodo.** 1997. Liposomal amphotericin B and recombinant human granulocyte-macrophage colony stimulating factor in the treatment of paediatric AIDS-related cryptococcosis. *Int. J. STD AIDS* **8:**406–408.

221. **Manfredi, R., G. Rezza, and V. G. Coronado.** 1995. Is AIDS-related cryptococcosis more frequent among men? *AIDS* **9:**397–398.

222. **Mansoor, G. A., and D. B. Ornt.** 1994. Cryptococcal peritonitis in peritoneal dialysis patients: a case report. *Clin. Nephrol.* **41:**230–232.

223. **Martinelli, C., C. E. Comin, S. Ambu, D. Bartolozzi, and F. Leoncini.** 1997. Solitary cutaneous cryptococcosis resembling chickenpox: a case report. *AIDS* **11:**260–261.

224. **Massa, M. C., and J. A. Doyle.** 1981. Cutaneous cryptococcosis simulating pyoderma gangrenosum. *J. Am. Acad. Dermatol.* **5:**32–36.

225. **Matsushita, T., and K. Suzuki.** 1985. Spastic paraparesis due to cryptococcal osteomyelitis. A case report. *Clin. Orthopaed. Rel. Res.* **196:**279–284.

226. **McAllister, C. K., C. E. Davis, Jr., A. J. Ognibene, and J. L. Carpenter.** 1984. Cryptococcal pleuro-pulmonary disease: infection of the pleural fluid in the absence of disseminated cryptococcosis. Case report. *Milit. Med.* **149:**684–686.

227. **McClusky, P. J., and D. Wakefield.** 1985. Ocular involvement in the acquired immune deficiency syndrome. *Aust. N.Z. J. Ophthalmol.* **13:**293–298.

228. **Menon, A., and R. Rajamani.** 1976. Giant "cryptococcoma" of the lung. *Br. J. Dis. Chest* **70:**269–272.

229. **Metroka, C. E., S. Cunningham-Rundles, M. S. Pollack, J. A. Sonnabend, J. M. Davis, B. Gordon, R. D. Fernandez, and J. Mouradian.** 1983. Generalized lymphadenopathy in homosexual men. *Ann. Intern. Med.* **99:**585–591.

230. **Meyohas, M. C., P. Roux, D. Bollens, C. Chouaid, W. Rozenbaum, J. L. Meynard, J. L. Poirot, J. Frottier, and C. Mayaud.** 1995. Pulmonary cryptococcosis: localized and disseminated infections in 27 patients with AIDS. *Clin. Infect. Dis.* **21:**628–633.

231. **Middleton, F. G., and W. H. Prioleau.** 1982. Pulmonary cryptococcosis: review of eleven cases. *J. S.C. Med. Assoc.* **78:**419–423.

232. **Miller, K. D., J. A. M. Mican, and R. T. Davey.** 1996. Asymptomatic solitary pulmonary modules due to *Cryptococcus neoformans* in patients infected with human immunodeficiency virus. *Clin. Infect. Dis.* **23:**810–812.

233. **Miller, M. J.** 1995. Fungal infections, p. 709–713. *In* J. S. Remington and J. O. Klein (ed.), *Infectious Diseases of the Fetus and Newborn Infant.* W.B. Saunders, Philadelphia.

234. **Miller, W. T., and J. M. Edelman.** 1990. Cryptococcal pulmonary infection in patients with AIDS: radiographic appearance. *Radiology* **175:**725–728.

235. **Mitchell, D. H., and T. C. Sorrell.** 1992. Pancoast's syndrome due to pulmonary infection with *Cryptococcus neoformans* variety *gattii*. *Clin. Infect. Dis.* **14:**1142–1144.

236. **Mitchell, D. H., T. C. Sorrell, A. M. Allworth, C. H. Heath, A. R. McGregor, K. Papahaoum, M. J. Richards, and T. Gottlieb.** 1995. Cryptococcal disease of the CNS in immunocompetent hosts: influence of cryptococcal variety on clinical manifestations and outcome. *Clin. Infect. Dis.* **20:**611–616.

237. **Mitchell, T. G., and J. R. Perfect.** 1995. Cryptococcosis in the era of AIDS—100 years after the discovery of *Cryptococcus neoformans*. *Clin. Microbiol. Rev.* **8:**515–548.

238. **Mohr, J. A., H. Long, B. A. McKown, and H. G. Muchmore.** 1972. *In vitro* susceptibility of *Cryptococcus neoformans* to steroids. *Sabouraudia* **10:**171–172.

239. **Molina, J. M., E. Oksenhendler, M. T. Daniel, and J. P. Clauvel.** 1988. Fine needle aspiration and cryptococcosis in the acquired immunodeficiency syndrome. *Ann. Intern. Med.* **108:**772.

240. **Molnar-Nadasdy, G., I. Haesly, J. Reed, and G. Altshuler.** 1994. Placental cryptococcosis in a mother with systemic lupus erythematosus. *Arch. Pathol. Lab. Med.* **118:**757–759.

241. **Moncino, M. D., and L. T. Gutman.** 1990. Severe systemic cryptococcal disease in a child: review of prognostic indicators predicting treatment failure and an approach to maintenance therapy with oral fluconazole. *Pediatr. Infect. Dis. J.* **9:**363–368.

242. **Monteil, R. A., P. Hofman, J. F. Michiels, and R. Loubiere.** 1997. Oral cryptococcosis: a case report of salivary gland involvement in an AIDS patient. *J. Oral Pathol. Med.* **26:**53–56.

243. **Moosa, M. Y. S., and V. M. Coovadia.** 1997. Cryptococcal meningitis in Durban, South Africa: a comparison of clinical features, laboratory findings, and outcome for human immunodeficiency virus (HIV)-positive and HIV-negative patients. *Clin. Infect. Dis.* **24:**131–134.

244. **Morris, E., and E. Wolinsky.** 1965. Localized osseous cryptococcosis. A case report. *J. Bone Joint Surg.* **47:**1027.

245. **Moser, S. A., L. Friedman, and A. R. Varraux.** 1978. Atypical isolate of *Cryptococcus neoformans* cultured from sputum of a patient with pulmonary cancer and blastomycosis. *J. Clin. Microbiol.* **7:**316–318.

246. **Muccioli, C., J. R. Belfort, R. Neves, and N. Rao.** 1995. Limbal and choroidal cryptococcus infection in the acquired immunodeficiency syndrome. *Am. J. Ophthalmol.* **120:**539–540.

247. **Mulanovich, V. E., W. E. Dismukes, and N. Markowitz.** 1995. Cryptococcal empyema: case report and review. *Clin. Infect. Dis.* **20:**1396–1398.

248. **Mundy, L. M., and W. G. Powderly.** 1997. Invasive fungal infections: cryptococcosis. *Semin. Respir. Crit. Care Med.* **18:**249–257.

249. **Munoz-Perez, M. A., M. A. Colmenero, A. Rodriquez-Pichardo, F. J. Rodriquez-Pinero, J. J. Ris, and F. Comacho.** 1996. Disseminated cryptococcosis presenting as molluscum-like lesions as the first manifestation of AIDS. *Int. J. Dermatol.* **35:**646–648.

250. **Murakawa, G. J., R. Kerschmann, and T. Berger.** 1996. Cutaneous Cryptococcus infection and AIDS—report of 12 cases and review of the literature. *Arch. Dermatol.* **132:**545–548.

251. **Murakawa, G. J., T. M. Mauro, and B. Egbert.** 1995. Disseminated cutaneous cryptococcus clinically mimicking basal cell carcinoma. *Dermatol. Surg.* **21:**992–993.

252. **Murata, K., A. Khan, and P. G. Herman.** 1989. Pulmonary parenchymal disease: evaluation with high-resolution CT. *Radiology* **170:**629–635.

253. **Murray, R. J., P. Becker, P. Furth, and G. J. Criner.** 1988. Recovery from cryptococcemia and the adult respiratory distress syndrome in the acquired immunodeficiency syndrome. *Chest* **93:**1304–1307.

254. **Myers, S. A., and O. H. Kamin.** 1996. Cutaneous cryptococcosis and histoplasmosis coinfection in a patient with AIDS. *J. Am. Acad. Dermatol.* **34:**898–900.

255. **Naka, W., M. Masuda, A. Konohana, T. Shinoda, and T. Nishikawa.** 1995. Primary cutaneous cryptococcosis and *Cryptococcus neoformans* serotype D. *Clin. Exp. Dermatol.* **20:**221–225.

256. **Nampoory, M. R., Z. V. Khan, K. V. Johny, J. N. Constandi, R. K. Gupta, I. Al-Muzqiri, M. Samhan, M. Mozavi, and T. D. Chugh.** 1996. Invasive fungal infections in renal transplant recipients. *J. Infect.* **33:**95–101.

257. **Neuhauser, E. B., and A. Tucker.** 1948. The roentgen changes produced by diffuse torulosis in the newborn. *Am. J. Roentgenol.* **59:**805–815.

258. **Newberry, W. M., Jr., J. E. Walter, J. W. Chandler, Jr., and F. E. Tosh.** 1967. Epidemiologic study of *Cryptococcus neoformans. Ann. Intern. Med.* **67:**724–732.

259. **Noble, R. C., and L. P. Fajardo.** 1972. Primary cutaneous cryptococcosis: a review and morphologic study. *Am. J. Clin. Pathol.* **57:**13–17.

260. **Ofori-Kwakye, S. K., A. M. Wang, J. H. Morris, G. V. O'Reilly, E. G. Fischer, and C. L. Rumbaugh.** 1986. Septation and focal dilatation of ventricles associated with cryptococcal meningoencephalitis. *Surg. Neurol.* **25:**253–260.

261. **Okun, E., and W. T. Butler.** 1964. Ophthalmologic complications of cryptococcal meningitis. *Arch. Ophthalmol.* **71:**52–57.

262. **Page, B., E. Thervet, C. Legendre, and H. Kreis.** 1995. Cryptococcosis after renal transplantation. *Transplant. Proc.* **27:**1732.

263. **Paillas, J., M. Sinico, A. Vieillefond, and N. Blaye.** 1982. Un cas de cryptococcose pulmonaire pseudo-tumorale. *Sem. Hop. Paris* **58:**2402–2404.

264. **Pankey, G. A., J. D. Key, and T. K. Setze.** 1976. Osseous cryptococcosis syndrome distinct from other forms of disseminated cryptococcosis, p. 84. *In* W. H. Holloway (ed.), *Infectious Disease Reviews IV.* W. H. Futura, New York.

265. **Pappas, P., J. Perfect, H. Henderson, C. Kauffman, M. Saccente, D. Haas, G. Pankey, D. Lancaster, M. Holloway, G. Cloud, and W. Dismukes.** 1997. Cryptococcosis in non-HIV infected patients: a multicenter survey, abstr. 128. 35th Annu. Meet. Infectious Disease Society of America.

266. **Patronas, N. J., and E. V. Makariou.** 1993. MRI of choroidal plexus involvement in intracranial cryptococcosis. *J. Comput. Assist. Tomogr.* **17:**547–550.

267. **Paya, C. V.** 1993. Fungal infections in solid-organ transplantation. *Clin. Infect. Dis.* **16:**677–688.

268. **Paya, C. V., P. E. Hermans, and J. A. Washington.** 1989. Incidence, distribution, and outcome of episodes of infection in 100 orthotopic liver transplantations. *Mayo Clin. Proc.* **64:**555–564.

269. **Pema, K., J. Diaz, L. G. Guerra, D. Nabhan, and A. Verghese.** 1994. Disseminated cutaneous cryptococcosis. Comparison of clinical manifestations in the pre-AIDS and AIDS eras. *Arch. Intern. Med.* **154:**1032–1034.

270. **Penar, P. L., J. Kim, and D. Chylatte.** 1988. Intraventricular cryptococcal granuloma. Report of two cases. *J. Neurosurg.* **68:**145–148.

271. **Pereira, C. A., O. Fischman, A. L. Colombo, A. L. Moron, and A. C. Pignatari.** 1993. Cryptococcal meningitis in pregnancy. Review of the literature. Report of two cases. *Rev. Inst. Med. Trop. Sao Paulo* **35:**367–371.

272. **Perfect, J. R.** 1989. Cryptococcosis. *Infect. Dis. Clin. North Am.* **3:**77–102.

272a. **Perfect, J. R.** Unpublished observation.

273. **Perfect, J. R., and M. L. Cameron.** 1994. Pulmonary cryptococcosis: pathophysiological and clinical characteristics, p. 249–279. *In* H. Friedman, M. Berdinelli, and H. Chmel (ed.), *Infectious Agents and Pathogenesis.* Plenum Press, New York.

274. **Perfect, J. R., D. T. Durack, and H. A. Gallis.** 1983. Cryptococcemia. *Medicine* **62:**98–109.

275. **Perfect, J. R., S. D. R. Lang, and D. T. Durack.** 1980. Chronic cryptococcal meningitis: a new experimental model in rabbits. *Am. J. Pathol.* **101:**177–194.

276. **Perfect, J. R., and B. Seaworth.** 1985. Penile cryptococcosis with a review of mycotic infections of the penis. *Urology* **25:**528–531.

277. **Perry, H. D., and E. D. Donnenfeld.** 1990. Cryptococcal keratitis after keratoplasty. *Am. J. Ophthalmol.* 110:320–321.

278. **Petrites-Murphy, M. B., L. A. Robbins, J. M. Donahue, and B. Smith.** 1996. Equine cryptococcal endometritis and placentitis with neonatal cryptococcal pneumonia. *J. Vet. Diagn. Invest.* **8:**383–386.

279. **Petrucco, O. M., R. F. Seamark, K. Holmes, I. J. Forbes, and R. G. Symons.** 1976. Changes in lymphocyte function during pregnancy. *Br. J. Obstet. Gynaecol.* **83:**245–250.

280. **Philpot, C. R., and D. Lo.** 1972. Cryptococcal meningitis in pregnancy. *Med. J. Aust.* **28:**1005–1007.

281. **Pierard, G. E., C. Pierard-Franchimont, J. A. Estrada, A. Rurangirwa, and F. L. Dosal.** 1990. Cutaneous mixed infections in AIDS. *Am. J. Dermatopathol.* **12:**63–66.

282. **Pincus, M. R., M. Silva-Hutner, G. Rebotta, and A. W. Branwood.** 1981. Disseminated cryptococcosis in an asymptomatic alcoholic man. *Arch. Intern. Med.* **141:**796–798.

283. **Pippard, M. J., A. Dagleish, P. Gibson, M. Malkovsky, and A. D. B. Webster.** 1986. Acquired immunodeficiency with disseminated cryptococcosis. *Arch. Dis. Child.* **61:**289–302.

284. **Plunkett, J. M., B. I. Turner, and M. B. Tallent.** 1981. Cryptococcal septicemia associated with urologic instrumentation in a renal allograft recipient. *J. Urol.* **125:**241–242.

285. **Poprich, M. J., R. H. Arthur, and E. Helmer.** 1990. CT of intracranial cryptococcosis. *Am. J. Roentgenol.* **154:**603–606.

286. **Pounden, W. D., J. M. Amberson, and R. F. Jaeger.** 1952. A severe mastitis problem associated with *Cryptococcus neoformans* in a large dairy herd. *Am. J. Vet. Res.* **13:**121–128.

287. **Powderly, W. G.** 1993. Cryptococcal meningitis and AIDS. *Clin. Infect. Dis.* **17:**837–842.

288. **Powers, C. N., G. M. Rupp, S. J. Maygarden, and W. J. Frable.** 1991. Fine-needle aspiration cytology of adrenal cryptococcosis: a case report. *Diagn. Cytopathol.* **7:**88–91.

289. **Proknow, J. J., J. R. Benfield, and J. W. Rippon.** 1965. Cryptococcal hepatitis presenting as a surgical emergency. *JAMA* **191:**269.

290. **Pursley, T. J., I. K. Blomquist, J. Abraham, H. F. Andersen, and J. A. Bartley.** 1996. Fluconazole-induced congenital anomalies in three infants. *Clin. Infect. Dis.* **22:**336–340.

291. **Ramirez-Ortiz, R., J. Rodriquez, Z. Soto, M. Rivas, and W. Rodriquez-Clintron.** 1997. Synchronous pulmonary cryptococcosis and histoplasmosis. *South. Med. J.* **90:**729–732.

292. **Randall, R. E., Jr., W. K. Stacy, and E. C. Toone.** 1968. Cryptococcal pyelonephritis. *N. Engl. J. Med.* **279:**60.

293. **Randhawa, H. S., and D. K. Paliwal.** 1979. Survey of *Cryptococcus neoformans* in the respiratory tract of patients with bronchopulmonary disorders and in the air. *Sabouraudia* **17:**399–404.

294. **Reiss, F., and G. Szilagyi.** 1965. Ecology of yeast-like fungi in a hospital population. *Arch. Dermatol.* **91:**611–14.

295. **Rex, J. H., R. A. Larsen, W. E. Dismukes, G. A. Cloud, and J. A. Bennett.** 1993. Catastrophic vision loss due to *Cryptococcus neoformans* meningitis. *Medicine* **72:**207–224.

296. **Ricciardi, D. D., D. U. Sepkowitz, and L. B. Berkowitz.** 1986. Cryptococcal arthritis in a patient with acquired immune deficiency syndrome. Case report and review of the literature. *J. Rheumatol.* **13:**455–458.

297. **Rigdon, R. H., and O. T. Kirksey.** 1952. Mycotic aneurysm (cryptococcosis) of the abdominal aorta. *Am. J. Surg.* **84:**486.

298. **Riley, E., and W. G. Cahan.** 1972. Pulmonary cryptocococcis followed by pulmonary tuberculosis. A case report. *Am. Rev. Respir. Dis.* **106:**594–599.

299. **Rinaldi, M. G., D. J. Drutz, A. Howell, M. A. Sande, C. B. Wofsky, and W. K. Hadley.** 1986. Serotypes of *Cryptococcus neoformans* in patients with AIDS. *J. Infect. Dis.* **153:**642.

300. **Rosenheim, S. H., and J. Schwarz.** 1975. Cavitary pulmonary cryptococcosis complicated by aspergilloma. *Am. Rev. Respir. Dis.* **111:**549–553.

301. **Rozenbaum, R., and A. J. Goncalves.** 1994. Clinical epidemiological study of 171 cases of cryptococcosis. *Clin. Infect. Dis.* **18:**369–380.

303. **Ruiz, A., M. J. Post, and C. C. Bundschu.** 1997. Dentate nuclei involvement in AIDS patients with CNS cryptococcosis: imaging findings with pathologic correlation. *J. Comput. Assist. Tomogr.* **21:**175–182.

304. **Sabesin, S. M., H. J. Fallon, and V. T. Andriole.** 1963. Hepatic failure as manifestation of cryptococcosis. *Arch. Intern. Med.* **111:**661.

305. **Sabetta, J. R., and V. T. Andriole.** 1985. Cryptococcal infection of the central nervous system. *Med. Clin. North Am.* **69:**333–344.

306. **Salyer, W. R., and D. C. Salyer.** 1974. Pleural involvement in cryptococcosis. *Chest* **66:**139.

307. **Salyer, W. R., D. C. Salyer, and R. D. Baker.** 1974. Primary complex of *Cryptococcus* and pulmonary lymph nodes. *J. Infect. Dis.* **130:**74–77.

308. **Sandler, B., T. S. Potter, and K. Hashimoto.** 1996. Cutaneous *Pneumocystis carinii* and *Cryptococcus neoformans* in AIDS. *Br. J. Dermatol.* **134:**159–163.

309. **Sarosi, G. A., P. M. Silberfarb, and F. E. Tosh.** 1971. Cutaneous cryptococcosis. *Arch. Dermatol.* **104:**1–3.

310. **Saul, A., P. Lavalle, and G. Rodriquez.** 1980. Cutaneous cryptococcosis. *Int. J. Dermatol.* **8:**457–458.

311. **Schhitzlein, H. N., F. R. Murtaugh, J. A. Arrington, and C. R. Martinez.** 1993. CT of multiple intracranial cryptococcoma. *Am. J. Nucl. Reson.* **5:**472–473.

312. **Schmidt, D. M., J. A. Sercarz, K. F. Kevorkian, and R. F. Canalis.** 1995. Cryptococcosis presenting as a neck mass. *Ann. Otol. Rhinol. Laryngol.* **104:**711–714.

313. **Schmidt-Westhausen, A., T. Grunewald, P. A. Reichart, and H. D. Pohle.** 1995. Oral cryptococcosis in a patient with AIDS. A case report. *Oral. Dis.* **1:**77–79.

314. **Schroter, G. P. L., M. Hoelscher, and C. W. Putnam.** 1977. Fungus infections after liver transplantation. *Ann. Surg.* **186:**115–122.

315. **Schroter, G. P. L., D. R. Temple, and B. Husberg.** 1976. Cryptococcosis after renal transplantation: report of ten cases. *Surgery* **79:**268–277.

316. **Schubert, S., A. Lebean, R. Zell, and F. Goebel.** 1997. Cavitary cryptococcoma of the lungs and meningitis by *Cryptococcus neoformans* in a patient with AIDS. *Eur. J. Med. Res.* **2:**173–176.

317. **Schupbach, C. W., C. E. Wheeler, and R. A. Briggaman.** 1976. Cutaneous manifestations of disseminated cryptococcosis. *Arch. Dermatol.* **112:**1734–1744.

318. **Seaton, R. A., N. Verma, S. Naragi, J. P. Wembri, and D. A. Warrell.** 1997. Visual loss in immunocompetent patients with *Cryptococcus neoformans* var. *gattii* meningitis. *Trans. R. Soc. Trop. Med. Hyg.* **91:**44–49.

319. **Shah, B., H. C. Taylor, and I. Pillay.** 1986. Adrenal insufficiency due to cryptococcosis. *JAMA* **256:**3247.

320. **Shaunak, S., W. A. Schell, and J. R. Perfect.** 1989. Cryptococcal meningitis with normal cerebrospinal fluid. *J. Infect. Dis.* **160:**912. (Letter.)

321. **Shijubo, N., T. Fujishima, K. Ooashi, S. Morita, K. Shigehara, H. Nakata, and S. Abe.** 1995. Pulmonary cryptococcal infection in an untreated patient with sarcoidosis. *Sarcoidosis* **12:**71–74.

322. **Shrader, S. K., J. C. Watts, J. A. Dancik, and J. D. Band.** 1986. Disseminated crypto-

coccosis presenting as cellulitis with necrotizing vasculitis. *J. Clin. Microbiol.* **24:**860–862.

323. **Sider, L., and M. A. Westcott.** 1994. Pulmonary manifestations of cryptococcosis in patients with AIDS: CT features. *J. Thoracic Imaging* **9:**78–84.

324. **Sieving, R. R., C. A. Kaufmann, and C. Watanakunakorn.** 1975. Deep fungal infection in systemic lupus erythematosus. *J. Rheumatol.* **2:**61–72.

325. **Siewers, C. M. F., and H. G. Cramblett.** 1964. Cryptococcosis (torulosis) in children. *Pediatrics* **34:**393–400.

326. **Silberfarb, P. M., G. A. Sarosi, and F. E. Tosh.** 1972. Cryptococcosis and pregnancy. *Am. J. Obstet. Gynecol.* **112:**714–720.

327. **Siman, J., R. E. Nichols, and E. V. Morse.** 1953. An outbreak of bovine cryptococcosis. *J. Am. Vet. Med. Assoc.* **122:**31–35.

328. **Singh, M., P. P. Gupta, J. S. Rana, and S. K. Jand.** 1994. Clinico-pathological studies on experimental cryptococcal mastitis in goats. *Mycopathologia* **126:**147–155.

329. **Singh, N., T. Gayowski, M. M. Wagener, and I. R. Marino.** 1997. Clinical spectrum of invasive cryptococcosis in liver transplant recipients receiving tacrolimus. *Clin. Transplant.* **11:**66–70.

330. **Singh, N., J. D. Rihs, T. Gayowski, and V. L. Yu.** 1994. Cutaneous cryptococcosis mimicking bacterial cellulitis in a liver transplant recipient: case report and review in solid organ transplant recipients. *Clin. Transplant.* **8:**365–368.

331. **Sinha, P., K. G. Naik, and C. P. Bhagwat.** 1978. Mediastinal cryptococcoma. *Thorax* **33:**657–659.

332. **Slobodniuk, R., and S. Naraqi.** 1980. Cryptococcal meningitis in the central province of Papua New Guinea. *New Guinea Med. J.* **23:**111.

333. **Smith, J. H., M. M. Nichols, A. S. Goldman, F. C. Schmalstieg, and R. M. Goldblum.** 1982. Disseminated cryptococcosis in an infant with severe combined immunodeficiency. *Hum. Pathol.* **13:**500–502.

334. **Song, M. M.** 1974. Experimental cryptococcosis of the skin. *Sabouraudia* **133:**137.

335. **Speed, B., and D. Dunt.** 1995. Clinical and host differences between infections with the two varieties of *Cryptococcus neoformans.* *Clin. Infect. Dis.* **21:**28–34.

336. **Speed, B. R., and J. Kaldor.** 1997. Rarity of cryptococcal infection in children. *Pediatr. Infect. Dis. J.* **16:**52–53.

337. **Speed, B. R., L. Strawbridge, and D. H. Ellis.** 1993. *Cryptococcus neoformans* var. *gattii* meningitis in an Australian patient with AIDS. *J. Med. Vet. Mycol.* **31:**395–399.

338. **Spickard, A.** 1973. Diagnosis and treatment of cryptococcal disease. *South. Med. J.* **66:**26–31.

339. **Sridoma, V., F. Pacini, S. L. Yang, A. Moawad, M. Reilly, and L. J. De Groot.** 1982. Decreased levels of help T-cells: a possible cause of immunodeficiency in pregnancy. *N. Engl. J. Med.* **307:**352–356.

340. **Stafford, C. R., J. F. Fisher, and H. E. Fadel.** 1983. Cryptococcal meningitis in pregnancy. *Obstet. Gynecol.* **62:**35–37.

341. **Staib, F., M. Seibold, M. L'age, W. Heise, J. Skorde, G. Grosse, F. Nurnberger, and G. Baver.** 1989. *Cryptococcus neoformans* in the seminal fluid of an AIDS patient. A contribution to the clinical course of cryptococcosis. *Mycoses* **32:**171–180.

342. **Stansell, J. D.** 1993. Pulmonary fungal infections in HIV-infected persons. *Semin. Respir. Infect.* **8:**116–123.

343. **Starr, J. C., H. Che, and J. Montgomery.** 1995. Cryptococcal pneumonia simulating chronic eosinophilic pneumonia. *South. Med. J.* **88:**845–846.

344. **Steiner, I., I. Polacheck, and E. Melamed.** 1984. Dementia and myoclonus in a case of cryptococcal encephalitis. *Arch. Neurol.* **14:**216.

345. **Stone, B. J., and J. G. Wheeler.** 1990. Disseminated cryptococcal infection in a patient with hyperimmunoglobulinemia E syndrome. *J. Pediatr.* **117:**92–95.

346. **Su, M. C., W. L. Ho, and J. H. Chen.** 1994. Intramedullary cryptococcal granuloma of spinal cord: a case report. *Chung-Hua I Hsueh Tsa Chih* **53:**58–61.

347. **Subramanian, S., S. S. Kherdekar, P. G. V. Babu, and C. S. Christianson.** 1982. Lipoid pneumonia with *Cryptococcus neoformans* colonisation. *Thorax* **37:**319–320.

348. **Sulica, R. L., J. Kelly, B. J. Berberian, and R. Glaun.** 1994. Cutaneous cryptococcosis with molluscum contagiosum coinfection in a patient with acquired immunodeficiency syndrome. *Cutis* **53:**88–90.

349. **Suster, B., M. Akerman, M. Orenstein, and M. R. Wax.** 1986. Pulmonary manifestations of AIDS: review of 106 episodes. *Radiology* **161:**86–93.

350. **Swenson, R. S., S. L. Kountz, N. Blank, and T. C. Merigan.** 1969. Successful renal allograft in a patient with pulmonary cryptococcus. *Arch. Intern. Med.* **124:**502–506.

351. **Symmers, W. S. C.** 1970. Curiosa et exotica. *Br. Med. J. Clin. Res.* **4:**763–767.

352. **Szporn, A. H., S. Tepper, and C. W. Watson.** 1984. Disseminated cryptococcosis presenting as thyroiditis: fine needle aspiration and autopsy findings. *Acta Cytol.* **29:**449–453.

353. **Tabone, M. D., G. Leverger, J. Landman, C. Asnar, L. Boccon-Gibod, and G. Lasfargues.** 1994. Disseminated lymphonodular cryptococcosis in a child with x-linked hyper-IgM immunodeficiency. *Pediatr. Infect. Dis. J.* **13:**77–79.

354. **Takos, M. J., and N. W. Elton.** 1953. Spontaneous cryptococcosis of marmoset monkeys in Panama. *Arch. Pathol. Lab. Med.* **55:**403–407.

355. **Tan, C. T.** 1988. Intracranial hypertension causing visual failure in cryptococcus meningitis. *J. Neurol. Neurosurg. Psychiatr.* **51:**944.

356. **Tan, C. T., and B. B. Kuan.** 1987. Cryptococcus meningitis, clinical-CT scan considerations. *Neuroradiology* **29:**43–46.

357. **Tenholder, M. F., F. W. Ewald, Jr., N. K. Khankhanian, and J. H. Crosby.** 1992. Complex cryptococcal empyema. *Chest* **101:**586–588.

358. **Tien, R. D., P. K. Chu, J. R. Hesselink, A. Duberg, and C. Wiley.** 1991. Intracranial cryptococcosis in immunocompromised patients: CT and MR findings in 29 cases. *Am. J. Roentgenol.* **156:**1245–1251.

359. **Ting, S. F., B. E. Glader, and C. G. Prober.** 1991. Cryptococcus infections in a nine-year old child with hemophilia and the acquired immunodeficiency syndrome. *Pediatr. Infect. Dis. J.* **1:**76–77.

360. **Tollemar, J., B. G. Ericzon, K. Holmberg, and J. Anderson.** 1990. The incidence and diagnosis of invasive fungal infections in liver transplant recipients. *Transplant. Proc.* **22:**242.

361. **Tomasini, C., V. Caliendo, P. Puiatii, and M. G. Bernengo.** 1997. Granulomatous-ulcerative vulvar cryptococcosis in a patient with advanced HIV disease. *J. Am. Acad. Dermatol.* **37:**116–117.

362. **Torre, D., R. Martegani, F. Speranza, C. Zeroli, and G. P. Fiori.** 1995. Pulmonary cryptococcosis presenting as pneumothorax in a patient with AIDS. *Clin. Infect. Dis.* **21:**1524–1525.

363. **Town, G. I.** 1985. Pulmonary cryptococcosis: a report of two cases and review of the literature. *N.Z. Med. J.* **98:**894–895.

364. **Trivino, A., C. de los Heros, I. Claros, and J. A. Onrubia.** 1988. Cryptococcal lymphadenitis: fine needle aspiration in acquired immunodeficiency syndrome. *Ann. Intern. Med.* **108:**772.

365. **Tynes, B., K. N. Mason, A. E. Jennings, and J. E. Bennett.** 1968. Variant forms of pulmonary cryptococcosis. *Ann. Intern. Med.* **69:**1117–1125.

366. **Vandersmissen, G., K. Meuleman, G. Tits, and A. Verhaeghe.** 1996. Cutaneous cryptococcosis in corticosteroid-treated patients without AIDS. *Acta Clin. Belg.* **51:**111–117.

367. **Van't Wout, J. W., E. R. de Graeff-Meeder, L. C. Paul, W. Kuis, and R. van Furth.**

1988. Treatment of two cases of cryptococcal meningitis with fluconazole. *Scand. J. Infect. Dis.* **20:**193–198.

368. **Vogelaers, D., M. Petrovic, M. Deroo, P. Verplancke, Y. Claessens, J. M. Naeyaert, and M. Afschrift.** 1997. A case of primary cutaneous cryptococcosis. *Eur. J. Clin. Microbiol.* **16:**150–152.

369. **Wade, J. J., N. Rolando, K. Hayllar, J. Philpott-Howard, M. W. Casewell, and R. Williams.** 1995. Bacterial and fungal infections after liver transplantation: an analysis of 284 patients. *Hepatology* **21:**1328.

370. **Wahab, J. A., M. J. Hanifah, and K. E. Choo.** 1995. Bruton's agammaglobinaemia in a child presenting with cryptococcal empyema thoracic and periauricular pyogenic abscess. *Singapore Med. J.* **36:**686–689.

371. **Warr, W., J. H. Bates, and A. Stone.** 1968. The spectrum of pulmonary cryptococcosis. *Ann. Intern. Med.* **69:**1109–1116.

372. **Wasser, L., and W. Talavera.** 1987. Pulmonary cryptococcosis in AIDS. *Chest* **92:**692–695.

373. **Watanabe, K., and H. Ikemoto.** 1973. Review of cases with cryptococcal meningitis encountered in Japan in the past ten years. *Jpn. J. Med. Mycol.* **14:**143.

374. **Watson, A. J., R. P. Russell, and R. F. Cabreja.** 1985. Cure of cryptococcal infection during continued immunosuppressive therapy. *Q. J. Med.* **55:**169–172.

375. **Watson, A. J., A. Whelton, and R. P. Russell.** 1984. Cure of cryptococcemia and preservation of graft function in a renal transplant recipient. *Arch. Intern. Med.* **14:**1877.

376. **Watson, N. E., and A. H. Johnson.** 1973. Cryptococcal peritonitis. *South. Med. J.* **66:**387.

377. **Wehn, S. M., R. Heinz, and P. C. Burger.** 1989. Dilated Virchow-Robin spaces in cryptococcal meningitis associated with AIDS: CT and MR findings. *J. Comput. Assist. Tomogr.* **13:**756–762.

378. **White, M., C. Cirrincione, A. Blevins, and D. Armstrong.** 1992. Cryptococcal meningitis with AIDS and patients with neoplastic disease. *J. Infect. Dis.* **165:**960–966.

379. **White, M. H., and D. Armstrong.** 1994. Cryptococcosis. *Infect. Dis. Clin. North Am.* **8:**383–398.

380. **Whitley, T. H., J. R. Graybill, and R. H. Alford.** 1976. Pulmonary cryptococcosis in chronic lymphocytic leukemia. *South. Med. J.* **69:**33–37.

381. **Witt, D., D. McKay, L. Schwam, D. Goldstein, and J. Gold.** 1987. Acquired immune deficiency syndrome presenting as bone marrow and mediastinal cryptococcosis. *Am. J. Med.* **82:**149–150.

382. **Wong, K. F., S. K. Ma, J. K. Chan, and K. W. Lam.** 1993. Acquired immunodeficiency syndrome presenting as marrow cryptococcosis. *Am. J. Hematol.* **42:**392–394.

383. **Worland, R. G.** 1994. Cryptococcal hand infection: a case report. *J. Hand Surg.* **19:**609–610.

384. **Yantsos, V. A., J. Carney, and D. L. Green.** 1994. Review of the morphological variations in cutaneous cryptococcosis with a new case resembling varicella. *Cutis* **54:**343–347.

385. **Yinnon, A. M., and R. F. Betts.** 1994. *Pneumocystis carinii* pneumonia associated with concurrent cryptococcosis or toxoplasmosis in patients with AIDS. *Clin. Infect. Dis.* **18:**113–114.

386. **Yinnon, A. M., B. Rudensky, E. Sagi, G. Brever, C. Brautbar, I. Polacheck, and J. Halevy.** 1997. Invasive cryptococcosis in a family with epidermodysplasia verruciformis and idiopathic CD4 cell depletion. *Clin. Infect. Dis.* **25:**1252–1253.

387. **Yinnon, A. M., A. Solages, and J. J. Treanor.** 1993. Cryptococcal peritonitis: report of a case developing during continuous ambulatory peritoneal dialysis and review of the literature. *Clin. Infect. Dis.* **17:**736–741.

388. **Young, E. J., D. D. Hirsh, V. Fainstein, and T. W. Williams.** 1980. Pleural effusions due to *Cryptococcus neoformans*: a review of the literature and report of two cases with cryptococcal antigen determinations. *Am. Rev. Respir. Dis.* **121:**743–746.

389. **Yu, F. C., W. C. Perng, C. P. Wu, C. Y. Shen, and H. S. Lee.** 1993. Adult respiratory distress syndrome caused by pulmonary cryptococcosis in an immunocompetent host: a case report. *Chin. Med. J.* **52:**120–124.

390. **Zimmerman, L. E., and H. Rappaport.** 1954. Occurrence of cryptococcosis in patients with malignant disease of reticuloendothelial system. *Am. J. Clin. Pathol.* **24:**1050–1072.

391. **Zlupko, G. M., F. J. Fochler, and Z. H. Goldschmidt.** 1980. Pulmonary cryptococcosis presenting with multiple pulmonary nodules. *Chest* **77:**575.

392. **Zuger, A., E. Louie, R. S. Holzman, M. S. Simberkoff, and J. J. Rahal.** 1986. Cryptococcal disease in patients with acquired immunodeficiency syndrome. Diagnostic features and outcome of treatment. *Ann. Intern. Med.* **104:**234–240.

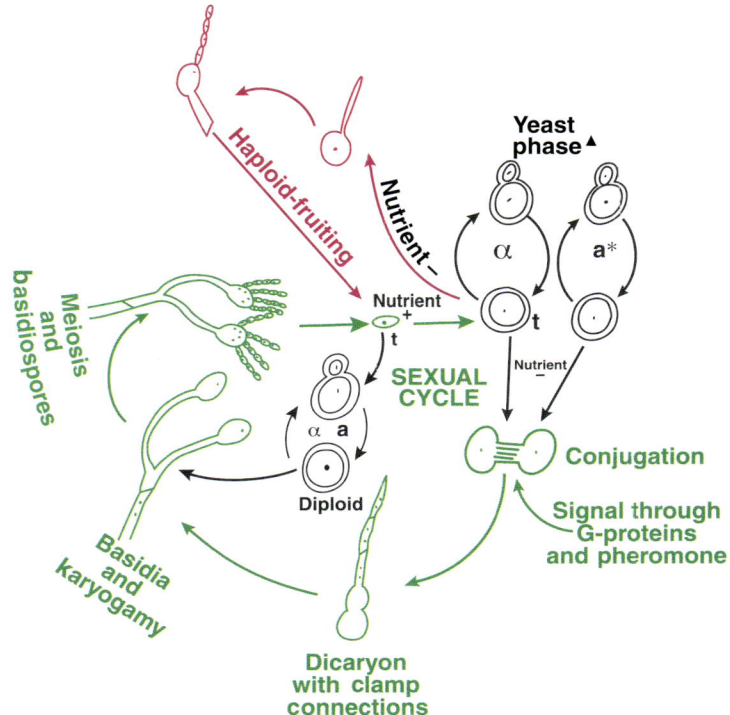

Plate 1. Life cycle of *Filobasidiella neoformans* (see chapter 2). Symbols: *, is *MAT***a** *C. neoformans* var. *neoformans* (serotype a) lost? t, infectious propagule(s)? Single outline, basidiospore; double outline, yeast. ▲, infection stage; α > **a**. Nutrient −, minimal nutrients with low nitrogen source and water content; Nutrient +, fully supplemented medium.

Plate 2. *Eucalyptus* sp., Vitoria, Brazil (see chapter 3).

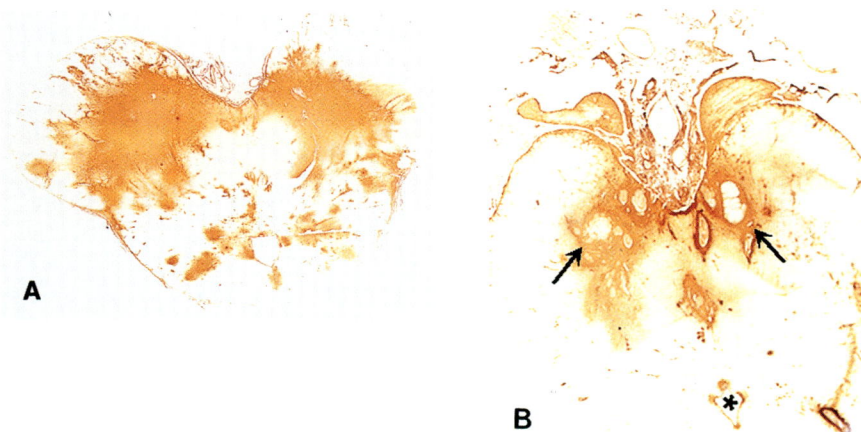

Plate 3. Immunohistochemical localization of *C. neoformans* glucuronoxylomannan (GXM) in human brain tissue (see chapter 9). GXM was detected using a specific monoclonal antibody (see references 157 and 158 in chapter 9). (A) Whole brain mount of midbrain section stained for GXM showing tissue deposits of cryptococcal antigen along the substantia nigra bilaterally and multiple perivascular foci in the tegmentum (original magnification, x1.5). (B) Another tissue shows parenchymal polysaccharide immunoreactivity around cryptococcal accumulations along the medial substantia nigra (arrows) as well as in tissue around large penetrating vessels. The cerebral aqueduct is indicated by the asterisk. The meninges, subpial, and periaqueductal regions are all staining for polysaccharide antigen (original magnification, x3.5). Photographs are courtesy of Sunhee Lee and reproduced here with permission.

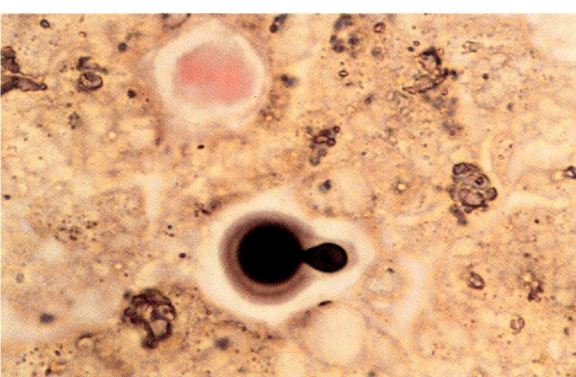

Plate 4. Gomori-methenamine silver stain of cytology from bronchoalveolar lavage (see chapter 12).

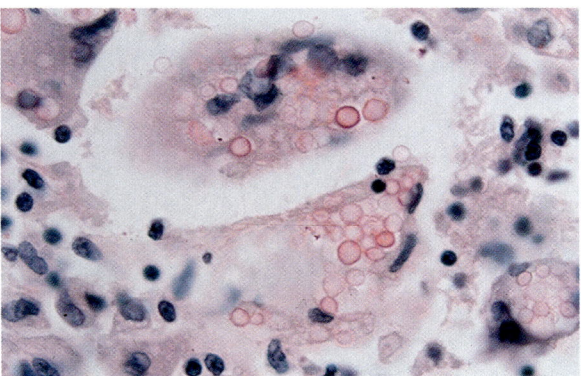

Plate 5. Mucicarmine stain of lung tissue (see chapter 12).

Plate 6. Alcian blue stain of lung tissue (see chapter 12).

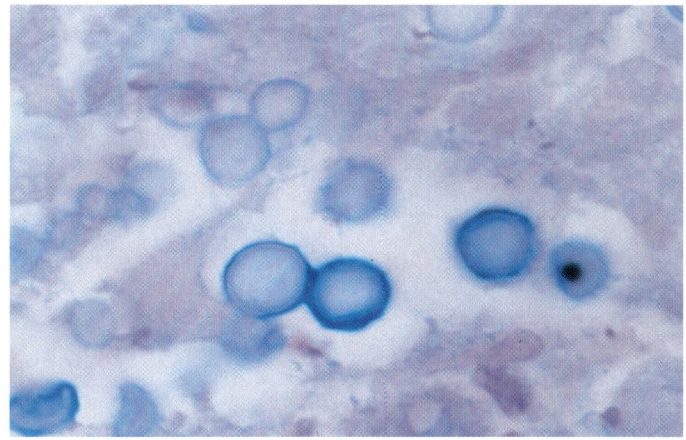

Plate 7. Fontana-Masson stain of lung tissue (see chapter 12).

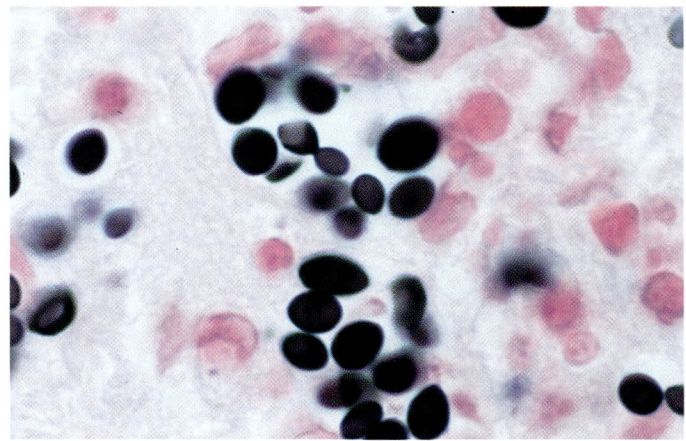

Plate 8. Hematoxylin-eosin (H&E) stain of tissue (see chapter 12).

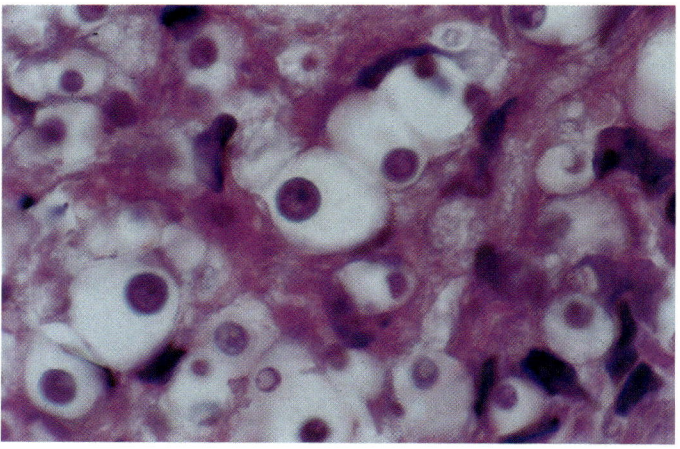

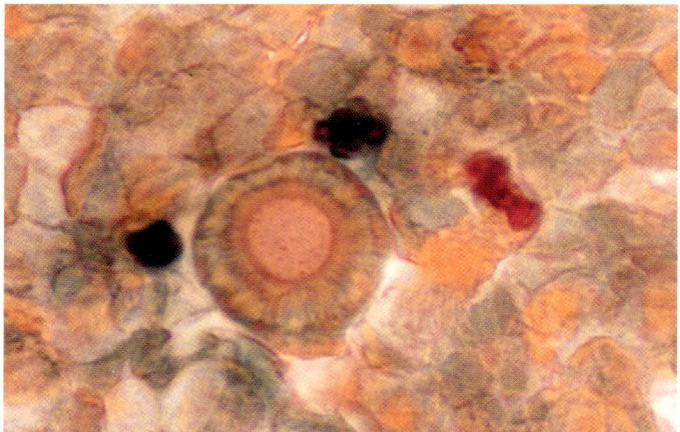

Plate 9. Papanicolaou stain of bronchoalveolar lavage (see chapter 12).

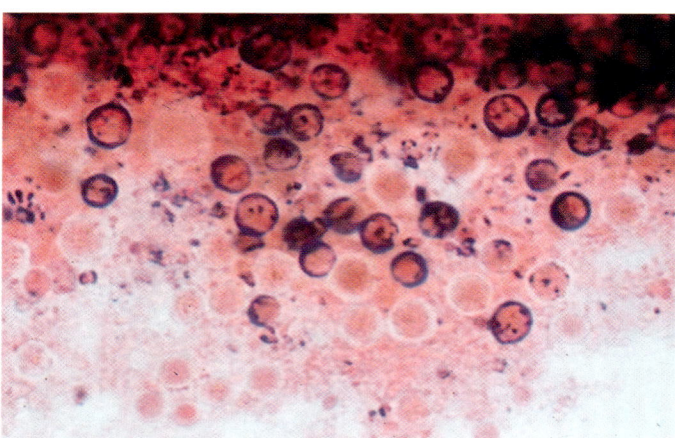

Plate 10. Gram stain of sputum (see chapter 12).

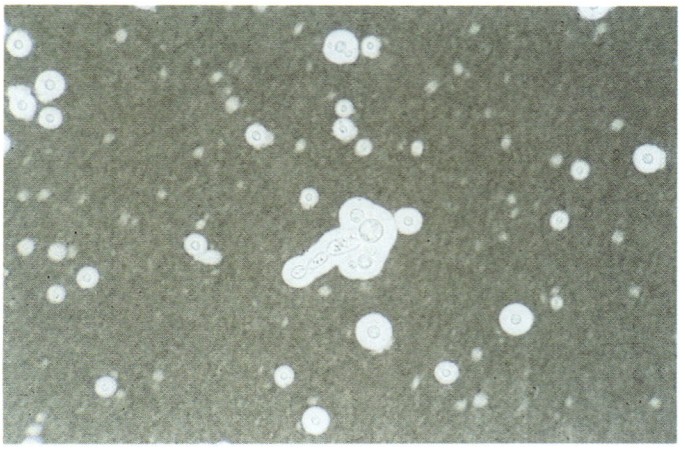

Plate 11. India ink stain of CSF from HIV-infected patient (see chapter 12).

Plate 12. Mucoid colonies, *C. neo-formans* (see chapter 12).

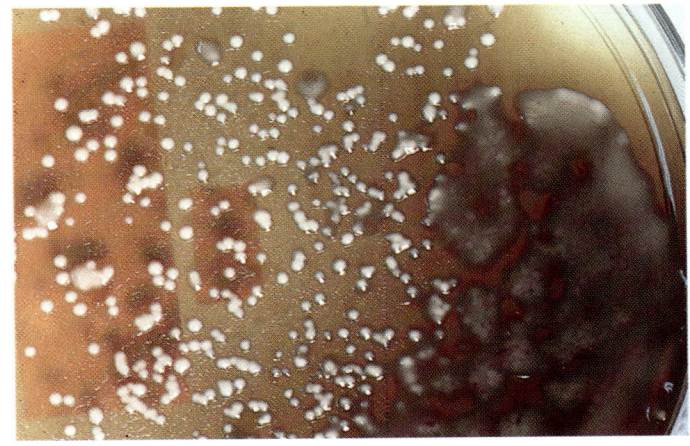

Plate 13. Haploid fruiting, *F. neo-formans* (see chapter 12).

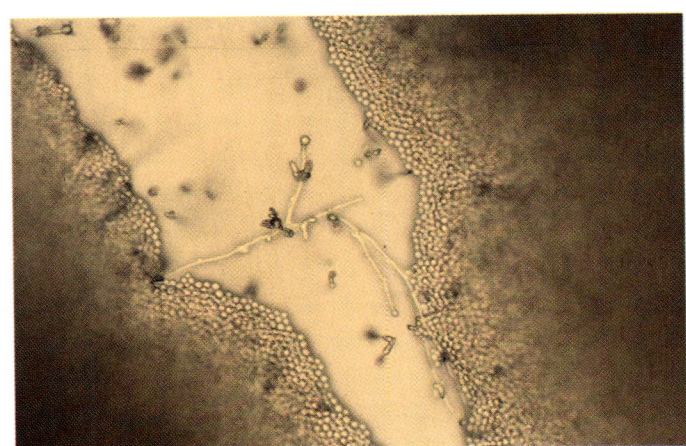

Plate 14. Basidium with basidio-spores (see chapter 12).

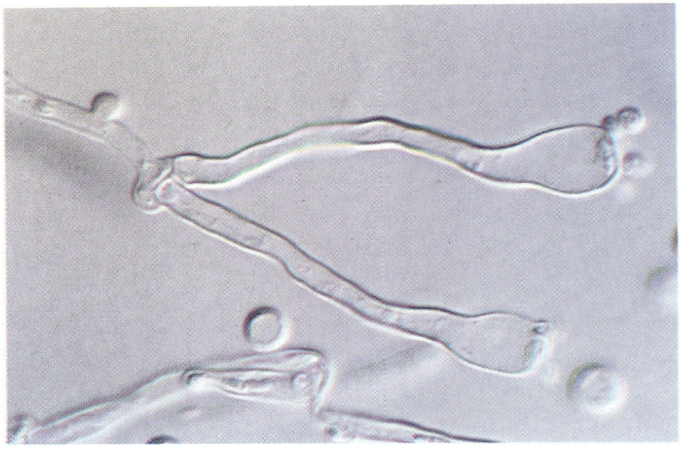

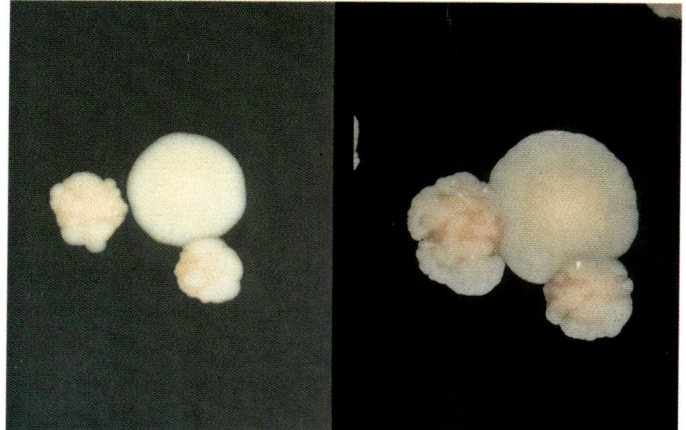

Plate 15. Variation in colony morphology (see chapter 12).

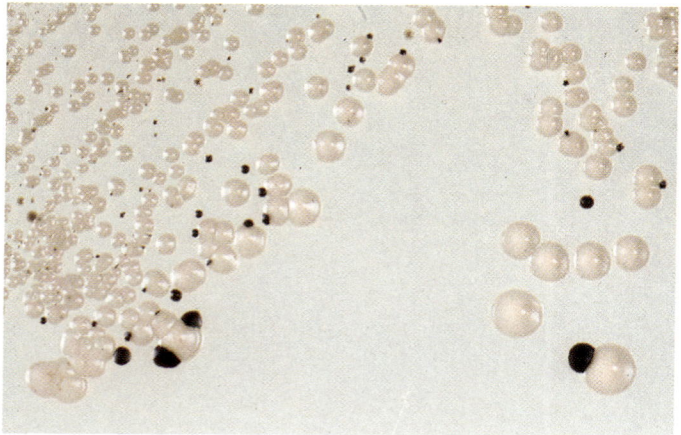

Plate 16. Mixture of *Candida albicans* (white) and *Cryptococcus neoformans* (dark) on a caffeic acid plate (see chapter 12).

Plate 17. Genetic cross between *C. neoformans* var. *neoformans* serotypes a and d on V-8 juice agar (see chapter 12).

Plate 18. Cryptococcal skin ulcer in HIV-infected patient (see chapter 13).

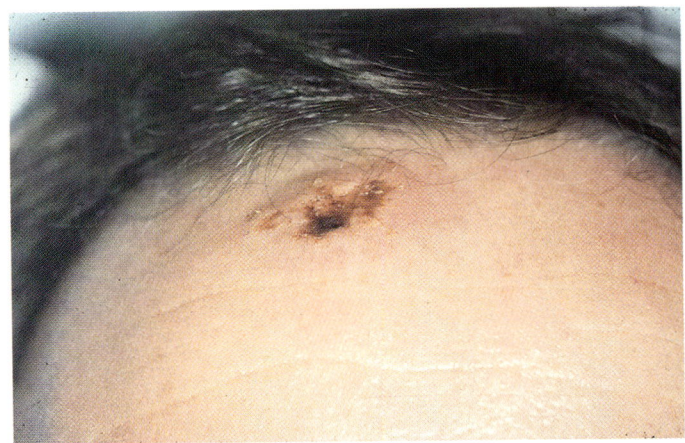

Plate 19. Cryptococcal cellulitis in patient receiving prednisone (see chapter 13).

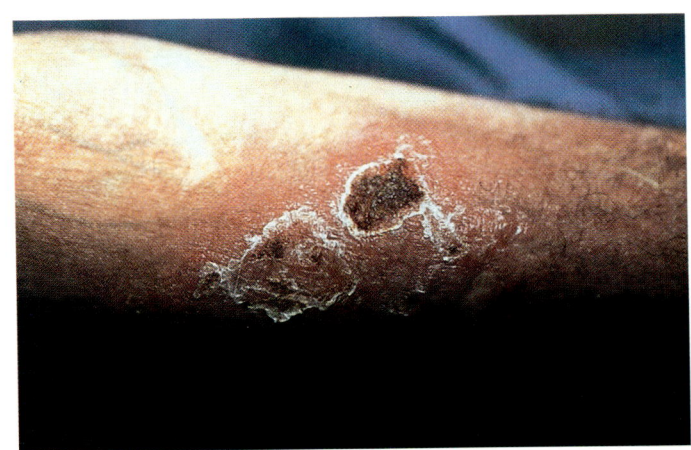

Plate 20. Cryptococcal lobar pneumonia (left lung) (see chapter 13).

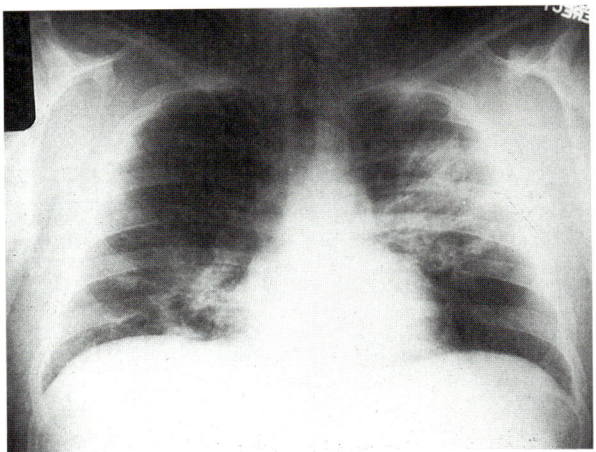

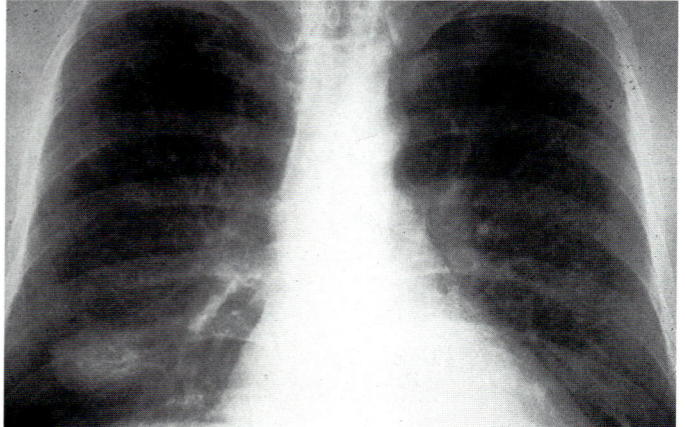

Plate 21. Cryptococcal nodule (right lower lobe) (see chapter 13).

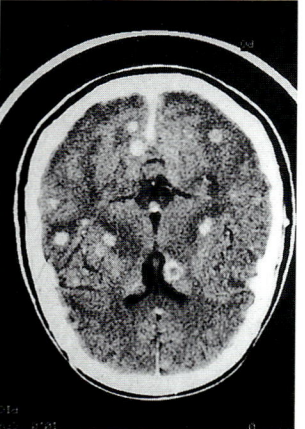

Plate 22. Multiple brain cryptococcomas (see chapter 13).

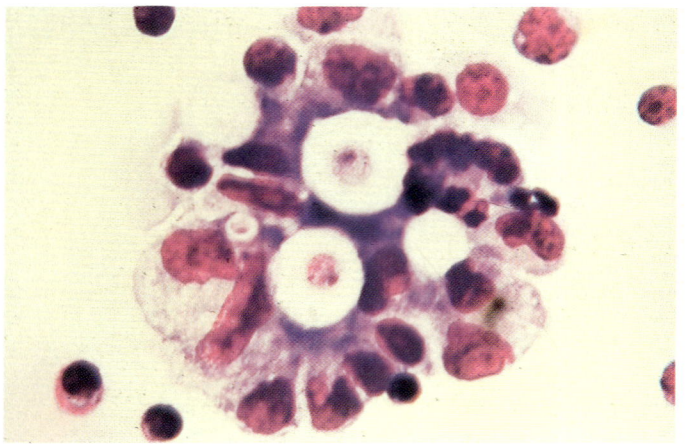

Plate 23. Cytological preparation of cerebrospinal fluid during cryptococcal meningitis in an immunocompetent host (hematoxylin and eosin stain) (see chapter 13).

14 | Therapy of Cryptococcosis

By the middle of the 20th century, *Cryptococcus neoformans* had finally been consistently identified in the clinical laboratories, and there were now the early beginnings of treatment strategies. For instance, by 1950 over 250 cases of cryptococcosis had been identified, and the natural history of the untreated infection was well documented. It appeared that 86% of patients with cryptococcal meningitis would die within 1 year and 70% would die within 3 months of the onset of symptoms (34). There were occasional cases of untreated cryptococcal meningitis that produced chronic meningitis for years. Beeson (15) carefully chronicled the remarkable 16-year history of untreated cryptococcal meningitis in one woman as it produced a series of relapses and ultimately led to her death. Despite approximately a half dozen reported cases of cryptococcal meningitis that have lasted for 9 to 22 years (22, 207, 238), it became generally appreciated that cryptococcal meningitis without treatment will be uniformly fatal to the host. Therefore, pulmonary cryptococcal infections might be controlled by the host, but once the yeast is able to disseminate from this initial site into the brain, death is inevitable. Unfortunately, in 1998, we have reports of the natural history of untreated HIV and cryptococcal meningitis from Africa, in which median survival from diagnosis to death is 14 days and only 22% of patients survived for >30 days without treatment (111a).

HISTORY

In the early attempts at treatment of cryptococcosis, the breadth of treatment modalities was quite broad. Both Carton (34) and Littman and Zimmerman (155) gave excellent descriptions of the variety of treatment strategies before 1956. For instance, these strategies included the unsuccessful use of antibiotics such as penicillin, tetracyclines, and streptomycin. Heavy metals were injected for treatment, from antimony/arsenical compounds to colloidal silver, copper, and gold thiosulfate. Although it is not certain that these inorganic compounds had any impact on cryptococcosis, we should not necessarily dismiss them too quickly. In a screen of a series of inorganic compounds against cryptococci, several, including those based on gold, were extremely potent in their anticryptococcal activity (184a).

Both active and passive immunological strategies were also attempted, including the use of an autogenous vaccine and administration of immune rabbit sera.

A series of chemical and hormonal agents, from intravenous alcohol to vitamin D, were tried without success. One of the interesting hormonal manipulations for possible therapies was the use of estrogens. The hypothesis that estrogen therapy might be useful in treatment was based on the observation that in both humans and animals, males appear to be more susceptible to cryptococcosis than females. In fact, Mohr et al. (169) performed two insightful studies on sex hormones. These studies showed that patients with cryptococcal meningitis who were given doses of the estrogenic hormone diethylstilbestrol increased their host-cell phagocytic capacities. This work, like serotherapy, was one of the first studies on immune modulation for treatment of cryptococcosis. Second, the studies of Mohr et al. showed that estrogens such as diethylstilbestrol (10 μg/ml) possessed direct anticryptococcal activity in vitro. In contrast, other hormones, such as progesterone and testosterone, had no anticryptococcal activity (168). It is unlikely that these specific estrogens will have any particular therapeutic use, but the pathophysiological principles behind these antifungal effects will be interesting to explore further.

Finally, antifungal substances derived from protoanemonin, a lactone found in higher plants (buttercups to garlic cloves), were found to be potent anticryptococcal substances. The initial clinical work with protoanemonin found it to be too toxic for human use. However, garlic (*Allium sativum*) continued to enjoy interest as a therapeutic agent for many decades, and at least one relatively favorable trial of garlic infusions in cryptococcal meningitis was published in China in 1980 (116).

Garlic extracts were shown by a series of investigators to have potent in vitro anticryptococcal activity (87, 170) and even to demonstrate anticryptococcal synergy with amphotericin B (53). Analysis of garlic extracts found that the active ingredient is allicin, which is a low-molecular-weight organosulfhydryl compound. This compound's biological activity is very unstable and is rapidly abolished by exposure to thiols, heat, and/or alkali (13). Once produced by physically breaking open a fresh garlic clove, the antifungal activity of the juice is rapidly lost, even at room temperature. However, because of the potent in vitro activity of this plant product, a clinical trial in the treatment of cryptococcal meningitis was performed. Patients were given either intramuscular or intravenous aqueous extracts of garlic in addition to oral garlic, and the success rate of this treatment was reported as 69% (116). Recent studies have not produced such a favorable scenario for garlic treatment. Volunteers given fresh garlic extracts had difficulty because a concentrated solution contained irritating substances for the oral mucosa. When tolerated doses of 10 to 25 mg were taken, only low antifungal activities in serum were found at 30 to 60 min after ingestion of the extract, and there was no detectable antifungal activity in urine (33). Louria et al. (157) examined the use of garlic extracts in a mouse model of cryptococcal meningitis. They found modest and inconsistent effects on reduction of yeast numbers in the brain but no impact on survival. In our own studies in the rabbit model of cryptococcal meningitis (184a), there was no effect of multiple intravenous garlic infusions on cerebrospinal fluid (CSF) yeast counts, and garlic was definitely not as active as standard drugs such as amphotericin B and azoles.

It is unlikely that garlic or its active ingredient, allicin, will become a therapeutic option, but this work emphasizes the potential importance of the discovery of antifungal drugs in plant products. The work on plant substances for anticryptococcal activity should be used as a prototype for further exploration in this discovery area of natural products. For example, compounds from a variety of plants, including cedar trees, Spanish moss, and tomatoes, have been shown to possess anticryptococcal activity (154). Further fractionation of these biological materials can specifically identify anticryptococcal compounds. For instance, an ethanolic extract of root bark from *Cleistopholis patens* contains 3-methoxysampangine, which exhibits potent anticryptococcal activity (156). Essential oils such as palmarosa and cinnamon also possess anticryptococcal activity. A variety of phenolic compounds, such as thymol and carvacrol, and terpenoids will kill cryptococci in vitro (250).

Early therapeutic strategies also used physical approaches such as radiotherapy or fever therapy for treatment of cryptococcosis. In our experience within the laboratory, radiotherapy does not appear justified, since in mutagenesis experiments with gamma irradiation, *C. neoformans* is resistant to killing by any radiation doses that could be safely delivered to a human host. For example, in vitro it requires exposure of 2 Gy for 5 min of gamma irradiation to kill 90% of a 10^8 inoculum of *C. neoformans* H99 in our laboratory. However, investigators have reported an inhibitory effect of treatment with X-rays on in vitro growth of *C. neoformans* in broth culture and spinal fluid (231). There have also been anecdotal clinical reports of regression of cryptococcal lesions with direct high-voltage irradiation to the skin, mouth, or lung (107, 124, 135). However, in our opinion there should be some caution for any enthusiasm regarding radiation therapy for *C. neoformans*, since it is unlikely that conventional radiotherapy will have any role in treatment of cryptococcosis, particularly in central nervous system (CNS) infections. Fever therapy uses the known sensitivity of *C. neoformans* to environmental temperatures between 37 and 42°C. The findings of Kligman et al. (136) on survival studies of chick embryos infected intravenously with *C. neoformans* when temperatures were raised to 42°C were an impetus to try thermal treatment in patients with cryptococcosis (136). Use of heat cabinets, typhoid vaccines, and malaria to raise body temperatures met with inconsistent results for treatment of cryptococcosis (153). Furthermore, it is clear from febrile patients with cryptococcosis that this yeast can still cause disease at 38 to 40°C in humans. Similarly, in experimental rabbits with body temperatures of 39 to 41°C and under immune suppression, this yeast can survive. These results demonstrate that some *C. neoformans* strains will adapt and survive in mammalian hosts at temperatures that would slow or even stop the growth of the yeast under similar temperatures in vitro (192a). Fever therapy is not likely to work clinically on a consistent basis, and its effects on other CNS physiological processes during infection make it no longer a therapeutic option. Whether fever should be controlled during treatment of cryptococcosis has not been conclusively answered, but antipyretics have not yet been determined to produce a negative impact on the therapeutic outcome of cryptococcosis.

In the mid-1950s the antibiotic actidone (cycloheximide), produced by *Streptomyces griseus*, had excellent in vitro inhibitory activity against *C. neoformans*, but unfortunately, it failed in treatment of experimental infections (137). This drug was the first antibiotic to be clinically studied on a substantial number of humans. Because of the in vitro susceptibility results and lack of effective therapies, the drug was evaluated in several dozen patients for treatment of cryptococcal meningitis by the intrathecal, intravenous, and intramuscular routes by multiple investigators (152). Actidone was even given in combination with the antibacterial agent polymyxin. Most of the treated patients died, and the several patients who did have some improvement were not completely followed up for final outcome measurements (262). It became clear to clinicians that this particular antifungal agent, which had shown such excellent anticryptococcal activity in vitro, was not particularly effective in the treatment of animals or humans with cryptococcosis. It was this clinical experience with actidone that first confirmed the usefulness of evaluating therapies for cryptococcosis in animal models and their excellent clinical predictive value.

It was not until a single case report by Appelbaum and Shtokalko (7) that the present era of successful therapy for cryptococcal meningitis was begun. In 1956, a 46-year-old female diabetic laboratory technician with cryptococcal meningitis was first treated with a new polyene antibiotic, amphotericin B. It was considered a safe compound that showed fungicidal activity against *C. neoformans* in vitro. Initially, the patient was treated orally with 1.6 to 8 g of amphotericin B per day, with no apparent clinical benefit after 5 weeks of treatment. She then received 100 mg of amphotericin B suspended in 500 ml of 5% dextrose in saline over 6 h. This treatment was given daily for 12 days, then repeated on alternate days for a total of 5 weeks. It is calculated that the first patient to be treated with amphotericin B for cryptococcal meningitis probably received approximately 3 g of intravenous amphotericin B. The patient significantly improved, and cultures of CSF became and remained negative. She was well 7 months after treatment. In her case report it was noted that a transitory impairment of renal function occurred during treatment. With this single successful case of treatment using intravenous amphotericin B, the era for studying the successful management of cryptococcosis, particularly cryptococcal meningitis, had begun. It was only with this introduction of amphotericin B in the 1950s that reported cure rates for cryptococcal meningitis rose to over 50% (75, 216, 225). From this humble beginning of a single patient and then open trials with amphotericin B, 40 years of investigation into improved therapies for this infection were undertaken. From these investigations, which are collated and summarized in Table 1, it is now possible to offer an improved prognosis for most patients with local and/or disseminated cryptococcal disease.

In many respects, the treatment of cryptococcosis is one of the best studied of all infectious diseases. Table 1 charts more than 25 studies into treatment of cryptococcal meningitis, and yet many therapeutic questions still need answers. In Table 1, we have attempted to review the studies and give some approximation of outcomes, but there should be a word of caution in the interpretation of the numbers within the table. First, the studies are not directly comparable. For example, endpoints such as success and toxicity are defined differently in the various studies. The success rates were calculated for each study by identifying the best group outcome reported by the investigators, and this varied between studies. Second, all drug study groups were combined, and thus details of specific therapeutic regimens and the important studies that support them are reviewed separately in the text. Third, in some respects, death rates are the most objective figures and possibly the best outcome measurement, but they may also primarily reflect the stage of the patient's underlying disease(s). Fourth, we did not necessarily consider cure the only criterion for success, but on the other hand, the table may not include quiescent disease as success if it was not included in the highest or best outcome group reported by the investigators.

Since 1963, rates of success in these studies, as judged by the clinical investigators, ranged from 36 to 100%, with most studies in the 55 to 75% success range.

Table 1 Treatment studies of cryptococcal meningitis and cryptococcosis[a]

Study	Year	Site	Underlying disease	No. of patients	Treatment regimen	Randomized	Length	Success (%)	Deaths (%)	Drug toxicity (%)
Mayanja-Kizza et al. (161a)	1998	CNS	AIDS	58	FLU 200 mg/day FLU 200 mg/day + FC 150 mg/kg/day	Yes	8 wk	38	46	0
Van der Horst et al. (246)	1997	CNS	AIDS	306	AmB 0.7 mg/kg/day vs. AmB 0.7 mg/kg/day + FC 100 mg/kg/day FLU 400 mg/day vs. ITZ 400 mg/day	Yes	2 wk 8 wk	69	9.4	39
Leenders et al. (147)	1997	CNS	AIDS	28	AmBisome 4 mg/kg/day AmB 0.7 mg/kg	Yes	10 wk	85	11	37
Yamaguchi et al. (267)	1996	CNS/lung/ other	Non-AIDS	44	FLU 200–400 mg/day	No	24 wk	48	0	2
Dromer et al. (73)	1996	CNS/lung/ skin/UTI	Non-AIDS	83	AmB (median total 1 g) FLU range (100–400 mg)	No	Variable	76	19	NR
Joly et al. (123)	1996	CNS	AIDS	90	AmB 0.5–0.7 mg/kg/day Intralipid/AmB 0.7–1.0 mg/kg/day	Yes	6 wk	72	30.9	58
Just-Nubling et al. (127)	1996	CNS	AIDS	42	AmB >0.3 mg/kg/day + FC 150 mg/kg/day + FLU 200–400 mg	No	1–8 wk	90	7.1	NR
Sharkey et al. (223)	1996	CNS	AIDS	55	AmB 0.7–1.2 mg/kg/day ABLC 1.2–5.0 mg/ kg/day	Yes	6 wk	46	22	95
De Lalla et al. (57)	1995	CNS disseminated	AIDS	31	AmB 1 mg/kg/day + FC 100–150 mg/kg/day	No	2 wk	94	0	23

(continued)

Table 1 (*Continued*)

Study	Year	Site	Underlying Disease	No. of patients	Treatment regimen	Randomized	Length	Success (%)	Deaths (%)	Drug toxicity (%)
Oppenheim et al. (180)	1995	CNS	AIDS	11	ABCD 0.5–4.0 mg/kg/day	No	Variable	45	NR	64
Haubrich et al. (111)	1994	CNS/blood	AIDS	8	FLU 800 mg/day	No	4–5 wk	87	0	12.5
Larsen et al. (145)	1994	CNS	AIDS	49	FLU 400 mg/day + FC 150 mg/kg/day	No	10 wk	75	8	28
Coker et al. (42)	1993	CNS	AIDS	23	AmB 3 mg/kg/day	No	Variable	61	13.0	23.6
Berry et al. (21)	1992	CNS	AIDS	8	FLU 800 mg/day	No	4.5 mo	50	37.5	12.5
DeGans et al. (55)	1992	CNS	AIDS	28	AmB 0.3 mg/kg/day + FC 150 mg/kg/day ITZ 200 mg/b.i.d.	Yes	6 wk	68	8.0	64
Laroche et al. (142)	1992	CNS	AIDS	41	FLU 400 mg/day	No	Variable	63	NR	7
Saag et al. (211)	1992	CNS	AIDS	194	AmB 0.3 mg/kg/day FLU 200–400 mg/day	Yes	10 wk	36	17.0	40
Denning et al. (63)	1990	CNS/UTI/lung	AIDS	57	ITZ 200 mg/b.i.d.	No	4 mo	64	19.3	42
Larsen et al. (146)	1990	CNS	AIDS	21	AmB 0.7 mg/kg/day + FC 150 mg/kg/day FLU 400 mg/day	Yes	10 wk	55	18.2	38
Staib and Seibold (227)	1988	CNS/lung	AIDS	15	AmB 0.3–0.5 mg/kg/day + FC 150 mg/kg/day	No	4–8 wk	64	15	NR

Stern et al. (230)	1988	CNS/blood/UTI	AIDS	7	FLU 100–400 mg	No	Variable	57	42	NR
Dismukes et al. (69)	1987	CNS	Non-AIDS	194	AmB 0.3 mg/kg/day FC 150 mg/kg/day	Yes	4 vs. 6 wk	66	16.8	53.1
Bennett et al. (19)	1979	CNS	Non-AIDS	66	AmB 0.4 mg/kg/day AmB 0.3 mg/kg/day + FC 150 mg/kg/day	Yes	10 wk 6 wk	39	19.8	30
Jimbow et al. (121)	1978	CNS	Non-AIDS	28	AmB 0.3–0.5 mg/kg FC 150 mg/kg/day	No	Variable	75	NR	20.8
Utz et al. (241)	1975	CNS	Non-AIDS	20	AmB 20 mg/day + FC 150 mg/kg	No	6 wk	80	15	65
Sarosi et al. (216)	1969	CNS	Non-AIDS	31	AmB (total dose, 570–5,750 mg)	No	Variable	65	29	NR
Drutz et al. (75)	1968	CNS/lung	Non-AIDS	7	AmB (mean total dose, 1,904 mg)	No	10 wk	100	0	100
Spickard et al. (225)	1963	CNS	Non-AIDS	30	AmB (mean total dose, 2,410 mg)	No	Variable	53	23.3	NR

[a]Abbreviations: ABCD, amphotericin B colloidal dispersion; ABLC, amphotericin B lipid complex; AmB, amphotericin B; FC, flucytosine; FLU, fluconazole; ITZ, itraconazole; NR, not reported; UTI, urinary tract infection.

Death occurred in 10 to 25% of patients during treatment or in the immediate follow-up period. For detailed comparisons of outcomes among treatment groups, the reader is encouraged to review each study in depth to identify the specific successes for each risk group population. It is essential to understand each patient's particular risk factor for infection, and the stage of infection at diagnosis, since these two aspects have a major influence on final outcomes for the treatment of cryptococcosis.

In the following discussion, we will examine the antifungal agents, the therapeutic trials of these agents in animals and humans, and the predicted prognosis for certain risk groups with cryptococcosis. Although we will compare the results of these studies, it is important to emphasize that the criteria for success and failure may vary among studies and that direct comparisons between these studies may not be wholly accurate. Nevertheless, by reporting the conclusions of these studies, we wish to impart to the reader a general appreciation for the modern history of treatment of cryptococcosis and an understanding of recent concepts in the treatment of this infection.

IN VITRO SUSCEPTIBILITY TESTING

McCullough et al. (163) in 1958 first found that MICs of a new polyene, amphotericin B, against 14 strains of *C. neoformans* ranged from 0.03 to 0.12 μg/ml. In 1966 Bennett (17) confirmed these results and suggested for the first time that treatment with amphotericin B might increase yeast resistance to the drug but that this concern might be balanced by the apparent reduced virulence of the yeast isolates exposed to this polyene. In 1968 Shadomy et al. (221) began the modern era of in vitro susceptibility testing for *C. neoformans*. In this early in vitro testing, Shadomy et al. examined 77 *C. neoformans* strains against amphotericin B, hamycin, and flucytosine, using a broth dilution technique, and found that all strains were very susceptible to amphotericin B at MICs of <0.2 μg/ml. In this work, the importance of media for these antifungal tests was first recognized. For instance, flucytosine had no anticryptococcal activity when grown in media containing peptones. On the other hand, there were measured MICs for the same strains in the 50-μg/ml range when synthetic media were used and cytosines in the media were not available to compete with the drug. A few years later, Block et al. (28) took in vitro susceptibility testing for *C. neoformans* even further. With the use of 65 isolates of *C. neoformans*, they identified the importance of inocula and temperatures for obtaining consistent and interpretable MICs. For example, when MICs were read at 48 h at 32°C with an initial inoculum of 6×10^2, all pretreatment isolates were susceptible to 10 μg/ml of flucytosine, but 9 of 16 yeast isolates recovered during or after therapy had MICs of 40 to >320 μg/ml. With this study it was clear that MICs, at least for flucytosine, could potentially help predict clinical outcome (28) and that flucytosine resistance would be clinically troublesome and detected by in vitro susceptibility testing (27).

After these early evaluations for in vitro susceptibility of drugs against *C. neoformans*, there were few studies for the next 20 years. Most studies were related to comparison of new drugs with standard therapies. In the 1990s during the AIDS epidemic and with the widespread use of fluconazole, it was clear that *C. albicans* isolates were developing azole resistance, both in vitro and in vivo. With the combination of more antifungal agents and the concern about antifungal drug resistance, there was a renewed focus on antifungal drug susceptibility testing.

Although there have been only a few well-documented drug-resistant *C. neoformans* infections, there has been a renewed emphasis on the use of in vitro susceptibility testing for screening antifungal compounds for anticryptococcal activity to guide therapeutic decisions (6, 93, 94, 184, 218, 220). Both macrobroth and microbroth methods for in vitro susceptibility testing of antifungal agents against pathogenic yeasts such as *C. neoformans* have now been standardized for medium, inoculum, and endpoint determinations (79, 88, 175).

Throughout the development of the NCCLS (National Committee for Clinical Laboratory Standards) M27-T standardized in vitro methodology (175), it was noted that this procedure might be suboptimal for the testing of *C. neoformans* strains since the use of RPMI broth for growth of some strains of *C. neoformans* was inadequate. Many *C. neoformans* strains grew slowly and required 72-h incubations to read endpoints, and some strains would not even grow in this medium. Therefore, Ghannoum et al. (94) proposed a modification of the NCCLS M27-T method. In this method, yeast nitrogen base (YNB) medium buffered to a pH of 7.0, an inoculum of 10^4 cells/ml, and incubation at 35°C for 48 h was used in a microtiter format, with an endpoint defined spectrophotometrically as 50% inhibition of a control well at 420 nm. A multicenter evaluation of this method showed excellent intralaboratory and interlaboratory agreement within one tube dilution (96%) for the assay and an overall agreement of 90% with the standard NCCLS M27-T microdilution method (214). This revised method was then used retrospectively by Witt et al. (263) to examine 76 AIDS patients who had received fluconazole with or without flucytosine so that they could determine if the susceptibility tests (MICs) for the *C. neoformans* isolates could help predict clinical outcome. The microtiter broth method showed a reasonable distribution of MICs, and in a multivariant logistic regression model, higher fluconazole MICs for *C. neoformans* strains were predictive of treatment failure in patients. However, there were also other important clinical factors associated with therapeutic failures in this study; these included positive blood and urine cultures, high serum and CSF antigen titers, and those patients in whom flucytosine was not used. From this clinical experience, it is clear that in vitro susceptibility testing does not completely predict patient outcome, but it can be used as one reasonable factor for clinical decisions in regard to treatment regimens and dosage selections. The microtiter broth method for *C. neoformans* in vitro susceptibility testing has also been improved for endpoint determination by agitation of the yeast cultures (6). Although at this time most of the emphasis is on broth susceptibility, semisolid agar has successfully been used for *C. neoformans* in vitro susceptibility testing (58), and eventually, for some antifungal drugs, this correlative testing may lead to the use of E-tests for *C. neoformans* drug susceptibility testing (44). All new tests will need to be compared and validated in relation to the values of the NCCLS M27-T standard.

Studies using in vitro susceptibility methods have detected resistance to flucytosine and fluconazole both in vitro and in vivo. For most initial clinical isolates of *C. neoformans* without antifungal exposure, the MICs for amphotericin B, flucytosine, and the azoles are relatively low. However, resistance to all these antifungal agents has developed and may be detected by in vitro susceptibility testing, which is the primary value of all microbial susceptibility testing. These tests do not necessarily predict success, but they should identify strains unlikely to respond to certain drugs. In this respect, in vitro susceptibility testing is very good for *C. neoformans.*

Amphotericin B is uniformly active against initial clinical isolates of *C. neoformans*, and by most methods MICs range from 0.05 to 0.2 μg/ml. In fact, with most standardized testing, the MICs for amphotericin B are generally low and exhibit very little variation in the MIC distribution between *C. neoformans* isolates. This finding makes it difficult to find a subset of yeast strains with relative resistance compared to more susceptible strains and thus makes it difficult to choose possible therapeutic MIC endpoints for polyene drug resistance. However, an occasional case of cryptococcosis has been found to involve a particular amphotericin B-resistant *C. neoformans* strain. It was detected by failure of therapy, and the relapse isolate was confirmed to be relatively resistant to amphotericin B by in vitro testing and comparison with the initial isolate prior to treatment (132). In fact, there are *C. neoformans* strains that have even developed cross-resistance to both amphotericin B and the azoles (125, 248). Further surveillance is necessary to determine if amphotericin B resistance in *C. neoformans* will be a significant clinical factor for therapeutic decisions or only a rare occurrence. The actual breakpoint for amphotericin B resistance has not been established and may vary according to the methods used, but it is likely that an MIC of >1 μg/ml by any standard in vitro method for a *C. neoformans* strain will likely be difficult to treat successfully with the amphotericin B deoxycholate preparation. It should be emphasized that at this time amphotericin B resistance is a relatively rare event in cryptococcosis, and most amphotericin B treatment failures are generally not related to the development of direct polyene resistance in the relapse isolate but are simply due to the failure of the host and the treatment regimen to eradicate a fully susceptible yeast strain by in vitro testing (31, 35).

On the other hand, resistance of *C. neoformans* to flucytosine treatment has been a potential clinical concern from the beginning, and this problem is related to direct drug resistance. Studies have reported initial drug-resistant isolates (primary resistance) to flucytosine in *C. neoformans* to be as low as <2% (27). However, in the early studies with this antifungal agent, flucytosine therapy used alone for cryptococcal meningitis showed a direct correlation with rising MICs for the relapse isolate, compared with the initial isolate and clinical relapses (27, 243). Further predictions of clinical flucytosine resistance were made on the basis of confirmed cases of clinical relapses or failures, which were also identified by in vitro susceptibility testing in pulmonary cryptococcal infections when flucytosine treatment alone was used (134, 243). It is likely that isolates with MICs for flucytosine of ≤32 μg/ml will respond to conventional doses of this agent when used alone, but those isolates with greater MICs or those that have increased their MIC four- to eightfold from the initial isolate prior to treatment will probably not benefit from continued flucytosine therapy alone. Furthermore, it is not yet clear whether or not combination therapy with both amphotericin B and flucytosine will be useful against those yeasts that are resistant to flucytosine. Clinical experience suggests that the combination of amphotericin B and flucytosine for initial treatment of cryptococcosis may actually reduce the incidence of flucytosine-resistant relapse isolates during or after therapy.

The systemic azoles miconazole (MCZ), ketoconazole (KTZ), fluconazole (FLU), and itraconazole (ITZ) have all been used in the treatment of cryptococcosis

(62, 68, 76, 211, 230, 234, 251, 252, 258). These azoles have potent in vitro anticryptococcal properties, with an activity rating on a weight basis that is ranked as follows: ITZ > MCZ > KTZ > FLU. However, this potency ranking of in vitro anticryptococcal activity may have little clinical relevance, because differences in pharmacokinetics among these compounds will likely require different susceptibility breakpoints for determination of in vitro susceptibility correlations among the various azoles.

We have the most experience with fluconazole in the treatment of cryptococcosis. Early results found that most *C. neoformans* relapse isolates studied by in vitro susceptibility testing were due to strains in which the MICs for fluconazole were similar to those for the original isolate and were relatively low (12, 31, 35). Molecular typing methods have confirmed that most of these failures were caused by true relapses of the original strain rather than by reinfections with new strains (31, 226). These relapses, which occurred primarily in AIDS patients who were receiving azole therapy, did not represent the emergence of secondary azole-resistant strains but demonstrated the essential feature of a severely damaged host defense that could not help in eradication of the yeast despite therapy or simply represented drug compliance problems. These early findings in the AIDS epidemic suggested that despite widespread use of azoles, the apparent incidence of direct azole resistance in cryptococcosis was low. These findings of low numbers of azole-resistant strains might have been predicted by the finding that some azole-resistant *C. neoformans* mutants appear to be less virulent than the wild-type azole-susceptible parent in animal studies (120). However, as the AIDS epidemic has continued, with prolonged and profound immunosuppression combined with the frequent use of azoles for prophylaxis and/or low doses for oropharyngeal thrush in these high-risk patients, there have been increasing case reports of clinical relapse isolates of *C. neoformans* with higher MICs to fluconazole than those in the original isolate. Initially, there were case reports of failures or relapses with *C. neoformans* strains found to have relatively high MICs ($\geq 32 \mu g/ml$) (8, 25, 183, 206, 249). Some of these cases had paired serial isolates that demonstrated more than a fourfold rise in MICs between initial and relapse *C. neoformans* isolates (25, 206). These fluconazole-resistant isolates were generally isolated during exposure to fluconazole during suppressive therapy, but several severe cases of cryptococcosis occurred when low doses of fluconazole were administered for antifungal prophylaxis (20, 249). Clinicians need to be aware that low doses of fluconazole for prophylaxis or suppression do not completely protect patients against infection with azole-resistant *C. neoformans* strains.

Koletar et al. (138) first spotted this trend toward rising azole resistance when they examined the *C. neoformans* in vitro susceptibility testing patterns for fluconazole in 1995. None of 11 strains had MICs of $\geq 32 \mu g/ml$ in 1991, in contrast to the 11 of 20 strains with MICs of $\geq 32 \mu g/ml$ for isolates observed in 1994. In 1994, 13 of 14 *C. neoformans* strains with MICs of $>16 \mu g/ml$ had prior fluconazole exposure compared with 6 of 6 strains with MICs of $<8 \mu g/ml$ in which there was no fluconazole exposure in these patients (138). Recently, Davey et al. (52), from a Reference Laboratory in the United Kingdom, found some upward shift in in vitro susceptibility testing patterns for *C. neoformans* between 1994 and 1996. They have now documented seven relapse cases of cryptococcosis in which sequential isolates correlated with a substantial rise in fluconazole MICs (52). Furthermore, reports in animal models demonstrate that those *C. neoformans* strains with higher flucona-zole MICs isolated from clinical infections have also been shown to be more resistant to treatment in experimental models of infection compared to those with lower MICs (178, 247). Therefore, it is likely that azole-resistant *C. neoformans* strains will continue to increase in severely immunosuppressed patients and that these strains can be identified by present in vitro susceptibility testing methods. There is now a reasonable clinical base for using susceptibility testing on *C. neofor-mans* strains for treatment decisions. For instance, strains with fluconazole MICs of ≥16 to 32 μg/ml by standard testing methods are not likely to respond consistently to this drug treatment alone with standard dosing schedules.

The extent and importance of azole cross-resistance among strains of *C. neofor-mans* remain uncertain and will require in vitro and in vivo correlation, but most of the reported fluconazole-resistant *C. neoformans* strains do remain very susceptible to low concentrations of itraconazole. Also, some of the newer-generation azoles display potent in vitro activity against strains of *C. neoformans* that are relatively resistant to fluconazole by in vitro susceptibility testing (45, 186). In fact, most available azoles appear to have fungistatic activity against *C. neoformans* in vitro, but the newer azoles and polyenes appear to demonstrate more direct fungicidal prop-erties with in vitro susceptibility testing. The success of these new azoles in the treatment of apparent fluconazole-resistant cryptococcosis remains to be examined.

Although in vitro susceptibility testing of antifungal combinations for *C. neo-formans* remains poorly standardized for the clinical arena, the continued examina-tion of drug combinations for synergy and improved fungicidal activity though research protocols may be important to improve the clinical use of drugs. In vitro susceptibility testing of *C. neoformans* has shown a positive interaction for the following drug combinations: amphotericin B–flucytosine (45), amphotericin B–azole (106), amphotericin B–rifampin (89), flucytosine–azole (177), and am-photericin B–flucytosine–azole. In general, no significant antagonism has been noted among the major groups of antifungal combinations during in vitro suscep-tibility testing studies against *C. neoformans*.

A new class of antifungal agents, the 1,3 β-glucan synthase inhibitors repre-sented by pneumocandins or echinocandin B congeners, have been particularly ac-tive in vitro against *Candida* species, *Pneumocystis carinii*, and some pathogenic molds. However, in vitro susceptibility testing shows that this class of agents has relatively poor in vitro activity against *C. neoformans*. The mechanism(s) for the lack of activity of these cell-wall active antifungal agents against *C. neoformans* remains uncertain, but it has been suggested that the 1,3 β-glucan linkage in the *C. neoformans* cell wall is less important than the 1,6 β-glucan linkage. Further molecular studies are needed to identify a specific mechanism(s) for this very wide difference in po-tency between these yeasts and *Candida* species. However, Franzot and Casadevall (84) have shown with in vitro susceptibility testing that a pneumocandin without significant activity by itself can significantly enhance the antifungal properties of both amphotericin B and fluconazole (84). These results suggest that these agents may have some antifungal potential for treatment against *C. neoformans* infections.

An important clinical principle that may be helpful in treatment strategies against *C. neoformans* infections and their known ability to relapse is that all isolates of *C. neoformans* from clinical specimens should be stored for at least 1 year by the clinical laboratory after diagnosis of the initial infection. With this policy, relapse isolate(s) can become a very important aspect of infection management. The relapse isolate can be compared reliably to the initial isolate with in vitro susceptibility testing to determine if secondary drug resistance has developed. In the comparison of primary and subsequent isolates, a significant increase in the MIC (fourfold or greater) for the treatment drugs would suggest that a component of the treatment failure is related to direct drug resistance. In most cases, a new or changed therapeutic regime is probably warranted for treatment of the clinical relapse.

Besides the various differences in in vitro susceptibility testing, such as inocula, pH, media, temperatures, and endpoint determinations that may influence test results, several concepts have emerged regarding *C. neoformans* and its direct interaction with antifungal agents. First, clinical and environmental *C. neoformans* isolates generally have the same antifungal susceptibility patterns (85, 86), but within a limited geographic area, there has been a study that found that clinical isolates were less susceptible to amphotericin B, but not to fluconazole, than were a group of environmental isolates (50). However, this susceptibility distribution between clinical and environmental isolates might change further with frequent high-level exposure to certain antifungal agents in very high risk patient populations. Second, the effect of melanin in the cells may influence certain drug susceptibility results (256). In fact, *C. neoformans* cells grown with L-dopa in the medium and thus fully melanized were less susceptible to amphotericin B than were lessmelanized cells grown in standard medium (255). Third, it is clear that strains exposed to azoles such as fluconazole can change their sterol composition, and in fact, the sterol composition of many strains can vary. A direct correlation between the changing sterol compositions of *C. neoformans* isolates and their measured susceptibilities to fluconazole has not yet been shown (95). Furthermore, subtle differences in sterol composition and polyene susceptibility can be detected as *C. neoformans* strains are passaged through animals (50). These observations, however, could identify a potential increase in drug resistance as more infections occur and more antifungal drugs are used. Fourth, in vitro studies with *C. neoformans* have shown that fluconazole inhibitory activity is enhanced in the presence of serum. Human serum has two components: a macromolecular component with the ability to inhibit *C. neoformans* and a low-molecular-size component ($>$10,000 Da) that synergistically enhances the anticryptococcal activity of fluconazole (174); this interaction is not related to the presence of iron (96). Fifth, in severely immunocompromised hosts, when the burden of yeasts for cryptococcal meningitis can reach $>$10^6 CFU/ml of CSF, it should be expected that secondary drug resistance may occur if mutation rates of the drug target are high, as is found for flucytosine. Finally, postantibiotic effects have been studied for *C. neoformans* strains in their exposure to amphotericin B, flucytosine, and imidazoles. Postantibiotic effects were found for amphotericin B and flucytosine and ranged from 2.8 to 10.6 h and 2.4 to 5.4 h, respectively. Postantibiotic effects were not found with the imidazoles (miconazole and ketoconazole), but decreased growth rates of *C. neoformans* were noted at concentrations as low as 1/1,000 of the MIC for these agents (240).

ANTIFUNGAL DRUG RESISTANCE AND ITS MECHANISMS

Data continue to accumulate in our understanding of drug resistance mechanism(s) with *C. neoformans*. There is both primary drug resistance, such as with the new echinocandin analogs for *C. neoformans*, in which there is no prior exposure to the drug before development of resistance, and secondary drug resistance after drug exposure, which is common with flucytosine treatment. Mechanisms of drug resistance focus on alterations in (i) drug entry, (ii) drug efflux, (iii) target activity, (iv) inactivation of the drug within cells (i.e., degradation), and (v) effects on other enzymes in the target pathway that may circumvent the need for the drug target. With *C. neoformans* some of these resistance mechanisms have been explored. Although few data have yet implicated the various drug efflux pumps, such as ABC transporters and major facilitator pumps, in *C. neoformans* drug resistance, Thornwell et al. (237) have recently identified a gene in *C. neoformans* that appears to be related to genes in other species for multidrug resistance protein pumps. Most of the studies of resistance mechanisms in *C. neoformans* have focused on pyrimidine metabolism and ergosterol synthesis. Pyrimidine metabolism is directly affected by flucytosine, and ergosterol synthesis is directly targeted by polyenes and azoles. Ergosterol is essential for the fluidity and integrity of the fungal membrane and for the proper function of many membrane-bound enzymes, such as chitin synthetase. It is clear that a mechanism(s) for resistance develops in ergosterol synthesis within *C. neoformans*.

Amphotericin B Resistance

Amphotericin B resistance in *C. neoformans* has been found to develop after treatment for cryptococcal meningitis in an AIDS patient. The altered site in this polyene-resistant *C. neoformans* relapse isolate was determined to be a defect in sterol $\Delta^{8\to7}$ isomerase, which did not affect the azole susceptibility of the strain (132). Mutations in sterol biosynthesis such as those at the $\Delta^{5,6}$ desaturase have the potential to produce cross-resistance for both polyenes and azoles. Joseph-Horne et al. (126) were able to isolate amphotericin B-resistant mutants of *C. neoformans* that were able to accumulate ergosterol. Thus, they suggested that at least three categories of amphotericin B-resistant strains can potentially arise: (i) sterol mutants, (ii) amphotericin B and azole cross-resistant mutants, and (iii) amphotericin B-resistant mutants with no azole cross-resistance (126).

Flucytosine Resistance

Primary flucytosine resistance in *C. neoformans* is not common. In fact, in early studies over 90% of pretreatment isolates of *C. neoformans* were relatively susceptible to therapeutic levels of flucytosine (28). However, Block et al. (27) carefully described the development of six cases of stable resistance in *C. neoformans* strains obtained from patients who had failed monotherapy with flucytosine. Pretreatment isolates had been initially susceptible to the drug, but significant changes in the MICs occurred with relapse isolates. It is known that flucytosine is taken up into the cell by a cytosine permease and deaminated into 5-fluorouracil by cytosine deaminase, which is a fungal-specific enzyme; mammalian cells lack this enzyme. 5-Fluorouracil is eventually converted by cellular pyrimidine processing enzymes into 5-fluorodeoxyuridine monophosphate, which inhibits thymidylate synthetase, and 5-fluorouridine triphosphate, which disrupts RNA/protein synthesis.

These cryptococcal isolates by Block et al. (27) had also acquired massive resistance to 5-fluorouracil, and this finding suggested a mutation in the enzymes uridine-5-monophosphate pyrophosphorylase or uracil phosphoribosyl transferase. One isolate appeared to have a mutational defect in the gene encoding for either the cytosine-specific permease or cytosine deaminase of the strain. Conclusions from this careful study were that flucytosine resistance is massive, stable, and nearly always associated with 5-fluorouracil resistance without development of intermediate MICs. The frequencies of appearance of drug resistance mutants in susceptible isolates of *C. neoformans* within the laboratory were found to be <0.001% and averaged 70 ± 17.9 per 10^7 cryptococci (27). These data suggest that resistance is generally caused by a single mutational event within the pyrimidine salvage pathway. Whelan (259) further supported this opinion and described flucytosine-resistant variants in *C. neoformans*. Resistance in *C. neoformans* was found to arise via de novo mutations rather than a recombinational process. These data demonstrated that there were likely to be two nonlinked genes (*FCY1* and *FCY2*) that mutated to result in resistance to flucytosine and that the resistance phenotype was inherited as a simple Mendelian characteristic (259). These genes have not yet been identified by cloning and sequencing, but it is likely that they will encode for enzymes in the pyrimidine salvage pathway or a cytosine permease/deaminase. Therefore, in many respects, our understanding of the mechanism(s) for the common occurrence of flucytosine resistance in *C. neoformans* is mature in that both genetic and biochemical studies have identified probable targets. Further studies in this area could benefit from more molecular biological investigations into specific mutational events and their frequency. Since flucytosine is now generally used in concert with amphotericin B for treatment of cryptococcosis, and this combination appears clinically to reduce the development of flucytosine-resistant strains, the area of flucytosine resistance mechanism(s) has not received much recent research attention.

Azole Resistance

Azole drug resistance has been documented to be increasing in the AIDS population during treatment for oropharyngeal candidiasis and as the use of azoles such as fluconazole has further expanded to prophylactic and suppression indications. Initial in vitro studies of relapse isolates in cryptococcal meningitis suggested that most *C. neoformans* relapse isolates did not show evidence of azole resistance (31, 226). These studies demonstrated that relapses can occur for a variety of reasons, including persistent immunosuppression, the fungistatic treatments employed, medical noncompliance, and/or early withdrawal from suppressive antifungal drug regimens. However, with further experience, there have been a number of relapsed cases associated with evidence of increased drug resistance for *C. neoformans* isolates (8, 20, 25, 138, 183, 206, 249). Studies on the azole drug resistance mechanism(s) for pathogenic fungi, including *C. neoformans*, have focused on several aspects of sterol metabolism. Azole-resistant fungi may have changes associated with a reduced cellular content of azole uptake through either an influx or efflux change, elevated P450 enzymes, or an altered P450 enzyme(s) such as the major inhibition target of fluconazole, the sterol 14α-demethylase. Recent examination of sterol metabolism in fluconazole-resistant strains of *C. neoformans* from relapse isolates in AIDS patients showed that one site of resistance in some strains is at the level of the drug target, P450 lanosterol 14α-demethylase (P45051A1), in which there are most likely alterations in the affinity of the azole for the active catalytic site of the enzyme in the resistant *C. neoformans* strain (141).

For other enzymes in the ergosterol-specific pathway, mutations leading to azole resistance are not as likely to occur in *C. neoformans*. For instance, the inhibitory effects on ketosteroid reduction, which precedes sterol $\Delta^{5,6}$ desaturation in the ergosterol biosynthetic pathway in *C. neoformans* (125, 245), would make mutations for azole resistance in this enzyme unlikely. Also, itraconazole blocks both the lanosterol 14α-demethylase and the NADPH-dependent 3-ketosteroid reductase in *C. neoformans* (245). This dual target activity may have implications both for its potent antifungal activity in vitro and for its reduced development of cross-resistance compared to other azoles. In other studies, a sterol $\Delta^{8\rightarrow7}$ isomerase *C. neoformans* mutant was found not to be resistant to azoles but was resistant to amphotericin B and had a reduced membrane sterol content (132).

On the other hand, *C. neoformans* has been shown by Joseph-Horne et al. (125) to possess the ability to develop cross-resistance to both polyenes and azoles. These investigators found that a reduced cellular content of the azoles accounted for the drug resistance. In addition, these isolates retained high levels of ergosterol and yet displayed cross-resistance to polyenes. These findings suggest the appearance of multidrug transport mechanisms.

The most comprehensive study of mechanisms for fluconazole resistance in *C. neoformans* was performed by Venkateswarlu et al. (248). In this study, seven clinical isolates of *C. neoformans* with in vitro fluconazole resistance were examined biochemically. The fluconazole MICs for three isolates were three- to sixfold higher than those for the initial susceptible isolates, and four isolates possessed high-level resistance in which MICs were 100- to 200-fold higher than those for the initial susceptible isolates. Although the levels of ergosterol in resistant isolates varied, those isolates that contained relatively low levels of ergosterol in their membrane were also resistant to amphotericin B. The differences in strains with low- and high-level resistance to fluconazole were also reflected in differences in their mechanisms for drug resistance. The strains with a low level of fluconazole resistance showed changes in the strains' affinity for the target enzyme (lanosterol 14α-demethylase). In contrast, resistance of the strains with a high level of fluconazole resistance correlated with decreases in their cellular content of the drug. For instance, the 10- to 20-fold reduction in the intracellular accumulation of fluconazole found in the high-level-resistant yeast isolates directly reflected their drug resistance to fluconazole. It is likely that development of multidrug resistance transporters is involved in the drug resistance mechanism(s) for these high-level fluconazole-resistant strains. At least one study has also identified simultaneous cross-resistance to both polyene and azole drugs in *C. neoformans* by the use of in vitro mutagenesis (125). The frequency (10^{-8}) of the event and the relative ease of making these cross-resistant mutants in *C. neoformans* without defects in ergosterol biosynthesis suggest that the multidrug resistance phenomena exist in *C. neoformans*. As previously mentioned, a multidrug resistance-like gene has already been cloned from *C. neoformans* (237).

Although *C. neoformans* strains may be found to have developed azole drug resistance by multiple mechanisms, such as target affinity, drug uptake defects, use of multidrug resistance genes that encode drug efflux pump proteins, or overproduction of the target enzymes (Table 2), it is likely that only one mechanism is present in a single azole-resistant strain.

ANTIFUNGAL AGENTS IN ANIMAL MODELS

Animal models of cryptococcosis have been important links in the development of strategies to treat cryptococcosis. Present models of cryptococcosis allow a focused evaluation of new and old antifungal agents under controlled biological circumstances with proper assessment of drug dosage, drug pharmacokinetics, established endpoints for efficacy, and ability to perform comparative drug studies. These models bridge an important gap between in vitro drug susceptibility screening and human treatment trials. For instance, the models have been used either to stimulate or to discourage specific drug development, and their validation for prediction of clinical efficacy has been shown to be reasonable. In Table 3 is a short synopsis of more than 30 research drug trials in animal models of cryptococcosis. These studies demonstrate the breadth of research evaluations for optimizing the treatment of cryptococcosis.

Although no experimental model can exactly reproduce the human disease of cryptococcosis, studies in guinea pigs, rats, mice, and rabbits have all been used to carefully study therapeutic questions for management of cryptococcosis. The two most developed animal models for studying the therapeutics of cryptococcosis are the mouse and the rabbit.

The mouse model has been the primary model for study of drug evaluations in cryptococcal infections. Fatal cryptococcal infections produced by intravenous, intraperitoneal, intracerebral, or intranasal inoculation of yeast cells have been used to evaluate the efficacy of antifungal chemotherapeutic agents in mice. Mice are inexpensive, immunologically well characterized, and genetically stable. Like immunocompromised humans with cryptococcosis, most strains of mice are relatively susceptible to cryptococcosis. In fact, with the use of mice with specific gene defects, such as $CD4^+$ gene knock-out mice or mice $CD4^+$-depleted through specific antibodies (see chapter 10), the immunological background of these models

Table 2 Known drug resistance mechanism(s) for *C. neoformans*

Year	No. of isolates	Target	Drug resistance (MIC)	Reference
1973	6	Defect cytosine permease/deaminase mutation in UMP pyrophosphorylase	Flucytosine (>64 µg/ml)	27
1994	1	Sterol $\Delta^{8 \to 7}$ isomerase	Amphotericin B (4 µg/ml)	132
1995	1	$P450_{14dm}$	Fluconazole (306 µg/ml)	141
1995	2	$\downarrow\downarrow$ Drug flux	Amphotericin B (5–10 µg/ml) Fluconazole (128–256 µg/ml)	125
1997	4	$\downarrow\downarrow$ Drug flux	Fluconazole (128–256 µg/ml)	248
1997	3	$P450_{14dm}$	Fluconazole (3–6 µg/ml)	248

Table 3 Efficacy of antifungal drugs against cryptococcosis in animal models[a]

Year	Drug(s)	Animals	Efficacy results	Comments	Reference
1973	Amphotericin B/flucytosine	Mice	AmB/FC > SA	Prevention of FC resistance	26
1975	Amphotericin B/flucytosine	Mice	AmB/FC > SA; AmB/FC < SA	Results depend on FC sensitivity	108
1978	Amphotericin B/miconazole	Mice	MCZ = control; AmB = AmB + MCZ	No MCZ benefit	105
1980	Amphotericin B/ketoconazole	Mice	AmB/KTZ > FC/KTZ; AmB/KTZ = SA	Modest benefit of combination	106
1980	Amphotericin B/minocycline	Mice	AmB = AmB + minocycline	No benefit of combination	104
1980	Ketoconazole	Mice	KTZ > control	Better effect in immunologically intact animals	261
1982	Ketoconazole	Mice	KTZ > control	Modest drug effects	47
1982	Liposomal amphotericin B	Mice	L-AmB > AmB	Higher doses increase the therapeutic benefit	101
1982	Amphotericin B/flucytosine; amphotericin B/ketoconazole	Mice	AmB/KTZ or AmB/FC > SA	Combination gives additive effects	200
1982	Amphotericin B, ketoconazole, flucytosine	Rabbits	AmB + KTZ > SA	Drug combination is additive	187
1983	Bay n7133, ICI 153,066, KTZ	Mice	ICI 153,066, KTZ > Bay n7133	Differences in azoles	102
1984	Flucytosine/ketoconazole	Mice	FC/KTZ > SA	Combination increases efficacy	46
1984	Itraconazole/ketoconazole	Mice	ITZ = KTZ	Modest effects	100
1985	Amphotericin B methyl ester	Rabbits	AmB = ↑AME	Higher AME dose successful	188
1986	Fluconazole/amphotericin B/ketoconazole	Mice	FLU or AmB > KTZ	Suppressive effects	54
1986	Itraconazole/fluconazole	Rabbits	ITZ = FLU	Two triazoles equal	193
1987	Itraconazole/flucytosine; fluconazole/flucytosine	Mice	ITZ/FC or FLU/FC = SA	No combination benefit	199
1989	SCH 39304/amphotericin B/fluconazole	Mice	SCH 39304 = AmB > FLU	New azole more potent	208
1989	SCH 39304/fluconazole	Rabbits	SCH 39304 = FLU	Two triazoles equal	196

474

Year	Drug	Species	Result	Finding	Reference
1990	Bay R 3783/fluconazole	Rabbits	Bay R 3783 = FLU	Two triazoles equal	265
1990	Fluconazole/amphotericin B	Rabbits	AmB > FLU	More rapid killing with AmB	185
1991	Flucytosine/fluconazole	Mice	FLU/FC > SA	Synergy with combination	4
1991	SCH 39304/amphotericin B	Mice	SCH 39304/AmB > SA	Positive interaction with azole + polyene	2
1992	SCH 39304/isomer RR 42427	Mice	Isomer RR 42427 > SCH 39304	Isomer improves therapy	5
1992	Amphotericin B colloidal dispersion	Mice	ABCD = AmB	Equal on mg/kg basis	114
1993	Fluconazole	Mice	FLU > control	In vitro–in vivo correlation	247
1993	Itraconazole	Mice	Hydroxylpropyl-β cyclodextrin ITZ > ITZ	Effect of drug vehicle	115
1994	Amphotericin B lipid complex	Rabbits	AmB > ABLC (same dose); AmB = ABLC (high dose)	Importance of drug dose	195
1995	D0870/fluconazole	Mice	D0870 > FLU	New azole treats FLU-resistant strain	45
1995	Itraconazole/flucytosine	Hamsters	ITZ > ITZ + FC	Possible azole + FC antagonism	119
1996	Flucytosine/fluconazole	Rabbits	FLU > FC	No additive effects or synergy with combination	129
1996	Flucytosine/fluconazole	Mice	FLU/FC > SA	Drug dose important	143
1996	SCH 56592/fluconazole	Rabbits	SCH 56592 = FLU	New triazole with less azole resistance	186
1996	ER-30346/fluconazole/itraconazole	Mice	ER-30346 = FLU > ITZ	New triazole with CNS antifungal activity	110
1997	Flucytosine/fluconazole	Mice	FLU/FC > SA	In vitro–in vivo correlation	178
1998	Amphotericin B colloidal dispersion/liposomal amphotericin B/amphotericin B lipid complex	Mice	ABCD, L-AmB > ABLC > control	May be a difference in lipid products for CNS infection	41

[a]Abbreviations: >, efficacy greater than; =, efficacy equal; /, combination; AME, amphotericin B methyl ester; KTZ, ketoconazole; L-AmB = liposomal amphotericin B; MCZ = miconazole; SA = single agent. For other abbreviations, see Table 1. Numbers for certain drugs are pharmaceutical company developmental numbers.

can mimic the cellular milieu found in AIDS patients with cryptococcosis. Another value of the mouse model is in the study of immune-based treatment regimens. Most cloned biological factors, such as colony-stimulating factors (granulocyte [G-CSF], granulocyte-macrophage [GM-CSF], and macrophage [M-CSF]), gamma interferon (IFN-γ), interleukin (IL)-12, and IL-2, that may be used in therapeutic trials in humans are also available in mice. These treatment models, which are established by introducing inocula through intrarespiratory, intravenous, or intracerebral routes, allow evaluation of both new and old antifungal agents and immune modulators in both acute and chronic cryptococcal infections.

The immunosuppressed rabbit model has also been used to study a variety of treatment regimens. It has several advantages: (i) small numbers of animals can be used to determine treatment endpoints, (ii) death as an endpoint is not necessary for drug evaluation, and (iii) the cortisone-treated rabbit with a profound CSF leukopenia during cryptococcal meningitis is similar immunologically to AIDS patients and corticosteroid-treated patients at this important site of infection.

There are several examples of this model's ability to determine the outcome of human treatment when in vitro susceptibility testing or pharmacokinetics could not have correctly predicted the results. First, amphotericin B treatment is rapidly fungicidal in the animal model, and yet this therapeutic outcome occurs despite extremely low CSF drug levels in treated rabbits. For instance, CSF drug levels in this model with a 1-mg/kg/day dose of amphotericin B range between 2 and 4 ng/ml (188). These extremely low levels of amphotericin B in CSF are similar to those found in humans during amphotericin B treatment (24). Furthermore, comparative trials with fluconazole and amphotericin B treatment in the rabbit model show that amphotericin B is even more rapid in its fungicidal activity than fluconazole despite fluconazole's high CSF concentrations during infection (185). A comparative human trial using either amphotericin B (0.5 mg/kg/day) or fluconazole (400 mg/day) supported this finding of an apparently slower fungicidal activity of the azole treatment compared with the polyene treatment (211). In patients receiving amphotericin B treatment, the median length of time to their first negative CSF culture after starting therapy is 42 days, compared with 64 days for patients receiving fluconazole. These findings suggest that amphotericin B for cryptococcal meningitis in rabbits and humans may work through similar mechanisms, such as meningeal membrane drug accumulation or its known immunomodulating properties or both, since the low amphotericin B levels in the CSF would not have directly predicted the rapid therapeutic response. A second observation in the rabbit model of cryptococcal meningitis also predicted a clinical outcome that was later found in humans. In the rabbit model, fluconazole and itraconazole treatments at the same doses have similar outcomes, as measured by a decrease in CSF yeast counts (193). This finding occurs despite high CSF concentrations of fluconazole and unmeasurable CSF levels of itraconazole. Humans show similar pharmacokinetics for these two azoles, and although direct comparative studies have not been performed between the two azoles for initial cryptococcal therapy, itraconazole has been successfully used in the treatment of cryptococcal meningitis despite these poor CSF pharmacokinetics (62, 251, 252). It is possible that itraconazole has some therapeutic advantages in that its lipophilic nature allows it to accumulate in large concentrations on and within host cells (194). This physical property may promote delivery of itraconazole to yeasts through migrating host cells within the

CSF or may simply result in accumulation within the meningeal membrane and exposure to yeast cells as they come in contact with the meninges.

The use of animal models for the study of cryptococcosis has permitted the careful examination of therapeutic strategies and drug regimens to help in the design of human trials and anticipate their outcomes. In Table 3, which summarizes the treatment studies of experimental cryptococcal meningitis, several observations relevant to therapeutic strategies can be identified. First, not all azole compounds are equally effective in treating cryptococcal meningitis (54, 102, 208). Second, in vitro susceptibility testing can make some reasonable predictions for clinical responses of *C. neoformans* isolates to therapy with fluconazole (178, 247). These experiments are essential to our validation of in vitro susceptibility testing. Controlled animal experiments may be more useful for confirming the ability of in vitro susceptibility tests to predict resistant isolates than is direct study of the human disease, with its many confounding factors for the determination of treatment outcomes. Third, drug vehicles to improve pharmacokinetics (115) and the change of drug structure (5) can have an impact on therapeutic outcome in experimental cryptococcosis. Fourth, the animal models have been used to identify several potentially important features of lipid preparations of amphotericin B in cryptococcal treatment. Use of these models has shown that the dose of these preparations is important and that the lipid products may not be equal on a weight basis to the efficacy of amphotericin B in deoxycholate. However, with these lipid preparations, the ability to give higher doses of amphotericin B with less toxicity may eventually give them a therapeutic advantage (195). There are also data that suggest that the potency of the individual lipid preparations for treatment may not be equal. For instance, amphotericin B lipid complex may be less potent in a milligram-per-kilogram dose than is liposomal amphotericin B or amphotericin B colloidal dispersion (41).

Drug combination therapies have been extensively explored in the animal models, and in general, the results have been encouraging for the use of multiple drugs for treatment. Amphotericin B in combination with flucytosine, ketoconazole, or newer triazoles showed at least additive effects (26, 106, 108, 187, 200). In only one study were there reports of antagonism with amphotericin B and flucytosine, and this finding occurred only when the *C. neoformans* strain was initially resistant to flucytosine (108). There was also no therapeutic benefit if amphotericin B was combined with a tetracycline or a first-generation azole (104, 105). Therefore, the consensus of opinion regarding amphotericin B-containing regimens in animal models is that most of them produce a positive response with a probable therapeutic advantage over other single agents, and no consistent findings of antagonism for these combinations have been found. The combination of azole with flucytosine was first explored with ketoconazole in a mouse model (46), and later a series of elegant studies with flucytosine and fluconazole (4, 129, 143, 178, 199) demonstrated that there is no antagonism with this combination; in fact, a therapeutic additive or synergistic outcome with the azole-plus-flucytosine combination was demonstrated in some but not all of these studies. Larsen et al. (143) showed that the doses of the drugs and the degree of infection within the host were the most important factors for the final prediction of outcome of the combination therapies in animal models. Both these factors will be essential to consider as further studies into combination therapies for cryptococcosis are performed in both animal models and humans. Finally, it has been shown in these models that newer antifungal agents, such as the advanced-generation triazoles voriconazole and SCH 56592, which possess excellent inhibitory activity against azole-resistant *C. neoformans* in vitro, also effectively treat experimental cryptococcosis, including meningitis (5, 45, 186, 196, 208).

The animal models of cryptococcosis have given us a window to carefully view our in vitro susceptibility tests for new and old antifungal agents and to predict their success in treating human infections. It is clear that in current animal models of cryptococcosis, we can carefully study effects of immunity, pharmacokinetics, and drug activity with a controlled, reproducible evaluation. The endpoint can be survival, which also allows us to study drug toxicity. However, quantitative yeast counts at the site of infection have been validated and are frequently used as a study endpoint that does not require host survival (186, 188, 193, 195, 196). With the use of several models and attention to drug dosing, these models continue to be a crucial step in the development of drugs for the treatment of cryptococcosis. It is also important to note that no currently available agent or combination of agents has yet been found to be rapidly fungicidal in these animal models. Therefore, the models will allow for detection of any new agents that have more fungicidal activity and potentially a better therapeutic outcome for cryptococcosis.

ANTIFUNGAL AGENTS AND THERAPEUTIC REGIMENS

Although many questions still remain regarding specific regimens and the optimal length of treatment for certain patients, cryptococcosis is the best studied of all disseminated mycoses and one of the most carefully studied of all infectious diseases. Prior to specific antifungal treatments, a uniformly fatal infection such as cryptococcal meningitis can now be cured or at least controlled in the vast majority of patients with our present-day treatment regimens.

Table 1 summarizes a series of comparative studies for the treatment of cryptococcal meningitis and reports their success rates. It should be emphasized that criteria for success varied within each study design, and thus study outcomes may not be directly comparable. In this synopsis, success has been determined by averaging the success rates of the study's highest success category for all the drug regimens. This success figure does not necessarily reflect cure in all cases. On the other hand, some patients were placed in a category of quiescent disease, which was acceptable clinically but not the highest category of the study and thus not considered success for the purpose of this table. We also attempted to identify a more concrete outcome during the study period, i.e., mortality. Although mortality is a substantial outcome, it does not always reflect death from cryptococcal infection but can also be caused by the underlying disease. Finally, an attempt was made to list the drug toxicities in each study. Some studies reported both minor and major toxicities, while others reported only major toxicities. However, this summary gives the breadth of the evaluations and a general appreciation for therapeutic outcomes. In reviewing these studies of approximately 1,500 patients, we can confirm the following. (i) Treatment was successful in at least 50 to 80% of patients with cryptococcosis; (ii) drugs still cause moderate toxicities; and (iii) approximately one in five treated patients will die during treatment. In the following discussion we will analyze the specific antifungal agents used in these studies and amplify and discuss specific treatment options.

Amphotericin B

With over 35 years of amphotericin B use, disseminated cryptococcosis has became a curable infection (19, 69, 75, 216, 225, 241). Initial studies for the treatment of cryptococcal meningitis reported success rates with amphotericin B that ranged between 60 and 70%. The first prospective landmark comparative study with amphotericin B as one of the treatment regimens was by Bennett and colleagues (19), who found that therapy at a dosage of 0.4 mg/kg/day for 10 weeks was successful in 68% of the patients. This study was completed prior to the AIDS epidemic and may have had patients at lower risk for failure, but patients with underlying immunosuppressive diseases were also included. In contrast, in patients with AIDS, the reported success rate of amphotericin B therapy for cryptococcal meningitis with a 10-week regimen of amphotericin B at 0.5 mg/kg/day was only 40% (211). An analysis of these two reports emphasizes the dependence of outcome on many factors, including specific study criteria for cure, the patient's underlying disease or stage of illness, and certain prognostic factors that put the patient at high or low risk for treatment failure. Although it is clear that improved single-drug regimens are needed, amphotericin B is still the most potent and consistently effective single antifungal agent for the treatment of cryptococcal meningitis in humans. Most comparative studies for amphotericin B treatment have been performed in the management of cryptococcal meningitis; however, this agent has been used successfully in treatment of all forms of invasive cryptococcosis, including pneumonia (134) and cryptococcosis of the skin, eye, bone, and prostate. Because of its toxicity and unusual pharmacokinetics, amphotericin B has frequently been used in combination with flucytosine; this successful drug combination will be discussed later. In an attempt to optimize antifungal activity at certain sites of cryptococcal infection, amphotericin B has also been administered directly into the CNS site of infection. Amphotericin B is known to penetrate poorly into the subarachnoid space (23), but intraventricular administration of amphotericin B (0.5 to 1.0 mg per dose) has been successfully employed in cases of severe cryptococcal meningitis to ensure a high concentration of drug at the CNS site of infection. In one retrospective report, five of six patients with cryptococcal meningitis survived treatment with both systemic and intrathecally administered amphotericin B, with an average total intrathecal dose of 20 mg (201). However, it should be emphasized that intraventricular administration of amphotericin B can lead to complications, such as drug-induced arachnoiditis and/or bacterial infection of the Ommaya reservoirs that are used for these injections. Intrathecal use of amphotericin B should be balanced by the fact that there are two other available antifungal agents, flucytosine and fluconazole, that can be administered systemically and produce high concentrations of antifungal activity within the subarachnoid space. Intraventricular administration of amphotericin B for treatment of cryptococcal meningitis has generally been used in severe cases, with a wide range of dosing frequencies and no comparative trials to ensure its actual benefit. In most cases it should probably be considered as a last option, only when standard systemic therapies have failed both mycologically and clinically to control infection. This clinical scenario is quite rare.

Lipid Preparations of Amphotericin B

Formulations of amphotericin B have been created that encase the active antifungal polyene in a variety of lipid vehicles. These new products reduce amphotericin B toxicity, improve its delivery, and potentially allow more amphotericin B to be given to a single patient. It is clear from clinical experience that these formulations have reduced the troublesome nephrotoxicity of the polyene and thus allowed much higher doses of amphotericin B to be given compared to the standard amphotericin B deoxycholate suspension (Fungizone). Human and animal cryptococcal meningitis has been treated with several of these lipid preparations, including liposomal amphotericin B (AmBisome) (42, 148, 217a), amphotericin B colloidal dispersion (ABCD, or Amphocet) (180), amphotericin B lipid complex (ABLC, or Abelcet) (39, 223), and amphotericin B injected into Intralipid (123). Generally, these studies demonstrate some efficacy for the lipid preparations of amphotericin B in treatment of cryptococcal meningitis, but there are no comparative studies between the various formulations in humans (195). The studies generally represent open trials for compassionate use or dose-range studies. In the studies of Joly et al. (123), the lack of an improved amphotericin B toxicity profile with the Intralipid preparation and with no quality control studies for the preparation make it unlikely that amphotericin B infused in Intralipid will become a therapeutic alternative for the treatment of cryptococcal meningitis (123).

These early studies with the lipid preparations of amphotericin B for treatment of cryptococcosis do suggest that both the lipid formulation and the total dose of amphotericin B may be important. For instance, recent comparative treatment studies in a mouse model of cryptococcosis found that AmBisome and ABCD were more active than ABLC on a weight basis (41). The rabbit model of cryptococcal meningitis also emphasizes the importance of careful studies of these new vehicles in respect to their optimal dosing regimen. For example, in this model at comparable doses of 1 mg/kg/day for 2 weeks of amphotericin B versus ABLC, amphotericin B was more effective than ABLC in reducing yeast counts within the CSF. On the other hand, the ability to use a 10-fold higher dose of ABLC (10 mg/kg/day) because of reduced toxicity with the ABLC preparation improved the fungicidal activity of this formulation in the model, and treatment with these higher doses of ABLC even approached a more fungicidal CSF response than did standard amphotericin B treatment (195).

The animal studies emphasize how important it is for clinicians to optimize the therapeutic-toxic ratio for each lipid formulation to obtain the best therapeutic results. A recent study in patients with AIDS that compared amphotericin B at 0.7 mg/kg/day with AmBisome at 4 mg/kg/day in the treatment of cryptococcal meningitis showed, as in the rabbit studies, that higher doses of the lipid preparation actually improved fungicidal activity within the CSF over that with amphotericin B. In this study, the median time to CSF sterilization with AmBisome was 14 days, compared with more than 21 days with amphotericin B (148). These results suggest that AmBisome can effectively help in the management of cryptococcal meningitis in humans. On the other hand, clinicians should await other comparative trials with each lipid product at several doses so they can determine successful outcomes and identify the best doses for an optimal outcome. However, until these studies are performed, it is likely that this group of antifungal formulations will only be used as secondary, expensive alternatives in the management of cryptococcal meningitis in certain high-risk patients and in those with particular concerns about drug toxicities.

Flucytosine

Flucytosine is a pyrimidine antagonist with established potent in vitro antifungal activity against a series of *Candida* species, some dematiaceous fungi, and most *C. neoformans* strains. In its initial antifungal evaluation, it was used to treat cryptococcal meningitis as a single agent. With its excellent CSF penetration in which CSF drug levels can approach 60 to 75% of a simultaneous serum concentration, flucytosine initially enjoyed moderate success in the treatment of cryptococcal meningitis and pneumonia (18). However, very early in its clinical use, a serious drawback to therapy was identified: development of drug-resistant isolates associated with relapses in treatment of cryptococcal meningitis (27, 242, 243). The numbers of yeasts in the CSF during meningitis may exceed 10^6 CFU/ml of CSF in some cases (191). Since mutations in the yeast pyrimidine pathway are frequent and occur at several steps in this pathway (see section on antifungal drug resistance), the possibility of selection for drug-resistant yeasts was found to be high in vitro, and early therapeutic studies proved this concern to be valid in the clinics. The relapse rate of cryptococcal meningitis during treatment with flucytosine alone, with flucytosine-resistant isolates detected by in vitro susceptibility testing, was found to be between 30 and 40%. On the other hand, flucytosine alone has been used successfully in the management of many cases of cryptococcal pneumonia (81, 134, 197, 242), but as Kerkering et al. (134) have pointed out, an occasional case of flucytosine-resistant cryptococcal meningitis has developed after treatment of cryptococcal pneumonia with flucytosine alone in severely immunocompromised hosts. This clinical experience calls for caution in the use of flucytosine as a single therapeutic agent in the treatment of invasive cryptococcosis in a clinical scenario where there is likely to be a high burden of yeasts. This could be during meningitis with a positive India ink examination or in any invasive cryptococcal infection occurring in a severely immunosuppressed patient.

Flucytosine has been very successfully combined with amphotericin B to take advantage of their synergistic or additive mechanisms of actions and their different but complementary pharmacokinetics. First, this combination has reduced concerns about the development of drug-resistant strains and, in fact, flucytosine resistance has not generally been reported during combination therapy. This finding suggests that rapid reduction in yeast cell counts with the combination of agents is an advantage in the prevention of drug resistance to either agent. Second, the combination therapy has allowed a reduction in the amount of amphotericin B needed to treat cryptococcosis, and this helps to limit its known toxicity (19). In fact, a flucytosine-containing regimen as initial therapy for cryptococcal meningitis in AIDS patients was shown to be an independent prognostic factor for reducing the risk of infection relapse during the first year after starting antifungal therapy (210).

As previously discussed, in vitro antifungal susceptibility testing can help in predicting clinical resistance to treatment, particularly with the use of flucytosine. Ideally, initial and relapse cryptococcal isolates will be available for comparison, and any increase in MICs of four- to eightfold would suggest development of drug resistance. Although specific standards for flucytosine breakpoints for cryptococcosis have not yet been confirmed, it is reasonable to use these guidelines for flucytosine: strains with MICs of ≤ 4 to 6.25 μg/ml are susceptible, those of 8 to 16 μg/ml are intermediate, and those of ≥ 32 μg/ml are resistant. Further studies may redefine specific breakpoints, but by using present NCCLS standards, these flucytosine breakpoints seem reasonable.

Use of flucytosine can cause troubling side effects during the treatment of cryptococcal meningitis. Intolerable side effects such as leukopenias (32, 130) and gastrointestinal disturbances (18) have dampened enthusiasm for its use by some clinicians. These side effects have been particularly troublesome to those caring for AIDS patients with cryptococcosis (38). The most serious side effect of flucytosine treatment is bone marrow suppression. In patients with cryptococcal meningitis who were treated with amphotericin B and flucytosine, bone marrow toxicity was correlated with a serum flucytosine level of >100 μg/ml 2 h after an oral dose (32, 130, 229). It was hypothesized that the levels of the direct marrow suppressant 5-fluorouracil would also correlate with toxicity (66), but in one study the level of the parent drug was actually the most predictive value for toxicity (229). Therefore, it is recommended that all patients who receive prolonged, high-dose flucytosine for cryptococcal meningitis should have their flucytosine drug levels monitored soon after starting therapy. Most bone marrow and gastrointestinal toxicity is discovered between days 4 and 14 of treatment (229). Although it is not certain that drug levels correlate as well with gastrointestinal side effects as with pancytopenia, it is probably reasonable to assume that gastrointestinal toxicity is also dose related. It should be pointed out that gastrointestinal toxicity can range from nausea and vomiting to intestinal colitis (18). The serum flucytosine level should also be checked if there is any significant change in renal function during therapy, since flucytosine is primarily excreted by the kidneys. Renal deterioration generally calls for a dose reduction. A reasonable guideline for flucytosine dosage is to reduce the dose by one-half for creatinine clearances between 40 and 60 ml/h and one-fourth for clearances between 20 and 40 ml/h (18). For lower renal function, doses will need to be correlated with frequent drug levels. The drug is dialyzed by both peritoneal and hemodialysis and thus will need to be supplemented after dialysis. The clinician should optimally strive for serum flucytosine levels of 30 to 80 μg/ml 2 h after the dose is given and to avoid levels over 100 μg/ml. In the early studies with flucytosine therapy for cryptococcal meningitis, a dosage of 150 mg/kg/day was used. This dosage is probably not necessary in most patients and may increase the risk for toxicity. In patients with normal renal function, our experience with an initial flucytosine dosage of 100 mg/kg/day divided into four doses will generally achieve therapeutic serum drug levels and may actually reduce the number of drug toxicity cases, even in high-risk AIDS patients. For instance, with a flucytosine dosage of 100 mg/kg/day, van der Horst et al. (246) found that in 202 AIDS patients with cryptococcal meningitis who were receiving flucytosine with amphotericin B or amphotericin B alone, the drug toxicity withdrawal rate was only 3%. The group receiving combination therapy was similar to the group treated with amphotericin B alone, and no bone marrow suppression was detected whether the baseline hematologic values were initially normal or abnormal. Finally, the drug is well absorbed orally, but there is no widely available commercial intravenous preparation.

Azoles

In the last decade, azoles (imidazoles and triazoles) have made a significant impact on the management of cryptococcosis. Currently available azoles are fungistatic against *C. neoformans* both in vitro and in vivo. However, despite the lack of a fungicidal response to azole treatment in the immunocompromised host, the excellent safety profile, pharmacokinetics, and the clinical response of these agents have combined to make them extremely important therapeutic options for the management of cryptococcosis at various body sites.

Miconazole was the first azole agent used for human cryptococcal infections, but it is rarely used for cryptococcosis today. It offers no advantages over the newer triazoles and must be given intravenously, tends to be more toxic, and has limited penetration into the CSF, which is an important clinical site for this yeast. The early limited experience with this azole produced both successes and failures in the treatment of cryptococcosis (234, 258). In fact, because of its limited CSF penetration, Graybill and Levine (103) used intraventricular miconazole successfully to treat at least one case of cryptococcal meningitis.

Ketoconazole was the next azole developed for treatment of human cryptococcosis, and it possesses excellent in vitro activity against *C. neoformans*. A few cases of pulmonary and nonmeningeal cryptococcosis have been reported to be successfully managed with ketoconazole (68). Unfortunately, there is no large clinical series of reported human cases treated with ketoconazole, although it is likely that this drug was used frequently for cryptococcosis prior to the development of the triazoles. Thus, it is likely that for nonmeningeal cryptococcosis, ketoconazole is a relatively safe and cost-effective oral option for treatment of mild cryptococcal infections at doses of 200 to 400 mg/day. On the other hand, standard doses of ketoconazole alone have not been successful in the treatment of cryptococcal meningitis (192). Its limited penetration into CSF and its reduced accumulation in host cells compared with other clinically available triazoles make this drug unlikely to be useful alone in the treatment of meningitis. If ketoconazole is used for treatment of cryptococcosis, it is recommended that the clinician be certain that concomitant CNS disease is not present. Ketoconazole also has been found to be generally less effective in the treatment of invasive mycoses in severely immunosuppressed patients; many cases of cryptococcosis occur in this high-risk group of patients. Finally, ketoconazole has hormonal toxicities and drug absorption concerns that make it less attractive as an alternative to the triazoles.

Itraconazole

Itraconazole, a first-generation triazole, has been used successfully in a substantial number of patients for initial therapy of cryptococcosis, including those with meningitis (62, 251–253). In vitro, *C. neoformans* is extremely susceptible to itraconazole, and itraconazole has been effectively used to treat animals with invasive cryptococcosis (193, 244). However, there has been no direct comparison between itraconazole and another triazole, fluconazole, in the initial treatment of human pulmonary or disseminated cryptococcosis. Therefore, a judgment on the best triazole for initial management of cryptococcosis cannot be made with any precision. On the other hand, several comments about the treatment of cryptococcal meningitis with itraconazole should be made. First, despite its poor penetration into CSF, it can be used to successfully treat cryptococcal meningitis. This fact has been documented in both human and animal infections (62, 193, 251). Part of the explanation for this efficacy, despite poor biological fluid levels, may reside in its lipophilic characteristics, which allow it to attach to and accumulate in host cells and tissue (194). Along with its potent direct anticryptococcal activity, the drug may reach yeast cells carried by the host cells. Second, there can be significant oral absorption difficulties with this drug, and some patients will simply not absorb the drug. This drug absorption problem may be improved with a new beta-cyclodextrin elixir formulation. However, itraconazole serum levels will probably need to be measured during any treatments of serious cryptococcal infections to ensure drug absorption. Although drug level guidelines for successful management of cryptococcosis with itraconazole are uncertain, it would be encouraging to have peak levels (1 to 2 h after dose) >1 to 2 μg/ml. Third, the use of 400 mg/day of itraconazole or fluconazole for consolidation therapy in cryptococcal meningitis from weeks 2 through 10 of infection in AIDS patients, after induction therapy with amphotericin B or amphotericin B plus flucytosine, showed a similar clinical efficacy as judged by elimination of symptoms in 70 and 68% of patients, respectively (246). Therefore, no difference in final outcome between the two drugs during this phase of treatment was found, although there was a slightly greater sterilization of CSF with fluconazole (97%) than with itraconazole (92%). Fourth, comparative data between itraconazole and fluconazole for the suppressive phase of treatment for cryptococcal meningitis in AIDS patients show convincingly that fluconazole is superior to itraconazole in this stage of management. In a study of 107 AIDS patients receiving 200 mg/day of itraconazole or fluconazole for suppressive therapy, there was a relapse rate of 23.6% versus 3.8%, respectively (210). Finally, in the sum total of studies, although there has been more experience with fluconazole treatment for management of cryptococcosis than with itraconazole treatment, it is clear that itraconazole can effectively help in the management of cryptococcal infections. It is also important that we gain more clinical experience with this antifungal agent in immunocompetent hosts.

Fluconazole

Fluconazole has been established as a primary drug in the management of cryptococcosis. With its excellent CSF pharmacokinetics (9, 190, 193) and favorable animal studies (193), fluconazole was initially examined for the treatment of cryptococcal meningitis in patients with AIDS (230). Early pharmacokinetic studies showed that in humans the area under the CSF concentration-time curve was 86% of the area under the plasma concentration-time curve, and levels in CSF generally exceeded MICs for *C. neoformans* over the entire 24-h period after dosing (9). Therefore, in a large comparative treatment trial in 191 AIDS patients with cryptococcal meningitis by Saag et al. (211), the complete success rates, which included sterile CSF cultures at 10 weeks, for initial therapy with amphotericin B (0.5

mg/kg/day) versus fluconazole (400 mg/day) were only 40% and 34%, respectively. However, if quiescent disease, which has been defined for cryptococcal meningitis outcome endpoints as clinically stable, progressive improvement, or resolution of symptoms but persistently positive cultures from any body site or fluid, is included in the success category, a positive overall response rate for this infection was raised to 67% in the amphotericin B group and to 60% in the fluconazole group. The mortality rates in the amphotericin B and fluconazole treatment groups at 10 weeks were 14% and 18%, respectively. However, this study did find a slight trend toward earlier deaths in the fluconazole treatment group and a median length of time to the first negative CSF culture of 42 days in the amphotericin B treatment group compared with 64 days in the fluconazole treatment group. This pivotal study supported the use of fluconazole as one therapeutic option for initial therapy in patients with cryptococcal meningitis and AIDS. Although this study used a dosage of 400 mg/day of fluconazole, higher dosages (800 mg to 1,600 mg/day) have occasionally been successful in salvage therapy of cryptococcal meningitis for prior therapeutic failures (21). In a study by Larsen et al. (145) that used dose escalation in an attempt to optimize the best dose of fluconazole for treatment in humans, it remains unclear what is the best dose for fluconazole treatment of cryptococcal meningitis; however, it probably is in the range of 400 to 800 mg/day in AIDS patients.

It should be emphasized that fluconazole acts fungistatically both in vivo and in vitro, and therefore consistent cures in AIDS patients are not considered a goal with this antifungal agent. Also, in two small studies of treatments for cryptococcal meningitis, both fluconazole and itraconazole used for initial therapy of cryptococcal meningitis were not as effective as a combination regimen with amphotericin B and flucytosine (55, 146). At present, amphotericin B-containing regimens are probably more frequently used than triazoles as induction therapy for cryptococcal meningitis in both AIDS and non-AIDS patients.

Despite extensive and comparative studies in AIDS patients, the use of fluconazole in the management of cryptococcosis in HIV-negative patients is less well studied. However, its safety and ease of administration make fluconazole an attractive therapeutic option. On the other hand, the goal in most non-HIV-infected patients with cryptococcosis is cure without the need for lifelong suppressive therapy, and its fungistatic properties may not be ideal for this treatment. One study by Dromer et al. (73) reported on 83 HIV-negative patients with meningeal or extrameningeal cryptococcosis. In 25 patients with meningitis, there was a 68% cure rate with fluconazole treatment alone. Similarly, in a group of 35 patients who received amphotericin B alone, there was a 74% cure rate. This retrospective study was not a direct comparison and should probably not be used to suggest that amphotericin B and fluconazole are equal in efficacy within this population. However, it does support the probability that fluconazole can be used effectively in some aspects of treatment for cryptococcal meningitis in non-AIDS patients. Similarly, Yamaguchi et al. (267) treated 19 HIV-negative patients with fluconazole alone (200 to 400 mg/day) for cryptococcal meningitis and reported that 89% were mycologically cured. However, proper dosing and comparative studies on outcome with other standard regimens will need to be performed before single triazole therapies with fluconazole for cryptococcal meningitis can be recommended as first-time initial therapy in HIV-negative patients.

In the study by Dromer and her colleagues (73), 14 patients without meningitis completed a treatment course of fluconazole, and all were considered cured of their cryptococcosis. In the Yamaguchi study (267), 25 patients with either pulmonary or other non-CNS cryptococcal infections had an 89% success rate with fluconazole therapy, and in a small study four patients with pulmonary cryptococcosis treated with 400 to 600 mg of fluconazole for 3 months were cured (269). Fluconazole is likely to be a major therapeutic option in HIV-negative patients without CNS involvement. The dose and duration for fluconazole therapy in HIV-negative patients with non-CNS cryptococcosis may need to be individualized depending on the patient's underlying disease, but we recommend a dosage for nonmeningeal cryptococcosis of at least 200 to 400 mg/day for 3 to 6 months, pending further clinical experience. It should also be mentioned that many trials for disseminated cryptococcosis developed defined lengths of therapy ranging from 4 to 10 weeks. With safer, oral antifungal drugs, the length of treatment for invasive cryptococcosis does not need to be so confining and probably should be longer.

Combination Therapy

The use of combination therapy for infectious diseases has been a cornerstone of many therapeutic strategies. For instance, combination therapy has been used for necessary synergistic microbial killing, such as the use of a β-lactam antibiotic and an aminoglycoside for treatment of enterococcal endocarditis, or for prevention of drug resistance with the multiple drug regimens used for treatment of mycobacteriosis or HIV infection. For almost 20 years, the combination of amphotericin B and flucytosine has been used in the treatment of cryptococcal meningitis. This combination of drugs, which have different mechanisms of antifungal action, was hypothesized to both increase fungicidal activity, and thus reduce the length of a treatment course with the toxic amphotericin B, and reduce the incidence of relapses by preventing drug resistance to flucytosine. In a series of excellent clinical studies, these actions with this drug combination were confirmed (19, 55,69, 121, 146, 227, 241, 246). After encouraging preliminary studies for amphotericin B and flucytosine, a pivotal randomized collaborative trial of 61 patients by Bennett et al. (19) showed that amphotericin B (0.3 mg/kg/day) and flucytosine (150 mg/kg/day) for 6 weeks versus amphotericin B (0.4 mg/kg/day) for 10 weeks produced similar outcomes prior to the AIDS epidemic. Although there were no differences in the final outcome of infection, the combination of amphotericin B and flucytosine was found to result in more cured or improved patients, fewer failures or relapses, and a statistically improved sterilization of the CSF after 2 weeks of treatment compared to amphotericin B alone. In fact, 100% of patients treated with this drug combination had sterile CSF at the end of 2 weeks. These results were achieved with less nephrotoxicity in the combination group compared to the higher doses of amphotericin B used in the other arm of the study.

These findings were examined further in a follow-up study by Dismukes et al. (69) that used the observation that rapid CSF sterilization by the drug combination may be exploited to reduce the length of therapy for cryptococcal meningitis. Amphotericin B (0.3 mg/kg/day) plus flucytosine (150 mg/kg/day) therapy was found to give comparable results when given for either 4 or 6 weeks of treatment. In fact, the rate of success in this 194-patient study when both cure and improvement were considered reached approximately 85% in each group. However, a subset of severely immunosuppressed patients, such as recipients of solid organ transplants, failed the short-course (4-week) regimen, and it was clear even prior to the AIDS epidemic that severely immunocompromised hosts generally required longer therapy than 4 weeks for cryptococcal meningitis. For example, four of the first five renal transplant patients randomized to the 4-week regimen relapsed, and randomization of this risk group to the 4-week regimen was discontinued. Even 6 weeks of treatment may not be long enough, since three patients relapsed after 6 weeks of treatment, and Watson et al. (257) found that one of nine transplant patients relapsed after 6 weeks of combination therapy. Thus, the insights from these large comparative studies included the importance of identifying prognostic factors in tailoring the length of therapy for low- and high-risk patients.

The combination regimen of amphotericin B and flucytosine can be effectively given in non-AIDS patients with cryptococcal meningitis. In fact, with the AIDS epidemic and the high frequency of corticosteroid-induced cryptococcal infections, along with new, safer oral agents, there has been less enthusiasm for continuing short-course (4-week) definitive combination therapy of cryptococcal meningitis in both immunosuppressed and nonimmunosuppressed patients. However, the groundwork was laid by these studies for development of new strategies in the management of cryptococcal meningitis. For example, if an objective endpoint to drug therapy for cryptococcal meningitis is examined, such as CSF sterilization, the combination of amphotericin B and flucytosine appears to be better than either agent alone. For example, in non-AIDS patients, the combination of amphotericin B and flucytosine showed rapid and consistent CSF sterilization within 2 weeks of treatment initiation (19, 69).

Two small studies by Larsen et al. (146) and DeGans et al. (55) with several dozen AIDS patients using amphotericin B and flucytosine demonstrated that this combination was very effective in the treatment of AIDS patients and, in fact, appeared to be better at CSF sterilization than treatment with single triazoles such as fluconazole or itraconazole. When a series of studies on treatment of cryptococcal meningitis in AIDS and non-AIDS patients are reviewed, more than 90% of patients without AIDS and 60 to 70% of patients with AIDS have sterile CSF cultures after 2 weeks of treatment with the combination of amphotericin B and flucytosine.

These findings created a basis for a comparison study of amphotericin B and amphotericin B plus flucytosine for 2 weeks as initial (induction) therapy for cryptococcal meningitis; patients were then switched to either fluconazole or itraconazole (consolidation) therapy for another 8 weeks in the initial treatment of cryptococcal meningitis in AIDS patients (246). This large pivotal study of induction and consolidation treatments in 381 AIDS patients is encouraging for this new treatment strategy for cryptococcal meningitis. In this study, the negative CSF culture rates at 2 weeks were generally lower than those in other reports; 60% of patients with the combination versus 51% of patients with amphotericin B alone were culture-negative. This result did not reach statistical significance ($P = 0.06$), nor did differences in clinical outcome features between groups. However, a recent review of another study on antifungal drug suppression after cryptococcal meningitis found that an initial flucytosine-containing regimen did correlate with a reduced risk for infection relapse (210). Finally, an overall mortality of <11% was recorded with this design, and during induction therapy there was only a 5.5% death rate (246).

Thus, this new strategy for the initial treatment of cryptococcal meningitis has been endorsed in AIDS patients. This regimen also includes an initial higher induction dose of amphotericin B (0.7 mg/kg/day) than that given in previous studies plus a lower dose of flucytosine (100 mg/kg/day) for the 2 weeks of induction therapy. De Lalla and colleagues (57) have confirmed the principle of high doses of amphotericin B for short induction periods in the treatment of cryptococcosis. In their study, 31 consecutive patients with AIDS and cryptococcal meningitis received amphotericin B at 1 mg/kg/day and flucytosine (100 to 150 mg/kg/day) for 14 days of induction therapy with a 93.5% success rate for negative CSF cultures and symptoms at 2 months. Thus, aggressive combination therapy for reducing CSF yeast counts appears to be a well-established goal of initial therapy. With a lower burden of yeasts, treatment with an azole such as fluconazole (400 to 800 mg/day) is given for 8 weeks. However, in the pivotal AIDS study there was no difference in outcome between treatment with 400 mg of fluconazole or 400 mg of itraconazole in the consolidation phase of this regimen.

After an initial 10-week treatment period, an AIDS patient is then treated with continuous fluconazole (200 mg/day) as suppression therapy. It is hoped that better drugs and improvement in the effect of antiretroviral agents on the immune system and on viral infection will eventually allow us to stop medication. On the other hand, the optimal use of the drug combination and treatment strategy in HIV-negative patients in the initial treatment of cryptococcal meningitis remains uncertain, since there are no definitive studies on the induction-consolidation regimen that was so successful in AIDS patients. However, it is our experience that a regimen of induction followed by consolidation similar to that used in AIDS patients is reasonable to follow in non-AIDS patients. A recent retrospective review of approximately 300 non-AIDS patients with cryptococcosis found that this strategy is now very commonly used in clinical practice (181). However, it should be emphasized that the present goal in non-AIDS patients with cryptococcal infection is cure of the infection, and this fact might incline clinicians to use more standard regimens. Also, since we commonly treat many invasive mycoses, such as the dimorphic fungi, for at least 6 months, it is likely that the length of therapy with follow-up triazoles may be extended from 3 to 6 months to a year, since relapse rates for standard regimens in HIV-negative patients are reported to be 15 to 30% and generally occur in the first year after treatment.

There has also been interest in an oral combination regimen for the treatment of cryptococcal meningitis. The concern about the development of flucytosine resistance and fluconazole's fungistatic activity when each is used alone supported investigations into combining these oral agents. There are excellent data from animal studies (Table 3) that favor the added potency of these two agents together. A human trial with a 75% success rate was reported by Larsen et al. (145) and supports the use of combined flucytosine (150 mg/kg/day) and fluconazole (400 mg/day) (145). Mayanja-Kizza et al. found that fluconazole (200 mg/day) and flucytosine (150 mg/kg/day) for 2 weeks improved survival at 6 months compared to monotherapy in an African study (161a). The agents have different mechanisms of antifungal action, and both penetrate very well into the CSF. However, a review of the experience to date does not yet support the advantages of this combination oral regimen over conventional treatments, and its use will likely be decided on a case-by-case basis. In non-AIDS patients, a clinical trial of this combination compared to amphotericin B plus flucytosine was attempted by the NIAID Mycoses Study Group but was not completed. Several of the patients with severe liver disease died with this drug combination. Since in our opinion this is a very high-risk population, it is not clear that the combination regimen increased hepatic toxicity in patients with severe liver disease. However, anyone with severe liver disease who is treated with flucytosine- or fluconazole-containing regimens probably needs very careful monitoring, and it may be better to use amphotericin B alone in these cases.

Azoles and polyenes are another potential combination of agents for the treatment of cryptococcosis. These classes of drugs have a potential antagonism, based on similar primary mechanisms of action at the sterol membrane. In cryptococcosis, however, this antagonism has not been found (233). In vitro susceptibility testing and animal models suggest that there may actually be an additive effect for azoles and polyenes in the treatment of cryptococcosis (Table 3). In humans, there has now been extensive experience with sequential exposure of these two classes of drugs, in which both drugs are present in the host at the same time without an apparent negative impact on infection. It is also likely that there have been individual patients treated with amphotericin B and azole together, but there have been no reports of clinical failures related to apparent antagonism with this drug combination. However, any potential benefit of amphotericin B combined with azoles such as ketoconazole, fluconazole, or itraconazole remains undetermined until prospective, randomized trials are performed.

Finally, the use of multiple drug combinations with different mechanisms of action against these eukaryotic pathogens, similar to multiple drug treatments in cancer chemotherapy, needs to be explored in an attempt to improve fungicidal activity. It is our belief that as future agents are discovered, most deep-seated mycoses will be treated with combined agents. Just-Nubling et al. (127) reported triple drug therapy with polyene-azole-flucytosine in an open trial of cryptococcal meningitis in 42 AIDS patients, with over 90% initial success rates in controlling infection. However, except for the well-studied combination of amphotericin B and flucytosine in cryptococcal meningitis, other combination drug regimens should probably be used only with careful monitoring in individual cases until proper dosing and defined success rates with long-term follow-up are known.

TREATMENT CONCEPTS

Criteria for Success

In cryptococcal meningitis, "cure" is a simple term: the eradication of the organism and elimination of symptoms. In non-AIDS patients in whom a treatment goal is to stop medication, the two most important factors that suggest relapse and require reinstitution of treatment are (i) a positive culture and (ii) development of new or persistent neurological symptoms. All other changes in laboratory and clinical parameters are considered within the individual's total clinical condition and are not by themselves criteria for failure or relapse. These parameters include positive India ink examinations and elevated serum or CSF polysaccharide antigen titers.

In AIDS patients, the criteria for success with present antifungal regimens has been modified. Although many patients with HIV infection may be cured with initial therapy, the high rate of relapse, which approached approximately 50% in the early studies of cryptococcosis in AIDS patients (38, 77, 139, 273), has led to the goal of infection suppression as a marker for success. A group of AIDS patients has also been classified for clinical trials as having quiescent disease (202). These patients are defined as clinically stable, progressive improvement or resolution of symptoms but persistently positive CSF cultures. However, it is reasonable that clinicians strive to eliminate quiescent disease when it is present. Further studies with new agents and disease markers should attempt to identify and predict a true cure of cryptococcosis in AIDS patients. Specifically, the impact of highly active antiretroviral therapy (HAART) on cures for cryptococcal meningitis will need to be assessed. The only data on this subject suggest that lack of zidovudine therapy prior to treatment of cryptococcal meningitis was a risk factor for failure (211).

Length of Treatment

In non-AIDS patients, the initial treatment regimens for cryptococcal meningitis have ranged from 4 to 10 weeks. Documented relapse rates can still approach 15 to 20% with these regimens (19). In fact, in the largest drug combination study in non-AIDS patients, it was found that the regimen failed in 63 of 194 patients (33%) (69). Prolonged consolidation treatment with triazoles for 3 to 6 months and possibly up to a year should be considered. Prognostic factors such as those described by Diamond and Bennett (65) and the NIAID Mycoses Study Group Trials (69), however, remain important guidelines in establishing the risk of relapse and thus identify those who may require prolonged therapy. Risk factors will play a large part in stratifying patients to short- or long-term initial therapies. However, all patients without AIDS and with cryptococcal meningitis should be closely followed for 1 year, since the vast majority of relapses will occur within 3 to 6 months after therapy is stopped; rarely will relapse occur more than 1 year after therapy has been discontinued. In fact, a normal lumbar puncture 1 year after cessation of treatment is an excellent indication that cure has been achieved. It should be mentioned that successfully managed non-AIDS patients with cryptococcal meningitis may actually have a low-grade CSF pleocytosis of <10 to 50 cells/mm^3 for up to 6 months after treatment, but generally the CSF formula is completely normal by 1 year after treatment. Lumbar punctures should be performed during the follow-up period of cryptococcal meningitis whenever indicated by a negative change in the clinical status of the patient. On the other hand, a minimal number of lumbar punctures for management in a clinically well-appearing non-AIDS patient could be performed at the end of 2 weeks of therapy to ensure sterilization of the CSF, at the end of therapy, and at the 6-month to 1-year follow-up.

In AIDS patients, the length of therapy and follow-up is indefinite. In the first decade of cryptococcal meningitis and HIV infection, the relapse rates soared when therapy was stopped. Relapse rates of 50 to 60% were reported, with a corresponding reduction in life expectancy (38, 77, 139, 273). These findings led to the concept of chronic suppressive therapy after the initial treatment. Crucial pivotal studies showed that oral fluconazole therapy at a dosage of 200 mg/day for 1 year could reduce relapse rates to less than 5% (29, 204). When directly compared to amphotericin B, this regimen was also shown to be superior to suppressive doses of amphotericin B at 1.0 mg/kg intravenously once per week (204). Itraconazole has also been shown to be successful for suppression of infection (56). However, in a comparative study at 200 mg/day doses, itraconazole with its less reliably absorbed capsules was found to be significantly less effective than similar doses of fluconazole (210). Although the concept of chronic indefinite suppressive therapy for cryptococcosis in AIDS has been established by the medical community, it has not actually been proved to be effective for longer than 1 year after treatment. The long-term impact of these suppressive regimens on late relapse rates, the effect on survival and relapse rates during improved immunity of the underlying disease with new antiretroviral regimens (HAART), and the influence on fungal resistance patterns in colonizing *Candida* flora and cryptococcal isolates will need further studies to ascertain their importance to management of these patients.

Primary isolates of *C. neoformans* for all patients should ideally be stored for at least 1 year after diagnosis. Although relapse isolates may remain susceptible in vitro to the treatment agents, drug-resistant isolates are increasing (see drug resistance section). It is important to compare by MIC testing the susceptibility of the relapse isolate to the original isolate in all cases of relapses. An isolate that shows several dilutions (four- to eightfold) or greater increases in MICs to the administered drug should be considered relatively resistant to this drug, and either the drug should not be used in the retreatment regimen or significantly higher doses should be considered. There remains no consensus on the best regimen for treatment of clinical relapses, and the clinical situation may dictate strategy. However, assuming the yeast strain has not acquired high-level drug resistance, most patients, with or without AIDS, will respond to the same initial induction regimens previously administered, but given for longer periods of induction and generally with higher drug doses.

Immune Status

Every attempt should be made to improve the immunity of the host during and after treatment. In AIDS patients, cryptococcosis occurs generally with $CD4^+$ counts below 100 cells/μl. It is reasonable to develop an aggressive antiretroviral regimen (HAART) to reduce viral burden and potentially improve $CD4^+$ counts and function.

Immunosuppressive agents should also be eliminated or reduced. If possible, corticosteroid treatment should be stopped, but if necessary for treatment of the underlying disease, it is reasonable to strive for a dose of prednisone of $\leq$20 mg/day. Cyclosporine and tacrolimus possess potent anticryptococcal activity in vitro and sometimes in vivo (166), but their immunosuppressive activities generally dominate the clinical impact on this infection, and they generally produce a negative effect on the clearance of *C. neoformans*. For instance, organ transplant recipients who take these immunosuppressive drugs are still at risk for cryptococcal infection (see chapter 13). Because organ transplants now include livers, lungs, and hearts, which, unlike kidneys, cannot be sacrificed during infection management, the adjustment of immunosuppression becomes even more crucial. Antifungal drugs such as amphotericin B and azoles used for treatment of cryptococcal meningitis in this transplant population can also have profound drug interactions and resulting toxicities. For instance, azoles raise levels of dilantin, and rifampin reduces the effect of azoles. The interaction of cyclosporine and tacrolimus with amphotericin B to produce further nephrotoxicity must always be considered and may make a lipid formulation of amphotericin B attractive in this clinical setting.

Site of Infection

Because *C. neoformans* has a unique propensity for invading the CNS, it remains a reasonable practice to exclude meningitis in all patients in whom the yeast is identified from another body site and who have defined risk factors for dissemination even if they are asymptomatic. Knowledge of CNS involvement may change prognosis, follow-up strategies, and the initial treatment regimen selected. Whereas the treatment regimens that are successful for meningitis will also be adequate for invasive disease in other organs, a negative CNS evaluation would make consideration of lower doses, shorter therapies, and non-amphotericin B-containing regimens a possibility. What remains uncertain is whether a low-risk, asymptomatic patient should have a rule-out lumbar puncture when *C. neoformans* is isolated from another site such as the lung. These diagnostic decisions will need to be made on an individual basis, but many asymptomatic patients without risk factors for cryptococcal dissemination probably do not have silent cryptococcal meningitis.

There are several body sites that merit special recognition. Unlike the CNS, the second most important site of infection, the lung, has had fewer well-controlled clinical studies to gain insight into treatment outcomes (62, 68, 73, 75, 81, 109, 134, 197, 242, 252). Whether treatment is indicated and, if so, the most appropriate therapy to be used is still controversial in the immunocompetent and asymptomatic host. For example, immunocompetent hosts with single asymptomatic pulmonary nodules or yeast isolated from sputum without an infiltrate may not need treatment. However, if the yeast is viable by culture in the biopsied lesion or sputum, there is also a reasonable argument for the use of a relatively nontoxic oral azole regimen to potentially eliminate the yeast from the host and thus help reduce the risk for reactivation and dissemination of infection with a future immunosuppressed event. Those patients with symptoms, multiple nodules, masses or pseudotumorlike lesions, pleural effusions, or extensive interstitial lung disease require antifungal therapy. Most cryptococcal empyemas or effusions (78, 212, 271) can be managed with antifungal chemotherapy without surgical drainage, but there can be complex cases where tube drainage is considered beneficial (236). Prior to the development of amphotericin B therapy, surgical resection of symptomatic pulmonary lesions was a main therapeutic option. Surgical resection of pulmonary lesions today is rarely needed as a therapeutic modality with our present antifungal agents, but occasionally large pseudotumorlike masses unresponsive to antifungal chemotherapy have been surgically removed and the patient has improved (109, 118, 150, 159). However, it is unlikely that surgery will completely remove the entire focus of infection in the hilar lymph node complex (10, 213), so antifungal chemotherapy should also be given if the goal is to eliminate the infection. Despite suggestions in the literature, there is no convincing evidence that surgery on the lung of a patient with pulmonary cryptococcosis predisposes to dissemination of this infection (159).

In immunocompromised patients, including those with HIV infection, those receiving corticosteroid therapy, and those with an underlying immunosuppressive disease such as lymphoma or a solid organ transplant, it is clear that treatment is required for any isolation of *C. neoformans* from the respiratory tract. These patients should also definitely be checked for evidence of CNS infection at presentation. It should be noted that not all immunocompromised patients, even those with HIV infection, will necessarily be symptomatic on isolation of this yeast from clinical specimens (165), but unlike immunocompetent patients, they require therapy irrespective of clinical status since the risk of disseminated infection is so high.

The optimal treatment regimen for pulmonary cryptococcosis is not well defined, but several principles should be followed. If the patient is severely ill, then amphotericin B with or without flucytosine should probably be used in doses comparable to those in meningitis treatment regimens. If there is concern about nephrotoxicity, a lipid preparation of amphotericin B might be substituted. For those infections that are limited to the lung and not life-threatening, consideration can be given to an oral regimen such as ketoconazole, itraconazole, or fluconazole at initial doses of 400 mg/day. The lack of toxicity, excellent pharmacokinetics, and experience in cryptococcal meningitis make fluconazole probably the first choice for pulmonary disease in these cases. Flucytosine alone at 75 to 100 mg/kg/day can probably be used if the burden of yeasts in tissue is low and the patient is not severely immunosuppressed (134). It is possible that these oral regimens (triazoles and flucytosine) will fail after 2 to 3 weeks of therapy in symptomatic patients and that the patient will then rapidly respond to the institution of amphotericin B therapy. The optimal length of therapy for life-threatening infections is unknown, but guidelines for meningitis probably should be followed. For less serious infections treated with the use of azoles, it is likely that a 3- to 6-month course is appropriate. However, patients with severe immunosuppression such as organ transplant recipients and AIDS patients may require longer treatments, and consideration of chronic suppressive therapy, depending on the risk group, may be necessary.

CNS Torulomas or Brain Abscesses

Most cases of torulomas or cryptococcal brain abscesses can be successfully managed with systemic antifungal agents alone. However, there are several aspects about treatment of these CNS lesions that need to be emphasized. First, in AIDS patients and other severely immunocompromised patients, concomitant infections with other microorganisms or tumors can simultaneously occur, and a biopsy of the lesion may be required to rule out another diagnosis. Second, cryptococcomas can actually enlarge or new lesions may appear on serial CT or NMR scans during therapy; generally, these patients are clinically improving, and this finding probably represents an improved inflammatory reaction at the site of infection during successful killing of the yeast and does not necessarily represent clinical failure. With continued treatment such as an oral triazole and close follow-up, these lesions will disappear or regress over 3 months to a year but may last even longer. At times, the persistence of CNS lesions will help extend the length of treatment, and it is common to have patients with cryptococcomas on triazole therapy for 2 years. Finally, some lesions larger than 3 cm may require surgical removal to be managed successfully (90), but each of these cases needs to be considered individually, and probably a therapeutic trial with antifungal chemotherapy should be given prior to a decision to surgically remove the lesion. The parenchymal lesions may also be associated with the development of hydrocephalus, which will require appropriate shunting procedures.

Prostate

There are clinical data suggesting that the prostate may serve as a protective sanctuary for *C. neoformans* during treatment (144, 228). This finding of cryptococcal persistence in prostatic fluid or urine after therapy has been best described in patients with AIDS. In careful follow-up of patients with HIV infection, cultures of routinely voided urine, urine obtained by postprostatic massage, and seminal fluid were frequently positive at the end of therapy (144, 228). The significance of this prostatic focus for disease relapse is actually much less important in this population, because at this time chronic suppressive therapy is generally recommended. On the other hand, in non-AIDS patients, when stopping medications is considered, a urine culture after prostatic massage may identify residual viable organisms in some patients with disseminated disease and thus demonstrate a need for a prolonged treatment course. However, rarely are patients with positive cultures symptomatic, with signs of prostatitis as a guideline for treatment length (30). In the non-AIDS population the rates of infection at this site during disseminated disease are not defined, and thus the need to evaluate the prostate in these patients with disseminated disease in other sites cannot be generally recommended. However, occasionally the bloodstream may be seeded during a prostatectomy, as viable yeasts are apparently released into the circulation with surgical trauma (198). Despite this potential sanctuary from treatment, apparent localized infection of the prostate is generally treated similarly to other mild extraneural infections, and drugs like fluconazole penetrate well into prostatic tissue. Even in AIDS patients, it is likely that most CNS relapses come from the brain rather than reseeding of the brain from the prostate.

Ocular Infections

As previously noted, *C. neoformans* can cause blindness by progressive pressure symptoms on the optic nerve (122, 209). When this symptom complex is found, optic fenestration or shunts to relieve CSF pressure may be helpful and should be considered. Otherwise, treatment of endophthalmitis needs to be made early, and treatments appropriate for meningitis should be instituted. Occasional cases of cryptococcal endophthalmitis can be successfully managed medically (60).

Role of Increased Intracranial Pressure Hypertension

Increased intracranial pressure is an important area in the management of cryptococcal meningitis that is frequently overlooked. First, patients may present before diagnosis and prior to therapy with subacute hydrocephalus or may actually develop hydrocephalus during therapy. Although exact recommendations about the placement of shunts and antifungal drug regimens cannot be given, impressions from the literature and clinical experience suggest the following therapeutic strategy. Most ventricular shunts in place prior to the administration of antifungal drugs will need to be removed and replaced under antifungal coverage to ensure consistent cure. For instance, with shunt removal, three of seven patients survived infection (48, 117, 161, 254). However, one case was reported to respond with drug therapy alone; this success without shunt removal may have occurred because the shunt was not initially placed for hydrocephalus due to an unsuspected fungal meningitis (266). In most cases, after antifungal therapy has been started, shunts can generally be inserted for increased intracranial pressure without exacerbating or actually making the infection more difficult to eradicate. In fact, recognition and treatment of hydrocephalus improves neurological status in most patients. Common misconceptions about hydrocephalus and its treatment in cryptococcal meningitis are that (i) neurosurgical intervention is necessary only after medical therapy has been tried, and (ii) shunting will disseminate infection and serve as a foreign body that will prevent cure (182). This correctable complication makes it important to radiographically image all patients who develop dementia before, during, or even after successful therapy for cryptococcal meningitis. In fact, it is probably wise to perform either a CT or an NMR scan on all patients with the initial diagnosis of cryptococcal meningitis. Neurosurgical shunting is simply necessary for treatment of hydrocephalus, and the foreign body does not hinder the ability to cure the infection.

With the institution of anticryptococcal therapy, two manifestations of increased intracranial pressure may also become a management problem. In the first 1 to 2 weeks of treatment, patients can rapidly develop symptoms such as systemic hypertension, decreased sensorium, cranial neuropathies, sudden visual loss, increasing headaches, papilledema, hearing loss, meningismus, and sudden death. These patients have increased lumbar intracranial pressure or hypertension ($\geq$350 mm H_2O) with a normal CT or NMR scan or a scan without focal processes. These symptoms and findings suggest cerebral edema and not an obstructing hydrocephalus. The precise pathophysiological mechanisms are uncertain, but it is reasonable to hypothesize that the elevated subarachnoid pressures are caused by reduced CSF outflow and a resultant increased outflow resistance (61). This elevated pressure may also exist as part of cerebral edema from a large burden of yeasts and their extracellular polysaccharide blocking of the CSF circulation through the arachnoid villi and lymphatic drainage. This phenomenon generally occurs in AIDS patients or other severely immunosuppressed patients in whom large burdens of yeasts (10^6 CFU/ml) and high CSF and serum cryptococcal antigen titers are present. There is no relationship to the status of extracerebral disease and no evidence of a host inflammatory contribution. Cerebral edema may be increased by antifungal treatment, which causes the death and clumping of yeasts that further plug the resorptive membranes of the subarachnoid space. Also, during treatment there may be more release of osmotic materials into the brain and CSF, such as polysaccharide antigen or mannitol produced by the yeast (164). Unfortunately, there has been no direct study of CSF osmolality and pressures in these patients. When these signs of rapidly increasing intracranial pressure or hypertension ($\geq$350 mm H_2O) are identified, emergency measures should be taken. For example, van der Horst et al. (246) found that elevated opening pressure was associated with reduced CSF culture negativity at 2 weeks of treatment; in 13 of 14 early deaths the most recent lumbar puncture showed opening pressures over 250 mm H_2O, and half the patients had definite physical signs of elevated intracranial pressure.

The best strategy for immediate and long-term management is not certain. However, a lumbar puncture may provide dramatic relief in some patients and can help identify the need for further management strategies. The next step in management of these symptoms is likely a further withdrawal of CSF to relieve the pressure. Repeated lumbar punctures may help, but a more physiological approach would be the placement of an external lumbar drain for several days. In many cases, the elevated pressure is transient, and after several days the drains can be removed before bacterial superinfections occur. Occasional patients may continue to have symptoms with elevated pressure and will require a permanent shunt, but that is unusual for this syndrome. The use of acetazolamide to reduce CSF production has been suggested for management (122), but it is likely to be less effective than physical measures. Corticosteroid therapy to reduce cerebral edema has also been proposed but not studied. The administration of corticosteroids will need to be carefully studied before any recommendation for their use can be made. In most cases, corticosteroids generally have a profound negative effect on the outcome of this infection. Increased intracranial pressure may be a major factor in early deaths from cryptococcal meningitis, and it must be recognized on an acute basis. However, elevated CSF pressure is also associated with reduced long-term survival (246).

Patients may also develop new neurological and ocular signs in a more chronic fashion during antifungal treatment, and these symptoms need to be recognized. When there is a subacute neurological deterioration of a patient with cryptococcal meningitis before, during, or after therapy, these patients should be checked for a standard noncommunicating or communicating hydrocephalus as a result of chronic inflammation. This condition can be successfully managed with a ventricular shunt.

Cryptococcal Polysaccharide Antigen Titers

As good as polysaccharide antigen detection is for diagnostic purposes ($\geq$90 to 95% sensitivity/specificity), its use for treatment decisions is less precise and at times is simply confusing for clinicians (203). The initial CSF antigen titers correlate directly with the number of yeasts within the subarachnoid space (264), and generally high CSF antigen titers reflect a poorer prognosis (65). On the other hand, after initiation of therapy, the kinetics of polysaccharide antigen elimination in both sera and CSF are not predictable. Antigen may actually bind to tissue such as brain and be slowly eliminated despite the lack of viable yeasts in tissue. On the other hand, this high-molecular-weight polysaccharide can be rapidly eliminated from the subarachnoid space in normal rabbits when purified antigen is directly inoculated into CSF (184a). Therefore, its distribution and production during the infectious disease process must be relevant to its prolonged elimination. For example, patients may have persistent elevation of CSF antigen titers for months after therapy has been successfully stopped. It is also clear that polysaccharide antigen within the serum does not cross the blood-brain barrier and contribute to CSF levels (77). Therefore, CSF titers reflect a balance of production versus elimination, and neither specific polysaccharide antigen titers nor a drop in the titer should be specifically used as a guide for the duration of treatment in individual cases. In fact, the imprecision in current antigen testing kits for accuracy of quantitative polysaccharide measurements stretches their use for treatment decisions even aside from the difficulty in understanding polysaccharide kinetics during infection. Thus, polysaccharide antigen tests should not be solely used to predict relapses (67, 202).

Despite the unpredictable variations in antigen titers among patients and the inaccuracy of using them to guide decisions about individual lengths of therapy, it is always an encouraging sign to detect no or much less polysaccharide antigen in the CSF at the end of therapy. For instance, if treatment studies in AIDS patients are reviewed, there is some correlation between unchanged or increased titers of polysaccharide antigens and a higher risk for clinical and/or microbiological failure. Similarly, a drop in CSF antigen titer is associated with mycological success at 10 weeks of therapy, and a high CSF titer ($\geq$1:1,024) at the end of therapy or during suppressive therapy is associated with relapse. However, serum cryptococcal antigen titers generally have no correlation with infection outcome in AIDS patients (203). In non-AIDS patients, high CSF and serum titers at baseline are associated with failure during therapy, and persistently high titers are predictive of relapse (65). The problem for the clinician is that there are no exact guidelines for making specific therapeutic decisions on the basis of antigen titer endpoints. This difficulty is not likely to change until we develop more precise tests and/or fully understand polysaccharide kinetics during infection. Thus, a single polysaccharide antigen titer should rarely influence clinical decisions. In contrast, positive cultures and/or new or persistent neurological symptoms will influence most changes in treatment strategies during therapy.

Another area in which polysaccharide antigen titers must be evaluated for treatment decisions is in isolated cryptococcal polysaccharidemia (176, 202). A common practice in the management of HIV-infected patients in areas where cryptococcosis is frequently diagnosed is to perform screens of blood for polysaccharide antigen during episodes of fever. In some areas of the world, this practice can be useful for early diagnosis (64), whereas in other areas the incidence of cryptococcal disease is simply too low to justify its use (113). However, with this strategy, patients have been identified with positive cryptococcal antigens but with negative cultures for the yeast from blood, urine, and CSF samples. The management of these patients is controversial. In 28 patients from three small series, HIV-infected patients who had repeated positive cryptococcal antigen titers of $\geq$1:4 and negative culture work-ups for cryptococcal infection were at high risk for progressing to clinical infection without treatment, and at times this progression had a rapid downhill course to death. Most patients who empirically received fluconazole therapy for the positive antigen titer did not develop cryptococcosis (82, 160, 272), Thus, until prospective studies are performed on this high-risk group with isolated polysaccharide antigen titers, it is probably wise to start empiric therapy with a triazole when the positive antigen titer is detected and confirmed. In our opinion, the possibility of treating cryptococcal infection at an earlier stage of illness outweighs the costs and low toxicities of the drugs in this very high risk population for cryptococcosis.

Cytokines and Antibodies

A present challenge for improvement in management of cryptococcosis is to actively enhance the depressed host immune response in most patients with disseminated disease. Invasive cryptococcosis is generally associated with immunosuppression, and there is a body of evidence that biological response modifiers or antibodies can have a positive effect on cryptococcosis (see chapter 8). GM-CSF has potent in vitro immunostimulatory activities against *C. neoformans* (36, 43, 149). Since GM-CSF can modulate enhancement of T-cell proliferation through synergy with IL-2 and stimulate the phagocytic and fungicidal properties of neutrophils and macrophages, it is reasonable to determine its therapeutic effects on human cryptococcosis. An open clinical trial of GM-CSF use with amphotericin B for cryptococcal meningitis in AIDS patients suggested that sterilization of the cerebrospinal fluid was accelerated in the presence of GM-CSF (239). Several case reports of patients with cryptococcosis treated with 300 to 400 μg/day of GM-CSF for 7 days showed no negative effects (205). However, it remains difficult to determine its relative effects on *C. neoformans* compared to the antifungal agents alone, and we need further trials to determine its true efficacy. IL-2 and IFN-γ can directly or indirectly stimulate host cells to activate them against *C. neoformans* (140), and since these cytokines are commercially available, they should probably be studied in the treatment of cryptococcosis. They would be expected to have a positive impact on the host immune response. However, these cytokines may have complex interactions with many different cell populations, and not all studies will show a positive effect with IFN-γ treatment. The treatment doses and schedules could become quite complex. As the number of identified cytokines increases, more of these products may become available for study in cryptococcosis. For instance, IL-12 is an interesting cytokine for up-regulating the immune system, and it has been shown to be of significant benefit in a mouse model of cryptococcosis when used to augment antifungal drug response (40). If given early in infection, IL-12 can prevent dissemination from a pulmonary source (131).

A previous area of therapeutics, serotherapy, has recently been revisited for the treatment of cryptococcosis by Casadevall and colleagues (34a, 197a). Although primarily a host cell-mediated immune response occurs in cryptococcosis, humoral immunity with antibodies and complement do play an integral part in immunity (see chapter 8). For instance, anticryptococcal antibodies may serve as useful adjuvants by (i) enhancing opsonization of the whole yeast, which results in improved host killing of viable cryptococci and (ii) assisting in the elimination of the immunosuppressive cryptococcal polysaccharide. More rapid clearance of polysaccharide may permit a more vigorous host response, which will aid antifungal chemotherapy. Specific therapeutic monoclonal antibodies have been determined to have a positive protective impact on infection (72, 171, 173) and have already been confirmed to help in the treatment of murine cryptococcosis with amphotericin B, fluconazole, and flucytosine (71, 83).

The concept of antibody therapy for human cryptococcosis actually began in 1925 when Shapiro and Neal (222) treated a 12-year-old boy with intrathecal administration of rabbit immune sera. In the 1950s, Littman (151) successfully used human immunoglobulin as adjunctive therapy with amphotericin B, and in the 1960s, Gordon and coworkers (97–99) revealed that heterologous passive antibody increased clearance of serum cryptococcal polysaccharide. The identification of specific protective and therapeutic isotypes and the variability of effectiveness of these antibodies for different cryptococcal strains emphasizes the precise nature of this therapeutic modality. The ability of passive antibody to mediate protection is dependent on the quantity (72), isotype (172, 215), fine specificity of the antibody reagent (172), and animal model used (74). However, the proof of concept for antibody use for treatment has been validated in several animal models, and although occasional concerns about passive immunity have been raised (217), the vast majority of work has been very positive. With new technology in antibody research and the understanding of the biological relevance of antibodies in crypto-coccosis, the use of specific monoclonal antibodies as adjuvants in treatment of cryptococcosis is beginning to be studied again in the treatment of human infections.

The potential of these immunomodulatory strategies with a variety of cytokines or specific antibodies has not yet been realized in cryptococcosis, because the basic science establishment for proof of concept has not yet been translated into carefully designed clinical trials. However, as we approach the new millennium, we will need to challenge ourselves to clinically define the value of these strategies and help translate our basic science understanding into clinical practice.

Investigational Antifungal Agents

Despite significant progress in the treatment of cryptococcal meningitis, there continue to be relapses and failures, and even the current use of chronic suppressive therapy in AIDS patients shows a lack of optimal fungicidal regimens. It is also likely that drug resistance to available agents will continue to increase. Therefore, it is essential that there be continued development of new antifungal agents for the treatment of cryptococcosis.

A series of new triazoles (SCH 56592 and voriconazole) are being developed for cryptococcosis (92, 112, 186). In fact, the in vitro susceptibility testing of these triazoles shows that they are more active than fluconazole on a weight basis and, in particular, these compounds have activity against fluconazole-resistant strains of *C. neoformans*.

Unfortunately, a new class of antifungal drugs, echinocandin B congeners or pneumocandins, which target the inhibition of 1,3 β-glucan synthesis within the fungal cell wall are not particularly active in vitro against *C. neoformans*. These compounds generally express only slightly inhibitory activity (MICs of 16 to 32 μg/ml) in vitro (14) and are ineffective in treatment of an animal model of crypto-coccosis (1). It has been postulated that the resistance to this class of compounds may be due to the fact that *C. neoformans* possesses 1,6 β-glucan or one of the non-1,3 β-D-glucans (i.e., 1,3 α or 1,6 α) in its cell wall. However, the variable responses to these agents against *C. neoformans* may also involve flux of drug to the metabolic target of this yeast or some other undefined resistance mechanism(s). It has been shown that this class of drugs may interact positively with azole compounds against *C. neoformans* (84). Further work and understanding in this important area of cell wall biology and its inhibitors in *C. neoformans* is needed.

Several benzimidazoles, which act by blocking polymerization of micro-tubules, have been shown to possess potent anticryptococcal activity in vitro (49). Pentamidine, a drug found to be effective in pneumocytosis, and multiple other aromatic dicationic compounds have fungicidal properties against *C. neoformans* strains (11, 59). Pradimicins, a cell wall-active group of antifungal drugs that bind mannans, have been further developed into the compound BMY-28864, which has anticryptococcal activity in vitro and in vivo (91), but development has been stopped until possible hepatotoxicity issues are addressed. Benanomicin A, a novel antifungal agent produced by *Actinomadura* species, has been found to possess potent antifungal activity against a variety of fungi both in vitro and in vivo, including *C. neoformans* through alterations in cell structure (268). A new group of potent anticryptococcal compounds are sordarin-derived molecules, which specifically target inhibition of translation through elongation factor 2. These molecules also have broad-spectrum antifungal activity (70, 224).

The use of signal transduction as a target for new anticryptococcal drugs continues to be studied. For instance, paradoxically, the immunophilins, such as cyclosporine and tacrolimus, have potent anticryptococcal activity in vitro and in an animal model (167), but generally their immunosuppressive activity outweighs the importance of the direct antifungal potential of these drugs (189). However, there are now congeners of these immunosuppressants that lack immunosuppressive activity but retain anticryptococcal activity (179). Also, aureobasidium A, which blocks pathways in signaling for sphingolipid biosynthesis, has been shown to possess potent anticryptococcal activity (235, 270).

There is always the possibility of old targets being reexamined. Terbinafine has excellent in vitro anticryptococcal activity (219), but in our experience it does not effectively treat the experimental rabbit model of cryptococcal meningitis (184a). Further work on new or improved polyenes with more potent anticryptococcal activity than amphotericin B continues (158, 232). Even drugs for other maladies, such as cholesterol-lowering agents, may give a positive interaction with known anticryptococcal agents (37).

Finally, host biological products have been found to have direct potent killing activity against *C. neoformans*. Chloroquine, which is a weak base and increases vacuolar pH of cells, can possess anticryptococcal activity when given to mice through enhancing anticryptococcal defense mechanisms (162). Defensin NP1, a small protein derived from azurophilic granules of neutrophils, is fungicidal against *C. neoformans* in vitro (3).

PROGNOSIS

Table 4 lists risk factors for treatment of cryptococcal meningitis from seven major studies. In our opinion, however, the single most important prognostic factor in determining the success or failure of treatment outcome for cryptococcal meningitis is the patient's underlying disease and the ability to control it. If conditions such as cancer (128), AIDS, high-dose corticosteroid therapy, or organ transplantation are difficult to control, it is likely that there will be limited success with the treatment of cryptococcal meningitis. An example of this concept is a comparative review of patients with AIDS or malignancy as underlying diseases during the presentation of cryptococcal meningitis by White et al. (260). The median overall survival for patients with AIDS was 9 months, compared with 2 months in those patients with neoplastic disease. This correlated with the finding that 78% of AIDS patients and 43% of patients with neoplastic disease were considered to have had their infection controlled. Furthermore, patients with neoplasia were older, which probably indicated a greater degree of immunosuppression as suggested by longer duration of underlying illness, previous extensive chemotherapy, and advanced stage of illness (128). A

Table 4 Studies that identify risk factors predictive of failure for treatment of cryptococcal meningitis[a]

Study	Regimen	Risk factors
Non-AIDS		
Diamond and Bennett (65)	AmB	Positive India ink examination
		High CSF pressure
		Low CSF glucose
		Low CSF leukocyte count
		Extraneural sites
		Absent antibody
		CSF antigen titer >1:32
		Steroid therapy/ lymphoreticular malignancy
Bennett et al. (19)	AmB vs. AmB/FC	Increased CSF antigen
Dismukes et al. (69)	AmB/FC	No headache
		Abnormal mental status
		Low CSF leukocyte count
Saag et al. (211)	AmB vs. FLU	Abnormal mental status
		CSF antigen titer >1:1,024
		Low CSF leukocyte count
AIDS		
van der Horst et al. (246)	AmB/FC + FLU or ITZ	Intravenous drug use
		Positive culture at 2 wk
		Treatment without FLU
Witt et al. (263)	FLU	Positive blood and urine cultures
		High serum or CSF antigen titer
		High MIC to FLU
		Did not receive FC
Darras-Joly et al. (51)	AmB; FLU; lipid formulation of AmB	Age >30 yr
		Low CSF glucose
		Admission to intensive care unit
		Mechanical ventilation

[a]For definition of abbreviations, see Table 1.

focus on the underlying disease will be an essential part of strategies to improve outcomes in cryptococcal meningitis. In fact, with new combinations of highly active antiretroviral therapies, it is possible that there will be a decreased incidence of cryptococcosis. In patients with AIDS and cryptococcosis, antiviral regimens that reduce viral burden may improve immune function to the point where cure of cryptococcosis with current antifungal agents may be considered and thus studied. Only further studies will answer these questions, but encouraging signs are already noted. A large study of 391 AIDS patients by van der Horst et al. (246) had only a 5.5% death rate in the first 2 weeks of treatment and 3.9% in the next 8 weeks.

Another major prognostic factor in outcome is the burden of organisms at initial presentation. Those patients with high numbers of yeasts ($\geq 10^5$–10^6 CFU/ml in the CSF), which is reflected by heavily positive India ink examinations, polysaccharide antigen titers of $\geq 1,024$, and poor local inflammatory responses (<20 host cells/μl), are more likely to develop increased intracranial pressure during early treatment or relapse after initial treatment. Finally, a third significant factor is the initial mental status of the patient. Most studies have emphasized that patients presenting with stupor or coma have a worse outcome than those with a lucid sensorium.

Therefore, the three major prognostic factors for determining outcomes with treatment of cryptococcal meningitis are underlying disease, burden of organisms, and mental status, but several studies have identified specific factors that have made an impact on outcome (19, 65, 69, 226, 246). Although these prognostic factors are determined from different treatments and patient populations, a review will likely help clinicians to categorize their own patients into high- and low-risk categories for treatment success, even though they may not precisely predict outcome.

In the 1970s, Diamond and Bennett (65) identified specific clinical features in non-AIDS patients that correlated with failure or relapse during treatment with amphotericin B alone (65). Patients who died during therapy were more likely to have the following: (i) an initial positive India ink examination, (ii) high CSF opening pressure, (iii) low CSF glucose, (iv) low CSF leukocytes (<2/μl), (v) cryptococci isolated from extraneural sites, (vi) absence of anticryptococcal antibody, (vii) initial CSF or serum cryptococcal antigen titer of 32, and (viii) corticosteroid therapy or lymphoreticular malignancy. Patients who relapsed after treatment were characterized by (i) abnormal CSF glucose concentration for ≥ 4 weeks during therapy, (ii) low initial CSF leukocyte count, (iii) cryptococci isolated from extraneural sites, (iv) absence of anticryptococcal antibody, (v) posttreatment CSF or serum cryptococcal antigen titer of ≥ 8, (vi) no significant decrease in CSF and serum antigen titer during therapy, and (vii) daily corticosteroid therapy equivalent to 20 mg of prednisone or more after completion of therapy. This study, which was completed prior to the explosion of cases in severely immunosuppressed patients in whom the underlying disease can so profoundly influence outcome, cogently identified the importance of a high burden of organism and poor host immune response. Although many patients have normal CSF glucose levels, in our opinion persistent hypoglycorrhachia does suggest that there is persistent, abnormal glucose transport that reflects remaining active infection. Patients with low CSF glucose levels at the end of therapy should be closely monitored for relapse.

Before the AIDS era, studies with amphotericin B and flucytosine as treatment for cryptococcal meningitis identified three positive factors for a good outcome through a multivariant analysis of outcomes (69). These prognostic factors included a normal mental status on presentation, presence of a headache, and a CSF leukocyte count of >20 cells/μl. On the other hand, in AIDS patients treated with either amphotericin B or fluconazole, significant pretreatment predictors of death during treatment included abnormal mental status, a CSF antigen titer of 1:1,024, and a CSF leukocyte count of cells/μl (226). Retrospectively, it was suggested that the development of diastolic systemic hypertension at the beginning of or during treatment was associated with death, and this finding was correlated with an increased CSF opening pressure (80). In our experience with patients with AIDS

Table 5 Identification and quantification of risk factors for failure of treatment of cryptococcal meningitis

Category	High risk	Low risk
Underlying disease	Neoplasia, AIDS	None
Burden of organisms	Positive India ink examination of $>10^6$ CFU/ml; antigen titer $\geq$1:1,024	Positive culture only
Host inflammatory response	CSF leukocyte count $\leq$20 cells/µl; high doses of prednisone (>20 mg/day)	CSF leukocyte count $\geq$20 cells/µl
Mental status	Stupor, coma	Lucid
Intracranial pressure	Increased	Normal

and cryptococcal meningitis, a similar theme for poor prognosis is apparent. Poor outcomes are associated with the development of systemic hypertension, new cranial neuropathies, worsening mental status, and recurrence of headaches (184a). These observations emphasize the potential importance of increased intracranial pressure in early treatment outcomes.

Because it is important to gauge the relative risk of failure or relapse in each patient with meningitis, we suggest that the clinician review five clinical categories. Table 5 separates the five major categories of treatment risk into those with high and low risk for failure of therapy. Patients with poor prognostic findings who are thus at high risk for failure can potentially have their therapies more carefully designed and individualized to improve outcome. On the other hand, in low-risk patients, therapy may be stopped earlier and less fungicidal regimens can be used. There are no specific guidelines for prognosis in nonmeningeal cryptococcosis, except for the observation that outcomes are generally better. Finally, all patients with identifiable immunosuppressive events should be treated when they have asymptomatic pulmonary or urological disease, since their prognosis for development of symptomatic disseminated disease is high.

REFERENCES

1. **Abruzzo, G. K., A. M. Flattery, C. J. Gill, L. Kong, J. G. Smith, V. B. Pikounis, J. M. Balkovec, A. F. Bouffard, J. F. Dropinski, H. Rosen, H. Kropp, and K. Bartizal.** 1997. Evaluation of the echinocandin antifungal MK-0991 (L-743,872): efficacies in mouse models of disseminated aspergillosis, candidiasis, and cryptococcosis. *Antimicrob. Agents Chemother.* **41:**2333–2338.
2. **Albert, M. M., J. R. Graybill, and M. G. Rinaldi.** 1991. Treatment of murine cryptococcal meningitis with an SCH 39304-amphotericin B combination. *Antimicrob. Agents Chemother.* **35:**1721–1725.
3. **Alcovlomire, M. S., M. A. Ghannoum, A. S. Ibrahim, M. E. Selsted, and J. E. Edwards, Jr.** 1993. Fungicidal properties of defensin NP-1 and activity against *Cryptococcus neoformans* in vitro. *Antimicrob. Agents Chemother.* **37:**2628–2632.
4. **Allendoerfer, R., A. J. Marquis, M. G. Rinaldi, and J. R. Graybill.** 1991. Combined therapy with fluconazole and flucytosine in murine cryptococcal meningitis. *Antimicrob. Agents Chemother.* **35:**726–729.
5. **Allendoerfer, R., R. R. Yates, A. J. Marquis, D. Loebenberg, M. G. Rinaldi, and J. R.**

Graybill. 1992. Comparison of SCH 39304 and its isomers, RR 42427 and SS 42426, for treatment of murine cryptococcal and coccidioidal meningitis. *Antimicrob. Agents Chemother.* **36:**217–219.

6. **Anaissie, E. J., V. L. Paetznick, L. G. Ensign, A. Espinel-Ingroff, J. N. Galgiani, C. A. Hitchcock, M. LaRocco, T. Patterson, M. A. Pfaller, J. H. Rex, and M. G. Rinaldi.** 1996. Microdilution antifungal susceptibility testing of *Candida albicans* and *Cryptococcus neoformans* with and without agitation: an eight-center collaborative study. *Antimicrob. Agents Chemother.* **40:**2387–2391.

7. **Appelbaum, E., and S. Shtokalko.** 1957. Cryptococcus meningitis arrested with amphotericin B. *Ann. Intern. Med.* **47:**346–351.

8. **Armengou, A., C. Porcar, J. Mascaro, and F. Garcia-Bragado.** 1996. Possible development of resistance to fluconazole during suppressive therapy for AIDS-associated cryptococcal meningitis. *Clin. Infect. Dis.* **23:**1337–1338.

9. **Arndt, C. A., T. J. Walsh, C. L. McCully, F. M. Balis, P. A. Pizzo, and D. E. Poplack.** 1988. Fluconazole penetration into cerebrospinal fluid: implications for treating fungal infections of the central nervous system. *J. Infect. Dis.* **157:**178–180.

10. **Baker, R. D.** 1976. The primary pulmonary lymph node complex of cryptococcosis. *Am. J. Clin. Pathol.* **65:**83–92.

11. **Barchiesi, F., M. Del Poeta, V. Morbiducei, F. Ancarani, and G. Scalise.** 1994. Effect of pentamidine on the growth of *Cryptococcus neoformans*. *J. Antimicrob. Chemother.* **33:**1229–1232.

12. **Barchiesi, F., R. J. Hollis, S. A. Messer, G. Scalise, M. G. Rinaldi, and M. A. Pfaller.** 1995. Electrophoretic karyotype and *in vitro* antifungal susceptibility of *Cryptococcus neoformans* isolates from AIDS patients. *Diagn. Microbiol. Infect. Dis.* **23:**99–103.

13. **Barone, F. E., and M. R. Tansey.** 1977. Isolation, purification, identification, synthesis, and kinetics of activity of the anticandida component of Allium sativum and a hypothesis for its mode of action. *Mycologia* **69:**793–825.

14. **Bartizal, K., C. J. Gill, G. K. Abruzzo, A. M. Flattery, L. Kong, P. M. Scott, J. G. Smith, C. E. Leighton, D. Bouffard, J. F. Dropinski, and J. M. Balkovec.** 1997. In vitro preclinical evaluation studies with the echinocandin antifungal MK-0991 (L-743,872). *Antimicrob. Agents Chemother.* **41:**2326–2332.

15. **Beeson, P. B.** 1952. Cryptococcic meningitis of nearly sixteen years' duration. *Arch. Intern. Med.* **89:**797–801.

17. **Bennett, J. E.** 1967. Susceptibility of *Cryptococcus neoformans* to amphotericin B, p. 405–410. *Antimicrob. Agents Chemother. 1966.*

18. **Bennett, J. E.** 1977. Flucytosine. *Ann. Intern. Med.* **86:**319–322.

19. **Bennett, J. E., W. Dismukes, R. J. Duma, G. Medoff, M. A. Sande, H. Gallis, J. Leonard, B. T. Fields, M. Bradshaw, H. Haywood, Z. A. McGee, T. R. Cate, C. G. Cobbs, J. F. Warner, and D. W. Alling.** 1979. A comparison of amphotericin B alone and combined with flucytosine in the treatment of cryptococcal meningitis. *N. Engl. J. Med.* **301:**126–131.

20. **Berg, J., C. J. Clancy, and M. H. Nguyen.** 1998. The hidden danger of primary fluconazole prophylaxis for patients with AIDS. *Clin. Infect. Dis.* **26:**186–187.

21. **Berry, A. J., M. G. Rinaldi, and J. R. Graybill.** 1992. Use of high-dose fluconazole as salvage therapy for cryptococcal meningitis in patients with AIDS. *Antimicrob. Agents Chemother.* **36:**690–692.

22. **Bindford, C. H.** 1941. Torulosis of the central nervous system; review of recent literature and report of a case. *Am. J. Clin. Pathol.* **11:**242–251.

23. **Bindschadler, D. D., and J. E. Bennett.** 1968. Serology of human cryptococcosis. *Ann. Intern. Med.* **69:**45–52.

24. **Bindschadler, D. D., and J. E. Bennett.** 1969. A pharmacologic guide to the clinical use of amphotericin B. *J. Infect. Dis.* **120:**427–436.

25. **Birley, H. D., E. M. Johnson, P. McDonald, C. Parry, P. B. Carey, and D. W. Warnock.** 1995. Azole drug resistance as a cause of clinical relapse in AIDS patients with cryptococcal meningitis. *Int. J. STD AIDS* **6**:353–355.

26. **Block, E. R., and J. E. Bennett.** 1973. The combined effect of 5-fluorocytosine and amphotericin B in the therapy of murine cryptococcosis. *Proc. Soc. Exp. Biol. Med.* **142**:476–480.

27. **Block, E. R., A. E. Jennings, and J. E. Bennett.** 1973. 5-Fluorocytosine resistance in *Cryptococcus neofomans. Antimicrob. Agents Chemother.* **3**:649–656.

28. **Block, E. R., A. E. Jennings, and J. E. Bennett.** 1973. Variables influencing susceptibility testing of *Cryptococcus neoformans* to 5-fluorocytosine. *Antimicrob. Agents Chemother.* **4**:392–395.

29. **Bozzette, S. A., R. A. Larsen, and J. Chin.** 1991. A placebo-controlled trial of maintenance therapy with fluconazole after treatment of cryptococcal meningitis in the acquired immunodeficiency syndrome. *N. Engl. J. Med.* **324**:580–584.

30. **Braman, R. T.** 1981. Cryptococcosis (torulopsis) of prostate. *Urology* **17**:284–286.

31. **Brandt, M. E., M. A. Pfaller, R. A. Hajjeh, E. A. Graviss, J. Rees, E. D. Spitzer, R. W. Pinner, L. W. Mayer, and Cryptococcal Disease Active Surveillance Group.** 1996. Molecular subtypes and antifungal susceptibilities of serial *Cryptococcus neoformans* isolates in human immunodeficiency virus-associated cryptococcosis. *J. Infect. Dis.* **174**:812–820.

32. **Bryan, C. S., and J. A. McFarland.** 1978. Cryptococcal meningitis: fatal marrow aplasia from combined therapy. *JAMA* **239**:1068.

33. **Caporaso, N., S. M. Smith, and R. H. K. Eng.** 1983. Antifungal activity in human urine and serum after ingestion of garlic (*Allium sativum*). *Antimicrob. Agents Chemother.* **23**:700–702.

34. **Carton, C. A.** 1952. Treatment of central nervous system cryptococcosis: a review and report of four cases treated with actidione. *Ann. Intern. Med.* **37**:123–154.

34a. Casadevall, A. 1996. Antibody-based therapies for emerging infectious diseases. *Emerg. Infect. Dis.* **2**:200–208.

35. **Casadevall, A., E. D. Spitzer, D. Webb, and M. G. Rinaldi.** 1993. Susceptibilities of serial *Cryptococcus neoformans* isolates from patients with recurrent cryptococcal meningitis to amphotericin B and fluconazole. *Antimicrob. Agents Chemother.* **37**:1383–1386.

36. **Chen, G.-H., J. L. Curtis, C. H. Mody, P. J. Christensen, L. R. Armstrong, and G. B. Toews.** 1994. Effect of granulocyte-macrophage colony-stimulating factor on rat alveolar macrophage anticryptococcal activity *in vitro. J. Immunol.* **152**:724–734.

37. **Chin, N. X., I. Weitzman, and P. Della-Latta.** 1997. In vitro activity of fluvastatin, a cholesterol-lowering agent, and synergy with fluconazole and itraconazole against *Candida* species and *Cryptococcus neoformans. Antimicrob. Agents Chemother.* **41**:850–852.

38. **Chuck, S. L., and M. A. Sande.** 1989. Infections with *Cryptococcus neoformans* in the acquired immunodeficiency syndrome. *N. Engl. J. Med.* **321**:794–799.

39. **Clark, J. M., R. R. Whitney, S. J. Olsen, R. J. George, M. R. Swerdel, L. Kunselman, and D. P. Bonner.** 1991. Amphotericin B lipid complex therapy of experimental fungal infections in mice. *Antimicrob. Agents Chemother.* **35**:615–621.

40. **Clemons, K. V., E. Brummer, and D. A. Stevens.** 1994. Cytokine treatment of central nervous system infection: efficacy of interleukin-12 alone and synergy with conventional antifungal therapy in experimental cryptococcosis. *Antimicrob. Agents Chemother.* **38**:460–464.

41. **Clemons, K. V., and D. A. Stevens.** 1998. Comparison of Fungizone, Amphotec, AmBisome, and Abelcet for treatment of systemic murine cryptococcosis. *Antimicrob. Agents Chemother.* **42**:899–902.

42. **Coker, R. J., M. Viviani, B. G. Gazzard, B. Du Pont, H. D. Pohle, S. M. Murphy, J. Atouguia, J. L. Champalimaud, and J. R. Harris.** 1993. Treatment of cryptococcosis

with liposomal amphotericin B (AmBisome) in 23 patients with AIDS. *AIDS* **7:**829–835.

43. **Collins, H. L., and G. J. Bancroft.** 1992. Cytokine enhancement of complement-dependent phagocytosis of macrophages: synergy of tumor necrosis factor alpha and granulocyte macrophage colony stimulating factor for phagocytosis of Cryptococcus neoformans. *Eur. J. Immunol.* **22:**1447–1454.

44. **Colombo, A. L., F. Barchiesi, D. A. McGough, and M. G. Rinaldi.** 1995. Comparison of E-test and National Committee for Clinical Laboratory Standards broth macrodilution method for azole antifungal susceptibility testing. *J. Clin. Microbiol.* **33:**535–540.

45. **Correa, A. L., G. Velez, M. Albert, M. Luther, M. G. Rinaldi, and J. R. Graybill.** 1995. Comparison of D0870 and fluconazole in the treatment of murine cryptococcal meningitis. *J. Med. Vet. Mycol.* **33:**367–374.

46. **Craven, P. C., and J. R. Graybill.** 1984. Combination of oral flucytosine and ketoconazole as therapy for experimental cryptococcal meningitis. *J. Infect. Dis.* **149:**584–590.

47. **Craven, P. C., J. R. Graybill, and J. H. Jorgensen.** 1982. Ketoconazole therapy of murine cryptococcal meningitis. *Am. Rev. Respir. Dis.* **125:**696–700.

48. **Crum, C. P., and P. S. Feldman.** 1981. Cryptococcal peritonitis complicating a ventriculoperitoneal shunt in unsuspected cryptococcal meningitis. *Hum. Pathol.* **12:**660–663.

49. **Cruz, M. C., M. S. Bartlett, and T. D. Edlind.** 1994. In vitro susceptibility of the opportunistic fungus *Cryptococcus neoformans* to anthelmintic benzimidazoles. *Antimicrob. Agents Chemother.* **38:**378–380.

50. **Currie, B., H. Sanati, A. S. Ibrahim, J. E. Edwards, A. Casadevall, and M. A. Ghannoum.** 1995. Sterol compositions and susceptibilities to amphotericin B of environmental *Cryptococcus neoformans* isolates are changed by murine passage. *Antimicrob. Agents Chemother.* **39:**1934–1937.

51. **Darras-Joly, C., S. Chevret, M. Wolff, P. Longuet, E. Casalino, V. Joly, C. Chochillon, and J. P. Bedos.** 1996. Cryptococcus neoformans infection in France: epidemiologic features of and early prognostic parameters for 76 patients who were infected with human immunodeficiency virus. *Clin. Infect. Dis.* **23:**369–376.

52. **Davey, K. G., E. M. Johnson, A. D. Holmes, A. Szekely, and D. W. Warnock.** 1997. Emergence of drug resistant *Cryptococcus neoformans* strains during maintenance fluconazole treatment in AIDS patients, abstr. P493. Proc. International Society of Human and Animal Mycology.

53. **Davis, L. E., J. Shen, and R. E. Roger.** 1994. *In vitro* synergism of concentrated allium satiuum extract and amphotericin B against *Cryptococcus neoformans*. *Plant. Med.* **60:**546–549.

54. **de Fernandez, P., M. M. Patino, J. R. Graybill, and M. H. Tarbit.** 1986. Treatment of cryptococcal meningitis in mice with fluconazole. *J. Antimicrob. Chemother.* **18:**261–270.

55. **DeGans, J., P. Portegies, and G. Tiessens.** 1992. Itraconazole compared with amphotericin B plus flucytosine in AIDS patients with cryptococcal meningitis. *AIDS* **6:**185–190.

56. **DeGans, J., J. K. E. Schattenkerk, and R. J. van Ketel.** 1988. Itraconazole as maintenance treatment for cryptococcal meningitis in the acquired immune deficiency syndrome. *Br. Med. J. Clin. Res.* **296:**339.

57. **de Lalla, F., G. Pellizzer, and A. Vaglia.** 1995. Amphotericin B as primary therapy for cryptococcosis in patients with AIDS: reliability of relatively high doses administered over a relatively short period. *Clin. Infect. Dis.* **20:**263–266.

58. **Del Poeta, M., F. Barchiesi, V. Morbiducci, D. Arzeni, G. Marinucci, F. Ancarani, and G. Scalise.** 1994. Comparison of broth dilution and semisolid agar dilution for *in vitro* susceptibility testing of *Cryptococcus neoformans*. *J. Chemother.* **6:**173–176.

59. **Del Poeta, M., W. A. Schell, C. C. Dykstra, R. Tidwell, D. Boykin, and J. R. Perfect.** 1997. The dicationic aromatic compounds: a new class of antifungal agents, abstr.

F-101. 37th Interscience Conference on Antimicrobial Agents and Chemotherapy, Toronto.

60. **Denning, D. W., R. W. Armstrong, M. Fishman, and D. A. Stevens.** 1991. Endophthalmitis in a patient with disseminated cryptococcosis and AIDS who was treated with itraconazole. *Rev. Infect. Dis.* **13:**1126–1130.

61. **Denning, D. W., R. W. Armstrong, B. H. Lewis, and D. A. Stevens.** 1991. Elevated cerebrospinal fluid pressures in patients with cryptococcal meningitis and acquired immunodeficiency syndrome. *Am. J. Med.* **91:**267–272.

62. **Denning, D. W., R. M. Tucker, L. H. Hanson, J. R. Hamilton, and D. A. Stevens.** 1989. Itraconazole therapy for cryptococcal meningitis and cryptococcosis. *Arch. Intern. Med.* **149:**2301–2308.

63. **Denning, D. W., R. M. Tucker, J. S. Hostetler, S. Gill, and D. A. Stevens.** 1990. Oral itraconazole therapy of cryptococcal meningitis and cryptococcosis in patients with AIDS, p. 305–324. *In* H. Vanden Bossche (ed.), *Mycoses in AIDS Patients.* Plenum Press, New York.

64. **Desmet, P., K. D. Kayembe, and C. DeVroey.** 1989. The value of cryptococcal serum antigen screening among HIV positive AIDS patients in Kinshasa, Zaire. *AIDS* **3:**77–78.

65. **Diamond, R. D., and J. E. Bennett.** 1974. Prognostic factors in cryptococcal meningitis. A study of 111 cases. *Ann. Intern. Med.* **80:**176–181.

66. **Diasio, R. B., D. E. Lakings, and J. E. Bennett.** 1978. Evidence for conversion of 5-fluorocytosine to 5-fluorouracil in humans: possible factor in 5-fluorocytosine clinical toxicity. *Antimicrob. Agents Chemother.* **14:**903–908.

67. **Dismukes, W. E.** 1993. Management of cryptococcosis. *Clin. Infect. Dis.* **17:**507–512.

68. **Dismukes, W. E., A. M. Stamm, J. R. Graybill, P. C. Craven, D. A. Stevens, R. L. Stiller, G. A. Sarosi, G. Medoff, C. R. Gregg, H. A. Gallis, B. T. Field, R. L. Marier, T. A. Kerkering, L. G. Kaplowitz, G. Cloud, C. Bowles, and S. Shadomy.** 1983. Treatment of systemic mycoses with ketoconazole: emphasis on toxicity and clinical response in 52 patients. National Institute of Allergy and Infectious Diseases Collaborative Antifungal Study. *Ann. Intern. Med.* **98:**13–20.

69. **Dismukes, W. E., G. Cloud, H. A. Gallis, T. M. Kerkering, G. Medoff, P. L. Craven, L. G. Kaplowitz, J. F. Fisher, C. R. Gregg, C. A. Bowles, S. Shadomy, A. M. Stamm, R. B. Diasio, L. Kaufman, S.-J. Soong, W. Blackwelder, and National Institute of Allergy and Infectious Diseases Mycoses Study Group.** 1987. Treatment of cryptococcal meningitis with combination amphotericin B and flucytosine for four as compared with six weeks. *N. Engl. J. Med.* **317:**334–341.

70. **Dominguez, J. M., L. Capa, M. J. Serramia, A. Mendoza, J. M. Viava, J. F. Garcia-Bustos, and J. J. Martin.** 1997. Translation elongation factor 2 (EF2) is the target for sordarin-derived antifungals, abstr. F-55. 37th Interscience Conference on Antimicrobial Agents and Chemotherapy, Toronto.

71. **Dromer, F., and J. Charreire.** 1991. Improved amphotericin B activity by a monoclonal anti-*Cryptococcus neoformans* antibody: study during murine cryptococcosis and mechanisms of action. *J. Infect. Dis.* **163:**1114–1120.

72. **Dromer, F., J. Charreire, A. Contrepois, C. Carbon, and P. Yeni.** 1987. Protection of mice against experimental cryptococcosis by anti-*Cryptococcus neoformans* monoclonal antibody. *Infect. Immun.* **55:**749–752.

73. **Dromer, F., S. Mathoulin, B. Dupont, O. Brugiere, and L. Letenneur.** 1996. Comparison of the efficacy of amphotericin B and fluconazole in the treatment of cryptococcosis in human immunodeficiency virus-negative patients: retrospective analysis of 83 cases. French Cryptococcosis Study Group. *Clin. Infect. Dis.* **22**(Suppl. 2):S154–S160.

74. **Dromer, F., P. Yeni, and J. Charreire.** 1988. Genetic control of the humoral response to cryptococcal capsular polysaccharide in mice. *Immunogenetics* **28:**417–424.

75. **Drutz, D. J., A. Spickard, D. E. Rogers, and M. G. Koenig.** 1968. Treatment of disseminated mycotic infections. *Am. J. Med.* **45:**405–418.

76. **Dupont, B., and E. Drouhet.** 1987. Cryptococcal meningitis and fluconazole. *Ann. Intern. Med.* **106:**778.

77. **Eng, R. H. K., E. Bishburg, S. M. Smith, and R. Kapila.** 1986. Cryptococcal infections in patients with acquired immune deficiency syndrome. *Am. J. Med.* **81:**19–23.

78. **Epstein, R., R. Cole, and K. K. Hunt.** 1972. Pleural effusion secondary to pulmonary cryptococcosis. *Chest* **61:**296.

79. **Espinel-Ingroff, A., C. W. Kish, Jr., T. M. Kerkering, R. A. Fromtling, K. Bartizal, J. N. Galgiani, K. Villareal, M. A. Pfaller, T. Gerarden, and M. G. Rinaldi.** 1992. Collaborative comparison of broth macrodilution and microdilution antifungal susceptibility tests. *J. Clin. Microbiol.* **30:**3138–3145.

80. **Fan-Havard, P., E. Yamaguchi, S. M. Smith, and R. H. Eng.** 1992. Diastolic hypertension in AIDS patients with cryptococcal meningitis. *Am. J. Med.* **93:**347–348.

81. **Fass, R. J., and R. L. Perkins.** 1971. 5-Fluorocytosine in the treatment of cryptococcal and candida mycoses. *Ann. Intern. Med.* **74:**535.

82. **Feldmesser, M., C. Harris, S. Reichberg, S. Khan, and A. Casadevall.** 1996. Serum cryptococcal antigen in patients with AIDS. *Clin. Infect. Dis.* **23:**827–830.

83. **Feldmesser, M., J. Mukherjee, and A. Casadevall.** 1996. Combination of 5-flucytosine and capsule-binding monoclonal antibody in the treatment of murine *Cryptococcus neoformans* infections and *in vitro*. *J. Antimicrob. Chemother.* **37:**617–622.

84. **Franzot, S. P., and A. Casadevall.** 1997. Pneumocandin L743,872 enhances the activities of amphotericin B and fluconazole against *Cryptococcus neoformans* in vitro. *Antimicrob. Agents Chemother.* **44:**331–336.

85. **Franzot, S. P., and J. S. Hamdan.** 1996. In vitro susceptibilities of clinical and environmental isolates of *Cryptococcus neoformans* to five antifungal drugs. *Antimicrob. Agents Chemother.* **40:**822–824.

86. **Fromtling, R. A., G. K. Abruzzo, and A. Ruiz.** 1989. Virulence and antifungal susceptibility of environmental and clinical isolates of *Cryptococcus neoformans* from Puerto Rico. *Mycopathologia* **106:**163–166.

87. **Fromtling, R. A., and G. S. Bulmer.** 1978. *In vitro* effect of aqueous extract of garlic (Allium sativum) on the growth and viability of *Cryptococcus neoformans*. *Mycologia* **70:**397–405.

88. **Fromtling, R. A., J. N. Galgiani, M. A. Pfaller, A. Espinel-Ingroff, K. F. Bartizal, M. S. Bartlett, B. A. Body, C. Frey, G. Hall, G. D. Roberts, F. B. Nolte, F. C. Odds, M. G. Rinaldi, A. M. Sugar, and K. Villareal.** 1993. Multicenter evaluation of a broth macrodilution antifungal susceptibility test for yeasts. *Antimicrob. Agents Chemother.* **37:**39–45.

89. **Fujita, N. K., and J. E. Edwards.** 1981. Combined in vitro effect of amphotericin B and rifampin on *Cryptococcus neoformans*. *Antimicrob. Agents Chemother.* **19:**196–198.

90. **Fujita, N. K., M. Reynard, F. L. Sapico, L. B. Guze, and J. E. Edwards, Jr.** 1981. Cryptococcal intracerebral mass lesions. *Ann. Intern. Med.* **94:**382–388.

91. **Fung-Tomc, J. C., B. Minassian, E. Huczko, B. Kolek, D. P. Bonner, and R. E. Kessler.** 1995. In vitro antifungal and fungicidal spectra of a new pradimicin derivative, BMS-0181184. *Antimicrob. Agents Chemother.* **39:**295–300.

92. **Galgiani, J. N., and M .L. Lewis.** 1997. In vitro studies of activities of the antifungal triazoles SCH 56592 and itraconazole against *Candida albicans*, *Cryptococcus neoformans,* and other pathogenic yeasts. *Antimicrob. Agents Chemother.* **41:**180–183.

93. **Ghannoum, M. A., Y. Fu, A. S. Ibrahim, L. A. Mortara, M. C. Shafig, J. E. Edwards, Jr., and R. S. Criddle.** 1995. In vitro determination of optimal antifungal combinations against *Cryptococcus neoformans* and *Candida albicans*. *Antimicrob. Agents Chemother.* **39:**2459–2465.

94. **Ghannoum, M. A., A. S. Ibrahim, Y. Fu, M. C. Shafig, J. E. Edwards, and R. S. Criddle.** 1992. Susceptibility testing of *Cryptococcus neoformans*: a microdilution technique. *J. Clin. Microbiol.* **30:**2881–2886.

95. **Ghannoum, M. A., B. J. Spellberg, A. S. Ibrahim, J. A. Ritchie, B. Currie, E. D. Spitzer, J. E. Edwards, Jr., and A. Casadevall.** 1994. Sterol composition of *Cryptococcus neoformans* in the presence and absence of fluconazole. *Antimicrob. Agents Chemother.* **38:**2029–2033.

96. **Glover, D. D., E. Brummer, and D. A. Stevens.** 1996. Study of the role of iron in the anticryptococcal activity of human serum and fluconazole. *Mycopathologia* **133:**71–77.

97. **Gordon, M. A.** 1963. Synergistic serum therapy of systemic mycoses. *Mycopathologia* **19:**150.

98. **Gordon, M. A., and A. Casadevall.** 1995. Serum therapy of cryptococcal meningitis. *Clin. Infect. Dis.* **21:**1477–1479.

99. **Gordon, M. A., and E. Lapa.** 1964. Serum protein enhancement of antibiotic therapy in cryptococcosis. *J. Infect. Dis.* **114:**373–378.

100. **Graybill, J. R., and J. Ahrens.** 1984. R51211 (itraconazole) therapy of murine cryptococcosis. *Sabouraudia* **22:**445–453.

101. **Graybill, J. R., P. C. Craven, R. L. Taylor, D. M. Williams, and W. E. Magee**. 1982. Treatment of murine cryptococcosis with liposome-associated amphotericin B. *J. Infect. Dis.* **145:**748–752.

102. **Graybill, J. R., S. R. Kaster, and D. J. Drutz.** 1983. Comparative activities of Bay n 7133, ICI 153,066, and ketoconazole in murine cryptococcosis. *Antimicrob. Agents Chemother.* **24:**829–834.

103. **Graybill, J. R., and H. B. Levine.** 1978. Successful treatment of cryptococcal meningitis with intraventricular miconazole. *Ann. Intern. Med.* **138:**814–816.

104. **Graybill, J. R., and L. Mitchell.** 1980. Treatment of murine cryptococcosis with minocycline and amphotericin B. *Sabouraudia* **18:**137–144.

105. **Graybill, J. R., L. Mitchell, and H. B. Levine.** 1978. Treatment of experimental murine cryptococcosis: a comparison of miconazole and amphotericin B. *Antimicrob. Agents Chemother.* **13:**277–283.

106. **Graybill, J. R., D. M. Williams, E. VanCutsem, and D. J. Drutz.** 1980. Combination therapy of experimental *Histoplasmosis* and *Cryptococcosis* with amphotericin B and ketoconazole. *Rev. Infect. Dis.* **2:**551–558.

107. **Greening, R. R., and L. J. Menville.** 1947. Roentgen findings in torulosis; report of four cases. *Radiology* **48:**381–388.

108. **Hamilton, J. D., and D. M. Elliott.** 1975. Combined activity of amphotericin B and 5-fluorocytosine against *Cryptococcus neoformans in vitro* and *in vivo* in mice. *J. Infect. Dis.* **131:**129–137.

109. **Hammerman, K. J., K. E. Powell, C. S. Christianson, P. M. Huggin, H. W. Larsh, J. R. Vivas, and F. E. Tosh.** 1973. Pulmonary cryptococcosis: clinical forms and treatment. A Center for Disease Control Cooperative Mycoses Study. *Am. Rev. Respir. Dis.* **108:**1116–1123.

110. **Hata, K., J. Kimura, H. Miki, T. Toyosawa, M. Moriyama, and K. Katso.** 1996. Efficacy of ER-30346, a novel oral triazole antifungal agent, in experimental models of aspergillosis, candidiasis, and cryptococcosis. *Antimicrob. Agents Chemother.* **40:**2243–2247.

111. **Haubrich, R. H., D. Haghighat, S. A. Bozzette, and J. A. McCutchan.** 1994. High dose fluconazole for treatment of cryptococcal disease in patients with human immunodeficiency virus infection. *J. Infect. Dis.* **170:**238–242.

111a. **Heyderman, R. S., I. T. Gangaidzo, J. G. Hakim, J. Mielke, A. Taziwa, P. Musvaire, V. J. Robertson, and P. R. Mason.** 1998. Cryptococcal meningitis in human immunodeficiency virus-infected patients in Harare, Zimbabwe. *Clin. Infect. Dis.* **26:**284–289.

112. **Hitchcock, C. A., G. W. Pye, G. P. Oliver, and P. F. Drake.** 1995. UK-109.496, a novel, wide-spectrum triazole derivative for the treatment of fungal infections: antifungal activity and selectivity in vitro, abstr. F72. 35th Interscience Conference on Antimicrobial Agents and Chemotherapy, San Francisco.

113. **Hoffmann, S., J. Stenderup, and L. R. Mathiesen.** 1991. Low yield of screening for cryptococcal antigen by latex agglutination assay on serum and cerebrospinal fluid from Danish patients with AIDS or ARC. *Scand. J. Infect. Dis.* **23:**697–702.

114. **Hostetler, J. S., K. V. Clemons, L. H. Hanson, and D. A. Stevens.** 1992. Efficacy and safety of amphotericin B colloidal dispersion compared with those of amphotericin B deoxycholate suspension for treatment of disseminated murine cryptococcosis. *Antibiot. Chemother.* **36:**2656–2660.

115. **Hostetler, J. S., L. H. Hanson, and D. A. Stevens.** 1993. Effect of hydroxypropyl-beta-cyclodextrin on efficacy of oral itraconazole in disseminated murine cryptococcosis. *J. Antimicrob. Chemother.* **32:**459–463.

116. **Hunan Medical College.** 1980. Garlic in cryptococcal meningitis: a preliminary report of 21 cases. *Chin. Med. J.* **93:**123–126.

117. **Ingram, C. W., H. B. Haywood, V. M. Morris, R. L. Allen, and J. R. Perfect.** 1993. Cryptococcal ventricular peritoneal shunt infection: clinical and epidemiological evaluation of two closely associated cases. *Infect. Immun.* **14:**719–722.

118. **Inoue, N., T. Uozumi, T. Yamamoto, T. Ohishi, Y. Murai, M. Murakami, and H. Yoshimatsu.** 1984. Treatment of cryptococcal meningitis with pulmonary granuloma. *J. Neurol.* **231:**109–111.

119. **Iovannitti, C., R. Negroni, J. Baua, J. Finguelievich, and M. Kral.** 1995. Itraconazole and flucytosine plus itraconazole combination in the treatment of experimental Cryptococcosis in hamsters. *Mycoses* **38:**449–452.

120. **Iwata, K., T. Yamashita, M. Ohsumi, M. Baba, N. Naito, A. Taki, and N. Yamada.** 1990. Comparative morphological and biological studies on itraconazole- and ketoconazole-resistant mutants of *Cryptococcus neoformans*. *J. Med. Vet. Mycol.* **28:**77–90.

121. **Jimbow, T., Y. Tejima, and H. Ikemoto.** 1978. Comparison between 5-fluorocytosine, amphotericin B, and the combined administration of these agents in the therapeutic effectiveness for cryptococcal meningitis. *Chemotherapy* **24:**374–389.

122. **Johnston, S. R., E. L. Corbett, O. Foster, S. Ash, and J. Cohen.** 1992. Raised intracranial pressure and visual complications in AIDs patients with cryptococcal meningitis. *J. Infect.* **24:**185–189.

123. **Joly, V., P. Aubry, A. Ndayiragide, I. Carriere, E. Kawa, N. Mlika-Cabanne, J. P. Aboulker, J. P. Coulaud, B. Larouze, and P. Yeni.** 1996. Randomized comparison of amphotericin B deoxycholate dissolved in dextrose or Intralipid for the treatment of AIDS-associated cryptococcal meningitis. *Clin. Infect. Dis.* **23:**556–562.

124. **Jones, E. L.** 1927. Torula infection of the nasopharynx. *South. Med. J.* **20:**120–126.

125. **Joseph-Horne, T., D. Hollomon, R. S. Loeffler, and S. L. Kelly.** 1995. Cross-resistance to polyene and azole drugs in *Cryptococcus neoformans*. *Antimicrob. Agents Chemother.* **39:**1526–1529.

126. **Joseph-Horne, T., R. S. T. Loefflin, D. W. Halloman, and S. L. Kelly.** 1996. Amphotericin B resistant isolates of *Cryptococcus neoformans* without alteration in sterol biosynthesis. *J. Med. Vet. Mycol.* **34:**223–225.

127. **Just-Nubling, G., W. Heise, G. Rieg, S. Dieckmann, W. Enzensberger, M. L. Lage, E. R. Heln, and W. Stille.** 1996. Triple combination of amphotericin B, flucytosine and fluconazole for treatment of acute cryptococcal meningitis in patients with AIDS, abstr. V5. 3rd International Conference on Cryptococcus and Cryptococcosis, Paris.

128. **Kaplan, M. H., P. P. Rosen, and D. Armstrong.** 1977. Cryptococcus in a cancer hospital. *Cancer* **39:**2265–2274.

129. **Kartalija, M., K. Kaye, J. H. Tureen, Q. X. Liu, M. G. Tauber, B. R. Elliott, and M. A. Sande.** 1996. Treatment of experimental Cryptococcal meningitis with fluconazole: impact of dose and addition of flucytosine on mycologic and pathophysiologic outcome. *J. Infect. Dis.* **173:**1216–1221.

130. **Kauffman, C.A., and P.T. Frame.** 1977. Bone marrow toxicity associated with 5-fluorocytosine therapy. *Antimicrob. Agents Chemother.* **11:**244–247.

131. **Kawakami, K., M. Tohyama, Q. Xie, and A. Saito.** 1996. IL12 protects mice against pulmonary and disseminated infection caused by *Cryptococcus neoformans. Clin. Exp. Immunol.* **104:**208–214.

132. **Kelly, S. L., D. C. Lamb, M. Taylor, A. J. Corran, B. C. Baldwin, and W. G. Powderly.** 1994. Resistance to amphotericin B associated with defective sterol Δ 8→7 isomerase in a *Cryptococcus neoformans* strain from an AIDS patient. *FEMS Microbiol. Lett.* **122:**39–42.

134. **Kerkering, T. M., R. J. Duma, and S. Shadomy.** 1981. The evolution of pulmonary cryptococcosis: clinical implications from a study of 41 patients with and without compromising host factors. *Ann. Intern. Med.* **94:**611–616.

135. **Kessel, J. F., and F. Holtzwart.** 1935. Experimental studies with torula from a knee infection in man. *Am. J. Trop. Med. Hyg.* **15:**467–483.

136. **Kligman, A. M., A. P. Crane, and R. F. Norris.** 1951. Effect of temperature on survival of chick embryos infected intravenously with *Cryptococcus neoformans* (*Torula histolytica*). *Am. J. Med. Sci.* **221:**273–278.

137. **Kligman, A. M., and F. D. Weidman.** 1949. Experimental studies on treatment of human torulosis. *Arch. Dermatol. Syph.* **60:**726–741.

138. **Koletar, S. L., W. J. Buesching, and R. J. Fass.** 1995. Emerging resistance to fluconazole among blood and CSF *Cryptococcus neoformans* isolates, abstr. E70. 35th Interscience Conference on Antimicrobial Agents and Chemotherapy, San Francisco.

139. **Kovacs, J. A., A. A. Kovacs, M. Polis, W. C. Wright, V. J. Gill, C. U. Tuazon, E. P. Gelmann, H. C. Lane, R. Longfield, G. Overturf, A. M. Macher, A. S. Fauci, J. E. Parrillo, J. E. Bennett, and H. Masur.** 1985. Cryptococcosis in the acquired immunodeficiency syndrome. *Ann. Intern. Med.* **103:**533–538.

140. **Kowakami, K., M. Tohyama, K. Teruya, N. Kudeken, Q. Xie, and A. Saito.** 1996. Contribution of interferon-gamma in protecting mice during pulmonary and disseminated infection with *Cryptococcus neoformans. FEMS Immunol. Med. Microbiol.* **13:**123–130.

141. **Lamb, D. C., A. Corran, B. C. Baldwin, J. Kwon-Chung, and S. L. Kelly.** 1995. Resistant P45051 A1 activity in azole antifungal tolerant *Cryptococcus neoformans* from AIDS patients. *FEBS Lett.* **368:**326–330.

142. **Laroche, R., B. Dupont, J. E. Touze, H. Taelman, J. Bogaerts, A. Kadio, P. M'Pele, A. Latif, J. Aubry, J. P. Durbec, and J. F. Sauniere.** 1992. Cryptococcal meningitis associated with acquired immunodeficiency syndrome (AIDS) in African patients: treatment with fluconazole. *J. Med. Vet. Mycol.* **30:**71–78.

143. **Larsen, R. A., M. Bauer, J. M. Weiner, D. M. Diamond, M. E. Leal, J. C. Ding, M. G. Rinald, and J. R. Graybill.** 1996. Effect of fluconazole on fungicidal activity of flucytosine in murine cryptococcal meningitis. *Antimicrob. Agents Chemother.* **40:**2178–2182.

144. **Larsen, R. A., S. Bozzette, A. McCutchan, J. Chiu, M. A. Leal, and D. D. Richman.** 1989. Persistent *Cryptococcus neoformans* infection of the prostate after successful treatment of meningitis. *Ann. Intern. Med.* **111:**125–128.

145. **Larsen, R. A., S. A. Bozzette, B. E. Jones, D. Haghighat, M. A. Leal, D. Forthal, M. Baver, J. G. Tilles, J. A. McCutchan, and J. M. Ludom.** 1994. Fluconazole combined with flucytosine for treatment of cryptococcal meningitis in patients with AIDS. *Clin. Infect. Dis.* **19:**741–745.

146. **Larsen, R. A., M. A. E. Leal, and L. S. Chan.** 1990. Fluconazole compared with amphotericin B plus flucytosine for cryptococcal meningitis in AIDS. *Ann. Intern. Med.* **113:**183–187.

147. **Leenders, A. C., P. Reiss, P. Portegies, K. Clezy, W. C. J. Hop, J. Hoy, J. C. C. Borleffs, T. Allworth, R. Kauffmann, P. Jones, F. P. Kroon, H. A. Verbrugh, and S. de Marie.** 1997. Liposomal amphotericin B (AmBisome) compared with amphotericin B both followed by oral fluconazole in the treatment of AIDS-associated cryptococcal meningitis. *AIDS* **11:**1463–1471.

148. **Leenders, A. C., P. Reiss, P. Portegies, K. Clezy, W. C. J. Hop, and S. de Marie.** 1996. A randomized trial of liposomal amphotericin B (AmBisome) 4 mg/kg vs amphotericin B 0.7 mg/kg for cryptococcal meningitis in HIV-infected patients, abstr. LM 35. 36th Interscience Conference on Antimicrobial Agents and Chemotherapy, New Orleans.

149. **Levitz, S. M.** 1991. Activation of human peripheral blood mononuclear cells by interleukin-2 and granulocyte-macrophage colony stimulating factor to inhibit *Cryptococcus neoformans*. *Infect. Immun.* **59:**3393–3397.

150. **Lewis, J. L., and S. Rabinovich.** 1972. The wide spectrum of cryptococcal infections. *Am. J. Med.* **53:**315–322.

151. **Littman, M. L.** 1959. Cryptococcosis (torulosis). *Am. J. Med.* **27:**976–998.

152. **Littman, M. L., and L. E. Zimmerman.** 1956. *Cryptococcosis, Torulosis or European Blastomycosis*, p. 139–143. Grune & Stratton, New York.

153. **Littman, M. L., and L. E. Zimmerman.** 1956. *Cryptococcosis, Torulosis or European Blastomycosis*, p. 125–128. Grune & Stratton, New York.

154. **Littman, M. L., and L. E. Zimmerman.** 1956. *Cryptococcosis, Torulosis or European Blastomycosis*, p. 143–144. Grune & Stratton, New York.

155. **Littman, M. L., and L. E. Zimmerman.** 1956. *Cryptococcosis, Torulosis or European Blastomycosis*, p. 121–146. Grune & Stratton, New York.

156. **Liu, S. C., B. Oguntimein, C. D. Hufford, and A. M. Clark.** 1990. 3-Methoxysampangine, a novel antifungal copyrine alkaloid from *Cleistopholis patens*. *Antimicrob. Agents Chemother.* **34:**529–533.

157. **Louria, D. B., M. Lavenhar, T. Kaminski, and R. H. Eng.** 1989. Garlic (Allium sativum) in the treatment of experimental cryptococcosis. *J. Med. Vet. Mycol.* **27:**253–257.

158. **Luna, T., R. Mazzolla, G. Romano, and E. Blasi.** 1997. Potent antifungal effects of a new derivative of partricin A in a murine model of cerebral cryptococcosis. *Antimicrob. Agents Chemother.* **41:**706–708.

159. **Majid, A. A.** 1989. Clinical notes. Surgical resection of pulmonary cryptococcomas in the presence of cryptococcal meningitis. *J. R. Coll. Surg. Edinb.* **34:**332–333.

160. **Manfredi, R., A. Moroni, A. Mazzoni, A. Nanetti, M. Donati, A. Mastroiannis, O. V. Coronado, and F. Chiodo.** 1996. Isolated detection of Cryptococcal polysaccloride antigen in cerebrospinal fluid samples from patients with AIDS. *Clin. Infect. Dis.* **23:**849–850.

161. **Mangham, D., D. N. Gerding, L. R. Peterson, and G. A. Sarosi.** 1983. Fungal meningitis manifesting in hydrocephalus. *Arch. Intern. Med.* **143:**728–731.

161a. **Mayanja-Kizza, H., K. Oishi, S. Mitarai, H. Yamashita, K. Nalongo, K. Watanabe, T. Izumi, O. Jungala, K. Augustine, R. Mugerwa, T. Nagatake, and K. Matsumo.** 1998. Combination therapy with fluconazole and flucytosine for cryptococcal meningitis in Ugandan patients with AIDS. *Clin. Infect. Dis.* **26:**1362–1366.

162. **Mazzolla, R., R. Barluzzi, A. Brozzetti, J. R. Boelaert, T. Luna, S. Saleppico, F. Bistoni, and E. Blasi.** 1997. Enhanced resistance to *Cryptococcus neoformans* infection induced by chloroquine in a murine model of meningoencephalitis. *Antimicrob. Agents Chemother.* **41:**802–807.

163. **McCullough, N. B., D. B. Louria, T. F. Hilbish, L. B. Thomas, and C. Emmons.** 1958. Cryptococcosis: clinical staff conference at the National Institutes of Health. *Ann. Intern. Med.* **49:**642–661.

164. **Megson, G. M., D. A. Stevens, J. R. Hamilton, and D. W. Denning.** 1996. D-Mannitol

in cerebrospinal fluid of patients with AIDS and cryptococcal meningitis. *J. Clin. Microbiol.* **34:**218–221.

165. **Miller, K. D., J. A. M. Mican, and R. T. Davey.** 1996. Asymptomatic solitary pulmonary modules due to *Cryptococcus neoformans* in patients infected with human immunodeficiency virus. *Clin. Infect. Dis.* **23:**810–812.

166. **Mody, C. H., G. B. Toews, and M. F. Lipscomb.** 1988. Cyclosporin A inhibits the growth of *Cryptococcus neoformans* in a murine model. *Infect. Immun.* **56:**7–12.

167. **Mody, C. H., G. B. Toews, and M. F. Lipscomb.** 1989. Treatment of murine cryptococcosis with cyclosporin-A in normal and athymic mice. *Am. Rev. Respir. Dis.* **139:**8–13.

168. **Mohr, J. A., H. Long, B. A. McKown, and H. G. Muchmore.** 1972. *In vitro* susceptibility of *Cryptococcus neoformans* to steroids. *Sabouraudia* **10:**171–172.

169. **Mohr, J. A., R. J. Tacker, R. J. Devlin, R. F. Felton, G. Francis, and E. R. Rhoades.** 1969. Estrogen stimulation of phagocytosis. *Am. Rev. Respir. Dis.* **99:**979–983.

170. **Moore, G. S., and R. D. Atkin.** 1977. The fungicidal and fungistatic effects of an aqueous garlic extract on medically important yeast-like fungi. *Mycologia* **69:**341–348.

171. **Mukherjee, J., L. A. Pirofski, M. D. Scharff, and A. Casadevall.** 1993. Antibody-mediated protection in mice with lethal intracerebral *Cryptococcus neoformans* infection. *Proc. Natl. Acad. Sci. USA* **90:**3636–3640.

172. **Mukherjee, J., M. D. Sharff, and A. Casadevall.** 1992. Protective murine monoclonal antibodies to *Cryptococcus neoformans. Infect. Immun.* **60:**4534–4541.

173. **Mukherjee, J., L. S. Zuckier, M. D. Scharff, and A. Casadevall.** 1994. Therapeutic efficacy of monoclonal antibodies to *Cryptococcus neoformans* glucuronoxylomannan alone and in combination with amphotericin B. *Antimicrob. Agents Chemother.* **38:**580–587.

174. **Nassar, F., E. Brummer, and D. A. Stevens.** 1995. Different components in human serum inhibit multiplication of *Cryptococcus neoformans* and enhance fluconazole activity. *Antimicrob. Agents Chemother.* **39:**2490–2493.

175. **National Committee for Clinical Laboratory Standards.** 1995. *Reference Method for Broth Dilution Antifungal Susceptibility Testing of Yeast; Tentative Standard M27-T.* National Committee for Clinical Laboratory Standards, Villanova, Pa.

176. **Nelson, M. R., M. Bower, D. Smith, C. Reed, D. Shanson, and B. Gazzard.** 1990. The value of serum cryptococcal antigen in the diagnosis of cryptococcal infection in patients infected with the human immunodeficiency virus. *J. Infect. Dis.* **21:**175–181.

177. **Nguyen, M. H., F. Barchiesi, D. A. McGough, V. L. Yu, and M. G. Rinaldi.** 1995. In vitro evaluation of combination of fluconazole and flucytosine against *Cryptococcus neoformans* var. *neoformans. Antimicrob. Agents Chemother.* **39:**1691–1695.

178. **Nguyen, M. H., L. K. Najvar, C. Yu, and J. R. Graybill.** 1997. Combination therapy with fluconazole and flucytosine in the murine model of cryptococcal meningitis. *Antimicrob. Agents Chemother.* **41:**1120–1123.

179. **Odom, A., M. Del Poeta, J. Perfect, and J. Heitman.** 1997. The immunosuppressant FK506 and its non-immunosuppressive analog L-685,818 are toxic to *Cryptococcus neoformans* by inhibition of a common target protein. *Antimicrob. Agents Chemother.* **41:**156–161.

180. **Oppenheim, B. A., R. Herbrecht, and S. Kusne.** 1995. The safety and efficacy of amphotericin B colloidal dispersion in the treatment of invasive mycoses. *Mycoses* **21:**1145–1153.

181. **Pappas, P., J. Perfect, H. Henderson, C. Kauffman, M. Saccente, D. Haas, G. Pankey, D. Lancaster, M. Holloway, G. Cloud, and W. Dismukes.** 1997. Cryptococcosis in non-HIV infected patients: a multicenter survey, abstr. 128. 35th Annu. Meet. Infectious Disease Society of America.

182. **Park, M. K., and J. E. Bennett.** 1996. Management of hydrocephalus in cryptococcal meningitis, abstr. VI.1. 3rd International Conference on Cryptococcus and Cryptococcosis, Paris.

183. **Peetermans, W., H. Bobbaers, J. Verhaegen, and J. Vanderpitte.** 1993. Fluconazole-resistant *Cryptococcus neoformans* var. *gattii* in an AIDS patient. *Acta Clin. Belg.* **48:**405–409.

184. **Peng, T., and J. N. Galgiani.** 1993. In vitro studies of a new antifungal triazole, D0870, against *Candida albicans, Cryptococcus neoformans,* and other pathogenic yeasts. *Antimicrob. Agents Chemother.* **37:**2126–2131.

184a. **Perfect, J. R.** Unpublished data.

185. **Perfect, J. R.** 1990. Fluconazole therapy for experimental cryptococcosis and candidiasis in the rabbit. *Rev. Infect. Dis.* **12**(Suppl. 3)**:**299–302.

186. **Perfect, J. R., G. M. Cox, R. K. Dodge, and W. A. Schell.** 1996. In vitro and in vivo efficacies of the azole SCH 56592 against *Cryptococcus neoformans. Antimicrob. Agents Chemother.* **40:**1910–1913.

187. **Perfect, J. R., and D. T. Durack.** 1982. Treatment of experimental cryptococcal meningitis with amphotericin B, 5-fluorocytosine and ketoconazole. *J. Infect. Dis.* **146:**429–435.

188. **Perfect, J. R., and D. T. Durack.** 1985. Comparison of amphotericin B and *N*-D-ornithyl amphotericin B methyl ester in experimental crytococcal meningitis and *Candida albicans* endocarditis with pyelonephritis. *Antimicrob. Agents Chemother.* **28:**751–755.

189. **Perfect, J. R., and D. T. Durack.** 1985. Effects of cyclosporine in experimental cryptococcal meningitis. *Infect. Immun.* **50:**22–26.

190. **Perfect, J. R., and D. T. Durack.** 1985. Penetration of imidazoles and triazoles into cerebrospinal fluid of rabbits. *Antimicrob. Agents Chemother.* **16:**81–86.

191. **Perfect, J. R., D. T. Durack, and H. A. Gallis.** 1983. Cryptococcemia. *Medicine* **62:**98–109.

192. **Perfect, J. R., D. T. Durack, J. D. Hamilton, and H. A. Gallis.** 1982. Failure of ketoconazole in cryptococcal meningitis. *JAMA* **247:**3349–3351.

192a. **Perfect, J. R., S. D. R. Lang, and D. T. Durack.** 1980. Chronic cryptococcal meningitis: a new experimental model. *Am. J. Pathol.* **101:**177–193.

193. **Perfect, J. R., D. V. Savani, and D. T. Durack.** 1986. Comparison of itraconazole and fluconazole in treatment of cryptococcal meningitis and *Candida* pyelonephritis in rabbits. *Antimicrob. Agents Chemother.* **29:**579–583.

194. **Perfect, J. R., D. V. Savani, and D. T. Durack.** 1993. Uptake of itraconazole by alveolar macrophages. *Antimicrob. Agents Chemother.* **37:**903–904.

195. **Perfect, J. R., and K. A. Wright.** 1994. Amphotericin B lipid complex in the treatment of experimental cryptococcal meningitis and disseminated candidiasis. *J. Antimicrob. Chemother.* **33:**73–81.

196. **Perfect, J. R., K. A. Wright, M. M. Hobbs, and D. T. Durack.** 1989. Treatment of experimental cryptococcal meningitis and disseminated candidiasis with SCH 39304. *Antimicrob. Agents Chemother.* **33:**1735–1740.

197. **Perkins, W.** 1969. Pulmonary cryptococcosis: report on the treatment of nine cases. *Dis. Chest* **56:**389–394.

197a. **Pirofski, L. A., and A. Casadevall.** 1996. *Cryptococcus neoformans*: paradigm for the role of antibody immunity against fungi? *Zentralbl. Bakteriol.* **284:**475–495.

198. **Plunkett, J. M., B. I. Turner, and M. B. Tallent.** 1981. Cryptococcal septicemia associated with urologic instrumentation in a renal allograft recipient. *J. Urol.* **125:**241–242.

199. **Polak, A.** 1987. Combination therapy of experimental candidiasis, cryptococcosis, aspergillosis and wangiellosis in mice. *Chemotherapy* **33:**381–395.

200. **Polak, A., H. J. Schroder, and M. Wall.** 1982. Combination therapy of experimental candidiasis, cryptococcosis, and aspergillosis in mice. *Chemotherapy* **28:**461–479.

201. **Polsky, B., M. R. Depman, J. W. Gold, J. H. Galicich, and D. Armstrong.** 1986. Intraventricular therapy of cryptococcal meningitis via a subcutaneous reservoir. *Am. J. Med.* **81:**24–28.

202. **Powderly, W. G.** 1993. Cryptococcal meningitis and AIDS. *Clin. Infect. Dis.* **17:**837–842.

203. **Powderly, W. G., G. A. Cloud, W. E. Dismukes, and M. S. Saag.** 1994. Measurement of cryptococcal antigen in serum and cerebrospinal fluid: value in the management of AIDS-associated cryptococcal meningitis. *Clin. Infect. Dis.* **18:**789–792.

204. **Powderly, W. G., M. S. Saag, G. A. Cloud, P. Robinson, R. Meyer, J. M. Jacobson, J. R. Graybill, A. M. Sugar, V. J. McAuliffe, S. E. Follansbee, C. U. Tuazon, J. J. Stern, J. Feinberg, R. Hafner, and W. E. Dismukes.** 1992. A controlled trial of fluconazole or amphotericin B to prevent relapse of cryptococcal meningitis in patients with the acquired immunodeficiency syndrome. *N. Engl. J. Med.* **326:**793–798.

205. **Price, D. A., J. L. Klein, M. Fisher, J. Main, J. S. Bingham, and R. J. Coker.** 1997. Potential role for granulocyte-macrophage colony-stimulated factor in the treatment of HIV-associated cryptococcal meningitis. *AIDS* **11:**693–694.

206. **Prugam, A., J. Dupouy-Camet, P. Blanche, J. P. Gangneux, C. Tourte-Schaefer, and D. Sicard.** 1994. Increased fluconazole resistance of *Cryptococcus neoformans* isolated from a patient with AIDS and recurrent meningitis. *Clin. Infect. Dis.* **19:**975–976.

207. **Reilly, E. B., and E. L. Artman.** 1948. Cryptococcosis. Report of a case and experimental studies. *Arch. Intern. Med.* **81:**1–8.

208. **Restrepo, B. I., J. Ahrens, and J. R. Graybill.** 1989. Efficacy of SCH 39304 in murine cryptococcosis. *Antimicrob. Agents Chemother.* **33:**1242–1246.

209. **Rex, J. H., R. A. Larsen, W. E. Dismukes, G. A. Cloud, and J. A. Bennett.** 1993. Catastrophic vision loss due to *Cryptococcus neoformans* meningitis. *Medicine* **72:**207–224.

210. **Saag, M. S., and NIAID Mycoses Study Group.** 1995. Comparison of fluconazole versus itraconazole as maintenance therapy of AIDS-associated cryptococcal meningitis, abstr. I218. 35th Interscience Conference on Antimicrobial Agents and Chemotherapy, San Francisco.

211. **Saag, M. S., W. G. Powderly, G. A. Cloud, P. Robinson, M. H. Grieco, P. K. Sharkey, S. E. Thompson, A. Sugar, C. U. Tuazon, J. F. Fisher, N. Hyslop, J. M. Jacobson, R. Hafner, and W. E. Dismukes.** 1992. Comparison of amphotericin B with fluconazole in the treatment of acute AIDS-associated cryptococcal meningitis. *N. Engl. J. Med.* **326:**83–89.

212. **Salyer, W. R., and D. C. Salyer.** 1974. Pleural involvement in cryptococcosis. *Chest* **66:**139.

213. **Salyer, W. R., D. C. Salyer, and R. D. Baker.** 1974. Primary complex of *Cryptococcus* and pulmonary lymph nodes. *J. Infect. Dis.* **130:**74–77.

214. **Sanati, H., S. A. Messer, M. Pfaller, M. Witt, R. Larsen, A. Espinel-Ingroff, and M. Ghannoum.** 1996. Multicenter evaluation of broth microdilution method for susceptibility testing of *Cryptococcus neoformans* against flucanozole. *J. Clin. Microbiol.* **34:**1280–1282.

215. **Sanford, J. E., D. M. Lupan, A. M. Schlageter, and T. R. Kozel.** 1990. Passive immunization against *Cryptococcus neoformans* with an isotype-switch family of monoclonal antibodies reactive with cryptococcal polysaccharide. *Infect. Immun.* **58:**1919–1923.

216. **Sarosi, G. A., J. D. Parker, I. L. Doto, and F. E. Tosh.** 1969. Amphotericin B in cryptococcal meningitis: long-term results of treatment. *Ann. Intern. Med.* **71:**1079–1087.

217. **Savoy, A. C., D. M. Lupan, P. B. Manalo, J. S. Roberts, A. M. Schlageter, L. C. Weinhold, and T. R. Kozel.** 1997. Acute lethal toxicity following passive immunization for treatment of murine cryptococcosis. *Infect. Immun.* **65:**1800–1807.

217a. **Schurmann, D., B. de Matos Marques, T. Grunewald, H. D. Pohle, H. Hahn, and B. Ruf.** 1991. Safety and efficacy of liposomal amphotericin B in treating AIDS-associated disseminated cryptococcosis. *J. Infect. Dis.* **164:**620–622.

218. **Sekhon, A. S., A. A. Padhye, A. K. Gary, and Z. Hamir.** 1992. *In vitro* activity of amphotericin B, namycin and their novel water-soluble compounds against pathogenic yeasts. *Chemotherapy* **38:**297–302.

219. **Shadomy, S., A. Espinel-Ingraff, and R. J. Gebhart.** 1985. *In vitro* studies with SF 86–327, a new orally active allylamine derivative. *Sabouraudia* **23:**125–132.

220. **Shadomy, S., and M. A. Pfaller.** 1991. Laboratory studies with antifungal agents: susceptibility tests and quantitation in body fluids, p. 1173–1183. *In* A. Balows, W. J. Hausler, Jr., K. L. Herrmann, H. D. Isenberg, and H. J. Shadomy (ed.), *Manual of Clinical Microbiology,* 5th ed. American Society for Microbiology, Washington, D.C.

221. **Shadomy, S., H. J. Shadomy, J. A. McCay, and J. P. Utz.** 1969. In vitro susceptibility of *Cryptococcus neoformans* to amphotericin B, hamycin, and 5-fluorocytosine, p. 452–460. *Antimicrob. Agents Chemother. 1968.*

222. **Shapiro, L. L., and J. B. Neal.** 1925. Torula meningitis. *Arch. Neurol. Psych.* **13:**174–190.

223. **Sharkey, P. K., J. R. Graybill, E. S. Johnson, S. G. Hausroth, R. B. Pollard, A. Kolokathis, D. Mildvan, P. Fan-Havard, R. H. K. Eng, T. F. Patterson, J. C. Pottage, M. S. Simberkoff, J. Wolf, R. D. Meyer, R. Gupton, L. N. Lee, and D. S. Gordon.** 1996. Amphotericin B lipid complex compared with amphotericin B in the treatment of Cryptococcal meningitis in patients with AIDS. *Clin. Infect. Dis.* **22:**315–321.

224. **Sousa-Sanchez, A., M. E. Alvarez, C. Parra, and F. Baguero.** 1997. Activity on clinical yeast isolates of a new antifungal agent, GM 237354, abstr. F-56. 37th Interscience Conference on Antimicrobial Agents and Chemotherapy, Toronto.

225. **Spickard, A., W. T. Butler, V. Andriole, and J. P. Utz.** 1963. The improved prognosis of cryptococcal meningitis with amphotericin B therapy. *Ann. Intern. Med.* **58:**66–83.

226. **Spitzer, E. D., S. G. Spitzer, L. F. Freundlich, and A. Casadevall.** 1993. Persistence of initial infection in recurrent *Cryptococcus neoformans* meningitis. *Lancet* **341:**595–596.

227. **Staib, F., and M. Seibold.** 1988. Mycological-diagnostic assessment of the efficacy of amphotericin B+ flucytosine to control *Cryptococcus neoformans* in AIDS patients. *Mycoses* **31:**175–186.

228. **Staib, F., M. Seibold, M. L'age, W. Heise, J. Skorde, G. Grosse, F. Nurnberger, and G. Baver.** 1989. *Cryptococcus neoformans* in the seminal fluid of an AIDS patient. A contribution to the clinical course of cryptococcosis. *Mycoses* **32:**171–180.

229. **Stamm, A. M., R. B. Diasio, W. E. Dismukes, S. Shadomy, G. A. Cloud, C. A. Bowles, G. H. Karam, and A. Espinel-Ingroff.** 1987. Toxicity of amphotericin B plus flucytosine in 194 patients with cryptococcal meningitis. *Am. J. Med.* **83:**236–242.

230. **Stern, J. J., B. J. Hartman, P. Sharkey, V. Rowland, K. E. Squires, H. W. Murray, and J. R. Graybill.** 1988. Oral fluconazole therapy for patients with the acquired immunodeficiency syndrome and cryptococcosis: experience with 22 patients. *Am. J. Med.* **85:**477–480.

231. **Stone, W. J., and B. F. Sturdivant.** 1929. Meningo-encephalitis due to *Torula histolytica*. *Arch. Intern. Med.* **44:**560–575.

232. **Strippoli, V., F. D. D'Auria, N. Simonetti, D. Basti, and T. Bruzzese.** 1997. *In vivo* and *in vitro* antifungal activity of the polyene derivative SPA-S-753 against encapsulated form of *Cryptococcus neoformans*. *Infection* **25:**27–31.

233. **Sugar, A. M.** 1995. Use of amphotericin B with azole antifungal drugs: what are we doing? *Antimicrob. Agents Chemother.* **39:**1907–1912.

234. **Sung, J. P., G. D. Campbell, and J. G. Grendahl.** 1978. Miconazole therapy for fungal meningitis. **Arch. Intern. Med.** **35:**443–447.

235. **Takesako, K., H. Kuroda, T. Inoue, F. Haruna, Y. Yoshikawa, I. Kato, K. Uchida, T. Hiratani, and H. Yamaguchi.** 1993. Biological properties of aureobasidin A, a cyclic depsipeptide antifungal antibiotic. *J. Antibiot.* **46:**1414–1420.

236. **Tenholder, M. F., F. W. Ewald, Jr., N. K. Khankhanian, and J. H. Crosby.** 1992. Complex cryptococcal empyema. *Chest* **101:**586–588.

237. **Thornwell, S. J., R. B. Peery, and P. C. Skatrud.** 1997. Cloning and characterization of CneMDRI: a *Cryptococcus neoformans* gene encoding a protein related to multidrug resistance proteins. *Gene* **201:**21–29.

238. **Toone, E. C.** 1941. Torula histolytica (Blastomycoides histolytica) meningitis: report of a case with recovery. *V.A. Med. Monthly* **68:**405–407.

239. **Torres, I. R., U. C. Villareal, R. M. Robles, P. Aparicio, and D. C. Cano.** 1993. Comparative study between two treatment schedules in AIDS patients with meningitis caused by *Cryptococcus neoformans:* GM-CSF plus amphotericin B versus amphotericin B alone, p. 64. *In Leucomax, Current Use and Future Applications.* Adelphi Communications, Lucerne.

240. **Turnidge, T. D., S. Gudmundsson, B. Vogelman, and W. A. Craig.** 1994. The postantibiotic effect of antifungal agents against common pathogenic yeasts. *J. Antimicrob. Chemother.* **34:**83–92.

241. **Utz, J. P., I. L. Garrigues, M. A. Sande, J. F. Warner, G. L. Mandell, R. F. McGhee, R. J. Duma, and S. Shadomy.** 1975. Therapy of cryptococcosis with a combination of flucytosine and amphotericin B. *J. Infect. Dis.* **132:**368–373.

242. **Utz, J. P., S. Shadomy, and R. F. McGehee.** 1969. 5-Flucytosine: experience in patients with pulmonary and other forms of cryptococcosis. *Am. Rev. Respir. Dis.* **99:**975.

243. **Utz, J. P., S. Shadomy, and R. F. McGehee.** 1972. Flucytosine. *N. Engl. J. Med.* **286:**777–778.

244. **Van Cutsem, J., F. Van Gerven, and P. A. Janssen.** 1987. Activity of orally, topically, and parenterally administered itraconazole in the treatment of superficial and deep mycoses: animal models. *Rev. Infect. Dis.* **9**(Suppl. 1):S15–S32.

245. **Vanden Bossche, H., P. Marichal, L. Le June, M. C. Coese, J. Gorrens, and W. Cools.** 1993. Effects of itraconazole and cytochrome P-450 sterol 14 alpha-demethylation and reduction of 3-ketosteroids in *Cryptococcus neoformans. Antimicrob. Agents Chemother.* **37:**2101–2105.

246. **van der Horst, C., M. S. Saag, G. A. Cloud, R. J. Hamill, J. R. Graybill, J. D. Sobel, P. C. Johnson, C. U. Tuazon, T. Kerkering, B. L. Moskovitz, W. G. Powderly, and W. E. Dismukes.** 1997. Treatment of cryptococcal meningitis associated with the acquired immunodeficiency syndrome. *N. Engl. J. Med.* **337:**15–21.

247. **Velez, J. D., R. Allendoerfer, M. Luther, M. G. Rinaldi, and J. R. Graybill.** 1993. Correlation of *in vitro* azole susceptibility with *in vivo* response in a murine model of cryptococcal meningitis. *J. Infect. Dis.* **168:**508–510.

248. **Venkateswarlu, K., M. Taylor, N. J. Manning, M. G. Rinaldi, and S. L. Kelly.** 1997. Fluconazole tolerance in clinical isolates of *Cryptococcus neoformans. Antimicrob. Agents Chemother.* **41:**748–751.

249. **Viard, J. P., C. Hennequin, N. Fortineau, N. Pertuiset, C. Rothschild, and H. Zylberberg.** 1995. Fulminant cryptococcal infections in HIV-infected patients on oral fluconazole. *Lancet* **346:**118.

250. **Viollon, C., and J. P. Chaumont.** 1994. Antifungal properties of essential oils and their main components upon *Cryptococcus neoformans. Mycopathologia* **128:**151–153.

251. **Viviani, M. A., A. M. Tortorano, P. C. Giani, C. Arici, A. Goglio, P. Crocchiolo, and M. Alma-viva.** 1987. Itraconazole for cryptococcal infection in the acquired immunodeficiency syndrome. *Ann. Intern. Med.* **106:**166. (Letter.)

252. **Viviani, M. A., A. M. Tortorano, M. Langer, M. Alma-viva, E. Negri, S. Christina, S. Soccia, R. De Maria, R. Fiocchi, and P. Ferrazzi.** 1989. Experience with itraconazole in cryptococcosis and aspergillosis. *J. Infect.* **18:**151–165.

253. **Viviani, M. A., A. M. Tortorano, R. Woenstenborghs, and G. Cauwenbergh.** 1987. Experience with itraconazole in deep mycoses in northern Italy. *Mykosen* **30:**233–244.

254. **Walsh, T. J., R. Schlegel, M. M. Moody, J. W. Costerton, and M. Saloman.** 1986. Ventriculoatrial shunt infection due to *Cryptococcus neoformans:* an ultrastructural and quantitative microbiological study. *Neurosurgery* **18:**373–375.

255. **Wang, Y., and A. Casadevall.** 1994. Growth of *Cryptococcus neoformans* in presence of

L-dopa decreases its susceptibility to amphotericin B. *Antimicrob. Agents Chemother.* **38:**2648–2650.

256. **Wang, Y., and A. Casadevall.** 1996. Susceptibility of melanized and nonmelanized *Cryptococcus neoformans* to the melanin-binding compounds trifluoperazine and chloroquine. *Antimicrob. Agents Chemother.* **40:**540–545.

257. **Watson, A. J., R. P. Russell, and R. F. Cabreja.** 1985. Cure of cryptococcal infection during continued immunosuppressive therapy. *Q. J. Med.* **55:**169–172.

258. **Weinstein, L., and J. Irving.** 1980. Successful treatment of cerebral cryptococcoma and meningitis with miconazole. *Ann. Intern. Med.* **93:**569–571.

259. **Whelan, W. L.** 1987. The genetic basis of resistance to 5-fluorocytosine in *Candida* species and *Cryptococcus neoformans*. *Crit. Rev. Microbiol.* **15:**45–56.

260. **White, M., C. Cirrincione, A. Blevins, and D. Armstrong.** 1992. Cryptococcal meningitis with AIDS and patients with neoplastic disease. *J. Infect. Dis.* **165:**960–966.

261. **Williams, D. M., J. R. Graybill, D. J. Drutz, and H. B. Levine.** 1980. Suppression of cryptococcosis and histoplasmosis by ketoconazole in athymic nude mice. *J. Infect. Dis.* **141:**76–80.

262. **Wilson, H. M., and A. W. Duryea.** 1951. Cryptococcus meningitis (torulosis) treated with a new antibiotic, actidione. *Arch. Neurol. Psych.* **66:**470–480.

263. **Witt, M. D., R. J. Lewis, R. A. Larsen, E. N. Milefchik, M. A. E. Leal, R. H. Haubrich, J. A. Richie, J. E. Edwards, Jr., and M. A. Ghannoum.** 1996. Identification of patients with acute AIDS-associated cryptococcal meningitis who can be effectively treated with fluconazole: the role of antifungal susceptibility testing. *Clin. Infect. Dis.* **22:**322–328.

264. **Wong, B., J. R. Perfect, S. Beggs, and K. A. Wright.** 1990. Production of the hexitol L-mannitol by *Cryptococcus neoformans* in vitro and in rabbits with experimental meningitis. *Infect. Immun.* **58:**1664–1670.

265. **Wright, K. A., J. R. Perfect, and W. Ritter.** 1990. The pharmacokinetics of BAY R3783 and its efficacy in the treatment of experimental cryptococcal meningitis. *J. Antimicrob. Chemother.* **26:**387–397.

266. **Yadav, Y. R., J. R. Perfect, and A. Friedman.** 1988. Successful treatment of cryptococcal ventriculo-atrial shunt infection with systemic therapy alone. *Neurosurgery* **23:**317–322.

267. **Yamaguchi, H., H. Ikemoto, K. Watanabe, A. Ito, K. Hara, and S. Kohno.** 1996. Fluconazole monotherapy for cryptococcosis in non-AIDS patients. *Eur. J. Clin. Microbiol. Infect. Dis.* **15:**787–792.

268. **Yamaguchi, H., S. Inouye, Y. Orikasa, H. Tohyama, K. Komuro, S. Gomi, S. Ohuchi, T. Matsumoto, M. Yamaguchi, T. Hiratani, K. Uchida, Y. Ohsumi, S. Kondo, and T. Takeuchi.** 1992. A novel antifungal antibiotic, Benanomicin A, p. 393–402. *In* H. Yamaguchi, G. S. Kobayashi, and H. Takahashi (ed.), *Recent Progress in Antifungal Chemotherapy*. Marcel Dekker, Inc., New York.

269. **Yew, W. W., P. C. Wong, C. F. Wong, J. Lee, and C. H. Chau.** 1996. Oral fluconazole in the treatment of pulmonary cryptococcosis in non-AIDS patients. *Drugs Exp. Clin. Res.* **22:**25–28.

270. **Yoshikawa, Y., K. Ikai, Y. Umeda, A. Ogawa, K. Takesako, I. Kato, and H. Naganawa.** 1993. Isolation, structures, and antifungal activities of new aureobasidins. *J. Antibiot.* **46:**1347–1354.

271. **Young, E. J., D. D. Hirsh, V. Fainstein, and T. W. Williams.** 1980. Pleural effusions due to *Cryptococcus neoformans*: a review of the literature and report of two cases with cryptococcal antigen determinations. *Am. Rev. Respir. Dis.* **121:**743–746.

272. **Yuen, C., A. Graziani, N. Pietroski, R. MacGregor, and M. Schuster.** 1994. Cryptococcal antigenemia in HIV-infected patients. *Clin. Infect. Dis.* **19:**579.

273. **Zuger, A., E. Louie, R. S. Holzman, M. S. Simberkoff, and J. J. Rahal.** 1986. Cryptococcal disease in patients with acquired immunodeficiency syndrome. Diagnostic features and outcome of treatment. *Ann. Intern. Med.* **104:**234–240.

15 | Prevention of Human Infection

Considering the devastating consequences of disseminated cryptococcal infection, prevention of infection is an important goal in the care of high-risk patients. The benefits of prevention are obvious given that cryptococcal infections require prolonged therapy, have high mortality and morbidity despite appropriate therapy, and are virtually incurable in patients with severely impaired immune function. The propensity of *Cryptococcus neoformans* to invade the central nervous system can produce severe lifelong complications, including blindness and cranial nerve deficits, in patients who survive the infection. The inadequacy of existing therapies is further exemplified by their inability to eradicate most cryptococcal infections in patients with severe immunological impairment, such as those with late-stage human immunodeficiency virus (HIV) infection (7, 57). In recent years there has been considerable interest in developing strategies for the prevention of cryptococcal infection in patients at risk (Table 1). The development and application of preventive strategies is made feasible by the fact that populations at risk for cryptococcal infection have been clearly defined (see chapter 11).

AVOIDANCE OF INFECTION

C. neoformans infection is acquired from the environment. There are anecdotal data linking some cases of human infection to exposure to heavily contaminated sites, such as areas soiled with avian excreta (see chapter 3). The identification of environmental sources of infection suggests that it may be possible to design a prevention strategy based on "avoidance" of infection. The avoidance strategy would seek to reduce the likelihood of infection by minimizing exposure to areas contaminated with *C. neoformans*. Unfortunately, an avoidance strategy may not be practical given the multitude of contaminated sources in the environment and the likelihood of widespread dispersal of infectious organisms by aerosols. Also complicating the formulation of an effective avoidance strategy is uncertainty about the pathogenesis, route, and source of human infection (see chapters 3 and 11 of this volume). For example, it is not known whether most cases of cryptococcal meningitis are the result of acute infection or reactivation of a latent focus of infection. If cryptococcal infections reflect reactivation of latent infection when the individual becomes immunosuppressed, then a preventive strategy based on avoidance may

Table 1 Strategies for the prevention of *C. neoformans* infection

Method	Aim	Advantage	Disadvantage	Efficacy
Avoidance	Reduce likelihood of infection by avoiding contaminated sites, not keeping pet birds, and cessation of smoking	Common sense and prudent steps that could potentially reduce risk of infection at low cost	Impractical. Avoidance may not be possible given the multitude of contaminated sites in the environment.	Unknown
Primary chemoprophylaxis	Prevent infection by administration of an antifungal agent to patients at risk for infection	Highly effective	Chronic administration of antifungal drugs is expensive, may select for resistant fungal strains, increases likelihood of deleterious drug interactions, and is unnecessary for the majority of patients who would not have developed infection	Proved
Secondary chemoprophylaxis	Prevent recurrence of infection in successfully treated patients by chronic administration of an antifungal therapy	Highly effective	None, apart from the cost and risks associated with chronic administration of antifungal drugs. Suppressive therapy is recommended in all patients who are expected to remain immunosuppressed following the completion of successful initial therapy for cryptococcosis.	Proved
Early screening	Detect early infections by serum antigen determinations in patients at risk for infection	Theoretical advantage is that early detection could lead to more effective therapy	Problems include high cost, low sensitivity, and the occurrence of false-positive and false-negative latex agglutination tests.	Unknown
Vaccine and antibodies	Prevent infections by providing opsonic antibodies through either vaccination or passive antibody administration	Vaccines offer the theoretical advantage of lifelong protection. Antibodies are natural products of the immune system.	Both vaccines and antibodies are experimental agents that are not available for routine use. The safety and efficacy of vaccines and antibody prophylaxis remain to be proved. Antibody responses are likely to be costly.	Unknown

be futile. Nevertheless, there are some common-sense recommendations for individuals at risk for cryptococcosis.

(i) Avoid sites that are traditionally associated with high concentrations of *C. neoformans*. Patients with immunological disorders should avoid areas contaminated with avian excreta, in particular pigeon excreta. For example, patients with impaired immunity should avoid cleaning pigeon excreta-contaminated air-conditioning wells and bird aviaries. If heavily contaminated sites are to be cleaned, it may be advisable to sterilize the site with an alkaline solution (59). Furthermore, the association of *C. neoformans* var. *gattii* with some species of eucalyptus trees such as *Eucalyptus camaldulensis* and *Eucalyptus tereticornis* (20, 22, 37, 49) suggests that individuals at risk would be prudent to avoid sites where these trees are found in high concentrations and to avoid activities that involve exposure to these trees or their products.

(ii) Avoid pet birds. *C. neoformans* has been isolated from the excreta of many species of birds, including household bird pets and those in pet shops (see chapter 3). Keeping a pet bird may pose a danger to a patient with impaired immunity if the bird excreta is infected with *C. neoformans*. Activities related to pet care such as cleaning cages and being in the birds' general proximity could theoretically result in the inhalation of aerosolized infectious *C. neoformans* forms.

(iii) Stop smoking cigarettes and avoid cigarette smoke. A small study documented a significantly greater risk for disseminated cryptococcosis in HIV-infected patients who smoke than among nonsmokers (46). The authors of the study speculated that smoking may promote inhalation and deposition of the organism in the airways in association with smoke or particulates (46). Alternatively, tobacco smoking increases the chance for reactivation of a latent pulmonary infection (46). Alveolar macrophages from smokers have been shown to have reduced antifungal activity against *C. neoformans* (55).

CHEMOPROPHYLAXIS

Chemoprophylaxis involves the administration of antifungal drugs to high-risk patients to reduce the probability of primary infection or recurrence of infection. The chemoprophylaxis strategy is based on the premise that chronic administration of antifungal agents to patients at risk interferes with the establishment of a symptomatic infection. The introduction of safe and effective oral azole drugs in the early 1990s made feasible the design of effective, low-toxicity regimens for preventing cryptococcal infections.

Primary Prophylaxis

Primary prophylaxis refers to the administration of antifungal drugs to prevent the occurrence of *C. neoformans* infections in patients at high risk for infection. Several studies have provided convincing evidence that administration of fluconazole to patients at risk for *C. neoformans* infections can significantly reduce the likelihood of disseminated cryptococcosis (50). Nightingale et al. (45) reported that AIDS patients receiving fluconazole at a dose of 100 mg/day had significantly lower rates of cryptococcal infection relative to untreated control subjects. Powderly et al. (53) carried out a large randomized trial of fluconazole versus clotrimazole troches to evaluate their relative efficacy in preventing fungal infections in patients with AIDS. There was a sevenfold reduction in the number of cryptococcal infections in patients receiving 200 mg of fluconazole daily relative to those receiving clotrimazole troches (53). A retrospective study of HIV-infected patients revealed that those who took fluconazole were less likely to develop cryptococcal meningitis (1). Another retrospective study showed a drop in the prevalence of cryptococcal infections from 10.3 to 3.8% in an AIDS clinic after the introduction of weekly fluconazole therapy (100 mg once per week) for suppression of candida thrush (43). Finally, an analysis of trends in the incidence of cryptococcal infection in the Multicenter AIDS Cohort Study from 1985 to 1992 revealed a reduction in the incidence rate of cryptococcal meningitis from 3.3 to 0.6 per 100 person-years among individuals receiving antifungal prophylaxis (4).

These studies provide convincing evidence that routine fluconazole administration lowers the risk of cryptococcal infection in HIV-infected patients. The benefits of fluconazole prophylaxis in reducing fungal infections are most apparent in HIV-infected patients who have $CD4^+$ lymphocyte counts of ≤ 50 cells/μl (53). The increased use of fluconazole for the therapy of candidal infections in patients with AIDS may have the secondary benefit of reducing the prevalence of symptomatic C. neoformans infections. There is some evidence that the number of cases of cryptococcosis in the United States has declined since fluconazole has come into wide usage (4, 38, 44).

Despite its efficacy, primary prophylaxis of cryptococcal infection with fluconazole is not a routine recommendation in the care of patients with AIDS (25, 48, 53). Factors that have mitigated against the widespread use of fluconazole for prophylaxis are: (i) high expense, (ii) the possibility of selecting fluconazole-resistant Candida albicans strains in patients who receive chronic fluconazole therapy (48, 50, 52), (iii) the relatively low prevalence of cryptococcal infection, and (iv) the possibility of fluconazole drug interactions with other medications taken by HIV-infected patients.

Primary prophylaxis is appropriate for the rare accidents when an individual suffers an inoculation with material contaminated with C. neoformans. Cutaneous C. neoformans infection has been reported after needle injuries with contaminated needles (10, 26). Needle injuries can occur during the care of patients with cryptococcosis and in research laboratories that study C. neoformans (10, 26). Other individuals at risk are mortuary attendants and microbiology laboratory personnel. In the five contaminated-needle-related injuries reported in the literature, two cases of cutaneous infection occurred in two individuals who did not receive antifungal prophylaxis, and no infections occurred in three individuals who were promptly treated with 200 mg of fluconazole per day for 2 weeks (10, 26).

In summary, primary prophylaxis with fluconazole is highly effective in preventing C. neoformans infections in populations at high risk for infection. Concerns about cost, selection of antifungal-resistant strains, and the possibility of interactions with other drugs frequently used in patients with HIV infection have diminished the enthusiasm for universal fluconazole prophylaxis of patients at risk for infection. Nevertheless, there is evidence that the widespread use of fluconazole for the treatment of candidal thrush has reduced the prevalence of C. neoformans infections in HIV-infected patients at high risk.

Secondary Prophylaxis

Secondary prophylaxis refers to the administration of antifungal drugs to prevent recurrence of *C. neoformans* infections in patients who have been successfully treated for cryptococcosis. The need for secondary prophylaxis was made apparent by the high frequency of clinical relapses among patients with AIDS who survived an episode of cryptococcal meningitis (33, 65, 66). The experience with *C. neoformans* infections in patients with AIDS is different from that in other patient groups in that cryptococcal infection in patients with AIDS is seldom curable. Molecular analysis of initial and relapse *C. neoformans* isolates from treated patients with recurrent infection has consistently shown that patients tend to recur with the same isolate recovered during the initial infection (7, 24, 57). This implies that recurrent infection is a result of the inability of existing antifungal drugs to fully eradicate the infection in patients with severe immune deficits. Presumably, the immunological deficits in patients with AIDS are so severe that antifungal therapy alone cannot eradicate the infection without help from the immune system. There is also evidence that infection may persist in cryptic sites such as the prostate despite antifungal therapy (3, 34).

The high recurrence rate of cryptococcal meningitis in treated patients with AIDS has led to the use of chronic antifungal therapy in all patients who survive the initial infection to reduce the probability of relapse (42, 50, 52, 58, 66). Several studies have documented that fluconazole doses of 200 to 400 mg/day are very effective for reducing the number of recurrences among patients with AIDS treated for cryptococcal meningitis (6, 42, 50, 52, 58). A small number of patients have experienced recurrence of infection despite fluconazole suppressive therapy. Recurrences of *C. neoformans* infections in patients on suppressive therapy can reflect poor compliance, further deterioration of immune function, and/or the development of drug resistance. Analysis of antifungal drug susceptibilities of isolates from patients who have suffered recurrences of infection have shown no evidence of the development of resistance as a result of chronic antifungal drug use in the majority of cases (7, 11). However, there have been some cases where relapse of infection was associated with the recovery of an isolate that was less susceptible to the drugs used in therapy (15, 47, 54).

In summary, secondary prophylaxis is very successful at reducing the likelihood of relapse and is recommended for all patients with AIDS who survive the initial bout of infection. Secondary prophylaxis with fluconazole is well tolerated and should be considered for all immunocompromised patients who survive an episode of cryptococcosis and are expected to remain in an immunocompromised state.

Possible Effects of Nonstandard Antifungal Drugs on the Prevalence of *C. neoformans*

When evaluating the benefits and risks of primary prophylaxis, it is noteworthy that other drugs have been shown to be active against *C. neoformans* in vitro and in animals. Hence, the risk benefit and cost-effectiveness calculation for the benefits of primary prophylaxis for cryptococcosis may be affected if the patient is receiving other medicines with potential activity against *C. neoformans*. For example, trifluoperazine (21, 60), chloroquine (36, 60), pentamidine (5), benzimidazole (14), diethylcarbamazine (30–32), and fluvastatin (12) have each been reported to have activity against *C. neoformans* in vitro. There is no experience with the clinical use of any of these agents against *C. neoformans*, but the possibility exists that their use modifies the risk for cryptococcosis in patients at high risk for infection. For example, the use of chronic pentamidine for prophylaxis against *Pneumocystis carinii* infection in patients with AIDS could conceivably reduce the likelihood of *C. neoformans* infection (5).

Some dietary components may also affect the risk for infection. Garlic contains powerful antifungal compounds (8, 63, 64), and garlic consumption could theoretically reduce the risk of infection. An uncontrolled study from China reported a 69% response rate in patients treated with oral raw garlic suspensions, with some patients being reportedly cured (2). Measurements of antifungal activity in urine and serum after oral ingestion of fresh garlic extract demonstrated anticryptococcal activity 0.5 to 1 h after ingestion (8). However, the effects were transient, and the anticryptococcal activity of serum was low and measurable only in undiluted serum (8). In another study, administration of a commercial garlic extract available in China for the treatment of systemic fungal infections produced increased anti-*C. neoformans* activity in serum and cerebrospinal fluid (16).

In summary, it is important to consider that non-antifungal drugs and/or dietary practices may conceivably reduce the risk of cryptococcal infection in some patients. However, it is also important to note that there is no conclusive evidence for prevention of cryptococcosis by drugs other than the standard antifungal agents.

PREVENTION BY ANTIGEN SCREENING AND EARLY INTERVENTION

The availability of a sensitive and specific diagnostic test for *C. neoformans* infection in the form of the cryptococcal antigen detection assay suggests the feasibility of screening high-risk populations for early signs of cryptococcosis. This screening strategy is based on the premise that early diagnosis would permit more effective therapy and possibly prevent progression of infection to a life-threatening condition. Several studies have shown that screening of high-risk patients for cryptococcal antigen can identify some cases of cryptococcal infection. In Central Africa, routine serum screening for cryptococcal antigen revealed a positive assay in 55 (12.2%) of 450 newly diagnosed HIV-seropositive patients (17). In Brazil, a study of antigen screening in Brazil reported that only 13 (6.7 %) of 193 patients were positive for serum cryptococcal antigen; three of these patients were found to have cryptococcosis, and another three developed cryptococcosis in the next few months (41). In Denmark, a study of 530 serum samples from 334 HIV-infected patients revealed only 3 cases of cryptococcal antigenemia as measured by the latex agglutination assay (29). The results from these three studies reveal large differences in the likelihood that serum antigen screening will identify a patient with early or indolent cryptococcal infection. Furthermore, these studies suggest that the usefulness of screening for serum antigen may depend on the prevalence of cryptococcosis in the population under study. In Central Africa, where the prevalence of cryptococcosis is high, serum screening is more likely to detect individuals with cryptococcal infection than in Denmark, where the prevalence of cryptococcosis is low.

Problems with screening for serum antigen include cost of the assay and the possibility of false-positive and false-negative results. For example, among the 242 HIV-positive patients studied by Negroni et al. (41), 10 individuals had positive antigen titer but never developed cryptococcosis, and 3 individuals developed cryptococcosis despite having an initial negative serum antigen level. At present there is insufficient information to judge the efficacy or cost-effectiveness of screening populations at risk for infection by using routine serum antigen determinations. Hence, universal serum antigen screening for patients at risk is not recommended.

Physicians sometimes measure serum antigen in patients at risk for infection to search for hidden cryptococcal infection. When serum antigen is detected and subsequent diagnostic investigations confirm the presence of cryptococcal infection, the choice is clear: treatment is warranted. However, there are instances when the serum is positive for cryptococcal antigen but a diagnostic investigation fails to reveal a site of cryptococcal infection (23). In these cases the physician faces the quandary of whether to institute costly, possibly lifelong therapy or wait for the infection to either resolve or declare itself (23). Studies in rats have shown that localized pulmonary infection does not produce appreciable serum levels of cryptococcal polysaccharide (27). Hence, the detection of serum antigen could be considered evidence for invasive infection unless proved otherwise (23, 27). The optimal management for asymptomatic patients with AIDS who have positive serological tests for cryptococcal antigen is uncertain (23). A retrospective analysis of a small number of AIDS patients with isolated cryptococcal antigenemia suggested a high likelihood of progression to symptomatic infection (23). The present recommendation is to take an individualized approach to the patient with persistent serum antigenemia while maintaining a low threshold for the administration of antifungal therapy (23).

IMMUNOPROPHYLAXIS: VACCINE AND ANTIBODY PREVENTATIVE STRATEGIES

Interest in vaccines against *C. neoformans* dates to the early and mid-20th century when the futility of pre-amphotericin B era therapies led to experimentation with therapeutic vaccines (13). In recent years the high prevalence of cryptococcal infections in patients with AIDS has again rekindled interest in the development of a vaccine to prevent *C. neoformans* infections.

The Conjugate Vaccine

The conjugate vaccine is based on the premise that a vaccine that elicits a strong antibody response will protect against *C. neoformans* infection in susceptible hosts. In 1991 Devi and collaborators (19) reported the synthesis of a highly immunogenic vaccine composed of glucuronoxylomannan (GXM) covalently linked to tetanus toxoid (TT). In mice this vaccine elicits antibodies with the same specificity as those elicited by experimental cryptococcal infection (9). Monoclonal antibodies generated from splenocytes harvested from mice immunized with the vaccine have been shown to be protective against experimental murine infection (39, 40). Administration of the GXM-TT vaccine to mice elicits strong antibody responses that protect against experimental cryptococcal infection (18). The GXM-TT vaccine was tested in normal human subjects and found to be highly immunogenic and to result in the formation of immunoglobulin M (IgM) and IgG antibodies to the *C. neoformans* capsular GXM (51, 62). IgG levels increased 70-fold in normal subjects receiving a 50-μg dose of GXM-TT, and the antibody levels persisted for at least 6 months (61). Monoclonal antibodies generated from human subjects have serological and structural similarities to mouse protective antibodies, and some are opsonic (51, 62). Despite encouraging results in preclinical studies, the efficacy of the GXM-TT vaccine has not been tested in patients at risk for infection.

Antibody Prophylaxis and Therapy

Another immunological alternative for the prevention of infection is the administration of protective antibodies. Antibodies administered as adjuncts of amphotericin B have been used in a small number of patients for therapy of cryptococcal infection (28). Administration of human pooled globulin is routinely used in pediatric AIDS cases for prevention of infection. Children with AIDS have a markedly lower incidence of cryptococcosis relative to adults (35, 56). Whether the low frequency of cryptococcal infection in children is related to immunoglobulin use is unknown, but the association of these phenomena is noteworthy.

In summary, vaccines and antibodies offer the possibility of novel strategies for the prevention of cryptococcal infection. Unfortunately, neither vaccines nor antibodies for immunoprophylaxis are currently available, and there are theoretical concerns about whether vaccines or antibodies can provide sufficient protection in patients with severely impaired immune function.

REFERENCES

1. **Ammassari, A., A. Linzalone, R. Murri, G. Marasca, G. Morace, and A. Antinori.** 1995. Fluconazole for primary prophylaxis of AIDS-associated cryptococcosis: a case-control study. *Scand. J. Infect. Dis.* **27:**235–237.
2. **Anonymous.** 1980. Garlic in cryptococcal meningitis. A preliminary report of 21 cases. *Chin. Med. J.* **93:**123–126.
3. **Anonymous.** 1994. Case 7–1994. *N. Engl. J. Med.* **330:**490–496.
4. **Bacellar, H., A. Munoz, E. N. Miller, B. A. Cohen, D. Besley, O. A. Selnes, J. T. Becker, and J. C. McArthur.** 1994. Temporal trends in the incidence of HIV-1-related neurologic diseases. *Neurology* **44:**1892–1900.
5. **Barchiesi, F., M. Del Poeta, V. Morbiducci, F. Ancarani, and G. Scalise.** 1994. Effect of pentamidine on the growth of *Cryptococcus neoformans. J. Antimicrob. Chemother.* **33:**1229–1232.
6. **Bozette, S. A., R. A. Larsen, J. Chie, M. A. Leal, J. Jacobsen, P. Rothman, P. Robinson, G. Gilbert, J. A. McCutchan, J. Tilles, J. M. Leedom, and D. D. Richman.** 1991. A placebo-controlled trial of maintenance therapy with fluconazole after treatment of cryptococcal meningitis in the acquired immunodeficiency syndrome. *N. Engl. J. Med.* **324:**580–584.
7. **Brandt, M. E., M. A. Pfaller, R. A. Hajjeh, E. A. Graviss, J. Rees, E. D. Spitzer, R. W. Pinner, and L. W. Mayer.** 1996. Molecular subtypes and antifungal susceptibilities of

serial *Cryptococcus neoformans* isolates in human immunodeficiency virus-associated cryptococcosis. *J. Infect. Dis.* **174:**812–820.

8. **Caporaso, N., S. M. Smith, and R. K. Eng.** 1983. Antifungal activity in human urine and serum after ingestion of garlic (*Allium sativum*). *Antimicrob. Agents Chemother.* **23:**700–702.

9. **Casadevall, A., J. Mukherjee, S. J. N. Devi, R. Schneerson, J. B. Robbins, and M. D. Scharff.** 1992. Antibodies elicited by a *Cryptococcus neoformans* glucuronoxylomannan-tetanus toxoid conjugate vaccine have the same specificity as those elicited in infection. *J. Infect. Dis.* **65:**1086–1093.

10. **Casadevall, A., J. Mukherjee, Y. RuiRong, and J. Perfect.** 1994. Management of *Cryptococcus neoformans* contaminated needle injuries. *Clin. Infect. Dis.* **19:**951–953.

11. **Casadevall, A., E. D. Spitzer, D. Webb, and M. G. Rinaldi.** 1993. Susceptibilities of serial *Cryptococcus neoformans* isolates from patients with recurrent cryptococcal meningitis to amphotericin B and fluconazole. *Antimicrob. Agents Chemother.* **37:**1383–1386.

12. **Chin, N.-X., I. Weitzman, and P. Della-Latta.** 1997. In vitro activity of fluvastatin, a cholesterol-lowering agent, and synergy with fluconazole and itraconazole against *Candida* species and *Cryptococcus neoformans*. *Antimicrob. Agents Chemother.* **41:**850–852.

13. **Cox, L. B., and J. C. Tolhurst.** 1946. *Human Torulosis*. Melbourne University Press, Melbourne.

14. **Cruz, M. C., M. S. Bartlett, and T. D. Edlind.** 1994. In vitro susceptibility of the opportunistic fungus *Cryptococcus neoformans* to anthelmintic benzimidazoles. *Antimicrob. Agents Chemother.* **38:**378–380.

15. **Currie, B. P., M. A. Ghannoum, and A. Casadevall.** 1995. Decreased fluconazole susceptibility of a relapse *Cryptococcus neoformans* isolate after fluconazole treatment. *Infect. Dis. Clin. Pract.* **4:**318–319.

16. **Davis, L. E., J.-K. Shen, and Y. Cai.** 1990. Antifungal activity in human cerebrospinal fluid and plasma after intravenous administration of *Allium sativum*. *Antimicrob. Agents Chemother.* **34:**651–653.

17. **Desmet, P., K. D. Kayembe, and C. De Vroey.** 1989. The value of cryptococcal antigen screening among HIV-positive/AIDS patients in Kinshasa, Zaire. *AIDS* **3:**77–78.

18. **Devi, S. J. N.** 1996. Preclinical efficacy of a glucuronoxylomannan-tetanus toxoid conjugate vaccine of *Cryptococcus neoformans* in a murine model. *Vaccine* **14:**841–842.

19. **Devi, S. J. N., R. Schneerson, W. Egan, T. J. Ulrich, D. Bryla, J. B. Robbins, and J. E. Bennett.** 1991. *Cryptococcus neoformans* serotype A glucuronoxylomannan-protein conjugate vaccines: synthesis, characterization, and immunogenicity. *Infect. Immun.* **59:**3700–3707.

20. **Duarte, A., N. Ordonez, and E. Castaneda.** 1994. Association de levaduras del genero *Cryptococcus* con especies de *Eucalyptus* en Santa Fe de Bogota. *Rev. Inst. Med. Trop. Sao Paulo* **36:**125–130.

21. **Eilam, Y., I. Polacheck, G. Ben-Gigi, and D. Chernichovsky.** 1987. Activity of phenothiazines against medically important yeasts. *Antimicrob. Agents Chemother.* **31:**834–836.

22. **Ellis, D. H., and T. J. Pfeiffer.** 1990. Natural habitat of *Cryptococcus neoformans* var. *gattii*. *J. Clin. Microbiol.* **28:**1642–1644.

23. **Feldmesser, M., C. Harris, S. Reichberg, S. Khan, and A. Casadevall.** 1996. Serum cryptococcal antigenemia in patients with AIDS. *Clin. Infect. Dis.* **23:**827–830.

24. **Fries, B. C., F. Chen, B. P. Currie, and A. Casadevall.** 1996. Karyotype instability in *Cryptococcus neoformans* infection. *J. Clin. Microbiol.* **34:**1531–1534.

25. **Gallani, J. E., R. D. Moore, and R. E. Chaisson.** 1994. Prophylaxis for opportunistic infections in patients with early HIV infection. *Ann. Intern. Med.* **120:**932–944.

26. **Glaser, J. B., and A. Garden.** 1985. Inoculation of cryptococcosis without transmission of the acquired immunodeficiency syndrome. *N. Engl. J. Med.* **313:**266.

27. **Goldman, D., S. C. Lee, and A. Casadevall.** 1994. Pathogenesis of pulmonary *Cryptococcus neoformans* infection in the rat. *Infect. Immun.* **62:**4755–4761.

28. **Gordon, M. A., and A. Casadevall.** 1995. Serum therapy of cryptococcal meningitis. *Clin. Infect. Dis.* **21:**1477–1479.

29. **Hoffmann, S., J. Stenderup, and L. R. Mathiesen.** 1991. Low yield of screening for cryptococcal antigen by latex agglutination assay on serum and cerebrospinal fluid from Danish patients with AIDS or ARC. *Scand. J. Immunol.* **23:**697–702.

30. **Kitchen, L. W.** 1996. Adjunctive immunologic therapy for *Cryptococcus neoformans* infections. *Clin. Infect. Dis.* **23:**209–210.

31. **Kitchen, L. W., J. A. Ross, J. E. Hernandez, A. L. Zarraga, and F. J. Mather.** 1992. Effect of administration of diethylcarbamazine on experimental bacterial and fungal infections in mice. *Int. J. Antimicrob. Agents* **1:**259–268.

32. **Kitchen, L. W., J. A. Ross, B. S. Turner, J. E. Hernandez, and F. J. Mather.** 1995. Diethylcarbamazine enhances blood microbicidal activity. *Adv. Ther.* **12:**22–29.

33. **Kovacs, J. A., A. A. Kovacs, M. Polis, W. C. Wright, V. J. Gill, C. U. Tuazon, E. P. Gelmann, H. C. Lane, R. Longfield, G. Overturf, A. M. Macher, A. S. Fauci, J. E. Parrillo, J. E. Bennett, and H. Masur.** 1985. Cryptococcosis in the acquired immunodeficiency syndrome. *Ann. Intern. Med.* **103:**533–538.

34. **Larsen, R. A., S. Bozette, A. McCutchan, J. Chiu, M. A. Leal, and D. D. Richman.** 1989. Persistent *Cryptococcus neoformans* infection of the prostate after successful treatment of meningitis. *Ann. Intern. Med.* **111:**125–128.

35. **Leggiadro, R. J., F. F. Barrett, and W. T. Hughes.** 1992. Extrapulmonary cryptococcosis in immunocompromised infants and children. *Pediatr. Infect. Dis. J.* **11:**43–47.

36. **Levitz, S. M., T. S. Harrison, A. Tabuni, and X. Liu.** 1997. Chloroquine induces human nononuclear phagocytes to inhibit and kill *Cryptococcus neoformans* by a mechanism independent of iron deprivation. *J. Clin. Invest.* **100:**1640–1646.

37. **Licea, B. A., D. G. Garza, and M. T. Zuniga.** 1996. Aislamiento de *Cryptococcus neoformans* var. *gattii* de *Eucalyptus tereticornis*. *Rev. Iber. Micol.* **13:**27–28.

38. **McNeill, J. I., and V. L. Kan.** 1995. Decline in the incidence of cryptococcosis among HIV-infected patients. *J. Acquired Immune Defic. Syndr. Hum. Retrovirol.* **206:**207.

39. **Mukherjee, J., L. Pirofski, M. D. Scharff, and A. Casadevall.** 1993. Antibody mediated protection in mice with lethal intracerebral *Cryptococcus neoformans* infection. *Proc. Natl. Acad. Sci. USA* **90:**3636–3640.

40. **Mukherjee, J., M. D. Scharff, and A. Casadevall.** 1992. Protective murine monoclonal antibodies to *Cryptococcus neoformans*. *Infect. Immun.* **60:**4534–4541.

41. **Negroni, R., C. Cendoya, A. I. Arechavala, A. M. Robles, M. Bianchi, A. J. Bava, and S. Helou.** 1995. Detection of *Cryptococcus neoformans* capsular polysaccharide antigen in asymptomatic HIV-infected patients. *Rev. Inst. Med. Trop. Sao Paulo* **37:**385–389.

42. **Nelson, M. R., M. Fisher, J. Cartledge, T. Rogers, and B. G. Gazzard.** 1994. The role of azoles in the treatment and prophylaxis of cryptococcal disease in HIV infection. *AIDS* **8:**651–654.

43. **Newton, J. A., P. Olson, S. A. Tasker, W. D. Bone, M. T. Nguyen, and E. C. Oldfield.** 1993. Weekly fluconazole for the suppression of recurrent thrush in HIV seropositive patients: impact on the incidence of disseminated cryptococcal infection. *Clin. Infect. Dis.* **17:**563. (Abstract 194.)

44. **Newton, J. A., S. A. Tasker, W. D. Bone, E. C. Oldfield III, P. E. Olson, and M. T. Nguyen.** 1995. Weekly fluconazole for the suppression of recurrent thrush in HIV-seropositive patients: impact on the incidence of disseminated cryptoccal infection. *AIDS* **9:**1286–1287.

45. **Nightingale, S. D., S. X. Cal, D. M. Peterson, S. D. Loss, B. A. Gamble, D. A. Watson, C. P. Manzone, J. E. Baker, and J. D. Jockusch.** 1992. Primary prophylaxis with fluconazole against systemic fungal infections in HIV-positive patients. *AIDS* **6:**191–194.

46. **Olson, P. E., K. C. Earhart, R. J. Rossetti, J. A. Newton, and M. R. Wallace.** 1997. Smoking and risk of cryptococcosis in patients with AIDS. *JAMA* **277**:629.

47. **Paugam, A., J. Dupouy-Camet, P. Blanche, J. P. Gangneux, C. T. Schaefer, and D. Sicard.** 1994. Decreased fluconazole resistance of *Cryptococcus neoformans* isolated from a patient with AIDS and recurrent meningitis. *Clin. Infect. Dis.* **19**:975–976.

48. **Perfect, J. R.** 1993. Antifungal prophylaxis: to prevent or not. *Am. J. Med.* **94**:233–234.

49. **Pfeiffer, T. J., and D. H. Ellis.** 1992. Environmental isolation of *Cryptococcus neoformans* var. *gattii* from *Eucalyptus tereticornis*. *J. Med. Vet. Mycol.* **30**:407–408.

50. **Pinner, R. W., R. A. Hajjeh, and W. G. Powderly.** 1995. Prospects for preventing cryptococcosis in persons infected with human immunodeficiency virus. *Clin. Infect. Dis.* **21**(Suppl. 1):S103-S107.

51. **Pirofski, L., R. Lui, M. DeShaw, A. B. Kressel, and Z. Zhong.** 1995. Analysis of human monoclonal antibodies elicited by vaccination with a *Cryptococcus neoformans* glucuronoxylomannan capsular polysaccharide vaccine. *Infect. Immun.* **63**:3005–3014.

52. **Powderly, W. G.** 1996. Recent advances in the management of cryptococcal meningitis in patients with AIDS. *Clin. Infect. Dis.* **22**(Suppl. 2):S119-S123.

53. **Powderly, W. G., D. M. Finkelstein, J. Feinberg, P. Frame, W. He, C. van der hosrt, S. L. Koletar, M. E. Eyster, J. Carey, H. Wskin, T. M. Hooton, N. Hyslop, S. A. Spector, and S. A. Bozette.** 1995. A randomized trial comparing fluconazole with clotrimazole troches for the prevention of fungal infections in patients with advanced human immunodeficiency virus infection. *N. Engl. J. Med.* **332**:700–705.

54. **Powderly, W. G., E. J. Keath, M. Sokol-Anderson, D. Kitz, J. Russell Little, and G. Kobayashi.** 1992. Amphotericin B-resistant *Cryptococcus neoformans* in a patient with AIDS. *Infect. Dis. Clin. Pract.* **1**:314–316.

55. **Reardon, C. C., S. J. Kim, R. P. Wagner, H. Koziel, and H. Kornfeld.** 1996. Phagocytosis and growth inhibition of *Cryptococcus neoformans* by human alveolar macrophages: effects of HIV-1 infection. *AIDS* **10**:613–618.

56. **Rogers, M. F., P. Thomas, E. T. Starcher, M. C. Noa, T. J. Bush, and H. W. Jaffe.** 1987. Acquired immunodeficiency syndrome in children: report of the Centers for Disease Control national surveillance, 1982 to 1985. *Pediatrics* **79**:1008–1014.

57. **Spitzer, E. D., S. G. Spitzer, L. F. Freundlich, and A. Casadevall.** 1993. Persistence of the initial infection in recurrent cryptococcal meningitis. *Lancet* **341**:595–596.

58. **Sugar, A. M., and C. Saunders.** 1988. Oral fluconazole as suppressive therapy of disseminated cryptococcosis in patients with acquired immunodeficiency syndrome. *Am. J. Med.* **85**:481–489.

59. **Walter, J. E., and E. G. Coffee.** 1968. Control of *Cryptococcus neoformans* in pigeon coops by alkalinization. *Am. J. Epidemiol.* **87**:173–178.

60. **Wang, Y., and A. Casadevall.** 1996. Susceptibility of melanized and non-melanized *Cryptococcus neoformans* to the melanin-binding compounds trifluoperazine and chloroquine. *Antimicrob. Agents Chemother.* **40**:541–545.

61. **Williamson, P. R., J. E. Bennett, M. A. Polis, J. B. Robbins, and R. Schneerson.** 1993. Immunogenicity and safety of a conjugate glucuronoxylomannan-tetanus conjugate vaccine in volunteers. *Clin. Infect. Dis.* **17**:540. (Abstract 56.)

62. **Williamson, P. R., J. E. Bennett, J. B. Robbins, and R. Schneerson.** 1993. Vaccination for prevention of cryptococcosis, abstr. L22, p. 60. 2nd International Conference on Cryptococcus and Cryptococcosis, Milan.

63. **Yamada, Y., and K. Azuma.** 1977. Evaluation of the in vitro antifungal activity of allicin. *Antimicrob. Agents Chemother.* **11**:743–749.

64. **Yoshida, S., S. Kausga, N. Hayashi, T. Ushiroguchi, H. Matssura, and S. Nakagawa.** 1987. Antifungal activity of ajoene derived from garlic. *Appl. Environ. Microbiol.* **53**:615–617.

65. **Zuger, A., E. Louie, R. S. Holzman, M. S. Simberkoff, and J. J. Rahal.** 1986. Crypto-

coccal disease in patients with the acquired immunodeficiency syndrome: diagnostic features and outcome of treatment. *Ann. Intern. Med.* **104:**234–240.

66. **Zuger, A., M. Shuster, M. S. Simberkoff, J. J. Rahal, and R. S. Holzman.** 1988. Maintenence amphotericin B for cryptococcal meningitis in the acquired immunodeficiency syndrome (AIDS). *Ann. Intern. Med.* **109:**592–593.

INDEX